Springer Texts in Statistics

Springer
New York
Berlin
Heidelberg
Barcelona
Hong Kong
London
Milan
Paris
Singapore
Tokyo

Springer Texts in Statistics

Alfred: Elements of Statistics for the Life and Social Sciences
Berger: An Introduction to Probability and Stochastic Processes
Bilodeau and Brenner: Theory of Multivariate Statistics
Blom: Probability and Statistics: Theory and Applications
Brockwell and Davis: An Introduction to Times Series and Forecasting
Chow and Teicher: Probability Theory: Independence, Interchangeability, Martingales, Third Edition
Christensen: Plane Answers to Complex Questions: The Theory of Linear Models, Second Edition
Christensen: Linear Models for Multivariate, Time Series, and Spatial Data
Christensen: Log-Linear Models and Logistic Regression, Second Edition
Creighton: A First Course in Probability Models and Statistical Inference
Dean and Voss: Design and Analysis of Experiments
du Toit, Steyn, and Stumpf: Graphical Exploratory Data Analysis
Durrett: Essentials of Stochastic Processes
Edwards: Introduction to Graphical Modelling
Finkelstein and Levin: Statistics for Lawyers
Flury: A First Course in Multivariate Statistics
Jobson: Applied Multivariate Data Analysis, Volume I: Regression and Experimental Design
Jobson: Applied Multivariate Data Analysis, Volume II: Categorical and Multivariate Methods
Kalbfleisch: Probability and Statistical Inference, Volume I: Probability, Second Edition
Kalbfleisch: Probability and Statistical Inference, Volume II: Statistical Inference, Second Edition
Karr: Probability
Keyfitz: Applied Mathematical Demography, Second Edition
Kiefer: Introduction to Statistical Inference
Kokoska and Nevison: Statistical Tables and Formulae
Kulkarni: Modeling, Analysis, Design, and Control of Stochastic Systems
Lehmann: Elements of Large-Sample Theory
Lehmann: Testing Statistical Hypotheses, Second Edition
Lehmann and Casella: Theory of Point Estimation, Second Edition
Lindman: Analysis of Variance in Experimental Design
Lindsey: Applying Generalized Linear Models
Madansky: Prescriptions for Working Statisticians
McPherson: Statistics in Scientific Investigation: Its Basis, Application, and Interpretation
Mueller: Basic Principles of Structural Equation Modeling: An Introduction to LISREL and EQS

(continued after index)

Bernard Flury

A First Course in Multivariate Statistics

With 141 Figures

Springer

Bernard Flury
Department of Mathematics
Rawles Hall 231
Indiana University
Bloomington, IN 47405
USA

Library of Congress Cataloging-in-Publication Data
Flury, Bernhard, 1951–
A first course in multivariate statistics / Bernard Flury.
p. cm. — (Springer texts in statistics)
Includes index.

1. Multivariate analysis. I. Title. II. Series.
QA278.F59 1997
519.5′35–dc21 97-6237

Printed on acid-free paper.

Production managed by Victoria Evarretta; manufacturing supervised by Johanna Tschebull.
Photocomposed pages prepared from the author's Plain Tex files by Bartlett Press.

9 8 7 6 5 4 3 2

ISBN 978-1-4419-3113-9

To Leah

Preface

My goal in writing this book has been to provide teachers and students of multivariate statistics with a unified treatment of both theoretical and practical aspects of this fascinating area. The text is designed for a broad readership, including advanced undergraduate students and graduate students in statistics, graduate students in biology, anthropology, life sciences, and other areas, and postgraduate students. The style of this book reflects my belief that the common distinction between multivariate statistical theory and multivariate methods is artificial and should be abandoned. I hope that readers who are mostly interested in practical applications will find the theory accessible and interesting. Similarly I hope to show to more mathematically interested students that multivariate statistical modelling is much more than applying formulas to data sets.

The text covers mostly parametric models, but gives brief introductions to computer-intensive methods such as the bootstrap and randomization tests as well. The selection of material reflects my own preferences and views. My principle in writing this text has been to restrict the presentation to relatively few topics, but cover these in detail. This should allow the student to study an area deeply enough to feel comfortable with it, and to start reading more advanced books or articles on the same topic. I have omitted topics that I believe are of questionable value, like Factor

Analysis, and replaced the ever popular cluster analysis theme by an introduction to finite mixtures, an area that I feel is much better founded. Another principle has been to make this text self-contained, with only few references to the literature, and as few "it can be shown that's " as possible. As a consequence, the book contains extensive background material, mostly in Chapter 2, which the advanced student may already be familiar with. Yet even advanced students will find some of the material in Chapter 2 new, particularly the section on finite mixtures.

Most sections contain many exercises, ranging in difficulty from trivial to advanced. The student who is exposed to multivariate statistics for the first time should not expect to be able to solve all problems. The easy problems are meant to test the students' understanding of the basic concepts. The more advanced exercises may serve as a challenge for advanced students.

This book can be used as a text for various types of courses on multivariate statistics. For a one-semester introduction to multivariate statistics I suggest covering Chapter 1, selected material from Chapter 2 (depending on the preparedness of the audience), Sections 3.1 and 3.2, Section 4.1, Chapter 5, Chapter 6 except for Section 6.6, and selected material from Chapter 7. In a one-semester course for students with a good background in mathematical statistics I would present selected material from Chapters 3 to 5, and all of Chapters 6 to 9. Much of Chapter 5 can be given as a reading assignment in such a course. All of the book can be covered in a two-semester course.

The book starts out at an elementary level but reaches an intermediate level in Chapters 3 and 4. Chapters 5 and 6 (except Section 6.6) are again more elementary, while Chapters 7 to 9 are intermediate to advanced. The reader who is already familiar with Principal Component Analysis may find Chapter 8 rather unusual, and perhaps initially even difficult to understand. I have found it better to introduce principal components using the powerful notion of self-consistency rather than the commonly accepted method of maximization of variance. Hopefully many readers will find the self-consistency principle as fascinating as I do.

Graphical illustrations are abundant in this text, to illustrate theoretical concepts as well as practical applications. Most students will find the graphs helpful to develop their intuition.

Familiarity with elementary matrix algebra is an essential prerequisite for any student using this book, irrespective of their degree of knowledge of statistics. In a class with weak background in linear algebra the teacher may wish to start out with a review of matrix algebra, including elementary concepts such as matrix multiplication and inversion. For selected topics of particular importance in multivariate statistics, mostly eigenvectors and spectral decomposition theory, an appendix provides the necessary material. In some parts of the book, mostly in Sections 5.6 and 6.5, familiarity with linear regression will be an advantage but is not absolutely necessary.

In the spirit of erasing the boundary between theoretical and applied statistics it is essential that the students learn to use appropriate computer software. I believe that using canned statistics packages only is not the proper way to learn the material.

Instead the student should be able to see intermediate results of matrix calculations, and to design and modify programs herself or himself. Several software products provide an ideal environment for this purpose. In particular I have found GAUSS and MATLAB to be efficient and easy to use, but students with a good background in statistical computing might find S-PLUS more convenient. The mathematicians' preferred software MAPLE and MATHEMATICA may be used as well but will probably not be efficient for iterative numerical computations. At a later stage, after the student has understood the concepts well enough, it will be more convenient to use black-box routines of S-PLUS, SAS, SPSS, or other packages. At the learning stage I discourage students to use the packaged software because it tends to make them believe that they can apply multivariate statistical techniques without understanding the mathematical theory. To help readers getting ready to use software, Marco Bee and I have written tutorials in S-PLUS and GAUSS for selected examples; rather than putting these tutorials into the book (which might render the book outdated the day it is published) we have made them available from an ftp-site; see the *Software and Data Files* instructions following the Table of Contents.

I am indebted to an uncountable number of people for their help. Springer Statistics Editors John Kimmel and Martin Gilchrist have provided me with expert guidance during the many years it took me to write this book. Several anonymous referees have written extensive reviews of parts of the manuscript; their advice has been highly valuable. Special thanks go to Victoria Evarretta and the production staff at Springer-Verlag New York for turning my manuscript into a real book. Several institutions have supported me generously while I was working on this book during my sabbatical leave from Indiana University. I would like to thank the Statistics Departments at the University of Trento, the University of Auckland, and ETH Zürich for providing me with office space, access to computer and library facilities, and most importantly, interaction with their own faculty and students. Many friends of mine have gone through the trouble of proofreading parts of the manuscript, or even using it for their own classes. Others have given me permission to use their original data in this book, have suggested examples, found mistakes, and suggested changes to the text. In alphabetical order I would like to thank Jean-Pierre Airoldi, Chris Albright, Marco Bee, Peter Castelluccio, Siva Chittajallu, Zhuan Dong, Karin Dorman, Tim Dunne, Leah Flury, M.N. Goria, Klaus Haagen, Christian Klingenberg, Ed Levri, Nicola Loperfido, Steve McKinley, Ambrogio Monetti, A. Narayanan, Daan Nel, Beat Neuenschwander, Pierluigi Novi Inverardi, John Randall, Annalene Sadie, and Thaddeus Tarpey. I apologize for any omissions. Students at Indiana University, the University of Trento, the Scuola Estiva di Matematica in Perugia, and at Indiana University—Purdue University in Indianapolis have served as guinea pigs for my manuscript; they have helped me to find many major and minor mistakes. Needless to say that

Bloomington (Indiana), Bernard Flury
January, 1997

Contents

Software and Data Files

For selected examples in this book software instructions have been written in S-PLUS by Marco Bee, and in GAUSS by the author. One or two files are available for each section of the book that requires computing. You can get the files by anonymous ftp to 129.79.94.6, in subdirectory `/pub/flury`. The files are named `multiN.spl` (S-PLUS files), and `multiN.gau` (GAUSS files), where `N` is the combined chapter and section number, possibly followed by the letter `a` or `b`. For instance, file `multi42b.gau` is the second file containing GAUSS instructions for Section 4.2. It is important to study the tutorials in their sequential order because there are frequent references to material covered in earlier sections. We are planning to update and enhance the software instruction files regularly, and to add a similar set of instructions for MATLAB. Any comments will be greatly appreciated. Please send your comments by e-mail to `flury@indiana.edu` or `mbee@gelso.unitn.it`.

From the same ftp-site and subdirectory all data files used in this book are available as well. The names of the data files are `tabN_M.dat`, where `N` is the combined chapter and section number and `M` is the number of the table within section. For instance the data of Table 5.4.1 are in file `tab54_1.dat`.

1 Why Multivariate Statistics?

There is no better way to arouse interest in multivariate statistics than to study some good examples. This chapter gives four examples, three of them based on real data. The fourth one is a brainteaser.

Example 1.1 Classification of midges

The biologists Grogan and Wirth (1981) described two newly discovered species of predaceous midges, *Amerohelea fasciata* (*Af*) and *A. pseudofasciata* (*Apf*). Midges are small gnatlike insects. Because the two species are similar in appearance, it is useful for the biologist to be able to classify a specimen as *Af* or *Apf* based on external characteristics that are easy to measure. Among many characteristics that distinguish *Apf* from *Af*, Grogan and Wirth reported measurements of antenna length and wing length, both in millimeters, of nine *Af* midges and six *Apf* midges. The data are listed in Table 1.1. The question is, can we find a rule that allows us to classify a given midge as *Af* or *Apf*, respectively, based only on measurements of antenna length and wing length? Such a rule might be of practical importance — for instance, one species might be a valuable pollinator, and the other species the carrier of some disease. Identifying a midge correctly by simple external characteristics would, then save both time and money.

Table 1.1. Midge Data

Species	Antenna Length (x) (mm)	Wing Length (y) (mm)
Af	1.38	1.64
Af	1.40	1.70
Af	1.24	1.72
Af	1.36	1.74
Af	1.38	1.82
Af	1.48	1.82
Af	1.54	1.82
Af	1.38	1.90
Af	1.56	2.08
Apf	1.14	1.78
Apf	1.20	1.86
Apf	1.18	1.96
Apf	1.30	1.96
Apf	1.26	2.00
Apf	1.28	2.00

Source: Grogan and Wirth, 1981.

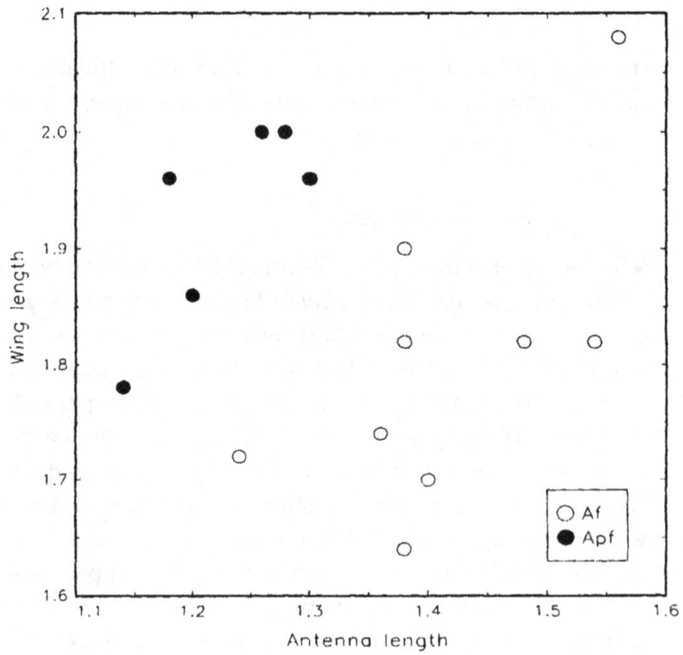

Figure 1-1
Scatterplot of the midge data.

We start out with a purely descriptive approach to the classification problem. For two variables measured on each object, the best way to analyze a small number of data points is to construct a scatterplot, as in Figure 1.1. The specimens of the two species appear well-separated in this graph. Even without any knowledge of statistical principles, one could formulate a classification rule as "*Apf* in the upper left part and *Af* in the lower right part of the graph." One could make this rule more precise by drawing some boundary curve in the graph. Figure 1.2 shows the same scatterplot as Figure 1.1, together with a boundary line

$$\text{wing length} - \text{antenna length} = 0.58$$

which is rather arbitrary, but seems to do the job. It will be simpler to use symbols, so let us write x = antenna length and y = wing length. Then the boundary line is $y - x = 0.58$, and two areas of classification can be defined as

$$C_1 = \{(x, y) : y - x < 0.58\} \qquad (\text{for } Af)$$

and

$$C_2 = \{(x, y) : y - x > 0.58\} \qquad (\text{for } Apf).$$

Thus, a decision on classification would be based solely on the value of the difference $y - x$ between antenna length and wing length. Grogan and Wirth (1981, p. 1286) actually suggested a different rule, based on the *ratio* x/y instead of the difference; see Exercise 1.

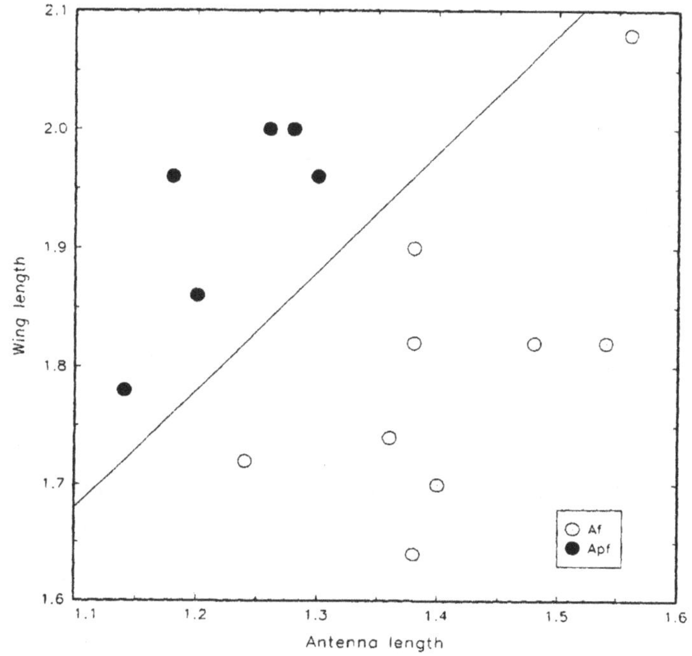

Figure 1-2
Scatterplot of the midge data with classification boundary $y - x = 0.58$.

What if three variables (or more) were to be analyzed? Obviously, things will get somewhat more tricky, because three-dimensional scatterplots are difficult to construct. Conversely, if there is only one variable on which to base a decision, then finding a proper classification rule should be simple. However, in our example it is clear (from Figure 1.1) that neither of the two variables by itself would allow us to construct a good classification rule, because there would be considerable overlap between the two groups.

Yet the idea of looking at one-dimensional distributions is promising if we allow for some more flexibility by changing to a new coordinate system. But first, let us illustrate graphically how a one-dimensional distribution can be obtained from a two-dimensional distribution. (This may sound a bit trivial, but you will understand the purpose very soon). Figure 1.3 shows the same scatterplot once again, this time with univariate distributions plotted along the coordinate axes. We can think of the distribution of "antenna length" as obtained by letting the dots in the scatterplot "drop" onto the horizontal axis. Similarly, a "horizontal rain" onto the vertical axis produces the distribution of "wing length."

The key idea, now, is to play with the coordinate system. Shifting the coordinate system to a different origin or changing the scale of each axis (for instance, from millimeters to miles) will produce the same graphs for the distributions along the two axes, so the crucial thing is to consider new coordinate directions. For instance, we could introduce a new coordinate system which is rotated by 45 degrees compared to

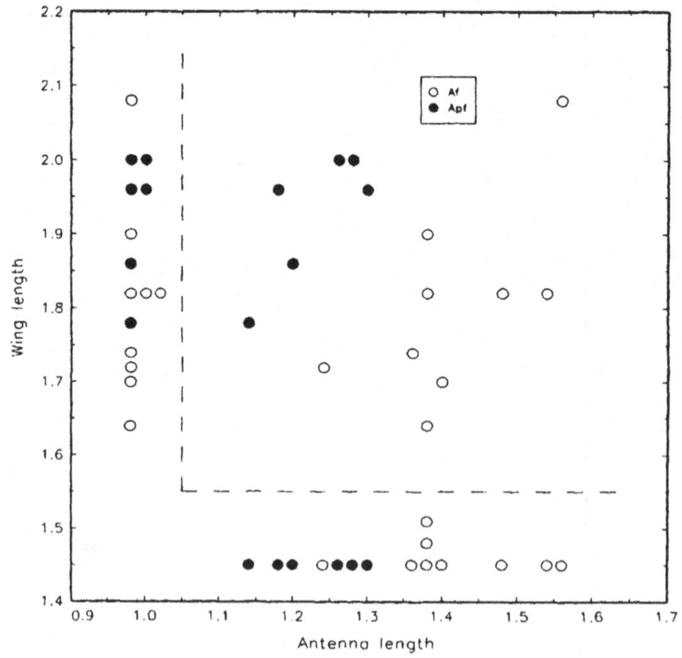

Figure 1-3
Univariate distributions of "antenna length" and "wing length" as projections of the data onto the coordinate axes.

the old one, as in Figure 1.4, and label the new coordinate axes as u and v. Without giving any numerical details at this point, the data can now be expressed in the (u, v)-coordinate system, and we can study the distribution of u or v by letting the data points "drop" onto the respective coordinate axes. As Figure 1.4 shows, we get an excellent separation of the two groups along the u-axis, and hardly any separation at all along the v-axis. Thus we might reasonably claim that a rule for classification should be based on the value of the u-coordinate of the data alone. Exercise 2 relates this idea to the boundary line discussed before.

Based on these ideas, we might now start to develop the mathematics of "descriptive multivariate discrimination," but the student should be aware that it will take quite some work to fully understand exactly how a new coordinate system should be defined or what constitutes an optimal boundary curve for classification. In particular, we will need to introduce important concepts related to bivariate and multivariate distributions, such as marginal and conditional distributions, expected values, moments, covariance and correlation, and transformations of random variables. Chapter 2 provides an adequate background. The student who finds the theoretical developments long and difficult should keep in mind that, ultimately, the theory will lead to methods for dealing with practical problems like the midge example. Two chapters of this text are devoted to problems of discrimination and classification (Chapters 5 and 7); the student who works his or her way through the theoretical foundation should find them rather rewarding.

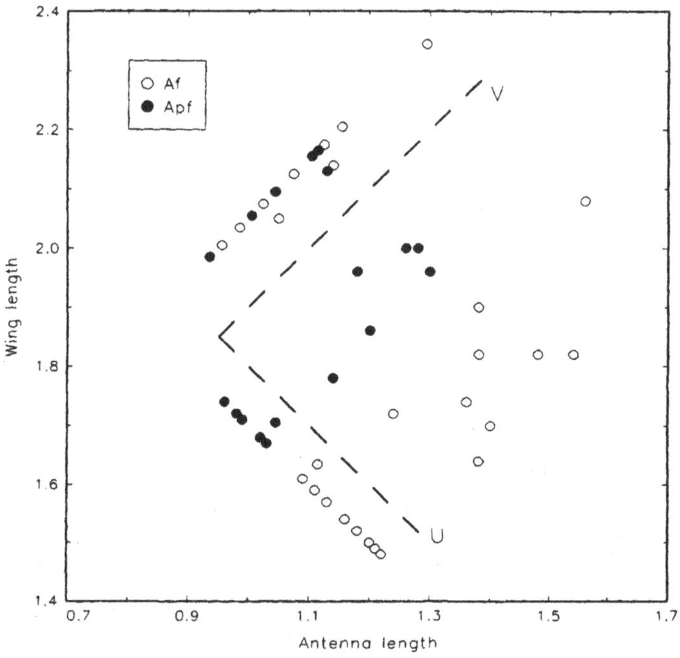

Figure 1-4
A new coordinate system for the midge data, rotated by 45 degrees.

Let us briefly discuss yet another aspect of the midge example. Although the data of the two groups appear well-separated in the graph of *antenna length* vs. *wing length*, one might doubt that the observed differences between the two species are more than just purely random. If the differences are, indeed, just random, then any classification rule based on this data set will itself be just a random procedure, with no potential benefit for the identification of future specimens. This remark points toward the need to use formal statistical *models*: The observed data are considered as a sample from some distribution (also referred to as "population"), and sample statistics serve as estimates of corresponding quantities called "parameters" in the statistical model. Thus, we will discuss parameter estimation and related testing procedures; this is done in Chapters 4 and 6.

The midge example is rather typical for discriminant analysis. The two samples of midges with known group membership would usually be referred to as the *training samples*. A classification rule based on the training samples would, then be applied later to midges with unknown group membership and lead to correct classification in most cases. □

Example 1.2 Head dimensions of 200 young men

In the mid-1980s, the Swiss government decided that new protection masks were needed for the members of the Swiss Army (the legendary Swiss Army knife still seemed ok for the next century). Approximately 900 members of the Swiss army were sampled, and 25 variables were measured on their heads. Figure 1.5 defines the six variables that the mask builders considered most important for the fit of protection masks; the variables are

$$MFB = \text{minimal frontal breadth},$$

$$BAM = \text{breadth of angulus mandibulae},$$

$$TFH = \text{true facial height},$$

$$LGAN = \text{length from glabella to apex nasi},$$

$$LTN = \text{length from tragion to nasion},$$

and

$$LTG = \text{length from tragion to gnathion}.$$

Table 1.2 displays the data of 200 male soldiers who were 20 years old at the time of the investigation; they may reasonably be thought of as a random sample of the "population" of all 20-year-old healthy Swiss males. Thus, we will study the joint distribution of $p = 6$ variables, based on $N = 200$ observations.

The first step in the analysis of this type of data is usually descriptive: histograms and scatterplots, and summary statistics, such as means, standard deviations, and correlations (see Exercise 3). One will observe, for instance, that all 15 correlations

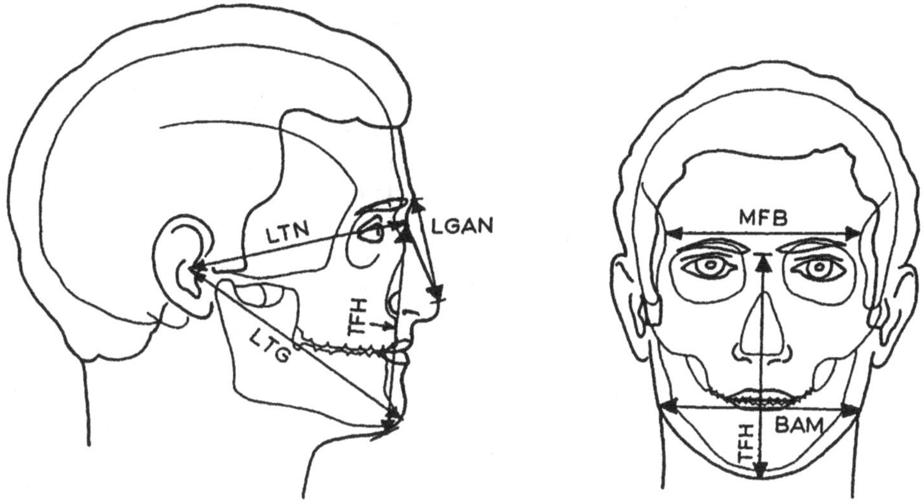

Figure 1-5
Definition of variables in the head dimension example. Reproduced with permission by Chapman and Hall. Flury and Riedwyl 1988, Figure 10.14, p.223

are positive, which is not surprising because all variables are head dimensions. Of course it would be nice if we were able to construct and understand six-dimensional scatterplots to detect structure in the data set, but, unfortunately, we are restricted to three-dimensional perception (and, for most practical purposes, even to two dimensions, although three-dimensional graphics are feasible with the proper computer hardware and software). For instance, we might detect that, in some directions of the six-dimensional space, the data show large spread, whereas in some other direction, there may be very little variability. This is fairly straightforward to illustrate in the two-dimensional case. Figure 1.6 shows a scatterplot of *LTG* vs. *LTN*, exhibiting a correlation of 0.656 and standard deviations of 5.64 for *LTG* and 3.92 for *LTN*, respectively. As in Example 1.1, let us introduce a new coordinate system. Without giving any theoretical reasons or computational details, Figure 1.7 shows the same data as Figure 1.6, with two new axes labeles u and v, centered at the mean of the scatterplot and rotated by 30 degrees compared to the original coordinate system. It can be verified (see Exercise 3) that in the (u, v)-coordinate system the two variables are almost exactly uncorrelated. We can also think of the u-axis, which follows the "main direction" of the scatterplot, as a good one-dimensional approximation to the two-dimensional data.

What we just discussed are aspects of a method called *Principal Component Analysis*, the topic of Chapter 8. In anticipation of a much more detailed discussion, let us briefly outline some ideas behind Principal Component Analysis. Suppose we deal with p-dimensional data (i.e., p variables measured on the same objects), and imagine a p-dimensional scatterplot constructed from the raw data. Now "play" with the coordinate system by shifting it around and rotating it. If we are lucky, we may find some rotated version of the coordinate system in which the data exhibits no or almost no variability in some of the coordinate directions. Then we might claim that the data

Table 1.2 Head Dimension Data in Millimeters

MFB	BAM	TFH	LGAN	LTN	LTG
113.2	111.7	119.6	53.9	127.4	143.6
117.6	117.3	121.2	47.7	124.7	143.9
112.3	124.7	131.6	56.7	123.4	149.3
116.2	110.5	114.2	57.9	121.6	140.9
112.9	111.3	114.3	51.5	119.9	133.5
104.2	114.3	116.5	49.9	122.9	136.7
110.7	116.9	128.5	56.8	118.1	134.7
105.0	119.2	121.1	52.2	117.3	131.4
115.9	118.5	120.4	60.2	123.0	146.8
96.8	108.4	109.5	51.9	120.1	132.2
110.7	117.5	115.4	55.2	125.0	140.6
108.4	113.7	122.2	56.2	124.5	146.3
104.1	116.0	124.3	49.8	121.8	138.1
107.9	115.2	129.4	62.2	121.6	137.9
106.4	109.0	114.9	56.8	120.1	129.5
112.7	118.0	117.4	53.0	128.3	141.6
109.9	105.2	122.2	56.6	122.2	137.8
116.6	119.5	130.6	53.0	124.0	135.3
109.9	113.5	125.7	62.8	122.7	139.5
107.1	110.7	121.7	52.1	118.6	141.6
113.3	117.8	120.7	53.5	121.6	138.6
108.1	116.3	123.9	55.5	125.4	146.1
111.5	111.1	127.1	57.9	115.8	135.1
115.7	117.3	123.0	50.8	122.2	143.1
112.2	120.6	119.6	61.3	126.7	141.1
118.7	122.9	126.7	59.8	125.7	138.3
118.9	118.4	127.7	64.6	125.6	144.3
114.2	109.4	119.3	58.7	121.1	136.2
113.8	113.6	135.8	54.3	119.5	130.9
122.4	117.2	122.2	56.4	123.3	142.9
110.4	110.8	122.1	51.2	115.6	132.7
114.9	108.6	122.9	56.3	122.7	140.3
108.4	118.7	117.8	50.0	113.7	131.0
105.3	107.2	116.0	52.5	117.4	133.2
110.5	124.9	122.4	62.2	123.1	137.0
110.3	113.2	123.9	62.9	122.3	139.8
115.1	116.4	118.1	51.9	121.5	133.8
119.6	120.2	120.0	59.7	123.9	143.7
119.7	125.2	124.5	57.8	125.3	142.7
110.2	116.8	120.6	54.3	123.6	140.1
118.9	126.6	128.2	63.8	125.7	151.1
112.3	114.7	127.7	59.4	125.2	137.5
113.7	111.4	122.6	63.3	121.6	146.8
108.1	116.4	115.5	55.2	123.5	134.1
105.6	111.4	121.8	61.4	117.7	132.6
111.1	111.9	125.2	56.1	119.9	139.5

Table 1.2 *Continued*

MFB	BAM	TFH	LGAN	LTN	LTG
111.3	117.6	129.3	63.7	124.3	142.8
119.4	114.6	125.0	62.5	129.5	147.7
113.4	120.5	121.1	61.5	118.1	137.2
114.7	113.8	137.7	59.8	124.5	143.3
115.1	113.9	118.6	59.5	119.4	141.6
114.6	112.4	122.2	54.5	121.2	126.3
115.2	117.2	122.2	60.1	123.9	135.7
115.4	119.5	132.8	60.3	127.8	140.3
119.3	120.6	116.6	55.8	121.5	143.0
112.8	119.3	129.6	61.0	121.1	139.4
116.6	109.6	125.4	54.6	120.2	122.6
106.5	116.0	123.2	52.8	121.7	134.9
112.1	117.4	128.2	59.9	120.3	131.5
112.8	113.0	125.4	64.8	119.4	136.6
114.6	119.0	116.8	57.4	123.8	140.0
110.9	116.5	125.8	53.5	124.8	142.9
109.1	117.0	123.7	60.0	120.1	137.7
111.7	117.3	121.0	51.5	119.7	135.5
106.4	111.1	124.4	59.1	122.4	138.4
121.2	122.5	117.8	54.8	121.5	143.9
115.2	121.2	117.4	54.9	121.9	144.0
123.2	124.2	120.0	57.9	119.4	138.4
113.1	114.5	118.9	56.9	121.8	135.0
110.3	108.9	115.2	55.9	119.0	138.0
115.0	114.7	123.5	66.7	120.3	133.6
111.9	111.1	122.3	63.8	117.1	131.6
117.2	117.5	120.2	60.5	119.5	129.6
113.8	112.5	123.2	62.0	113.5	132.4
112.8	113.5	114.3	53.8	128.4	143.8
113.3	118.4	123.8	51.6	122.7	141.7
123.9	120.5	118.3	54.3	122.0	133.8
119.8	119.6	126.1	57.5	124.7	130.9
110.9	113.9	123.7	62.7	124.8	143.5
111.9	125.1	121.8	58.1	112.1	134.8
114.0	120.8	131.2	61.0	124.7	152.6
113.6	110.4	130.9	60.2	118.5	132.5
118.9	126.1	121.9	56.1	127.3	145.6
119.4	127.8	128.0	61.8	120.6	141.3
121.0	121.1	116.8	56.3	124.2	140.9
109.0	105.7	126.2	59.4	121.2	143.2
117.9	125.1	122.6	58.2	128.4	151.1
124.8	123.8	128.3	60.4	129.1	147.2
120.6	124.3	120.0	59.5	123.4	144.1
115.6	115.9	117.2	54.0	119.9	135.3
116.6	119.1	131.0	58.0	123.3	136.4
118.7	118.9	129.6	68.6	123.0	141.4

Table 1.2 *Continued*

MFB	BAM	TFH	LGAN	LTN	LTG
114.3	117.1	127.1	55.7	119.1	139.8
110.9	113.1	124.1	60.6	115.7	132.1
119.2	120.0	136.9	55.1	129.5	142.0
117.1	123.7	108.7	53.2	125.6	136.6
109.3	110.2	129.3	58.5	121.0	136.8
108.8	119.3	118.7	58.9	118.5	132.7
109.0	127.5	124.6	61.1	117.6	131.5
101.2	110.6	124.3	62.9	124.3	138.9
117.8	109.0	127.1	53.9	117.9	135.8
112.4	115.6	135.3	55.8	125.0	136.1
105.3	109.8	115.4	59.6	116.6	137.4
117.7	122.4	127.1	74.2	125.5	144.5
110.9	113.7	126.8	62.7	121.4	142.7
115.6	117.5	114.2	55.0	113.2	136.6
115.4	118.1	116.6	62.5	125.4	142.1
113.6	116.7	130.1	58.5	120.8	140.3
116.1	117.6	132.3	59.6	122.0	139.1
120.5	115.4	120.2	53.5	118.6	139.4
119.0	124.1	124.3	73.6	126.3	141.6
122.7	109.0	116.3	55.8	121.8	139.4
117.8	108.2	133.9	61.3	120.6	141.3
122.3	114.2	137.4	61.7	125.8	143.2
114.4	117.8	128.1	54.9	126.5	140.6
110.6	111.8	128.4	56.7	121.7	147.5
123.3	119.1	117.0	51.7	119.9	137.9
118.0	118.0	131.5	61.2	125.0	140.5
122.0	114.6	126.2	55.5	121.2	143.4
113.4	104.1	128.3	58.7	124.1	142.8
117.0	111.3	129.8	55.6	119.5	136.1
116.6	108.3	123.7	61.0	123.4	134.0
120.1	116.7	122.8	57.4	123.2	145.2
119.8	125.0	124.1	61.8	126.9	141.2
123.5	123.0	121.6	59.2	115.3	138.4
114.9	126.7	131.3	57.3	122.7	139.2
120.6	110.8	129.6	58.1	122.7	134.7
113.0	114.8	120.7	54.1	119.7	140.9
111.8	110.2	121.0	56.4	121.4	132.1
110.8	114.9	120.5	58.7	113.4	131.6
114.8	118.8	120.9	58.4	119.7	135.9
122.5	122.3	116.7	57.4	128.1	147.3
105.9	105.6	129.3	69.5	123.6	136.6
108.0	111.3	116.9	53.8	117.8	129.6
114.4	111.7	116.3	54.3	120.2	130.1
117.9	112.9	119.1	54.2	117.9	134.8
110.7	113.9	114.5	53.0	120.1	124.5
112.3	110.4	116.8	52.0	121.0	133.4

Table 1.2 *Continued*

MFB	BAM	TFH	LGAN	LTN	LTG
110.9	110.0	116.7	53.4	115.4	133.0
126.6	127.0	135.2	60.6	128.6	149.6
116.2	115.2	117.8	60.8	123.1	136.8
117.2	117.8	123.1	61.8	122.1	140.8
114.5	113.2	119.8	50.3	120.6	135.1
126.2	118.7	114.6	55.1	126.3	146.7
118.7	123.1	131.6	61.8	123.9	139.7
116.2	111.5	112.9	54.0	114.7	134.2
113.9	100.6	124.0	60.3	118.7	140.7
114.4	113.7	123.3	63.2	125.5	145.5
114.5	119.3	130.6	61.7	123.6	138.5
113.3	115.9	116.1	53.5	127.2	136.5
120.7	114.6	124.1	53.2	127.5	139.1
119.1	115.3	116.6	53.5	128.2	142.6
113.2	107.7	122.0	60.6	119.4	124.2
113.7	110.0	131.0	63.5	117.3	134.6
116.3	119.3	116.6	57.3	122.0	141.6
117.6	117.8	122.5	59.9	119.4	136.3
114.8	115.0	115.2	58.9	122.5	135.2
127.3	123.9	130.3	59.8	128.3	138.7
130.5	125.5	127.4	62.1	130.1	153.3
110.4	105.4	122.1	56.2	114.6	122.8
108.5	105.4	119.1	59.4	120.4	134.7
121.6	112.1	126.5	60.6	122.7	142.9
117.9	115.2	139.1	59.6	125.5	141.3
112.7	111.5	114.9	53.5	113.9	132.6
121.8	119.0	116.9	56.5	120.1	139.2
118.5	120.0	129.8	59.5	127.8	150.5
118.3	120.0	127.5	56.6	122.0	139.4
117.9	114.4	116.4	56.7	123.1	136.3
114.2	110.0	121.9	57.5	116.1	126.5
122.4	122.7	128.4	58.3	131.7	148.1
114.1	109.3	124.4	62.8	120.8	133.4
114.6	118.0	112.8	55.6	118.5	135.6
113.6	114.6	127.1	60.8	123.8	143.1
111.3	116.7	117.7	51.2	125.7	141.9
111.4	120.4	112.1	56.4	120.3	137.1
119.9	114.4	128.8	69.1	124.9	144.3
116.1	118.9	128.3	55.8	123.7	139.7
119.7	118.2	113.5	59.5	127.0	146.5
105.8	106.7	131.2	61.3	123.7	144.3
116.7	118.7	128.2	55.8	121.2	143.9
106.4	107.3	122.9	57.6	122.3	132.9
112.2	121.3	130.1	65.3	120.3	137.9
114.8	117.3	130.3	60.9	125.6	137.4
110.0	117.4	114.1	54.8	124.8	135.1

Table 1.2 *Continued*

MFB	BAM	TFH	LGAN	LTN	LTG
121.5	121.6	125.4	59.5	128.5	144.7
119.8	119.4	119.6	53.9	122.3	143.6
107.7	108.4	125.1	62.3	122.7	137.2
118.4	115.7	121.1	57.8	124.9	140.5
119.8	113.9	132.0	60.8	122.4	137.6
114.1	112.8	119.3	52.7	114.2	136.9
117.7	121.8	120.0	59.1	122.6	138.3
111.1	117.7	117.7	60.2	124.6	139.2
111.1	117.7	117.7	59.1	124.7	141.9
128.1	118.3	129.4	61.0	134.7	148.6
120.4	118.7	126.4	59.4	133.1	147.1
112.9	112.0	123.5	57.2	121.3	133.3
118.2	114.4	114.8	55.3	126.1	149.1
119.0	112.7	129.1	62.0	127.6	146.6
111.8	116.0	117.8	60.9	114.4	128.7
116.6	111.4	115.6	60.9	117.8	137.4

Source: Flury & Riedwyl, 1988.

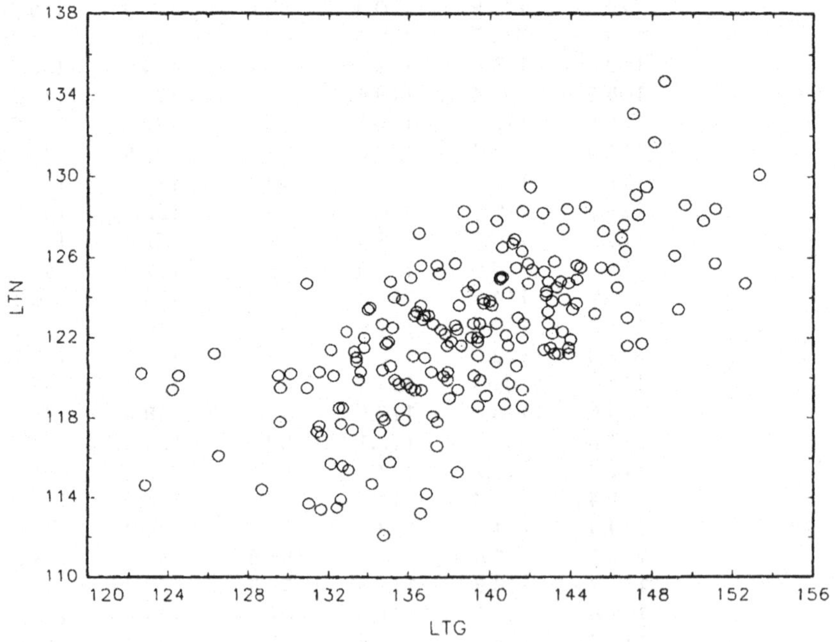

Figure 1-6
Scatterplot of *LTN* vs. *LTG* in the head dimension example.

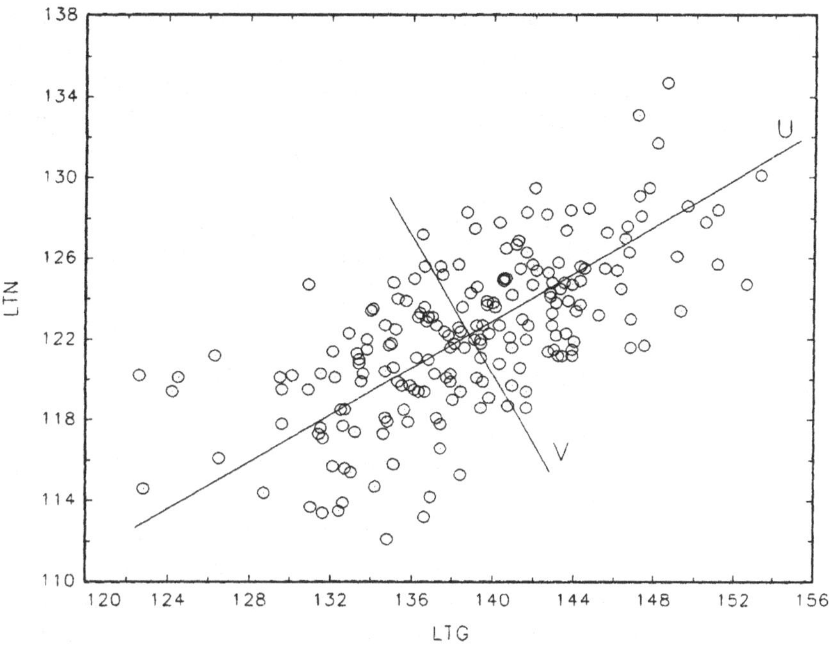

Figure 1-7
The same data as in Figure 1.6 with a new coordinate system rotated by 60 degrees.

is well-represented by fewer than p coordinate directions, and thus, approximate the p-dimensional data in a subspace of lower dimension. For instance, if we have $p = 3$ variables, we might find that, in the three-dimensional scatterplot, all data points fall almost exactly in a plane. Then we would find a new coordinate system, such that two of the coordinates span this plane, and ignore the third coordinate direction. Or the three-dimensional data might fall almost exactly on a straight line, in which case we would try to find a one-dimensional approximation. This may sound a bit abstract at this moment, but Exercise 4 gives a real data example, the well-known *turtle carapace data*, which illustrates the point well.

In the head dimensions example, and particularly if only two variables are analyzed as in Figure 1.6, it is perhaps not clear why approximating the six-dimensional data by a smaller number of variables would be interesting or beneficial, so let us discuss some further aspects of the study. As mentioned earlier, the purpose of the project was to give an empirical basis to the construction of new gas masks. After having completed the data collection, the investigators were to determine k "typical heads" after which to model the new masks, k being a number between two and six. But how exactly would such "typical heads" be defined?

If the number k of types were one, then we might just take the average value of each variable. If k is larger than one, then we would probably want the "typical heads" to represent the dominant directions of variability in the data. For instance, in the preceding (hypothetical) example of three-dimensional data falling almost exactly onto a straight line, one would want to choose the representatives along this line. For $p = 6$ variables (or even $p = 25$ as in the original data set), it would be helpful if the data could be approximated reasonably well in a subspace of lower dimension. In particular, if we were able to find a good one-dimensional approximation, then it should be relatively straightforward to select k good representatives along the approximating line.

This example actually goes beyond the material of this text, but we will return to it in Chapter 8, and continue the discussion of approximating high-dimensional data in a subspace of lower dimension. □

Example 1.3 Wing length of male and female birds

Capturing and tagging birds is an important method for studying their migration. When capturing an animal, the biologist would like to obtain certain measurements, without disturbing the bird too much. Flury, Airoldi, and Biber (1992) report such an experiment, in which water pipits (*Anthus spinoletta*) were caught and tagged at Bretolet in Valais, Switzerland. The raw data obtained from $N = 381$ birds are summarized in Table 1.3. There is a single variable: *wing length*, in millimeters. Perhaps, it seems a bit odd that this example, with only *one* variable reported, should serve as an introductory example to multivariate statistics, but we will justify it in a moment.

A striking feature of the data is that there are many birds with wing length in the range from 85 to 87 mm and again many in the range from 91 to 93 mm, with a distinct "valley" of low frequency in between, as illustrated by Figure 1.8. It is likely that most of the larger birds are male and most of the smaller birds female, but the gender of the birds was not identified at the time of the investigation to avoid hurting them. Thus the data set is a mixture of an unknown number of male birds and an unknown number of female birds. The situation is similar to Example 1.1 in the sense that there are data from two well-defined groups, and we would like to establish a classification rule which allows us to assign each specimen to one of the two groups, keeping the probability of misclassification as small as possible. But there is a crucial difference to the midge example: group membership is unknown, that is, no training samples are available. Nevertheless, it would be useful to partition the data set into subsets labeled "likely male" and "likely female," respectively.

Methods for partitioning data sets into homogeneous subgroups are commonly referred to as *cluster analysis*; they enjoy considerable popularity but will be treated only marginally in this book, for theoretical reasons and to limit the amount of material. In the bird example, a cluster analysis would typically find a "cutoff point" or "critical value" c for wing length, declare all birds with wing length less than c as one group, and those with wing length larger than c as the second group. Although such an ad-hoc partition would identify the two genders fairly well, it would fail to reflect

Table 1.3 Wing Length in Millimeters of 381 Water Pipits

Wing Length	Frequency
82	5
83	3
84	12
85	36
86	55
87	45
88	21
89	13
90	15
91	34
92	59
93	48
94	16
95	12
96	6
97	0
98	1

Source: Flury, Airoldi, and Biber, 1992.

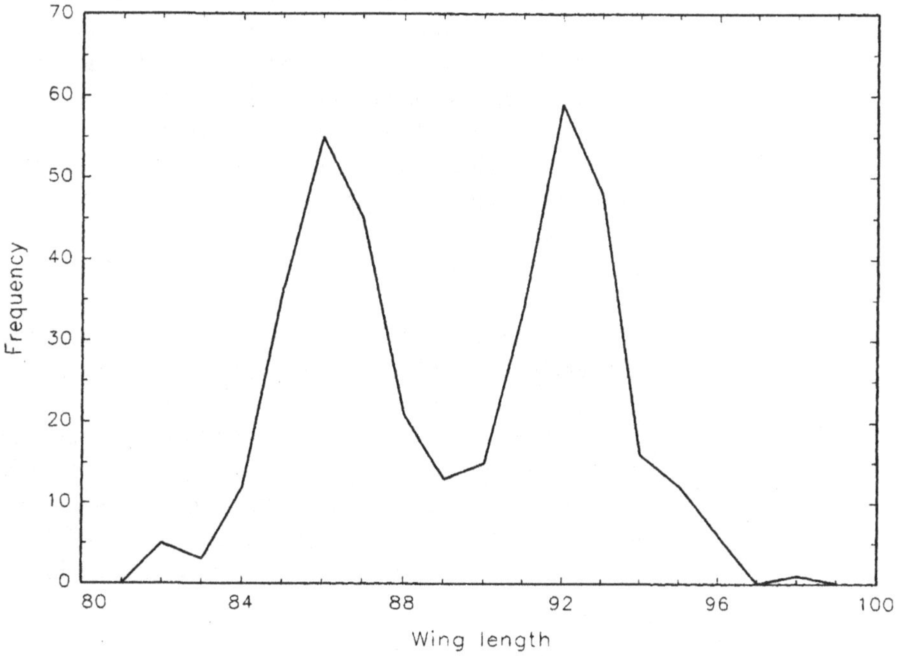

Figure 1-8
Frequency polygon for wing length of water pipits.

the uncertainty inherent in the allocation of any given specimen to the male or the female group, particularly for birds near the dip at wing length 89 to 90 millimeters.

A more sophisticated method for dealing with this classification problem is *mixture analysis*, an area of statistics which has made tremendous progress during the last 20 or so years. Mixture analysis assumes that the data come from two or more homogeneous populations (or rather, models), but it is not known from which of the populations each given observation originated. In the bird example, the populations would correspond to males and females of the species *Anthus Spinoletta*; it is a rather characteristic example for mixture analysis. In the midge example, if we were given only the values of wing length and antenna length, but *not* the species, we would again be in a mixture situation (but we might not be sure that there are really only two species represented in the data). Thus, mixture analysis is quite closely related to discriminant analysis, the crucial difference being that the information on group membership is missing in the mixture situation. We will later see that a convenient way to handle this situation mathematically is to think of group membership as a random variable. In the bird example, this means that we are really analyzing *two* variables, namely, X = gender and Y = wing length. We are given observations of Y only, but it is not possible to understand the distribution of Y without some knowledge of how Y depends on X. Thus, we "secretly" deal with a bivariate distribution, which justifies the inclusion of the bird example in this chapter. In the midge example, we could think of an analogous setup with *three* variables, namely, X = species, Y_1 = antenna length, and Y_2 = wing length.

As mentioned before, mixture analysis is more sophisticated than cluster analysis and other purely data-oriented techniques, more sophisticated in the sense that it requires a stronger theoretical background. In particular, mixture analysis is always based on some parametric model. In the bird example, this means that we do not get around specifying the distributional form for "wing length." For instance, we might assume that wing length of male water pipits follows a normal distribution, and wing length of female water pipits follows some other normal distribution, with different parameters. The problem, then is to estimate the parameters of *both* normal distributions without actually knowing for sure which observation is from which group. □

As outlined in the three examples discussed so far, the ultimate goal of this course is to enable the student to perform good multivariate statistical analyses. To achieve this goal, a fair amount of statistical theory will be needed. Of course, everyone who knows how to invoke a statistical software package can perform various kinds of multivariate tests, classification analyses, or principal component analyses, but, without a proper background, such a user of statistical software will hardly ever understand what the results mean. Thus, being familiar with the elementary statistical concepts underlying multivariate methods is crucial.

The student who puts enough effort into the theoretical parts of this course will also discover that the theory is much more than just a necessary evil. Rather, it should be taken as a stimulating challenge. Indeed, avoiding the theory means missing much of the fun. Many notions of multivariate statistics, even basic ones like marginal and conditional distributions, are full of surprises, and common sense is often misleading or plain wrong. To illustrate this point, let us discuss a final example, inspired by an article by D.J. Hand (1992). The example is purely fictitious but, nevertheless, relevant for real-life situations.

Example 1.4 A statistical paradox

A drug company has developed a new sleeping pill, and wishes to compare it to a standard medication. Three hundred human subjects are tested with the old and new drugs. After all data have been collected, two statisticians analyze the results independently. The data consist of number of hours slept (rounded to the nearest integer) for each of the 300 subjects. Statistician Marge reports as follows:

Old drug

Hours slept	# Subjects
0	100
2	100
4	100

New drug

Hours slept	# Subjects
1	100
3	100
5	100

Under the old drug, 100 subjects slept 0 hours, 100 slept 2 hours, and 100 slept 4 hours. Under the new drug, the frequencies were 100 each for 1, 3, and 5 hours of sleep. Marge concludes that the new drug is better. She argues for this conclusion in two ways: (i) the average duration of sleep was two hours under the old drug, and three hours under the new drug; (ii) if we give a randomly selected person the old drug, and another randomly selected person the new drug, the probability is 2/3 that the person with the new drug will sleep longer than the person taking the old drug.

Marge's fellow statistician Joyce reaches exactly the opposite conclusion. Joyce has compiled the following table of frequencies, based on the fact that both drugs were tested on all 300 subjects.

Hours slept under old drug	Hours slept under new drug	# Subjects
0	5	100
2	1	100
4	3	100

The same subjects who slept 0 hours under the old drug slept 5 hours under the new drug, and similarly for the second and third rows in the table of frequencies. Clearly, 200 of the 300 subjects slept *longer* with the old drug than with the new drug, and therefore Joyce concludes that the old drug is better. (It may be remarked at this point that several weeks passed between the two tests of each subject to eliminate possible carryover effects from the first to the second experiment).

Who is right, Marge or Joyce? It may be hard to accept that indeed *both* argue correctly. Let us repeat their arguments in slightly different words. Joyce argues that two out of three subjects sleep longer under the old drug than under the new drug. Marge argues that if we compare a subject who uses the old drug to a subject who uses the new drug, the odds are 2 to 1 that the person with the new drug will sleep longer. There is no contradiction between the two statements.

Interestingly, this seeming paradox and many other paradoxes arising in statistics are easy to understand once the proper theoretical tools have been developed. In this particular case, the proper tools are the notions of joint and marginal distributions, and stochastic independence, as outlined in Sections 2.2 to 2.4. We will return to Example 1.4 in Exercise 15 of Section 2.3, and in Exercise 14 of Section 2.4. □

Exercises for Chapter 1

1. This exercise is based on the midge data of Table 1.1.

   (a) With X = antenna length, Y = wing length, compute the value of $Z = X/Y$ for all 16 midges, and verify that the variable Z separates the two groups completely.

   (b) Consider the classification rule based on the two areas of classification

   $$D_1 = \{(x, y) : x/y > 0.7\} \qquad \text{(for } Af\text{)}$$

   and

   $$D_2 = \{(x, y) : x/y < 0.7\} \qquad \text{(for } Apf\text{).}$$

   Graph the boundary between the two areas of classification in a scatterplot of x vs. y (like Figure 1.1), and compare it to Figure 1.2.

2. Suppose the (x, y)-coordinate system in the Euclidean plane is rotated by an angle α into a new coordinate system (u, v). Then the point (x^*, y^*) in the old coordinate

system has coordinates

$$u^* = cx^* + sy^*$$

$$v^* = -sx^* + cy^*$$

in the new system, where $c = \cos(\alpha)$, $s = \sin(\alpha)$. In the midge example, consider a rotation by $-\pi/2 = -45$ degrees. Since $\cos(\pi/2) = \sin(\pi/2) = 1/\sqrt{2}$, the transformation is expressed by

$$u = (x - y)/\sqrt{2}$$

and

$$v = (x + y)/\sqrt{2}.$$

(a) Compute the coordinates u and v of the data of all 16 midges in the rotated coordinate system.

(b) Consider the (u, v)-coordinates of the 16 midges as two variables, and graph U vs V. Thus, verify (graphically) that V contains very little information about group differences, whereas variable U separates the two groups well.

(c) Can you relate the classification rule given in Example 1.1 to the (u, v)-coordinate system? In your plot of U vs V, what would the classification regions C_1 and C_2 look like?

3. This exercise is based on the head dimension data of Example 1.2.

(a) Get started with your favorite software by computing means and standard deviations for all six variables, as well as all 15 pairwise correlations between variables.

(b) Now consider only the two variables $X = LTG$ and $Y = LTN$. Rotate the (x, y)-coordinate system by 30 degrees, i.e., compute

$$u = cx + sy$$

and

$$v = -sx + cy,$$

where $c = \cos(\pi/6) \approx 0.866$, $s = \sin(\pi/6) = 0.5$, for all 200 observations.

(c) Compute means, variances, and standard deviations for the two variables U and V. Verify that the correlation between U and V is close to zero. Verify (numerically) also the following interesting fact: The sum of the variances of LTG and LTN is equal to the sum of the variances of U and V.

4. This exercise is based on the well-known *turtle carapace data* of Jolicoeur and Mosimann (1960). The data are given in Table 1.4. There are three variables,

L = carapace length,

W = carapace width,

and

H = carapace height,

all measured in *mm*, of 24 male and 24 female painted turtles (*Chrysemys picta marginata*).

(a) For both groups individually, compute means and standard deviations for all three variables, as well as all three correlations between pairs of variables. Based on these descriptive statistics, how could you describe the differences between carapaces of male and female turtles?

(b) For female turtles only, plot L vs. W.

(c) In part (b), the strong relationship between L and W suggests approximating the scatterplot by a straight line. You might consider (i) the regression line of W on L and (ii) the regression line of L on W. Compute these two lines, using software for least squares regression, and insert them into the scatterplot from part (b). Where do the two lines intersect? (*Note*: This part of the exercise assumes, of course, that you are familiar with simple linear regression.)

(d) Which of the two lines in part (c) would you prefer for the purpose of approximating the two-dimensional scatterplot by a straight line? Recall that the two regression lines result from least squares approximations where distance from the line is measured either horizontally or vertically. Can you think of a third way to measure distance from the line that would lead to a "compromise" between the two regression lines? How would you tackle the analogous problem with a three-dimensional scatterplot, i.e., if you want to approximate three-dimensional data by a single straight line? (*Note*: At this point, no "perfect" solution is expected. The purpose of part (d) is to get you started thinking about problems that we will solve later in Chapter 8).

(e) Construct a scatterplot of L vs. H for all 48 turtles, using different symbols for males and females. Find a rule (by simply drawing a straight boundary line in the scatterplot) that would allow you to classify a turtle as male or female, based on the measurements of L and H. How well would you expect your rule to perform? Do you think that classification based on both variables would be better than classification based on either L or H alone? (*Note*: Again, you are not expected to give a perfect solution. This type of problem will be discussed extensively in Chapters 5 and 7).

5. This exercise is based on the water pipit data of Example 1.3.

(a) Divide the data into two groups according to wing length ≤ 89, and wing length ≥ 90. Separately, for each group, compute the mean and the standard deviation of wing length.

(b) In part (a), one might argue that most birds in the first group are female, and most birds in the second group are male. Therefore, the means and variances computed in (a) should be valid estimates of the means and variances of wing length of female and male birds of the species *Anthus spinoletta*. Can you think of a reason why this argument is flawed? (*Hint*: Think of a situation where the distributions of the two groups overlap more than in this particular example.)

6. Suppose you play a game of chance with a friend (who may no longer be your friend after the game): Take three identical balls and label them as follows. The first ball is

Table 1.4 Turtle Carapace Data

Gender	Length	Width	Height
M	93	74	37
M	94	78	35
M	96	80	35
M	101	84	39
M	102	85	38
M	103	81	37
M	104	83	39
M	106	83	39
M	107	82	38
M	112	89	40
M	113	88	40
M	114	86	40
M	116	90	43
M	117	90	41
M	117	91	41
M	119	93	41
M	120	89	40
M	120	93	44
M	121	95	42
M	125	93	45
M	127	96	45
M	128	95	45
M	131	95	46
M	135	106	47
F	98	81	38
F	103	84	38
F	103	86	42
F	105	86	40
F	109	88	44
F	123	92	50
F	123	95	46
F	133	99	51
F	133	102	51
F	133	102	51
F	134	100	48
F	136	102	49
F	137	98	51
F	138	99	51
F	141	105	53
F	147	108	57
F	149	107	55
F	153	107	56
F	155	115	63
F	155	117	60
F	158	115	62
F	159	118	63

Table 1.4 *Continued*

Gender	Length	Width	Height
F	162	124	61
F	177	132	67

Source: Jolicoeur and Mosimann, 1960.

marked with the number '0' in red, and '5' in yellow. The second ball is marked '2' in red, and '1' in yellow. Finally, the third ball is marked '4' in red, and '3' in yellow. Put all three balls into an urn and mix well. Then you pick a color (red or yellow), and your ex-friend takes the other color. The person with the larger number will win, so you want to pick the color wisely. Which color should you choose under each of the rules (a) to (c)?

(a) The rules are that a single ball is taken at random from the urn, determining both the red and the yellow number.

(b) You take a ball at random and read the number of the color you chose. Then you put the ball back into the urn, mix well, and your friend chooses a ball at random and reads the number written in her color.

(c) Like (b), but your ball is not put back into the urn (and your former friend chooses only from among the two remaining balls).

(d) How is this exercise related to the contents of Chapter 1?

2 Joint Distribution of Several Random Variables

2.1 Introductory Remarks

Multivariate statistical methods are useful when they offer some advantage over a variable by variable approach. In Example 1.1, we saw that considering one variable at a time may not be optimal for classification purposes. Hence, it is important that we establish some terminology and acquire a basic knowledge of bivariate and multivariate distribution theory. In particular, we shall discuss notions such as independence of random variables, covariance and correlation, marginal and conditional distributions, and linear transformations of random variables. In the spirit of keeping this chapter on an elementary level, we shall restrict ourselves to the case of two variables most of the time, and outline the general case of $p \geq 2$ jointly distributed random variables only toward the end in Section 2.10.

It is assumed that the student who follows this course knows the basic notions of univariate statistics and probability. For instance, the first four chapters in Larson (1974) or the first five chapters in Ross (1994) would provide a good background. In particular, it is assumed that students know about discrete and continuous random variables, expected values, probability functions, density functions, and (cumulative) distribution functions.

Much of this chapter covers standard material that can be found in introductory statistics textbooks. An unusual feature of the chapter is Section 2.8, where finite mixtures of distributions are introduced as marginal distributions of a mixed, discrete - continuous, bivariate random variable.

For Sections 2.2 to 2.8 of this chapter, we shall use the notation (X,Y) to denote a two-dimensional (or bivariate) random variable, meaning that both X and Y are simultaneously measured on the same objects. For instance, we could have

$$X = \text{height}, \qquad Y = \text{weight}$$

of a person, or

$$X = \text{today's minimum temperature}$$
$$Y = \text{today's maximum temperature},$$

or

$$(X,Y) = \text{coordinates of the point where a dart hits the target.}$$

In contrast to regression analysis, the notation (X,Y) is not to imply an asymmetric treatment of the two variables. In regression, we distinguish between a dependent variable, usually Y, and an independent (or regressor) variable, usually X. In the current context, the labels X and Y are arbitrary, and we shall mostly concentrate on the *joint* distribution of X and Y rather than the prediction of Y, given X. Of course, there are strong relationships between bivariate distribution theory and regression analysis, which we will briefly treat in Section 2.7.

A little more formally than above, a bivariate random variable (X,Y) is defined as an assignment of ordered pairs $(x,y) \in \mathbb{R}^2$ to the elements of the sample space S of an experiment. As an example, suppose you draw a circle of radius one meter on your driveway (assuming the decimal system has finally taken over) and wait for rain. The experiment consists of marking the spot where the first raindrop falls inside the circle. Hence the sample space consists of all points inside the circle. Now suppose you paint a rectangular coordinate system on your driveway and label the axes as x and y. Then the coordinates of the first raindrop inside the circle constitute a two-dimensional random variable.

Another example is the simultaneous measurement of wing length (X) and antenna length (Y) of midges, as in Example 1.1. In this example it is crucial that both measurements are taken on the *same* objects. In this example and in most practical examples to follow, we will not write a sample space explicitly, but rather describe the pair of random variables (X,Y) directly in terms of its distribution.

As in the case of one-dimensional random variables, bivariate random variables are described by

- a *domain* D of values that (X,Y) can take, and
- either a probability function (for discrete variables) or a density function (for continuous variables).

The domain D consists of ordered pairs $(x,y) \in \mathbb{R}^2$. The number of elements in D may be finite or infinite, or D may even coincide with $\mathbb{R}^2$, the two-dimensional Euclidean space. Probability functions and density functions will be summarized under the notion *probability density function*, or *pdf*, for short.

2.2 Probability Density Function and Distribution Function of a Bivariate Random Variable

Consider first, bivariate random variables (X,Y) where both components are discrete. The *probability function* of (X,Y) is a function $f_{XY}(x,y)$ which assigns probabilities to all possible outcomes, i.e., to the ordered pairs (x,y) in the domain D. That is,

$$f_{XY}(x,y) = \Pr[X = x, Y = y],$$

where "Pr" is our acronym for *Probability*.
A probability function must satisfy the following properties:

(i) $f_{XY}(x,y) \geq 0$ for all $(x,y) \in D$

(ii) $\sum_{(x,y)\in D} f_{XY}(x,y) = 1.$

Let's consider some examples.

Example 2.2.1 Suppose you choose an integer at random between 0 and 2. (At random means here that each of the three numbers is chosen with equal probability). Denote the number by X. Then your friend chooses an integer Y at random between 0 and $X + 1$. This procedure defines a bivariate random variable (X, Y) with domain

$$D = \{(x, y) \in \mathbb{Z}^2 : x = 0, 1, 2;\ y = 0, \dots, x + 1\},$$

where $\mathbb{Z}$ is the set of integers. The joint probability function of X and Y can be represented by the table

		Y			
		0	1	2	3
	0	1/6	1/6	0	0
X	1	1/9	1/9	1/9	0
	2	1/12	1/12	1/12	1/12

Actually at this point the student who is not familiar with marginal and conditional distributions may not know how the joint probabilities can be obtained from the setup of the example; we will give a full justification in Section 2.6. However, for the moment we may take the above table as the definition of a probability function. It is then straightforward to show that conditions (i) and (ii) are satisfied. From the table of joint probabilities we can also calculate probabilities of certain events. For instance,

$$\Pr[X < Y] = \frac{1}{6} + \frac{1}{9} + \frac{1}{12} = \frac{13}{36},$$

$$\Pr[X > Y] = \frac{1}{9} + \frac{1}{12} + \frac{1}{12} = \frac{10}{36},$$

and the probability of a tie between X and Y is

$$\Pr[X = Y] = \frac{1}{6} + \frac{1}{9} + \frac{1}{12} = \frac{13}{36}.$$

Therefore, if this example represents a game in which the person with the higher number wins, then Y (the second player) has a higher probability to win. □

Example 2.2.2 Roll a red and a green die simultaneously, and set X = number shown by the red die, Y = number shown by the green die. Then conventional wisdom suggests the probability function

$$f_{XY}(x,y) = \begin{cases} 1/36 & \text{if } 1 \le x \le 6,\ 1 \le y \le 6, \\ 0 & \text{otherwise,} \end{cases}$$

where the domain D_{XY} consists of all ordered pairs of integers between 1 and 6. The reason we write D_{XY} instead of just D for the domain will be apparent from the next example. □

Example 2.2.3 In the same setup as the preceding example, let $Z = X + Y$ denote the sum of the numbers shown, and study the joint distribution of X and Z. The domain for (X,Z) is $D_{XZ} = \{(x,z) \in \mathbb{Z}^2 : 1 \le x \le 6,\ x+1 \le z \le x+6\}$. The joint probability function of (X,Z) is obtained as follows: for each $(x,z) \in D_{XZ}$,

$$f_{XZ}(x,z) = \Pr[X = x,\ Z = z]$$

$$= \Pr[X = x,\ X + Y = z]$$
$$= \Pr[X = x,\ Y = z - x]$$
$$= f_{XY}(x, z - x)$$
$$= 1/36.$$

Hence,

$$f_{XZ}(x,z) = \begin{cases} 1/36 & \text{if } 1 \leq x \leq 6,\ x + 1 \leq z \leq x + 6, \\ 0, & \text{otherwise.} \end{cases}$$

Although this function looks similar to f_{XY}, it is quite different from it because the domains are different. Perhaps, this is more evident if we write f_{XZ} in form of a table:

Z =	2	3	4	5	6	7	8	9	10	11	12
X = 1	c	c	c	c	c	c	0	0	0	0	0
= 2	0	c	c	c	c	c	c	0	0	0	0
= 3	0	0	c	c	c	c	c	c	0	0	0
= 4	0	0	0	c	c	c	c	c	c	0	0
= 5	0	0	0	0	c	c	c	c	c	c	0
= 6	0	0	0	0	0	c	c	c	c	c	c

Here, $c = 1/36$. Again, it is a simple exercise to check that conditions (i) and (ii) are satisfied. □

Now, we proceed to bivariate random variables where both components are continuous. Such random variables are described by a joint *density function* f_{XY} defined on the domain D, which satisfies the following properties:

(i) $f_{XY}(x,y) \geq 0$ for all $(x,y) \in D$.

(ii) $\displaystyle\int_{(x,y)\in D} f_{XY}(x,y)\,dy\,dx = 1.$

(iii) For any intervals $[a, b]$ and $[c, d]$, such that the rectangle $[a, b] \times [c, d]$ is inside D,

$$\Pr[a \leq X \leq b,\ c \leq Y \leq d] = \int_{x=a}^{b} \int_{y=c}^{d} f_{XY}(x,y)\,dy\,dx.$$

Condition (iii) can be somewhat simplified if we set $f_{XY}(x,y) = 0$ for all (x,y) outside the domain D. Then (iii) will hold for all rectangles in $\mathbb{R}^2$. More generally, it is required that, for any subset $A \subset \mathbb{R}^2$,

$$\Pr[(X,Y) \in A] = \int_{(x,y)\in A} f_{XY}(x,y)\,dy\,dx.$$

However, without giving a formal proof, we note that condition (iii) is sufficient to guarantee that $\Pr[(X,Y) \in A]$ can be computed by integration over A for any subset $A \subset \mathbb{R}^2$.

As in the univariate case, values of the density function are not to be interpreted as probabilities. In fact, each single point (x,y) has probability zero, because the integral over a rectangle of length zero vanishes. Thus, continuous random variables require a higher degree of mathematical sophistication, but they also offer the advantage of great simplicity, as will be clear from the examples.

Example 2.2.4 Let us return to the coordinate of the first raindrop inside the unit circle, as introduced in Section 2.1. An intuitively reasonable assumption would be that every point (x,y) inside the circle will be hit with the same probability – but what does that mean if there are infinitely many (actually even uncountably many) points inside the circle? What we mean by "equal probability" is the following. Let D = set of all points inside the circle, and denote by $A \subset D$ a subset of D. Then the probability that the first raindrop falls inside A should be proportional to the area of A. In turn, this implies that the density function must be constant in D. More formally, assume that the coordinate system has its origin at the center of the circle, then

$$D = \{(x,y) \in \mathbb{R}^2 : x^2 + y^2 \leq 1\},$$

and the density function can be defined as

$$f_{XY}(x,y) = \begin{cases} c & \text{if } x^2 + y^2 \leq 1, \\ 0 & \text{otherwise,} \end{cases}$$

where c is a positive constant. From condition (ii), it follows that $c = 1/\pi$. We shall say that (x,y) has a uniform distribution inside the circle.

Now, we can ask for probabilities of certain events, for instance: what is the probability that the first raindrop falls onto a spot at most $\frac{1}{2}$ meter from the center of the circle? Hence, let $A = \{(x,y) : x^2 + y^2 \leq \frac{1}{4}\}$, then the probability to be calculated is

$$\begin{aligned} \Pr[(X,Y) \in A] &= \Pr\left[X^2 + Y^2 \leq \frac{1}{4}\right] \\ &= \int_{(x,y)\in A} f_{XY}(x,y)\,dy\,dx \\ &= \int_{(x,y)\in A} \pi^{-1}\,dy\,dx = \frac{\pi/4}{\pi} = \frac{1}{4}. \end{aligned}$$

□

Example 2.2.5 This is an example where, perhaps, it is not obvious that all conditions imposed on the density function are satisfied. Consider the function

$$f_{XY}(x,y) = \begin{cases} \frac{1}{\sqrt{2\pi}} \exp[-\frac{x^2}{2} + x - y] & \text{if } x \leq y, \\ 0 & \text{if } x > y. \end{cases} \tag{1}$$

This density function (if it is actually a proper density) is positive in the area to the left and above the line $x = y$. Figure 2.2.1 illustrates the density function f_{XY} in the form of a contour plot. The curves in Figure 2.2.1 represent sets of points where the density takes a fixed constant value, like curves of equal altitude on a geographical map. It is often useful to think of bivariate densities as "mountains" rising out of a plain and to represent them either in the form of a contour plot or in a three-dimensional view. In this particular case, the "mountain" has a rather unusual shape, with a vertical wall along the line $x = y$. The maximum (i.e., the "top of the mountain") is reached at $x = y = 0$, and the volume of the "mountain" is 1, as required by condition (ii).

This is an example without any practical purpose, but it will be useful in later sections to illustrate concepts, such as conditional distribution and conditional expectation. □

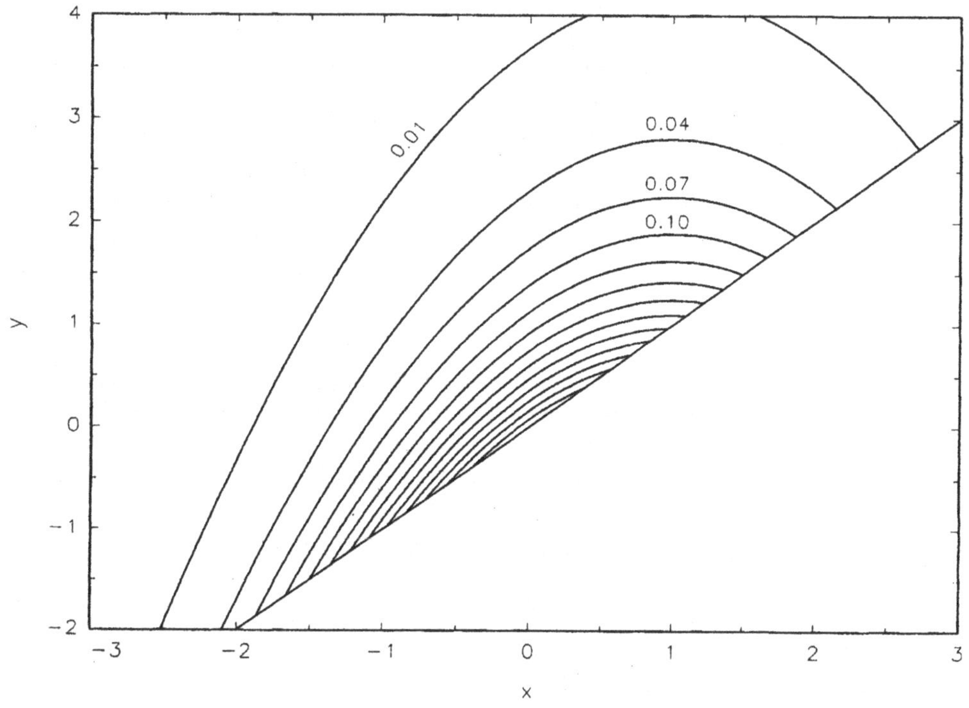

Figure 2-2-1 Contour plot of the density function in Example 2.2.5. Curves are sets of points on which the density function takes a constant value c, for $c = 0.01, 0.03, 0.05, 0.1, 0.15, 0.2$, and 0.3.

Example 2.2.6 Perhaps the simplest of all bivariate random variables are those with a uniform distribution inside a rectangle. Let $a < b,\ c < d$ denote four real constants, and define

$$f_{XY}(x,y) = \begin{cases} \dfrac{1}{(b-a)(d-c)} & \text{if } a \le x \le b, c \le y \le d \\ 0 & \text{otherwise.} \end{cases} \tag{2}$$

It is easy to check that f_{XY} is, indeed, a density function. In the simplest case, $a = c = 0$, $b = d = 1$, that is, a uniform distribution on the unit square. Then

$$f_{XY}(x,y) = \begin{cases} 1 & \text{if } 0 \le x \le 1, 0 \le y \le 1, \\ 0 & \text{otherwise} \end{cases}$$

which constitutes a generalization of the one-dimensional uniform distribution on the interval [0,1]. □

Example 2.2.7 In the spirit of a course on parametric multivariate statistics, we do not get around the good old normal distribution. Without (for the moment) any proofs—they are postponed to Chapter 3—consider the following density function:

$$f_{XY}(x, y) = \frac{1}{2\pi\sigma_1\sigma_2(1-\rho^2)^{1/2}} \times$$
$$\exp\left\{-\frac{1}{2(1-\rho^2)}\left[\left(\frac{x-\mu_1}{\sigma_1}\right)^2 - 2\rho\frac{(x-\mu_1)(y-\mu_2)}{\sigma_1\sigma_2} + \left(\frac{y-\mu_2}{\sigma_2}\right)^2\right]\right\}, \quad (x,y) \in \mathbb{R}^2, \tag{3}$$

where $\mu_1,\ \mu_2,\ \sigma_1,\ \sigma_2$, and ρ are fixed numbers, or parameters. This frightening formula is the density function of a bivariate normal distribution, and one may wonder how complicated things will get in the case of $p > 2$ random variables. Do not fear. First we shall introduce appropriate tools from matrix algebra, which will make things much simpler. For the moment, let us just remark that three of the parameters are subject to constraints, namely, $\sigma_1 > 0$, $\sigma_2 > 0$, and $|\rho| < 1$. A particularly simple form of (3) results if $\mu_1 = \mu_2 = \rho = 0$ and $\sigma_1 = \sigma_2 = 1$. Then

$$f_{XY}(x,y) = (2\pi)^{-1}\exp\left[-(x^2+y^2)/2\right], \tag{4}$$

which is recognized as the product of two univariate standard normal densities. To understand (3) better, we will have to come up with appropriate interpretations for all parameters involved.

Figure 2.2.2 gives two graphical representations of the bivariate normal density with parameter values $\mu_1 = \mu_2 = 0,\ \sigma_1 = \sigma_2 = 1$, and $\rho = 0.7$. Part (a) is a contour plot, illustrating contours of equal density

$$f_{XY}(x,y) = c$$

for various values of the constant c. Part (b) is a three-dimensional view of the "normal mountain," showing the well-known bell shape. □

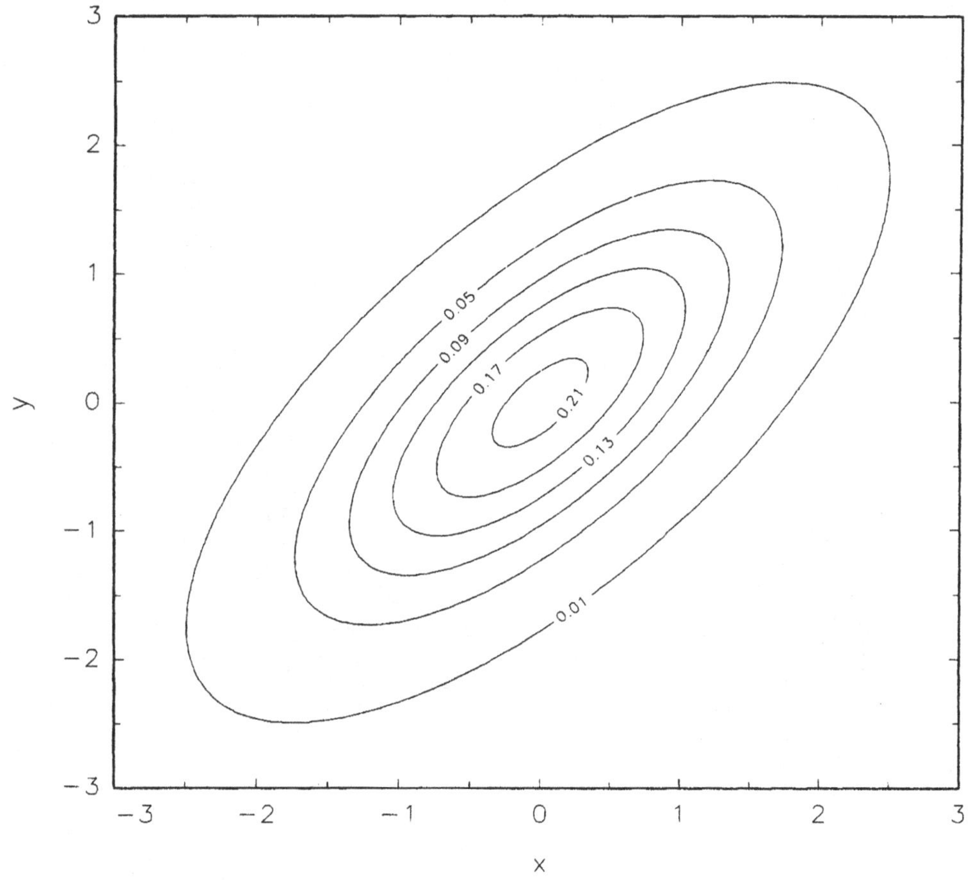

Figure 2-2-2A

The *distribution function* (sometimes called *cumulative distribution function*, or *cdf*, for short) of a bivariate random variable (X,Y), is defined as

$$F_{XY}(x,y) = \Pr[X \leq x,\ Y \leq y] \qquad \text{for all } (x,y) \in \mathbb{R}^2.$$

Hence, for discrete random variables,

$$F_{XY}(x,y) = \sum_{a \leq x} \sum_{b \leq y} f_{XY}(a,b). \tag{5}$$

Typically, this is quite a complicated step function, as you may verify by computing it for various values of (x,y) in any of the discrete examples. Hence, for all practical purposes, the probability function is usually preferred. However, the distribution function is useful for deriving marginal and conditional probability functions, as we shall see later, and it is a very important theoretical concept.

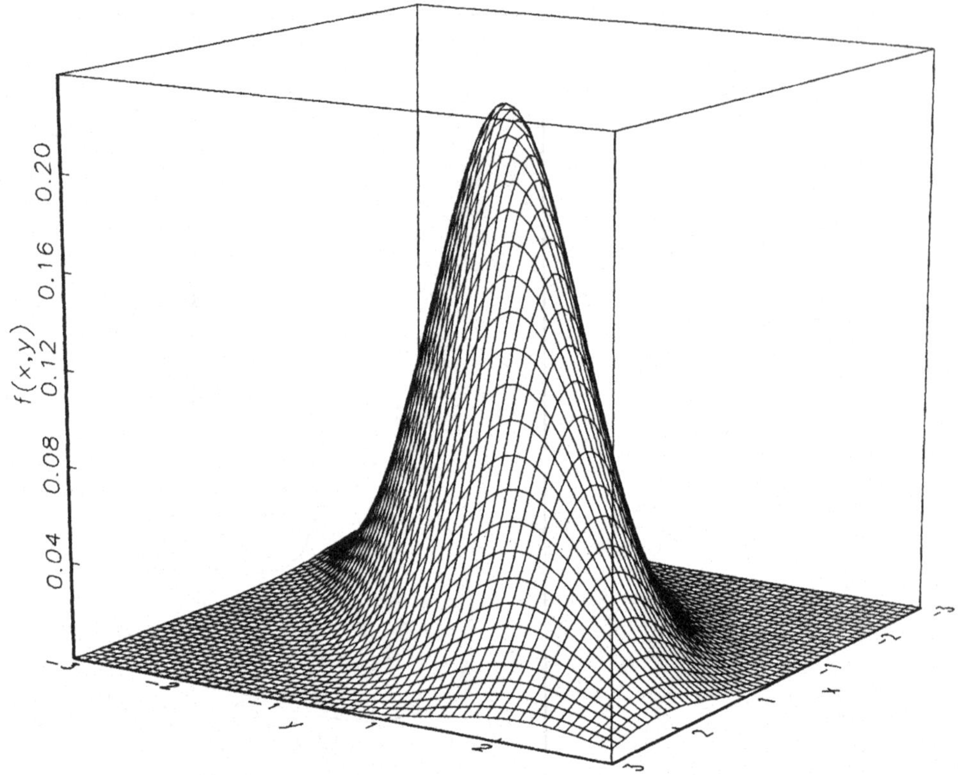

Figure 2-2-2B
Density function of a bivariate normal distribution; (a) contour plot, (b) three-dimensional view. The parameters of the normal distribution are $\mu_1 = \mu_2 = 0$, $\sigma_1 = \sigma_2 = 1$, and $\rho = 0.7$.

In the continuous case, the analog of (5) is

$$F_{XY}(x,y) = \int_{u=-\infty}^{x} \int_{v=-\infty}^{y} f_{XY}(u,v)\, dv\, du. \tag{6}$$

As in the discrete case, one usually prefers to work with density functions, but it may be instructive, anyway, to consider a simple example.

Example 2.2.8 continuation of Example 2.2.4

Recall the density function

$$f_{XY}(x,y) = \begin{cases} 1/\pi & \text{if } x^2 + y^2 \le 1, \\ 0 & \text{otherwise.} \end{cases}$$

Figure 2.2.3 shows two particular choices of (x,y). In each case, the value of the distribution function is given by

$$F_{XY}(x,y) = \frac{1}{\pi} \times \text{(shaded area)}.$$

In particular, if $x \geq 1$ and $y \geq 1$, then the entire unit circle is to the left and below (x,y), and, for all such points, the value of F_{XY} is 1. We will return to this example in Exercise 14. □

Example 2.2.9 continuation of Example 2.2.6

Consider, again, the uniform distribution inside the unit square, given by the density function

$$f_{XY}(x,y) = \begin{cases} 1 & \text{if } 0 \leq x \leq 1,\ 0 \leq y \leq 1, \\ 0 & \text{otherwise.} \end{cases}$$

We leave it to the student (see Exercise 13) to verify that the distribution function F_{XY} is given as in Figure 2.2.4, where the Euclidean plane is partitioned into several areas. In each of the areas, the value of F_{XY} is indicated. Note also that this distribution function is continuous everywhere. For the student who has difficulty understanding this example, it may be useful to construct a contour plot of F_{XY}. □

Given the distribution function of a random variable, it is always possible to reconstruct its probability density function. In the discrete case, the nonzero probabilities "sit" at points where F_{XY} jumps from one value to the next. In the continuous

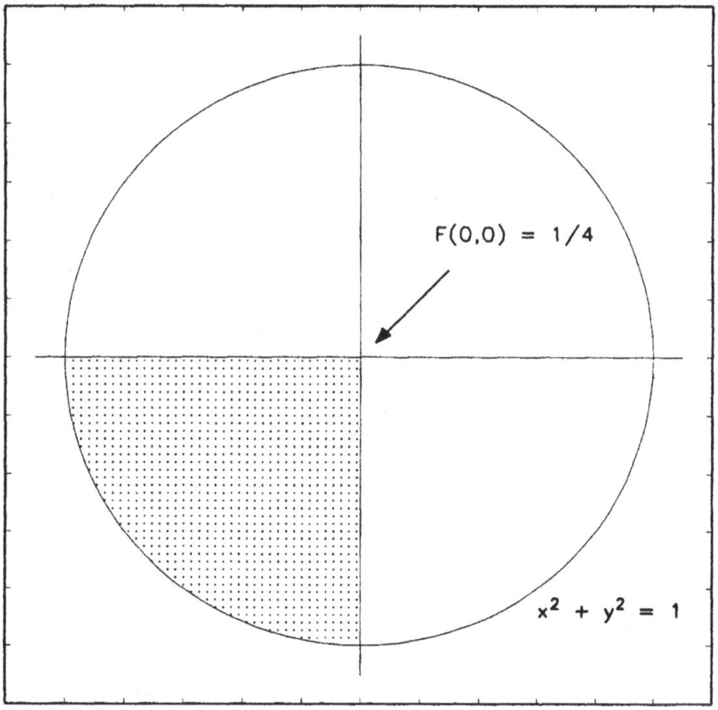

Figure 2-2-3A

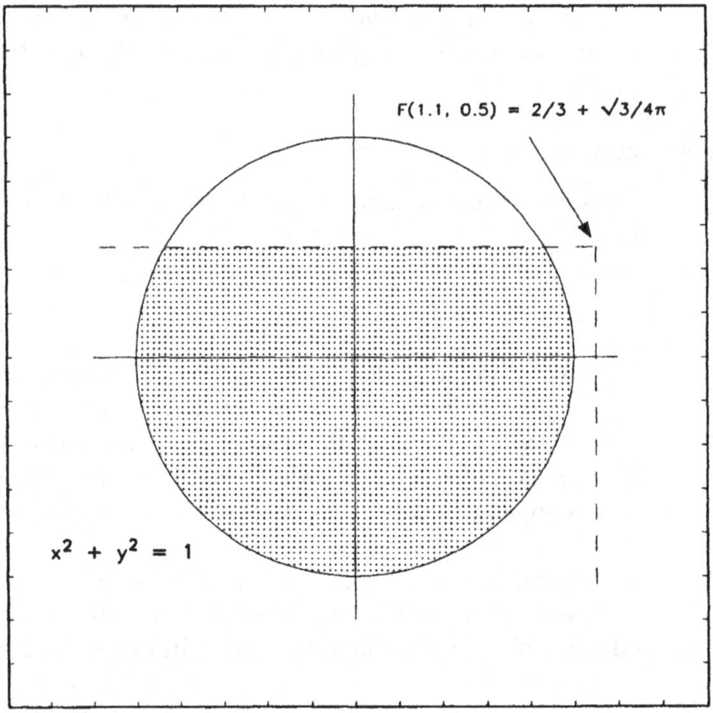

Figure 2-2-3B Values of the distribution function for two choices of (x, y) in the raindrop example.

case, it follows from (6) that

$$\frac{\partial^2 F_{XY}}{\partial x \partial y} = f_{XY}(x, y).$$

It also follows from the definition of the distribution function that

(i) $F_{XY}(-\infty, -\infty) = 0$,

(ii) $F_{XY}(\infty, \infty) = 1$,

and

(iii) $F_{XY}(x, y)$ is a nondecreasing function of x and y.

We shall use distribution functions only rarely in this book. However, as the preceding short discussion of distribution functions has pointed out, they are, indeed, a powerful concept for handling random variables, making the distinction into the discrete and the continuous case somewhat redundant.

Exercises for Section 2.2

1. Verify that the table in Example 2.2.1 defines a proper probability function.

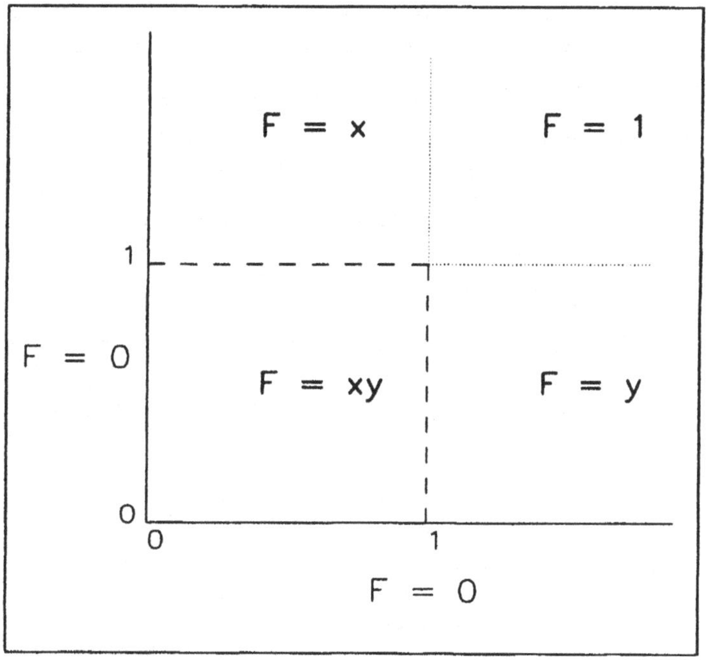

Figure 2-2-4 Distribution function $F_{XY}(x, y)$ of the uniform distribution inside the unit square.

2. Verify that the table in Example 2.2.3 defines a proper probability function.

3. Using the same technique as in Example 2.2.3, find the joint probability function of (X,U), where $U = X - Y$, and the joint probability function of (X,V), where $V = X + 2Y$.

4. Using the probability function in Example 2.2.2, find the joint probability function of X and Z, where $Z = |X - Y|$.

5. Suppose that the continuous bivariate random variable (X,Y) has density function

$$f_{XY}(x,y) = \begin{cases} c & \text{if } x \geq 0;\ y \geq 0;\ x + y \leq 1, \\ 0 & \text{otherwise.} \end{cases}$$

What is the value of the constant c?

6. In Example 2.2.4, find
   (a) $\Pr[X \leq 0, Y \leq 0]$,
   (b) $\Pr[\,|XY| > 0]$, and
   (c) $\Pr[X = Y]$.

7. In Example 2.2.4, define a function $h(u)$ for $u \in \mathbb{R}$ as

$$h(u) = \Pr[X^2 + Y^2 \leq u].$$

Find an analytical expression (a formula) for $h(u)$ and graph $h(u)$ in the range $-1 \leq u \leq 2$.

8. Find and graph the function $h(u) = \Pr[X^2 + Y^2 \leq u]$ for the bivariate uniform distribution on the unit square (see Example 2.2.6).

9. Show that the function f_{XY} in Example 2.2.5 is a proper density function.

10. Show that equation (4) defines a proper bivariate density function, using a transformation to polar coordinates $[x = r\cos(\theta),\ y = r\sin(\theta)]$.

11. For the special case $\rho = 0$, prove that (3) is a joint density function of two continuous random variables.

12. Find the distribution function for the bivariate random variable of Example 2.2.1.

13. Find the distribution function for the bivariate uniform distribution inside the unit square, and, thus, verify the results given in Figure 2.2.4. Construct a contour plot of the *cdf* $F_{XY}(x, y)$ for $-0.5 \leq x, y \leq 1.5$, using contour levels 0.1, 0.3, 0.5, 0.7, and 0.9.

14. Find the distribution function for the uniform distribution inside the unit circle (see Example 2.2.8).

15. Verify that the function

$$f_{XY}(x,y) = \begin{cases} (x+y)/36 & \text{if } x = 1, 2, 3;\ y = 1, 2, 3, \\ 0 & \text{otherwise} \end{cases}$$

is a proper probability function of a bivariate discrete random variable. Represent the probability function in form of a table.

16. In Exercise 15, find the distribution function.

17. Verify that the function

$$f_{XY}(x,y) = \begin{cases} x+y & \text{if } 0 \leq x \leq 1,\ 0 \leq y \leq 1, \\ 0 & \text{otherwise} \end{cases}$$

is a proper density function of a bivariate continuous random variable.

18. Find the distribution function for the density in Exercise 17.

19. A bivariate *exponential* random variable. Let $\lambda > 0$ and $\mu > 0$ denote two real constants, and define a bivariate continuous random variable (X, Y) by the density function

$$f_{XY}(x,y) = \begin{cases} \dfrac{1}{\lambda\mu} \exp\left[-\left(\dfrac{x}{\lambda} + \dfrac{y}{\mu}\right)\right] & \text{if } x \geq 0;\ y \geq 0, \\ 0 & \text{otherwise.} \end{cases}$$

Show that f_{XY} is a proper density function, and graph it as a contour plot.

20. A bivariate *double-exponential* random variable. Show that the function

$$f_{XY}(x,y) = \frac{1}{4}\exp[-(|x| + |y|)] \quad (x,y) \in \mathbb{R}^2$$

is a density function of a bivariate random variable, and graph f_{XY} as a contour plot.

21. Figure 2.2.5 shows contours of constant values of the function

$$f_{XY}(x,y) = \begin{cases} \frac{1}{8}(x^2 - y^2)e^{-x} & \text{if } x > 0,\ -x \le y \le x, \\ 0 & \text{otherwise.} \end{cases}$$

Show that f_{XY} is a proper density function. *Note*: Verifying condition (ii) requires knowledge of partial integration. Being familiar with gamma functions would help.

22. Show that the function

$$f_{XY}(x,y) = \begin{cases} \frac{1}{2}e^{-x} & \text{if } x > 0, \quad -x \le y \le x, \\ 0 & \text{otherwise} \end{cases}$$

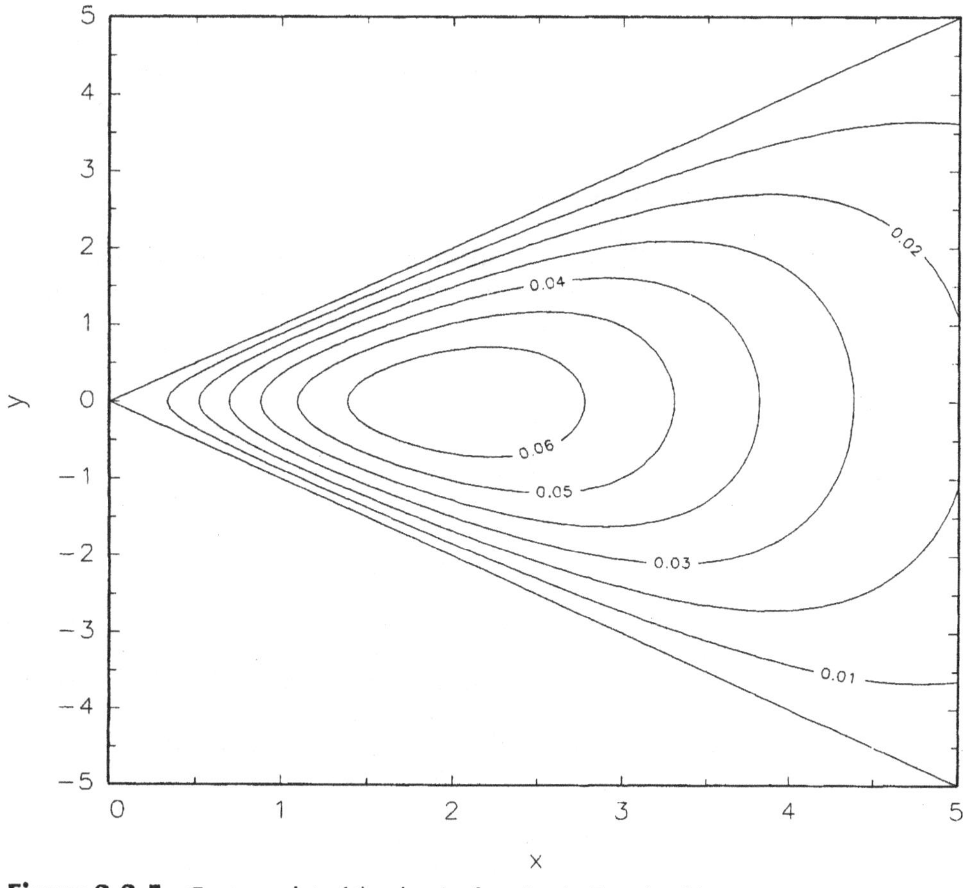

Figure 2-2-5 Contour plot of the density function in Exercise 21.

is a density function. Graph the density function as a contour plot in the range $0 \leq x \leq 5$, $-5 \leq y \leq 5$. Repeat the exercise for

$$f_{XY}(x, y) = \begin{cases} \frac{1}{2x}e^{-x} & \text{if } x > 0, \quad -x \leq y \leq x, \\ 0 & \text{otherwise} \end{cases}$$

(but notice that the latter function has an infinite peak at $x = y = 0$).

23. A bivariate *Pareto* variable. Show that the function

$$f_{XY}(x, y) = \begin{cases} \frac{2}{(1 + x + y)^3} & \text{if } x, y \geq 0, \\ 0 & \text{otherwise} \end{cases}$$

is a density function. Graph the density as a contour plot for $0 \leq x, y \leq 5$.

24. Show that the function

$$f_{XY}(x,y) = \frac{1}{2\sqrt{2\pi}} \exp\left[-\frac{x^2}{2} - |y - x|\right], \qquad (x, y) \in \mathbb{R}^2$$

is a density function of a bivariate continuous random variable and graph the density as a contour plot for $-5 \leq x, y \leq 5$. *Hint:* Use Example 2.2.5.

2.3 Marginal Distributions

Given the distribution of a bivariate random variable (X,Y), we shall often be interested in finding the distribution of X or the distribution of Y, ignoring the other component. These are called the *marginal* distributions—perhaps, because the probability functions in the discrete case appear in the margins of the table of probabilities, as we shall see very soon.

Based on the joint distribution function F_{XY} of (X,Y), the marginal distribution function of X is given by

$$\begin{aligned} F_X(a) &= \Pr[X \leq a] \\ &= \Pr[X \leq a, \ Y < \infty] \qquad (1) \\ &= F_{XY}(a, \infty). \end{aligned}$$

If (X,Y) is continuous, with density function f_{XY}, then

$$F_X(a) = \int_{-\infty}^{a} \int_{-\infty}^{\infty} f_{XY}(x,y)\, dy\, dx. \qquad (2)$$

Setting

$$f_X(x) = \int_{-\infty}^{\infty} f_{XY}(x,y)\, dy, \qquad (3)$$

it follows that f_X is the density function of X. We leave it as an exercise to verify that f_X satisfies all the properties of a density function.

Similarly, it follows that the marginal distribution function and density function of Y are given by

$$F_Y(a) = F_{XY}(\infty, a)$$

and

$$f_Y(y) = \int_{-\infty}^{\infty} f_{XY}(x,y)\,dx,$$

respectively. Thus, marginal densities are obtained by integrating out the "other" variable.

The analogous formulas for the discrete case are found either by the distribution function or simply by an enumeration of all elements in the sample space that give the same value of X, say. Thus, the marginal probability functions are

$$f_X(x) = \Pr[X = x] = \sum_y f_{XY}(x,y)$$

and

$$f_Y(y) = \Pr[Y = y] = \sum_x f_{XY}(x,y), \tag{4}$$

where the summation is over all values that X or Y can take. In the continuous case, integration, as in equation (3), takes the place of the sum in (4).

Example 2.3.1 continuation of Example 2.2.1

The marginal probability functions are obtained by computing row sums and column sums in the table of joint probabilities. Hence, the marginal probability functions are

$$f_X(x) = \begin{cases} 1/3 & \text{if } x = 0, 1, 2 \\ 0 & \text{otherwise,} \end{cases}$$

and

$$f_Y(y) = \begin{cases} 13/36 & \text{if } y = 0 \text{ or } 1, \\ 7/36 & \text{if } y = 2, \\ 3/36 & \text{if } y = 3, \\ 0 & \text{otherwise.} \end{cases}$$

□

Example 2.3.2 continuation of Example 2.2.3

The marginal probability function of X assigns probability $1/6$ to each number from 1 to 6. The marginal probability function of Z is given in Figure 2.3.1; it has a symmetric triangular shape. Note that this is the probability function for the sum of two dice, assuming "fair play." See also Exercise 6. □

Example 2.3.3 continuation of Example 2.2.7

Again, take the density function of the bivariate normal distribution, in its simplest form with parameters $\mu_1 = \mu_2 = 0$, $\sigma_1 = \sigma_2 = 1$, and $\rho = 0$. Then

$$f_{XY}(x,y) = (2\pi)^{-1}\exp[-(x^2 + y^2)/2]$$
$$= \frac{1}{\sqrt{2\pi}}e^{-x^2/2} \cdot \frac{1}{\sqrt{2\pi}}e^{-y^2/2} = f_X(x) \cdot f_Y(y),$$

where both f_X and f_Y denote standard normal density functions. From integrating the product of two univariate standard normal densities over either of the variables, it follows that both marginals correspond to standard normal distributions. □

Deriving the marginals of the general bivariate normal (with five parameters) is technically much more complicated; see Exercise 11.

In the continuous case, things also tend to become a little involved when the domain on which the density function is positive has finite bounds, as in our Examples 2.2.4 and 2.2.6. This is illustrated now.

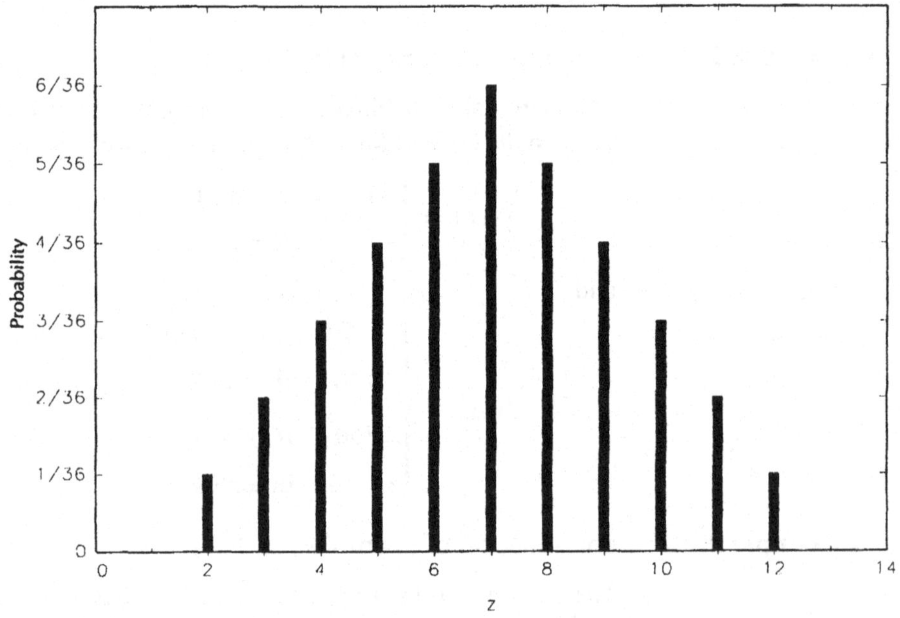

Figure 2-3-1 Probability function of the sum of numbers shown by two dice.

Example 2.3.4 continuation of Example 2.2.4

Recall that the density function of the uniform distribution inside the unit circle is given by

$$f_{XY}(x,y) = \begin{cases} 1/\pi & \text{if } x^2+y^2 \le 1, \\ 0 & \text{otherwise.} \end{cases}$$

To find the marginal density f_X, we need to integrate f_{XY} with respect to y. To make the following argument as clear as possible, suppose that we take a fixed value of x, say x_0, and find the value of the marginal density function f_X at x_0. We need to distinguish between two cases.

Case 1: $|x_0| > 1$. Then $f_{XY}(x_0, y) = 0$ for all $y \in \mathbb{R}$, and hence,

$$f_X(x_0) = \int_{-\infty}^{\infty} 0\, dy = 0.$$

Case 2: $|x_0| \le 1$. Then

$$f_{XY}(x_0,y) = \begin{cases} \dfrac{1}{\pi} & \text{for } -\sqrt{1-x_0^2} \le y \le \sqrt{1-x_0^2}, \\ 0 & \text{for } |y| > \sqrt{1-x_0^2}. \end{cases}$$

This is illustrated in Figure 2.3.2. In this graph, integrating the joint density f_{XY} over y at $x = x_0$ can be thought of as "adding" along the vertical line at $x = x_0$. Hence,

$$\begin{aligned} f_X(x_0) &= \int_{-\infty}^{\infty} f_{XY}(x_0,y)dy \\ &= \int_{y=-\infty}^{-\sqrt{1-x_0^2}} f_{XY}(x_0, y)\, dy + \int_{y=-\sqrt{1-x_0^2}}^{\sqrt{1-x_0^2}} f_{XY}(x_0, y)\, dy + \int_{y=\sqrt{1-x_0^2}}^{\infty} f_{XY}(x_0, y)\, dy \\ &= \int_{y=-\infty}^{-\sqrt{1-x_0^2}} 0\, dy + \int_{y=-\sqrt{1-x_0^2}}^{\sqrt{1-x_0^2}} \frac{1}{\pi} dy + \int_{y=\sqrt{1-x_0^2}}^{\infty} 0\, dy \\ &= \frac{2}{\pi}\sqrt{1-x_0^2}. \end{aligned}$$

This holds for all x_0 in $[-1, 1]$, and hence, the marginal density of X is given by

$$f_X(x) = \begin{cases} \dfrac{2}{\pi}\sqrt{1-x^2} & \text{if } |x| \le 1, \\ 0 & \text{otherwise.} \end{cases}$$

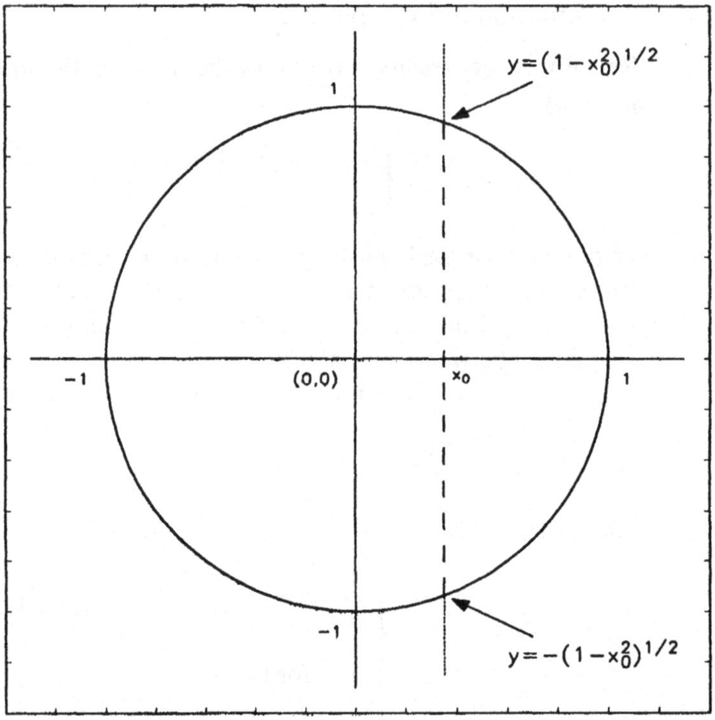

Figure 2-3-2
Finding the marginal density of X in the raindrop example.

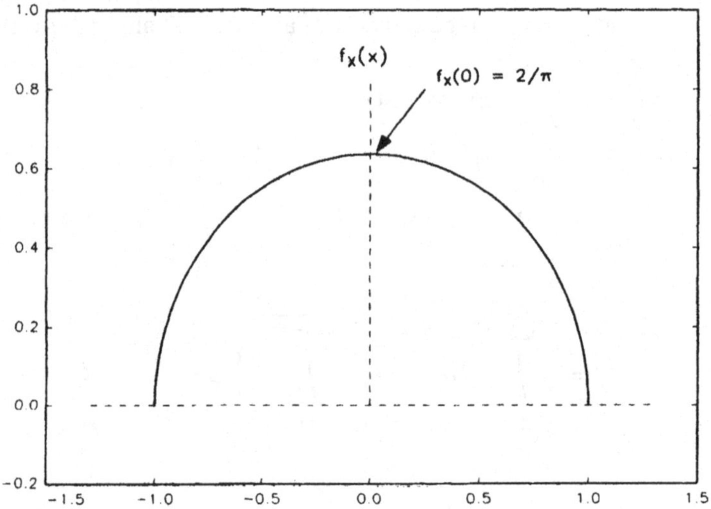

Figure 2-3-3
Marginal density function of one coordinate in the raindrop example.

This density is graphed in Figure 2.3.3. A similar result holds for the marginal distribution of Y. □

Example 2.3.5 Consider a pair of continuous random variables with joint *pdf*

$$f_{XY}(x, y) = \begin{cases} \frac{1}{8}(x^2 - y^2)e^{-x} & \text{if } x \geq 0, |y| \leq x \\ 0 & \text{otherwise,} \end{cases}$$

which has been graphically illustrated in Figure 2.2.5 (p. 40). To find the marginal *pdf* of X, notice that $f_X(x) = 0$ for $x < 0$. For $x \geq 0$, we use the same technique as in the preceding example to obtain

$$\begin{aligned} f_X(x) &= \int_{-\infty}^{\infty} f_{XY}(x, y)\, dy \\ &= \int_{-\infty}^{-x} f_{XY}(x, y)\, dy \quad + \int_{-x}^{x} f_{XY}(x, y)\, dy \quad + \int_{x}^{\infty} f_{XY}(x, y)\, dy \\ &= \int_{y=-x}^{x} \frac{1}{8}(x^2 - y^2)e^{-x}\, dy \\ &= \frac{1}{8}e^{-x} \int_{-x}^{x} (x^2 - y^2)\, dy \\ &= \frac{1}{6}x^3 e^{-x}. \end{aligned}$$

Hence, we obtain the marginal *pdf* of X as

$$f_X(x) = \begin{cases} \frac{1}{6}x^3 e^{-x} & \text{if } x \geq 0 \\ 0 & \text{otherwise.} \end{cases}$$

The advanced student will recognize this as a so-called *gamma* density function. The density f_X is zero at $x = 0$, increases monotonically to a maximum at $x = 3$, and then decreases again; see Exercise 16. □

Exercises for Section 2.3

1. Find the marginal probability functions $f_X(x)$ and $f_Y(y)$ for the bivariate discrete random variable (X,Y) with joint probability function

$$f_{XY}(x,y) = \begin{cases} (x + y)/36 & \text{if } x = 1, 2, 3 \;\; y = 1, 2, 3, \\ 0 & \text{otherwise,} \end{cases}$$

(see Exercise 15 in Section 2.2, p. 36).

2. Find the marginal density functions $f_X(x)$ and $f_Y(y)$ for the continuous random variables X and Y with joint density

$$f_{XY}(x,y) = \begin{cases} x+y & \text{if } 0 \le x \le 1,\, 0 \le y \le 1, \\ 0 & \text{otherwise.} \end{cases}$$

3. Define a joint probability function of two discrete random variables by

$$f_{XY}(x,y) = \begin{cases} \dfrac{1}{36} & \text{if } x = 1,\ldots,6,\ y = 1,\ldots,6, \\ 0 & \text{otherwise,} \end{cases}$$

as in Example 2.2.2 and in Exercise 3 of Section 2.2. Let $U = X - Y$ and $V = X + 2Y$. Find the marginal probability functions of U and V.

4. In the same setup as Exercise 3, find the marginal probability function of the random variable $Z = |X - Y|$.

5. If (X,Y) has density function

$$f_{XY}(x,y) = \begin{cases} c & \text{if } x \ge 0;\ y \ge 0;\ x + y \le 1, \\ 0 & \text{otherwise,} \end{cases}$$

find the constant c and the marginal density functions f_X and f_Y.

6. Max the Magician has a pair of dice (X,Y) that show the numbers 1 to 6 with the following joint probabilities (entries in the table are multiples of $1/36$):

		X=					
		1	2	3	4	5	6
	1	1	2	0	3	0	0
	2	0	0	0	0	3	3
Y =	3	3	0	1	1	0	1
	4	1	3	1	1	0	0
	5	1	1	1	0	2	1
	6	0	0	3	1	1	1

You are allowed to watch Max's right hand roll die X, and your friend is allowed to watch the left hand roll die Y. Each of you tallies the frequencies with which the respective die shows the numbers 1 to 6, but you are not allowed to combine the results. Will you or your friend be able to detect that something is "magic" about the dice? (In plain English, this exercise asks you to find the marginal probability functions f_X and f_Y, and compare them to the marginal probability functions obtained from Example 2.2.2).

7. Kopy the Kat has figured out that something is fishy about Max's dice (see Exercise 6) and wants to construct her own set of dice with the properties that the marginal distributions of X and Y are those of regular dice, but the joint probabilities are different from $1/36$. Help her find an example.

8. Returning to Max the Magician (Exercise 6), a third friend is joining your efforts by tallying the *sum* of the numbers shown by the two dice, i.e., by studying the distribution of $Z = X + Y$. Does the distribution of the sum help you to detect that something is "magic" about the dice? (Translation: Find the marginal probability function of Z, and compare it to the probability function of Z in Example 2.3.2).

9. In Exercise 8, if the third friend were to observe the difference $U = X - Y$, would this help you to detect the magic property of the dice?

10. Kopy the Kat (see Exercise 7) would like to construct her dice such that even a *fourth* friend who is allowed to study $U = X - Y$ cannot find out that something is magic. Help her construct such a pair of dice, or prove that she is asking for too much. (In formal language: Find a joint probability function f_{XY} of two discrete variables such that the marginal probability functions of $X, Y, Z = X + Y$, and $U = X - Y$ behave as if f_{XY} assigned probability $1/36$ to each pair of values (x, y), $x, y = 1, \ldots, 6$, but f_{XY} is *not* a discrete uniform probability function).

11. For the bivariate normal distribution with $\mu_1 = \mu_2 = 0$, $\sigma_1 = \sigma_2 = 1$, and $-1 < \rho < 1$ [see equation (3) of Section 2.2], prove that both marginal distributions are standard normal. *Hint*: Verify that

$$\begin{aligned} f_{XY}(x,y) &= \frac{1}{2\pi(1-\rho^2)^{1/2}} \exp\left[-\frac{1}{2(1-\rho^2)}(x^2 - 2\rho xy + y^2)\right] \\ &= \frac{1}{\sqrt{2\pi}} \exp[-x^2/2] \cdot \frac{1}{\sqrt{2\pi}\sqrt{1-\rho^2}} \exp\left[-\frac{1}{2} \cdot \frac{(y-\rho x)^2}{1-\rho^2}\right], \end{aligned}$$

and integrate over y to find the marginal density of X.

12. Let X denote a discrete random variable that takes values $-10, -9, \ldots, 9, 10$ with probability $1/21$ each, and let c denote a fixed number, $0 \le c \le 10$. (For instance, take $c = 7.5$). Define a new random variable

$$Y = \begin{cases} X & \text{if } |X| \le c, \\ -X & \text{if } |X| > c. \end{cases}$$

Find the joint probability function of (X, Y) and the marginal probability function of Y.

13. This is a continuous version of Exercise 12. Let X denote a random variable with uniform distribution in $[-1, 1]$, i.e., with density

$$f_X(x) = \begin{cases} \frac{1}{2} & \text{if } -1 \le x \le 1, \\ 0 & \text{otherwise.} \end{cases}$$

Let c denote a real constant ($0 \le c \le 1$), and define

$$Y = \begin{cases} X & \text{if } |X| \le c, \\ -X & \text{if } |X| > c. \end{cases}$$

Find the density function of Y. (*Note*: This is somewhat trickier than Exercise 12, because the pair of random variables (X,Y) does *not* have a density function. Hence, try to find the distribution function F_Y of Y directly. A graph of the function

$$h(x) = \begin{cases} x & \text{if } |x| \le c, \\ -x & \text{if } |x| > c \end{cases}$$

should help you to understand this example.)

14. Same as Exercise 13, but under the much weaker condition that X has a density function symmetric to 0, i.e.,

$$f_X(x) = f_X(-x) \text{ for all } x \in \mathbb{R},$$

and $c > 0$ is an arbitrary constant. (For instance, a normal random variable with mean 0 satisfies this condition.) Show that Y has the same distribution as X, whatever the constant c is.

15. Consider the following joint probability function of two discrete random variables X and Y:

		X = 1	X = 3	X = 5
Y =	0	0	0	1/3
	2	1/3	0	0
	4	0	1/3	0

(a) Compute the marginal probability functions of X and Y.

(b) Compute $\Pr[X > Y]$.

(c) How is this problem related to Example 1.4?

16. Consider the joint density function

$$f_{XY}(x, y) = \begin{cases} \frac{1}{8}(x^2 - y^2)e^{-x} & \text{if } x > 0,\ |y| \le x, \\ 0 & \text{otherwise} \end{cases}$$

from Figure 2.2.5 (p. 37), and the corresponding marginal density function f_X from Example 2.3.5. Show that the joint *pdf* has its maximum at $(x, y) = (2, 0)$, and the marginal *pdf* of X has its maximum at $x = 3$. Graph the marginal *pdf*, and verify that $\int_0^\infty f_X(x)\,dx = 1$. *Hint*: Use partial integration repeatedly.

17. If the pair of continuous random variables (X, Y) has joint *pdf*

$$f_{XY}(x, y) = \begin{cases} \frac{1}{2}e^{-x} & \text{if } x > 0,\ |y| \le x, \\ 0 & \text{otherwise,} \end{cases}$$

find the marginal *pdf* of X. Similarly, for

$$f_{XY}(x,y) = \begin{cases} \dfrac{1}{2x}e^{-x} & \text{if } x > 0,\ |y| \le x, \\ 0 & \text{otherwise.} \end{cases}$$

18. Find the marginal *pdf* of X if (X, Y) is a bivariate continuous random variable with joint *pdf*

$$f_{XY}(x,y) = \begin{cases} \dfrac{2}{(1+x+y)^3} & \text{if } 0 \le x, y, \\ 0 & \text{otherwise.} \end{cases}$$

(This is the bivariate Pareto density of Exercise 23 in Section 2.2).

2.4 Independence of Random Variables

In probability theory, two events A and B are called *independent* if $\Pr[A,B] = \Pr[A] \cdot \Pr[B]$. For jointly distributed random variables (X, Y) and considering events such as $\{X \le x, Y \le y\}$, this suggests the following definition.

Definition The random variables X and Y with joint distribution function F_{XY} and marginal distribution functions F_X and F_Y, respectively, are *stochastically independent* if

$$F_{XY}(x,y) = F_X(x) \cdot F_Y(y) \quad \text{for all } (x,y) \in \mathbb{R}^2, \tag{1}$$

i.e., if the joint *cdf* is the product of the two marginal distribution functions. □

The term "statistical independence" is sometimes used instead of "stochastic independence." If it is clear from the context that the objects under consideration are random variables, it is acceptable to simply use "independence." However, some caution is in order, because the word independence is also used in other meanings, particularly in regression analysis (regressors are often called "independent variables") and in linear algebra (linear independence of vectors).

Whether (X,Y) is discrete or continuous, it follows from (1) that stochastic independence can be defined equivalently in terms of the probability density functions. In the continuous case, if (1) holds, then

$$\begin{aligned} f_{XY}(x,y) &= \frac{\partial^2 F_{XY}(x,y)}{\partial x \partial y} \\ &= \frac{\partial^2 F_X(x) F_Y(y)}{\partial x \partial y} \\ &= \frac{\partial F_X(x)}{\partial x} \cdot \frac{\partial F_Y(y)}{\partial y} \\ &= f_X(x) \cdot f_Y(y). \end{aligned} \tag{2}$$

Conversely, if $f_{XY}(x,y) = f_X(x) f_Y(y)$ for all (x, y), then

$$\begin{aligned} F_{XY}(a,b) &= \int_{-\infty}^{a} \int_{-\infty}^{b} f_{XY}(x,y)\, dy\, dx \\ &= \int_{-\infty}^{a} \int_{-\infty}^{b} f_X(x) f_Y(y)\, dy\, dx \\ &= \int_{-\infty}^{a} f_X(x)\, dx \cdot \int_{-\infty}^{b} f_Y(y)\, dy \\ &= F_X(a) \cdot F_Y(b). \end{aligned}$$

A similar proof for the discrete case is left as an exercise.

For practical purposes, it is often more convenient to check independence using the factorization of the probability density function rather than the factorization of the distribution function. Note that to show independence it is necessary to prove that (1) or (2) holds for *all* (x,y). Conversely, to verify that X and Y are not independent it suffices to find one single event for which the factorization does not hold.

In our examples considered so far, independence holds in the following cases: Examples 2.2.2, 2.2.6, and 2.2.7 if $\rho = 0$. Let us briefly consider some of the remaining ones.

Example 2.4.1 continuation of Examples 2.2.1 and 2.3.1

Take, e.g., $f_{XY}(0,3) = 0 \neq f_X(0) \cdot f_Y(3)$. Hence, X and Y are dependent. □

Example 2.4.2 continuation of Examples 2.2.4 and 2.3.4

Take any point (x,y) outside the unit circle but inside the square of side length 2 centered at the origin. For such a point, $f_X(x) > 0$ and $f_Y(y) > 0$, but $f_{XY}(x,y) = 0$. Hence, independence is rejected. For the more picky student: Of course, we could always change the density function f_{XY} on a set of probability zero without changing the distribution function. Hence, take any small rectangle inside the square but outside the circle to verify that independence does not hold. □

In Example 2.4.2, the lack of independence is not as evident as in Example 2.4.1, and it is somewhat counterintuitive. After all, one might be tempted to say that "the x-coordinate and the y-coordinate of the first raindrop have nothing to do with each other and hence, are independent." It is the geometrical shape of the domain of positive density (the circle) that makes this argument wrong. This will become more evident when we talk about conditional distributions, but some informal reasoning may be useful at this point. For instance, suppose you are told that the x-coordinate of the raindrop is $X = 0.5$. Then you know that Y is between $-\sqrt{3}/2$ and $\sqrt{3}/2$, whereas if $X = 0$, then Y could be anywhere between -1 and 1. Or, to take an even more extreme case, if you happen to know that $X = 1$, then you can tell with certainty that $Y = 0$.

Thus, knowledge of X provides you with some information on the distribution of Y. The notion of statistical independence characterizes the lack of such information. In any case, the example shows that independence is a rather restrictive notion and not as trivial as one might think at first.

For a similar reason as in Example 2.4.2, the two variables in Example 2.3.5 are dependent, as is seen from Figure 2.2.5 (p. 37). Actually independence of two continuous random variables can often be ruled out immediately by just studying the joint support of X and Y, i.e., the set of (x, y) for which the joint density function is strictly positive. See Exercise 4.

Here are some further remarks for the theoretically inclined student. Both the uniform distribution inside the circle and the bivariate standard normal distribution given in (2.4) are examples of *spherically symmetric* distributions. If we center the distribution at the origin and rotate the coordinate system by any angle α, the distribution in the rotated coordinate system has exactly the same form as in the original (x,y) coordinate system. Equivalently, if X and Y have a joint density function, spherical symmetry means that the contours of equal density are concentric circles centered at the origin. That means the joint density function depends on x and y only through the value of $x^2 + y^2$, as in the bivariate standard normal density

$$f_{XY}(x, y) = \frac{1}{2\pi} \exp\left[-\frac{1}{2}(x^2 + y^2)\right], \quad (x, y) \in \mathbb{R}^2.$$

Interestingly, the following result holds: If the distribution of (X,Y) is spherically symmetric and if X and Y are independent, then X and Y are normal with the same variance. For a proof of this remarkable fact, which singles out the normal distribution among all spherical ones, the reader is referred to Muirhead (1982, Chapter 1.5), or to Ross (1994, Chapter 6.2).

Example 2.4.3 Let X and Y denote two independent continuous random variables with identical distributions, where

$$f_X(x) = \begin{cases} 3(1 - x^2)/4 & \text{if } |x| \le 1, \\ 0 & \text{otherwise.} \end{cases}$$

By independence, the joint *pdf* of X and Y is

$$f_{XY}(x, y) = \begin{cases} 9(1 - x^2)(1 - y^2)/16 & \text{if } |x| \le 1;\ |y| \le 1, \\ 0 & \text{otherwise.} \end{cases}$$

This joint *pdf* is shown in Figure 2.4.1 in the form of a contour plot (a) and in a three-dimensional view (b). The contours of equal density are almost circles at the center but approach the shape of the square for small values of f_{XY}. Thus, we have independence and symmetric marginal distributions, but the joint distribution of X and Y is not spherically symmetric.

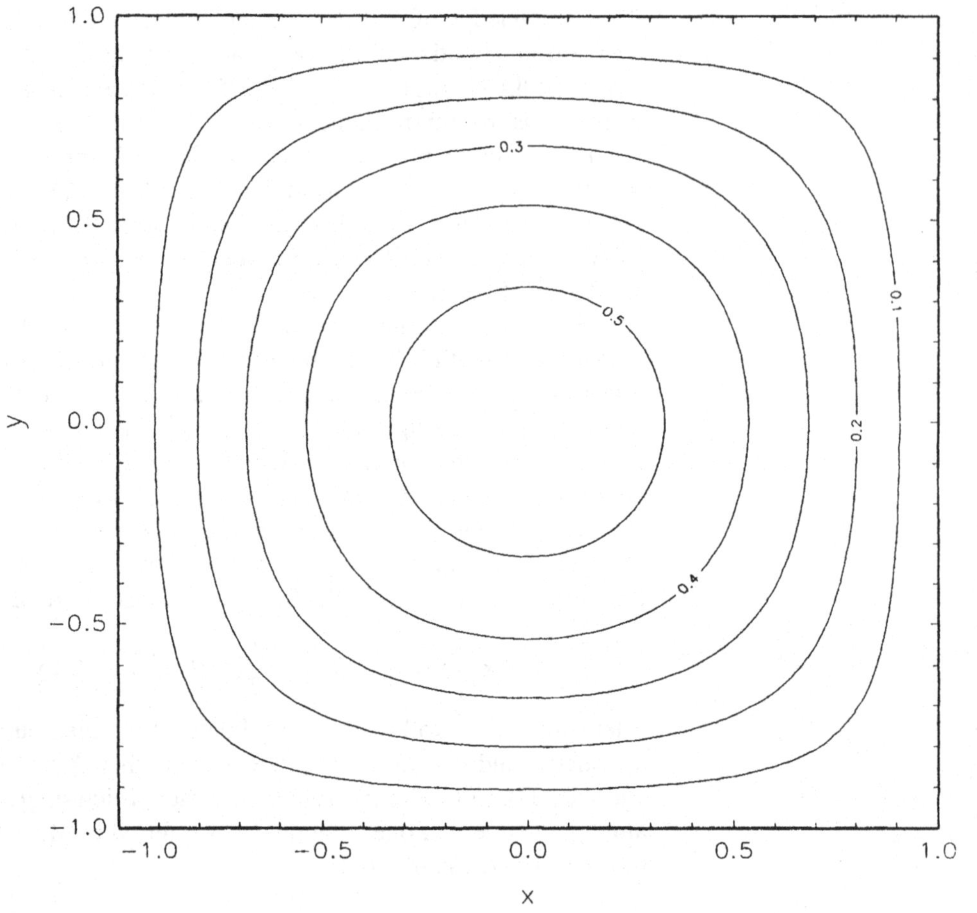

Figure 2-4-1A

Next, consider two random variables U and V with joint *pdf*

$$g_{UV}(u, v) = \begin{cases} 2\left[1 - (u^2 + v^2)\right]/\pi & \text{if } u^2 + v^2 \le 1, \\ 0 & \text{otherwise.} \end{cases}$$

This density function, which looks similar to f_{XY}, is shown in Figure 2.4.2, again (a) as a contour plot and (b) in a three-dimensional view. Note that the contours are exact circles and g_{UV} depends on u and v only through $u^2 + v^2$. The same argument, as in Example 2.4.2, shows that U and V are dependent. □

Loosely speaking, the message from Example 2.4.3 and the text preceding it is the following: You can ask for either independence of two random variables or a joint spherical distribution. If you ask for *both* at the same time, then you have automatically specified the normal distribution.

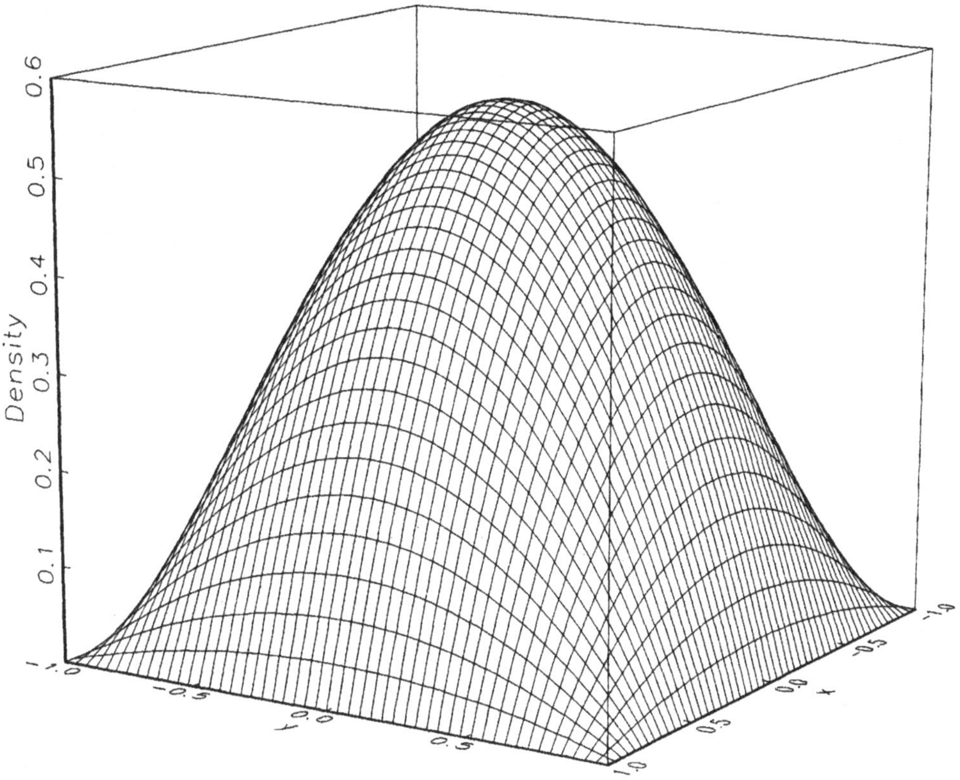

Figure 2-4-1B
Joint density function of X and Y in Example 2.4.3 (a) as a contour plot and (b) in a three-dimensional view.

Example 2.4.4 continuation of Example 2.4.2

Although the x- and y-coordinates in the raindrop example are not independent, it is surprisingly easy to transform the pair of random variables (X, Y) with uniform distribution inside the unit circle into a a new pair of random variables for which independence holds. The trick is to use *polar coordinates*: Each point $(x, y) \neq (0, 0)$ in the Euclidean plane has a unique representation as

$$x = r\cos(a)$$

and

$$y = r\sin(a),$$

where the angle a is allowed to vary in the interval $[0, 2\pi)$ and $r > 0$ is the Euclidean distance of (x, y) from the origin. The numbers r and a are the polar coordinates of (x, y). If we consider only points inside the unit circle, then we have the further constraint $r \leq 1$. Hence, we can write the random variables X and Y in the raindrop example as

$$X = R\cos(A)$$

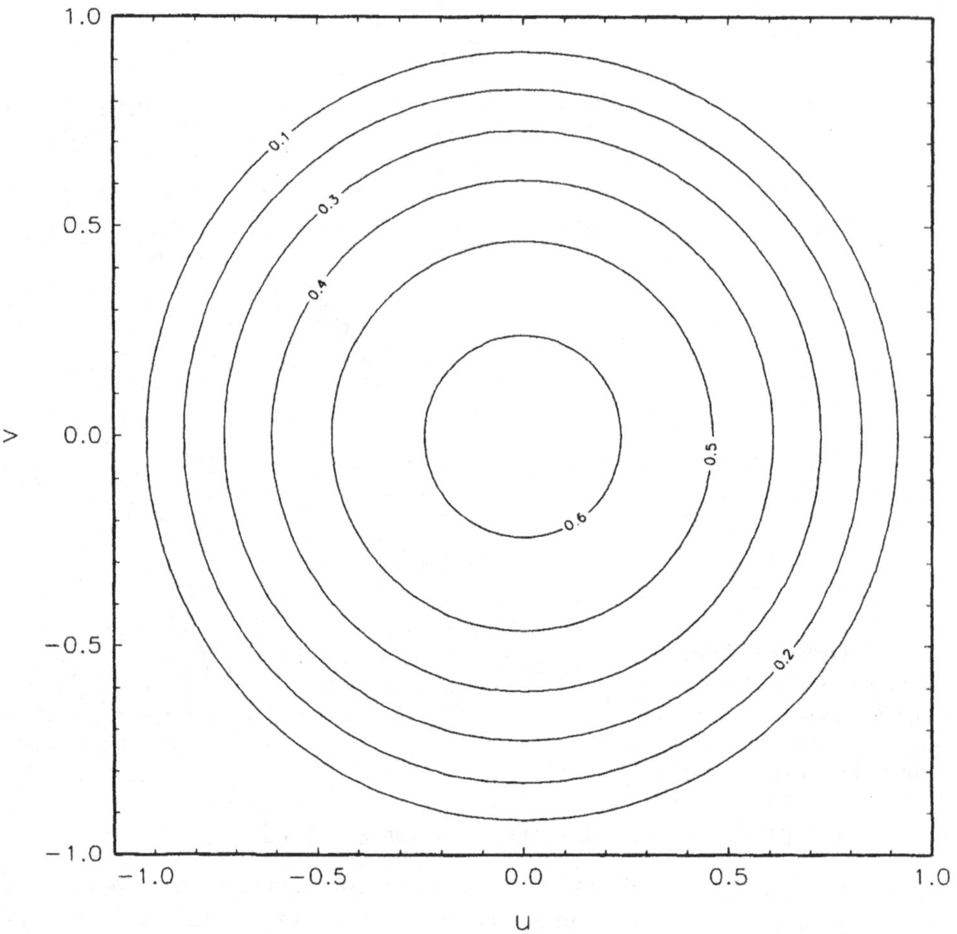

Figure 2-4-2A

and

$$Y = R\ \sin(A),$$

and thus implicitly define two new random variables R and A, the polar coordinates corresponding to X and Y. Now, we are going to show that R and A are independent.

Figure 2.4.3 shows a graph of the unit circle along with an (arbitrary) fixed point (x, y) inside the circle, with associated polar coordinates (r, a). Consider the event $\{R \leq r,\ A \leq a\}$. This event is equivalent to $(X, Y) \in \mathcal{S}$, where $\mathcal{S}$ is the shaded area in Figure 2.4.3. But $\mathcal{S}$ is a sector of a circle with radius r and angle a, and the area of this sector is given by

$$r^2\pi \cdot \frac{a}{2\pi} = \frac{r^2 a}{2}.$$

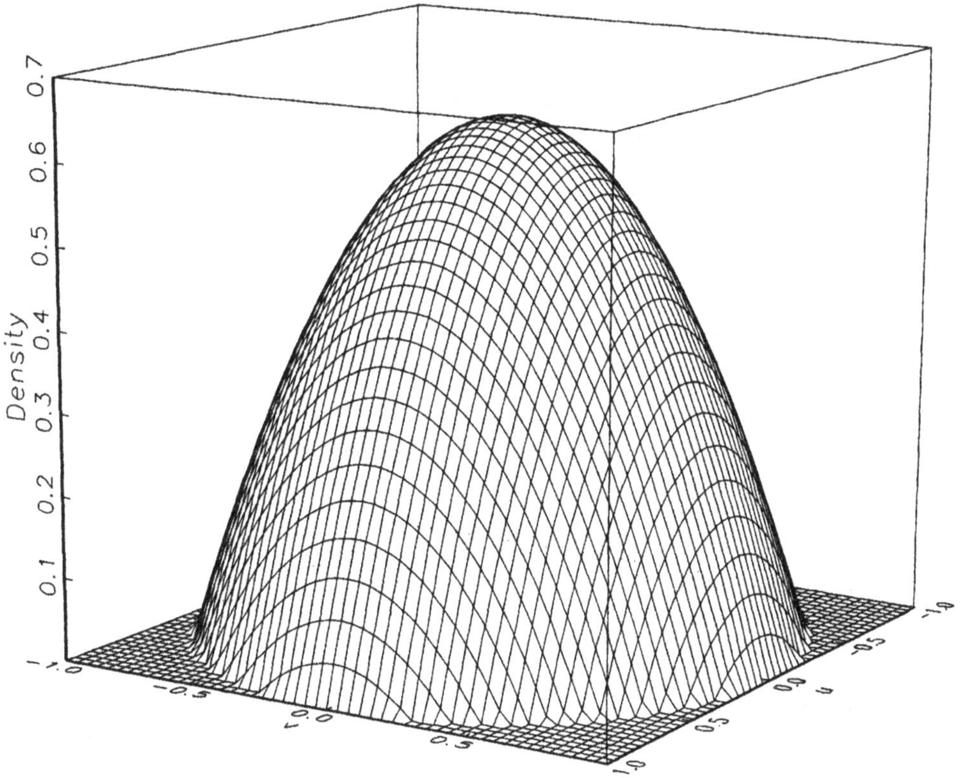

Figure 2-4-2B
Joint density function of U and V in Example 2.4.3, (a) as a contour plot, and (b) in a three-dimensional view.

Since (X, Y) is uniform, we obtain

$$\begin{aligned} F_{RA}(r, a) &= \Pr[R \le r,\ A \le a] \\ &= \Pr[(X, Y) \in \mathcal{S}] \\ &= \frac{r^2 a}{2} \cdot \frac{1}{\pi} \\ &= r^2 \cdot \frac{a}{2\pi} \,. \end{aligned}$$

Because this holds for all values of r in the interval $(0, 1]$ and for all values of a in the interval $[0, 2\pi)$, the last expression can be viewed as the product of two distribution functions:

$$F_R(r) = \begin{cases} 0 & \text{if } r < 0, \\ r^2 & \text{if } 0 \le r < 1, \\ 1 & \text{if } 1 \le r, \end{cases}$$

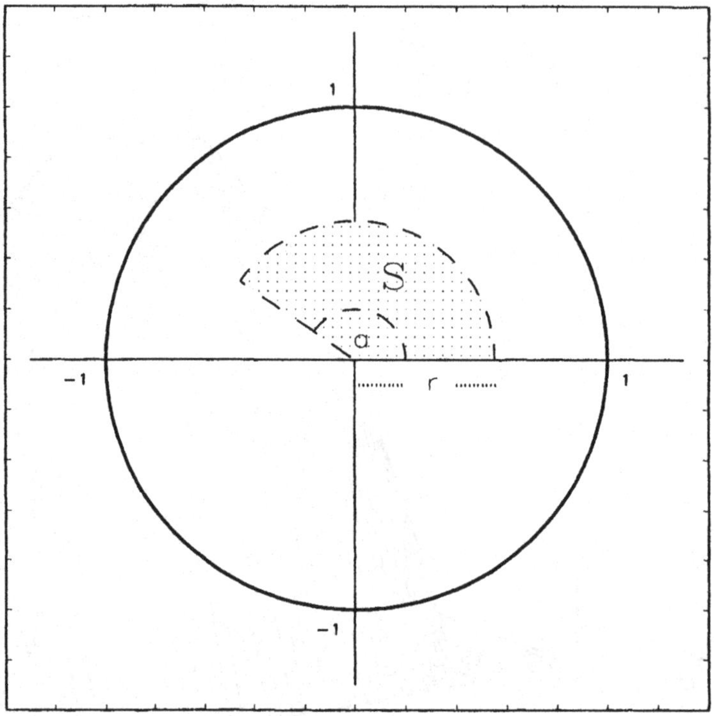

Figure 2-4-3 Calculation of the distribution function of the polar coordinates R and A in the raindrop example.

and

$$F_A(a) = \begin{cases} 0 & \text{if } a < 0, \\ a/2\pi & \text{if } 0 \le a < 2\pi, \\ 1 & \text{if } 2\pi \le a. \end{cases}$$

Therefore, the polar coordinates R and A are independent random variables. The angle A has a uniform distribution in the interval $(0, 2\pi)$, and the distance R has density $f_R(r) = 2r$ in the interval $(0, 1)$. □

Exercises for Section 2.4

1. Suppose that X and Y are two discrete random variables with marginal probability functions f_X and f_Y and distribution functions F_X and F_Y, respectively. Show that the following two statements are equivalent:

   (a) The joint probability function can be factorized, i.e.,

$$f_{XY}(x, y) = f_X(x) \cdot f_Y(y) \text{ for all } (x, y) \in \mathbb{R}^2.$$

(b) The joint distribution function can be factorized, i.e.,

$$F_{XY}(a,b) = F_X(a) \cdot F_Y(b) \text{ for all } (a,b) \in \mathbb{R}^2.$$

2. Verify independence in Examples 2.2.2 and 2.2.6.

3. Prove the following result. If (X,Y) is bivariate normal with $\rho = 0$, then X and Y are stochastically independent.

4. Prove the following theorem. Suppose that $\mathcal{A}$ is a subset of $\mathbb{R}^2$ with finite area and (X,Y) is a continuous random variable with joint *pdf* f_{XY}, such that $f_{XY}(x,y)$ is strictly positive for all $(x,y) \in \mathcal{A}$ and 0 otherwise. Suppose that the marginal densities f_X and f_Y are strictly positive in some intervals (x_0, x_1) and (y_0, y_1), respectively, and zero outside these intervals. If X and Y are independent, then $\mathcal{A}$ is a rectangle with sides parallel to the x- and y-coordinate axes. *Hint*: Study Example 2.4.2.

5. Suppose that X and Y are discrete random variables with joint *pdf*

$$f_{XY}(x,y) = \begin{cases} \dfrac{x+y}{36} & \text{if } x = 1,2,3,\ y = 1,2,3, \\ 0 & \text{otherwise.} \end{cases}$$

Show that X and Y are not independent. (See also Exercise 15 in section 2.2.)

6. If X and Y are continuous random variables with joint *pdf*

$$f_{XY}(x,y) = \begin{cases} x+y & \text{if } 0 \le x \le y,\ 0 \le y \le 1, \\ 0 & \text{otherwise,} \end{cases}$$

show that X and Y are not independent. (See also Exercise 17 in Section 2.2).

7. If $\lambda > 0$ and $\mu > 0$ are fixed numbers (parameters) and the pair of continuous random variables (X,Y) has joint *pdf*

$$f_{XY}(x,y) = \begin{cases} \dfrac{1}{\lambda\mu} \exp\left[-\left(\dfrac{x}{\lambda} + \dfrac{y}{\mu}\right)\right] & \text{if } x \ge 0,\ y \ge 0, \\ 0 & \text{otherwise,} \end{cases}$$

find the marginal densities of X and Y, and show independence. (See also Exercise 19 in Section 2.2).

8. Show that the continous random variables X and Y with joint *pdf*

$$f_{XY}(x,y) = \frac{1}{4} \exp\left[-(|x| + |y|)\right], \quad (x,y) \in \mathbb{R}^2,$$

are independent. (Both X and Y follow so-called *double-exponential* distributions; see also Exercise 20 in Section 2.2).

9. Let $k \ge 1$ denote a fixed integer, and define a bivariate continuous random variable (X,Y) by the density function

$$f_{XY}(x,y) = \begin{cases} \dfrac{1}{k} & \text{if } 0 \le x < k,\ 0 \le y < k, \text{ and } \lfloor x \rfloor = \lfloor y \rfloor, \\ 0 & \text{otherwise} \end{cases}$$

Here the symbol $\lfloor\ \rfloor$ is the "floor" function, i.e., $\lfloor z \rfloor$ is the largest integer less or equal to z or z truncated to its integral part. For some value of k, sketch the area in the (X, Y) plane where the density is positive, and find the marginal densities of X and Y. For which value(s) of k are X and Y stochastically independent?

10. Consider, again, Max's magic dice, as given by the table of joint probabilities in Exercise 6 of Section 2.3. Show that X and Y are not independent.

11. Let X denote a discrete random variable taking values 1 and 2 with probabilities p_1 and p_2, respectively, $p_1 + p_2 = 1$. Let Y denote a discrete random variable, independent of X, taking values 1 and 2 with probabilities q_1 and q_2, respectively. Define $Z = X + Y$, and show that the probability function of Z is given by $f_Z(2) = p_1 q_1$, $f_Z(3) = p_1 q_2 + p_2 q_1$, and $f_Z(4) = p_2 q_2$.

12. Suppose that (X, Y) is a bivariate random variable with marginal probability functions as in Exercise 11, and assume that $Z = X + Y$ has the same probability function as in Exercise 11. Prove that X and Y are independent.
Hint: In the table of joint probabilities

	$X = 1$	$X = 2$	
$Y = 1$			q_1
$Y = 2$			q_2
	p_1	p_2	

fill in the empty spaces, subject to the constraints imposed by the probability function of $Z = X + Y$.

13. Let X and Y denote two discrete random variables with probability functions

$$f_X(i) = \Pr[X = i] = p_i \quad i = 1, 2, 3$$

and

$$f_Y(i) = \Pr[Y = i] = q_i \quad i = 1, 2, 3.$$

(a) If X and Y are independent, find the joint distribution (i.e., the joint probability function) of (X, Y).
(b) If X and Y are independent, find the probability function of $Z = X + Y$.
(c) If $Z = X + Y$ has the probability function as in (b), are X and Y necessarily independent? (*Hint*: Consider a simple case where $p_1 = p_2 = p_3 = q_1 = q_2 = q_3 = \frac{1}{3}$, and let Exercise 10 inspire you).

14. (This is the continuation of Exercise 15 in Section 2.3). Let X^* and Y^* denote two *independent* discrete random variables such that X^* has the same marginal distribution as X and Y^* has the same marginal distribution as Y.

(a) Find the joint *pdf* of X^* and Y^*.

(b) Find $\Pr[X^* > Y^*]$.

(c) How is this exercise related to Example 1.4?

15. Abe, Bill, and Charlie like to play darts. Abe always scores a 2. Bill scores 0 with probability 0.6 and 4 with probability 0.4. Charlie scores 1 or 3 with probability 0.5 each. Assume that the scores reached by Bill and Charlie are independent.

(a) If Abe and Bill throw one dart each, who is more likely to have the higher score?

(b) If Abe, Bill, and Charlie throw one dart each, who is most likely to have the highest score?

Exercise 15 has a rather amusing anecdotal version as follows. A statistician eats daily at the same restaurant, where apple pie and blueberry pie are offered for dessert, sometimes also cherry pie. On the statistician's taste scale, apple pie always gets a score of 2, whereas 60 percent of the time blueberry pie reaches a score of 0 and 40 percent of the time a score of 4. Finally, whenever available, cherry pie scores 1 or 3 fifty percent of the time each. Since the statistician is a rational person, she always chooses apple pie when she is asked to decide between apple pie and blueberry pie, as seen from part (a) of Exercise 15. However, a few days ago I overheard the following conversation.

Statistician: *I'd like apple pie for dessert.*
Waiter: *Today, we also have Cherry pie.*
Statistician: *Then bring me blueberry pie, please.*

According to part (b) of Exercise 15, the statistician's reaction to the information that Cherry pie is available is perfectly rational, however illogical it may sound, because it maximizes the probability of choosing the best of the three pies.

2.5 Expected Values, Moments, Covariance, and Correlation

The notion of mathematical expectation is one of the fundamental concepts in statistics. Unless the purpose of a statistical analysis is purely descriptive, expected values appear in form of parameters of distributions, in testing and in estimation problems. Most of the "practical" material in the later chapters of this book is based on the notions of mean vectors and covariance matrices, which play a central role in multivariate statistics. It is assumed that the reader of this book has been introduced to the notion of the expected value of a random variable before, so we will give only a brief review of the basic results, and then focus on expected values based on joint distributions of random variables.

The *expected value*, or *mean*, of a random variable X is defined as

$$E[X] = \sum_x x\, f_X(x) \tag{1}$$

if X is discrete, and

$$E[X] = \int_{\mathbb{R}} x\, f_X(x)\, dx \tag{2}$$

in the continuous case, provided that the sum or integral exists. Thus, the expected value of X can be understood as a weighted average of the values that X can take, the weights being given by the probability function or the density function, respectively. It is not too difficult to construct examples of random variables which do not have an expected value, but we will not be concerned about this case in this book except in exercises 1 and 2. Table 2.5.1 lists expected values for a few of the standard distributions encountered in statistics. The reader who is not familiar with these results is encouraged to review the material on expectation in some introductory statistics book.

Example 2.5.1 If X is the number shown by a fair die, then

$$E[X] = \sum_{x=1}^{6} x \Pr[X = x] = \frac{1}{6} \sum_{x=1}^{6} x = 3.5.$$

For two independent fair dice X and Y, the probability function of $Z = X + Y$ has been found in Example 2.3.2. Thus,

$$E[X + Y] = E[Z] = \sum_{z=3}^{12} z \Pr[Z = z] = 2 \cdot \frac{1}{36} + 3 \cdot \frac{2}{36} + \ldots + 11 \cdot \frac{2}{36} + 12 \cdot \frac{1}{36} = 7.$$

Note that, in this calculation, we used only the definition of the expected value of a random variable but not any results on sums of random variables with which the reader may already be familiar. □

Example 2.5.2 Consider, again, the coordinates of the first raindrop in the unit circle, previously discussed in Examples 2.2.4, 2.3.4, and 2.4.2. Since the marginal *pdf* of X is

$$f_X(x) = \begin{cases} \dfrac{2}{\pi}\sqrt{1 - x^2} & \text{if } |x| \le 1, \\ 0 & \text{otherwise,} \end{cases}$$

$$E[X] = \int_{\mathbb{R}} x\, f_X(x)\, dx = \frac{2}{\pi} \int_{-1}^{1} x \sqrt{1 - x^2}\, dx = 0.$$

Table 2.5.1 Some standard probability laws and their means and variances

Distribution	*pdf*	Mean	Variance
Binomial(n, p)	$f(x) = \binom{n}{x} p^x (1-p)^{n-x}, \quad x = 0, 1, \ldots, n$	np	$np(1-p)$
Poisson(λ)	$f(x) = e^{-\lambda} \frac{\lambda^x}{x}, \quad x = 0, 1, \ldots$	λ	λ
Uniform(a, b)	$f(x) = \frac{1}{b-a}$ if $a \le x \le b$, $f(x) = 0$ otherwise	$\frac{a+b}{2}$	$\frac{(b-a)^2}{12}$
Exponential(λ)	$f(x) = \frac{1}{\lambda} e^{-x/\lambda}$ if $x \ge 0$, $f(x) = 0$ otherwise	λ	λ^2
Normal(μ, σ^2)	$f(x) = \frac{1}{\sqrt{2\pi}\sigma} \exp\left[-\frac{1}{2}\left(\frac{x-\mu}{\sigma}\right)^2 \right], \quad x \in \mathbb{R}$	μ	σ^2

Similarly, $E[Y] = 0$. Now, consider the distance from the origin $R = (X^2 + Y^2)^{1/2}$. From Example 2.4.4, we know that the *pdf* of R is

$$f_R(r) = \begin{cases} 2r & \text{if } 0 \le r \le 1, \\ 0 & \text{otherwise,} \end{cases}$$

and hence,

$$E[\sqrt{X^2 + Y^2}] = E[R] = \int_0^1 2r^2\, dr = \frac{2}{3}. \qquad \square$$

In both examples discussed so far, we have found the expected value of some function of two jointly distributed random variables, the function being $X + Y$ in the first example and $(X^2 + Y^2)^{1/2}$ in the second. The expected value of $R = (X^2 + Y^2)^{1/2}$ was obtained by finding the *pdf* of R, first, and then applying the definition (2) to the random variable R. Let us clarify this procedure with yet another example, even simpler than the preceding ones.

Example 2.5.3 Suppose that X is the number shown by a regular die, and let $S = X^2$, $V = (X - 3.5)^2$. Then S takes values 1, 4, 9, 16, 25, and 36 with probability 1/6 each, and hence, $E[S] = \sum_{j=1}^{6} j^2/6 = 91/6$. The *pdf* of V is

$$f_V(v) = \begin{cases} \dfrac{1}{3} & \text{if } v \in \{\dfrac{1}{4}, \dfrac{9}{4}, \dfrac{25}{4}\}, \\ 0 & \text{otherwise,} \end{cases}$$

and hence, from equation (1), $E[V] = E[(X - 3.5)^2] = 35/12$.

It is sometimes difficult, or even impossible, to find the distribution of a function of a random variable, particularly in the continuous case, but also in the discrete case, if the number of values which the random variable can take is large or infinite. For instance, suppose that X has a uniform distribution in the interval (0, 1), and we are (for whatever reason) interested in the expected value of $Z = X^3 - \sqrt{X}$. Finding the distribution of Z might be a nightmare, but, fortunately, it is not necessary, as the following theorem shows.

Theorem 2.5.1 (Expected Value Theorem, version for a single variable) *Let X denote a random variable with pdf $f_X(x)$, and let $G(x)$ denote a real-valued function defined on the domain of X. Then the expected value of the random variable $Z = G(X)$ is*

$$E[G(X)] = \sum_x G(x)\, f_X(x) \tag{3}$$

in the discrete case and

$$E[G(X)] = \int_{\mathbb{R}} G(x)\, f_X(x)\, dx \tag{4}$$

in the continuous case.

Many textbooks give (3) and (4) as the *definition* of the expected value of $G(X)$, instead of stating it as a result, thus, implicitly assuming that $E[G(X)]$, as in (3) or (4), is the same as $E[Z]$, as in (1) or (2), where $Z = G(X)$. For this reason, Ross (1994) calls Theorem 2.5.1 the "law of the unconscious statistician." In the discrete case, if the function $G(x)$ is one-to-one in the domain of X, the proof of the result is simple, but in the continuous case it is not trivial.

Because we will focus mostly on the joint distribution of two (or several) variables, we give a more general version of the theorem. A proof of the theorem in the discrete case is given at the end of this section.

Theorem 2.5.2 (Expected Value Theorem, bivariate version) *Let (X,Y) denote a bivariate random variable with pdf $f_{XY}(x,y)$, and let $G(x,y)$ denote a real-valued function defined on the domain of (X,Y). Then the expected value of the random variable $Z = G(X,Y)$ is given by*

$$E[G(X,Y)] = \sum_{(x,y)} G(x,y)\, f_{XY}(x,y) \tag{5}$$

in the discrete case, where the sum is to be taken over all (x,y) in the domain of (X,Y), and by

$$E[G(X,Y)] = \int_{x=-\infty}^{\infty} \int_{y=-\infty}^{\infty} G(x,y)\, f_{XY}(x,y)\, dy\, dx \tag{6}$$

in the continuous case.

Example 2.5.4 continuation of Example 2.5.1

With X and Y denoting the numbers shown by two independent regular dice,

$$E[X+Y] = \sum_{x=1}^{6}\sum_{y=1}^{6}(x+y)\Pr[X=x, Y=y] = \frac{1}{36}\sum_{x=1}^{6}\sum_{y=1}^{6}(x+y) = 7.$$

This confirms the result obtained earlier. □

Example 2.5.5 continuation of Example 2.5.2

If the *pdf* of (X, Y) is uniform in the unit circle, then

$$\begin{aligned} E[\sqrt{X^2+Y^2}] &= \frac{1}{\pi} \int\limits_{x^2+y^2\le 1} \sqrt{x^2+y^2}\, dy\, dx \\ &= \frac{1}{\pi} \int\limits_{x=-1}^{1} \int\limits_{y=-\sqrt{1-x^2}}^{\sqrt{1-x^2}} \sqrt{x^2+y^2}\, dy\, dx, \end{aligned}$$

which can be shown (see Exercise 4) to be equal to 2/3. However, in this case, it is easier to calculate the expected value of $R = \sqrt{X^2 + Y^2}$ using the definition of the expected value as in equation (2), given that the *pdf* of R is known from Example 2.2.4. □

Now, we turn to special choices of the function $G(x, y)$. For simplicity, we will give most formulas only for the continuous case. For instance, if we set $G(x, y) = x$, then we obtain

$$\begin{aligned} E[G(X, Y)] = E[X] &= \int_{\mathbb{R}^2} x\, f_{XY}(x, y)\, dy\, dx \\ &= \int_{\mathbb{R}} x \int_{\mathbb{R}} f_{XY}(x, y)\, dy\, dx \\ &= \int_{\mathbb{R}} x\, f_X(x)\, dx, \end{aligned}$$

consistent with our former definition of $E[X]$. Next, writing $\mu_X = E[X]$ and choosing the function $G(x, y) = (x - \mu_X)^2$, we obtain the expected squared deviation of X from its mean, called the *variance*:

$$\text{var}[X] = E\big[(X - \mu_X)^2\big] = \int_{\mathbb{R}} (x - \mu_X)^2\, f_X(x)\, dx, \tag{7}$$

with which the reader is likely to be familiar. The square root of the variance is called the *standard deviation*.

Other useful choices of the function G are the integral powers of X: The kth *moment* of X is defined as

$$E[X^k] = \int_{\mathbb{R}} x^k\, f_X(x)\, dx.$$

Both choices of the function $G(x, y)$, just studied, share the property that G depends only on x but not on y. In such cases, $E[G(X, Y)]$ can be computed using the marginal distribution of X; see Exercise 5. Before turning to more interesting functions $G(x, y)$, recall the following set of rules on expected values. Here, a and b denote arbitrary real constants, and G, H are real-valued functions defined on the domain of X.

(i) $$E[G(X) + H(X)] = E[G(X)] + E[H(X)], \tag{8}$$

provided both sides of (8) exist.

(ii) $$E[a + bX] = a + b\, E[X]. \tag{9}$$

(iii) $$\text{var}[X] = E[X^2] - \big(E[X]\big)^2. \tag{10}$$

(iv) $$\text{var}[a \pm bX] = b^2 \text{var}[X]. \tag{11}$$

Proving these four rules is left to the reader; see Exercise 6.

Writing $\mu_X = E[X]$ and $\mu_Y = E[Y]$, next, consider the function

$$G(x,y) = (x - \mu_X)(y - \mu_Y).$$

This special function leads to the following definition:

Definition 2.5.1 For two jointly distributed random variables X and Y, the expectation of $(X-\mu_X)(Y-\mu_Y)$ is called the *covariance* between X and Y. □

Because of its importance, we introduce a new symbol cov$[X,Y]$ for the covariance and give the formula explicitly for both the discrete and the continuous case, that is,

$$\text{cov}[X,Y] = E[(X - \mu_X)(Y - \mu_Y)] = \sum_{(x,y)}(x - \mu_X)(y - \mu_Y) f_{XY}(x,y) \quad (12)$$

if (X,Y) is discrete, and

$$\text{cov}[X,Y] = \int_{x=-\infty}^{\infty}\int_{y=-\infty}^{\infty}(x - \mu_X)(y - \mu_Y)\, f_{XY}(x, y)\, dy\, dx \quad (13)$$

if (X,Y) is continuous.

Moments of bivariate random variables are expected values of products of powers: Consider the functions $G(x,y) = x^j y^h$, where j and h are nonnegative integers. Then define

$$\gamma_{jh} = E[X^j Y^h].$$

Thus, $\gamma_{10} = \mu_X$, $\gamma_{20} = E[X^2]$, $\gamma_{01} = \mu_Y$, and $\gamma_{02} = E[Y^2]$. Of particular interest is the *mixed moment* $\gamma_{11} = E[XY]$ because of its connection to the covariance. It turns out that

$$\text{cov}[X,Y] = E[XY] - \mu_X\mu_Y = E[XY] - E[X]\, E[Y], \quad (14)$$

as we shall be able to show shortly. For computational purposes, formula (14) is often more convenient than (12) or (13) because

$$E[XY] = \sum_{(x,y)} xy\, f_{XY}(x,y)$$

in the discrete case and

$$E[XY] = \int_{x=-\infty}^{\infty}\int_{y=-\infty}^{\infty} xy\, f_{XY}(x,y)\, dy\, dx$$

in the continuous case. Later, we will illustrate this with examples.

To prove (14), first, we state another useful result: For a bivariate random variable (X,Y) and real constants a and b,

$$E[aX + bY] = a\, E[X] + b\, E[Y]. \quad (15)$$

The proof of (15) is straightforward (and given here for the continuous case):

$$\begin{aligned} E[aX+bY] &= \int_{\mathbb{R}^2} (ax+by)\, f_{XY}(x,y)\, dy\, dx \\ &= a \int_{\mathbb{R}^2} x\, f_{XY}(x,y)\, dy\, dx + b \int_{\mathbb{R}^2} y\, f_{XY}(x,y)\, dy\, dx \\ &= a\, E[X] + b\, E[Y]. \end{aligned}$$

A similar proof of the result

$$E[XY+aX+bY+c] = E[XY] + a\,E[X] + b\,E[Y] + c \tag{16}$$

is left as an exercise. Returning to the notion of covariance and using (16), we see that

$$\begin{aligned} \text{cov}[X,Y] &= E[(X-\mu_X)(Y-\mu_Y)] \\ &= E[XY - \mu_X Y - \mu_Y X + \mu_X\mu_Y] \\ &= E[XY] - \mu_X E[Y] - \mu_Y E[X] + \mu_X\mu_Y, \end{aligned}$$

and the desired result (14) follows.

Some further useful rules, the proof of which is left to the student (see Exercise 8), are as follows.

(i) $$\text{cov}[X,X] = \text{var}[X]. \tag{17}$$

(ii) $$\text{cov}[aX,bY] = ab\,\text{cov}[X,Y]. \tag{18}$$

(iii) $$\text{cov}[X+a,\; Y+b] = \text{cov}[X,Y]. \tag{19}$$

(iv) $$\text{var}[aX+bY] = a^2\text{var}[X] + b^2\text{var}[Y] + 2ab\,\text{cov}[X,\; Y]. \tag{20}$$

Property (i) shows that the covariance is a generalization of the notion of variance. Properties (ii) and (iii) parallel closely the analogous results for the variance, namely; $\text{var}[aX] = a^2\text{var}[X]$, and $\text{var}[a+X] = \text{var}[X]$. Property (iv) is of fundamental importance and will be generalized to a formula involving an arbitrary number of variables in Section 2.11. As a special case of (iv),

$$\text{var}[aX+bY] = a^2\,\text{var}[X] + b^2\,\text{var}[Y] \tag{21}$$

exactly if $\text{cov}[X,Y] = 0$. Specifically, if $\text{cov}[X,Y] = 0$, then $\text{var}[X+Y] = \text{var}[X] + \text{var}[Y]$.

The following result is formulated as a theorem because of its importance.

Theorem 2.5.3 *If the random variables X and Y are stochastically independent, then* $\text{cov}[X,Y] = 0$.

Proof (continuous case): Consider

$$E[XY] = \int_{x=-\infty}^{\infty} \int_{y=-\infty}^{\infty} xy\, f_{XY}(x,y)\, dy\, dx.$$

By definition of independence, f_{XY} factorizes, and hence,

$$E[XY] = \int_{x=-\infty}^{\infty} x\, f_X(x)\, dx \int_{y=-\infty}^{\infty} y\, f_Y(y)\, dy$$
$$= E[X] \cdot E[Y],$$

and the theorem follows from equation (14). ■

We will see from examples that the inverse implication is not true; that is, if the covariance between X and Y vanishes, the random variables are not necessarily independent.

Example 2.5.6 continuation of Example 2.5.5

If (X, Y) follows a uniform distribution inside the unit circle, then we know already that $\mu_X = \mu_Y = 0$. Next,

$$E[XY] = \int\limits_{x^2+y^2 \le 1} xy \frac{1}{\pi}\, dy\, dx = \frac{1}{\pi} \int\limits_{x=-1}^{1} x \int\limits_{y=-\sqrt{1-x^2}}^{\sqrt{1-x^2}} y\, dy\, dx.$$

The inner integral is zero for all values of x, and hence, $E[XY] = 0$, which implies that the covariance is zero. However, X and Y are dependent. □

Before turning to further examples, let us introduce yet another notion; the correlation coefficient.

Definition 2.5.2 The correlation coefficient (or simply *correlation*) between two random variables is defined as

$$\text{corr}[X,Y] = \rho_{XY} = \frac{\text{cov}[X, Y]}{\sqrt{\text{var}[X]}\sqrt{\text{var}[Y]}}.$$ □

It will be convenient to use the symbols σ_X^2 and σ_Y^2 for var$[X]$ and var$[Y]$, respectively, and hence, σ_X and σ_Y for the standard deviations. Then

$$\rho_{XY} = \frac{\text{cov}[X, Y]}{\sigma_X \sigma_Y}, \tag{22}$$

showing that the correlation can be considered as a "standardized covariance." The correlation vanishes exactly if the covariance vanishes, and Theorem 2.5.3 could, therefore, equivalently be formulated in terms of ρ_{XY}, namely; stochastic independence implies uncorrelatedness, but the reverse implication is not true.

The correlation coefficient has the following well-known property.

Theorem 2.5.4 *The correlation coefficient ρ_{XY} is between -1 and 1, i.e., $|\rho_{XY}| \leq 1$, with equality exactly if $Y = a + bX$ for some constants a and b ($b \neq 0$).*

Proof Let (X,Y) denote a bivariate random variable with $E[X] = \mu_1$, $E[Y] = \mu_2$, $\text{var}[X] = \sigma_1^2$, $\text{var}[Y] = \sigma_2^2$, and $\text{cov}[X,Y] = \sigma_{12}$. Define two new random variables

$$U = \frac{X - \mu_1}{\sigma_1},$$

and

$$V = \frac{Y - \mu_2}{\sigma_2}.$$

Then $E[U] = E[V] = 0$, $E[U^2] = E[V^2] = 1$, and

$$\begin{aligned}\rho_{XY} = \frac{\sigma_{12}}{\sigma_1\sigma_2} &= \frac{E[(X - \mu_1)(Y - \mu_2)]}{\sigma_1\sigma_2}\\ &= E\left[\frac{(X - \mu_1)}{\sigma_1} \cdot \frac{(Y - \mu_2)}{\sigma_2}\right] = E[UV].\end{aligned}$$

Next, consider $E[(U - V)^2]$. This expectation is positive, and is equal to zero exactly if $\Pr[U = V] = 1$. Hence,

$$\begin{aligned}0 \leq E[(U - V)^2] &= E[U^2 + V^2 - 2UV]\\ &= E[U^2] + E[V^2] - 2E[UV]\\ &= 2 - 2\rho_{XY},\end{aligned}$$

which implies $\rho_{XY} \leq 1$. Similarly, $E[(U + V)^2] \geq 0$ implies $\rho_{XY} \geq -1$, with equality exactly if $\Pr[U = -V] = 1$. Hence, $\rho_{XY} = 1$ implies that

$$\frac{X - \mu_1}{\sigma_1} = \frac{Y - \mu_2}{\sigma_2}$$

or, equivalently,

$$Y = \mu_2 + \frac{\sigma_2}{\sigma_1}(X - \mu_1)$$

with probability one. Thus the constants are $a = \mu_2 + (\sigma_2/\sigma_1)\mu_1$ and $b = (\sigma_2/\sigma_1)$. Note that the slope is positive in this case. Similarly, if $\rho_{XY} = -1$, then $Y = \mu_2 - \frac{\sigma_2}{\sigma_1}(X - \mu_1)$. The proof of the fact that $Y = a + bX$ ($b \neq 0$) implies $|\rho_{XY}| = 1$ is left as an exercise. ■

The foregoing proof is based on a simple but powerful idea, namely, transform the variables U and V into new variables $U - V$ and $U + V$. The result, then follows from the fact that variances are always nonnegative. See also Exercise 10.

The following property of the correlation coefficient is very appealing and not too difficult to prove (see Exercise 11). If we transform the bivariate random variable (X,Y) into (X^*, Y^*) by

$$X^* = a + bX \qquad (b \neq 0),$$
$$Y^* = c + dY \qquad (d \neq 0),$$

then $\rho_{X^*Y^*} = \pm\rho_{XY}$, with a sign change exactly if $bd < 0$. This means that the correlation coefficient, up to possible changes in sign, is *scale-invariant* under linear transformations of each component. For instance, the correlation remains unchanged if we switch from Fahrenheit to Celsius or from inches to light-years.

Now, let us turn to examples. In all examples with independence (2.2.2, 2.2.6, and 2.2.7 if $\rho = 0$), the covariance and correlation are zero, so there is no need to reconsider them.

Example 2.5.7 continuation of Example 2.2.1

Calculations give $E[X] = E[Y] = 1$, $E[X^2] = 5/3$, $E[Y^2] = 17/9$, and $\sigma_X^2 = E[X^2] - (E[X])^2 = 2/3$, $\sigma_Y^2 = E[Y^2] - (E[Y])^2 = 8/9$. Note that the means are equal, although $\Pr[X < Y] \neq \Pr[X > Y]$, as found earlier. Next, from the joint probabilities given in Example 2.2.1,

$$E[XY] = \sum_{x=0}^{2}\sum_{y=0}^{3} xy\, f_{XY}(x,y) = \frac{4}{3}.$$

Thus, $\text{cov}[X,Y] = E[XY] - E[X]E[Y] = 1/3$, and the correlation is $\rho_{XY} = \text{cov}[X,Y]/(\sigma_X\sigma_Y) = \sqrt{3/13} \approx 0.433$. This indicates a positive relationship between the two variables, which makes sense in view of the setup of the example. □

Example 2.5.8 continuation of Example 2.2.3

Let X and Y denote the numbers shown by two independent regular dice, and let $Z = X + Y$. By the bivariate probability function of (X,Z) and the marginal probability functions, we obtain $E[X] = 7/2$, $E[X^2] = 91/6$, $\sigma_X^2 = 35/12$, and $\sigma_X \approx 1.71$ for X = number shown by the red die, and $E[Z] = 7$, $E[Z^2] = 329/6$, $\sigma_Z^2 = 35/6$, and $\sigma_Z \approx 2.42$ for Z = sum of the numbers shown by the two dice. Moreover, $E[XZ] = 329/12$, and hence, $\text{cov}[X, Z] = E[XZ] - E[X]E[Z] = 35/12$. It follows that $\rho_{XY} = 1/\sqrt{2} \approx 0.707$. □

In Example 2.5.8, we found the numerical result the hard way, doing the actual computations instead of thinking about the setup of the example. So let us consider a more general version.

Example 2.5.9 Suppose X and Y are two uncorrelated (or more strongly, independent) random variables with identical distributions. For instance, X and Y may denote the numbers shown by two independent regular dice. Let $Z = X + Y$. Then what is the correlation between X and Z? Let $E[X] = E[Y] = \mu$ and $\text{var}[X] = \text{var}[Y] = \sigma^2$. Then uncorrelatedness implies that

$$\text{var}[Z] = \text{var}[X + Y] = \text{var}[X] + \text{var}[Y] = 2\sigma^2.$$

Moreover,

$$\begin{aligned}\text{cov}[X, Z] &= \text{cov}[X, X + Y] \\ &= E\big[(X - \mu)(X + Y - 2\mu)\big] \\ &= E\big[(X - \mu)^2 + (X - \mu)(Y - \mu)\big] \\ &= \text{var}[X] + \text{cov}[X, Y] \\ &= \sigma^2.\end{aligned}$$

Hence,

$$\rho_{XZ} = \frac{\text{cov}[X, Z]}{\sqrt{\text{var}[X]\,\text{var}[Z]}} = \frac{\sigma^2}{\sqrt{2\sigma^4}} = \frac{1}{\sqrt{2}}.$$ □

Example 2.5.9 can be generalized as follows. If $X_1, \ldots, X_n$ denote independent, identically distributed, random variables, and

$$S_k = \sum_{i=1}^{k} X_i \qquad (k = 1, \ldots, n)$$

denotes the partial sum of the first k variables, then the correlation between S_k and S_n is $\sqrt{k/n}$. For instance, if we roll a die $n = 9$ times, then the correlation between the sum of the first four rolls (S_4) and the sum of all nine (S_9) is $\sqrt{4/9} = 2/3$. A proof of this more general result (which follows the technique used in Example 2.5.9) is left to the student; see Exercise 12.

The square of the correlation coefficient is sometimes called the *coefficient of determination* and is usually introduced in the context of regression analysis as a numerical measure of the fit of a regression line. The preceding example and the discussion following it offer a better reason for this terminology, namely; the fact that $\rho_{XZ}^2 = 1/2$ shows that Z is "half determined" by X. A similar argument can be applied to the more general situation just mentioned: A proportion of $\text{corr}^2[S_k, S_n] = k/n$ of the sum $X_1 + \ldots + X_n$ is determined by the partial sum of the first k variables.

Finding the correlations between X and Y in Examples 2.2.5 and 2.2.7 (for the general case with five parameters) is left as an exercise to the technically more inclined student; see problems 13 and 14.

Example 2.5.10 A measurement error model

Measuring is often affected by random errors, and repeated measurements may be taken to improve precision. For instance, if two different nurses measure the blood pressure of the same individual, they may not arrive at exactly the same result. In situations like Example 1.3 (wing length of birds), the "objects" to be measured may themselves resent the experiment and introduce errors of measurement. Or two different observatories may get somewhat different readings on the brightness of a star due to differences in atmospheric conditions that are impossible to control.

The following is a simple model which accounts for errors of measurement. Suppose that Z is the random variable of interest, and two measurements are to be made. Write the measurements as

$$X_1 = Z + e_1$$

and

$$X_2 = Z + e_2, \tag{23}$$

where e_1 and e_2 are "errors of measurement." Assume that e_1 and e_2 are independent random variables, independent also of Z, with $E[e_1] = E[e_2] = 0$ and $\text{var}[e_1] = \text{var}[e_2] = \tau^2$. Let $\text{var}[Z] = \sigma^2$. Then $\text{var}[X_1] = \text{var}[X_2] = \sigma^2 + \tau^2$, $E[X_1X_2] = E[Z^2]$, $\text{cov}[X_1, X_2] = \sigma^2$, and hence,

$$\text{corr}[X_1, X_2] = \frac{\sigma^2}{\sigma^2 + \tau^2}.$$

This correlation is always nonnegative and 0 only if Z is a constant. If the error terms have no variability (i.e., if $\tau^2 = 0$), then the correlation is one. In fact, in this simple model with no systematic error of measurement the two values would be identical with probability one.

What about the improvement in precision? Suppose that we take the average of the two measurements, $\bar{X} = (X_1 + X_2)/2$, then

$$\text{var}[\bar{X}] = \sigma^2 + \frac{1}{2}\tau^2,$$

which should be compared to the variance of a single measurement, $\text{var}[X_1] = \sigma^2 + \tau^2$. It follows that taking repeated measurements and averaging them is beneficial when the variance of the error of measurement (τ^2) is relatively large, compared to the variance σ^2 of the variable of interest. See also Exercise 15. □

Example 2.5.11 Suppose that Z_1 and Z_2 are two uncorrelated random variables, both with mean zero and variance one. Consider the transformation to two new variables

$$Y_1 = bZ_1 + \sqrt{1 - b^2}\, Z_2$$

and

$$Y_2 = bZ_1 - \sqrt{1-b^2}\, Z_2, \tag{24}$$

where b is a fixed number in $[0, 1]$. Then $E[Y_1] = E[Y_2] = 0$, $\text{var}[Y_1] = \text{var}[Y_2] = 1$, and $\rho = \text{corr}[Y_1, Y_2] = 2b^2 - 1$. For a given value of ρ, we can thus choose the coefficient b as $b = \sqrt{(1+\rho)/2}$. Then the transformation (24) is

$$\begin{aligned} Y_1 &= \sqrt{\frac{1+\rho}{2}}\, Z_1 + \sqrt{\frac{1-\rho}{2}}\, Z_2, \\ Y_2 &= \sqrt{\frac{1+\rho}{2}}\, Z_1 - \sqrt{\frac{1-\rho}{2}}\, Z_2, \end{aligned} \tag{25}$$

and the correlation between Y_1 and Y_2 is ρ. This is possible for any value of ρ in the interval $[-1, 1]$. □

Transformations like (25) are used in computer simulations to generate pairs of random variables with a given, specified correlation coefficient. See also Exercise 16.

Example 2.5.12 Sample correlation

Suppose that $(x_1, y_1), \ldots, (x_N, y_N)$ are N pairs of real numbers (in a statistical context, they would typically be regarded as a sample from some bivariate distribution, but this assumption is not needed here). The sample correlation coefficient for the N pairs of numbers is defined as

$$r = \frac{\sum_{i=1}^{N}(x_i - \bar{x})(y_i - \bar{y})}{\sqrt{\sum_{i=1}^{N}(x_i - \bar{x})^2}\sqrt{\sum_{i=1}^{N}(y_i - \bar{y})^2}},$$

where $\bar{x}$ and $\bar{y}$ are the usual averages. Is it true that the sample correlation coefficient is always between -1 and 1? Fortunately it is not necessary to establish a special proof of this property, as the following construction shows. Define a pair of discrete random variables (U, V) such that each of the N data points (x_i, y_i) has probability $1/N$, i.e.,

$$\Pr[(U, V) = (x_i, y_i)] = \frac{1}{N} \quad \text{for } i = 1, \ldots, N.$$

If two pairs (x_i, y_i) and (x_j, y_j) are identical, we can either distinguish them formally or assign probability $2/N$, and similarly for multiple points. Thus, the pair of random variables (U, V) is an exact representation of the N data points. Then it is straightforward to verify that $E[U] = \bar{x}$, $\text{var}[U] = \sum_{i=1}^{N}(x_i - \bar{x})^2/N$, similarly for V, and $\text{cov}[U, V] = \sum_{i=1}^{N}(x_i - \bar{x})(y_i - \bar{y})/N$. It follows that the sample correlation is the same as the correlation between the random variables U and V as constructed in this example. Therefore any property shown for correlation coefficients, in general, holds for sample correlation coefficients, including the facts that they are in the interval $[-1, 1]$ and that a correlation of ± 1 implies an exact linear relationship. □

To conclude this section, we give a proof of the "expected value theorem" (or the "law of the unconscious statistician"). The proof holds for the bivariate version but only for the discrete case. Recall the setup: Suppose that (X, Y) is a pair of discrete random variables, with joint probability function f_{XY}, and $G(x, y)$ is a real-valued function defined on the domain of (X, Y). Define a new random variable $Z = G(X, Y)$. Then the theorem states that

$$E[Z] = \sum_{(x,y)} G(x, y)\, f_{XY}(x, y).$$

Notice that Z is a discrete random variable and that its probability function $f_Z(z)$ is obtained from the joint probability function of X and Y by

$$\begin{aligned} f_Z(z) &= \Pr[Z = z] \\ &= \sum_{\{(x,y):G(x,y)=z\}} f_{XY}(x,y) \qquad \text{for all } z. \end{aligned}$$

Hence, defining $A(z) = \{(x,y) : G(x,y) = z\}$ for each z in the domain of Z,

$$\begin{aligned} E[Z] &= \sum_z z\, f_Z(z) \\ &= \sum_z \sum_{A(z)} G(x,y)\, f_{XY}(x, y). \end{aligned}$$

But this double sum is a sum over all possible values of (x,y), and hence,

$$E[Z] = \sum_{(x,y)} G(x,y) \cdot f_{XY}(x,y),$$

as stated in the theorem. A proof for the continuous case can be constructed following the same idea, but it is technically more difficult. A rather elegant proof which covers both the discrete and the continuous case can be found in Ross (1994). ■

Exercises for Section 2.5

1. Let X denote a discrete random variable such that

$$\Pr[X = 2^k] = 2^{-k} \quad \text{for } k = 1, 2, \ldots$$

Show that $E[X]$ does not exist for this random variable.

2. Suppose X is a *Cauchy* random variable, i.e., a continuous random variable with *pdf*

$$f_X(x) = \frac{1}{\pi(1 + x^2)}, \qquad x \in \mathbb{R}.$$

Show that $E[X]$ does not exist for this random variable.

3. Suppose that the (discrete or continuous) random variable X with *pdf* $f_X(x)$ is symmetric to m, i.e.,

$$f_X(m-x) = f_X(m+x) \quad \text{for all } x \in \mathbb{R}.$$

Show that $E[X] = m$, provided the expected value exists.

4. Verify the calculations in Example 2.5.5, i.e., show that

$$\int_{x^2+y^2\le 1} \sqrt{x^2+y^2}\, dy\, dx = \frac{2}{3}\pi.$$

5. Let (X, Y) denote a bivariate random variable, and let $G(x, y)$ denote a real-valued function which, in the domain of (X, Y), does not depend on y. That is, there exists a function H such that $G(x, y) = H(x)$ for all $(x, y) \in \mathbb{R}^2$ for which $f_{XY}(x, y) > 0$. Show that $E[G(X, Y)]$ can be computed from the marginal distribution of X.

6. Prove equations (8) to (11).

7. Prove equation (16), assuming that (X, Y) is continuous and has a density function f_{XY}.

8. Prove equations (17) to (20).

9. Suppose that X is a random variable with $\text{var}[X] > 0$, and let $Y = a + bX$ for some constants a and b, $b \neq 0$. Show that $|\rho_{XY}| = 1$.

10. Suppose that X and Y are jointly distributed random variables with $\text{var}[X] = \text{var}[Y] = \sigma^2$ and $\text{corr}[X, Y] = \rho$. Let $U = X + Y$ and $V = X - Y$.

(a) Find $\text{var}[U]$ and $\text{var}[V]$. Under what conditions are these variances zero?

(b) Show that U and V are uncorrelated.

11. Prove the following result: Let (X,Y) denote a bivariate random variable, and define

$$X^* = a + bX \qquad (b \neq 0),$$

and

$$Y^* = c + dY \qquad (d \neq 0),$$

for some real constants a, b, c, d. Then $\text{corr}[X^*, Y^*] = \pm\text{corr}[X, Y]$, with a sign change exactly if $bd < 0$.

12. Prove the following result: Let $X_1, \ldots, X_n$ denote n independent, identically distributed, random variables, and choose an integer k between 1 and n. Let

$$S_n = \sum_{i=1}^{n} X_i, \text{ and } S_k = \sum_{i=1}^{k} X_i.$$

Then the squared correlation between S_k and S_n is k/n.

13. Suppose that (X, Y) is a continuous random variable with *pdf*

$$f_{XY}(x, y) = \begin{cases} \frac{1}{\sqrt{2\pi}} \exp\left[-\frac{x^2}{2} + x - y\right] & \text{if } x \le y, \\ 0 & \text{otherwise,} \end{cases}$$

(see Example 2.2.5). Find the means and variances of both variables and the correlation.

14. For the bivariate normal distribution with density function given by equation (2.3) (see Example 2.2.7), find $E[X]$, $E[Y]$, var$[X]$, and var$[Y]$. Then find the correlation between X and Y, assuming $\mu_1 = \mu_2 = 0$, and $\sigma_1 = \sigma_2 = 1$.

15. Suppose that $Z, e_1, \ldots, e_N$ are pairwise uncorrelated random variables, such that var$[Z] = \sigma^2$ and var$[e_i] = \tau^2$ for $i = 1, \ldots, N$. Let $X_i = Z + e_i$, $i = 1, \ldots, N$. Find corr$[X_i, X_j]$ $(i \neq j)$ and var$[\bar{X}]$, where $\bar{X} = (X_1 + \ldots + X_N)/N$.

16. This exercise provides an alternative to the method outlined in Example 2.5.11 for constructing a pair of random variables with a given correlation coefficient ρ. Suppose that Z_1 and Z_2 are uncorrelated, both with mean 0 and variance 1. Define

$$Y_1 = Z_1,$$

and

$$Y_2 = \rho Z_1 + \sqrt{1 - \rho^2} Z_2.$$

Show that both Y_i have mean 0 and variance 1 and that corr$[Y_1, Y_2] = \rho$.

17. Find $E[X]$, $E[Y]$, var$[X]$, var$[Y]$, cov$[X,Y]$, and ρ_{XY} for the bivariate random variable with probability function

$$f_{XY}(x,y) = \begin{cases} (x + y)/36 & \text{if } x = 1, 2, 3 \;\; y = 1, 2, 3, \\ 0 & \text{otherwise.} \end{cases}$$

18. Find $E[X]$, $E[Y]$, var$[X]$, var$[Y]$, cov$[X,Y]$, and ρ_{XY} for the bivariate random variable with density function

$$f_{XY}(x,y) = \begin{cases} x + y & \text{if } 0 \le x \le 1,\ 0 \le y \le 1, \\ 0 & \text{otherwise.} \end{cases}$$

19. Find ρ_{XY} for the bivariate random variable with density function

$$f_{XY}(x,y) = \begin{cases} 2 & \text{if } x \ge 0;\ y \ge 0,\ x + y \le 1, \\ 0 & \text{otherwise.} \end{cases}$$

20. Find ρ_{XY} for the magic dice in Exercise 6 of Section 2.3.

21. Let X denote a discrete random variable that takes values $-10, -9, \ldots, 9, 10$ with probability 1/21 each, and let c denote a fixed number, $0 \le c \le 10$. Define a new random variable

$$Y = \begin{cases} X & \text{if } |X| \leq c, \\ -X & \text{if } |X| > c. \end{cases}$$

(see Exercise 12 in Section 2.3). Find ρ_{XY} if $c = 7.5$.

22. A continuous version of Exercise 21: Let X denote a continuous random variable with a uniform distribution in the interval $[-1, 1]$, and let $0 \leq c \leq 1$ denote a real constant. Define

$$Y = \begin{cases} X & \text{if } |X| \leq c \\ -X & \text{if } |X| > c, \end{cases}$$

(a) Find ρ_{XY} as a function of c. [*Hint*: Since (X,Y) do not have a joint density, compute $E[XY]$ as $E[G(X)]$, where

$$G(x) = \begin{cases} x & \text{if } |x| \leq c, \\ -x & \text{if } |x| > c. \end{cases}$$

(b) Show that ρ_{XY} is a continuous, monotonically increasing function of c for $0 \leq c \leq 1$.

(c) Find the value of c such that $\rho_{XY} = 0$.
Note: This is a particularly surprising exercise because it gives a pair of random variables (X,Y) that are completely dependent (in the sense that either X or Y determines both variables), yet their correlation is zero.

23. Let $X \sim \mathcal{N}(0,1)$ (a standard normal random variable), and let

$$Y = \begin{cases} X & \text{if } |X| \leq c, \\ -X & \text{if } |X| > c, \end{cases}$$

where $c > 0$ is a constant. From Exercise 14 in Section 2.3, we know that $Y \sim N(0,1)$.

(a) Show that $\rho_{XY} = 4[\Phi(c) - c \cdot \varphi(c)] - 3$, where φ and Φ are the density function and the cumulative distribution function, respectively, of the standard normal distribution. *Hint*: Compute $E[XY]$ as $E[G(X)]$, as in Exercise 14, using partial integration and the fact that $\partial\varphi(z)/\partial z = -z\varphi(z)$.

(b) Using (a), show that $\rho_{XY} = \rho_{XY}(c)$ is a monotonically increasing function of c for $c \in [0, \infty)$. What are the values of ρ_{XY} for $c = 0$ and $c = \infty$? That is, argue that there exists a unique value c_0 such that $\rho_{XY}(c_0) = 0$.

(c) Using numerical approximation or a table of the normal distribution function Φ, find c_0 to at least two decimal digits.

24. In Exercise 9 of Section 2.4, show that $\rho_{XY} = 1 - 1/k^2$. *Hint*: Use $\sum_{i=1}^{k} i^2 = k(k+1)(2k+1)/6$ to show that $E[XY] = (4k^2 - 1)/12$.

25. Let X and Y denote two discrete random variables taking values $1, 2, 3$ with probability $1/3$ each. Suppose that the probability function of $Z = X + Y$ is given by the following table:

z	2	3	4	5	6
Pr[Z=z]	$\frac{1}{9}$	$\frac{2}{9}$	$\frac{1}{3}$	$\frac{2}{9}$	$\frac{1}{9}$

Show that $\rho_{XY} = 0$, although X and Y need not be independent (See Exercise 13 in Section 2.4.)

26. This exercise serves to illustrate that the "expected value theorem" is not trivial. We demonstrate this by showing that, for other functions of the *pdf* of a random variable, a similar result does not necessarily hold. So suppose that, for a continuous random variable X with *pdf* $f(x)$, the *unexpected value* of X is defined as

$$\mathcal{U}[X] = \int_{x=-\infty}^{\infty} x\sqrt{f(x)}\,dx.$$

(a) If X is uniform in $[0, 1]$, show that $\mathcal{U}[X] = 1/2$.

(b) Define $Y = 4X$. Find the *pdf* of Y and its unexpected value $\mathcal{U}[Y]$. *Hint*: If it is not immediately clear what the *pdf* of Y is, first, find its *cdf*.

(c) Show that $\mathcal{U}[Y] \neq \int_{x=-\infty}^{\infty} 4x\sqrt{f(x)}\,dx$. This shows that, in general,

$$\mathcal{U}[G(X)] \neq \int_{x=-\infty}^{\infty} G(x)\sqrt{f(x)}\,dx.$$

27. This exercise is for the student who is familiar with moment generating functions of univariate random variables. The *moment generating function* of a bivariate random variable (X, Y) is defined as the function

$$m_{XY}(s, t) = E\left[e^{sX+tY}\right]$$

for (s, t) in an open circle around the origin, provided this expectation exists. Throughout this exercise, assume that the moment generating function is continuously differentiable and the order of summation (or integration) and differentiation can be interchanged, wherever necessary.

(a) If X and Y are independent, show that $m_{XY}(s, t)$ is the product of the moment generating functions of X and Y.

(b) Show that

$$\left.\frac{\partial m_{XY}}{\partial s}\right|_{(s,t)=(0,0)} = E[X].$$

(c) Show that

$$\left.\frac{\partial^2 m_{XY}}{\partial s\partial t}\right|_{(s,t)=(0,0)} = E[XY].$$

(d) How can higher order moments, i.e., moments $\gamma_{hj} = E\left[X^h Y^j\right]$, be obtained from the moment generating function?

2.6 Conditional Distributions

Again, suppose that we are given the joint *pdf* f_{XY} of two random variables X and Y. When studying *marginal* distributions, we asked the question "What is the distribution of Y, ignoring X?" Now we are going to ask for the univariate distribution of Y in a different setup. Suppose that the value of X has been observed. Then how does this knowledge affect our assessment of the distribution of Y? For instance, if (X, Y) denotes the high temperatures on two randomly selected consecutive days of the year, then knowing the high temperature on the first day (X) certainly affects our knowledge of the possible values which the high temperature can take on the second day (Y), and our assessment of their probabilities.

Recall from probability theory that, for two events A and B with $\Pr[B] > 0$, the conditional probability of A, given B, is defined as

$$\Pr[A|B] = \Pr[A, B]/\Pr[B]. \tag{1}$$

This leads immediately to the following definition of conditional distributions.

Definition 2.6.1 For a bivariate random variable (X,Y) with probability density function $f_{XY}(x,y)$, the *conditional pdf* of Y, given $X = x$, is given by

$$f_{Y|X}(y|x) = \frac{f_{XY}(x,y)}{f_X(x)}, \tag{2}$$

where f_X is the marginal *pdf* of X. This definition holds for all x such that $f_X(x) > 0$. □

Several remarks need to be made to clarify this definition. In the discrete case, the values of f_{XY} and of f_X are probabilities, and therefore,

$$\begin{aligned} \frac{f_{XY}(x, y)}{f_X(x)} &= \frac{\Pr[X = x, Y = y]}{\Pr[X = x]} \\ &= \Pr[Y = y|X = x]. \end{aligned}$$

So the conditional *pdf* of Y, given $X = x$, is nothing but the definition of conditional probability. It follows also that $f_{Y|X}$ is, indeed, a proper probability function for all values of x such that $f_X(x) > 0$. To see this, note that $f_{Y|X}(y|x) \geq 0$ and

$$\sum_y f_{Y|X}(y|x) = \frac{1}{f_X(x)} \sum_y f_{XY}(x,y) = 1$$

by the definition of the marginal distribution of X (see Section 2.3). Note also that (2) defines a conditional distribution for each fixed value of x.

Before attempting a discussion of the somewhat trickier continuous case, let us return to some of the earlier examples. Sometimes, we will use the notation

$f_{Y|X}(y|X=x)$ instead of $f_{Y|X}(y|x)$ to better express the fact that we are studying the conditional distribution of Y, given $X = x$.

Example 2.6.1 continuation of Example 2.2.1

First, consider the conditional distribution of Y, given $X = x$, for various values of x. The values of the conditional probability functions are obtained by dividing each entry in the table of joint probabilities by the marginal total. For instance,

$$f_{Y|X}(y|X=2) = \begin{cases} \dfrac{1/12}{1/3} = 1/4 & y = 0, 1, 2, 3, \\ 0 & \text{otherwise.} \end{cases}$$

Generally, for $x = 0, 1, 2$, we obtain

$$f_{Y|X}(y|x) = \begin{cases} \dfrac{1}{x+2} & \text{for } y = 0, 1, \ldots, x+1, \\ 0 & \text{otherwise.} \end{cases}$$

Thus, for each value of x, the conditional distribution of Y, given $X = x$, is discrete uniform with equal weights on the integers from 0 to $x + 1$. The conditional distributions of X for given values of Y are obtained similarly. For instance,

$$f_{X|Y}(x|Y=2) = \begin{cases} 4/7 & x = 1, \\ 3/7 & x = 2, \\ 0 & \text{otherwise.} \end{cases}$$

The conditional probability function of X, given $Y = 3$, puts weight 1 on $x = 2$, i.e., $\Pr[X = 2|Y = 3] = 1$. This is also clear from the fact that Y can take the value 3 only if $X = 2$. □

In Example 2.2.2 (two independent dice) it turns out that

$$f_{Y|X}(y|x) = f_Y(y) = \begin{cases} 1/6 & y = 1, \ldots, 6, \\ 0 & \text{otherwise} \end{cases}$$

for all values of x, that is, the conditional distribution of Y, given $X = x$, does not depend on x. This is not surprising. Under independence, $f_{XY}(x,y) = f_X(x) \cdot f_Y(y)$, and thus,

$$f_{Y|X}(y|x) = \frac{f_{XY}(x,y)}{f_X(x)} = \frac{f_X(x) \cdot f_Y(y)}{f_X(x)} = f_Y(y).$$

Hence, independence implies that the conditional distribution of Y, given $X = x$, does not depend on x and, therefore, is equal to the marginal distribution of Y. Conversely, suppose that $f_{Y|X}(y|x) = h(y)$, a function that does not depend on x. Then

$$h(y) = f_{XY}(x,y)/f_X(x),$$

and therefore,

$$f_{XY}(x,y) = f_X(x) \cdot h(y). \tag{3}$$

Hence, by the definition of independence, X and Y must be independent, and summation of (3) over x shows that $h(y)$ is the probability function of Y. The same argument carries over to the continuous case, with integration replacing summation.

In Example 2.2.3, all conditional probability functions are discrete uniform; details are left to the student (Exercise 1). Let us now turn to the continuous case. The difficulty here is that the event on which we condition has probability zero, and it is not possible to write $\Pr[Y \le y|X = x]$ in the form $\Pr[Y \le y, X = x]/\Pr[X = x]$. Instead, we shall look at well-defined probabilities of the form

$$\Pr[a \le Y \le b|x \le X \le x + h]$$

for $h > 0$, and then let h tend to zero. Hence, define

$$\Pr[a \le Y \le b|X = x] = \lim_{h\to 0} \Pr[a \le Y \le b|x \le X \le x + h]. \tag{4}$$

The probability on the right-hand side, for fixed $h > 0$, is given by

$$\frac{\Pr[x \le X \le x + h, a \le Y \le b]}{\Pr[x \le X \le x + h]} = \frac{\int_{y=a}^{b} \int_{u=x}^{x+h} f_{XY}(u,y)\,du\,dy}{\int_{u=x}^{x+h} f_X(u)\,du}.$$

Now suppose that $f_X(u)$ and $f_{XY}(u,y)$ are continuous functions of u in a neighborhood of $u = x$. Then by the mean value theorem of calculus,

$$\int_{u=x}^{x+h} f_X(u)du \approx f_X(x) \cdot h,$$

and

$$\int_{u=x}^{x+h} f_{XY}(u,y)\,du \approx f_{XY}(x,y) \cdot h$$

for h small. More precisely, there exist numbers ξ_1 and ξ_2 in the interval $(x, x+h)$ such that $\int_{u=x}^{x+h} f_X(u)\,du = h\, f_X(\xi_1)$ and $\int_{u=x}^{x+h} f_{XY}(u, y)\,du = h\, f_{XY}(\xi_2, y)$. Therefore, the limit (4) is obtained as

$$\begin{aligned} \lim_{h\to 0} \Pr[a \le Y \le b|x \le X \le x + h] &= \lim_{h\to 0} \frac{h \int_{y=a}^{b} f_{XY}(x,y)\,dy}{h\, f_X(x)} \\ &= \int_a^b \frac{f_{XY}(x,y)}{f_X(x)}\,dy. \end{aligned} \tag{5}$$

Since this holds for all a, b, the integrand in (5) is the conditional density function.

Example 2.6.2 continuation of Example 2.2.4

For the raindrop example, we obtain

$$f_{Y|X}(y|x) = \begin{cases} \frac{1}{\pi} / \frac{2\sqrt{1-x^2}}{\pi} = \frac{1}{2\sqrt{1-x^2}} & \text{if } |y| \le \sqrt{1-x^2}, \\ 0 & \text{otherwise.} \end{cases}$$

Hence, the conditional distribution of Y, given $X = x$, is uniform in the interval $(-\sqrt{1-x^2}, \sqrt{1-x^2})$. It is widest for $x = 0$ and narrows to a degenerate distribution putting probability 1 on the point $y = 0$ if $x \to \pm 1$. □

Example 2.6.3 continuation of Example 2.2.5

Recall the bivariate density

$$f_{XY}(x,y) = \begin{cases} \exp[-\frac{x^2}{2} + x - y]/\sqrt{2\pi} & \text{if } x \le y, \\ 0 & \text{otherwise.} \end{cases} \tag{6}$$

Leaving the technical details to the student (Exercise 2), the marginal distribution of X turns out to have density

$$f_X(x) = \frac{1}{\sqrt{2\pi}} e^{-x^2/2} \quad x \in \mathbb{R},$$

that is, X is standard normal. The density of the conditional distribution of Y, given $X = x$, is

$$\begin{aligned} f_{Y|X}(y|x) &= \frac{\exp[-\frac{x^2}{2} + x - y]/\sqrt{2\pi}}{\exp[-x^2/2]/\sqrt{2\pi}} \\ &= e^{x-y}, \text{ if } y \ge x \end{aligned}$$

and

$$f_{Y|X}(y|x) = 0, \text{ otherwise.}$$

This is a "shifted" exponential distribution with positive density for all $y \ge x$. □

Example 2.6.3 points immediately towards a useful representation of bivariate probability density functions. For all $y \ge x$,

$$\begin{aligned} f_{XY}(x,y) &= \frac{1}{\sqrt{2\pi}} \exp[-\frac{x^2}{2} + x - y] \\ &= \frac{1}{\sqrt{2\pi}} \exp[-\frac{x^2}{2}] \cdot \exp[x - y] \\ &= f_X(x) \cdot f_{Y|X}(y|x). \end{aligned}$$

However, this is not specific to the particular example. From the definition (2) of the conditional *pdf*, we can always write

$$\begin{aligned} f_{XY}(x,y) &= f_{Y|X}(y|x) \cdot f_X(x) \\ &= f_{X|Y}(x|y) \cdot f_Y(y). \end{aligned} \tag{7}$$

Hence, in Example 2.6.3, just looking at the bivariate density (6) might have told us that the conditional density of Y has the form e^{x-y}.

Example 2.6.4 Suppose that the joint *pdf* of (X, Y) is given by

$$f_{XY}(x,y) = \begin{cases} \dfrac{1}{10\sqrt{2\pi}} \exp\left[-\dfrac{1}{2}(y-x)^2 \right] & \text{if } 0 \le x \le 10,\ y \in \mathbb{R}, \\ 0 & \text{otherwise.} \end{cases}$$

For fixed $x \in [0, 10]$,

$$\begin{aligned} f_X(x) &= \frac{1}{10} \int_{-\infty}^{\infty} \frac{1}{\sqrt{2\pi}} \exp\left[-\frac{1}{2}(y-x)^2 \right] dy \\ &= \frac{1}{10}, \end{aligned}$$

and therefore, X is uniform in the interval $[0, 10]$. It follows that

$$f_{Y|X}(y|x) = \frac{1}{\sqrt{2\pi}} \exp\left[-\frac{1}{2}(y-x)^2 \right], \quad y \in \mathbb{R},$$

that is, the conditional distribution of Y, given $X = x$, is normal with mean x and variance 1. □

Actually, equation (7) is useful for constructing bivariate distributions as follows. First, take a *pdf* f_X for the marginal distribution of X. Then choose a *pdf* $f_{Y|X}$ for the conditional distribution of Y, given $X = x$, such that $f_{Y|X}$ depends in some way on x, and define a joint *pdf* f_{XY} according to (7). For instance, the joint density in Example 2.6.4 was constructed by choosing X to be uniform and $f_{Y|X}$ to be the density function of a normal with mean x and variance 1. Example 2.2.1 was constructed this way and now can finally be fully justified.

Example 2.6.5 Let X denote an exponential random variable with *pdf* $f_X(x) = e^{-x}, x > 0$. Suppose that the conditional distribution of Y, given $X = x$, is normal with mean x and variance x^2, i.e.,

$$f_{Y|X}(y|x) = \frac{1}{\sqrt{2\pi}\, x} \exp\left[-\frac{1}{2}\left(\frac{y-x}{x}\right)^2 \right], \quad y \in \mathbb{R},$$

which is well defined for all $x > 0$. Then the joint *pdf* of (X, Y) is given by

$$
\begin{aligned}
f_{XY}(x, y) &= f_X(x)\, f_{Y|X}(y|x) \\
&= \begin{cases} \dfrac{1}{\sqrt{2\pi}\, x} \exp\left[-\dfrac{y^2}{2x^2} + \dfrac{y}{x} - \dfrac{1}{2} - x \right] & \text{if } x > 0,\ y \in \mathbb{R}, \\ 0 & \text{otherwise.} \end{cases}
\end{aligned}
$$

This might be difficult to recognize as a joint density function unless one is aware of the way it was constructed. Figure 2.6.1 shows a contour plot of f_{XY}; with an infinite peak at $x = y = 0$. □

A particularly useful application of equation (7) occurs in the theory of *finite mixtures* of distributions to be introduced in Section 2.8.

Students, who have not been exposed before to marginal and conditional distributions, sometimes find them difficult to understand and to distinguish, so we conclude this section by explaining these concepts once more, on an intuitive level. The purpose is to illustrate the principles because, once the concepts are clear, the mathematical formulation is a direct consequence.

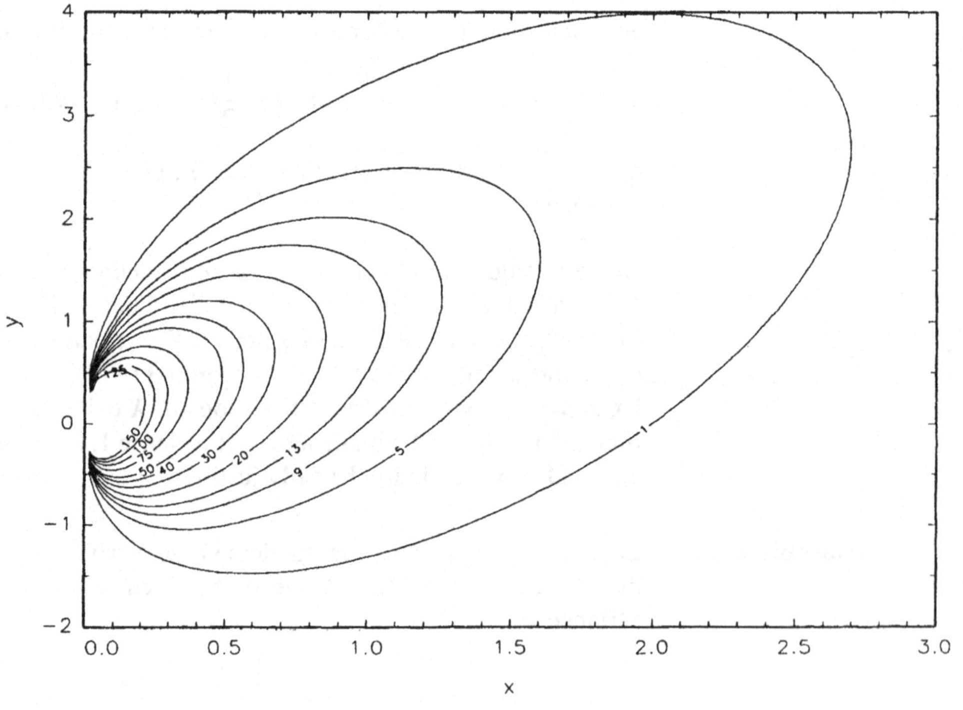

Figure 2-6-1
Contour plot of the joint *pdf* in Example 2.6.5. Contours are marked with the value of $100 \times f_{XY}(x, y)$.

Consider the discrete case first. Suppose the joint *pdf* of two random variables X and Y is given by

$$f_{XY}(i,j) = Pr[X=i, Y=j] = p_{ij} \qquad \text{for } i=0,1, \quad j=0,1,2, \tag{8}$$

or, perhaps, in less abstract form as a table:

	$Y=0$	$Y=1$	$Y=2$
$X=0$	p_{00}	p_{01}	p_{02}
$X=1$	p_{10}	p_{11}	p_{12}

The *marginal* distribution of Y is the distribution of Y, ignoring X. "Ignoring X" means, mathematically, considering events of the form $\{Y=j, X=\text{anything}\}$. According to the rules of probability, $f_Y(j) = Pr[Y=j] = \sum_i Pr[X=i, Y=j]$. In words, we have to add all probabilities in the column for $Y=j$ and write the sum as the marginal probability in the margin of the table. This is perhaps even more intuitive if we recall the frequency definition of probability: Suppose that the table above represents percentages in a statistic of car accidents compiled by an insurance company, involving all policyholders during a given year. Here, X could denote gender (0 for females, 1 for males), and Y could denote the number of accidents, ranging from 0 to 2. Thus, if we are interested in the distribution of the number of accidents per policyholder, irrespective of gender, we just have to add up the two numbers in each column of the table. Conversely, if we are interested in the distribution of gender among policyholders, we would want to ignore variable Y and, thus, add the percentages in each row.

Conditional distributions take a *slice* out of the joint distribution. For instance, if we want to find the conditional distribution of Y, given $X=0$, then we have to take the "slice" (or row in the table above) corresponding to $X=0$. In plain English, we are interested in the distribution of the number of accidents for females only. The probabilities, or relative frequencies, in the row labeled $X=0$ are p_{00}, p_{01}, and p_{02}, respectively. Since these three probabilities do not add up to one, they need to be rescaled. Thus, we want three percentages, or weights, that reflect the same proportions as those in the selected row of the table of probabilities, but adding up to one. The proper choice (for the slice $X=0$) is to define

$$q_0 = p_{00}/(p_{00}+p_{01}+p_{02}),$$

$$q_1 = p_{01}/(p_{00}+p_{01}+p_{02}),$$

and

$$q_2 = p_{02}/(p_{00}+p_{01}+p_{02}) \tag{9}$$

as the conditional probabilities for the three values of Y, given $X=0$. Hence, the conditional *pdf* of Y, given $X=x$, is nothing but a properly rescaled version of the probabilities associated with the chosen slice of the joint distribution. It is also apparent from (9) that this rescaling corresponds exactly to a division by the value of the *marginal* probability function of X at $X = 0$. Thus, to find the conditional *pdf* of Y, given $X = i$, it is useful to know the value of the marginal *pdf* of X at $X = i$, i.e., $f_X(i) = Pr[X = i]$.

In a purely theoretical context, the distinction between marginal and conditional distributions is usually clear. In real-world applications, however, one often has to be careful. If the above example refers to the joint distribution of gender and number of accidents, then this joint distribution is itself the bivariate marginal distribution obtained from a three-dimensional distribution, in which the probabilities would be further broken down by age group. At the same time, it is also a conditional distribution, because it refers to the accident statistics of one particular insurance company, and therefore, we have taken a slice out of a three-dimensional distribution in which the third variable is "company." The bivariate probabilities in the table above may be regarded as sums of probabilities obtained by ignoring the value of one or several other variables (marginal distribution) or by taking a slice (conditional distribution). Thus, it is important, in any practical situation, that we define clearly what is the set of variables to be considered. Sometimes this is trivial. If we study an experiment that consists of tossing coins or rolling dice, then we would never even consider the distributions as marginal distributions obtained by ignoring other variables, such as room temperature, at the time of the experiment. However, if the experiment consists of asking people questions about their eating habits, then we might be well advised to consider the reported frequencies as marginal, ignoring age group, or as conditional on age group, depending on the design of the study.

Although the continuous case may look more frightening to the novice, the concepts of marginal and conditional distributions are really exactly the same. Marginal distribution: ignore one of the variables. Conditional distribution: take a slice out of the joint distribution. The difficulty with "ignoring one variable" is that finding the marginal *pdf* is no longer as simple as adding discrete probabilities, but if we recall that an integral is actually the limiting case of a sum, the basic idea, nevertheless, is the same. The concept of a conditional density, fortunately, is straightforward to illustrate: If we think of the bivariate density function as a mountain, then finding a conditional distribution may be thought of as "cutting the mountain." Suppose that we are asking for the conditional distribution of Y, given $X = x_0$. Then we cut the mountain along a line parallel to the y-axis at $x = x_0$. The shape of the cut, then corresponds to the conditional density of Y, given $X = x_0$. The conditional density is the same as the joint density $f_{XY}(x_0, y)$ with x_0 considered as a constant, but since the density must integrate to 1, we have to multiply it by a proper constant. In mathematical terms,

$$f_{Y|X}(y|X = x_0) = c \cdot f_{XY}(x_0, y), \tag{10}$$

where the (positive) constant c has to be chosen such that

$$\int_{y=-\infty}^{\infty} c \cdot f_{XY}(x_0, y)dy = 1. \tag{11}$$

Therefore the constant c is nothing but a normalization, which typically depends on the value x_0.

Example 2.6.6 continuation of Example 2.6.2

Figure 2.6.2(a) shows a three-dimensional representation of the density function

$$f_{XY}(x, y) = \begin{cases} 1/\pi & \text{if } x^2 + y^2 \le 1, \\ 0 & \text{otherwise.} \end{cases} \tag{12}$$

The density function appears as a cylinder with radius 1 and height $1/\pi$. Suppose that we wish to find the marginal density function of X. Then we have to "add up the probabilities" over all values of Y. At $X = 0$, we will get a larger value than if X is near 1 or near -1, and the result is the density function illustrated in Figure 2.3.3 (p. 44). Now, suppose that we know that the x-coordinate of the raindrop is 1/2 and we would, therefore, like to know the density function of Y, given $X = 1/2$. According

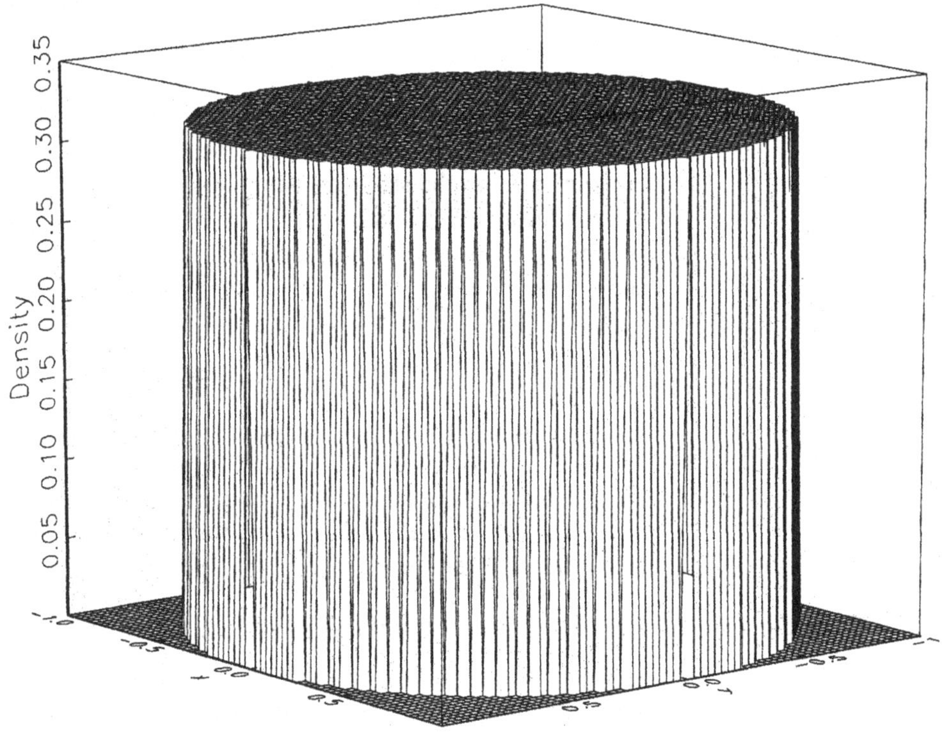

Figure 2-6-2A

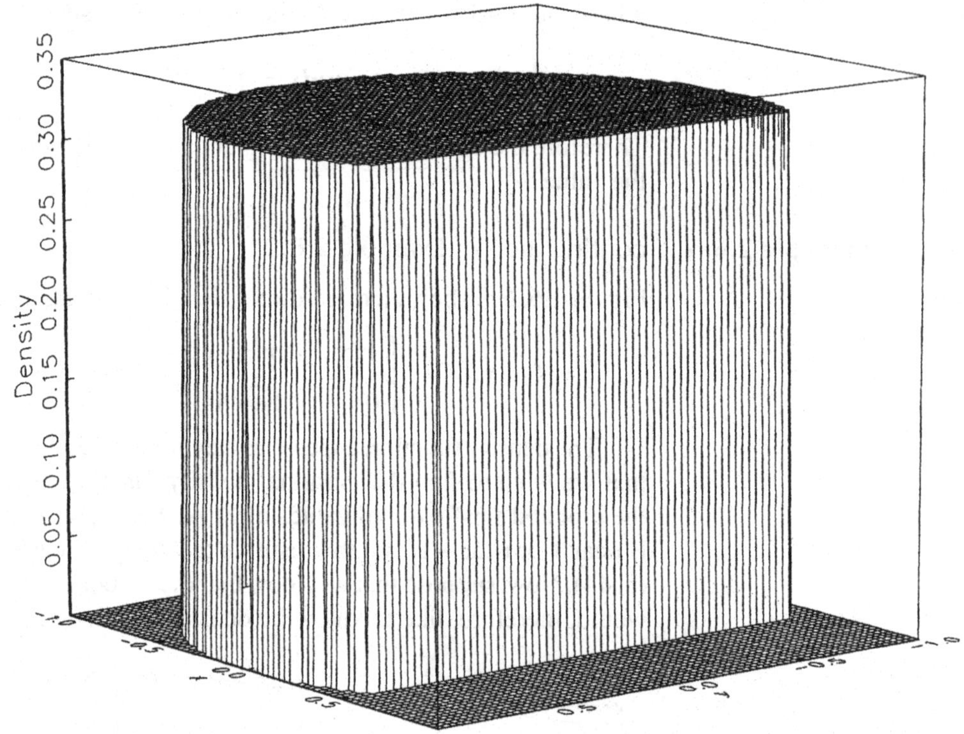

Figure 2-6-2B
(a) Density function of the uniform distribution inside the unit circle and (b) conditional *pdf* of Y, given $X = 1/2$, illustrated as the face created by a cut at $x = 1/2$.

to the above heuristic explanation, we can cut the cylinder at $x = 1/2$, as illustrated in Figure 2.6.2(b). The shape of the cut, then will determine the conditional *pdf*. Because the joint *pdf* has the shape of a cylinder, the cut has an exact rectangular shape, and the conditional distribution of Y, given $X = 1/2$, must be uniform in an interval symmetric about zero. More precisely, this interval is $\left[-\sqrt{1-(1/2)^2}, \sqrt{1-(1/2)^2}\right]$, and hence, the exact form of the conditional density function is given by

$$f_{Y|X}(y|X = 1/2) = \begin{cases} 1/\sqrt{3} & \text{if } |y| \le \sqrt{3}/2, \\ 0 & \text{otherwise.} \end{cases} \tag{13}$$

More generally, for each fixed value x of X between -1 and 1, the conditional distribution of Y, given $X = x$, will be uniform in an interval symmetric about zero. For $x = 0$, it is the interval $[-1, 1]$; for x approaching ± 1, the interval narrows down to $y = 0$, and the conditional density function (in the limit) becomes an infinite peak at $y = 0$; see Exercise 7. □

Example 2.6.7 The bivariate normal distribution

In Figure 2.2.2 (p. 33), we saw the *pdf* $f_{XY}(x, y)$ of a bivariate normal distribution with parameters $\mu_1 = \mu_2 = 0$, $\sigma_2 = \sigma_2 = 1$, and $\rho = 0.7$. Finding the conditional distribution of Y, given $X = 1.5$, corresponds to slicing the mountain at $x = 1.5$.

This is illustrated in Figure 2.6.3. The cut appears to be nicely bell-shaped. Exercise 6 shows that, indeed, this is the case, i.e., the conditional distribution of Y, given $X = x$, is normal for all values of x. This is a remarkable property that we will study in more detail in Chapter 3. □

Example 2.6.8 Mixtures of bivariate normal distributions

Figure 2.6.4 gives yet another example of "slicing a mountain," this time for a much more complicated distribution, which is actually a so-called *mixture* of two bivariate normals, a type of distribution we will treat in Section 2.8. Part (a) of Figure 2.6.4 shows the complete bivariate density function. Part (b) shows a cut at $x = 1$, and the conditional *pdf* of Y, given $X = 1$, appears as a nice bimodal curve. Actually, this curve itself corresponds to a mixture of two univariate normal densities, but we are not yet equipped with the proper mathematical tools to give a formal treatment of this example. □

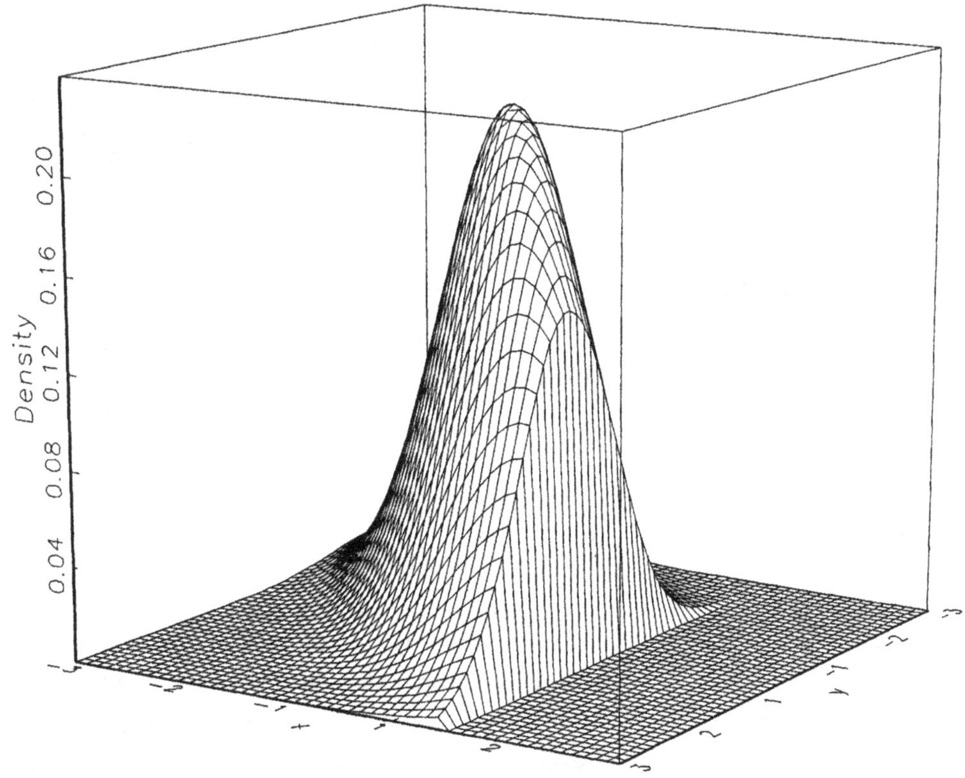

Figure 2-6-3 The conditional *pdf* of Y, given $X = 1.5$, for a bivariate normal distribution.

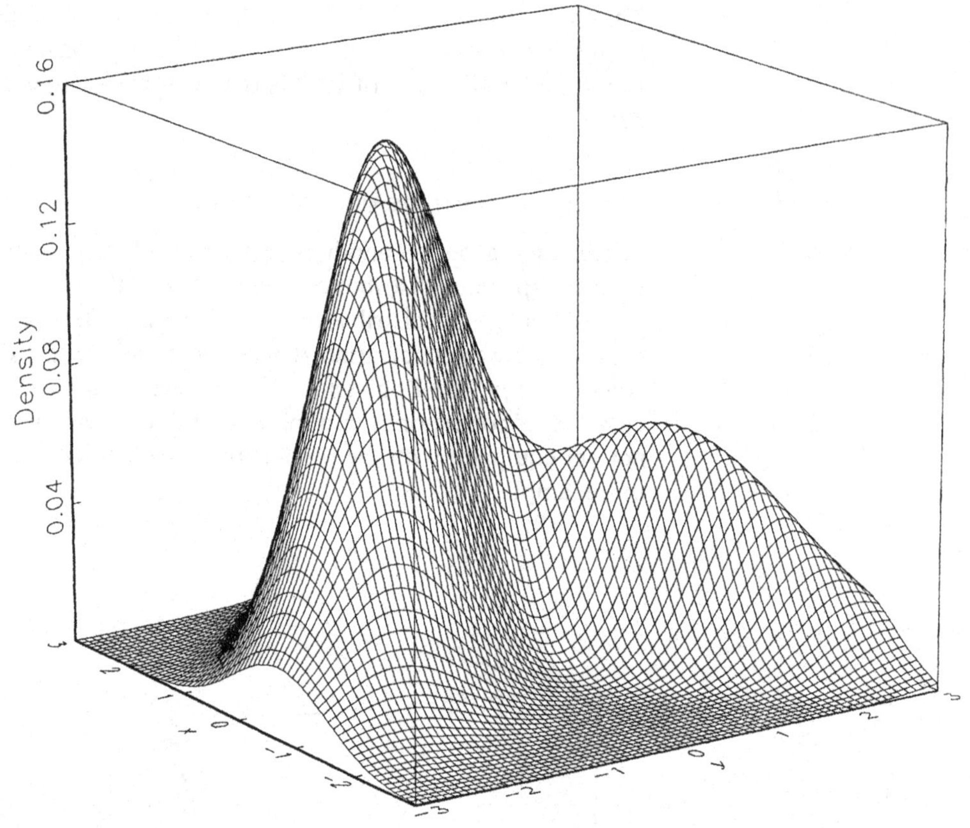

Figure 2-6-4A

Example 2.6.9 continuation of Example 2.6.5

Slicing the mountain, shown in Figure 2.6.1, by a line parallel to the y-axis will produce a normal density curve for all values of $x > 0$. For x close to zero, the density has a high peak at x, which goes to infinity as x tends to 0. □

Exercises for Section 2.6

1. This exercise refers to Example 2.2.3, where $Z = X + Y$ denotes the sum of the numbers shown by two independent regular dice.
   (a) Find all conditional distributions of Z, given $X = x$, for $x = 1, \ldots, 6$.
   (b) Find all conditional distributions of X, given $Z = z$, for $z = 2, \ldots, 12$.
2. Verify that the marginal distribution of X in Example 2.6.3 is standard normal, and hence, find the conditional density of Y, given $X = x$.

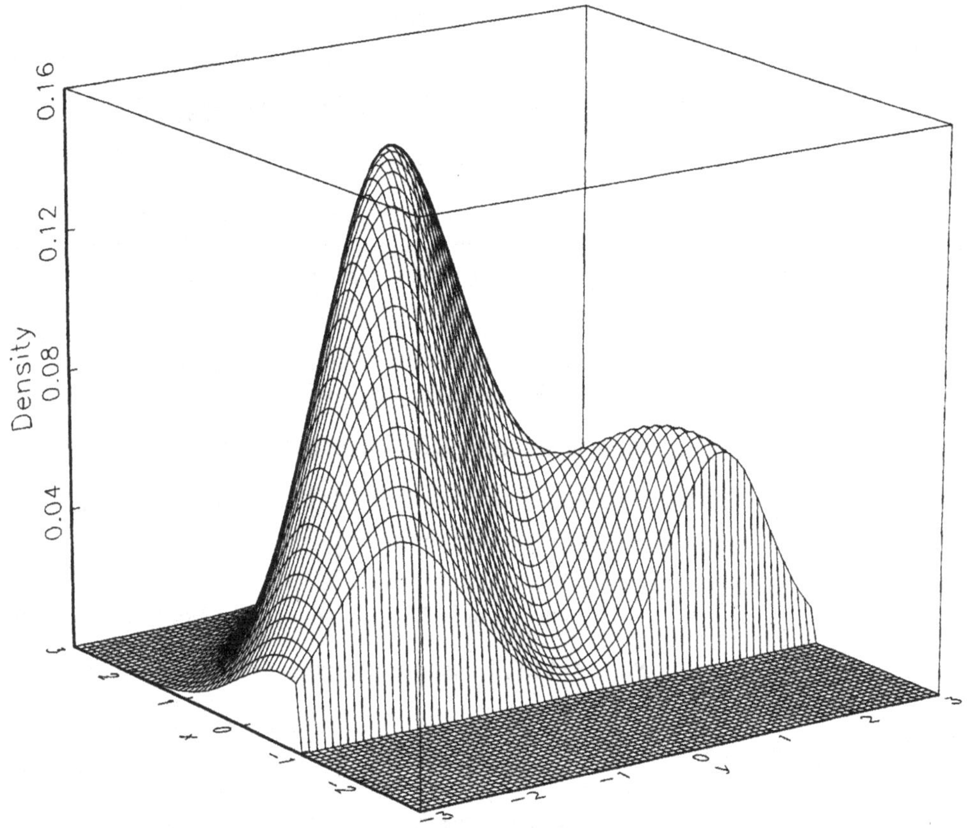

Figure 2-6-4B (a) Density function of a mixture of two bivariate normal distributions and (b) conditional *pdf* of Y, given $X = -1$, illustrated as the face created by a cut at $x = -1$.

3. Let (X, Y) denote a pair of discrete random variables with joint probability function

$$f_{XY}(x, y) = \begin{cases} \dfrac{x+y}{36} & \text{if } x = 1, 2, 3,\ y = 1, 2, 3, \\ 0 & \text{otherwise.} \end{cases}$$

Find the probability function of the conditional distribution of Y, given $X = x$, for $x = 1, 2, 3$.

4. Let (X, Y) denote a pair of continuous random variables with joint *pdf*

$$f_{XY}(x, y) = \begin{cases} x + y & \text{if } 0 \le x, y \le 1, \\ 0 & \text{otherwise.} \end{cases}$$

Find $f_{Y|X}(y|x)$ for $0 \le x \le 1$.

5. If (X, Y) has density function

$$f_{XY}(x, y) = \begin{cases} 2 & \text{if } x \geq 0,\ y \geq 0,\ x + y \leq 1, \\ 0 & \text{otherwise,} \end{cases}$$

find the conditional *pdf* of Y, given $X = x$.

6. For the bivariate normal distribution with $\mu_1 = \mu_2 = 0$, $\sigma_1 = \sigma_2 = 1$, and $-1 < \rho < 1$, prove that the conditional distribution of Y, given $X = x$, is normal with mean ρx and variance $1 - \rho^2$. *Hint*: Use Exercise 11 in Section 2.3.

7. For the uniform distribution inside the unit circle as given in equation (12), find the conditional *pdf* of Y, given $X = x$, for $x = 0, \pm 0.5, \pm 0.75$, and ± 0.95, and graph it. What will this conditional *pdf* look like if x approaches ± 1 ?

8. Consider a pair of continuous random variables with joint *pdf*

$$f_{XY}(x, y) = \begin{cases} \frac{1}{8}(x^2 - y^2)e^{-x} & \text{if } x \geq 0,\ |y| \leq x \\ 0, & \text{otherwise} \end{cases}$$

(see Example 2.3.5). Find the conditional *pdf* of Y, given $X = x$, for $x > 0$.

9. Let (X, Y) denote a pair of continuous random variables with joint *pdf*

$$f_{XY}(x, y) = \begin{cases} 2[1 - (x^2 + y^2)]/\pi & \text{if } x^2 + y^2 \leq 1, \\ 0 & \text{otherwise} \end{cases}$$

(see Example 2.4.3). Find the conditional *pdf* of Y, given $X = x$.

10. Suppose that (X, Y) follows a spherically symmetric distribution centered at the origin (see Section 2.4). Show that the conditional *pdf* $f_{Y|X}(y|x)$ is symmetric about zero.

11. Let the joint *pdf* of (X, Y) be given by

$$f_{XY}(x, y) = \begin{cases} \frac{1}{20}e^{-|y-x|} & \text{if } 0 \leq x \leq 20,\ y \in \mathbb{R}, \\ 0 & \text{otherwise.} \end{cases}$$

Find the conditional *pdf* of Y, given $X = x$, for $x \in [0, 10]$.

12. Let $k \geq 1$ denote a fixed integer, and define a bivariate continuous random variable (X, Y) by the joint density function

$$f_{XY}(x, y) = \begin{cases} \frac{1}{k} & \text{if } 0 \leq x < k,\ 0 \leq y < k, \text{ and } \lfloor x \rfloor = \lfloor y \rfloor, \\ 0 & \text{otherwise} \end{cases}$$

(see Exercise 9 in Section 2.4). Show that all conditional distributions of Y, given $X = x$, are uniform.

13. Suppose that the joint *pdf* of (X, Y) is given by

$$f_{XY}(x, y) = \begin{cases} \dfrac{1}{\sqrt{2\pi(1+x)}} \exp\left[-\dfrac{y^2 + 2x^2(1-y) + 2x + x^4}{2(1+x)}\right] & \text{if } x \geq 0,\ y \in \mathbb{R}, \\ 0 & \text{otherwise.} \end{cases}$$

Graph the joint *pdf* as a contour plot, and find the conditional *pdf* of Y, given $X = x$. *Hint*: Read Example 2.6.5.

14. Let the joint *pdf* of (X, Y) be given by

$$f_{XY}(x, y) = \begin{cases} (x+1)y^x & \text{if } 0 \leq x, y \leq 1, \\ 0 & \text{otherwise.} \end{cases}$$

Graph the joint *pdf* as a contour plot, and find the conditional *pdf* of Y, given $X = x$.

15. Suppose that the joint *pdf* of two random variables (X, Y) is given by

$$f_{XY}(x, y) = \begin{cases} 1 & \text{if } 0 \leq x \leq 1,\ |y| \leq x, \\ 0 & \text{otherwise.} \end{cases}$$

(a) Find the conditional *pdf* of Y, given $X = x$.

(b) Find the conditional *pdf* of X, given $Y = y$.

16. A stick of fixed length $L > 0$ is broken at a random point. Let X denote the length of the remaining piece, i.e., X is uniform in the interval $(0, L)$. The piece of length X is, again, broken at a random point, and the length of the remaining piece is Y. That is, conditionally on $X = x$, the distribution of Y is uniform in the interval $(0, x)$.

(a) Find the joint *pdf* of X and Y.

(b) Find the marginal *pdf* of Y.

(c) Find the conditional *pdf* of X, given $Y = y$.

2.7 Conditional Expectation and Regression

Conditional expectations are expected values obtained from conditional distributions. Using the setup and notation of the preceding sections, we define the conditional expectation of Y, given $X = x$, as

$$E[Y|X = x] = \begin{cases} \displaystyle\sum_y y\, f_{Y|X}(y|x) & \text{if } Y \text{ is discrete,} \\ \displaystyle\int_{-\infty}^{\infty} y\, f_{Y|X}(y|x)\, dy & \text{if } Y \text{ is continuous.} \end{cases} \tag{1}$$

More generally, as justified by Theorem 2.5.1, we can define the conditional expectation of a real-valued function $G(Y)$, given $X = x$, as

$$E[G(Y)|X = x] = \begin{cases} \sum_y G(y)\, f_{Y|X}(y|x) & \text{if } Y \text{ is discrete,} \\ \int_{-\infty}^{\infty} G(y)\, f_{Y|X}(y|x)\, dy & \text{if } Y \text{ is continuous.} \end{cases} \tag{2}$$

For instance, the conditional m^{th} moment of Y, given $X = x$, is given by

$$E[Y^m|X = x] = \int_{y=-\infty}^{\infty} y^m\, f_{Y|X}(y|x)\, dy,$$

written here for the continuous case. We will be interested mostly in conditional means and variances, and thus, introduce the symbols

$$\mu_{Y|x} = E[Y|X = x] \tag{3}$$

and

$$\begin{aligned} \sigma^2_{Y|x} &= \text{var}[Y|X = x] \\ &= E\big[(Y - \mu_{Y|x})^2 | X = x\big]. \end{aligned} \tag{4}$$

We leave it to the student (Exercise 1) to show that

$$\text{var}[Y|X = x] = E[Y^2|X = x] - \big(E[Y|X = x]\big)^2. \tag{5}$$

If X and Y are independent, then the conditional distribution of Y, given $X = x$, does not depend on x and, therefore is the same as the marginal distribution. Then $\mu_{Y|x}$ and $\sigma^2_{Y|x}$ will be the mean and the variance, respectively, of the marginal distribution of Y. In general, however, the conditional mean $\mu_{Y|x}$ and the conditional variance $\sigma^2_{Y|x}$ will depend on x. In fact, conditional means are perhaps the single most important topic in statistics and, hence, deserve a special name.

Definition 2.7.1 For a bivariate random variable (X, Y), the conditional mean $\mu_{Y|x}$, considered as a function of x, is called the *regression* of Y on X. □

The student who has taken a regression course before may be somewhat surprised by this definition, because regression analysis is often viewed as a collection of techniques for fitting functions to observed data. Yet, this definition expresses perfectly the central goal of regression analysis, namely, to estimate the (conditional) mean of a variable, knowing the values of one or several other variables. Classical regression analysis is also, usually, connected to least squares estimation. Later in this section, we are going to provide some theoretical background to this, but first let us illustrate the concepts introduced, so far, with some examples.

Example 2.7.1 continuation of Example 2.6.1

From the conditional probability functions, we obtain the following numerical results:

x	$\mu_{Y\|x}$	$E[Y^2\|X = x]$	$\sigma^2_{Y\|x}$	$\sigma_{Y\|x}$
0	1/2	1/2	1/4	0.5
1	1	5/3	2/3	0.816
2	3/2	7/2	5/4	1.118

It appears that both $\mu_{Y|x}$ and $\sigma^2_{Y|x}$ are increasing functions of x; in addition, $\mu_{Y|x}$ is a linear function of x. More precisely,

$$E[Y|X = x] = \frac{1}{2} + \frac{1}{2}x$$

for all x in the domain of X. We leave it to the student (Exercise 2) to calculate $E[X|Y = y]$ and $\text{var}[X|Y = y]$ in this example. □

Example 2.7.2 continuation of Example 2.2.3

Consider, again, the pair of random variables (X, Z), where X = number shown by the red die and Z = sum of numbers shown. Using the conditional probability functions (Exercise 1 in Section 2.6), we find that

$$\mu_{Z|x} = 3.5 + x, \quad x = 1, \ldots, 6$$

and

$$\sigma^2_{Z|x} = 35/12, \quad x = 1, \ldots, 6,$$

that is, the conditional mean, again, is a linear function of x, but the conditional variance does not depend on x. This is an almost classical regression situation (see the discussion later in this section), except that the deviations of Z from the conditional mean are discrete uniform rather than normal. Conversely, the conditional mean of X, given $Z = z$, is $\mu_{X|z} = z/2$. The conditional variance of X, given $Z = z$, is the variance of a discrete uniform distribution on m integers, m ranging from 1 (if $Z = 2$ or $Z = 12$) to 6 (if $Z = 7$). It turns out that the conditional variance can be written as

$$\text{var}[X|Z = z] = \frac{m^2 - 1}{12}, \quad z = 2, \ldots, 12,$$

where

$$m = \min\{z - 1,\ 13 - z\};$$

see Exercise 3. □

Example 2.7.3 In the "raindrop in a circle" example, the conditional distribution of Y, given $X = x$, has been found to be uniform in the interval $(-\sqrt{1 - x^2}, \sqrt{1 - x^2})$, see equation (2) in

Section 2.6. Hence, $\mu_{Y|x} = 0$, and $\sigma^2_{Y|x} = (1-x^2)/3$. In this example, the conditional mean does not depend on x, but the conditional variance does. □

Example 2.7.4 continuation of Example 2.2.7

If (X,Y) is bivariate normal with parameters μ_X, μ_Y, σ^2_X, σ^2_Y, and ρ, the density function of the conditional distribution of Y, given $X = x$, is

$$f_{Y|X}(y|x) = \frac{1}{\sqrt{2\pi}\,\sigma_{Y|x}} \exp\left[-\frac{1}{2}\left(\frac{y-\mu_{Y|x}}{\sigma_{Y|x}}\right)^2\right], \tag{6}$$

where

$$\mu_{Y|x} = \mu_Y + \frac{\rho\sigma_Y}{\sigma_X}(x-\mu_X) \tag{7}$$

and

$$\sigma^2_{Y|x} = (1-\rho^2)\sigma^2_Y. \tag{8}$$

In words, the conditional distribution of Y is normal, the regression function is a linear function of x, and the conditional variance is a constant, independent of x. Technical details are left to the student; see Exercise 4. Figure 2.7.1 shows a contour plot of a bivariate normal, with parameters $\mu_X = 4$, $\mu_Y = 2$, $\sigma_X = 4$, $\sigma_Y = 3$, and $\rho = 0.7$. The two conditional means, or *regression functions*

$$r(x) := E[Y|X=x] = \mu_Y + \rho\frac{\sigma_Y}{\sigma_X}(x-\mu_X)$$

and

$$r^*(y) := E[X|Y=y] = \mu_X + \rho\frac{\sigma_X}{\sigma_Y}(y-\mu_Y)$$

are shown as a solid and a broken line, respectively. Notice that the two regression lines intersect the ellipses of constant density exactly at those points where tangent lines to the ellipses are horizontal or vertical. We leave it to the reader to explain this fact; see Exercise 5. Most important, however, it should be noted that the two regression lines are not identical, contrary to what one might naively assume. Only if the correlation ρ approaches ± 1 will the two regression lines coincide. □

The bivariate normal distribution is an example where the classical assumptions of the simple linear regression model, namely,

(i) $E[Y|X=x] = \alpha + \beta x$,

(ii) $\text{var}[Y|X=x]$ is a constant, independent of x, and

(iii) the conditional distribution of Y, given $X = x$, is normal for all x,

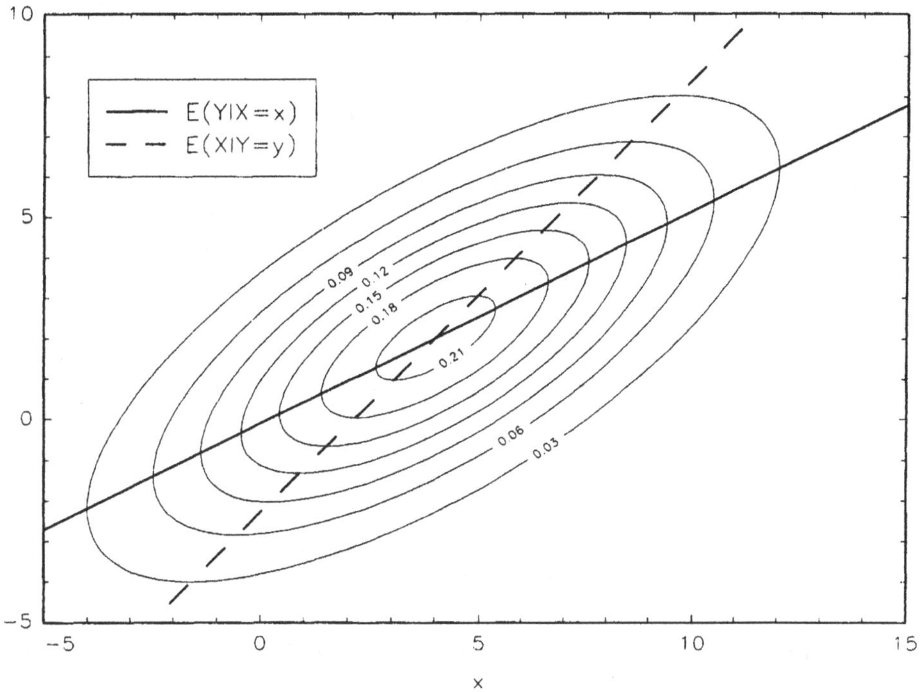

Figure 2-7-1 Contour plot of a bivariate normal distribution with $\mu_X = 4$, $\mu_Y = 2$, $\sigma_X = 4$, $\sigma_Y = 3$, and $\rho = 0.7$. The solid line and the broken line are graphs of the regression of Y on X and the regression of X on Y, respectively.

are all satisfied. However, the bivariate normal is by no means the only such example, as the following construction shows. Define the conditional distribution of Y, given $X = x$, as normal with mean $\alpha + \beta x$ and variance σ^2, and choose any probability density function f_X for X, discrete or continuous. Then the joint *pdf*

$$f_{XY}(x,y) = f_{Y|x}(y|x) \cdot f_X(x)$$

satisfies the three conditions. Note that the case where X is discrete is actually an example of a mixed distribution, which we will discuss in Section 2.8.

Example 2.7.5 continuation of Examples 2.6.4 and 2.6.5

Suppose that the joint *pdf* of (X, Y) is given by

$$f_{XY}(x, y) = \begin{cases} \frac{1}{10\sqrt{2\pi}} \exp\left[-\frac{1}{2}(y-x)^2\right] & \text{if } 0 \le x \le 10;\ y \in \mathbb{R} \\ 0 & \text{otherwise.} \end{cases}$$

Then $E[Y|X=x]=x$, and $\text{var}[Y|X=x]=1$. Similarly, if

$$f_{XY}(x,y)=\begin{cases}\dfrac{1}{\sqrt{2\pi}\,x}\exp\left[-\dfrac{y^2}{2x}+y-\dfrac{3}{2}x\right] & \text{if } x>0,\ y\in\mathbb{R},\\ 0 & \text{otherwise,}\end{cases}$$

then $E[Y|X=x]=x$, and $\text{var}[Y|X=x]=x^2$. In Figure 2.6.1 (p. 80), the regression of Y on X can, again, be obtained by connecting all points on the ellipses with vertical tangents. □

Example 2.7.6 This example illustrates a situation where the regression of Y on X is not linear. Suppose that X has *pdf*

$$f_X(x)=\begin{cases}2x & \text{if } 0\le x\le 1,\\ 0 & \text{otherwise,}\end{cases}$$

and the conditional distribution of Y, given $X=x$, is normal with mean $2x^2$ and variance 1. Then the joint *pdf* is

$$f_{XY}(x,y)=\begin{cases}\sqrt{\dfrac{2}{\pi x}}\exp\left[-\dfrac{1}{2}(y-2x^2)^2\right] & \text{if } 0\le x\le 1,\ y\in\mathbb{R},\\ 0 & \text{otherwise.}\end{cases}$$

This joint *pdf*, along with the regression function $r(x)=2x^2$, is shown in Figure 2.7.2 in form of a contour plot. □

Although the preceding examples have been constructed such that the conditional distributions of Y, given $X=x$, are all normal, it is just as easy to construct examples where the conditional distribution has some other specified form, with a given regression function; see Exercises 6 and 7.

Let us return briefly to Example 2.7.4, the bivariate normal. If we write the regression function as

$$r(x)=E[Y|X=x]=\alpha+\beta x,$$

then $\beta=\rho\sigma_Y/\sigma_X$, and $\alpha=\mu_Y-\beta\mu_X$. Since $\text{cov}[X,Y]=\rho\sigma_X\sigma_Y$, the slope of the regression function is

$$\beta=\frac{\rho\sigma_X\sigma_Y}{\sigma_X^2}=\frac{\text{cov}[X,Y]}{\text{var}[X]}.$$

Similarly, in both Examples 2.7.1 and 2.7.2, one can verify that

$$E[Y|X=x]=\mu_Y+\frac{\text{cov}[X,Y]}{\text{var}[X]}(x-\mu_X).$$

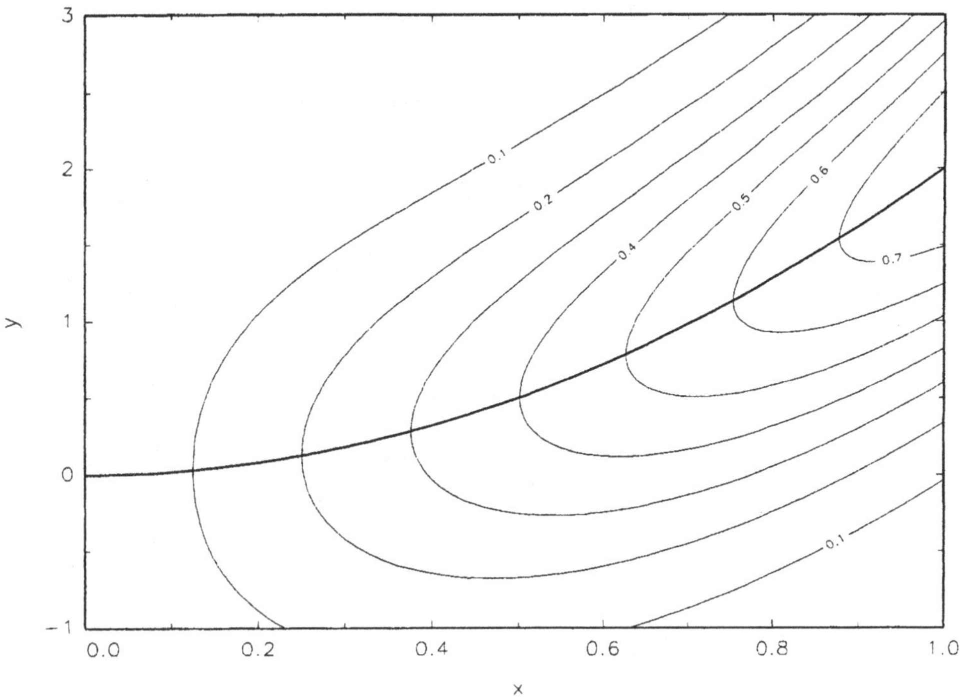

Figure 2-7-2 Contour plot of the joint density function in Example 2.7.6. The solid curve is a graph of the regression function of Y on X, $r(x) = 2x^2$.

This is the same expression for the intercept and slope as in the bivariate normal case. The same result holds even for the "raindrop" example, where the slope is zero. Lucky coincidence? Not really, as the following theorem shows.

Theorem 2.7.1 *Let (X,Y) denote a bivariate random variable such that $E[Y|X = x] = \alpha + \beta x$, i.e., the regression of Y on X is a straight line with some intercept α and some slope β. Then*

$$\beta = cov[X,Y]/var[X], \tag{9}$$

and

$$\alpha = E[Y] - \beta \cdot E[X]. \tag{10}$$

Proof We give a proof for the continuous case, the discrete case being analogous. The assumption of the theorem says that

$$E[Y|X = x] = \int_{y=-\infty}^{\infty} y\, f_{Y|X}(y|x)\, dy = \alpha + \beta x \quad \text{for all } x.$$

But $f_{Y|X}(y|x) = f_{XY}(x,y)/f_X(x)$, which implies

$$\int_{y=-\infty}^{\infty} y\, f_{XY}(x,y)\,dy = (\alpha + \beta x) f_X(x) \qquad \text{for all } x. \tag{11}$$

Then integrating (11) over x gives

$$\int_{x=-\infty}^{\infty}\int_{y=-\infty}^{\infty} y\, f_{XY}(x,y)\,dy\,dx = \alpha \int_{x=-\infty}^{\infty} f_X(x)\,dx + \beta \int_{x=-\infty}^{\infty} x\, f_X(x)\,dx,$$

or

$$E[Y] = \alpha + \beta\, E[X],$$

which immediately yields $\alpha = E[Y] - \beta\, E[X]$. Next, multiplying (11) by x and integrating, again, over x gives

$$\int_{x=-\infty}^{\infty}\int_{y=-\infty}^{\infty} xy\, f_{XY}(x,y)\,dy\,dx = \alpha \int_{x=-\infty}^{\infty} x\, f_X(x)\,dx + \beta \int_{y=-\infty}^{\infty} x^2 f_X(x)\,dx,$$

or

$$E[XY] = \alpha\, E[X] + \beta\, E[X^2].$$

Using α from above gives

$$E[XY] = E[X]E[Y] - \beta \cdot (E[X])^2 + \beta \cdot E[X^2],$$

which, in turn, implies

$$\beta = \frac{E[XY] - E[X]E[Y]}{E[X^2] - (E[X])^2} = \frac{\text{cov}[X,Y]}{\text{var}[X]}.$$

This completes the proof. ■

In applied regression analysis, with (x_i, y_i), $i = 1, \ldots, n$, denoting observed data pairs, the least squares estimators of the parameters α and β are

$$\hat{\beta} = S_{xy}/S_{xx},$$

and

$$\hat{\alpha} = \bar{y} - \hat{\beta}\bar{x}, \tag{12}$$

where $S_{xy} = \sum_{i=1}^{n}(x_i - \bar{x})(y_i - \bar{y})$ and $S_{xx} = \sum_{i=1}^{n}(x_i - \bar{x})^2$. Using S_{xy}/n and S_{xx}/n to estimate cov$[X,Y]$ and var$[X]$, respectively, $\hat{\alpha}$ and $\hat{\beta}$ may be viewed as valid estimators of a regression function under the very general setup of this theorem, without invoking any normality assumptions.

We conclude this section with two results that are useful for computing means and variances in seemingly intractable situations. The "conditional mean theorem" will be of considerable importance in the derivation of principal components in Chapter 8. Suppose we know the joint distribution of X and Y, and we are interested in the mean and variance of Y, but these are difficult to obtain from the joint distribution. Yet, if the conditional mean and variance of Y, given $X = x$, are known for all x, then it is often easy to find $E[Y]$ and $\text{var}[Y]$, as Theorems 2.7.2 and 2.7.3 show.

We will use special notation. For a pair of random variables (X, Y) with joint *pdf* $f_{XY}(x, y)$, we will write

$$r(x) = E[Y|X = x] \tag{13}$$

for the *regression function*, and

$$v(x) = \text{var}[Y|X = x], \tag{14}$$

which we will call the *conditional variance function*. By applying the functions r and v to the random variable X, we obtain two random variables $r(X)$ and $v(X)$, which will be useful in the following two theorems.

Theorem 2.7.2 (Conditional Mean Theorem) *For jointly distributed random variables* (X,Y) *with regresssion function* $r(x)$ *as in equation (13),*

$$E[Y] = E\big[r(X)\big]. \tag{15}$$

Proof We give the proof for the continuous case, the discrete case being analogous with summation replacing integration. By definition,

$$\begin{aligned} E[Y] &= \int_{x=-\infty}^{\infty} \int_{y=-\infty}^{\infty} y\, f_{XY}(x, y)\, dy\, dx \\ &= \int_{x=-\infty}^{\infty} \left[\int_{y=-\infty}^{\infty} y\, f_{Y|X}(y|x)\, dy \right] f_X(x)\, dx. \end{aligned}$$

But the inner integral is $r(x)$, and therefore,

$$\begin{aligned} E[Y] &= \int_{x=-\infty}^{\infty} r(x)\, f_X(x)\, dx \\ &= E\big[r(X)\big], \end{aligned}$$

as was to be shown. ■

Example 2.7.7 continuation of Example 2.7.2

Consider, again, the joint distribution of X and Z, where $Z = X + Y$ and X, Y are the numbers shown by two independent regular dice. Then $r(x) = x + 3.5$, and, by the conditional mean theorem,

$$E[Y] = E[r(X)] = E[X + 3.5] = E[X] + 3.5 = 7. \qquad \square$$

Example 2.7.8 continuation of Example 2.7.6

If (X, Y) have joint *pdf*

$$f_{XY}(x, y) = \begin{cases} \sqrt{\dfrac{2}{\pi}} \exp\left[-\dfrac{1}{2}(y - 2x^2)^2 \right] & \text{if } 0 \le x \le 1;\ y \in \mathbb{R} \\ 0 & \text{otherwise,} \end{cases}$$

then it may seem impossible to compute $E[Y]$ because finding the marginal *pdf* of Y is not easy. However, we know that

$$f_X(x) = \begin{cases} 2x & \text{if } 0 \le x \le 1, \\ 0 & \text{otherwise} \end{cases}$$

and

$$r(x) = 2x^2.$$

Hence,

$$\begin{aligned} E[Y] &= E[r(X)] \\ &= \int_{-\infty}^{\infty} r(x)\, f_X(x)\, dx \\ &= \int_0^1 4x^3\, dx \\ &= 1. \end{aligned}$$

$\square$

The next theorem shows how to obtain the variance of Y from the conditional mean and variance of Y, given $X = x$.

Theorem 2.7.3 (Conditional Variance Theorem) *For two jointly distributed random variables (X,Y) and with $r(x)$ and $v(x)$ defined as in equations (13) and (14),*

$$\text{var}[Y] = E\big[v(X)\big] + \text{var}\big[r(X)\big]. \tag{16}$$

Proof First, notice that

$$v(X) = \text{var}[Y|X] = E[Y^2|X] - (E[Y|X])^2 = E[Y^2|X] - r^2(X).$$

By the conditional mean theorem applied to Y^2,

$$E[v(X)] = E\Big[\mathrm{var}[Y|X]\Big] = E\left[E[Y^2|X]\right] - E\left[r^2(X)\right]$$
$$= E[Y^2] - E\left[r^2(X)\right].$$

Moreover,

$$\mathrm{var}[r(X)] = \mathrm{var}\Big[E[Y|X]\Big] = \mathrm{var}[r(X)]$$
$$= E\left[r^2(X)\right] - \{E[r(X)]\}^2$$
$$= E\left[r^2(X)\right] - \{E[Y]\}^2.$$

Then adding the two previous equations gives

$$E[v(X)] + \mathrm{var}[r(X)] = E[Y^2] - \{E[Y]\}^2 = \mathrm{var}[Y]. \quad ■$$

Example 2.7.9 continuation of Example 2.7.7

In the dice example, the conditional mean and conditional variance of Z, given $X = x$, are given by

$$r(x) = 3.5 + x, \quad x = 1, \ldots, 6,$$

and

$$v(x) = 35/12, \quad x = 1, \ldots, 6.$$

Since $v(x)$ is constant, $E\big[v(X)\big] = 35/12$. Moreover, $\mathrm{var}\big[r(X)\big] = \mathrm{var}[3.5 + X] = \mathrm{var}[X] = 35/12$. Hence $\mathrm{var}[Y] = 35/12 + 35/12 = 35/6$. (This result, of course, is easier to obtain using the fact that Z is the sum of two independent random variables, each of them with variance $35/12$). □

Example 2.7.10 This particularly simple example is meant to further clarify the two preceding theorems. Suppose that the joint probability function of X and Y is given by the following table.

	$X = 1$	$X = 2$	$X = 3$
$Y = 1$	1/9	1/9	1/9
$Y = 2$	1/9	1/9	1/9
$Y = 3$	1/9	1/9	0
$Y = 4$	1/9	0	0
$f_X(x)$	4/9	3/9	2/9
$r(x)$	2.5	2	1.5
$v(x)$	5/4	2/3	1/4

The table also displays the marginal probability function of X, the values of the regression function $r(x)$, and the conditional variance function $v(x)$ for $x =$

1, 2, 3. Thus, $r(X)$ is a discrete random variable, taking values 2.5, 2, and 1.5 with probabilities 4/9, 3/9, and 2/9, respectively, and $v(X)$ is a discrete random variable taking values 5/4, 2/3, and 1/4 with the same probabilities. By the conditional mean theorem,

$$E[Y] = E\big[r(X)\big] = \frac{4}{9} \cdot 2.5 + \frac{3}{9} \cdot 2 + \frac{2}{9} \cdot 2 = \frac{19}{9}.$$

Similarly, we can calculate $\text{var}\big[r(X)\big] = E\big[r^2(X)\big] - (19/9)^2 = 25/162$ and $E\big[v(X)\big] = 5/6$, and use the conditional variance theorem to obtain

$$\text{var}[Y] = E\big[v(X)\big] + \text{var}\big[r(X)\big] = \frac{80}{81}.$$ □

In Example 2.7.10, we could have obtained var$[Y]$ more easily from the marginal distribution of Y, so let us proceed to a more challenging one.

Example 2.7.11 continuation of Example 2.6.5

Suppose that the random variable X is exponential with parameter $\lambda = 1$, i.e.,

$$f_X(x) = \begin{cases} e^{-x} & \text{if } x > 0 \\ 0 & \text{otherwise,} \end{cases}$$

and the conditional distribution of Y, given $X = x$, is normal with mean x and variance x^2. The joint *pdf* of X and Y has been illustrated as a contour plot in Figure 2.6.1 (p. 80). From the joint *pdf* of (X, Y), it seems difficult to calculate the mean and variance of Y, but using the two preceding theorems it turns out that it is actually straightforward. First, note that $r(x) = x$ and $v(x) = x^2$. Then the conditional mean theorem gives

$$E[Y] = E\big[r(X)\big] = E[X] = 1.$$

Next, $\text{var}\big[r(X)\big] = \text{var}[X] = 1$, and $E\big[v(X)\big] = E[X^2] = 2$. Therefore, the conditional variance theorem gives

$$\text{var}[Y] = \text{var}\big[r(X)\big] + E\big[v(X)\big] = 3.$$ □

Particularly useful applications of the two theorems occur in the theory of mixtures of distributions, which will be introduced in the next section.

Here are some final remarks on notation. Most textbooks use the symbol $E[Y|X]$ instead of $r(X)$ and $\text{var}[Y|X]$ instead of $v(X)$, that is, $E[Y|X]$ is a random variable which takes the value $E[Y|X = x]$ when X takes the value x, and $\text{var}[Y|X]$ is a random variable taking the value $\text{var}[Y|X = x]$ when $X = x$. Then equations (15) and (16) read

$$E[Y] = E\Big[E[Y|X]\Big] \tag{17}$$

and

$$\text{var}[Y] = E\Big[\text{var}[Y|X]\Big] + \text{var}\Big[E[Y|X]\Big], \tag{18}$$

respectively. This notation will be used extensively in Chapter 8.

Exercises for Section 2.7

1. Prove equation (5).

2. In Example 2.7.1, find the conditional mean and the conditional variance of X, given $Y = y$, for $y = 0, \ldots, 4$.

3. In Example 2.7.2, find $E[X|Z = z]$ and $\text{var}[X|Z = z]$ for $z = 2, \ldots, 12$, and thus verify the results given in the text.

4. Verify the results of Example 2.7.4, i.e., verify that, for the bivariate normal distribution with *pdf* as in equation (3) of Section 2.2, the conditional distribution of Y, given $X = x$, is normal with mean and variance given by (7) and (8). *Hint*: Find f_{XY} as the product of $f_{Y|X}$ and f_X, where f_X is the *pdf* of a normal random variable with mean μ_X and variance σ_X^2.

5. Suppose that (X, Y) is bivariate normal. Argue that the regression of Y on X can be thought of as the set of all points where the ellipses of constant density have a vertical tangent; see also Figure 2.7.1.

6. Consider a joint *pdf* of two random variables such that X is exponential with parameter $\lambda = 1$, i.e.,

$$f_X(x) = \begin{cases} e^{-x} & \text{if } x \geq 0, \\ 0 & \text{otherwise,} \end{cases}$$

and the conditional distribution of Y, given $X = x$, is double-exponential with mean x, i.e.,

$$f_{Y|X}(y|x) = \frac{1}{2}e^{-|y-x|}, \quad y \in \mathbb{R}.$$

(a) Find the joint *pdf* of X and Y.

(b) Find the regression function $r(x) = E[Y|X = x]$ and the conditional variance function $v(x) = \text{var}[Y|X = x]$.

(c) Graph the joint *pdf* as a contour plot, and graph the regression function in the same plot. Do the "vertical tangents" of problem 5 work in this case?

(d) Find $E[Y]$ and $\text{var}[Y]$ using the conditional mean and conditional variance theorems.

7. Consider the joint distribution of two random variables (X, Y) such that X has *pdf*

$$f_X(x) = \begin{cases} 2x & \text{if } 0 \leq x \leq 1, \\ 0 & \text{otherwise,} \end{cases}$$

and the *pdf* of the conditional distribution of Y, given $X = x$, is given by

$$f_{Y|X}(y|x) = \begin{cases} 3[1 - (y - \sqrt{x})^2]/4 & \text{if } |y - \sqrt{x}| \leq 1, \\ 0 & \text{otherwise.} \end{cases}$$

(a) Find the joint *pdf* of X and Y.

(b) Find the regression function $r(x) = E[Y|X = x]$ and the conditional variance function $v(x) = \text{var}[Y|X = x]$.

(c) Graph the joint *pdf* as a contour plot, and graph the regression function in the same plot. Do the "vertical tangents" of problem 5 work in this case?

(d) Find $E[Y]$ and $\text{var}[Y]$ using the conditional mean and conditional variance theorems.

8. Find the conditional mean and variance of Y, given $X = x$, if X and Y have joint *pdf*

$$f_{XY}(x, y) = \begin{cases} \exp[-\frac{x^2}{2} + x - y]/\sqrt{2\pi} & \text{if } x \leq y \\ 0 & \text{otherwise;} \end{cases}$$

see Example 2.6.2.

9. Find the regression of Y on X and the conditional variance of Y, given $X = x$, if the joint probability function of X and Y is expressed by

$$f_{XY}(x, y) = \begin{cases} (x + y)/36 & \text{if } x = 1, 2, 3;\ y = 1, 2, 3 \\ 0 & \text{otherwise.} \end{cases}$$

See also Exercise 15 in Section 2.2.

10. If X and Y have joint *pdf*

$$f_{XY}(x, y) = \begin{cases} x + y & \text{if } 0 \leq x, y \leq 1 \\ 0 & \text{otherwise,} \end{cases}$$

find the conditonal mean and the conditional variance of Y, given $X = x$, for $0 \leq x \leq 1$. See also Exercise 17 in Section 2.2.

11. Suppose that X is uniform in the interval $[-1, 1]$ and

$$Y = \begin{cases} X & \text{if } |X| \leq c \\ -X & \text{if } |X| > c \end{cases}$$

where c is a fixed constant in the interval $[0, 1]$. Find the regression of Y on X and the conditional variance of Y, given $X = x$. See also Exercises 12 and 13 of Section 2.3.

12. For a fixed integer $k \geq 1$, define the joint *pdf* of two random variables X and Y by

$$f_{XY}(x, y) = \begin{cases} \frac{1}{k} & \text{if } 0 \leq x < k,\ 0 \leq y < k, \text{ and } \lfloor x \rfloor = \lfloor y \rfloor \\ 0 & \text{otherwise,} \end{cases}$$

as in Exercise 9 of Section 2.4. Show that the regression of Y on X is a step function, i.e., piecewise constant.

13. Find the regression of Y on X for the random variable (X,Y) with density function

$$f_{XY}(x,y) = \begin{cases} 2 & \text{if } x \geq 0,\ y \geq 0,\ x+y \leq 1, \\ 0 & \text{otherwise.} \end{cases}$$

Also find the conditional variance of Y, given $X = x$.

14. Let (X,Y) denote a bivariate continuous random variable with a uniform distribution inside the triangle with vertices (0,0), (1,1), and (2,0).

(a) Show that the regression of X on Y is a linear function of Y.

(b) Show that the regression of Y on X is a piecewise linear function of X but is not overall linear.

15. If the density function of the bivariate random variable (X,Y) is given by

$$f_{XY}(x,y) = \begin{cases} \dfrac{1}{x}\exp[-(x+\dfrac{y}{x})] & x > 0,\ y > 0, \\ 0 & \text{otherwise,} \end{cases}$$

find the regression of Y on X.

16. Let the joint density function of (X,Y) be given by

$$f_{XY}(x,y) = \begin{cases} \dfrac{1}{20}\exp[-|y-2x|] & 0 \leq x \leq 10,\ \ y \in \mathbb{R}, \\ 0 & \text{otherwise.} \end{cases}$$

(a) Find the marginal distribution of X.

(b) Find the conditional distribution of Y, given $X = x$.

(c) Find the regression of Y on X.

17. Consider the following game. You roll a fair die (X), then your friend tosses a fair coin. Suppose that the die shows x, then

- you get from your friend $(x-3)^2/2$ dollars if the coin shows "heads," or
- you pay your friend 2 dollars if the coin shows "tails."

Call your gain (in one game) Y.

(a) Find the joint distribution of X and Y.

(b) Is the game fair, that is, is $E[Y] = 0$?

(c) Find the regression of Y on X.

18. Mort, the statistician in the Insurance Company "Dewey, Cheatam, and Howe," figures that the number of people who will buy life insurance in 1999 is a Poisson random variable N with parameter λ, i.e.,

$$\Pr[N = n] = e^{-\lambda}\frac{\lambda^n}{n!}, \qquad n = 0, 1, \ldots$$

Each of the insured people will produce a claim with probability p and fail to do so with probability $1-p$. Furthermore, it is assumed that claims are independent, so, conditionally on $N = n$, the number of claims Y is a binomial random variable with parameters n and p.

(a) Find the joint probability function of N and Y.

(b) Find $E[Y]$ and $\mathrm{var}[Y]$.

19. Suppose that X and Y are two jointly distributed random variables such that $E[Y|X = x] = \alpha + \beta x$ (i.e., the regression of Y on X is linear). If the conditional variance of Y, given $X = x$, does not depend on x, show that

$$\mathrm{var}[Y|X = x] = (1 - \rho^2)\mathrm{var}[Y],$$

where ρ is the correlation between X and Y.

20. Suppose that (X, Y) are two jointly distributed, random variables, such that $E[X] = \mu$, $\mathrm{var}[X] = \sigma^2$, $E[Y|X = x] = \alpha + \beta x$ for some real numbers α and β and $\mathrm{var}[Y|X = x] = \tau^2$ for all x. Find $E[Y]$ and $\mathrm{var}[Y]$.

21. In Exercise 16 of Section 2.6, use the conditional mean and conditional variance theorems to find $E[Y]$ and $\mathrm{var}[Y]$.

22. Prove equations (9) and (10) using the conditional mean theorem and by calculating $E[X\,r(X)]$, where $r(X)$ is the conditional mean.

23. This exercise generalizes Examples 2.2.1 and 2.7.1. Let $N \geq 1$ be a fixed integer, and denote by X a discrete random variable taking values $0, 1, \ldots, N$ with probability $1/(N + 1)$ each. Let Y denote a discrete random variable which, conditionally on $X = x$, takes values $0, 1, \ldots, x$ with equal probabilities.

(a) Find the joint *pdf* of X and Y.

(b) Find $E[Y|X = x]$.

(c) Using the conditional mean theorem, find $E[Y]$

(d) Using the conditional variance theorem, find $\mathrm{var}[Y]$.
Hint: $\sum_{i=0}^{N} i^2 = N(N + 1)(2N + 1)/6$.

24. Let $X_1, \ldots, X_N$ be identically distributed random variables. Show that

$$E\left[X_1 \middle| \sum_{i=1}^{N} X_i = c\right] = \frac{c}{N}.$$

Hint: Find $\sum_{i=1}^{N} E\left[X_i \middle| \sum_{h=1}^{N} X_h = c\right]$.

2.8 Mixed Discrete-Continuous Distributions and Finite Mixtures

This section is somewhat unusual for a course in multivariate statistics. Perhaps surprisingly so, because finite mixtures of distributions are an interesting topic with great potential for applications, as we shall see soon. Yet the notions of joint, marginal and conditional distributions are all we need to understand finite mixtures.

Let us start out with a simple example. Suppose the probability that a randomly selected man is a smoker is q_1, and, for a woman, it is q_2. Let $Y = 0$ for a nonsmoker, and $Y = 1$ for a smoker. Then for men,

$$\Pr[Y = y] = f_1(y) = q_1^y(1 - q_1)^{1-y} \qquad y = 0, 1,$$

and, for women,

$$\Pr[Y = y] = f_2(y) = q_2^y(1 - q_2)^{1-y} \qquad y = 0, 1.$$

Next, assume that we cannot choose to interview a fixed number of men and women, but rather ask randomly selected people. Thus, set $X = 1$ for men, and $X = 2$ for women. Assume $\Pr[X = 1] = p_1$, and $\Pr[X = 2] = p_2 = 1 - p_1$. Then the joint probability function of (X, Y) and the marginal probability functions of X and Y are given in the following table:

	Y=0	Y=1	
X=1	$p_1(1 - q_1)$	p_1q_1	p_1
X=2	$p_2(1 - q_2)$	p_2q_2	p_2
	$1 - (p_1q_1 + p_2q_2)$	$p_1q_1 + p_2q_2$	

The marginal probability function of Y can be written as

$$\begin{aligned} f_Y(y) &= (p_1q_1 + p_2q_2)^y[1 - (p_1q_1 + p_2q_2)]^{1-y}, \qquad y = 0, 1 \\ &= p_1 f_1(y) + p_2 f_2(y). \end{aligned}$$

Conditionally on $X = j$ ($j = 1, 2$), the distribution of Y is binary with success probability q_j. The marginal distribution of Y is itself binary, with parameter $p_1q_1 + p_2q_2$. The marginal distribution of Y is called a *mixture* because its probability function can be written in the form $p_1 f_1 + p_2 f_2$, where f_i are the conditional probability functions, which are "mixed" in proportions p_1 and p_2.

In the preceding example, the conditional distributions and the marginal distribution of Y are all of the same type, namely, binary. In general, however, this will not be the case. Suppose, for instance, that the number of traffic accidents per year caused by a member of risk group j is a Poisson random variable with parameter λ_j. For the jth risk group,

$$\Pr[Y = y] = f_j(y) = \frac{\lambda_j^y}{y!}e^{-\lambda_j}, \qquad y = 0, 1, 2, \ldots$$

Now, suppose that there are k risk groups, and a customer falls into the jth group with probability p_j ($p_1 + \ldots + p_k = 1$). Then again we can define a random variable X such that $\Pr[X = j] = p_j$ ($j = 1, \ldots, k$), find the joint distribution of X and Y, and the marginal distribution of Y. The latter is given by the probability function

$$\Pr[Y = y] = f_Y(y) = \sum_{j=1}^{k} p_j f_j(y)$$

$$= \frac{1}{y!}\sum_{j=1}^{k} p_j \lambda_j^y e^{-\lambda_j}, \qquad y = 0, 1, 2, \ldots$$

This marginal distribution is not a Poisson distribution, except for special cases.

In both examples just discussed, variable X plays the role of a "group code" and Y is, in a sense, the variable in which we are interested. In both cases, the joint probability function was defined in terms of the *marginal* probability function of X and the *conditional* probability function of Y, given $X = j$; see Section 2.6. Typically, however, mixture analysis is concerned with cases where Y is a continuous random variable. Thus, first, we need to discuss bivariate distributions with one discrete (X) and one continuous (Y) component.

To simplify notation, we restrict ourselves to the case where X takes values $1, \ldots, k$ with probabilities $p_1, \ldots, p_k$, where $\sum_{j=1}^{k} p_j = 1$. Let the conditional distribution of Y, given $X = j$, be given by the density function $f_j(y)$. Then we can use equation (7) of Section 2.6 to define the joint *pdf* of X and Y as

$$f_{XY}(j,y) = p_j\, f_j(y), \qquad j = 1, \ldots, k;\ y \in \mathbb{R}\,. \tag{1}$$

This is neither a probability function nor a density function, but it is a valid *probability density function:* for each $j = 1, \ldots, k$, and for any real numbers a, b ($a < b$),

$$\Pr[X = j, a \le Y \le b] = p_j \int_a^b f_j(y)\, dy. \tag{2}$$

Moreover,

$$\sum_{j=1}^{k} \int_{y=-\infty}^{\infty} f_{XY}(j, y)\, dy = \sum_{j=1}^{k} p_j \int_{y=-\infty}^{\infty} f_j(y)\, dy$$

$$= \sum_{j=1}^{k} p_j$$

$$= 1,$$

and $f_{XY}(j, y) \ge 0$ for all (j, y). Hence, all conditions for a probability density function are satisfied. In working with the mixed discrete-continuous random variable (X, Y), we need to remember only to use summation for all operations involving X and integration for all operations involving Y.

Using this rule, we can now find all marginal and conditional distributions:

(i) *marginal distribution* of X:

$$f_X(j) = \int_{y=-\infty}^{\infty} f_{XY}(j, y)\, dy = p_j \int_{y=-\infty}^{\infty} f_j(y)\, dy \\ = p_j \qquad j = 1, \ldots, k, \tag{3}$$

as was to be anticipated since we started out with this marginal probability function.

(ii) *conditional distribution* of Y, given $X = j$:

$$f_{Y|X}(y|X = j) = \frac{f_{XY}(j, y)}{f_X(j)} \\ = \frac{p_j\, f_j(y)}{p_j} \\ = f_j(y) \qquad j = 1 \ldots, k, \tag{4}$$

which, again, is only what we started with.

(iii) *marginal distribution* of Y:

$$f_Y(y) = \sum_{j=1}^{k} f_{XY}(j, y) \\ = \sum_{j=1}^{k} p_j\, f_j(y) \qquad y \in \mathbb{R}. \tag{5}$$

(iv) *conditional distribution* of X, given $Y = y$:

$$f_{X|Y}(j|Y = y) = \frac{f_{XY}(j, y)}{f_Y(y)} \\ = \frac{p_j\, f_j(y)}{\sum_{h=1}^{k} p_h\, f_h(y)} \\ = \Pr[X = j|Y = y] \qquad i = 1, \ldots, k. \tag{6}$$

The marginal distribution of Y, given in equation (5), is the main topic of this section. Therefore, we give a formal definition.

Definition 2.8.1 Suppose that the random variable Y has a distribution that can be represented by a *pdf* of the form

$$f_Y(y) = p_1 f_1(y) + \ldots + p_k f_k(y),$$

where all p_j are positive, $p_1 + \ldots + p_k = 1$, and all $f_j(y)$ are probability density functions. Then Y is said to have a *finite mixture distribution* with k components. □

The word "finite" in Definition 2.8.1 refers to the fact that the number k of groups, or components, is assumed to be finite. A good way to think of finite mixtures is as follows. There are k different possible experiments that can be performed. The outcome of the jth experiment can be described by a density function $f_j(y)$. First, the investigator chooses an experiment, such that the jth experiment has probability p_j of being selected. Then the chosen experiment is performed, and the value of Y is measured. The random variable Y then follows the mixture *pdf* $f_Y(y) = \sum_{j=1}^{k} p_j f_j(y)$.

Example 2.8.1 This example makes no practical sense whatsoever, but is simple enough to illustrate the notions introduced so far. Suppose we first toss a fair coin. If the coin shows "heads," then we measure the time (in years) until an earthquake hits Los Angeles. If the coin shows "tails," then we measure the time until an earthquake shakes Hoople, North Dakota. Suppose the waiting time for an earthquake in LA is an exponential random variable with *pdf*

$$f_1(y) = \begin{cases} e^{-y} & \text{if } y \geq 0, \\ 0 & \text{otherwise.} \end{cases}$$

Similarly, suppose the waiting time in Hoople has *pdf*

$$f_2(y) = \begin{cases} \frac{1}{2}y^2e^{-y} & \text{if } y \geq 0, \\ 0 & \text{otherwise.} \end{cases}$$

Then the mixture density, with $p_1 = p_2 = 1/2$, is given by

$$f_Y(y) = \begin{cases} \frac{1}{2}[e^{-y}(1 + \frac{1}{2}y^2)] & \text{if } y \geq 0, \\ 0 & \text{otherwise.} \end{cases}$$

This mixture density, along with its components $p_1 f_1(y)$ and $p_2 f_2(y)$, is shown in Figure 2.8.1. The mixture density looks similar to $f_1(y)$ for small values of y and similar to $f_2(y)$ for larger values of y, but quite different from both components for values of y around 2. Note that if the experimenter observes *only* the waiting time Y, but not the location, then we do not know whether the observed waiting time refers to LA or to Hoople. If the measured value y happens to be close to 0 (say below 1), then we would be reasonably sure that the earthquake shook LA. Conversely, if the waiting time is more than about 3, chances are much higher that the quake shook Hoople. More formally, what we just tried to do is to "guess" the value of X, given $Y = y$, a concept that we will later formalize by introducing the notion of posterior probabilities. □

Here is some more terminology. In the finite mixture setup, the p_j are often called the *mixing weights*, or *mixing proportions*, and the conditional densities $f_j(y)$ are called the *components* of the mixture. If we focus on classification, the p_j are also

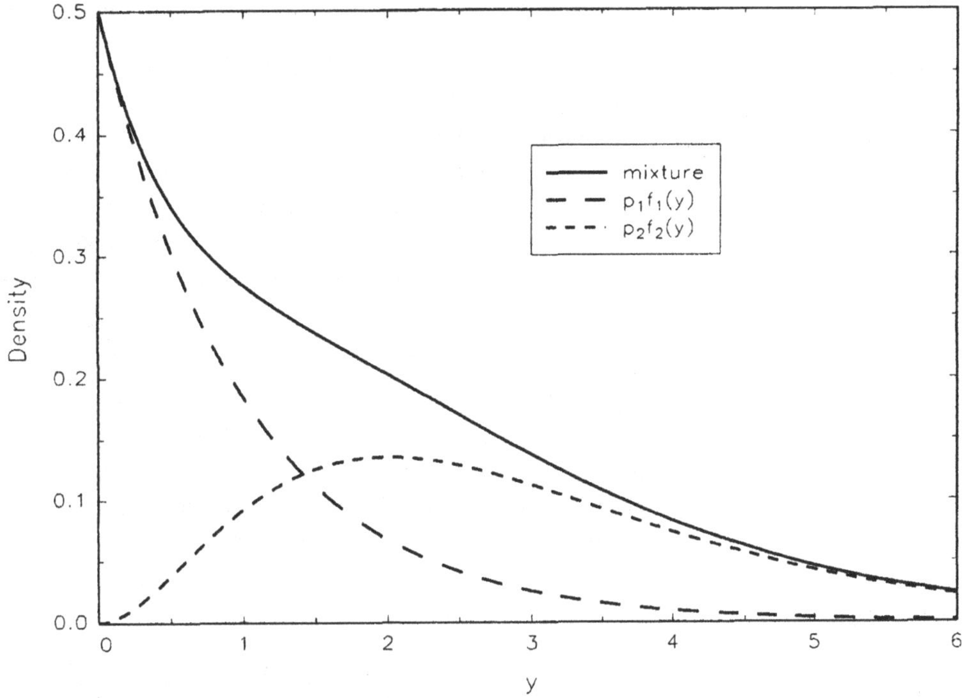

Figure 2-8-1 Construction of the mixture density in Example 2.8.1. The two dashed curves correspond to $p_1 f_1(y)$ and $p_2 f_2(y)$, respectively, with $p_1 = p_2 = 1/2$. The solid line is the graph of the mixture density $p_1 f_1(y) + p_2 f_2(y)$.

referred to as *prior probabilities*. The word "prior" refers to the fact that the p_j give the probability that an observation is from the jth group before any measurements have been taken. After having measured the value of Y, we may wish to reassess the probability of group membership. We would then use the conditional probability function of X, given $Y = y$, and define the *posterior probabilities* of membership in group j as

$$p_{jy} = \Pr[X = j | Y = y] = \frac{p_j f_j(y)}{f_Y(y)} \qquad j = 1, \ldots, k. \tag{7}$$

This is the same as the conditional *pdf* of X given in equation (6). The values of p_{jy}, then can be compared for $j = 1, \ldots, k$, and the category with the highest posterior probability can be identified to classify an observation. The posterior probabilities use the information obtained by measuring Y and are therefore "informed guesses," as opposed to the "uninformed guesses" expressed by the prior probabilities.

Let us introduce a few more notions, restricted for simplicity to the case of $k = 2$ components. The ratio of the posterior probabilities is called the *posterior odds*.

Formally, the posterior odds for component 1, given $Y = y$, is defined as

$$\frac{p_{1y}}{p_{2y}} = \frac{p_{1y}}{1 - p_{1y}} = \frac{p_1 f_1(y)}{p_2 f_2(y)}. \tag{8}$$

The posterior odds tells how many times as likely it is that the observation y came from component 1 as from component 2. Whereas posterior probabilities vary in the interval $[0, 1]$, posterior odds can take all values between 0 and infinity. Posterior odds of 1 means that both components have the same posterior probability.

Occasionally, it is useful to study the *posterior log-odds*, defined as

$$\ell(y) = \log \frac{p_{1y}}{p_{2y}} = \log \frac{p_1 f_1(y)}{p_2 f_2(y)} = \log \frac{p_{1y}}{1 - p_{1y}}. \tag{9}$$

where "log" is the natural logarithm. This function can take any value on the real line. In particular, $\ell(y) = 0$ means indifference, i.e., both components are equally likely. For all y such that $\ell(y) > 0$, the first component is more likely than the second component. Note also that transformation (9) from posterior probabilities to posterior log odds is one-to-one. More precisely, as some straightforward algebra on (9) shows,

$$p_{1y} = \frac{e^{\ell(y)}}{1 + e^{\ell(y)}}, \tag{10}$$

meaning that posterior probabilities are uniquely determined by posterior log odds. The transformation (10) maps the real line into the interval $[0, 1]$ which is appropriate for probabilities. We will spend considerable time and effort on transformations (9) and (10) in normal theory classification and in logistic regression.

Example 2.8.2 continuation of Example 2.8.1

In the earthquake example, the posterior probability that the earthquake is in LA, given that the waiting time is y, is

$$\begin{aligned} p_{1y} &= \Pr[X = 1 | Y = y] \\ &= \frac{\frac{1}{2}e^{-y}}{\frac{1}{2}\left[e^{-y} + \frac{1}{2}y^2 e^{-y}\right]} \\ &= \frac{1}{1 + \frac{1}{2}y^2} \qquad \text{for all } y \geq 0. \end{aligned}$$

Similarly, the posterior probability for Hoople is given by

$$p_{2y} = \Pr[X = 2 | Y = y] = 1 - p_{1y} = \frac{y^2}{y^2 + 2}.$$

These posterior probabilities are illustrated in Figure 2.8.2. The posterior probability for component 1 is a monotonically decreasing function of y, starting at 1 for $y = 0$ and approaching 0 as y tends to infinity. Similarly, the curve of posterior probabilities for component 2 is monotonically increasing and approaches 1, as y grows large. The two curves intersect at $y = \sqrt{2}$, where the posterior probability for both components is 1/2. The posterior odds for LA is $p_{1y}/p_{2y} = 2/y^2$, and the posterior log odds is $\ell(y) = \log(2) - 2\log(y)$. □

Example 2.8.3 continuation of Example 1.3

The "wing length of birds" example is a rather typical situation where mixture analysis is applicable. There are two distinct groups, females and males, which we can represent by a discrete variable X taking values 1 and 2 for females and males, respectively. Recall, however, that gender was not registered in this example, so we are in a genuine mixture situation where sampling is from the mixture density rather from the joint *pdf* of X and Y. Wing length (Y) is reasonably assumed to be a continous random variable, although, in the particular example, it was recorded to the nearest integer only. Moreover, since all birds are fully grown at the time of their migration, all differences between the two groups can be ascribed to sexual dimorphism rather than to age. As is the case with many morphometric variables, we may also be willing to assume that wing length within each group is approximately normal, an assumption that appears reasonable in view of the frequency plot in Figure 1.8 (p. 15). Hence,

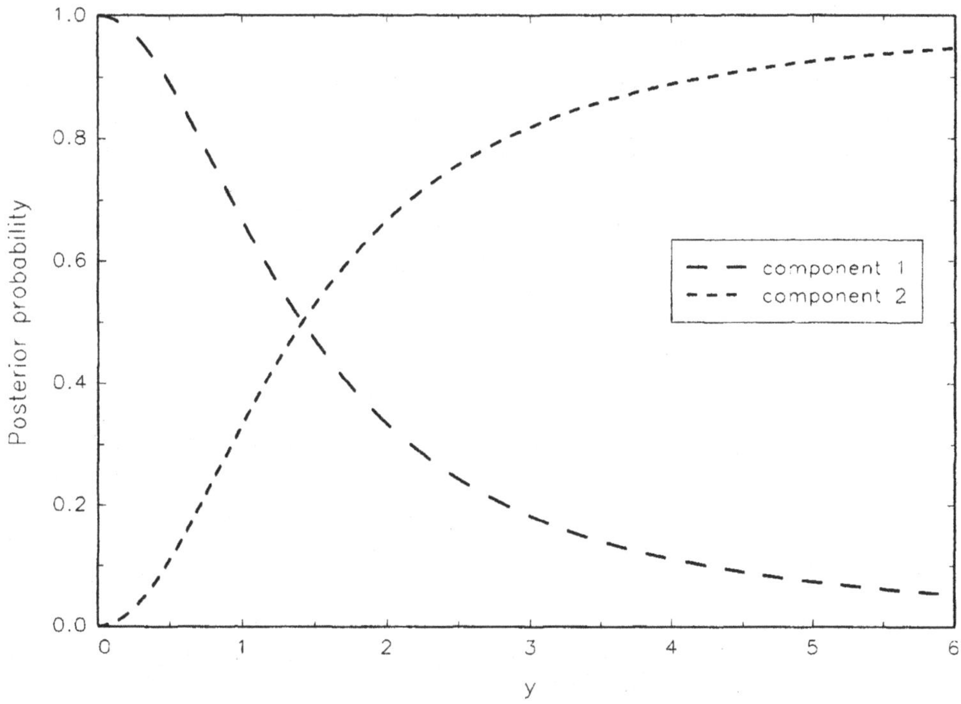

Figure 2-8-2 Graph of posterior probabilities in Examples 2.8.1 and 2.8.2.

now we will develop some theory of normal mixtures and then return to the wing length example as a practical application. □

Mixtures of normal distributions, or *normal mixtures*, for short, are the most prominent and most studied example of finite mixtures. If the jth conditional distribution is normal with mean μ_j, variance σ_j^2, and mixing proportions p_j $(j = 1, \ldots, k)$, then the mixture density is given by

$$f_Y(y) = \frac{1}{\sqrt{2\pi}} \sum_{j=1}^{k} \frac{p_j}{\sigma_j} \exp\left[-\frac{1}{2}\left(\frac{y-\mu_j}{\sigma_j}\right)^2\right]. \tag{11}$$

A normal mixture with k components is determined by $3k - 1$ parameters: k means μ_j, k standard deviations σ_j, and (considering that they add up to 1) $k - 1$ mixing proportions p_j. We shall return to the problem of parameter estimation in Chapter 9; for the moment let us just look at some particular examples, as shown in Figure 2.8.3. A mixture of two normal densities can exhibit a wide variety of shapes: unimodal, bimodal, almost flat in an interval, heavy-tailed, and so on. For three or more components, the family of normal mixtures becomes even more flexible; see Exercise 1.

Two special cases of normal mixtures are of particular interest for both theoretical and practical reasons. We discuss them in turn.

(a) *Normal mixtures with equal variances.* Consider the mixture density

$$f_Y(y) = p_1\, g(y; \mu_1, \sigma^2) + p_2\, g(y; \mu_2, \sigma^2), \tag{12}$$

where $g(y; \mu, \sigma^2)$ is the normal density with parameters as indicated. Parts (a), (b), and (f) of Figure 2.8.3 represent normal mixtures with equal variances. Besides its practical appeal, as will become evident in the continuation of the bird example, this model has nice theoretical properties. The posterior probability for component j, given $Y = y$, is given by

$$p_{jy} = \frac{p_j\, g(y; \mu_j, \sigma^2)}{f_Y(y)} \qquad j = 1, 2. \tag{13}$$

In a moment, we shall see that p_{jy} is a monotonic function of y, increasing for one component and decreasing for the other one. The posterior odds is expressed by

$$\begin{aligned}\frac{p_{1y}}{p_{2y}} &= \frac{p_1}{p_2} \exp\left[-\frac{1}{2}\left(\frac{y-\mu_1}{\sigma}\right)^2 + \frac{1}{2}\left(\frac{y-\mu_2}{\sigma}\right)^2\right] \\ &= \frac{p_1}{p_2} \exp\left\{\frac{1}{2\sigma^2}\left[2(\mu_1 - \mu_2)y - (\mu_1^2 - \mu_2^2)\right]\right\}.\end{aligned} \tag{14}$$

Hence, p_{1y}/p_{2y} is monotonically increasing, if $\mu_1 > \mu_2$, and decreasing if $\mu_1 < \mu_2$. Since $p_{1y} + p_{2y} = 1$, the posterior probability p_{1y} has the same monotonicity property. Note that the monotonicity property fails to hold if the two variances are not equal, because the quadratic term in the exponent of (14) does not vanish.

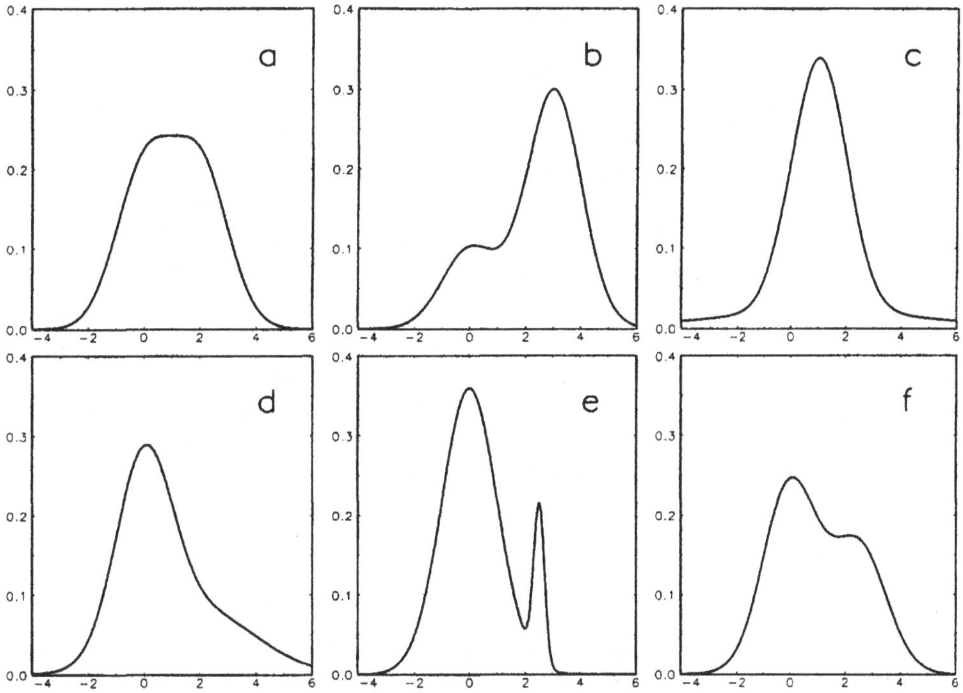

Figure 2-8-3 Six examples of normal mixture densities with two components

(a) $p_1 = 0.5$, $\mu_1 = 0$, $\sigma_1 = 1$, $p_2 = 0.5$, $\mu_2 = 2$, $\sigma_2 = 1$
(b) $p_1 = 0.25$, $\mu_1 = 0$, $\sigma_1 = 1$, $p_2 = 0.75$, $\mu_2 = 3$, $\sigma_2 = 1$
(c) $p_1 = 0.8$, $\mu_1 = 1$, $\sigma_1 = 1$, $p_2 = 0.2$, $\mu_2 = 1$, $\sigma_2 = 4$
(d) $p_1 = 0.6$, $\mu_1 = 0$, $\sigma_1 = 1$, $p_2 = 0.4$, $\mu_2 = 2$, $\sigma_2 = 2$
(e) $p_1 = 0.9$, $\mu_1 = 0$, $\sigma_1 = 1$, $p_2 = 0.1$, $\mu_2 = 2.5$, $\sigma_2 = 0.2$
(f) $p_1 = 0.6$, $\mu_1 = 0$, $\sigma_1 = 1$, $p_2 = 0.4$, $\mu_2 = 2.5$, $\sigma_2 = 1$

The log odds, that is, the natural logarithm of (14), is particularly simple:

$$\ell(y) = \log \frac{p_{1y}}{p_{2y}} = \log \frac{p_1}{p_2} - \frac{1}{2\sigma^2}(\mu_1^2 - \mu_2^2) + \frac{\mu_1 - \mu_2}{\sigma^2}\, y, \tag{15}$$

that is, $\ell(y)$ is a linear function of y — a function closely related to the *linear discriminant function* discussed in Chapter 5.

In practical applications like the bird example, one often wishes to find a cutoff point for classification, that is, a number c such that an object would be classified as coming from component 1, if $y > c$, and from component 2, if $y < c$, or vice versa, depending on which mean is larger. An intuitively reasonable rule is to choose c such that the posterior probabilities at $y = c$ are both $1/2$. This is equivalent to finding c

such that the log odds is zero, i.e.,

$$\log \frac{p_{1c}}{p_{2c}} = 0.$$

Some straightforward algebra shows that the cutoff point in the normal mixture with equal variances is given by

$$c = \frac{1}{2}(\mu_1 + \mu_2) - \frac{\sigma^2}{\mu_1 - \mu_2} \log \frac{p_1}{p_2}. \tag{16}$$

Therefore, if $\mu_1 > \mu_2$, then an observation with value y has higher posterior probability for component 1, if $y > c$, and otherwise, lower posterior probability.

Mixtures of two normal distributions with equal variances have led us quite deeply into ideas of classification, on which we shall elaborate in more detail in Chapters 5 and 7. The continuation of the bird example will serve as an illustration. For now, however, let us turn to a second, interesting, normal mixture model.

(b) *Scale mixtures of two normal distributions.* In the notation just introduced, let $p_1 = 1 - \varepsilon$, $p_2 = \varepsilon$, and

$$f_Y(y) = (1 - \varepsilon)\, g(y; 0, 1) + \varepsilon\, g(y; 0, \sigma^2). \tag{17}$$

Both components are centered at the same mean, and the density is symmetric. Typically, ε is small (0.05 or 0.01), and σ is much larger than 1. Figure 2.8.3(c) gives an example of a scale mixture. The same graph also illustrates the typical purpose of scale mixtures as a model for "contaminated" distributions. The second component, which occurs only relatively infrequently, is a generator of, possibly, very distant observations, or outliers. This explains the popularity of this model in studies of the robustness of normal-theory-based statistical procedures. We leave it to the student to study posterior probabilities for scale mixtures of normals; see Exercise 2.

Example 2.8.4 continuation of Examples 1.3 and 2.8.3

Without giving any details on parameter estimation in normal mixtures, let us return to the wing lengths of water pipits. A normal mixture model was fitted to the data, assuming equality of variances in both components. Using the *EM*-algorithm explained in Chapter 9, the following parameter estimates were obtained:

Component j	Mixing proportion $\hat{p}_j$	Mean $\hat{\mu}_j$	Standard dev. $\hat{\sigma}$
1 (female)	0.49	86.17	1.53
2 (male)	0.51	92.35	1.53

The mixture density

$$\hat{f}_Y(y) = \hat{p}_1\, g(y; \hat{\mu}_1, \hat{\sigma}^2) + \hat{p}_2\, g(y; \hat{\mu}_2, \hat{\sigma}^2)$$

is shown in Figure 2.8.4, together with a histogram of the data. Figure 2.8.5 gives a plot of estimated posterior probabilities

$$\hat{p}_{jy} = \hat{p}_j\, g(y; \hat{\mu}_j, \hat{\sigma}^2)/\hat{f}_Y(y)$$

for $87 \le y \le 91$. The optimal cutoff point obtained using the estimated parameters in equation (16) is $\hat{c} = 89.25$, visible in Figure 2.8.5 as the point where the two curves intersect. For each given value of y, a table of estimated posterior probabilities, or a graph like Figure 2.8.5, allows the investigator to assess the confidence in the assignment of an object to one of two or several components. In our example, there would be very little doubt how to classify birds with wing length smaller than 88 or larger than 90.5. For wing lengths in between these limits, however, there would be considerable uncertainty, particularly, if the measured value y is near the cutoff point $\hat{c} = 89.25$. □

Example 2.8.5 This example illustrates the use of mixtures in a rather unexpected way. Consider the following setup: Beta Airlines offers hourly flights from Chicago to Atlanta, leaving Chicago on the hour between 7A.M. and 11P.M. During the night hours from 11P.M. to

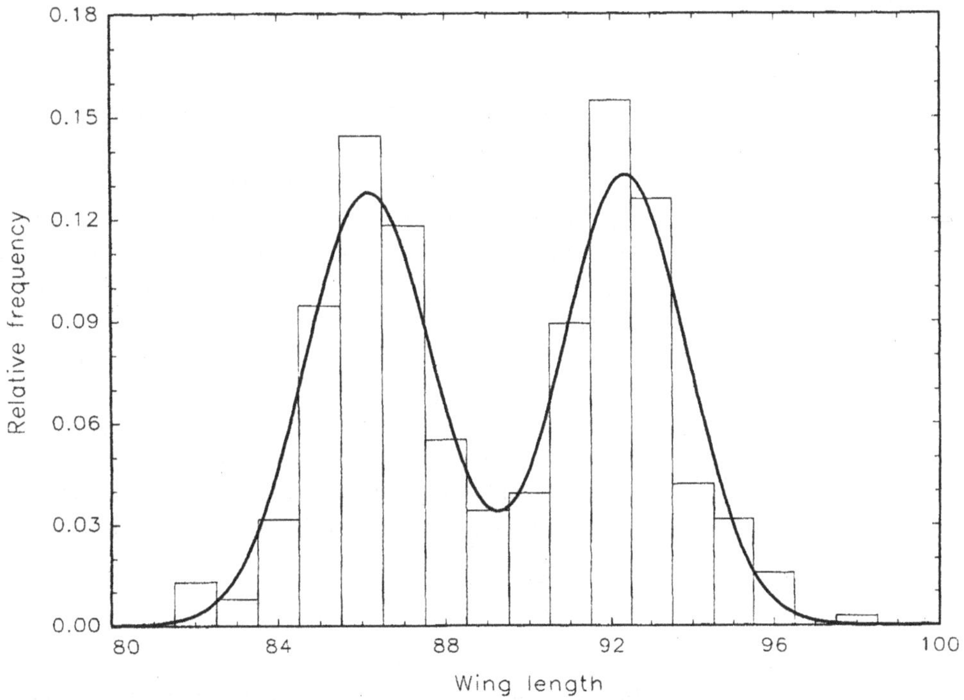

Figure 2-8-4 Histogram of wing length and estimated normal mixture density in the wing length of birds example.

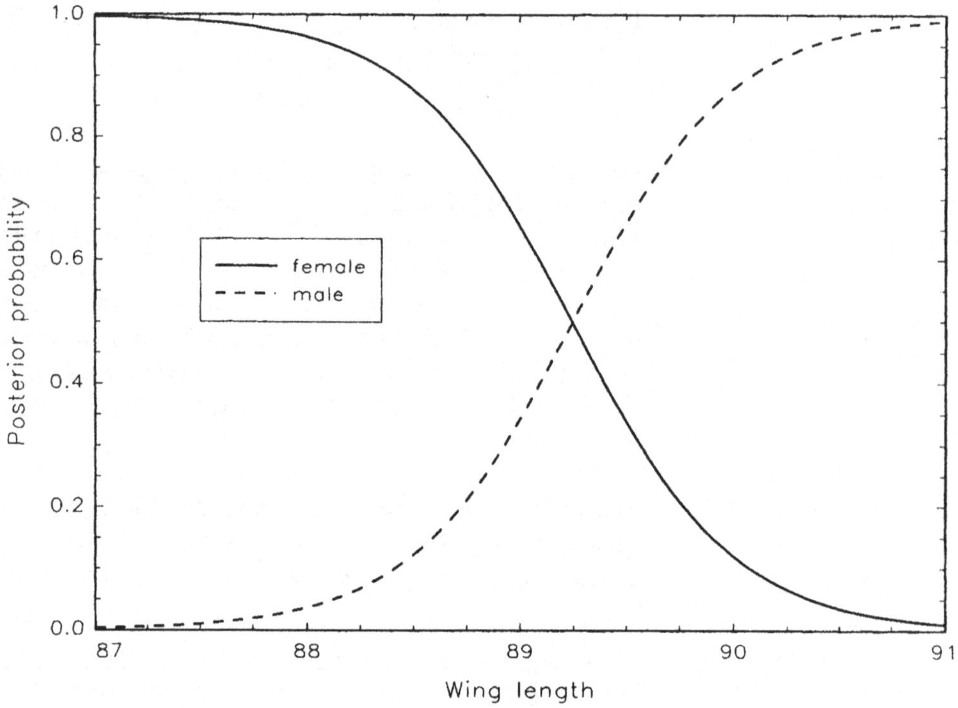

Figure 2-8-5
Estimated posterior probabilities for female and male birds.

7A.M. there are no flights. Furthermore, Beta Airlines is proud of its on-time record. No departure is ever late.

One week in March, a spaceship from Mars lands at Chicago O'Hare International Airport. The little green men quickly discover their mistake — they actually wanted to invade Atlanta, not Chicago — and walk straight to the gate where the next Beta flight for Atlanta is to leave. The problem is what is the distribution of their waiting time?

Since Mars and Earth operate on different time scales, it is reasonable to assume a uniform arrival time for the little green men at Chicago. If they get to the gate between 7A.M. and 11pm (which happens with probability 2/3), their waiting time is uniform in the interval (0, 1), and we do not worry whether this interval is open or closed. If they arrive between 11P.M. and 7A.M. (with probability 1/3), their waiting time is uniform in the interval (0, 8). Thus, define a discrete variable X with

$$\Pr[X = 1] = p_1 = 2/3,$$

$$\Pr[X = 2] = p_2 = 1/3.$$

Let Y = waiting time. Then the conditional densities of Y are

$$f_1(y) = \begin{cases} 1 & \text{if } 0 < y \le 1, \\ 0 & \text{otherwise,} \end{cases}$$

and

$$f_2(y) = \begin{cases} 1/8 & \text{if } 0 < y \le 8, \\ 0 & \text{otherwise.} \end{cases}$$

The unconditional (marginal) distribution of the waiting time is given by the piecewise constant density

$$\begin{aligned} f_Y(y) &= p_1 f_1(y) + p_2 f_2(y) \\ &= \begin{cases} 17/24 & \text{if } 0 < y \le 1, \\ 1/24 & \text{if } 1 < y \le 8, \\ 0 & \text{otherwise.} \end{cases} \end{aligned}$$

Although correct, this derivation of the mixture density is not completely rigorous; see Exercise 3. □

Any piecewise constant density can be modeled as a mixture of uniform distributions, if we allow both endpoints of the uniform components to vary. At the same time, the wealth of densities that can be generated this way leads to some trouble because different bivariate distributions with uniform conditional densities $f_j(y)$ may produce the same marginal mixture density $f_Y(y)$. To make this clear, denote by $u(y; \alpha, \beta)$ the uniform density between α and β, i.e.,

$$u(y; \alpha, \beta) = \begin{cases} \dfrac{1}{\beta - \alpha} & \text{if} \quad \alpha < y \le \beta, \\ 0 & \text{otherwise.} \end{cases}$$

In the extraterrestrial example,

$$\begin{aligned} f_Y(y) &= \frac{2}{3} u(y; 0, 1) + \frac{1}{3} u(y; 0, 8) \\ &= \frac{17}{24} u(y; 0, 1) + \frac{7}{24} u(y; 1, 8), \end{aligned}$$

as is easily verified. This problem is commonly referred to as *nonidentifiability*: different conditional distributions of Y, all from the same family, lead to the same mixture. An even more confusing example is given by the density

$$f_Y(y) = \begin{cases} \dfrac{1}{2} & \text{if} \quad 1 \le y < 2, \\ \dfrac{1}{4} & \text{if} \quad 0 \le y < 1 \text{ or } 2 \le y \le 3, \\ 0 & \text{otherwise,} \end{cases}$$

which can be represented in two different ways as a mixture of two uniform distributions; see Exercise 4.

Example 2.8.6 Stochastic simulation is the art of generating random variables and stochastic processes using computers; it is commonly used to find approximate numerical solutions to problems that are mathematically intractable. It is relatively easy to generate so-called *pseudorandom numbers* on a computer, and many software packages actually have built-in random number generators. Pseudorandom numbers mimic independent random variables with a uniform distribution in the interval $[0, 1)$; the word "pseudo" refers to the fact that their generation is actually entirely deterministic, based on relatively simple recursive algorithms. Yet, for most practical purposes, good pseudorandom numbers behave as if they were genuinely uniform random variables.

One of the basic problems in stochastic simulation is to generate random variables Y with a given *pdf* $f_Y(y)$, using pseudorandom numbers as the only building block. One of the popular methods used, the *composition* method, is actually based on mixtures. The basic idea is as follows. Suppose the density $f_Y(y)$ can be represented as a mixture of two densities, say $f_Y(y) = p_1 f_1(y) + p_2 f_2(y)$. Also suppose that it is difficult to generate the random variable Y with *pdf* f_Y, but it is easy to generate random variables with the densities f_1 and f_2. Then the composition method generates Y from *pdf* f_j with probability p_j, $j = 1, 2$. Then by construction of the mixture, Y has the mixture *pdf* f_Y. Actually, this method reflects exactly the idea of "doing experiment j with probability p_j" introduced in the text following Definition 2.8.1.

A simple example is as follows. Suppose that we wish to generate Y with the "house density"

$$f_Y(y) = \begin{cases} \dfrac{2+y}{5} & \text{if } 0 \le y < 1, \\ \dfrac{4-y}{5} & \text{if } 1 \le y < 2, \\ 0 & \text{otherwise} \end{cases}$$

Then f_Y can be written as $f_Y(y) = 0.8 f_1(y) + 0.2 f_2(y)$, where

$$f_1(y) = \begin{cases} \dfrac{1}{2} & \text{if } 0 \le y < 2, \\ 0 & \text{otherwise} \end{cases}$$

and

$$f_2(y) = \begin{cases} y & \text{if } 0 \le y < 1, \\ 2-y & \text{if } 1 \le y < 2, \\ 0 & \text{otherwise.} \end{cases}$$

We leave it to the reader to graph these two densities, and the mixture density, to verify the decomposition; see Exercise 5. Generating a random variable with *pdf* f_1 is easy: if V is uniform in $[0, 1)$, then $2V$ is uniform in $[0, 2)$. Similarly, generating a random variable with *pdf* f_2 is easy because $f_2(y)$ is the density function of the sum of two independent, uniform random variables (see Section 2.9). Hence, we can generate Y with the house density f_Y as follows: With probability 0.8, generate a uniform random variable and multiply it by 2; with probability 0.2, generate two independent, uniform random variables, and add them. But how will the computer know which of the two possibilities to choose? Again, the answer is easy: Generate a pseudorandom number, and decide if it is smaller or larger than 0.8. A formal description of the algorithm follows.

Algorithm for generating the house distribution:

- Step 1: generate a pseudorandom number U.
- Step 2: If $U \le 0.8$, generate a pseudorandom number V, and return $Y = 2V$. If $U > 0.8$, generate two pseudorandom numbers V_1 and V_2, and return $Y = V_1 + V_2$.

The reader who has access to a random number generator is encouraged to implement this algorithm on a computer, generate a large sample (say 10^4 observations), and graph the data in the form of a histogram to verify empirically that the algorithm works properly. □

Another useful application of mixtures to a problem of transforming continuous random variables is given in Exercise 22.

All mixtures discussed so far are univariate mixtures. The bivariate setup with a group indicator variable X helped to motivate and understand them. However, mixtures of bivariate distributions, or mixtures of multivariate distributions, in general, can be constructed in the same way. Suppose Y and Z are two jointly distributed random variables, whose joint *pdf* depends on the value of a discrete group indicator variable X. As before, suppose X takes values $1, \ldots, k$ with probabilities $p_1, \ldots, p_k$, and denote the joint *pdf* of (Y, Z), given $X = j$, as $f_j(y, z)$. Then the joint unconditional distribution of Y and Z is given by the mixture density

$$f_{YZ}(y, z) = p_1 f_1(y, z) + \ldots + p_k f_k(y, z).$$

Example 2.8.7 Mixtures of k bivariate normal distributions can be used to generate a rich class of bivariate densities with up to k distinct peaks. Figure 2.8.6 shows a bivariate normal mixture with two components; the parameters are not given here because they are of no particular interest. As in the univariate case, there are regions where the mixture density is dominated by one of the components, and regions of transition where the mixture *pdf* looks very "unnormal." Exercise 6 explores this area further. The same bivariate normal mixture was actually used before in Figure 2.6.4 (p. 86) to illustrate the notion of conditional distributions. Bivariate normal mixtures have the property that all conditional and marginal distributions are univariate normal mixtures; see Exercise 7. □

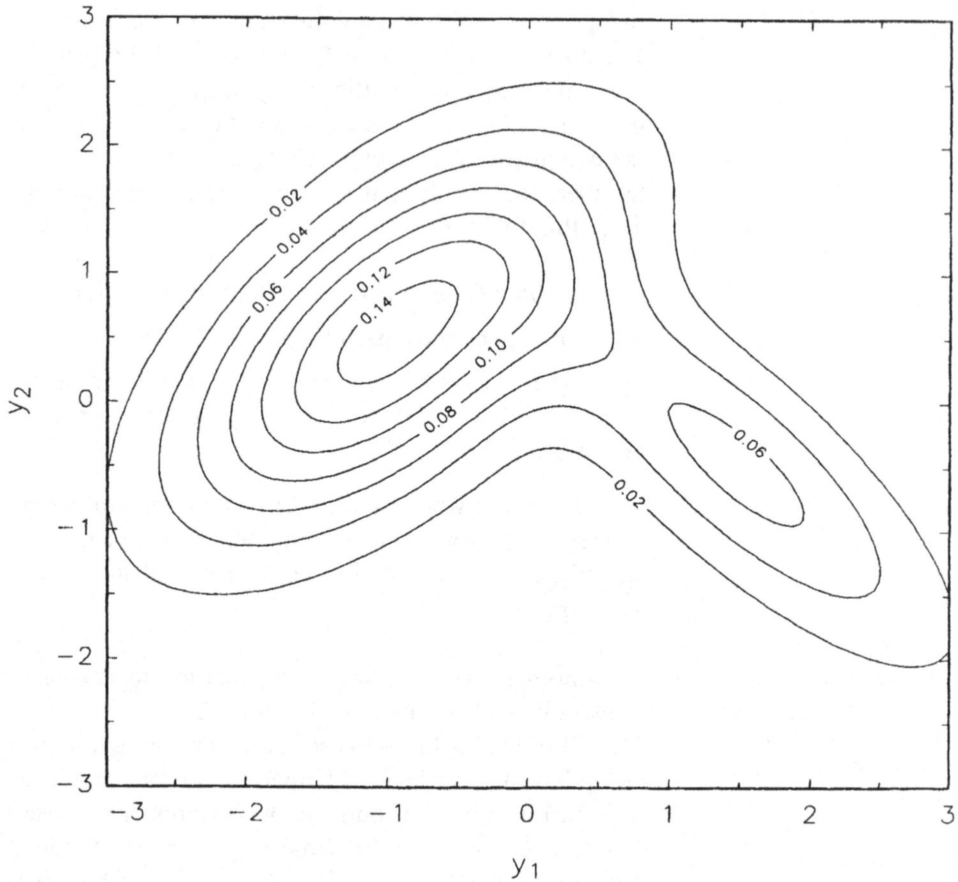

Figure 2-8-6 A bivariate normal mixture shown as a contour plot. See Figure 2.6.4(a), p. 2.6-7, for a surface plot of the same distribution.

The mean and variance of a mixture can be obtained from the means and variances of its components in a relatively straightforward way, as we are going to show in the following theorem. Again, we use the mixture setup of equations (1) to (5), where X is a discrete group indicator taking values $1, \ldots, k$ with probabilities $p_1, \ldots, p_k$, the conditional *pdf* of Y, given $X = j$, is $f_j(y)$, and the marginal density of Y is the mixture *pdf* $f_Y(y) = \sum_{j=1}^k p_j f_j(y)$.

Theorem 2.8.1 *Let $\mu_j = E[Y|X = j]$ and $\sigma_j^2 = \text{var}[Y|X = j]$ denote the mean and variance of Y in the jth component ($j = 1, \ldots, k$), and let*

$$\bar{\mu} = E[Y] = \int_{\mathbb{R}} y \, f_Y(y) \, dy,$$

and

$$\bar{\sigma}^2 = var[Y] = \int_{\mathbb{R}} (y - \bar{\mu})^2 f_Y(y)\, dy,$$

denote the mean and variance of the mixture. Then

$$\bar{\mu} = \sum_{j=1}^{k} p_j \mu_j, \tag{18}$$

and

$$\bar{\sigma}^2 = \sum_{j=1}^{k} p_j \sigma_j^2 + \sum_{j=1}^{k} p_j (\mu_j - \bar{\mu})^2. \tag{19}$$

Proof By the conditional mean theorem in Section 2.7, $E[Y] = E\big[r(X)\big]$, where $r(x) = E[Y|X = x]$. In the mixture setup, $r(X)$ is a discrete random variable which takes value μ_j with probability p_j, $j = 1, \ldots, k$. Hence, (18) follows. The conditional variance theorem states that $\text{var}[Y] = E\big[v(X)\big] + \text{var}\big[r(X)\big]$, where $v(x) = \text{var}[Y|X = x]$. Here, $v(X)$ is a discrete random variable taking values σ_j^2 with probability p_j, $j = 1, \ldots, k$. Hence, $E\big[v(X)\big] = \sum_{j=1}^{k} p_j \sigma_j^2$ and $\text{var}\big[r(X)\big] = \sum_{j=1}^{k} p_j(\mu_j - \bar{\mu})^2$, which implies (19). ■

This theorem allows for a remarkably simple computation of the mean and variance of a mixture, provided that the conditional means and variances are known. For the normal distribution examples, the calculations are straightforward. In Example 2.8.5, the expected value and variance of the Martians' waiting time can be obtained directly from the means and variances of uniform distributions; see Exercise 8.

Example 2.8.8 The *double exponential* distribution, in its simplest form, has *pdf*

$$f_Y(y) = \frac{1}{2} e^{-|y|} \qquad y \in \mathbb{R}.$$

This *pdf* can be written as $f_Y(y) = \frac{1}{2} f_1(y) + \frac{1}{2} f_2(y)$, where

$$f_1(y) = \begin{cases} e^{-y} & \text{if } y > 0, \\ 0 & \text{otherwise,} \end{cases}$$

and

$$f_2(y) = \begin{cases} e^{y} & \text{if } y \le 0, \\ 0 & \text{otherwise.} \end{cases}$$

This decomposition actually considers the double exponential distribution as a mixture of a standard exponential and the negative of a standard exponential random variable, with equal mixing weights. Each of the components has variance 1, and the means are $\mu_1 = 1$, $\mu_2 = -1$. Hence, by Theorem 2.8.1, the mean of the standard double exponential distribution is 0, and its variance is 2. This result would be straightforward to obtain by direct integration, but in more complex cases it is often easier to find means and variances of the components, rather than calculating $E[Y]$ and var$[Y]$ by using the *pdf* $f_Y(y)$. See Exercise 9. □

Equation (19) is actually the theoretical analog of a celebrated equation usually referred to as the *analysis of variance* equality. Its main appeal is that it allows one to split the variance of the mixture (or of the marginal distribution of Y) into a part involving only the conditional variances $\sum p_j \sigma_j^2$ and a part involving only the means $\sum p_j (\mu_j - \bar{\mu})^2$. In words, the variance of Y is the sum of a weighted average of the conditional variances and a weighted average of the the squared deviations of the conditional means μ_j from the "overall mean" $\bar{\mu}$, the weights given by the mixing proportions p_j. This is often symbolically represented as

$$\text{total variance} = \text{variance within} + \text{variance between}.$$

In the analysis of variance models, it is often assumed that all conditional variances σ_j^2 are equal, say σ^2. Equation (19) then reads

$$\text{var}[Y] = \sigma^2 + \sum_{j=1}^{k} p_j (\mu_j - \bar{\mu})^2. \tag{20}$$

Since the second term in (20) is nonnegative, and equals zero exactly if all μ_j are identical, this means that var$[Y] = \sigma^2$ exactly if all μ_j are equal. This fact is used in the analysis of variance model to construct the classical F-test for equality of k means, based on "variance within" and "variance between." Exercise 10 gives some more details how equation (19) can be interpreted in a sample setup.

As outlined in this section, mixture distributions are closely connected to methods commonly referred to as *classification analysis*. Three types of methods of classification frequently used are *discriminant analysis*, *logistic regression*, and *cluster analysis*. We comment on them briefly in some final remarks.

1. *Discriminant analysis* is a classical multivariate technique, based on samples with known group membership, called the training samples, from each of k different populations. Based on the observed data, the investigator may want to describe and test for differences between the groups, and formulate rules for optimal allocation of future observations with unknown group membership. Example 1.1 (two species of midges) is a rather typical case. Much of Chapters 5 to 7 of this book is concerned with discriminant analysis. In the terminology of the current section, discriminant analysis, primarily, compares the conditional distributions of Y for the various values of the group indicator X.

2. *Logistic regression* focuses on estimation of posterior probabilities. In a way, logistic regression is a reversal of discriminant analysis since it studies the conditional distribution of X, given $Y = y$. Suppose, for simplicity, that the group indicator variable X takes only the two values 1 and 2. Then logistic regression models the distribution of X as a function of y. More precisely, let $p(y) = Pr[X = 1|Y = y]$ as in equation (7). Then logistic regression assumes that the posterior log odds function has the form

$$\ell(y) = \log \frac{p(y)}{1 - p(y)} = \alpha + \beta y \tag{21}$$

for some real parameters α and β. Likelihood methods are commonly used to estimate these parameters. No distributional assumptions are made on Y. We will discuss logistic regression in Chapter 7. Note that particular cases where (21) holds are normal mixtures with identical variances; see equation (15).

3. *Cluster analysis* is a collection of mostly heuristic techniques for partitioning multivariate data into homogeneous subgroups, called clusters. As in the mixture setup, group membership is not known for each individual observation. Cluster analysis is usually meant to provide the user with an objective assessment of how many different subgroups the data contain. This is an ambitious goal, but (not too surprisingly) cluster analysis often fails to give convincing results other than those the investigator anticipated anyway. In contrast to finite mixture analysis, most cluster analytical methods are not explicitly based on a probabilistic model, which is bound to lead the naive investigator into believing that he or she did not make any assumptions at all, and that the results therefore are "objective" in some sense. The unpleasant fact is that there are so many different options and possibilities to choose from that the claim of objectiveness is hard to justify. Moreover, due to the lack of a probabilistic foundation, most methods of cluster analysis fail to assess the uncertainty inherent in classifying objects into groups. In conclusion, many statisticians (including the author of this text) consider cluster analysis as the poor brother of finite mixture analysis, and no attempt is made here to introduce the topic. Good references for the interested reader are Hartigan (1975), Kaufman and Rousseeuw (1990), and Everitt (1993); see also Exercise 11.

Exercises for Section 2.8

1. Graph the following normal mixture densities with k components.
   (a) $k = 3$; $p_1 = p_2 = p_3 = 1/3$; $\mu_1 = 1$, $\mu_2 = 2$, $\mu_3 = 3$; $\sigma_1 = \sigma_2 = \sigma_3 = 1$.
   (b) $k = 3$; $p_1 = p_2 = p_3 = 1/3$; $\mu_1 = 1$, $\mu_2 = 4$, $\mu_3 = 7$; $\sigma_1 = \sigma_2 = \sigma_3 = 1$.
   (c) $k = 3$; $p_1 = 0.25$, $p_2 = 0.5$, $p_3 = 0.25$; $\mu_1 = 1$, $\mu_2 = 3$, $\mu_3 = 5$; $\sigma_1 = 4$, $\sigma_2 = 1$, $\sigma_3 = 4$.

(d) $k = 10$; $p_1 = \cdots = p_{10} = 0.1$; $\mu_i = 1.5i$ $(i = 1, \ldots, 10)$; $\sigma_1 = \cdots = \sigma_{10} = 1$.
In each case, comment on the shape of the mixture density.

2. Consider a scale mixture of two normals,

$$f_Y(y) = \frac{1}{2}g(y; 0, 1) + \frac{1}{2}g(y; 0, 2^2),$$

where $g(y; \mu, \sigma^2)$ is the *pdf* of a normal random variable with parameters as indicated.

(a) Find the posterior probabilities p_{1y} and p_{2y} as functions of y.

(b) Find the posterior odds p_{1y}/p_{2y} and the log odds.

(c) Graph the component densities, the mixture density f_Y, the posterior probabilities, the posterior odds, and the log odds in the range $-5 \le y \le 5$.

(d) Find the value(s) of y such that $p_{1y} = p_{2y} = 1/2$.

3. Give a rigorous derivation of the mixture *pdf* in Example 2.8.5 along the following lines. Assume for simplicity that the Beta flights leave exactly at hours 1, 2, 3, ..., 16, and 24.

(a) Let Z denote the arrival time of the Martians at the gate, i.e., Z is a continuous random variable with a uniform distribution in $[0, 24]$. Let X denote a discrete random variable taking values 1 to 17 such that $X = j$ exactly if $j - 1 < Z \le j$, $j = 1, \ldots, 16$, and $X = 17$ if $16 < Z \le 24$. Find the probability function of X.

(b) Let Y denote the waiting time for the next flight. Find the conditional *cdf* of Y, given $X = j$, i.e., find $F_j(a) = \Pr[Y \le a | X = j]$, $j = 1, \ldots, 17$.

(c) Find the conditional densities $f_j(y)$ as the derivatives of the F_j, and find the joint *pdf* of X and Y.

4. Find two different ways to represent the density function

$$f_Y(y) = \begin{cases} \frac{1}{2} & \text{if } 1 \le y \le 2, \\ \frac{1}{4} & \text{if } 0 \le y < 1 \text{ or } 2 < y \le 3, \\ 0 & \text{otherwise} \end{cases}$$

as a mixture of two uniform densities.

5. This exercise is based on Example 2.8.6.

(a) Graph the mixture *pdf* $f_Y(y)$ and the component densities $f_1(y)$ and $f_2(y)$.

(b) Write a computer program to generate random variables with the house density $f_Y(y)$.

(c) Generate a sample of size 10,000, and graph the results as a histogram with bin width 0.1.

(d) For each bin in (c), compute the expected relative frequency. For instance, for the first bin this is $\int_0^{0.1} f_Y(y)\,dy$. Compare observed and expected relative frequencies, and check how well the observed frequencies match the expected frequencies.

6. Let $g(y, z; \rho)$ denote the bivariate normal density with $\mu_1 = \mu_2 = 0, \sigma_1 = \sigma_2 = 1$, and correlation ρ, i.e.,

$$g(y, z; \rho) = \frac{1}{2\pi(1-\rho^2)^{1/2}} \exp\left[-\frac{1}{2(1-\rho^2)}\left(y^2 - 2\rho yz + z^2\right)\right] \quad (y,z) \in \mathbb{R}^2.$$

Consider mixtures of two such densities,

$$f_{YZ}(y,z) = p_1\, g(y, z; \rho_1) + p_2\, g(y, z; \rho_2),$$

where $0 < p_1 < 1,\ 0 < p_2 < 1,\ p_1 + p_2 = 1,\ |\rho_1| < 1$, and $|\rho_2| < 1$.

(a) Show that f_{YZ} is a bivariate density function.

(b) Show that f_{YZ} is not a bivariate normal density, unless $\rho_1 = \rho_2$.

(c) Show that the marginal distributions of Y and Z are both standard normal.

(d) Find the correlation $\rho(Y, Z)$ for the mixture density as a function of p_1, p_2, ρ_1, and ρ_2.

(e) Find parameter combinations (i.e., values of p_1, p_2, ρ_1, ρ_2) for which $\rho(Y, Z) = 0$.

(f) Graph f_{YZ} as a contour plot for $p_1 = p_2 = \frac{1}{2}$, $\rho_1 = -\rho_2 = \frac{1}{2}$. (This is a case where $\rho_{YZ} = 0$).

7. Suppose that the joint distribution of Y and Z is a bivariate normal mixture with k components.

(a) Show that Y and Z follow univariate normal mixture distributions.

(b) Show that the conditional distribution of Z, given $Y = y$, is a normal mixture for all y. *Hint*: Use results from Section 2.6.

(c) Find a bivariate normal mixture with $k = 2$ components such that Y follows a univariate normal mixture distribution with two nonidentical components and Z is standard normal.

8. In the mixture setup of equations (1) to (5), suppose that $f_Y(y) = \sum_{j=1}^{k} p_j f_j(y)$ is a mixture of k uniform densities such that

$$f_j(y) = \begin{cases} \dfrac{1}{b_j - a_j} & \text{if } a_j \le y \le b_j, \\ 0 & \text{otherwise.} \end{cases}$$

(a) Show that $\mu_j = E[Y|X = j] = (a_j + b_j)/2$, and $\sigma_j^2 = \text{var}[Y|X = j] = (b_j - a_j)^2/12$.

(b) Find $E[Y]$ and $\text{var}[Y]$.

(c) Using the results from part (b), find the mean and the variance of the Martians' waiting time in Example 2.8.5. Verify your result by computing the mean and variance directly from the mixture *pdf*.

9. Use the technique of Example 2.8.8 to find the mean and variance of Y if Y has the following *pdf*.

(a)

$$f_Y(y) = \frac{1}{2\lambda} e^{-|y-\alpha|/\lambda} \qquad y \in \mathbb{R},$$

where $\lambda > 0$ and α are parameters;

(b)

$$f_Y(y) = \begin{cases} \frac{1}{4} + \frac{3}{4}e^{-y} & \text{if } 0 \le y < 1, \\ \frac{3}{4}e^{-y} & \text{if } 1 \le y, \\ 0 & \text{otherwise;} \end{cases}$$

(c)

$$f_Y(y) = \begin{cases} \frac{1}{2}[e^{-y}(1 + \frac{1}{2}y^2)] & \text{if } y \ge 0, \\ 0 & \text{otherwise.} \end{cases}$$

Hint: Read Example 2.8.1, and use the identity $\int_0^\infty y^m e^{-y}\,dy = m!$ for all nonnegative integers m.

10. In *one-way analysis of variance* with k treatment groups, y_{ji} ($j = 1, \ldots, k$; $i = 1, \ldots, n_j$) denote measurements taken in k samples. Set $n = \sum_{j=1}^{k} n_j$, and let

$$\bar{y}_j = \frac{1}{n_j} \sum_{i=1}^{n_j} y_{ji} \qquad j = 1, \ldots, k$$

and

$$\bar{y} = \frac{1}{n} \sum_{j=1}^{k} n_j \bar{y}_j$$

denote the groupwise averages and the overall average, respectively. The analysis of variance equality in this setup is given by

$$\sum_{j=1}^{k} \sum_{i=1}^{n_j} (y_{ji} - \bar{y})^2 = \sum_{j=1}^{k} \sum_{i=1}^{n_j} (y_{ji} - \bar{y}_j)^2 + \sum_{j=1}^{k} n_j (\bar{y}_j - \bar{y})^2 .$$

Prove this equality. *Hint*: The analysis of variance equality can be shown directly with some tedious algebra, but try to prove it by defining a random variable Y^* which, conditionally on $X = j$, takes values $y_{j1}, \ldots, y_{jn_j}$ with probability $1/n_j$ each. Use equation (19).

11. Consider a normal mixture density

$$f_Y(y) = \frac{1}{2} g(y; \mu_1, \sigma^2) + \frac{1}{2} g(y; \mu_2, \sigma^2)$$

with equal variances in both components. Suppose that we define a cutoff point for clustering as $c = (\mu_1 + \mu_2)/2$. Assuming $\mu_1 > \mu_2$, let $\mu_1^* = E[Y|Y > c]$, $\mu_2^* =$

$E[Y|Y < c]$, and $\sigma_*^2 = \text{var}[Y|Y > c] = \text{var}[Y|Y < c]$. Show that $\mu_1^* > \mu_1, \mu_2^* < \mu_2$, and $\sigma_*^2 < \sigma^2$. *Note*: This exercise shows that partitioning the data into two groups according to a fairly reasonable criterion and computing means and variances "within groups" will lead to an erroneous assessment of the parameters of the components of the distribution. See the wing length of birds example for some practical motivation.

12. Show that the density function of the uniform distribution in $[0,1]$ can be written as a mixture of two densities of the form

$$f_1(y) = \begin{cases} 2y & \text{if } 0 \le y \le 1, \\ 0 & \text{otherwise,} \end{cases}$$

and

$$f_2(y) = \begin{cases} 2(1-y) & \text{if } 0 \le y \le 1, \\ 0 & \text{otherwise.} \end{cases}$$

What are the mixing proportions?

13. Consider a mixture of two exponential distributions given by the mixture density

$$f_Y(y) = \begin{cases} p_1 \frac{1}{\lambda_1} e^{-y/\lambda_1} + p_2 \frac{1}{\lambda_2} e^{-y/\lambda_2} & \text{if } y \ge 0, \\ 0 & \text{otherwise,} \end{cases}$$

where $\lambda_1 > 0, \lambda_2 > 0$ are fixed parameters and p_1, p_2 are the mixing proportions $(p_1 + p_2 = 1)$.

(a) Find the posterior probabilities p_{1y} and p_{2y} as functions of y.

(b) Find the posterior odds p_{1y}/p_{2y}, and the log odds.

(c) For the values $p_1 = 0.7$, $p_2 = 0.3$, $\lambda_1 = 1$, and $\lambda_2 = 3$, graph the component densities, the mixture density f_Y, the posterior probabilities, the posterior odds and the log odds in the range $0 \le y \le 7$.

(d) Find the value(s) of y such that $p_{1y} = p_{2y} = \frac{1}{2}$.

14. Consider normal mixture models of the form

$$f_Y(y) = p\, g(y; 1, \sigma^2) + (1-p)\, g(y; 0, \sigma^2),$$

where $g(y; \mu, \sigma^2)$ is the density of a normal random variable with mean μ and variance σ^2. What does the mixture density look like if

(a) σ is very large (say $\sigma > 10$)?

(b) σ is very small (say $\sigma < 0.1$)?

Note: A more precise formulation of this problem is as follows for the student who has been exposed to limit theorems.

(a) Find the limiting distribution as $\sigma \to \infty$ of the random variable Y/σ, where Y has the mixture density given above.

(b) Find the limiting distributions as $\sigma \to 0$ of Y with the above mixture density.

15. *Nonparametric density estimation* is a branch of statistics which studies methods of estimating density functions from data without making explicit distributional assumptions. Histograms may be regarded as the simplest method of density estimation. *Kernel density* estimates work as follows. Let $y_1, \ldots, y_n$ denote observed data, and let $K(z)$ denote a density function, usually symmetric to 0, such that a random variable with density K has mean 0 and variance τ^2 (which depends on n, but we are not going to go into much detail here). The kernel density estimate based on the observed data is defined, for all $y \in \mathbb{R}$, as

$$\hat{f}(y) = \frac{1}{n} \sum_{i=1}^{n} K(y - y_i).$$

(a) Show that $\hat{f}$ is a mixture *pdf*.

(b) If Y^* is a random variable with *pdf* $\hat{f}$, find $E[Y^*]$ and $\text{var}[Y^*]$. How do these quantities differ from the usual sample mean and variance? *Hint*: Use equations (18) and (19).

16. Let Z denote a continuous random variable with a uniform distribution in the interval $[-1, 1]$, and let X denote a discrete random variable, independent of Z, taking values 1 and 2 with probabilities p_1 and p_2, respectively ($p_1 + p_2 = 1$). Define a random variable Y as

$$Y = \begin{cases} Z & \text{if } X = 1, \\ -Z & \text{if } X = 2. \end{cases}$$

(a) Find the conditional *pdf* of Y, given $X = j$, for $j = 1, 2$.

(b) Find the marginal *pdf* of Y as a mixture *pdf*, and show that it is the same as the *pdf* of Z.

(c) What are the possible values (y, z) that the pair of random variables (Y, Z) can take?

(d) Show that the correlation between Y and Z is $p_1 - p_2$. Hence, show that, for $p_1 = p_2$, Y and Z are uncorrelated but not independent. *Hint*: Find $E[YZ|X = j]$, $j = 1, 2$.

17. The same as Exercise 16, for the student who hates continuous distributions: Instead of Z uniform in the interval $[-1, 1]$, suppose that Z is a discrete random variable that takes values -2, -1, 0, 1, and 2 with probability $1/5$ each. In part (c), find the joint probability function of Y and Z.

18. On their way to the Beta airlines gate at Chicago O'Hare, the Martians of Example 2.8.5 drop a post card to their friends at home into a mailbox. The mail from this particular box is collected exactly at 10A.M., 2P.M., and 7P.M.. Find the *pdf* of the waiting time of the letter in the mailbox.

19. Space Shuttle *Recover* circles the earth once in exactly 72 minutes. During 24 minutes of each cycle, the space shuttle is visible from your house, and you can't overlook it because it shines so brightly in the sky, even during daytime and when the sky is cloudy. For the remaining 48 minutes of each cycle, the shuttle is invisible. Being

unaware of the schedule of the shuttle, you step out of your house, look up in the sky, and measure the time Y until you discover *Recover*. (This is to be interpreted as: you step out of the house at a time which is uniformly and continuously distributed in the 72-minute cycle).

(a) Find the *cdf* of Y.

(a) Find $E[Y]$ and $\text{var}[Y]$, using equations (18) and (19). *Note*: This is correct although Y is neither continuous nor discrete.

20. Prove equations (18) and (19) along the following lines, without using the conditional mean and variance theorems.

(a) Show that $E[Y^m] = \sum_{j=1}^k p_j E[Y^m | X = j]$, which implies (18) as the special case $m = 1$.

(b) Show that $E[Y^2] = \sum_{j=1}^k p_j(\sigma_j^2 + \mu_j^2)$.

(c) Using parts (a) and (b), show (19).

21. Table 2.8.1 displays data from a study by M. Oses and J. Paul (1997) on wing length of angels. There are two species of angels, ***Angelus angelus*** and ***A. diabolicus***, which occur in proportions p_1 and p_2 ($p_1 + p_2 = 1$). The purpose of the study was to estimate the unknown proportions. According to an infallible source, the wing length of ***A. angelus*** follows a so-called *half-halo* (HH) distribution with *pdf*

$$f_1(y) = \begin{cases} \frac{2}{\pi}\sqrt{1-(y-2)^2} & \text{if } 1 \le y \le 3, \\ 0 & \text{otherwise,} \end{cases}$$

and the wing length of *A. diabolicus* follows a *hoof* (H) distribution with *pdf*

$$f_2(y) = \begin{cases} 1 - \frac{2}{\pi}\sqrt{1-(y-2)^2} & \text{if } 1 \le y \le 3, \\ 0 & \text{otherwise.} \end{cases}$$

(a) Graph the HH and H densities and the mixture *pdf* $f_Y(y) = p_1 f_1(y) + p_2 f_2(y)$ for $p_1 = 0.1, 0.3, 0.5, 0.7$, and 0.9.

(b) Graph the data of Table 2.8.1, and compare the graph to the mixture densities from (a). Guess a reasonable value for the mixing proportion p_1.

(c) Based on your assessment of the mixing proportions from part (b), estimate the posterior probability that an angel is from species *A. diabolicus*, given that the wing length is 1.75.

22. This exercise is for the student who is familiar with the transformation theorem for continuous random variables, which states the following. Let U denote a continuous random variable with density function $f_U(u)$, and let $h(u)$ denote a strictly monotonic function, continuously differentiable in the domain D of U. Let $V = h(U)$, or equivalently, $U = h^{-1}(V)$. Then the *pdf* of V is given by

$$f_V(v) = f_U\left(h^{-1}(v)\right) \left| \frac{\partial h^{-1}}{\partial v} \right|.$$

Table 2.8.1 Wing length of 130 angels of the species *Angelus angelus* and *A. diabolicus.*

Wing Length	Frequency
1.0 – 1.2	20
1.2 – 1.4	14
1.4 – 1.6	13
1.6 – 1.8	10
1.8 – 2.0	8
2.0 – 2.2	7
2.2 – 2.4	11
2.4 – 2.6	14
2.6 – 2.8	14
2.8 – 3.0	19

Reprinted with permission from M. Oses and J. Paul (1997). Morphometric discrimination between two species of angels and estimation of their relative frequency. *J. Celest. Morph.* **1011**, 747–767. The unit of measurement is *mm* (miracle-meters). Rather than raw data, the authors gave only the observed frequencies in 10 intervals.

See, e.g., Ross (1994). Prove the following generalization. Suppose that the function h is not monotonic, but there exists a partition $D_1, \ldots, D_k$ of the domain D such that, for $u \in D_j$, $h(u)$ is strictly monotonic and continuously differentiable. Let $h_j(u)$ denote the function $h(u)$ constrained to D_j. Let $p_j = \Pr[U \in D_j]$, and

$$f_j(u) = \begin{cases} \dfrac{1}{p_j} f_U(u) & \text{if } u \in D_j, \\ 0 & \text{otherwise.} \end{cases}$$

Then the *pdf* of $V = h(U)$ is given by

$$f_V(v) = \sum_{j=1}^{k} p_j f_j\left[h_j^{-1}(v)\right]\left|\frac{\partial h_j^{-1}}{\partial v}\right| = \sum_{j=1}^{k} f\left[h_j^{-1}(v)\right]\left|\frac{\partial h_j^{-1}}{\partial v}\right|.$$

Hint: Consider the *pdf* of U as a mixture with k components $f_U(u) = \sum_{j=1}^{k} p_j f_j(u)$, where each of the component densities is positive in a different interval.

2.9 Sums of Random Variables

In this section we give a method for finding the distribution of $X + Y$ when X and Y are two jointly distributed random variables. It is not always the quickest or most elegant method, but it is interesting and relatively easy to understand because of its similarity to the method of finding marginal distributions outlined in Section 2.3. Actually, the discrete case has already been discussed to some extent in Section 2.3, but we are now going to present it more formally.

Suppose that X and Y are two jointly distributed, random variables, taking values (x, y) in some domain D with positive probabilities. Let $Z = X + Y$. Then for a fixed value $z \in \mathbb{R}$,

$$\begin{aligned} \Pr[Z = z] &= \Pr[X + Y = z] \\ &= \sum_{x+y=z} \Pr[X = x, Y = y] \\ &= \sum_{x} \Pr[X = x, Y = z - x]. \end{aligned} \tag{1}$$

This is to be understood as follows. The second line in equation (1) represents a sum over all pairs $(x, y) \in D$ whose sum is z. It follows from the first line because the event $\{X + Y = z\}$ is composed of all pairs $(x, y) \in D$ such that $x + y = z$. The third line in equation (1) may be formally regarded as a sum over all x in the domain of X, but many of the terms in the sum will typically be zero.

In the usual notation for probability functions thus, we can write

$$f_Z(z) = \sum_x f_{XY}(x, z - x) \tag{2}$$

or

$$f_Z(z) = \sum_y f_{XY}(z - y, y). \tag{3}$$

Note the similarity of equations (2) and (3) to the formulas for marginal probability functions given in Section 2.3. In the current case, for a fixed value z, we have to add the probabilities of all pairs (x, y) whose sum is z. To find the marginal distribution of Y, we would have to add the probabilities of all pairs (x, y) where y is a given fixed value, i.e., $f_Y(y) = \sum_x f_{XY}(x, y)$.

Example 2.9.1 Let (x, y) denote a bivariate, discrete random variable whose *pdf* is given by the following table:

	$y = 1$	$y = 2$	$y = 3$
$x = 1$	1/9	0	2/9
$x = 2$	2/9	1/9	0
$x = 3$	0	2/9	1/9

Of course, it is easy to just "read" the probability function of $Z = X + Y$ off the table of joint probabilities, but a graph will prepare us for later. Figure 2.9.1 represents the points (x, y), $x = 1, 2, 3$, $y = 1, 2, 3$, in the Euclidean plane, marked with their respective probabilities. If we want to find, e.g., $\Pr[Z = 3]$, then we have to add the probabilities of all points falling on the line $x + y = 3$, shown in Figure 2.9.1 as a dashed line. Similarly, the dotted line in Figure 2.9.1 contains all points whose sum is 4. The probability function of Z thus is obtained as $f_z(2) = f_z(6) = 1/9$, $f_z(3) = f_z(5) = 2/9$, and $f_z(4) = 1/3$. See also Exercise 1. □

Example 2.9.2 This is a wellknown and classical example. Suppose X and Y are independent Poisson random variables with parameters $\lambda > 0$ and $\theta > 0$, respectively. Then the joint *pdf* of X and Y is given by

$$f_{XY}(x, y) = \begin{cases} e^{-(\lambda+\theta)} \dfrac{\lambda^x}{x!} \dfrac{\theta^y}{y!} & \text{if } x, y = 0, 1, \ldots, \\ 0 & \text{otherwise.} \end{cases}$$

Let $Z = X + Y$. Then the *pdf* of Z, for $z = 0, 1, \ldots$, is given by

$$f_Z(z) = \sum_x f_{XY}(x, z - x) = \sum_{x=0}^{z} f_{XY}(x, z - x)$$

$$= e^{-(\lambda+\theta)} \sum_{x=0}^{z} \frac{\lambda^x \theta^{z-x}}{x!(z-x)!} .$$

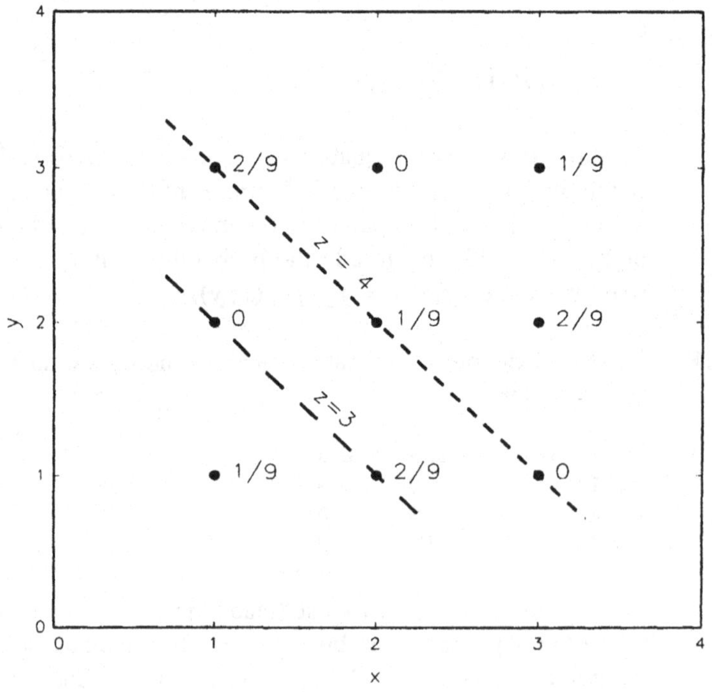

Figure 2-9-1 Illustration of Example 2.9.1.

But, for any $z \geq 0$,

$$1 = \left(\frac{\lambda}{\lambda+\theta} + \frac{\theta}{\lambda+\theta}\right)^z = \sum_{x=0}^{z} \binom{z}{x} \left(\frac{\lambda}{\lambda+\theta}\right)^x \left(\frac{\theta}{\lambda+\theta}\right)^{z-x}$$
$$= \frac{z!}{(\lambda+\theta)^z} \sum_{x=0}^{z} \frac{\lambda^x \theta^{z-x}}{x!(z-x)!},$$

and therefore we obtain

$$f_Z(z) = e^{-(\lambda+\theta)} \frac{(\lambda+\theta)^z}{z!}, \qquad z = 0, 1, \ldots,$$

that is, Z follows itself a Poisson distribution with parameter $\lambda + \theta$. See Exercise 3 for a continuation of this example. □

Now, we turn to the continuous case, formulating the main result as a theorem.

Theorem 2.9.1 (Convolution Theorem) *Suppose X and Y are two jointly distributed continuous random variables with density function $f_{XY}(x, y)$. Let $Z = X + Y$. Then the pdf of Z is given by*

$$\begin{aligned} f_Z(z) &= \int_{x=-\infty}^{\infty} f_{XY}(x, z-x)\, dx \\ &= \int_{y=-\infty}^{\infty} f_{XY}(z-y, y)\, dy. \end{aligned} \tag{4}$$

Proof Let $F_Z(z)$ denote the *cdf* of Z. Then for fixed $z \in \mathbb{R}$,

$$\begin{aligned} F_Z(z) &= \Pr[Z \leq z] \\ &= \Pr[X + Y \leq z] \\ &= \int_{x=-\infty}^{\infty} \int_{y=-\infty}^{z-x} f_{XY}(x, y)\, dy\, dx. \end{aligned} \tag{5}$$

The last line in equation (5) follows because all points $(x, y) \in \mathbb{R}^2$, such that $x+y \leq z$, form a set $\{(x, y) : x \in \mathbb{R}, y \leq z - x\}$. Then differentiating $F_Z(z)$ with respect to z gives the *pdf* of Z:

$$\begin{aligned} f_Z(z) &= \frac{\partial}{\partial z} F_Z(z) \\ &= \int_{x=-\infty}^{\infty} f_{XY}(x, z-x)\, dx. \end{aligned}$$

The second line of equation (4) follows similarly. ■

Equation (4) is a straightforward, continuous analog of (2) and (3). It means that if we want to find the value of the density of Z at point z, we have to integrate over the joint *pdf* of X and Y along a straight line $x + y = z$. This is very much like adding probabilities along straight lines as illustrated in Example 2.9.1, and parallels the computation of marginal densities in Section 2.3.

To illustrate Theorem 2.9.1, we start with a classical example.

Example 2.9.3 Sum of two independent normal random variables

Suppose X and Y are independent standard normal random variables. Then their joint density is expressed by $f_{XY}(x, y) = (2\pi)^{-1} \exp\left[-\frac{1}{2}(x^2 + y^2)\right]$, and the *pdf* of $Z = X + Y$ is given by

$$\begin{aligned} f_Z(z) &= \int_{x=-\infty}^{\infty} f_{XY}(x, z - x)\, dx \\ &= \frac{1}{2\pi} \int_{x=-\infty}^{\infty} \exp\left\{-\frac{1}{2}\left[x^2 + (z - x)^2\right]\right\} dx \\ &= \frac{1}{\sqrt{2\pi}\sqrt{2}} \exp\left[-\frac{1}{2}\left(z/\sqrt{2}\right)^2\right] \cdot \int_{x=-\infty}^{\infty} \frac{1}{\sqrt{2\pi}\sqrt{1/2}} \exp\left[-\frac{1}{2}\left(\frac{x - z/2}{\sqrt{1/2}}\right)^2\right] dx. \end{aligned}$$

The function in the integral in the last expression is the *pdf* of a normal random variable with mean $z/2$ and variance $1/2$, and hence, the integral is unity. Thus, Z is normal with mean 0 and variance 2. □

Often, the main difficulty in using Theorem 2.9.1 is to find the correct boundaries for integration. The next example is a rather typical case.

Example 2.9.4 Sums of uniform random variables

Suppose that X and Y are independent random variables, both uniform in $[0, 1]$. Then their joint *pdf* is given by

$$f_{XY}(x, y) = \begin{cases} 1 & \text{if } 0 \le x \le 1,\ 0 \le y \le 1, \\ 0 & \text{otherwise,} \end{cases}$$

that is, the joint *pdf* takes the value 1 inside the unit square and is zero outside. Let $Z = X + Y$. To find the *pdf* $f_Z(z)$, it is useful to graph, for fixed $z \in \mathbb{R}$, the line $x + y = z$ along which integration takes place. This is illustrated in Figure 2.9.2. We can distinguish among three cases.

Case 1: $z < 0$ or $z \ge 2$. Then the line $x + y = z$ does not intersect the unit square, and therefore, $f_Z(z) = 0$.

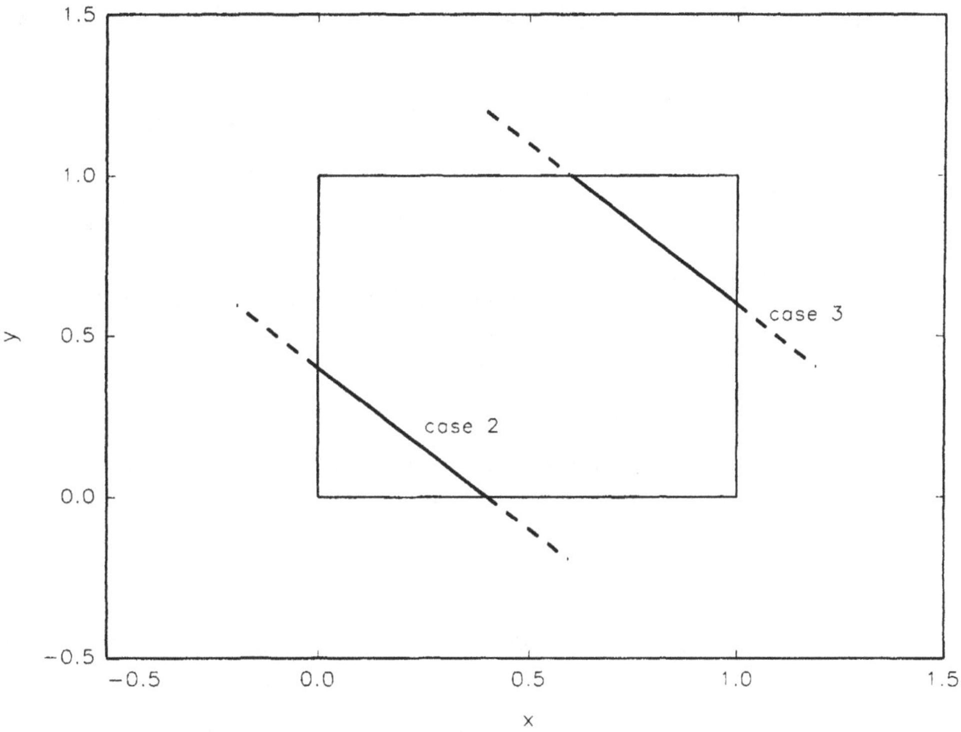

Figure 2-9-2 Computation of the density of $X + Y$ in Example 2.9.4. The two lines correspond to cases 2 and 3.

Case 2: $0 \le z < 1$. The line $x + y = z$ intersects the x-axis at $x = z$. Therefore, for $0 \le z < 1$, write

$$\begin{aligned} f_Z(z) &= \int_{x=-\infty}^{\infty} f_{XY}(x, z - x)\, dx \\ &= \int_{x=-\infty}^{0} f_{XY}(x, z - x)\, dx + \int_{x=0}^{z} f_{XY}(x, z - x)\, dx + \int_{x=z}^{\infty} f_{XY}(x, z - x)\, dx. \end{aligned}$$

Only in the second of the three integrals is the density positive, and therefore,

$$f_Z(z) = \int_0^z 1\, dx = z, \qquad 0 \le z < 1.$$

Case 3: $1 \le z < 2$. In this case, the line $x + y = z$ intersects the unit square for $z - 1 \le x \le 1$, and hence,

$$f_Z(z) = \int_{z-1}^{1} 1\, dx = 2 - z, \qquad 1 \le z < 2.$$

Summarizing this, we have found the *pdf* of $Z = X + Y$ as

$$f_Z(z) = \begin{cases} z & \text{if } 0 \le z < 1, \\ 2 - z & \text{if } 1 \le z < 2, \\ 0 & \text{otherwise.} \end{cases} \tag{6}$$

This is a triangular density function with a peak at $z = 1$. □

What about adding a third uniform random variable? This can be done using the same technique, as outlined in the next example.

Example 2.9.5 continuation of Example 2.9.4

Suppose that X follows the distribution of the sum of two independent, uniform [0, 1] random variables, and Y is uniform in [0, 1], independent of X. Then by (6), the joint *pdf* of X and Y is given by

$$f_{XY}(x, y) = \begin{cases} x & \text{if } 0 \le x < 1,\ 0 \le y \le 1, \\ 2 - x & \text{if } 1 \le x < 2,\ 0 \le y \le 1, \\ 0 & \text{otherwise.} \end{cases}$$

Again, it is useful to graph lines of the form $x + y = z$ for fixed values of z, as illustrated in Figure 2.9.3. But this time the density is positive in a rectangle, taking values x for $0 \le x < 1$ and $2 - x$ for $1 \le x < 2$. We distinguish four cases.

Case 1: $z < 0$ or $z \ge 3$. Then $f_Z(z) = 0$.

Case 2: $0 \le z < 1$. Then

$$f_Z(z) = \int_0^z x\,dx = \frac{1}{2}z^2.$$

Case 3: $1 \le z < 2$. Then

$$f_Z(z) = \int_{z-1}^{z} f_{XY}(x, z - x)\,dx = \int_{z-1}^{1} x\,dx + \int_1^z (2 - x)\,dx$$
$$= -\frac{3}{2} + 3z - z^2.$$

Case 4: $2 \le z < 3$. Then

$$f_Z(z) = \int_{z-1}^{2} (2 - x)\,dx = \frac{9}{2} - 3z + \frac{1}{2}z^2.$$

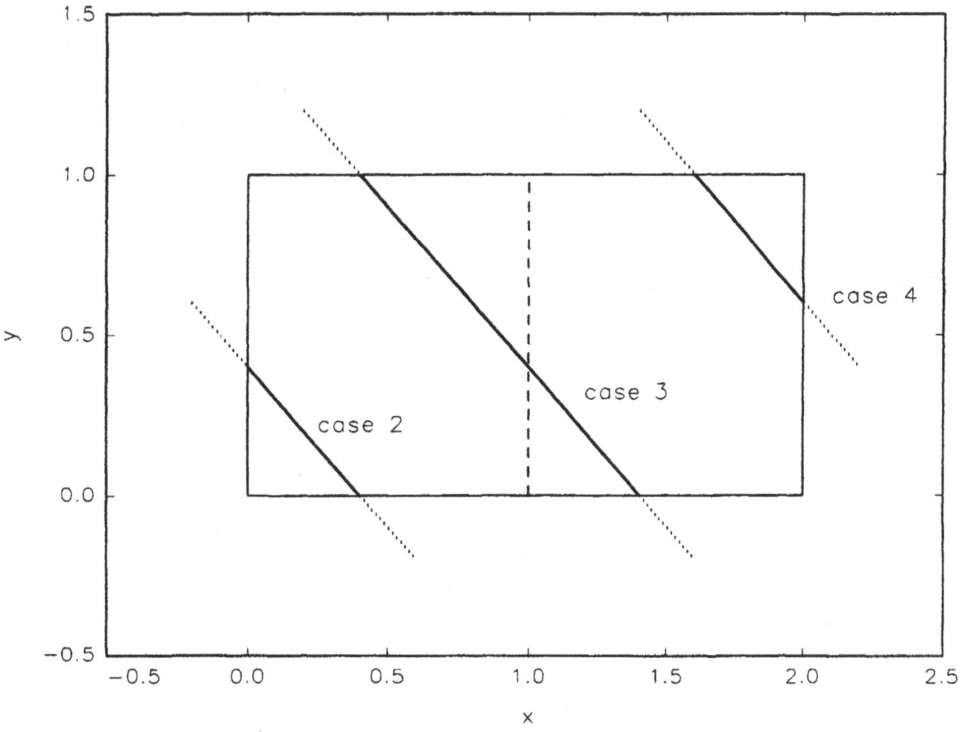

Figure 2-9-3 Computation of the density in Example 2.9.5. The three lines correspond to cases 2, 3, and 4.

Thus, we have found the *pdf* of the sum of three independent random variables, all uniform in [0, 1], as

$$f_Z(z) = \begin{cases} \frac{1}{2}z^2 & \text{if } 0 \le z < 1, \\ -\frac{3}{2} + 3z - z^2 & \text{if } 1 \le z < 2, \\ \frac{9}{2} - 3z + \frac{1}{2}z^2 & \text{if } 2 \le z < 3, \\ 0 & \text{otherwise.} \end{cases} \tag{7}$$

We leave it to the reader (Exercise 4) to investigate some properties of this rather interesting distribution. □

Theorem 2.9.1 has many useful applications, among them the derivation of the sum of independent chi-square random variables (Exercise 7). We have applied it only to examples where the two terms in the sum are independent, but it works just as well in the dependent case. Exercises 8 and 9 are examples of this kind.

Using the same technique as in the proof of Theorem 2.9.1, it is possible to find a formula for a general linear combination $Z = aX + bY$ of two jointly distributed, continuous random variables X and Y; see Exercise 10. However, the same goal, usually, can be achieved by first finding the joint distribution of aX and bY, using the transformation theorem of Section 2.11 and then applying Theorem 2.9.1. Therefore, the generalization is not presented here.

Exercises for Section 2.9

1. This exercise is based on Example 2.9.1.
   (a) Find the marginal probability functions $f_X(x)$ and $f_Y(y)$.
   (b) Let X^* and Y^* denote two independent discrete random variables with the probability functions $f_X(x)$ and $f_Y(y)$ from part (a). Find the joint distribution of X^* and Y^*.
   (c) Find the *pdf* of $Z^* = X^* + Y^*$, and compare it to the *pdf* of $Z = X + Y$ from Example 2.9.1.

2. Let X and Y denote two independent random variables such that $X \sim$ binomial(n_1, p) and $Y \sim$ binomial(n_2, p). Find the *pdf* of $Z = X + Y$.

3. Let X denote a standard exponential random variable with *pdf* $f_X(x) = e^{-x},\ x \geq 0$. Let Y denote another standard exponential random variable, independent of X. Show that the *pdf* of $Z = X + Y$ is given by

$$f_Z(z) = \begin{cases} ze^{-z} & \text{if } z \geq 0, \\ 0 & \text{otherwise.} \end{cases}$$

4. Suppose that the random variable Z has the density given in equation (7), i.e., the density of the sum of three independent, identically distributed, uniform random variables.
   (a) Graph the density, and verify that it is continuous, differentiable, and that the first derivative is itself continuous.
   (b) Show, by integration or otherwise, that $E[Z] = 3/2$ and $\text{var}[Z] = 1/4$.
   (c) In the same graph as in part (a), plot the density function of a normal random variable with mean 3/2 and variance 1/4, and comment.

5. This exercise gives a generalization of Example 2.9.5. It requires some work. Suppose that $Z = X_1 + \cdots + X_N$, where N is fixed, and all X_i are independent, identically distributed $U(0, 1)$ random variables. Show, by induction, that the *pdf* of Z is given by

$$f_Z(z) = \begin{cases} \dfrac{1}{(N-1)!} \displaystyle\sum_{j=0}^{k} \binom{N}{j} (-1)^j (z-j)^{N-1} & \text{if } k \leq z < k+1,\ k = 0, 1, \ldots, N-1, \\ 0 & \text{otherwise.} \end{cases}$$

This is called the *Irwin-Hall* distribution; see Johnson, Kotz, and Balakrishnan (1995, p. 296).

6. Find the distribution of $Z = X + Y$, where X is $U(0, 1)$ and Y is $U(0, b)$ for some $b > 1$, X and Y independent. *Hint*: A graph like Figure 2.9.2 will help.

7. For an integer $m \geq 1$, the random variable X has a chi-square distribution with m degrees of freedom if the *pdf* of X is given by

$$f_X(x) = \begin{cases} \dfrac{1}{\Gamma(\frac{1}{2}m)2^{m/2}}\, x^{\frac{m}{2}-1}\, e^{-x/2} & \text{if } x < 0, \\ 0 & \text{otherwise.} \end{cases}$$

Here, $\Gamma(\alpha)$ is the gamma function, i.e.,

$$\Gamma(\alpha) = \int_{u=0}^{\infty} u^{\alpha-1}\, e^{-u}\, du, \ \alpha > 0.$$

If X and Y are independent chi-square random variables with m and n degrees of freedom, respectively, show that $Z = X + Y$ has a chi-square distribution with $m + n$ degrees of freedom.

8. Suppose that random variables X and Y have the joint *pdf*

$$f_{XY}(x, y) = \begin{cases} x + y & \text{if } 0 \leq x \leq 1,\ 1 \leq y \leq 1, \\ 0 & \text{otherwise.} \end{cases}$$

Find the *pdf* of $Z = X + Y$.

9. Suppose that (X, Y) is bivariate normal with means $\mu_x = \mu_y = 0$, variances $\sigma_x^2 = \sigma_y^2 = 1$, and correlation ρ ($|\rho| < 1$); see equation (3) in Section 2.2. Use Theorem 2.9.1 to show that $Z = X + Y$ is normal with mean 0 and variance $2(1 + \rho)$.

10. Generalize Theorem 2.9.1 to the *pdf* of $Z = aX + bY$, where a and b are fixed, nonzero constants. First, consider the case $a > 0$, $b > 0$ and then the case $a > 0,\ b < 0$.

11. Consider a pair of continuous random variables (X, Y) with the density function

$$f_{XY}(x, y) = \begin{cases} 1 & \text{if } (x, y) \in S_1, \\ 2 & \text{if } (x, y) \in S_2, \\ 0 & \text{otherwise,} \end{cases}$$

where $S_1 = [0, \frac{1}{3}] \times [0, \frac{1}{3}] \cup [\frac{1}{3}, \frac{2}{3}] \times [\frac{1}{3}, \frac{2}{3}] \cup [\frac{2}{3}, 1] \times [\frac{2}{3}, 1]$ and $S_2 = [0, \frac{1}{3}] \times [\frac{2}{3}, 1] \cup [\frac{1}{3}, \frac{2}{3}] \times [0, \frac{1}{3}] \cup [\frac{2}{3}, 1] \times [\frac{1}{3}, \frac{2}{3}]$.

(a) Show that both X and Y are uniform in $(0, 1)$.

(b) Show that X and Y are dependent.

(c) Show that the *pdf* of $X + Y$ is given by equation (6). *Hint*: Draw the unit square in the (x, y)-plane and partition it into nine subsquares of side length 1/3 each. Mark each subsquare with the proper value of the joint density function. Then draw lines of the form $x + y = z$ for various fixed values of z.

12. The continuous random variable X has a gamma(α, β) distribution, for parameters $\alpha > 0$ and $\beta > 0$, if its *pdf* is expressed by

$$f_X(x; \alpha, \beta) = \begin{cases} \dfrac{1}{\Gamma(\alpha)\beta^\alpha} x^{\alpha-1} e^{-x/\beta} & \text{if } x > 0, \\ 0 & \text{otherwise.} \end{cases}$$

Here, $\Gamma(\alpha)$ is the gamma function as in Exercise 7. Show the following result: If $X \sim$ gamma(α_1, β) and $Y \sim$ gamma(α_2, β), X and Y independent, then $X + Y \sim$ gamma$(\alpha_1 + \alpha_2, \beta)$.

13. Let X and Y be independent Poisson random variables with parameters $\lambda > 0$ and $\theta > 0$, respectively. Show that the conditional distribution of X, given $X + Y = z$, is binomial with z trials and success probability $\lambda/(\lambda + \theta)$.

14. Let $X_1, \ldots, X_N$ be independent, identically distributed, Poisson random variables with parameter $\lambda > 0$. Let $Z = \sum_{i=1}^N X_i$. Show that $E[X_1|Z] = Z/N$ and $\text{var}[X_1|Z] = (N-1)Z/N^2$. *Hint*: See Exercise 24 in Section 2.7.

15. Let X and Y be two jointly distributed, random variables such that both X and Y take values 1, 2, and 3 with probability 1/3 each. Find a joint probability distribution of X and Y such that $X + Y$ takes values 2 to 6 with probability 1/5 each. *Hint*: Fill the entries in a 3×3 table with probabilities that have the desired properties, i.e, the row sums and the column sums are all 1/3, and the sums of all entries corresponding to a fixed value of $X + Y$ are all 1/5.

16. Generalize Exercise 15 to the case where X and Y take values 1 to 4 with probability 1/4 each, and $X + Y$ takes values $2, 3, \ldots, 8$ with probability 1/7 each.

17. Suppose that X and Y are continuous random variables, both uniform in the interval $(0, 1)$. If $X + Y$ is uniform in the interval $(0, 2)$, show that $\Pr[X = Y] = 1$. *Hint*: Use equation (20) from Section 2.5 to show that the correlation between X and Y is 1.

2.10 Notions and Concepts of *p*-variate Distributions

In this section, we shall introduce the p–dimensional analogs of the notions and results introduced in Sections 2.1 to 2.7. From now on, vector and matrix notation will be used extensively. Having understood the bivariate concepts, it should not be overly difficult to follow the p-variate generalization. Again, we shall distinguish between the discrete and the continuous case, being aware that this distinction is somewhat artificial. Later, we shall concentrate, more and more, on the continuous case because it is easier to handle. It will often be convenient to write p–variate random variables as random column vectors

$$\mathbf{X} = \begin{pmatrix} X_1 \\ \vdots \\ X_p \end{pmatrix}$$

of p components. In the discrete case, $\mathbf{X}$ is described by a probability function

$$f_{\mathbf{X}}(x_1, \ldots, x_r) = \Pr[X_1 = x_1, \ldots, X_p = x_p],$$

such that $f_{\mathbf{X}}$ is nonnegative and

$$\sum_{(x_1,\ldots,x_p)} f_{\mathbf{X}}(x_1, \ldots, x_p) = 1.$$

In the continuous case, $f_{\mathbf{X}}$ is a density function satisfying the conditions

(i) $f_{\mathbf{X}}(x_1, \ldots, x_p) \geq 0$ for all $\mathbf{x} = (x_1, \ldots, x_p)' \in \mathbb{R}^p$,
(ii) $\int_{x_1=-\infty}^{\infty} \cdots \int_{x_p=-\infty}^{\infty} f_{\mathbf{X}}(x_1, \ldots, x_p) dx_p \cdots dx_1 = 1$, and
(iii) for any rectangular area $[a_1, b_1] \times \cdots \times [a_p, b_p]$,

$$\Pr[a_1 \leq X_1 \leq b_1, \ldots, a_p \leq X_p \leq b_p] = \int_{x_1=a_1}^{b_1} \cdots \int_{x_p=a_p}^{b_p} f_{\mathbf{X}}(x_1, \ldots, x_p)\, dx_p \cdots dx_1.$$

For dimension $p > 2$, it is often inconvenient to write explicit probability functions. Let us try it, anyway, for two selected examples.

Example 2.10.1 Let (X_1, X_2, X_3) denote the numbers shown by three independent fair dice. Assuming that no cheating occurs, the joint probability function is given by

$$f_{\mathbf{X}}(x_1, x_2, x_3) = \begin{cases} 1/216 & \text{if } x_i = 1, 2, \ldots, 6,\ i = 1, 2, 3, \\ 0 & \text{otherwise.} \end{cases}$$

Now, suppose that we consider, instead, the joint distribution of (Y_1, Y_2, Y_3), where

$$Y_1 = X_1,$$
$$Y_2 = X_1 + X_2,$$

and

$$Y_3 = X_1 + X_2 + X_3.$$

By a similar argument as in Example 2.2.3,

$$f_{\mathbf{Y}}(y_1, y_2, y_3) = \begin{cases} 1/216 & \text{if } y_1 = 1, \ldots, 6, \\ & \quad y_2 = y_1 + 1, \ldots, y_1 + 6, \\ & \quad y_3 = y_2 + 1, \ldots, y_2 + 6, \\ 0 & \text{otherwise.} \end{cases}$$

Hence, the joint distribution of $\mathbf{Y}$ is uniform on 216 points in a rather complicated set. □

Example 2.10.2 Let **X** denote a three-dimensional random variable given by its probability function

$$f_{\mathbf{X}}(x_1, x_2, x_3) = \begin{cases} \dfrac{1}{4} & \text{if } x_1 = 0,\ x_2 = 0,\ x_3 = 0, \\ & \text{or } x_1 = 0,\ x_2 = 1,\ x_3 = 1, \\ & \text{or } x_1 = 1,\ x_2 = 0,\ x_3 = 1, \\ & \text{or } x_1 = 1,\ x_2 = 1,\ x_3 = 0, \\ 0 & \text{otherwise.} \end{cases}$$

This is a simple but interesting probability function, which puts probability mass $\frac{1}{4}$ onto four of the eight corners of the unit cube. We will return to it later. □

Example 2.10.3 It is easy to construct discrete or continuous multivariate examples in dimension p by taking the product of p univariate densities. For instance, the joint density of p variables

$$f_{\mathbf{Y}}(y_1, \ldots, y_p) = (2\pi)^{-p/2} \exp\left(-\frac{1}{2}\sum_{i=1}^{p} y_i^2\right) \qquad y_i \in \mathbb{R},\ i = 1, \ldots, p, \quad (1)$$

is the product of p univariate standard normal densities, and is not very interesting by itself. However, it can serve as a prototype of the general, multivariate normal distribution, just as the univariate standard normal may be used to generate any univariate normal distribution. □

Starting with the joint distribution of p variables, we can select a subset of q variables ($1 \le q < p$) and study their joint distribution. Any marginal distribution is obtained either by summation or integration of the joint probability function over those variables not selected. For instance, take the subset of the first q variables ($X_1, X_2, \ldots, X_q$ in our notation). Their joint marginal *pdf* is given by

$$f_{X_1, X_2, \ldots, X_q}(x_1, \ldots, x_q) = \sum_{x_{q+1}} \cdots \sum_{x_p} f_{\mathbf{X}}(x_1, \ldots, x_q,\ x_{q+1}, \ldots, x_p)$$

in the discrete case, and by

$$f_{X_1, \ldots, X_q}(x_1, \ldots, x_q) = \int_{x_{q+1}=-\infty}^{\infty} \cdots \int_{x_p=-\infty}^{\infty} f_{\mathbf{X}}(x_1, \ldots, x_q,\ x_{q+1}, \ldots, x_p)\, dx_p \cdots dx_{q+1}.$$

in the continuous case. The case of mixed discrete - continuous variables can be handled similarly by using summation or integration, whenever appropriate.

Independence and conditional distributions follow along the lines of Sections 2.4 and 2.6, although we have to be a little more careful about independence, as we shall see. Variables $X_1, \ldots, X_p$ are called independent (or better, mutually independent), if their joint probability density function factors into the p marginals, i.e.,

$$f_{\mathbf{X}}(x_1, \ldots, x_p) = \prod_{i=1}^{p} f_{X_i}(x_i).$$

We may also look at independence of subsets of variables. For instance, the subset $\{X_1, \ldots, X_q\}$ and the subset $\{X_{q+1}, \ldots, X_p\}$ are independent if

$$f_{\mathbf{X}}(x_1, \ldots, x_p) = f_{X_1,\ldots,X_q}(x_1, \ldots, x_q) \cdot f_{X_{q+1},\ldots,X_p}(x_{q+1}, \ldots, x_p).$$

Let us illustrate with an example why the notion of independence is somewhat tricky.

Example 2.10.4 continuation of Example 2.10.2

The marginal probability function of (X_1, X_2) is given by

$$f_{X_1,X_2}(x_1, x_2) = \begin{cases} \dfrac{1}{4} & x_1 = 0, 1;\ x_2 = 0, 1, \\ 0 & \text{otherwise.} \end{cases}$$

Hence, X_1 and X_2 are stochastically independent binary variables with probability $1/2$ on values 0 and 1. Exactly the same is true for the pairs (X_1, X_3) and (X_2, X_3). Yet, the random variables (X_1, X_2, X_3) are not mutually independent because the joint probability function does not factor. In fact, whenever the values of two of the three components are given, the remaining one is uniquely determined. The lesson is that pairwise independence does not, in general, imply mutual independence. □

Example 2.10.5 continuation of Example 2.10.1

Finding the marginal probability function of Y_3 (i.e., of the sum of the numbers shown by three independent dice) seems a bit tricky because of the strange shape of the set of points with positive probability. However, we know the marginal distribution of Y_2 from Example 3.2, and the joint probability function of (Y_2, Y_3) can be written in a table like the one in Example 2.2. Details are left to the student; see Exercise 1. □

Example 2.10.6 A generalization of our favorite raindrop example to dimension p is the uniform distribution inside the unit sphere, given by its *pdf*

$$f_X(x_1, \ldots, x_p) = \begin{cases} c & \text{if } x_1^2 + \ldots + x_p^2 \leq 1, \\ 0 & \text{otherwise,} \end{cases} \tag{2}$$

where c is the inverse of the volume of the unit sphere; see Exercises 2 and 3. In dimension $p = 3$, we obtain

$$f_{\mathbf{X}}(x_1, x_2, x_3) = \begin{cases} \dfrac{3}{4\pi} & \text{if } x_1^2 + x_2^2 + x_3^2 \leq 1, \\ 0 & \text{otherwise.} \end{cases}$$

Finding the marginal distribution of any of the three components of this distribution is somewhat tricky; see exercise 5. However, independence can be ruled out immediately by the following argument: Each one of the marginal densities is positive in the interval $(-1, 1)$. Now, take the point $(x_1, x_2, x_3) = (0.6, 0.6, 0.6)$. Then

$$f_{X_1}(0.6) \cdot f_{X_2}(0.6) \cdot f_3(0.6) > 0 = f_X(0.6, 0.6, 0.6).$$

Actually, we could take any point inside the cube of side length 2, centered at the origin, but outside the unit circle. □

The uniform distribution inside a sphere, and the product of p univariate, standard normals in Example 2.10.3 provide illustrations of *spherical distributions*, characterized by the fact that the density depends on $x_1, \ldots, x_p$ only through the Euclidean distance $(x_1^2 + \ldots + x_p^2)^{1/2}$ from the origin. Spherical distributions will be treated in Section 3.4.

Example 2.10.7 In Section 2.8, we considered a "group code" variable X, taking values 1 to k with probabilities $p_1, \ldots, p_k$. For such a variable, notions like mean and standard deviation are generally meaningless concepts, because the assignment of numbers ("group codes") to subpopulations is arbitrary, and, typically, no order relationship is given between them. A somewhat more complicated, but powerful, alternative is provided by the multinomial distribution, or rather, a special case of it. The idea is to introduce indicator variables, one for each group. Hence, we shall define a k-variate random vector $\mathbf{X} = (X_1, X_2, \ldots, X_k)'$, such that exactly one of the k components can take the value 1, and all others take value 0. In other words, we want

$$\Pr[X_i = 1] = p_i$$

and

$$\Pr[X_i = 0] = 1 - p_i,$$

and whenever $X_i = 1$, then all other X_j must be 0. Such a model is provided by the probability function

$$f_{\mathbf{X}}(x_1, \ldots, x_k) = \begin{cases} \prod_{i=1}^{k} p_i^{x_i} & x_i = 0 \text{ or } 1, \quad \sum_{j=1}^{k} x_j = 1, \\ 0 & \text{otherwise,} \end{cases} \tag{3}$$

where $p_1 + \cdots + p_k = 1$. For instance, for $k = 3$,

$$f_{\mathbf{X}}(x_1, x_2, x_3) = \begin{cases} p_1^{x_1} \cdot p_2^{x_2} \cdot p_3^{x_3} & x_1, x_2, x_3 = 0 \text{ or } 1, \quad x_1 + x_2 + x_3 = 1, \\ 0 & \text{otherwise.} \end{cases}$$

Then $\Pr[X_1 = 1, X_2 = 0, X_3 = 0] = p_1^1 \cdot p_2^0 \cdot p_3^0 = p_1$, and similarly for X_2 and X_3.

Although this distribution appears somewhat cumbersome for the simple purpose of modeling group membership, it has great theoretical and practical appeal, as you may recall from the theory of regression using indicator variables. In its simplest form with $k = 2$ groups, it reduces to

$$f_{\mathbf{X}}(x_1, x_2) = \begin{cases} p_1^{x_1} p_2^{x_2} & x_1, x_2 = 0 \text{ or } 1, \quad x_1 + x_2 = 1, \\ 0 & \text{otherwise,} \end{cases}$$

where $p_1 + p_2 = 1$. Due to the exact relationship between X_1 and X_2 ($X_1 + X_2 = 1$), it is usually preferable to consider only the marginal probability function of X_1, given by

$$\Pr(X_1 = x_1) = p_1^{x_1}(1 - p_1)^{1-x_1}, \quad x_1 = 0, 1,$$

which is just the model for a binary variable. In the general setup of equation (3), the marginal distribution of each of the components X_i is given by the probability function

$$\Pr[X_i = x_i] = p_i^{x_i}(1 - p_i)^{1-x_i}, \quad x_i = 0, 1;$$

see Exercise 4. The general multinomial distribution is treated in Example 2.10.9. □

The concept of conditional distributions, again, is analogous to the one developed in Section 2.6. For both the discrete and continuous case, as well as any mixture of the two, the probability density function of a conditional distribution is given by the ratio of the joint to the marginal *pdf*s. More precisely,

$$\begin{aligned} & f_{X_1,\ldots,X_q|X_{q+1}\ldots X_p}(x_1, \ldots, x_q | X_{q+1} = x_{q+1}, \ldots, X_p = x_p) \\ &= \frac{f_{\mathbf{X}}(x_1, \ldots, x_p)}{f_{X_{q+1}\ldots X_p}(x_{q+1}, \ldots, x_p)}. \end{aligned}$$

At this point, it is more elegant to write a p-variate, random vector in the form $\begin{bmatrix} \mathbf{X} \\ \mathbf{Y} \end{bmatrix}$, where $\mathbf{X}$ has q components and $\mathbf{Y}$ has $m = p - q$ components. Then we can write the joint probability density function in the form $f_{\mathbf{XY}}(\mathbf{x}, \mathbf{y})$, the marginals as $f_{\mathbf{X}}(\mathbf{x})$ and $f_{\mathbf{Y}}(\mathbf{y})$, respectively, and the *pdf* of the conditional distribution of $\mathbf{Y}$, given $\mathbf{X} = \mathbf{x}$, as $f_{\mathbf{Y}|\mathbf{X}}(\mathbf{y}|\mathbf{x}) = f_{\mathbf{XY}}(\mathbf{x}, \mathbf{y})/f_{\mathbf{X}}(\mathbf{x})$. Thus the notation used in Sections 2.2 to 2.8 carries over to the case where $\mathbf{X}$ and $\mathbf{Y}$ are themselves multivariate.

Example 2.10.8 continuation of Example 2.10.6

Again, consider the uniform distribution inside the unit sphere in three-dimensional space, as given by the density function

$$f_{\mathbf{X}}(x_1, x_2, x_3) = \begin{cases} \dfrac{3}{4\pi} & \text{if } x_1^2 + x_2^2 + x_3^2 \le 1, \\ 0 & \text{otherwise.} \end{cases}$$

The joint density of X_1 and X_2, given $X_3 = x_3$, for any x_3 in the interval $(-1, 1)$, is expressed by

$$f_{X_1X_2|X_3}(x_1, x_2|x_3) = \frac{f_{\mathbf{X}}(x_1, x_2, x_3)}{f_{X_3}(x_3)}.$$

Now, recall that if x_3 is a fixed value, then $f_{X_3}(x_3)$ is a constant, say $f_{X_3}(x_3) = c$. But

$$f_{\mathbf{X}}(x_1, x_2, x_3) = \begin{cases} \dfrac{3}{4\pi} & \text{if } x_1^2 + x_2^2 \le 1 - x_3^2, \\ 0 & \text{otherwise.} \end{cases}$$

Hence, the conditional density has the value $3/(4\pi c)$ for all (x_1, x_2) inside the circle with radius $\sqrt{1 - x_3^2}$, that is,

$$f_{X_1X_2|X_3}(x_1, x_2|x_3) = \begin{cases} \dfrac{3}{4\pi c} & \text{if } x_1^2 + x_2^2 \le 1 - x_3^2, \\ 0 & \text{otherwise,} \end{cases}$$

meaning that the conditional density is uniform inside the circle of radius $\sqrt{1 - x_3^2}$.

This result makes good intuitive sense if we think of the conditional density as a "slice" cut across the three-dimensional density along the plane of constant x_3. The intersection of this plane with the sphere has circular shape, and, for each point inside the circle, the density function is constant. As a by-product, we get the marginal distribution of X_3 "almost free". Since the conditional density of (X_1, X_2), given $X_3 = x_3$, must integrate to 1 and the area of the circle of radius r is $r^2\pi$,

$$(1 - x_3^2)\pi = 4\pi c/3,$$

or $c = \frac{3}{4}(1 - x_3^2)$. Recalling that $c = f_{X_3}(x_3)$, we have, therefore, obtained the marginal density of any component as

$$f_{X_i}(x_i) = \begin{cases} \dfrac{3}{4}(1 - x_i^2) & \text{if } |x_i| \le 1, \\ 0 & \text{otherwise.} \end{cases}$$

It is interesting to compare this density with the marginal density of a component obtained from the uniform distribution inside the unit circle; see Example 2.3.4 and Exercise 5.

Next, consider the conditional distribution of X_1, given $X_2 = x_2$ and $X_3 = x_3$. By the same argument, we obtain

$$f_{X_1|X_2X_3}(x_1|x_2, x_3) = \begin{cases} \dfrac{3}{4\pi} / f_{X_2X_3}(x_2, x_3) & \text{if } x_1^2 \le 1 - (x_2^2 + x_3^2), \\ 0 & \text{otherwise.} \end{cases}$$

Thus the conditional density of X_1, given $(X_2, X_3) = (x_2, x_3)$, is uniform in the interval $(-t, t)$, where $t = \sqrt{1 - x_2^2 - x_3^2}$. It follows that the marginal distribution of (X_2, X_3) has *pdf*

$$f_{X_2X_3}(x_2, x_3) = \begin{cases} \frac{3}{4}(1 - x_2^2 - x_3^2)^{1/2} & \text{if } x_2^2 + x_3^2 \leq 1, \\ 0 & \text{otherwise.} \end{cases}$$

Hence, this marginal distribution has the unit circle as the domain of positive density, but it is not uniform inside this domain.

All conditional and marginal densities found in this example are of the spherical type. This is no coincidence, as it is possible to show that conditional and marginal distributions obtained from spherical random variables are themselves always spherical; see Muirhead (1982, chapter 1.5). □

Now, we turn to a brief discussion of expectations of p-variate random vectors. Analogously with the theory of Section 2.5, define

$$E[G(X_1, \ldots, X_p)] = \int_{x_1=-\infty}^{\infty} \cdots \int_{x_p=-\infty}^{\infty} G(x_1, \ldots, x_p) f_{\mathbf{X}}(x_1, \ldots, x_p)\, dx_p \cdots dx_1$$

in the continuous case, and

$$E[G(X_1, \ldots, X_p)] = \sum_{(x_1, \ldots, x_p)} G(x_1, \ldots, x_p) f_{\mathbf{X}}(x_1, \ldots, x_p) \tag{4}$$

in the discrete case. Here, $G(x_1, \ldots, x_p)$ is a real-valued function of p variables. The most important special cases, together with some convenient notation, are as follows:

(a) $G(x_1, \ldots, x_p) = x_i$ (i fixed). Then we obtain the mean of the ith variable,

$$\mu_i = E[X_i].$$

(b) $G(x_1, \ldots, x_p) = x_i^2$ to obtain $E[X_i^2]$. Similarly, $G(x_1, \ldots, x_p) = x_i x_j$ (for fixed $(i, j), i \neq j$) yields $E[X_i X_j]$.

(c) $G(x_1, \ldots, x_p) = (x_i - \mu_i)^2$ to get

$$\sigma_i^2 = \text{var}[X_i] = E[(X_i - \mu_i)^2] = E[X_i^2] - \mu_i^2.$$

(d) $G(x_1, \ldots, x_p) = (x_i - \mu_i)(x_j - \mu_j)$. Then

$$\sigma_{ij} = \text{cov}[X_i, X_j] = E[(X_i - \mu_i)(X_j - \mu_j)] = E[X_i X_j] - \mu_i \mu_j.$$

Thus, these second-order moments add nothing to those defined in Section 2.5.

The notions of mean vector and covariance matrix, introduced in the following definitions, will be important throughout this course.

Definition 2.10.1 Let $\mathbf{X} = \begin{pmatrix} X_1 \\ \vdots \\ X_p \end{pmatrix}$ denote a random vector of dimension p. Then

$$E[\mathbf{X}] = \begin{pmatrix} E[X_1] \\ \vdots \\ E[X_p] \end{pmatrix} = \begin{pmatrix} \mu_1 \\ \vdots \\ \mu_p \end{pmatrix}$$

is the *mean vector* of $\mathbf{X}$. □

We will often use the symbol $\boldsymbol{\mu} = E[\mathbf{X}]$ to indicate a mean vector. More generally, if

$$\mathbf{Y} = \begin{pmatrix} Y_{11} & \cdots & Y_{1q} \\ \vdots & & \vdots \\ Y_{p1} & \cdots & Y_{pq} \end{pmatrix} \tag{5}$$

denotes a $p \times q$ matrix of random variables (or a random matrix, for short), then we define its expectation by

$$E[\mathbf{Y}] = \begin{pmatrix} E[Y_{11}] & \cdots & E[Y_{1q}] \\ \vdots & & \vdots \\ E[Y_{p1}] & \cdots & E[Y_{pq}] \end{pmatrix}. \tag{6}$$

This is immediately useful in the following definition.

Definition 2.10.2 The *covariance matrix* of a p-variate, random vector $\mathbf{X}$ is the $p \times p$ matrix

$$\begin{aligned}
\text{Cov}[\mathbf{X}] &= E[(\mathbf{X} - \boldsymbol{\mu})(\mathbf{X} - \boldsymbol{\mu})'] \\
&= E\left\{\begin{bmatrix} (X_1 - \mu_1)^2 & (X_1 - \mu_1)(X_2 - \mu_2) & \cdots & (X_1 - \mu_1)(X_p - \mu_p) \\ (X_2 - \mu_2)(X_1 - \mu_1) & (X_1 - \mu_1)^2 & \cdots & (X_2 - \mu_2)(X_p - \mu_p) \\ \vdots & \vdots & \ddots & \vdots \\ (X_p - \mu_p)(X_1 - \mu_1) & (X_p - \mu_p)(X_2 - \mu_2) & \cdots & (X_p - \mu_p)^2 \end{bmatrix}\right\} \\
&= \begin{pmatrix} \text{var}[X_1] & \text{cov}[X_1, X_2] & \cdots & \text{cov}[X_1, X_p] \\ \text{cov}[X_2, X_2] & \text{var}[X_2] & \cdots & \text{cov}[X_2, X_p] \\ \vdots & \vdots & \ddots & \vdots \\ \text{cov}[X_p, X_1] & \text{cov}[X_p, X_2] & \cdots & \text{var}[X_p] \end{pmatrix}.
\end{aligned} \tag{7}$$

□

We shall often write

$$\boldsymbol{\psi} = \begin{pmatrix} \sigma_1^2 & \sigma_{12} & \cdots & \sigma_{1p} \\ \sigma_{21} & \sigma_2^2 & \cdots & \sigma_{2p} \\ \vdots & \vdots & \ddots & \vdots \\ \sigma_{p1} & \sigma_{p2} & \cdots & \sigma_p^2 \end{pmatrix}$$

for a covariance matrix. The covariance matrix of a random vector contains the variances as diagonal elements and the covariances as off-diagonal elements. Covariance matrices are symmetric because $\text{cov}[U, V] = \text{cov}[V, U]$ for any two random variables U and V. Note the fine distinction in the notation: We use the symbol $\text{Cov}[\mathbf{X}]$ to denote the covariance matrix of the random vector $\mathbf{X}$, whereas $\text{cov}[U, V]$ stands for the covariance between two variables.

A useful result for the computation of a covariance matrix, generalizing equation (14) of Section 2.5, is

$$\text{Cov}[\mathbf{X}] = E[\mathbf{X}\mathbf{X}'] - \boldsymbol{\mu}\boldsymbol{\mu}'. \tag{8}$$

See Exercise 7 for a proof.

If ψ is the covariance matrix of a p–variate random vector $\mathbf{X}$, then it contains $p(p-1)/2$ distinct off-diagonal elements, or covariances σ_{ij}. Each of them can be transformed into a correlation coefficient by the transformation

$$\rho_{ij} = \text{corr}[X_i, X_j] = \frac{\sigma_{ij}}{\sigma_i \sigma_j}.$$

The $p \times p$ symmetric matrix with 1's on the diagonal and correlation coefficients ρ_{ij} off-diagonal is called the *correlation matrix* of $\mathbf{X}$:

$$\mathbf{R} = \begin{pmatrix} 1 & \rho_{12} & \rho_{13} & \cdots & \rho_{1p} \\ \rho_{21} & 1 & \rho_{23} & \cdots & \rho_{2p} \\ \rho_{31} & \rho_{32} & 1 & \cdots & \rho_{3p} \\ \vdots & \vdots & \vdots & \ddots & \vdots \\ \rho_{p1} & \rho_{p2} & \rho_{p3} & \cdots & 1 \end{pmatrix}.$$

There are some straightforward matrix manipulations relating the covariance matrix ψ and the correlation matrix $\mathbf{R}$, based on multiplication of a matrix by a diagonal matrix (see Exercise 8). Let $\text{diag}\psi$ denote a diagonal matrix with elements $\sigma_1^2, \ldots, \sigma_p^2$ on the diagonal, i.e.,

$$\text{diag}\psi = \text{diag}(\sigma_1^2, \ldots, \sigma_p^2) = \begin{pmatrix} \sigma_1^2 & 0 & \cdots & 0 \\ 0 & \sigma_2^2 & \cdots & 0 \\ \vdots & \vdots & \ddots & \vdots \\ 0 & 0 & \cdots & \sigma_p^2 \end{pmatrix}.$$

Then we can define its square root by taking square roots of all diagonal elements:

$$\left[\text{diag}\psi\right]^{1/2} = \text{diag}(\sigma_1, \ldots, \sigma_p) = \begin{pmatrix} \sigma_1 & 0 & \cdots & 0 \\ 0 & \sigma_2 & \cdots & 0 \\ \vdots & \vdots & \ddots & \vdots \\ 0 & 0 & \cdots & \sigma_p \end{pmatrix}.$$

The inverse of $[\text{diag}\psi]^{1/2}$ is

$$[\text{diag}\psi]^{-1/2} = \left\{[\text{diag}\psi]^{1/2}\right\}^{-1} = \begin{pmatrix} 1/\sigma_1 & 0 & \cdots & 0 \\ 0 & 1/\sigma_2 & \cdots & 0 \\ \vdots & \vdots & \ddots & \vdots \\ 0 & 0 & \cdots & 1/\sigma_p \end{pmatrix}.$$

With this notation, it can, then be verified (see Exercise 9) that

$$\mathbf{R} = [\text{diag}\psi]^{-1/2}\,\psi\,[\text{diag}\psi]^{-1/2}. \tag{9}$$

Conversely, we can write the covariance matrix ψ in terms of the standard deviations and correlations as

$$\psi = [\text{diag}\psi]^{1/2}\,\mathbf{R}\,[\text{diag}\psi]^{1/2}. \tag{10}$$

Example 2.10.9 The general multinomial distribution

Suppose that

$$\mathbf{X}_1 = \begin{pmatrix} X_{11} \\ \vdots \\ X_{1k} \end{pmatrix}, \mathbf{X}_2 = \begin{pmatrix} X_{21} \\ \vdots \\ X_{2k} \end{pmatrix}, \ldots, \mathbf{X}_N = \begin{pmatrix} X_{N1} \\ \vdots \\ X_{Nk} \end{pmatrix}$$

are N independent, identically distributed, random vectors, each following the multinomial distribution of Example 2.10.7 with probabilities $p_1, \ldots, p_k$. Let

$$\mathbf{Y} = \begin{pmatrix} Y_1 \\ \vdots \\ Y_k \end{pmatrix} = \sum_{j=1}^{N} \mathbf{X}_j = \begin{pmatrix} \sum_{j=1}^{N} X_{j1} \\ \vdots \\ \sum_{j=1}^{N} X_{jk} \end{pmatrix}.$$

Then $\mathbf{Y}$ is said to have a *multinomial distribution* with N trials and probabilities $p_1, \ldots, p_k$. We will write

$$\mathbf{Y} \sim \text{mult}(N;\ p_1, \ldots, p_k),$$

for short. Exercise 18 shows that the *pdf* of $\mathbf{Y}$ is given by

$$\begin{aligned} f_{\mathbf{Y}}(y_1, \ldots, y_k) &= \Pr[Y_1 = y_1,\ Y_2 = y_2, \ldots, Y_k = y_k] \\ &= \binom{N}{y_1\ y_2\ \cdots\ y_k} p_1^{y_1}\, p_2^{y_2} \cdots p_k^{y_k}, \end{aligned} \tag{11}$$

where all y_i are nonnegative integers, and $\sum_{i=1}^{k} y_i = N$. Here,

$$\binom{N}{y_1\ y_2\ \cdots\ y_k} = \frac{N!}{y_1!\, y_2! \cdots y_k!} \tag{12}$$

is the *multinomial coefficient*, i.e., the number of ways to partition a set of N objects into subsets of size $y_1, \ldots, y_k$.

A convenient way to think of multinomial distributions is to consider an experiment where N balls are thrown at a collection of k bins; each ball lands in the ith bin with probability p_i, independently of all other balls. Then the Y_i are the numbers of balls in the k bins.

We leave it to the reader to show that if $\mathbf{Y} \sim \text{mult}(N;\ p_1, \ldots, p_k)$, then the mean vector and the covariance matrix of $\mathbf{Y}$ are given by

$$E[\mathbf{Y}] = \begin{pmatrix} Np_1 \\ Np_2 \\ \vdots \\ Np_k \end{pmatrix} \quad \text{and} \quad \text{Cov}[\mathbf{Y}] = N \begin{bmatrix} p_1(1-p_1) & -p_1p_2 & \cdots & -p_1p_k \\ -p_2p_1 & p_2(1-p_2) & \cdots & -p_2p_k \\ \vdots & \vdots & \ddots & \vdots \\ -p_kp_1 & -p_kp_2 & \cdots & p_k(1-p_k) \end{bmatrix}; \qquad (13)$$

see Exercise 19. □

Exercises for Section 2.10

1. In Example 2.10.1, find the marginal probability function of Y_3, i.e., find the distribution of the sum of the numbers shown by three independent fair dice. *Hint*: The probability function of $Y_2 = X_1 + X_2$ is known from Section 2.2. Write the joint probablity function of Y_2 and X_3 as a table.

2. Show that the volume of the unit sphere in dimension $p = 3$ is $4\pi/3$, and thus, verify the constant value of the density function in Example 2.10.6. *Hint*: Evaluate the integral

$$\int\int\int_{x^2+y^2+z^2\le 1} dz\,dy\,dx = \int_{x=-1}^{1}\int_{y=-\sqrt{1-x^2}}^{\sqrt{1-x^2}}\int_{z=-\sqrt{1-x^2-y^2}}^{\sqrt{1-x^2-y^2}} dz\,dy\,dx.$$

3. Show that the volume of the unit sphere in dimension $p \ge 1$ is given by

$$V_p = \frac{\left[\Gamma\left(\frac{1}{2}\right)\right]^p}{\Gamma\left(\frac{p}{2}+1\right)} = \frac{2\pi^{p/2}}{p\cdot\Gamma(p/2)},$$

where $\Gamma(\cdot)$ is the gamma function, i.e., $\Gamma(\alpha) = \int_0^\infty u^{\alpha-1}e^{-u}\,du$, defined for $\alpha > 0$. *Note*: this is an exercise for the mathematically more advanced student. A proof by induction is suggested.

4. Show that the marginal distribution of X_i in the multinomial model (3) is given by $\Pr[X_i = 1] = p_i$, $\Pr[X_i = 0] = 1 - p_i$.

5. Again, consider the uniform distribution inside the p-dimensional sphere, as given in equation (2). Note that, for $p = 1$, this is the uniform distribution in the interval $[-1, 1]$. Graph the marginal density function of X_i for $p = 1$, $p = 2$, and $p = 3$. See Example 2.10.6.

6. Extend Exercise 5 to the cases $p = 4$ and $p = 5$. *Hint*: Use the same technique, as in Example 2.10.7, to find the marginal density of X_1, and use Exercise 3.

7. Prove equation (8) either by writing the (i, j)th entry of Cov[**X**] in scalar form or, directly, by using matrix operations on $E[(\mathbf{X} - \mu)(\mathbf{X} - \mu)']$ and results on linear transformations from Section 2.11.

8. Let **A** denote a matrix of dimension $p \times q$, and let $\mathbf{B} = \text{diag}(b_1, \ldots, b_p)$ and $\mathbf{C} = \text{diag}(c_1, \ldots, c_q)$ denote diagonal matrices. Compute **BA** and **AC**, and thus, verify that
   (a) the ith row of **BA** is the ith row of **A** multiplied by b_i and
   (b) the jth column of **AC** is the jth column of **A** multiplied by c_j.

9. Prove equations (9) and (10).

10. Let **X** denote a multinomial random vector of dimension k, as defined in equation (3). Find $E[\mathbf{X}]$ and Cov[**X**]. *Hint*: Use Exercise 4, and find the joint marginal distribution for a pair (X_i, X_j).

11. Let $\mathbf{X} = (X_1, \ldots, X_p)$ denote a p-variate random vector with pdf $f_{\mathbf{X}}(\mathbf{x})$. Show that

$$f_{\mathbf{X}}(\mathbf{x}) = f_{X_1|X_2\ldots X_p}(x_1|x_2, \ldots, x_p) \times f_{X_2|X_3\ldots X_p}(x_2|x_3, \ldots, x_p) \times \cdots \times f_{X_{p-1}|X_p}(x_{p-1}|x_p) \times f_{X_p}(x_p),$$

provided all the conditional *pdf's* exist. *Hint*: Write

$$f_{\mathbf{X}}(\mathbf{x}) = \frac{f_{X_1\ldots X_p}(x_1, \ldots, x_p)}{f_{X_2\ldots X_p}(x_2, \ldots, x_p)} \times \frac{f_{X_2\ldots X_p}(x_2, \ldots, x_p)}{f_{X_3\ldots X_p}(x_3, \ldots, x_p)} \times \cdots \times \frac{f_{X_{p-1}X_p}(x_{p-1}, x_p)}{f_{X_p}(x_p)} \times f_{X_p}(x_p).$$

12. For $\mathbf{Y} = (Y_1, Y_2, Y_3)$ in Example 2.10.1, find $E[\mathbf{Y}]$ and Cov[**Y**],
    (a) using the *pdf* of **Y**, and
    (b) using the transformation $Y_1 = X_1$, $Y_2 = X_1 + X_2$, and $Y_3 = X_1 + X_2 + X_3$.

13. Let $X_1, \ldots, X_n$ denote n independent identically distributed random variables, with $E[X_i] = \mu$, and $\text{var}[X_i] = \sigma^2$. Let $Y_k = \sum_{i=1}^{k} X_i$ for $k = 1, \ldots, n$, that is, $Y_1 = X_1$, $Y_2 = X_1 + X_2$, etc. Let $\mathbf{Y} = (Y_1, \ldots, Y_n)'$. Find $E[\mathbf{Y}]$, Cov[**Y**], and the correlation matrix of **Y**. *Hint*: Find the typical elements of $E[\mathbf{Y}]$ and Cov[**Y**]. Also see Exercise 6 in Section 2.5.

14. Let (X_1, X_2, Y) denote a three-dimensional random variable with density function

$$f_{X_1X_2Y}(x_1, x_2, y) = \begin{cases} \dfrac{1}{100\sqrt{2\pi}} \exp[-\frac{1}{2}(y - 1 - 2x_1 - 3x_2)^2] & \text{if } 0 \le x_1 \le 10, 0 \le x_2 \le 10, y \in \mathbb{R}, \\ 0 & \text{otherwise.} \end{cases}$$

   (a) Find the marginal distribution of (X_1, X_2).
   (b) Find the conditional distribution of Y, given X_1 and X_2.

(c) Find the regression of Y on X_1 and X_2, i.e., $E[Y|X_1 = x_1, X_2 = x_2]$.

15. Let the joint distribution of three discrete random variables X_1, X_2, and Y be given by the following table:

x_1	x_2	y	$\Pr[X_1 = x_1, X_2 = x_2, Y = y]$
0	0	0	0.28
0	0	1	0.12
0	1	0	0.09
0	1	1	0.01
1	0	0	0.01
1	0	1	0.09
1	1	0	0.12
1	1	1	0.28

(a) Find the conditional distribution of (X_2, Y), given $X_1 = 0$.
(b) Find the conditional distribution of (X_2, Y), given $X_1 = 1$.
(c) Show that $E[Y|X_1 = 0, X_2 = 0] > E[Y|X_1 = 0, X_2 = 1]$.
(d) Show that $E[Y|X_1 = 1, X_2 = 0] > E[Y|X_1 = 1, X_2 = 1]$.
(e) Find the joint marginal distribution of (X_2, Y).
(f) Show that $E[Y|X_2 = 0] < E[Y|X_2 = 1]$.

Note: Parts (c) and (d) show that, for both values of X_1, the conditional mean of Y, given $X_2 = x_2$, is a decreasing function of x_2. Part (f) shows that the conditional mean of Y, given $X_2 = x_2$, is an increasing function of x_2. This is an example of *Simpson's paradox*, which states that a relationship found between two variables (here, Y and X_2) in each of two subpopulations (here, conditionally on $X_1 = 0$ and conditionally on $X_1 = 1$) does not necessarily hold overall.

16. This is a data example to illustrate Simpson's paradox (Exercise 15). Suppose that, in a company with two departments A and B, the following data was collected on job applicants.

Department A

Gender	# Applicants	# Accepted	# Rejected
M	800	560	240
F	200	180	20

Department B

Gender	# Applicants	# Accepted	# Rejected
M	200	20	180
F	800	240	560

(a) Show that men were discriminated against in department A (i.e., a higher percentage of female applicants was hired).

(b) Show that men were also discriminated against in department B.

(c) Show that overall (i.e., for the data of both departments combined), women were discriminated against.

(d) Relate this data example to Exercise 15. What are X_1, X_2, and Y?

17. Consider the joint distribution of three random variables X, Y, Z, given by the following table.

x	y	z	$\Pr[X = x, Y = y, Z = z]$
0	0	0	0.04
0	0	1	0.36
0	1	0	0.01
0	1	1	0.09
1	0	0	0.12
1	0	1	0.03
1	1	0	0.28
1	1	1	0.14

(a) Find the conditional distribution of (Y, Z), given $X = 0$. Show that conditionally on $X = 0$, Y and Z are independent.

(b) Find the conditional distribution of (Y, Z), given $X = 1$. Show that conditionally on $X = 1$, Y and Z are independent.

(c) Find the (joint) marginal distribution of (Y, Z), and show that they are not independent.

Note: This is an example of a *conditional independence* model, in which independence between Y and Z holds conditionally on all values of X, yet Y and Z are not independent.

18. Prove equation (11) along the following lines:

(a) For fixed nonnegative integers $y_1, \ldots, y_k$, with $\sum_{i=1}^{k} y_i = N$, define $\alpha_1 = \Pr[Y_1 = y_1]$ and

$$\alpha_m = \Pr[Y_m = y_m | Y_1 = y_1, \ldots, Y_{m-1} = y_{m-1}], \quad m = 2, \ldots, k.$$

Show that $f_{\mathbf{Y}}(y_1, \ldots, y_k) = \alpha_1 \alpha_2 \cdots \alpha_k$.

(b) Show that

$$\alpha_1 = \binom{N}{y_1} p_1^{y_1} (1 - p_1)^{N - y_1}$$

(this follows from first principles) and (for $m > 1$)

$$\alpha_m = \binom{N - y_1 - \ldots - y_{m-1}}{y_m} \frac{p_m^{y_m} (1 - p_1 - \ldots - p_m)^{N - y_1 - \ldots - y_m}}{(1 - p_1 - \ldots - p_{m-1})^{N - y_1 - \ldots - y_{m-1}}}.$$

(c) Combine parts (a) and (b) to prove equation (11).

19. If $\mathbf{Y} \sim \text{mult}(N;\ p_1, \ldots, p_k)$, find the mean vector and the covariance matrix of $\mathbf{Y}$. *Hint*: The covariance matrix of a multinomial random vector with $N = 1$ is calculated in Example 2.11.3; use the fact that $\mathbf{Y}$ is the sum of N independent, identically distributed, random variables.

2.11 Transformations of Random Vectors

In this section, we shall discuss mostly how bivariate and more generally p-variate random variables behave under linear transformations. For instance, as outlined in Example 2.10.1, we might want to find the distribution of $X_1 + \cdots + X_p$, where each X_i denotes the number shown in rolling a die. We will start out by studying first- and second-order moments, i.e., mean vectors and covariance matrices, but not the actual distributions of linearly transformed random vectors. Next, we will discuss linear combinations of p variables. These transformations play a particularly important role, as we shall see in Chapters 4 and 5. Accordingly, the first part of Section 2.11 is written on a relatively low level of technical difficulty, and the case of $p = 2$ variables is stressed once again. It is essential that the student understand the concept of a linear combination to be able to follow the derivation of discriminant functions. A more rigorous treatment of both linear and nonlinear transformations of random vectors is given at the end of this section. Although the transformation theorem (or rather, its application to multivariate normal distributions) is crucial for the development of multivariate methods, the mathematically less inclined reader may choose to skip it at the first reading and return to it later.

Consider a p-variate, random vector $\mathbf{X} = (X_1, \ldots, X_p)'$ and a matrix

$$\mathbf{A} = (a_{ij}) = \begin{pmatrix} a_{11} & \cdots & a_{1p} \\ \vdots & \ddots & \vdots \\ a_{k1} & \cdots & a_{kp} \end{pmatrix}$$

of dimension $k \times p$. Define a new random vector

$$\mathbf{Y} = \mathbf{A}\mathbf{X} = \begin{pmatrix} Y_1 \\ \vdots \\ Y_k \end{pmatrix}, \tag{1}$$

with

$$Y_j = \sum_{h=1}^{p} a_{jh} X_h$$

as its jth component. Here, the elements of the matrix $\mathbf{A}$ are fixed real numbers. Hence, every variable Y_j is a *linear combination* of $X_1, \ldots, X_p$. Now, we may ask for the relationship between the moments of $\mathbf{X}$ and those of $\mathbf{Y}$, assuming that all means and variances exist.

Theorem 2.11.1 *If* $\mathbf{Y} = \mathbf{AX} + \mathbf{c}$, *where* $\mathbf{X}$ *is a* p*-dimensional random vector,* $\mathbf{A}$ *is a fixed matrix of dimension* $k \times p$, *and* $\mathbf{c} \in \mathbb{R}^k$ *is a fixed vector, then*

$$E[\mathbf{Y}] = \mathbf{A} \cdot E[\mathbf{X}] + \mathbf{c}, \tag{2}$$

and

$$\text{Cov}[\mathbf{Y}] = \mathbf{A} \cdot \text{Cov}[\mathbf{X}] \cdot \mathbf{A}'. \tag{3}$$

Proof The jth element of $E[\mathbf{Y}]$ is $E[Y_j] = E\left[\sum_{h=1}^{p} a_{jh}X_h + c_j\right] = \sum_{h=1}^{p} a_{jh}E[X_h] + c_j$. Hence, (2) follows. To prove (3), first note that (2) holds in a more general setup, namely, if $\mathbf{U}$ is a random matrix of dimension $p \times q$ and $\mathbf{A}$, $\mathbf{B}$ are fixed matrices such that the product $\mathbf{AUB}$ is defined, then

$$E[\mathbf{AUB}] = \mathbf{A} \cdot E[\mathbf{U}] \cdot \mathbf{B}, \tag{4}$$

see Exercise 1. Thus, writing μ_X and μ_Y for $E[\mathbf{X}]$ and $E[\mathbf{Y}]$, respectively, we get

$$\begin{aligned}
\text{Cov}[\mathbf{Y}] &= E[(\mathbf{Y} - \mu_Y)(\mathbf{Y} - \mu_Y)'] \\
&= E[(\mathbf{AX} - \mathbf{A}\mu_X)(\mathbf{AX} - \mathbf{A}\mu_X)'] \\
&= E[\mathbf{A}(\mathbf{X} - \mu_X)(\mathbf{X} - \mu_X)'\mathbf{A}'] \\
&= \mathbf{A} \cdot E[(\mathbf{X} - \mu_X)(\mathbf{X} - \mu_X)']\mathbf{A}' \\
&= \mathbf{A} \cdot \text{Cov}[\mathbf{X}] \cdot \mathbf{A}',
\end{aligned}$$

as was to be shown. Notice that this matrix has dimension $k \times k$, as required. ∎

The student who hesitates to use matrix algebra may try to prove equation (3) in scalar notation: Let

$$\text{Cov}[\mathbf{X}] = \psi = (\sigma_{ij}).$$

Then it is to be shown that

$$\text{cov}[Y_h, Y_m] = \sum_{i=1}^{p}\sum_{j=1}^{p} a_{hi}a_{mj}\sigma_{ij},$$

which is the (h, m)-entry of $\text{Cov}[\mathbf{Y}]$. The benefit of using matrix algebra will be quite obvious.

We will often be interested in a special case of Theorem 2.11.1, where $\mathbf{A}$ has dimension $1 \times p$. Then we will typically use a lowercase letter (like $\mathbf{a}$) to write the coefficients in the form of a column vector, that is, we will study the linear combination

$$Y = \mathbf{a}'\mathbf{X} = a_1X_1 + a_2X_2 + \ldots a_pX_p, \tag{5}$$

where $\mathbf{a} = (a_1, \ldots, a_p)'$. The covariance matrix of Y has dimension 1×1 in this case, that is, it collapses to $\text{var}[Y]$. Again, writing $\psi = (\sigma_{ij})$ for the covariance matrix of

$\mathbf{X}$, $\sigma_i^2 = \sigma_{ii}$ for var$[X_i]$, and $\rho_{ij} = \sigma_{ij}/\sigma_i\sigma_j$ for the correlation between X_i and X_j,

$$\begin{aligned}\text{var}[Y] &= \mathbf{a}'\psi\mathbf{a} \\ &= \sum_{i=1}^{p}\sum_{j=1}^{p} a_i a_j \sigma_{ij} \\ &= \sum_{i=1}^{p} a_i^2\sigma_i^2 + 2\sum_{i<j} a_i a_j \sigma_{ij} \\ &= \sum_{i=1}^{p} a_i^2\sigma_i^2 + 2\sum_{i<j} a_i\sigma_i a_j\sigma_j\rho_{ij}. \end{aligned} \tag{6}$$

Here, the symbol $\sum_{i<j}$ is to be read as the sum over all pairs of indexes (i, j) such that $1 \le i < j \le p$. In other words,

$$\sum_{i<j} m_{ij} = \sum_{i=1}^{p-1}\sum_{j=i+1}^{p} m_{ij}.$$

Functions of the form $q(\mathbf{v}) = \mathbf{v}'\mathbf{H}\mathbf{v}$, as in equation (6) above, where $\mathbf{v} \in \mathbb{R}^p$ and $\mathbf{H}$ is a real $p \times p$ matrix, are called *quadratic forms*. They play an important role in multivariate statistics, as we will see in the chapter on multivariate normal theory and in later chapters.

In the case where all covariances between X_i and X_j are zero,

$$\text{Cov}[\mathbf{X}] = \text{diag}(\sigma_1^2, \ldots, \sigma_p^2) = \begin{pmatrix} \sigma_1^2 & 0 & \cdots & 0 \\ 0 & \sigma_2^2 & \cdots & 0 \\ \vdots & \vdots & \ddots & \vdots \\ 0 & 0 & \cdots & \sigma_p^2 \end{pmatrix}.$$

Then the following result holds.

Corollary 2.11.2 *If $X_1, \ldots, X_p$ are pairwise uncorrelated, then*

$$var\left[\sum_{i=1}^{p} a_i X_i\right] = \sum_{i=1}^{p} a_i^2 var[X_i]. \tag{7}$$

In particular, recall that the condition of uncorrelatedness is satisfied when $(X_1, \ldots, X_p)$ are mutually independent.

Example 2.11.1 continuation of Examples 2.2.3 and 2.10.1

Let $\mathbf{X} = (X_1, X_2, \ldots, X_p)'$ denote the numbers shown in rolling a die p times. Then

$$\boldsymbol{\mu}_X = E[\mathbf{X}] = \begin{bmatrix} 3.5 \\ \vdots \\ 3.5 \end{bmatrix} = 3.5 \cdot \mathbf{1}_p,$$

where $\mathbf{1}_p$ is a vector of length p with 1 in each position. Also,

$$\psi_X = \text{cov}[\mathbf{X}] = \text{diag}\left(\frac{35}{12}, \ldots, \frac{35}{12}\right) = \frac{35}{12} \cdot \mathbf{I}_p.$$

Now, consider partial sums

$$Y_1 = X_1$$

$$Y_2 = X_1 + X_2$$

$$\vdots$$

and

$$Y_p = X_1 + X_2 + \ldots + X_p.$$

We can write $\mathbf{Y} = (Y_1, \ldots, Y_p)' = \mathbf{AX}$, where

$$\mathbf{A} = \begin{pmatrix} 1 & 0 & 0 & \cdots & 0 \\ 1 & 1 & 0 & \cdots & 0 \\ 1 & 1 & 1 & \cdots & 0 \\ \vdots & \vdots & \vdots & \ddots & \vdots \\ 1 & 1 & 1 & \cdots & 1 \end{pmatrix}$$

is a lower triangular matrix. Then

$$\begin{aligned} \psi_Y = \text{Cov}[\mathbf{Y}] &= \mathbf{A}\psi_X\mathbf{A}' \\ &= \mathbf{A} \cdot \frac{35}{12}\mathbf{I}_p \cdot \mathbf{A}' \\ &= \frac{35}{12}\mathbf{AA}' \\ &= \frac{35}{12} \cdot \begin{pmatrix} 1 & 1 & 1 & \cdots & 1 \\ 1 & 2 & 2 & \cdots & 2 \\ 1 & 2 & 3 & \cdots & 3 \\ \vdots & \vdots & \vdots & \ddots & \vdots \\ 1 & 2 & 3 & \cdots & p \end{pmatrix}. \end{aligned}$$

The squared correlation between two partial sums is given by $\rho^2(Y_i, Y_j) = i/j \ (i \leq j)$. See also Exercise 6 in Section 2.5. □

Example 2.11.2 continuation of Example 2.10.2

It can be verified directly that $E[\mathbf{X}] = \mu = (\frac{1}{2}, \frac{1}{2}, \frac{1}{2})'$ and $\psi = \text{cov}[\mathbf{X}] = \frac{1}{4} \cdot \mathbf{I}_3$. Diagonality of ψ follows from pairwise independence. Suppose that we are interested in $Y = X_1 + X_2 + X_3$. It follows from Theorem 2.11.1 and Corollary 2.11.2 that

$$E[Y] = \frac{3}{2}$$

and

$$\text{var}[Y] = \text{var}[X_1] + \text{var}[X_2] + \text{var}[X_3] = \frac{3}{4}.$$

Remarkably, this is the same result as if (X_1, X_2, X_3) were mutually independent. Variances and covariances fail to detect "pathological" cases of mutual dependence when all pairs of variables exhibit pairwise independence. This result can also be obtained from the exact distribution of Y, see Exercise 2. □

Example 2.11.3 continuation of Example 2.10.7

Again, consider the multinomial distribution with probability function

$$f_{\mathbf{X}}(x_1, \ldots, x_k) = \begin{cases} p_1^{x_1} p_2^{x_2} \ldots p_k^{x_k} & x_i = 0, 1, \ \sum_{j=1}^{k} x_j = 1, \\ 0 & \text{otherwise.} \end{cases}$$

The marginal probability function of each single component has been given in Example 2.10.7; it follows that $E[X_i] = p_i$ and $\text{var}[X_i] = p_i(1 - p_i)$. The joint marginal probability function of X_i and X_j $(i \neq j)$ is (see Exercise 10 in Section 2.10) expressed by

$$f_{X_i X_j}(x_i, x_j) = \begin{cases} p_i & \text{if } x_i = 1, x_j = 0, \\ p_j & \text{if } x_i = 0, x_j = 1, \\ 1 - (p_i + p_j) & \text{if } x_i = x_j = 0, \\ 0 & \text{otherwise.} \end{cases}$$

Hence, $E[X_i X_j] = 0$, $\text{cov}[X_i, X_j] = -p_i p_j$,

$$E[\mathbf{X}] = \mathbf{p} = \begin{pmatrix} p_1 \\ p_2 \\ \vdots \\ p_k \end{pmatrix},$$

and

$$\text{Cov}[\mathbf{X}] = \begin{pmatrix} p_1(1-p_1) & -p_1 p_2 & \cdots & -p_1 p_k \\ -p_2 p_1 & p_2(1-p_2) & \cdots & -p_2 p_k \\ \vdots & \vdots & \ddots & \vdots \\ -p_k p_1 & -p_k p_2 & \cdots & p_k(1-p_k) \end{pmatrix}.$$

Now, consider $Y = \sum_{i=1}^{k} X_i$. By construction of the multinomial distribution, the X_i add up to 1, and hence, $Y = 1$ with probability 1. Let us see if this can be confirmed by the calculation of the mean and variance of Y. Setting $\mathbf{1}_k = (1, \ldots, 1)'$, a vector of length k with 1 in each position,

$$E[Y] = \mathbf{1}_k' E[\mathbf{X}] = \sum_{i=1}^{k} p_i = 1,$$

and

$$\begin{aligned}\operatorname{var}[Y] &= \sum_{i=1}^{k} p_i(1-p_i) - \sum_{\substack{i=1 \\ i\neq j}}^{k}\sum_{j=1}^{k} p_i p_j \\ &= \sum_{i=1}^{k}\left[p_i\left(1 - p_i - \sum_{\substack{j=1 \\ j\neq i}}^{k} p_j\right)\right] \\ &= 0.\end{aligned}$$

Hence, the result is confirmed, albeit in a rather complicated way. Y is a "degenerated" random variable, centered at 1, with variability zero. □

In all examples considered so far, the linear transformations were from a p-variate random vector to a new random vector of dimension p or lower. However, the theorem covers other cases as well.

Example 2.11.4 continuation of Example 2.2.2

Suppose that we are interested in the joint distribution of X_1, X_2, and $Y = X_1 - X_2$, where X_1 and X_2 are the numbers shown in two independent rolls of a die. The joint probability function is not difficult to obtain, but, for the moment let us concentrate on the mean vector and covariance matrix. Formally, let

$$U_1 = X_1,$$

$$U_2 = X_2,$$

and

$$U_3 = X_1 - X_2.$$

Then we can write $\mathbf{U} = \mathbf{AX}$, where $\mathbf{A} = \begin{bmatrix} 1 & 0 \\ 0 & 1 \\ 1 & -1 \end{bmatrix}$. It follows from Example 2.11.1 that

$$E[\mathbf{U}] = \mathbf{A} \cdot E[\mathbf{X}] = \begin{pmatrix} 3.5 \\ 3.5 \\ 0 \end{pmatrix}$$

and

$$\text{cov}[\mathbf{U}] = \mathbf{A} \cdot \text{Cov}[\mathbf{X}] \cdot \mathbf{A}' = \frac{35}{12} \cdot \mathbf{AA}' = \frac{35}{12} \cdot \begin{pmatrix} 1 & 0 & 1 \\ 0 & 1 & -1 \\ 1 & -1 & 2 \end{pmatrix}.$$

This covariance matrix is *singular*, i.e., its determinant is zero. Later, we shall see that singularity of a covariance matrix means that there are exact linear relationships between some variables. In our example, with $U_3 = U_1 + U_2$, this is obviously the case. □

In the discrete case, singularity does not imply that we cannot find a joint probability function for all variables involved. In the continuous case, things are a little more tricky, as we can see from the following example.

Example 2.11.5 Suppose that X follows a uniform distribution in the interval $[0, 1]$. Consider the joint distribution of X and $2X$. Using the transformation

$$\mathbf{Y} = \begin{pmatrix} Y_1 \\ Y_2 \end{pmatrix} = \begin{pmatrix} 1 \\ 2 \end{pmatrix} X,$$

we find that

$$E[\mathbf{Y}] = \begin{pmatrix} 1/2 \\ 1 \end{pmatrix}$$

and

$$\text{Cov}[\mathbf{Y}] = \begin{pmatrix} 1 \\ 2 \end{pmatrix} \cdot \frac{1}{12} \cdot (1, 2) = \begin{pmatrix} \frac{1}{12} & \frac{1}{6} \\ \frac{1}{6} & \frac{1}{3} \end{pmatrix}.$$

Again, this covariance matrix is singular, reflecting the exact linear relationship $Y_2 = 2Y_1$ between the two variables. Variable Y_2 is uniform in $[0, 2]$, with density function

$$f_{Y_2}(y_2) = \begin{cases} \dfrac{1}{2} & \text{if } 0 \leq y_2 \leq 2, \\ 0 & \text{otherwise,} \end{cases}$$

but there is no such thing as a joint density function of Y_1 and Y_2 because all of the probability mass is concentrated on the line $y_2 = 2y_1$, which has area 0 in $\mathbb{R}^2$. See also Exercise 6. □

We finish this section with a highly useful and well-known result, the transformation theorem.

Theorem 2.11.3 (Transformation theorem for continuous random vectors) *Let* $\mathbf{X} = (X_1, \ldots, X_p)'$ *denote a* p*-variate continuous random vector, with density function* $f_{\mathbf{X}}(\mathbf{x})$ *positive in a domain* $D_{\mathbf{X}} \subset \mathbb{R}^p$. *Let*

$$y_i = y_i(x_1, \ldots, x_p) \qquad i = 1, \ldots, p \tag{8}$$

denote p real-valued functions of p variables, and assume that the transformation (8) is one-to-one, i.e., the x_j are functions of the y_i,

$$x_j = x_j(y_1, \ldots, y_p) \qquad j = 1, \ldots, p. \tag{9}$$

Define p jointly distributed random variables $\mathbf{Y} = (Y_1, \ldots, Y_p)'$ by

$$Y_i = y_i(X_1, \ldots, X_p) \qquad i = 1, \ldots, p. \tag{10}$$

Assume that the functions (9) are differentiable, i.e., $\frac{\partial x_j}{\partial y_i}$ exists for all (i, j). Then the pdf of $\mathbf{Y}$ is given by

$$f_{\mathbf{Y}}(y_1, \ldots, y_p) = f_X\left[x_1(y_1, \ldots, y_p), \ldots, x_p(y_1, \ldots, y_p)\right] \cdot J(y_1, \ldots, y_p). \tag{11}$$

Here, $J(y_1, \ldots, y_p)$ is the absolute value of the Jacobian of the transformation, that is,

$$J(y_1, \ldots, y_p) = |\det \mathbf{J}|, \tag{12}$$

and $\mathbf{J}$ is the matrix of derivatives of the x variables with respect to the y variables,

$$\mathbf{J} = \begin{pmatrix} \frac{\partial x_1}{\partial y_1} & \frac{\partial x_1}{\partial y_2} & \cdots & \frac{\partial x_1}{\partial y_p} \\ \frac{\partial x_2}{\partial y_1} & \frac{\partial x_2}{\partial y_2} & \cdots & \frac{\partial x_2}{\partial y_p} \\ \vdots & \vdots & \ddots & \vdots \\ \frac{\partial x_p}{\partial y_1} & \frac{\partial x_p}{\partial y_2} & \cdots & \frac{\partial x_p}{\partial y_p} \end{pmatrix}. \tag{13}$$

Proof Suppose that we are interested in

$$Pr[\mathbf{X} \in \mathcal{A}] = \int_{\mathcal{A}} f_{\mathbf{X}}(\mathbf{x})\, d\mathbf{x} \tag{14}$$

for some set $\mathcal{A} \subset \mathbb{R}^p$. Let $\mathcal{B}$ be the transform of $\mathcal{A}$ under the transformation (8), that is, each $\mathbf{x} \in \mathcal{A}$ is mapped into a $\mathbf{y} \in \mathcal{B}$ by (8), and each $\mathbf{y} \in \mathcal{B}$ is mapped into a $\mathbf{x} \in \mathcal{A}$ by (9). Then the theory of multiple integrals (see, e.g., Khuri (1993, Theorem 7.11.1) shows that

$$\int_{\mathcal{A}} f_{\mathbf{X}}(\mathbf{x})\, d\mathbf{x} = \int_{\mathcal{B}} f_X\left[x_1(y_1, \ldots, y_p), \ldots, x_p(y_1, \ldots, y_p)\right] \cdot J(y_1, \ldots, y_p)\, d\mathbf{y}. \tag{15}$$

Since this holds for all $\mathcal{A} \subset \mathbb{R}^p$, the integrand on the right hand side in equation (15) is the density function of $\mathbf{Y}$. ∎

In the univariate case, the transformation theorem simplifies somewhat because the determinant of a number is the number itself; see Exercise 12. The theorem works only for one-to-one transformations in $\mathbb{R}^p$. For instance, it cannot be used to find the distribution of $X_1 + X_2$ for a bivariate random vector $\mathbf{X} = (X_1, X_2)'$. However, it often provides an intermediate step for finding the distribution of a sum or of some other function of several random variables, as we will see in Example 2.11.7.

The class of nonsingular linear transformations is particularly important and also relatively simple to handle. Suppose that we wish to find the distribution of $\mathbf{Y} = \mathbf{AX} + \mathbf{b}$, where $\mathbf{X}$ is a p-variate continuous random vector with density function $f_{\mathbf{X}}$, $\mathbf{A}$ is a nonsingular $p \times p$ matrix, and $\mathbf{b} \in \mathbb{R}^p$. Then the Jacobian is given by

$$J(y_1, \ldots, y_p) = |\det \mathbf{A}^{-1}| = 1/|\det \mathbf{A}|$$

and does not depend on the y_i; see Exercise 13.

Example 2.11.6 Suppose that X_1 and X_2 are jointly normal with means 0, variances 1, and correlation ρ, $|\rho| < 1$. Then the *pdf* of $\mathbf{X} = (X_1, X_2)$ is given by (see Example 2.2.7)

$$f_{\mathbf{X}}(\mathbf{x}) = \frac{1}{2\pi\left(1-\rho^2\right)^{1/2}} \exp\left[-\frac{1}{2\left(1-\rho^2\right)}\left(x_1^2 - 2\rho x_1 x_2 + x_2^2\right)\right], \quad \mathbf{x} \in \mathbb{R}^2.$$

Let $Y_1 = X_1$ and $Y_2 = X_2 - \rho X_1$, that is,

$$\begin{pmatrix} X_1 \\ X_2 \end{pmatrix} = \begin{pmatrix} 1 & 0 \\ \rho & 1 \end{pmatrix} \begin{pmatrix} Y_1 \\ Y_2 \end{pmatrix},$$

and the Jacobian is $J(y_1, y_2) = 1$. Then Theorem 2.11.3 gives

$$\begin{aligned}
f_{\mathbf{Y}}(\mathbf{y}) &= f_{\mathbf{X}}(y_1, y_2 + \rho y_1) \\
&= \frac{1}{2\pi(1-\rho^2)^{1/2}} \exp\left\{-\frac{1}{2(1-\rho^2)}\left[y_1^2 - 2\rho y_1(y_2 + \rho y_1) + (y_2 + \rho y_1)^2\right]\right\} \\
&= \frac{1}{2\pi(1-\rho^2)^{1/2}} \exp\left\{-\frac{1}{2(1-\rho^2)}\left[(1-\rho^2)y_1^2 + y_2^2\right]\right\} \\
&= \frac{1}{\sqrt{2\pi}} \exp\left(-\frac{1}{2}y_1^2\right) \cdot \frac{1}{\sqrt{2\pi}\left(1-\rho^2\right)^{1/2}} \exp\left[-\frac{1}{2(1-\rho^2)}y_2^2\right].
\end{aligned}$$

This is the joint *pdf* of two independent normals. Variable Y_1 is standard normal; variable Y_2 has mean zero and variance $1 - \rho^2$. In Theorem 3.3.1, we will discuss a more general version of this result and give a more elegant proof. □

Example 2.11.7 In some cases, the main difficulty in applications of the transformation theorem is to identify the domain of positive density of the transformed variables correctly. For instance, suppose the joint *pdf* of X_1 and X_2 is given as

$$f_{\mathbf{X}}(x_1, x_2) = \begin{cases} x_1 + x_2 & \text{if } 0 \le x_1 \le 1, 0 \le x_2 \le 1, \\ 0 & \text{otherwise,} \end{cases}$$

as in Exercise 8 of Section 2.9. Suppose that we wish to find the *pdf* of $X_1 + X_2$. This can be achieved by first computing the joint *pdf* of $Y_1 = X_1$ and $Y_2 = X_1 + X_2$, and then integrating over Y_1 to find the *pdf* of the sum. Since the inverse transformation is given by

$$X_1 = Y_1$$

and

$$X_2 = Y_2 - Y_1,$$

the domain of positive density of the joint *pdf* of Y_1 and Y_2 is the set $\{(y_1, y_2) \in \mathbb{R}^2 : 0 \le y_1 \le 1,\ y_1 \le y_2 \le y_1 + 1\}$, as illustrated in Figure 2.11.1. The Jacobian is given by $J(y_1, y_2) = 1$, and therefore,

$$f_{\mathbf{Y}}(y_1, y_2) = f_{\mathbf{X}}(y_1, y_2 - y_1) = \begin{cases} y_2 & \text{if } 0 \le y_1 \le 1,\ y_1 \le y_2 \le y_1 + 1, \\ 0 & \text{otherwise.} \end{cases}$$

The marginal *pdf* of $Y_2 = X_1 + X_2$ is now obtained directly by integrating over Y_1 and observing the boundaries in Figure 2.11.1; it follows that

$$f_{Y_2}(y_2) = \begin{cases} y_2^2 & \text{if } 0 \le y_2 < 1 \\ y_2(2 - y_2) & \text{if } 1 \le y_2 < 2 \\ 0 & \text{otherwise.} \end{cases}$$ □

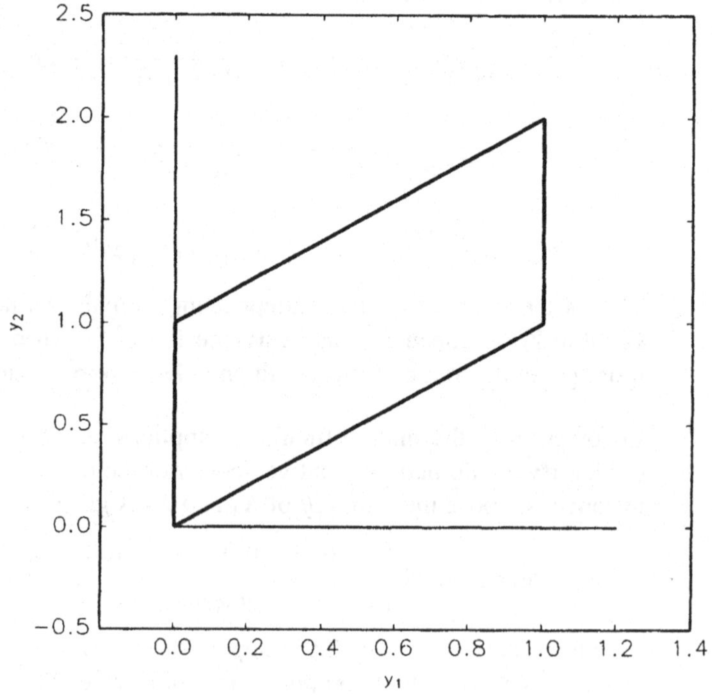

Figure 2-11-1
The domain of positive density of the joint distribution of Y_1 and Y_2 in Example 2.11.7.

Example 2.11.8 In Example 2.4.4, we saw that if (X_1, X_2) has a uniform distribution inside the unit circle, then the polar coordinates R and A, where $X_1 = R \cos A$ and $X_2 = R \sin R$, are independent. The transformation from (x_1, x_2) in the unit circle to (r, a) in the rectangle $0 \le r \le 1, 0 \le a < 2\pi$ is one-to-one, the matrix of derivatives is given by

$$\mathbf{J} = \begin{pmatrix} \dfrac{\partial x_1}{\partial r} & \dfrac{\partial x_1}{\partial a} \\ \dfrac{\partial x_2}{\partial r} & \dfrac{\partial x_2}{\partial a} \end{pmatrix} = \begin{pmatrix} \cos(a) & -r \sin(a) \\ \sin(a) & r \cos(a) \end{pmatrix},$$

and hence, the Jacobian is expressed by $J(r, a) = \left| r \left[\cos^2(a) + \sin^2(a)\right] \right| = r \ (\ge 0)$. The joint *pdf* of R and A, therefore, is given by

$$\begin{aligned} f_{RA}(r, a) &= f_{\mathbf{X}}\left[r \cos(a),\ r \sin(a)\right] \cdot r \\ &= \begin{cases} r/\pi & \text{if } 0 \le r \le 1, 0 \le a < 2\pi, \\ 0 & \text{otherwise,} \end{cases} \end{aligned}$$

that is, the angle A is uniform in the interval $[0, 2\pi)$, and the radius R has density $2r$ in $[0, 1]$. This confirms the result found in Example 2.4.4. Transformations to polar coordinates will be important in our discussion of spherical distributions in Section 3.4. □

Exercises for Section 2.11

1. Prove the following result. If $\mathbf{X}$ is a $p \times q$ matrix of random variables and $\mathbf{A}, \mathbf{B}$ are matrices of fixed numbers ($\mathbf{A}, \mathbf{B}$ of proper dimensions such that the product $\mathbf{AXB}$ is defined), then $E[\mathbf{AXB}] = \mathbf{A} \cdot E[\mathbf{X}] \cdot \mathbf{B}$. See equation (4).

2. For the random variables X_1, X_2, and X_3 used in Examples 2.10.2 and 2.11.2, find the probability function of $Y = X_1 + X_2 + X_3$. Then compute $E[Y]$ and var$[Y]$ using this *pdf*, and thus, verify the result of Example 2.11.2.

3. For the multinomial model used in Example 2.11.3, show that
$$\text{corr}[X_i, X_j] = \rho_{ij} = -\sqrt{p_i/(1-p_i)} \cdot \sqrt{p_j/(1-p_j)}\,.$$

4. In Example 2.11.4, find the correlation matrix of $\mathbf{U}$, and show that it is singular.

5. In Section 2.5, we saw that the correlation coefficent ρ_{XY} for two random variables X and Y is always in the interval $[-1, 1]$. Show that the covariance matrix of $\binom{X}{Y}$ is singular exactly if $|\rho_{XY}| = 1$. *Hint*: Write $\text{Cov}\begin{bmatrix} X \\ Y \end{bmatrix} = \begin{bmatrix} \sigma_{11} & \sigma_{12} \\ \sigma_{21} & \sigma_{22} \end{bmatrix}$, and recall that $\rho = \sigma_{12}/\sqrt{\sigma_{11}\sigma_{22}}$.

6. This exercise serves to further illustrate problems of continuous random variables where bivariate densities do not exist; see Example 2.11.5. Suppose that X is uniform

in the interval [0, 1] and $\varepsilon > 0$ is an arbitrary (but fixed) real number. Define a random variable Y such that, conditionally on $X = x$, Y is uniform in the interval $[x, x + \varepsilon]$, that is,

$$f_{Y|X}(y|x) = \begin{cases} \dfrac{1}{\varepsilon} & \text{if } x \leq y \leq x + \varepsilon, \\ 0 & \text{otherwise.} \end{cases}$$

(a) Find the joint density of (X, Y). For $\varepsilon = \frac{1}{2}$, graph the area where f_{XY} is positive.

(b) Find the marginal density of Y.

(c) Now let ε become smaller and tend to 0. What happens to the joint density f_{XY}?

(d) Show that, as ε tends to 0, the density of Y tends to a uniform density in [0, 1].

7. Same as Exercise 6, but for X with a standard exponential distribution, i.e.,

$$f_X(x) = \begin{cases} e^{-x} & x \geq 0, \\ 0 & \text{otherwise.} \end{cases}$$

Hint: In part (d), you will need to use

$$\lim_{\varepsilon \to 0} \frac{1 - e^{-\varepsilon}}{\varepsilon} = 1$$

to show that the density of Y tends to a standard exponential.

8. Let T denote a uniform random variable in the interval $[0, 2\pi]$, i.e.

$$f_T(t) = \begin{cases} \dfrac{1}{2\pi} & \text{if } 0 \leq t \leq 2\pi, \\ 0 & \text{otherwise.} \end{cases}$$

Let $X = \cos T$ and $Y = \sin T$.

(a) Find the distribution of X. *Hint*: First, assume that T is uniform in the interval $[0, \pi]$, and then argue that this gives the correct solution for T uniform in $[0, 2\pi]$. See also Exercise 22 in Section 2.8.

(b) Do X and Y have a joint (bivariate) density function? How can you adequately describe the joint distribution of X and Y?

(c) Show that $\text{corr}[X, Y] = 0$.

9. Four guests attend a party. As they arrive, they deposit their hats. Two hours and many Molsons later they leave, and each of them picks up a (randomly) chosen hat. We will be interested in the random variable Y = # of matches, i.e., number of hats picked up by the correct owners.

(a) Write out all 24 permutations of the numbers 1, 2, 3, 4, and assign probability 1/24 to each of them. Each permutation gives a value of Y, according to how many of the four numbers appear in the right position. For instance, the permutation

$$1 \quad 4 \quad 3 \quad 2$$

gives $Y = 2$, corresponding to guests 1 and 3 getting the right hats back. Hence, find the probability function of Y.

(b) Show that $E[Y] = 1$ and $\text{var}[Y] = 1$.

(c) Define four binary random variables X_1, X_2, X_3, and X_4, such that

$$X_i = \begin{cases} 1 & \text{if the } i\text{th guest gets the correct hat} \\ 0 & \text{otherwise.} \end{cases}$$

Show that $\Pr[X_1 = 1] = \frac{1}{4}$, $\Pr[X_1 = 0] = \frac{3}{4}$, and argue that the same law must hold for X_2, X_3, and X_4. Hence, show that $E[X_i] = \frac{1}{4}$ and $\text{var}[X_i] = \frac{3}{16}$ $(i = 1, 2, 3, 4)$.

(d) Show that $\Pr[X_1 = 1, X_2 = 1] = \frac{1}{12}$, and argue that the same result must hold for any pair (X_i, X_j), whenever $i \neq j$.

(e) Show that $\text{cov}[X_i, X_j] = \frac{1}{48}$, for $i \neq j$. *Hint*: Recall that $\text{cov}[X_i, X_j] = E[X_i X_j] - E[X_i]E[X_j]$. Furthermore, $X_i X_j = 1$ exactly if both variables are 1, and $X_i X_j = 0$ otherwise. Hence, to calculate $E[X_i X_j]$, you do not need to know the exact joint distribution of X_i and X_j.

(f) From the setup of the problem, it follows that $Y = X_1 + X_2 + X_3 + X_4$. Use equations (2) and (3) or (6) to show that $E[Y] = 1$ and $\text{var}[Y] = 1$.

10. The neighbors of our friends in Exercise 9 also throw a party, with $k \geq 2$ guests who (two hours and many Budweisers after arrival) pick up random hats. Let Y = number of guests who get the right hats. Finding the exact distribution of Y, as in part (a) of Exercise 10, turns out to be a job for Sisyphus for a party with as few as $k = 10$ guests $(10! = 3,628,800)$. Nevertheless, finding $E[Y]$ and $\text{var}[Y]$ is possible, using the same technique as in parts (c) through (f) of Exercise 9.

(a) Define k random variables $X_1, \ldots, X_k$ such that

$$X_i = \begin{cases} 1 & \text{if the } i\text{th guest gets the correct hat} \\ 0 & \text{otherwise.} \end{cases}$$

Show that $E[X_i] = 1/k$ and $\text{var}[X_i] = (k-1)/k^2$.

(b) Show that $\text{cov}[X_i, X_j] = 1/[k^2(k-1)]$ $(i \neq j)$.

(c) Let $Y = \sum_{i=1}^k X_i$, and show that $E[Y] = 1$, $\text{var}[Y] = 1$.

Note: This result is quite remarkable because it tells that, for any number of guests, the expected number of matches is always 1 and the variance is always 1. This holds despite the fact that, for large k, the exact distribution of Y is almost "infinitely complicated" to obtain.

11. Let $\mathbf{X} = (X_1, \ldots, X_p)'$ denote a p-variate random vector, and $Y = X_1 + X_2 + \ldots + X_p = \mathbf{1}_p'\mathbf{X}$, where $\mathbf{1}_p$ is a vector in $\mathbb{R}^p$ with 1 in each position. Show that

(a) $E[Y] = \sum_{i=1}^k E[X_i]$, and

(b) $\text{var}[Y] = \sum_{i=1}^k \text{var}[X_i] + \sum_{i<j} \text{cov}[X_i, X_j]$, using equations (2) and (3).

12. Show that, for dimension $p = 1$, the transformation theorem for one-to-one transformations of a continuous random variable can be written as

$$f_Y(y) = f_X\left[x(y)\right] \left|\frac{\partial x}{\partial y}\right|.$$

13. For a transformation $\mathbf{Y} = \mathbf{AX} + \mathbf{b}$, where $\mathbf{A}$ is nonsingular, show that the Jacobian is given by $|\det(\mathbf{A}^{-1})|$.

14. Suppose that (X_1, X_2) is bivariate normal with means zero, variances one, and correlation ρ ($|\rho| < 1$). Use the transformation theorem to show that $X_1 + X_2$ and $X_1 - X_2$ are independent normal random variables with means zero and variances $2(1 + \rho)$ and $2(1 - \rho)$, respectively. See also Exercise 9 in Section 2.9.

15. Use the technique of Example 2.11.7 to find the distribution of the sum of two independent random variables, both uniform in the interval (0, 1).

16. Let $\mathbf{X}$ denote a p-variate, random vector with *pdf* $f_{\mathbf{X}}$, and define p new variables as $Y_i = b_i + a_i X_i$, $i = 1, \ldots, p$. Here, the a_i and the b_i are constants, all a_i different from zero. Show that the joint *pdf* of $Y_1, \ldots, Y_p$ is expressed by

$$f_{\mathbf{Y}}(\mathbf{y}) = f_{\mathbf{X}}\left(\frac{y_1 - b_1}{a_1}, \ldots, \frac{y_p - b_p}{a_p}\right)\left|\prod_{i=1}^{p} \frac{1}{a_i}\right|,$$

(a) directly, i.e., via the cumulative distribution function of $\mathbf{Y}$, and

(b) using the transformation theorem.

16. Use the transformation theorem to show that the sum of two independent exponential random variables, both with parameter 1, has *pdf* $f_Z(z) = z \exp[-z]$ for $z > 0$. See also Exercise 3 in Section 2.9.

17. Solve Exercise 7 of Section 2.9 on the sum of independent chi-square random variables, using the transformation theorem.

18. Suppose that (X_1, X_2) has a uniform distribution in the unit circle. Find the joint *pdf* of $Y_1 = X_1^2$ and $Y_2 = X_2^2$. *Hint*: The transformation from (X_1, X_2) to (Y_1, Y_2) is not one-to-one. First, find the joint distribution of $|X_1|$ and $|X_2|$.

19. This exercise connects the transformation theorem to the theory of finite mixtures.

(a) Suppose that the p-variate random vector $\mathbf{X}$ has a finite mixture density

$$f_{\mathbf{X}}(\mathbf{x}) = \sum_{j=1}^{k} p_j f_j(\mathbf{x}).$$

Conditionally on $\mathbf{X}$ being from the jth component, let $\mathbf{Y} = h_j(\mathbf{X})$ be a one-to-one transformation, such that the conditions of the transformation theorem are satisfied, that is, conditionally on the jth component, the *pdf* of $\mathbf{Y}$ is given by

$$g_j(\mathbf{y}) = f_j\left(h_j^{-1}(\mathbf{y})\right) \cdot J_j(\mathbf{y}),$$

where $J_j(\mathbf{y}) = |\det \mathbf{J}_j|$, and $\mathbf{J}_j$ is the matrix of derivatives of the p functions $\mathbf{x} = h_j^{-1}(\mathbf{y})$. Show that the (unconditional) distribution of $\mathbf{Y}$ has *pdf*

$$g(\mathbf{y}) \sum_{j=1}^{k} p_j g_j(\mathbf{y}) = \sum_{j=1}^{k} p_j f_j\left(h_j^{-1}(\mathbf{y})\right) J_j(\mathbf{y}).$$

(b) Suppose that the p-variate random vector $\mathbf{X}$ has *pdf* $f_{\mathbf{X}}$ and the transformation $\mathbf{y} = h(\mathbf{x})$ is not one-to-one in the domain D of $\mathbf{X}$, but D can be partitioned into subsets $D_1, \ldots, D_k$ such that within each D_j the transformation is one-to-one. Let h_j be the function h restricted to D_j, such that the transformation $\mathbf{y} = h_j(\mathbf{x})$ is invertible for $\mathbf{x} \in D_j$. Using part (a), show that the *pdf* of $\mathbf{Y} = h(\mathbf{X})$ is given by

$$f_{\mathbf{Y}} = \sum_{j=1}^{k} f_{\mathbf{X}} \left[h_j^{-1}(\mathbf{y}) \right] J_j(\mathbf{y}),$$

where J_j is the Jacobian of the transformation constrained to D_j. *Hint*: Consider $f_{\mathbf{X}}$ as a mixture density with k components.

20. This exercise is based on Example 2.2.5. Suppose that X and Y have joint *pdf*

$$f_{XY}(x, y) = \begin{cases} \exp\left(-\frac{1}{2}x^2 + x - y\right) / \sqrt{2\pi} & \text{if } x \le y, \\ 0 & \text{otherwise.} \end{cases}$$

Find the *pdf* of $Y - X$. *Hint*: Find the joint *pdf* of $U = X + Y$ and $V = Y - X$.

21. If $\mathbf{X}_1$ and $\mathbf{X}_2$ are independent p-variate random vectors, show that $\text{Cov}[\mathbf{X}_1 + \mathbf{X}_2] = \text{Cov}[\mathbf{X}_1] + \text{Cov}[\mathbf{X}_2]$. More generally, if $\mathbf{X}_1, \ldots, \mathbf{X}_N$ are mutually independent p-variate random vectors, show that $\text{Cov}\left[\sum_{i=1}^N \mathbf{X}_i\right] = \sum_{i=1}^N \text{Cov}[\mathbf{X}_i]$.

3 The Multivariate Normal Distribution

3.1 Review of the Univariate Normal Distribution

Before introducing the multivariate normal distribution, let us briefly review some important results about the univariate normal.

Definition 3.1.1 A continuous random variable X has a (univariate) normal distribution if it has a density function of the form

$$f_X(x) = \frac{1}{\sqrt{2\pi}\sigma} \exp\left[-\tfrac{1}{2}\left(\frac{x-\mu}{\sigma}\right)^2 \right], \qquad x \in \mathbb{R}, \tag{1}$$

for some real numbers μ and $\sigma > 0$. □

If X has density (1), then μ is the mean of X, and σ is the standard deviation (Exercise 1). We will write $X \sim \mathcal{N}(\mu, \sigma^2)$ to indicate that X follows a normal distribution with parameters as indicated.

Among all possible choices of the parameters μ and σ, the *standard normal distribution* with $\mu = 0$ and $\sigma = 1$ plays a particularly important role because its distribution function is tabulated in most statistics texts. We leave it to the student to show that if $X \sim \mathcal{N}(\mu, \sigma^2)$, and $Z = (X - \mu)/\sigma$, then $Z \sim \mathcal{N}(0, 1)$, i.e., Z follows a standard normal distribution. This is actually a special case of the result that if $X \sim N(\mu, \sigma^2)$, and $Y = aX + b$ for some real constants a and b, then $Y \sim \mathcal{N}(a\mu + b, a^2\sigma^2)$; see Exercise 2.

The cumulative distribution function (cdf) of the normal distribution does not exist in closed form. Let

$$\Phi(z) = \frac{1}{\sqrt{2\pi}} \int_{u=-\infty}^{z} e^{-u^2/2}\, du \,, \qquad z \in \mathbb{R}, \tag{2}$$

denote the cdf of the standard normal. For $X \sim \mathcal{N}(\mu, \sigma^2)$,

$$F_X(x) = \Pr[X \leq x] = \Phi\left(\frac{x-\mu}{\sigma}\right) ; \tag{3}$$

see Exercise 3.

As mentioned before, the parameter σ is assumed to be positive. If we take a fixed value of μ and compare different normal density functions with σ taking smaller and smaller values, then we see higher and higher peaks centered at μ. For the limiting case $\sigma = 0$, no density exists. Yet the distribution function of such a sequence of random variables converges to a limiting distribution function which is zero for all values less than μ and 1 for all values larger than μ. (This is formulated more precisely in Exercise 4). Thus, in the limit, which we will call the *degenerate case*, a normal random variable with mean μ and variance zero takes the value μ with probability one. In other words, we may consider fixed numbers as special cases of normal random variables with variance 0. This may look a bit far-fetched, but, in Section 3.2, we will see why it is useful to admit such degenerate normal distributions.

The following two theorems are well known and of considerable importance; proofs are left to the student (Exercises 5 and 6).

Theorem 3.1.1 *If X and Y are two stochastically independent random variables, both normal, then $X + Y$ is normal.*

Theorem 3.1.2 *If $Z \sim \mathcal{N}(0, 1)$, then Z^2 follows a chi-square distribution on one degree of freedom.*

An immediate consequence of the first result is that if $X_1, \ldots, X_k$ are k mutually independent, normal random variables, then their sum follows a normal distribution. Theorem 3.1.1 is also one of the most frequently misstated results in introductory textbooks, which sometimes produce blunders like "if X and Y are both normally distributed random variables, then $X + Y$ is normal." With an example, we show why this statement is wrong. At the same time, the example serves the purpose of waking the reader who has found this review boring because there is nothing new in it. In anticipation of things to be discussed, independence of X and Y in Theorem 3.1.1 is not a necessary condition for the sum to follow a normal distribution; see Exercises 7 and 8.

Example 3.1.1 Let $Z \sim \mathcal{N}(0, 1)$, and let U denote a discrete random variable, independent of Z and taking values 1 and 2 with probabilities p_1 and p_2, respectively ($p_1 + p_2 = 1$). Define a random variable Y as

$$Y = \begin{cases} Z & \text{if } U = 1, \\ -Z & \text{if } U = 2. \end{cases}$$

This definition of Y determines its conditional distribution, given the value of U. The marginal distribution of Y is a mixture distribution, as outlined in Section 2.8. However, conditionally on both $U = 1$ and $U = 2$, Y is standard normal, and hence, $Y \sim \mathcal{N}(0, 1)$, irrespective of the values of p_1 and p_2. Therefore, both Y and Z are standard normal random variables. Yet, they do not have a joint pdf, because only values (y, z) with $y = z$ or $y = -z$ can be taken. For the sum $Y + Z$,

$$Y + Z = \begin{cases} Z + Z & \text{if } U = 1, \\ (-Z) + Z & \text{if } U = 2, \end{cases}$$

that is,

$$Y + Z = \begin{cases} 2Z & \text{with probability } p_1, \\ 0 & \text{with probability } p_2. \end{cases}$$

That means, with probability p_1, the sum $X + Y$ follows a $\mathcal{N}(0, 2^2)$ distribution, and with probability p_2, it is zero. This is an extremely "unnormal" distribution, which is neither discrete nor continuous, except in the limiting cases $p_1 = 0$ and $p_1 = 1$. It is also interesting to find the correlation $\rho(Y, Z)$ between Y and Z. Since $\text{var}[Y] = \text{var}[Z] = 1$,

$$\begin{aligned} \rho(Y, Z) = E[YZ] &= p_1 E[YZ|U = 1] + p_2 E[YZ|U = 2] \\ &= p_1 E[Z^2] + p_2 E[-Z^2] \\ &= p_1 - p_2 \,. \end{aligned}$$

Thus, the correlation can take all values in $[-1, 1]$. In particular, if $p_1 = \frac{1}{2}$, then $\rho(Y, Z) = 0$, despite the fact that Y and Z are highly dependent. See also Exercises 21 to 23 in Section 2.5, and Exercises 16 and 17 in Section 2.8. □

Exercises for Section 3.1

1. If X has pdf (1), show that $E[X] = \mu$ and $\text{var}[X] = \sigma^2$.
2. If $X \sim \mathcal{N}(\mu, \sigma^2)$ and $Y = aX + b$ for some constants a and b, show that $Y \sim \mathcal{N}(a\mu + b, a^2\sigma^2)$. What happens if $a = 0$?
3. Prove equation (3).

4. Let $F_n(x)$ denote the distribution function of a $\mathcal{N}(\mu, 1/n)$ random variable. Show that the sequence F_n converges to a limiting function of the form

$$F_\infty(x) = \begin{cases} 0 & \text{if } x < \mu, \\ 1/2 & \text{if } x = \mu, \\ 1 & \text{if } x > \mu. \end{cases}$$

(Note that F_∞ is not a distribution function, but needs to be "fixed" only in one point to make it a proper distribution function).

5. Prove Theorem 3.1.1, and its extension to k mutually independent, normal random variables.

6. Prove Theorem 3.1.2. *Note*: The *pdf* of a chi-square, random variable U with one degree of freedom has the form

$$f_U(u) = (2\pi u)^{-1/2} \exp(-u/2), \qquad u > 0.$$

7. Define a bivariate random variable (X, Y) as follows. Divide the (x, y)-plane into four sectors according to $\{x > 0, y > 0\}$, $\{x > 0, y < 0\}$, $\{x < 0, y < 0\}$, and $\{x < 0, y > 0\}$. Shade the first and third sectors, and leave the other two sectors blank. Let the joint *pdf* of (X, Y) be

$$f_{XY}(x, y) = \begin{cases} \pi^{-1} \exp\left[-\frac{1}{2}(x^2 + y^2)\right] & \text{if } (x, y) \text{ is in shaded area} \\ 0 & \text{if } (x, y) \text{ is in blank area.} \end{cases}$$

(a) Show that X and Y both are $\mathcal{N}(0, 1)$ random variables.

(b) Show that $X + Y$ is not normal, without finding the exact distribution of the sum.

(c) Show that the covariance between X and Y is $2/\pi$.

(d) Find the *pdf* of $X + Y$.

8. Define a bivariate random variable (X, Y), as in Exercise 7, but divide the (x, y)-plane into *eight* sectors separated by the lines $x = 0$, $y = 0$, $x = y$, and $x = -y$. Starting in one of the sectors and going around the origin, shade each second sector, and leave the remaining four sectors blank. Then define a joint density, as in problem 7.

(a) Show that $X \sim \mathcal{N}(0, 1)$, $Y \sim \mathcal{N}(0, 1)$, $X+Y \sim \mathcal{N}(0, 2)$, and $X-Y \sim \mathcal{N}(0, 2)$.

(b) Show that X and Y are dependent.

9. Prove the following result by mathematical induction: Let $Z_1, \ldots, Z_k$ denote mutually independent standard normal random variables. Then $Z_1^2 + \ldots + Z_k^2$ follows a chi-square distribution on k degrees of freedom. *Note*: The *pdf* of a chi-square random variable with k degrees of freedom has the form

$$f_U(u) = \frac{1}{2^{k/2}\Gamma(\frac{1}{2}k)} u^{\frac{1}{2}k-1} e^{-\frac{1}{2}u}, \quad u > 0,$$

where $\Gamma(t) = \int_0^\infty y^{t-1} e^{-y}\, dy$ is the gamma function.

10. The moment-generating function of a random variable X is defined as

$$M_X(t) = E\left[e^{tX}\right], \qquad t \in \mathbb{R}, \tag{4}$$

provided this expectation exists.

(a) If $X \sim \mathcal{N}(0, 1)$, find $M_X(t)$.

(b) If $X \sim \mathcal{N}(\mu, \sigma^2)$, show that

$$M_X(t) = \exp\left(t\mu + \frac{1}{2}t^2\sigma^2\right). \tag{5}$$

Hint: Write $X = \mu + \sigma Z$, where $Z \sim \mathcal{N}(0, 1)$.

(c) With $M_X'(t)$ and $M_X''(t)$ denoting the first and second derivatives of M_X with respect to t, show that $M_X'(0) = \mu$, $M_X''(0) = \mu^2 + \sigma^2$.

3.2 Definition and Properties of the Multivariate Normal Distribution

Multivariate normal theory is an old and well-known topic, but a rather fascinating one. The reader who has been exposed to the multivariate normal distribution before may be a bit surprised to find the theory being developed in a different way from what most low-level or medium-level textbooks would do. Indeed, in this section we follow a modern approach which is more abstract than the traditional way of introducing the multivariate normal by a density function. The modern approach has the advantage that many results follow from the definition "almost free." The price to be paid is that we will have to skip the proof of one of the central theorems because it is beyond the level of this book. However, this is more than compensated for by the elegance and the generality of the approach followed here.

Definition 3.2.1 The p-variate random vector $\mathbf{X} = \begin{pmatrix} X_1 \\ \vdots \\ X_p \end{pmatrix}$ is multivariate normal (or p-variate normal) if, for all vectors $\mathbf{b} \in \mathbb{R}^p$, $U = \mathbf{b}'\mathbf{X}$ is either univariate normal or constant (i.e., degenerate, univariate normal). □

This definition requires immediate comments. What the definition tells is that if we want to verify multivariate normality of a p-variate random vector $\mathbf{X}$, we need to look at univariate distributions of linear combinations $U = b_1X_1 + \ldots + b_pX_p$ and verify that all such linear combinations are univariate normal. This seems to be an impossible job, and the question immediately arises whether we could limit the search to a finite number of linear combinations. We will deal with this question in Exercise 10 at the end of this section.

Even more importantly, the degenerate case requires some explanation. Definition 3.2.1 admits the case that some linear combinations of $\mathbf{X}$ yield a constant rather than a genuine, normally distributed random variable. In this full generality, even a fixed vector $\xi \in \mathbb{R}^p$ could be considered a multivariate normal random variable — a very special (and uninteresting) one, because all linear combinations of ξ will inevitably be constants.

The immediate question is do multivariate normal random variables exist, besides the trivial case of a constant vector? The following two examples show that this is, indeed, the case.

Example 3.2.1 Let $Z_1, \ldots, Z_p$ denote p independent univariate standard normal random variables, and set $\mathbf{Z} = \begin{pmatrix} Z_1 \\ \vdots \\ Z_p \end{pmatrix}$. Then the joint density function of $\mathbf{Z}$ is the product of the univariate densities, i.e.,

$$\begin{aligned} f_{\mathbf{Z}}(\mathbf{z}) &= \prod_{j=1}^{p} \frac{1}{\sqrt{2\pi}} e^{-z_j^2/2} \\ &= (2\pi)^{-p/2} \exp(-\tfrac{1}{2}\mathbf{z}'\mathbf{z}), \qquad \mathbf{z} \in \mathbb{R}^p. \end{aligned} \tag{1}$$

Now, consider a linear combination $U = \mathbf{b}'\mathbf{Z} = b_1 Z_1 + \ldots + b_p Z_p$. All terms $b_j Z_j$ in this sum are independent normal random variables, with means 0 and variances b_j^2, and therefore, normality of U follows from the fact that sums of independent normal random variables are normal; see Section 3.1. □

In fact, the particular p-variate random vector $\mathbf{Z}$ of the preceding example is going to play an important role: it serves as a prototype of a p-dimensional normal distribution and will be referred to as the *p-variate standard normal distribution*. It follows immediately that $E[\mathbf{Z}] = \mathbf{0}$, and independence implies $\mathrm{Cov}[\mathbf{Z}] = \mathbf{I}_p$.

Example 3.2.2 Let $X_1 \sim \mathcal{N}(0, 1)$, set $X_2 = X_1 + 1$, and let $\mathbf{X} = \begin{pmatrix} X_1 \\ X_2 \end{pmatrix}$. Then $\mathbf{X}$ does not have a density function (see Exercise 1). Consider linear combinations $U = b_1 X_1 + b_2 X_2 = (b_1 + b_2) X_1 + b_2$. Then it follows that $U \sim \mathcal{N}(b_2,\ (b_1 + b_2)^2)$ for all choices of b_1 and b_2. In the case $b_2 = -b_1$, U is degenerate. Hence, $\mathbf{X}$ is bivariate normal, but it takes values only on the line $x_2 = x_1 + 1$. Therefore, in a way, $\mathbf{X}$ is not truly bivariate, but our definition of multivariate normality admits this case. □

Theorem 3.2.1 *If $\mathbf{X}$ is p-variate normal, and $\mathbf{Y} = \mathbf{A}\mathbf{X} + \mathbf{c}$ for a matrix $\mathbf{A}$ of dimension $k \times p$ and a vector $\mathbf{c} \in \mathbb{R}^k$, then $\mathbf{Y}$ is k-variate normal.*

Proof We need to show that $U = \mathbf{b}'\mathbf{Y}$ is univariate normal for all $\mathbf{b} \in \mathbb{R}^k$. First, notice that $\mathbf{b}'\mathbf{Y} = \mathbf{b}'(\mathbf{A}\mathbf{X} + \mathbf{c}) = (\mathbf{b}'\mathbf{A})\mathbf{X} + \mathbf{b}'\mathbf{c}$. But $\mathbf{h} := \mathbf{A}'\mathbf{b}$ is a vector in $\mathbb{R}^p$, and $c^* := \mathbf{b}'\mathbf{c}$ is a constant, so $U = \mathbf{h}'\mathbf{X} + c^*$, which means that any linear combination of $\mathbf{Y}$ is a linear combination of **X** plus some constant. By assumption, all linear combinations of **X** are normal, and therefore, the same holds for **Y**. ■

In its shortness, the preceding proof is quite remarkable because it does not make any assumptions about the dimension of the matrix **A**. Indeed, k may be smaller than p, equal to p, or even larger than p. Thus, we can freely "jump" from one dimension to the other, but (not surprisingly) if $k > p$, then the new random vector **Y** will always take values only in a subspace of dimension smaller than k, and some linear combinations of **Y** will be constants. Example 3.2.2 was actually an example of this. Exercise 2 gives some more details and examples. Exercise 7 gives a straightforward but important application: The sample average $\bar{\mathbf{X}}_N$ of N independent observations from the $\mathcal{N}(\boldsymbol{\mu}, \boldsymbol{\psi})$ distribution is itself multivariate normal with mean vector $\boldsymbol{\mu}$ and covariance matrix $\boldsymbol{\psi}/N$.

Theorem 3.2.1 also has an immediate consequence.

Corollary 3.2.2 *If* $\mathbf{Z}$ *is* p*-variate standard normal, with density function as in equation (1) and if* $\mathbf{Y} = \mathbf{AZ} + \mathbf{c}$ *for a* $k \times p$ *matrix* $\mathbf{A}$ *and a vector* $\mathbf{c} \in \mathbb{R}^k$, *then* $\mathbf{Y}$ *is* k*-variate normal with mean vector* $\mathbf{c}$ *and covariance matrix* $\mathbf{AA}'$.

Proof Multivariate normality follows from Theorem 3.2.1. The mean vector and the covariance matrix of **Y** follow from Theorem 2.11.1. ■

Corollary 3.2.2 allows us to construct multivariate normal random vectors with any mean vector, and (as we shall see shortly) any "legal" covariance matrix $\boldsymbol{\psi}$. It will be convenient to introduce new notation: If **X** is p-variate normal with mean vector $\boldsymbol{\mu}$ and covariance matrix $\boldsymbol{\psi}$, then we shall write

$$\mathbf{X} \sim \mathcal{N}_p(\boldsymbol{\mu}, \boldsymbol{\psi}). \tag{2}$$

At this point, equation (2) is rather dangerous to write because we do not know yet whether there are different multivariate normal distributions with the same mean vector and covariance matrix, or if the mean vector and the covariance matrix together determine the distribution uniquely. We will return to this point in Theorem 3.2.4. Note that, if **Z** has a p-variate standard normal distribution as in Example 3.2.1, then we can write $\mathbf{Z} \sim \mathcal{N}_p(\mathbf{0}, \mathbf{I}_p)$.

Theorem 3.2.1 has yet another interesting and immediate consequence.

Corollary 3.2.3 *Suppose that* $\mathbf{X} \sim \mathcal{N}_p(\boldsymbol{\mu}, \boldsymbol{\psi})$ *and* $\mathbf{X}$ *is partitioned into* q *and* $p - q$ *components,* $1 \le q < p$. *Thus, write* $\mathbf{X} = \begin{pmatrix} \mathbf{X}_1 \\ \mathbf{X}_2 \end{pmatrix}$, *where* $\mathbf{X}_1$ *has dimension* q *and* $\mathbf{X}_2$ *has dimension* $p - q$. *Partition* $\boldsymbol{\mu}$ *and* $\boldsymbol{\psi}$ *analogously as*

$$\boldsymbol{\mu} = \begin{pmatrix} \boldsymbol{\mu}_1 \\ \boldsymbol{\mu}_2 \end{pmatrix}$$

and

$$\psi = \begin{pmatrix} \psi_{11} & \psi_{12} \\ \psi_{21} & \psi_{22} \end{pmatrix}, \tag{3}$$

with dimensions of the subvectors and submatrices as induced by the partition of $\mathbf{X}$*. Then*

$$\mathbf{X}_1 \sim \mathcal{N}_q(\mu_1, \psi_{11}),$$

and

$$\mathbf{X}_2 \sim \mathcal{N}_{p-q}(\mu_2, \psi_{22}). \tag{4}$$

Proof $E[\mathbf{X}_1] = \mu_1$ and $\text{Cov}[\mathbf{X}_1] = \psi_{11}$ follow directly from the definition of mean vector and covariance matrix. Normality of $\mathbf{X}_1$ follows from the fact that $\mathbf{X}_1 = (\mathbf{I}_q \quad \mathbf{O}) \begin{pmatrix} \mathbf{X}_1 \\ \mathbf{X}_2 \end{pmatrix}$, where $\mathbf{O}$ is a matrix of dimension $q \times p - q$ with all entries equal to zero. The proof for $\mathbf{X}_2$ is analogous. ∎

Corollary 3.2.3 is actually quite a remarkable result: Every marginal distribution, of whatever dimension, of a multivariate normal is itself multivariate or univariate normal. For other classes of distributions, similar results do typically not hold. For instance, in the raindrop example (uniform distribution inside a circle, see Example 2.3.4), the marginal distributions are far from being uniform.

To get ready for the next theorem, we need to recall the notions of positive definiteness and positive semidefiniteness; see the appendix on matrix algebra. A matrix ψ of dimension $p \times p$ is *positive definite* if $\mathbf{b}'\psi\mathbf{b} > 0$ for all $\mathbf{b} \in \mathbb{R}^p$ $(\mathbf{b} \neq \mathbf{0})$ and *positive semidefinite* if $\mathbf{b}'\psi\mathbf{b} \geq \mathbf{0}$ for all $\mathbf{b} \in \mathbb{R}^p$. In our current applications of these notions, ψ will be a symmetric covariance matrix. Positive semidefiniteness of ψ ensures that ψ is "legal": Suppose that $\text{Cov}[\mathbf{X}] = \psi$. Then by the results of Section 2.11, $\text{var}[\mathbf{b}'\mathbf{X}] = E\{(\mathbf{b}'\mathbf{X} - E[\mathbf{b}'\mathbf{X}])^2\} = \mathbf{b}'\psi\mathbf{b}$, which must always be nonnegative. Positive definiteness means that *all* nontrivial linear combinations of $\mathbf{X}$ have strictly positive variance. For dimension $p = 1$, the notions of positive definiteness and positive semidefiniteness reduce to the relationships "$>$" and "$\geq$"; see Exercise 3.

Theorem 3.2.4 *For any vector* $\mu \in \mathbb{R}^p$ *and positive semidefinite symmetric matrix* ψ *of dimension* $p \times p$*, there exists a unique multivariate normal distribution with mean vector* μ *and covariance matrix* ψ*.*

The proof of this theorem requires characteristic functions and is beyond the level of this introduction. The interested reader is referred to the book of Muirhead (1982).

Theorem 3.2.4 implies that mean vector and covariance matrix, together, determine a multivariate, normal distribution uniquely, just as a mean and a variance determine a univariate normal. Thus, it is justified to speak of *the* $N_p(\mu, \psi)$-distribution; see the remark following equation (2).

The next theorem will show how a random variable with a $\mathcal{N}_p(\mu, \psi)$-distribution can be generated from a p-variate, standard normal distribution. But first, we need a lemma from linear algebra whose proof is given in Appendix A.3.

Lemma 3.2.5 *If ψ is positive semidefinite symmetric of dimension $p \times p$, then there exists a matrix* **M** *of dimension $p \times p$ such that*

$$\psi = \mathbf{M}\mathbf{M}'. \tag{5}$$

Note that the matrix **M** in Lemma 3.2.5 is not unique. All the lemma states is the existence of such a matrix. Even in the univariate case, uniqueness does not hold because any positive number c can be written as $c = (\sqrt{c})^2 = (-\sqrt{c})^2$. Popular choices of **M** are the *Cholesky decomposition* (if **M** is positive definite), and the *symmetric square root matrix*, see Appendix A.7.

Theorem 3.2.6 *For any vector $\mu \in \mathbb{R}^p$ and positive semidefinite, symmetric matrix ψ of dimension $p \times p$, a random vector* **Y** *with a $\mathcal{N}_p(\mu, \psi)$-distribution can be generated as*

$$\mathbf{Y} = \mu + \mathbf{M}\mathbf{Z}, \tag{6}$$

where $\mathbf{Z} \sim \mathcal{N}_p(\mathbf{0}, \mathbf{I}_p)$, and $\mathbf{M}\mathbf{M}' = \psi$.

Proof This follows from Corollary 3.2.2 and Lemma 3.2.5. ∎

Example 3.2.3 Suppose we wish to construct a three-dimensional normal random vector with

$$\mu = \begin{pmatrix} 1 \\ 2 \\ 3 \end{pmatrix}$$

and

$$\psi = \begin{pmatrix} 1 & 1 & 1 \\ 1 & 2 & 2 \\ 1 & 2 & 3 \end{pmatrix}.$$

Without first verifying positive definiteness of ψ, we note that $\psi = \mathbf{M}\mathbf{M}'$, where

$$\mathbf{M} = \begin{pmatrix} 1 & 0 & 0 \\ 1 & 1 & 0 \\ 1 & 1 & 1 \end{pmatrix}.$$

Actually, this choice of **M** is the Cholesky decomposition. Thus, Theorem 3.2.6 tells us that $\mathbf{Y} = \boldsymbol{\mu} + \mathbf{MZ}$, or

$$\begin{pmatrix} Y_1 \\ Y_2 \\ Y_3 \end{pmatrix} = \begin{pmatrix} 1 \\ 2 \\ 3 \end{pmatrix} + \begin{pmatrix} 1 & 0 & 0 \\ 1 & 1 & 0 \\ 1 & 1 & 1 \end{pmatrix} \begin{pmatrix} Z_1 \\ Z_2 \\ Z_3 \end{pmatrix}$$

is $\mathcal{N}_3(\boldsymbol{\mu}, \boldsymbol{\psi})$ as desired, with Z_1, Z_2, and Z_3 being three independent standard normal random variables. Note that we were able to construct this three-dimensional normal random variable without knowing its density function and without even knowing whether or not **Y** has a density function. □

Further examples are given in Exercise 4.

Theorem 3.2.7 *Let* $\mathbf{X} \sim \mathcal{N}_p(\boldsymbol{\mu}, \boldsymbol{\psi})$, *where* $\boldsymbol{\psi}$ *is positive definite and symmetric. Then* **X** *has a density function*

$$f_{\mathbf{X}}(\mathbf{x}) = (2\pi)^{-p/2}(\det \boldsymbol{\psi})^{-1/2} \exp[-\tfrac{1}{2}\Delta^2(\mathbf{x})], \qquad \mathbf{x} \in \mathbb{R}^p, \tag{7}$$

where

$$\Delta^2(\mathbf{x}) = (\mathbf{x} - \boldsymbol{\mu})'\boldsymbol{\psi}^{-1}(\mathbf{x} - \boldsymbol{\mu}). \tag{8}$$

Proof The proof is essentially an application of the transformation Theorem (see Theorem 2.11.3) and uses the following rules from linear algebra:

(i) For a square matrix **A**, $\det(\mathbf{A}) = \det(\mathbf{A}')$.

(ii) For square matrices **A** and **B** of the same dimension, $\det(\mathbf{AB}) = \det(\mathbf{A})\det(\mathbf{B})$.

(iii) For nonsingular matrices **A** and **B**, $(\mathbf{AB})^{-1} = \mathbf{B}^{-1}\mathbf{A}^{-1}$, and $(\mathbf{A}')^{-1} = (\mathbf{A}^{-1})'$.

First notice that if $\boldsymbol{\psi}$ is positive definite, then it is nonsingular. Thus, for any decomposition $\boldsymbol{\psi} = \mathbf{MM}'$, **M** must be nonsingular, too, and

$$\boldsymbol{\psi}^{-1} = (\mathbf{M}^{-1})'\mathbf{M}^{-1}. \tag{9}$$

Now, let $\mathbf{Z} \sim \mathcal{N}_p(\mathbf{0}, \mathbf{I}_p)$, and set $\mathbf{X} = \boldsymbol{\mu} + \mathbf{MZ}$. Then by Theorem 3.2.6, $\mathbf{X} \sim \mathcal{N}_p(\boldsymbol{\mu}, \boldsymbol{\psi})$ as desired. The Jacobian of the transformation is given by $J = (\det \mathbf{M})^{-1}$, and hence,

$$\begin{aligned} f_{\mathbf{X}}(\mathbf{x}) &= |J|\, f_{\mathbf{Z}}\big(\mathbf{M}^{-1}(\mathbf{x} - \boldsymbol{\mu})\big) \\ &= (2\pi)^{-p/2}|(\det \mathbf{M})^{-1}| \exp\big\{-\tfrac{1}{2}[\mathbf{M}^{-1}(\mathbf{x} - \boldsymbol{\mu})]'[\mathbf{M}^{-1}(\mathbf{x} - \boldsymbol{\mu})]\big\} \\ &= (2\pi)^{-p/2}\big[(\det \mathbf{M})(\det \mathbf{M}')\big]^{-1/2} \exp\big[-\tfrac{1}{2}(\mathbf{x} - \boldsymbol{\mu})'(\mathbf{M}^{-1})'\mathbf{M}^{-1}(\mathbf{x} - \boldsymbol{\mu})\big] \qquad (10) \\ &= (2\pi)^{-p/2}\big[\det(\mathbf{MM}')\big]^{-1/2} \exp\big[(\mathbf{x} - \boldsymbol{\mu})'\boldsymbol{\psi}^{-1}(\mathbf{x} - \boldsymbol{\mu})\big] \\ &= (2\pi)^{-p/2}(\det \boldsymbol{\psi})^{-1/2} \exp[-\tfrac{1}{2}\Delta^2(\mathbf{x})], \end{aligned}$$

with $\Delta^2(\mathbf{x})$ as in equation (8). ∎

In Chapter 5, we will study the function $\Delta^2(\mathbf{x})$ in great detail. For the moment, let us just remark that curves of the form

$$\Delta(\mathbf{x}) = \text{constant} > 0$$

are ellipses, or ellipsoids in higher dimensions. Thus, contours of equal density of the multivariate normal distribution are constant on ellipses or ellipsoids, always provided that a density exists. Figure 3.2.1 shows the density function of a bivariate normal distribution graphically.

We conclude this section with a last theorem, the proof of which is left to the student in Exercises 5 and 6.

Theorem 3.2.8 *Let* $\mathbf{X} \sim \mathcal{N}_p(\boldsymbol{\mu}, \psi)$. *Partition* $\mathbf{X}$ *and the parameters into* q *and* $p - q$ *components* $(1 \le q < p)$, *such that*

$$\mathbf{X} = \begin{pmatrix} \mathbf{X}_1 \\ \mathbf{X}_2 \end{pmatrix},$$

$$\boldsymbol{\mu} = \begin{pmatrix} \boldsymbol{\mu}_1 \\ \boldsymbol{\mu}_2 \end{pmatrix},$$

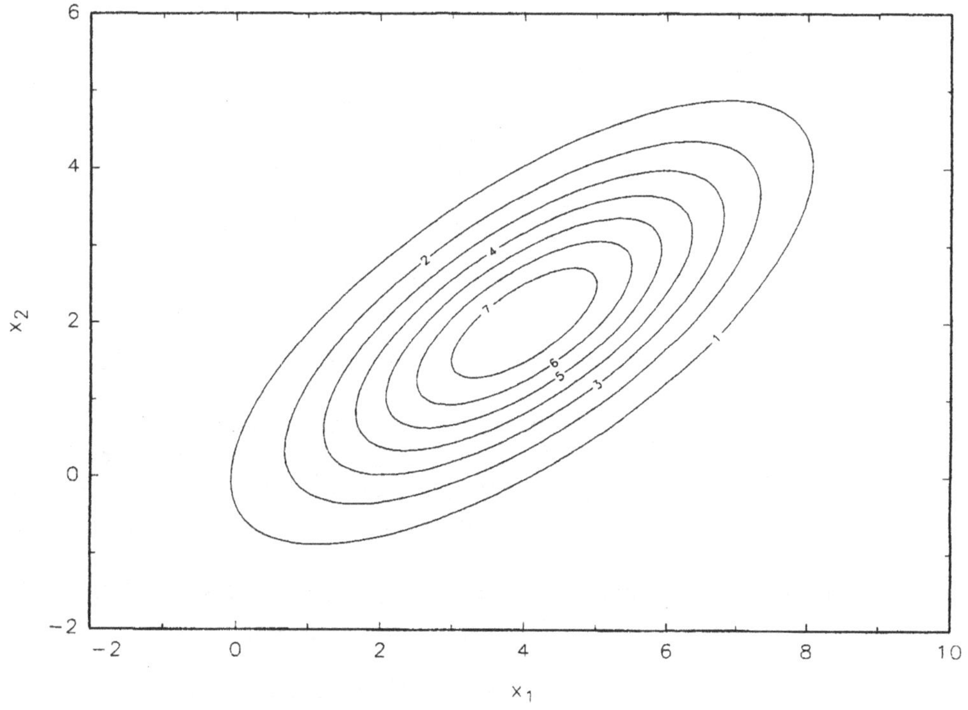

Figure 3-2-1A

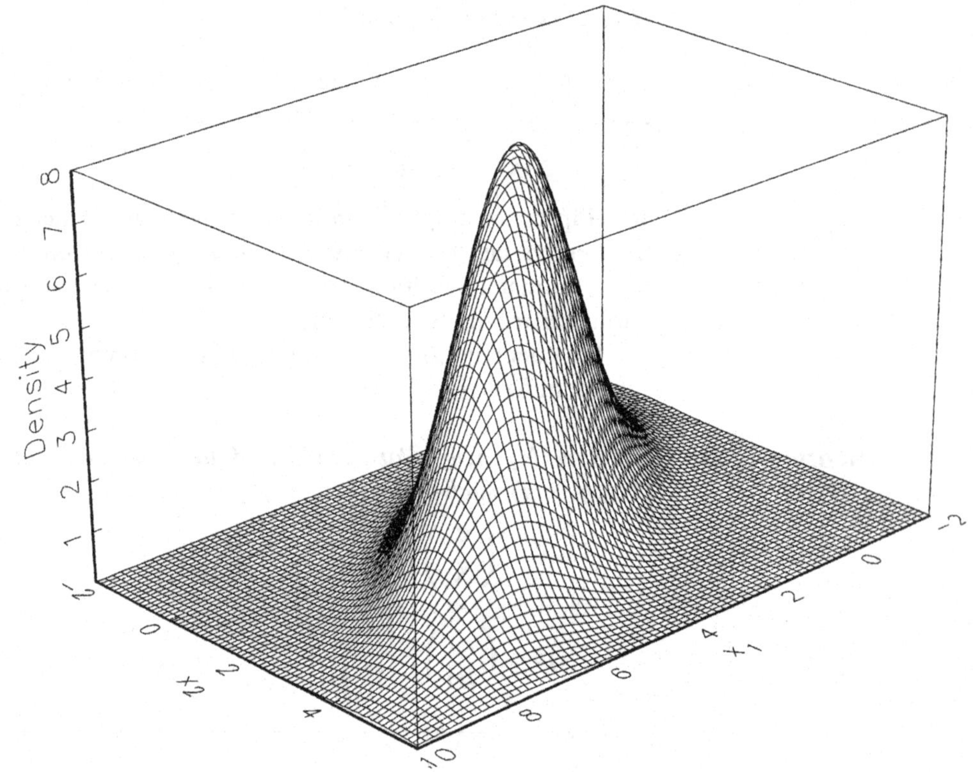

Figure 3-2-1B (a) contour plot and (b) 3D-view of the density of a bivariate normal distribution with $\mu = \begin{pmatrix} 4 \\ 2 \end{pmatrix}$ and $\psi = \begin{pmatrix} 4 & 2 \\ 2 & 2 \end{pmatrix}$ scaled by 100.

and

$$\psi = \begin{pmatrix} \psi_{11} & \psi_{12} \\ \psi_{21} & \psi_{22} \end{pmatrix}. \tag{11}$$

Then the following two conditions are equivalent:

(i) $\psi_{12} = \mathbf{O}$, *i.e., all covariances between variables in* $\mathbf{X}_1$ *and variables in* $\mathbf{X}_2$ *are zero.*

(ii) $\mathbf{X}_1$ *and* $\mathbf{X}_2$ *are stochastically independent.*

Theorem 3.2.8 is one of the most frequently misstated ones. For instance, a statement such as "If **X** and **Y** are multivariate normal random vectors and if all covariances between **X** and **Y** are zero, then they are independent" is wrong, as we show with a final example. More examples are provided in the exercises.

Example 3.2.4 For a positive integer k, let $\mathbf{X} \sim \mathcal{N}_k(\mathbf{0}, \mathbf{I}_k)$, and define **Y** by

$$\mathbf{Y} = \begin{cases} \mathbf{X} & \text{with probability } p_1 \\ -\mathbf{X} & \text{with probability } p_2, \end{cases}$$

where $p_1 + p_2 = 1$. Using the same technique as in Example 3.1.1, we find that $\mathbf{Y} \sim \mathcal{N}_k(\mathbf{0}, \mathbf{I}_k)$ and

$$\text{Cov}\begin{pmatrix} \mathbf{X} \\ \mathbf{Y} \end{pmatrix} = \begin{bmatrix} \mathbf{I}_k & (p_1 - p_2)\mathbf{I}_k \\ (p_1 - p_2)\mathbf{I}_k & \mathbf{I}_k \end{bmatrix}.$$

Therefore, if $p_1 = p_2 = \frac{1}{2}$, all covariances between $\mathbf{X}$ and $\mathbf{Y}$ are zero, yet $\mathbf{X}$ and $\mathbf{Y}$ are not independent because $\mathbf{Y}$ is always $\pm\mathbf{X}$. The point is that the joint $2k$-dimensional distribution of $\mathbf{X}$ and $\mathbf{Y}$ is *not* multivariate normal, and therefore, Theorem 3.2.8 does not apply. □

For related examples, see Exercise 23 in Section 2.5 and Exercise 8 in Section 3.1.

Exercises for Section 3.2

1. In Example 3.2.2, show that X_1 and X_2 do not have a joint bivariate density function.

2. This exercise deals with the situation of a linear transformation $\mathbf{Y} = \mathbf{AX} + \mathbf{c}$, where $\mathbf{A}$ has dimension $k \times p$, $k > p$.
   (a) If $p = 1$ and $k = 2$, show that there exist nonzero vectors $\mathbf{b} \in \mathbb{R}^2$ such that $\mathbf{b}'\mathbf{Y}$ is constant. How can such vectors $\mathbf{b}$ be found?
   (b) Let $X \sim \mathcal{N}(0, 1)$, and $\mathbf{Y} = \begin{pmatrix} a_1 \\ a_2 \end{pmatrix} X + \begin{pmatrix} c_1 \\ c_2 \end{pmatrix}$. Find the covariance matrix of $\mathbf{Y}$ and the correlation between Y_1 and Y_2. Find a linear combination of Y_1 and Y_2 which is constant.
   (c) In the general case $k > p$, show that there are at least $(k - p)$ different linear combinations of $\mathbf{Y}$ which are constant. (Two linear combinations are different if the two vectors of coefficients are linearly independent).
   Hint: The rank of $\mathbf{A}$ is at most p.

3. Show that for dimension $p = 1$, the set of positive definite (semidefinite) matrices reduces to the set of positive (nonnegative) real numbers.

4. This exercise shows how to generate a two-dimensional normal random vector with an arbitrary mean vector $\boldsymbol{\mu}$ and positive semidefinite covariance matrix $\boldsymbol{\psi}$. We will write

$$\boldsymbol{\psi} = \begin{pmatrix} \sigma_1^2 & \rho\sigma_1\sigma_2 \\ \rho\sigma_1\sigma_2 & \sigma_2^2 \end{pmatrix},$$

   where σ_1 and σ_2 are both positive.
   (a) Show that, if $|\rho| > 1$, then $\boldsymbol{\psi}$ is not positive semidefinite. *Hint*: If you find this difficult, first, try the special case $\sigma_1 = \sigma_2 = 1$.
   (b) If $|\rho| < 1$, show that there exists a lower triangular matrix $\mathbf{M} = \begin{pmatrix} m_{11} & 0 \\ m_{21} & m_{22} \end{pmatrix}$ with positive diagonal elements such that $\mathbf{MM}' = \boldsymbol{\psi}$. Find $m_{11} > 0$, m_{21}, and

$m_{22} > 0$ explicitly as functions of the parameters σ_1, σ_2, and ρ. *Note*: This is the *Cholesky decomposition* for dimension $p = 2$.

(c) Find the matrix $\mathbf{M}$ numerically for $\psi_1 = \begin{pmatrix} 2 & 3 \\ 3 & 9 \end{pmatrix}$.

5. Prove Theorem 3.2.8, assuming that the covariance matrix ψ is positive definite.
Hint: If a nonsingular square matrix $\mathbf{A}$ is block-diagonal, i.e.,

$$\mathbf{A} = \begin{pmatrix} \mathbf{A}_{11} & \mathbf{O} \\ \mathbf{O} & \mathbf{A}_{22} \end{pmatrix},$$

then

$$\mathbf{A}^{-1} = \begin{pmatrix} \mathbf{A}_{11}^{-1} & \mathbf{O} \\ \mathbf{O} & \mathbf{A}_{22}^{-1} \end{pmatrix},$$

and

$$\det(\mathbf{A}) = \det(\mathbf{A}_{11})\det(\mathbf{A}_{22}).$$

Use these two equalities to show that the joint *pdf* of $\mathbf{X}_1$ and $\mathbf{X}_2$ can be factorized as $f_1(\mathbf{x}_1)f_2(\mathbf{x}_2)$, where f_1 and f_2 are the marginal densities of $\mathbf{X}_1$ and $\mathbf{X}_2$, respectively.

6. Prove Theorem 3.2.8 without assuming positive definiteness of ψ.
Hint: Find a decomposition

$$\begin{pmatrix} \psi_{11} & \mathbf{O} \\ \mathbf{O} & \psi_{22} \end{pmatrix} = \begin{pmatrix} \mathbf{M}_1 & \mathbf{O} \\ \mathbf{O} & \mathbf{M}_2 \end{pmatrix} \begin{pmatrix} \mathbf{M}_1' & \mathbf{O} \\ \mathbf{O} & \mathbf{M}_2' \end{pmatrix},$$

and write $\mathbf{X}_1 = \mu_1 + \mathbf{M}_1\mathbf{Z}_1$, $\mathbf{X}_2 = \mu_2 + \mathbf{M}_2\mathbf{Z}_2$, where $\mathbf{Z}_1$ and $\mathbf{Z}_2$ are multivariate standard normal random vectors.

7. Suppose that $\mathbf{X}_1, \ldots, \mathbf{X}_N$ are independent, identically distributed, random vectors, all p-variate normal with mean vector μ and covariance matrix ψ (ψ is not necessarily positive definite). Let $\mathbf{U} = \sum_{j=1}^{N} \mathbf{X}_j$ and $\bar{\mathbf{X}}_N = \mathbf{U}/N$.

(a) Show that $\mathbf{U} \sim \mathcal{N}_p(N\mu, N\psi)$.

(b) Show that $\bar{\mathbf{X}}_N \sim \mathcal{N}_p(\mu, \psi/N)$.
Hint: Since the $\mathbf{X}_j$ are independent, their joint distribution is (Np)-variate normal with a mean vector consisting of N "copies" of μ stacked on top of each other and a block-diagonal covariance matrix with ψ in all blocks along the main diagonal. Then use Theorem 3.2.1.

8. Prove the following result: Suppose that $\mathbf{X}_1, \ldots, \mathbf{X}_N$ are independent random vectors, such that $\mathbf{X}_i \sim \mathcal{N}_p(\mu_i, \psi_i)$, $i = 1, \ldots, N$. Then for fixed constants $a_1, \ldots, a_N$,

$$\sum_{i=1}^{N} a_i \mathbf{X}_i \sim \mathcal{N}_p\left(\sum_{i=1}^{N} a_i \mu_i, \sum_{i=1}^{N} a_i^2 \psi_i\right).$$

Also show that this result implies Exercise 7.

9. Suppose that $\mathbf{X} = \begin{pmatrix} X_1 \\ X_2 \end{pmatrix}$ is bivariate normal with mean vector $\mu = \begin{pmatrix} \mu_1 \\ \mu_2 \end{pmatrix}$ and covariance matrix $\psi = \begin{pmatrix} \sigma_1^2 & \rho\sigma_1\sigma_2 \\ \rho\sigma_1\sigma_2 & \sigma_2^2 \end{pmatrix}$, where $\sigma_1 > 0$, $\sigma_2 > 0$, and $|\rho| < 1$. Use

Theorem 3.2.7 to show that the joint *pdf* of X_1 and X_2 is given by

$$f_{\mathbf{X}}(x_1, x_2) = \frac{1}{2\pi\sigma_1\sigma_2\sqrt{1-\rho^2}} \exp\left[-\frac{1}{2}\Delta^2(x_1, x_2)\right], \tag{12}$$

where

$$\Delta^2(x_1, x_2) = \frac{1}{1-\rho^2}\left[\left(\frac{x_1-\mu_1}{\sigma_1}\right)^2 - 2\rho\left(\frac{x_1-\mu_1}{\sigma_1}\right)\left(\frac{x_2-\mu_2}{\sigma_2}\right) + \left(\frac{x_2-\mu_2}{\sigma_2}\right)^2\right], \quad x_1, x_2 \in \mathbb{R}. \tag{13}$$

10. This exercise deals with the question asked after Definition 3.2.1, namely, if we know that a finite number (say k) of different linear combinations of two jointly distributed random variables X and Y are univariate normal, can we conclude (for some large k) that the joint distribution of X and Y is bivariate normal?

(a) Suppose that U and S are two independent random variables, such that U is uniform in the interval $[0, 2\pi)$, and S follows a chi-square distribution on two degrees of freedom, i.e., has *pdf*

$$f_S(s) = \tfrac{1}{2}e^{-s/2}, \quad s > 0 .$$

Let $X = \sqrt{S}\cos(U)$ and $Y = \sqrt{S}\sin(U)$, and show that the joint distribution of X and Y is bivariate standard normal.

(b) Let $U_1, U_2, \ldots$ denote a sequence of random variables such that U_k has a uniform distribution in the set

$$\mathcal{A}_k = \bigcup_{j=0}^{k-1}\left[\frac{2j}{k}\pi, \frac{2j+1}{k}\pi\right),$$

i.e., has *pdf*

$$f_k(u) = \begin{cases} 1/\pi & \text{if } u \in \mathcal{A}_k, \\ 0 & \text{otherwise.} \end{cases}$$

Show that the distribution of the random variable $\cos(U_k)$ is the same as the distribution of U, with U as in part (a), for any positive integer k. More generally, show that the distribution of the random variables

$$\cos\left(U_k - \frac{j}{k}\pi\right), \quad j = 0, \ldots, k-1,$$

is the same as the distribution of $\cos(U)$, for all positive integers k.

(c) Find the joint distribution of $X = \sqrt{S}\cos(U_k)$ and $Y = \sqrt{S}\sin(U_k)$, where S and U_k are independent, with distributions as defined above. Show that the joint distribution is not bivariate normal.
Hint: Study Exercises 7 and 8 in Section 3.1.

(d) Show that exactly k different linear combinations of X and Y, as defined in part (c), are univariate normal. (Two linear combinations are different if they are not multiples of each other). Thus, show that, for any finite integer k, normality of k different linear combinations does not imply joint normality.

3.3 Further Properties of the Multivariate Normal Distribution

In this section, we present some additional classical results on the multivariate normal distribution. Some proofs are omitted since they require technical tools not otherwise used in this text. The interested reader is referred to the books by Anderson (1984), Muirhead (1982), and Seber (1984). The student, to whom multivariate normal theory is new, may choose to skip this section at the first reading, since it is somewhat less important for the remainder of the book.

The first result concerns conditional distributions in the multivariate normal family.

Theorem 3.3.1 *Suppose that* $\mathbf{X} \sim \mathcal{N}_p(\boldsymbol{\mu}, \boldsymbol{\psi})$, *partitioned into* q *and* $(p-q)$ *components as*

$$\mathbf{X} = \begin{pmatrix} \mathbf{X}_1 \\ \mathbf{X}_2 \end{pmatrix},$$

$$\boldsymbol{\mu} = \begin{pmatrix} \boldsymbol{\mu}_1 \\ \boldsymbol{\mu}_2 \end{pmatrix},$$

and

$$\boldsymbol{\psi} = \begin{pmatrix} \boldsymbol{\psi}_{11} & \boldsymbol{\psi}_{12} \\ \boldsymbol{\psi}_{21} & \boldsymbol{\psi}_{22} \end{pmatrix}.$$

Assume that $\boldsymbol{\psi}_{11}$ *is positive definite. Then the conditional distribution of* $\mathbf{X}_2$, *given* $\mathbf{X}_1 = \mathbf{x}_1$, *is* $(p-q)$*-variate normal with the following parameters:*

$$\boldsymbol{\mu}_{2.1} := E[\mathbf{X}_2 | \mathbf{X}_1 = \mathbf{x}_1] = \boldsymbol{\mu}_2 + \boldsymbol{\psi}_{21}\boldsymbol{\psi}_{11}^{-1}(\mathbf{x}_1 - \boldsymbol{\mu}_1) \tag{1}$$

and

$$\boldsymbol{\psi}_{22.1} := \mathrm{Cov}[\mathbf{X}_2 | \mathbf{X}_1 = \mathbf{x}_1] = \boldsymbol{\psi}_{22} - \boldsymbol{\psi}_{21}\boldsymbol{\psi}_{11}^{-1}\boldsymbol{\psi}_{12}. \tag{2}$$

Proof

$$\begin{aligned} \mathbf{Y} = \begin{pmatrix} \mathbf{Y}_1 \\ \mathbf{Y}_2 \end{pmatrix} &= \begin{pmatrix} \mathbf{X}_1 \\ \mathbf{X}_2 - \boldsymbol{\psi}_{21}\boldsymbol{\psi}_{11}^{-1}\mathbf{X}_1 \end{pmatrix} \\ &= \begin{pmatrix} \mathbf{I}_q & \mathbf{O} \\ -\boldsymbol{\psi}_{21}\boldsymbol{\psi}_{11}^{-1} & \mathbf{I}_{p-q} \end{pmatrix} \begin{pmatrix} \mathbf{X}_1 \\ \mathbf{X}_2 \end{pmatrix}. \end{aligned} \tag{3}$$

Then $\mathbf{Y}$ is multivariate normal with mean vector

$$E[\mathbf{Y}] = \begin{pmatrix} \mathbf{I}_q & \mathbf{O} \\ -\boldsymbol{\psi}_{21}\boldsymbol{\psi}_{11}^{-1} & \mathbf{I}_{p-q} \end{pmatrix} \begin{pmatrix} \boldsymbol{\mu}_1 \\ \boldsymbol{\mu}_2 \end{pmatrix} = \begin{pmatrix} \boldsymbol{\mu}_1 \\ \boldsymbol{\mu}_2 - \boldsymbol{\psi}_{21}\boldsymbol{\psi}_{11}^{-1}\boldsymbol{\mu}_1 \end{pmatrix} \tag{4}$$

and covariance matrix

$$\begin{aligned}\text{Cov}[\mathbf{Y}] &= \begin{pmatrix} \mathbf{I}_q & \mathbf{O} \\ -\psi_{21}\psi_{11}^{-1} & \mathbf{I}_{p-q} \end{pmatrix} \begin{pmatrix} \psi_{11} & \psi_{12} \\ \psi_{21} & \psi_{22} \end{pmatrix} \begin{pmatrix} \mathbf{I}_q & -\psi_{11}^{-1}\psi_{12} \\ \mathbf{O} & \mathbf{I}_{p-q} \end{pmatrix} \\ &= \begin{pmatrix} \psi_{11} & \mathbf{O} \\ \mathbf{O} & \psi_{22} - \psi_{21}\psi_{11}^{-1}\psi_{12} \end{pmatrix}. \end{aligned} \tag{5}$$

By Corollary 3.2.3 and Theorem 3.2.8, both $\mathbf{Y}_1$ and $\mathbf{Y}_2$ are multivariate normal and independent. Hence, for any $\mathbf{y}_1$, the conditional distribution of $\mathbf{Y}_2$, given $\mathbf{Y}_1 = \mathbf{y}_1$, is the same as the unconditional (marginal) distribution of $\mathbf{Y}_2$. In terms of original variables, this means that the conditional distribution of $\mathbf{X}_2 - \psi_{21}\psi_{11}^{-1}\mathbf{X}_1$, given $\mathbf{X}_1 = \mathbf{x}_1$, is the same as the unconditional distribution of $\mathbf{Y}_2$, i.e., $(p-q)$-variate normal with mean vector $\boldsymbol{\mu}_2 - \psi_{21}\psi_{11}^{-1}\boldsymbol{\mu}_1$ and covariance matrix $\psi_{22.1} = \psi_{22} - \psi_{21}\psi_{11}^{-1}\psi_{12}$. Thus, conditionally on $\mathbf{X}_1 = \mathbf{x}_1$,

$$\mathbf{X}_2 - \psi_{21}\psi_{11}^{-1}\mathbf{x}_1 \sim \mathcal{N}(\boldsymbol{\mu}_2 - \psi_{21}\psi_{11}^{-1}\boldsymbol{\mu}_1,\ \psi_{22.1}). \tag{6}$$

Since $\mathbf{x}_1$ is fixed, the conditional distribution of $\mathbf{X}_2$ is normal with the same covariance matrix $\psi_{22.1}$ and with mean vector

$$E[\mathbf{X}_2|\mathbf{X}_1 = \mathbf{x}_1] = \boldsymbol{\mu}_2 - \psi_{21}\psi_{11}^{-1}\boldsymbol{\mu}_1 + \psi_{21}\psi_{11}^{-1}\mathbf{x}_1 = \boldsymbol{\mu}_2 + \psi_{21}\psi_{11}^{-1}(\mathbf{x}_1 - \boldsymbol{\mu}_1), \tag{7}$$

which concludes the proof. ∎

A more general version of Theorem 3.3.1 can be formulated for the case where ψ_{11} is not necessarily positive definite, but is not given here (see Muirhead, 1982). Instead, let us make some comments. First, the theorem says that *all* conditional distributions of a multivariate normal are themselves (univariate or multivariate) normal. In the case $p = 2$ and $q = 1$, this can be visualized by "slicing the normal mountain;" see Figure 2.6.3. Second, the conditional mean vector of $\mathbf{X}_2$, given $\mathbf{X}_1 = \mathbf{x}_1$, is a linear (or more precisely, affine) function of $\mathbf{x}_1$, which, by itself, would have allowed determining $\boldsymbol{\mu}_{2.1}$, using Theorem 2.7.1 and its multivariate generalization. Perhaps, most remarkably, the conditional covariance matrix $\psi_{22.1}$ does not depend on $\mathbf{x}_1$. In particular this means that, if we choose any of the components of a p-variate normal distribution as the dependent variable and the remaining $(p-1)$ components as regressors, then all assumptions of the classical, multiple linear regression model are automatically satisfied. However, multivariate normality is a sufficient, not a necessary, condition for these assumptions to hold; see Examples 2.7.5 and 2.7.6, and Exercise 7 in the current section. Exercise 17 gives another interesting result on conditional distributions of the multivariate normal.

If $X \sim \mathcal{N}(\mu, \sigma^2)$, $\sigma^2 > 0$, then $\left[(x-\mu)/\sigma\right]^2$ has a chi-square distribution with one degree of freedom (see Theorem 3.1.2). Our next theorem gives a generalization of this result.

Theorem 3.3.2 *Suppose that* $\mathbf{X} \sim \mathcal{N}_p(\boldsymbol{\mu}, \boldsymbol{\psi})$, *where* $\boldsymbol{\psi}$ *is positive definite. Let* $\Delta^2(\mathbf{X}) = (\mathbf{X} - \boldsymbol{\mu})'\boldsymbol{\psi}^{-1}(\mathbf{X} - \boldsymbol{\mu})$. *Then* $\Delta^2(\mathbf{X})$ *follows a chi-square distribution on p degrees of freedom.*

Proof Since $\boldsymbol{\psi}$ is positive definite symmetric, there exists a nonsingular matrix $\mathbf{M}$ such that $\boldsymbol{\psi} = \mathbf{M}\mathbf{M}'$ (Lemma 3.2.5). Let $\mathbf{Z} = \mathbf{M}^{-1}(\mathbf{X} - \boldsymbol{\mu})$. Then $\mathbf{Z} \sim \mathcal{N}_p(\mathbf{0}, \mathbf{I}_p)$ by Theorem 3.2.1. Moreover,

$$\Delta^2(\mathbf{X}) = (\mathbf{X} - \boldsymbol{\mu})'(\mathbf{M}\mathbf{M}')^{-1}(\mathbf{X} - \boldsymbol{\mu}) = \mathbf{Z}'\mathbf{Z} = \sum_{j=1}^{p} Z_j^2. \tag{8}$$

Since all Z_j are independent standard normal variables, the theorem follows from the fact that sums of independent chi-square random variables are themselves chi-square; see Exercise 9 in Section 3.1. ∎

As seen in Section 3.2, contours of constant density of a nonsingular, multivariate normal distribution are concentric ellipses or ellipsoids centered at the mean $\boldsymbol{\mu}$. Now, Theorem 3.3.2 says that the probability of a multivariate normal random vector $\mathbf{X}$ falling inside an ellipsoid of constant density can be computed using the chi-square distribution. More precisely, suppose that $\mathbf{X} \sim \mathcal{N}_p(\boldsymbol{\mu}, \boldsymbol{\psi})$, with $\boldsymbol{\psi}$ positive definite, and consider contours of equal density given by

$$\Delta^2(\mathbf{x}) = (\mathbf{x} - \boldsymbol{\mu})'\boldsymbol{\psi}^{-1}(\mathbf{x} - \boldsymbol{\mu}) = c^2 \qquad \text{(constant)} \tag{9}$$

or, equivalently, by

$$f_{\mathbf{X}}(\mathbf{x}) = (2\pi)^{-p/2}(\det \boldsymbol{\psi})^{-1/2} \exp\left(-\frac{1}{2}c^2\right). \tag{10}$$

Then

$$\Pr[\Delta^2(\mathbf{X}) \le c^2] = \Pr[\mathbf{Y}_p \le c^2], \tag{11}$$

where $\mathbf{Y}_p$ is a chi-square random variable on p degrees of freedom. That is, the probability that $\mathbf{X}$ falls inside the ellipsoid (9) can be obtained from a table of the chi-square distribution. Most often, this result is used to find confidence regions: Let $\chi^2_\alpha(p)$ denote the α-quantile of the chi-square distribution on p degrees of freedom. Then equation (9) with $c^2 = \chi^2_\alpha(p)$ gives an ellipse (or ellipsoid) within which $\mathbf{X}$ falls with probability α. This is particularly useful in the two-dimensional case, as illustrated by the following example.

Example 3.3.1 Suppose that $\mathbf{X} = \begin{pmatrix} X_1 \\ X_2 \end{pmatrix} \sim \mathcal{N}_2(\mu, \psi)$, where $\psi = \begin{pmatrix} \sigma_1^2 & \rho\sigma_1\sigma_2 \\ \rho\sigma_1\sigma_2 & \sigma_2^2 \end{pmatrix}$, $|\rho| < 1$. Then Exercise 4 shows that

$$\Delta^2(\mathbf{x}) = (\mathbf{x} - \mu)'\psi^{-1}(\mathbf{x} - \mu)$$

$$= \frac{1}{1-\rho^2}\left[\left(\frac{x_1 - \mu_1}{\sigma_1}\right)^2 - 2\rho\left(\frac{x_1 - \mu_1}{\sigma_1}\right)\left(\frac{x_2 - \mu_2}{\sigma_2}\right) + \left(\frac{x_2 - \mu_2}{\sigma_2}\right)^2\right]. \tag{12}$$

To find an ellipse within which $\mathbf{X}$ falls with probability α, we need to set (12) equal to the α-quantile of the chi-square distribution on two degrees of freedom. If U is chi-square on two degrees of freedom, then its distribution function is given by

$$F(u) = \Pr\left[U \le u\right] = \begin{cases} 0 & u \le 0 \\ 1 - e^{-\frac{1}{2}u} & u > 0 \end{cases}, \tag{13}$$

and therefore, the α-quantile is given by the value of u for which $F(u) = \alpha$, i.e.,

$$u = -2\log(1-\alpha). \tag{14}$$

Then choosing different values for α in the interval (0, 1) and plugging them into (14) will generate values of $u = c^2$ to be used in (12). For each value of c^2, the equation $\Delta^2(\mathbf{x}) = c^2$ defines a quadratic curve in the (x_1, x_2)-plane, corresponding to the constant value $(2\pi)^{-1}(\det\psi)^{-\frac{1}{2}}\exp(-\frac{1}{2}c^2)$ of the density function.

For instance, for $\mu = \begin{pmatrix} 3 \\ 4 \end{pmatrix}$ and $\psi = \begin{pmatrix} 2 & 1 \\ 1 & 1 \end{pmatrix}$, contours of equal density have the form

$$2\left[\frac{1}{2}(x_1 - 3)^2 - (x_1 - 3)(x_2 - 4) + (x_2 - 4)^2\right] = c^2.$$

Figure 3.3.1 shows such contours for various values of $c^2 = -2\log(1-\alpha)$ and associated contour levels $d = (2\pi)^{-1}(\det\psi)^{-\frac{1}{2}}\exp(-\frac{1}{2}c^2)$, according to the following table:

α	0.1	0.2	0.3	0.4	0.5	0.6	0.7	0.8	0.9
c^2	0.2107	0.4463	0.7133	1.0217	1.3863	1.8326	2.4079	3.2189	4.6052
d	0.1432	0.1273	0.1114	0.0955	0.0796	0.0637	0.0477	0.0318	0.0159

Thus, the innermost ellipse in Figure 3.3.1 gives a region within which $\mathbf{X}$ falls with probability 0.1; the outermost ellipse is associated with probability 0.9. □

To graph ellipses as in Figure 3.3.1, one might solve the equation $\Delta^2(\mathbf{x}) = c^2$ for one of the variables (say x_2) and evaluate the resulting formula for all values of x_1 that lead to a real solution. However, a much more elegant method to graph ellipses, as explained in the software instructions to Chapter 3, was actually used to generate the ellipses in Figure 3.3.1. In Chapter 6, we will encounter the graphing of ellipses, again, in the context of confidence regions for parameters.

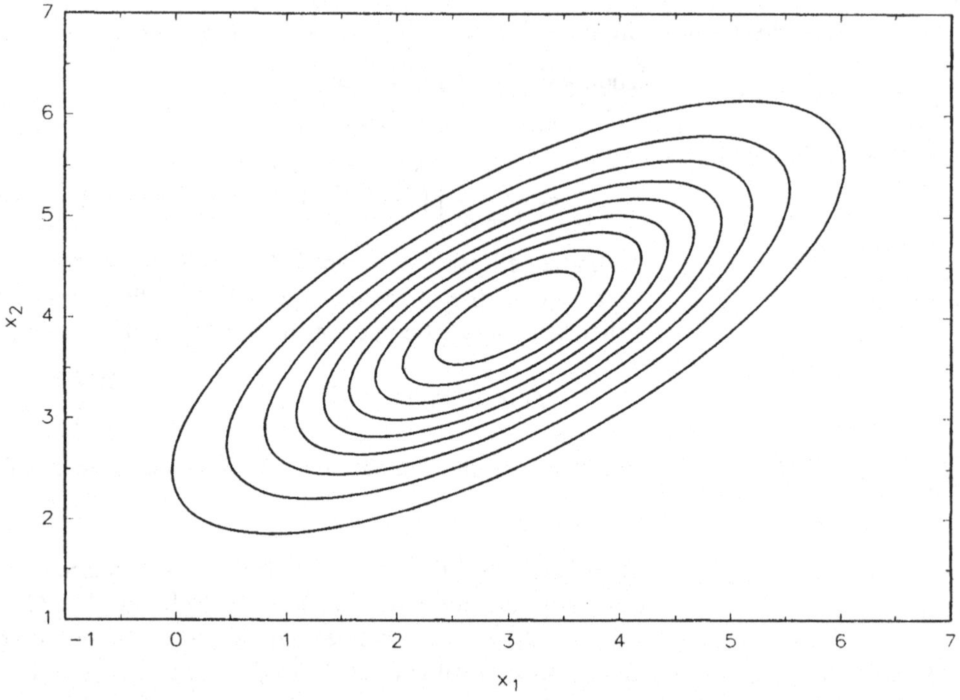

Figure 3-3-1 Ellipses of constant density of a bivariate normal distribution. The ellipses represent regions within which **X** falls with probability $0.1, 0.2, \ldots, 0.9$.

The next theorem is given without proof, despite its fundamental importance. The reason for omitting the proof is that it requires characteristic functions, which are not treated in this text. In Exercise 7 of Section 3.2, we have seen that the average $\bar{\mathbf{X}}_N$ of N independent random vectors, all distributed as $\mathcal{N}_p(\boldsymbol{\mu}, \boldsymbol{\psi})$, is itself multivariate normal. The multivariate central limit theorem tells that this is approximately true under more general conditions.

Theorem 3.3.3 (Multivariate Central Limit Theorem) *Suppose $\mathbf{X}_1, \mathbf{X}_2, \ldots, \mathbf{X}_N$ are independent, identically distributed, p-variate random vectors, with mean vectors $\boldsymbol{\mu} = E[\mathbf{X}_i]$ and covariance matrices $\boldsymbol{\psi} = \mathrm{Cov}[\mathbf{X}_i]$. Let $\bar{\mathbf{X}}_N = \frac{1}{N}\sum_{i=1}^N \mathbf{X}_i$. Then the asymptotic distribution (as $N \to \infty$) of $\sqrt{N}(\bar{\mathbf{X}}_N - \boldsymbol{\mu})$ or, equivalently, of $\frac{1}{\sqrt{N}}\sum_{i=1}^N (\mathbf{X}_i - \boldsymbol{\mu})$ is $\mathcal{N}_p(\mathbf{0}, \boldsymbol{\psi})$.*

For a proof of Theorem 3.3.3 the reader is referred again to Muirhead (1982). However, it is worth noticing that the proof of the multivariate Central Limit Theorem, in fact, is based on the univariate version, in the sense that it suffices to show that all linear combinations of $\sqrt{N}(\bar{\mathbf{X}}_N - \boldsymbol{\mu})$ converge to (univariate) normal distributions. Also see Exercise 18.

The most important application of the multivariate Central Limit Theorem is that, irrespective of the distribution from which a sample $\mathbf{X}, \ldots, \mathbf{X}_N$ is obtained, the distribution of $\bar{\mathbf{X}}_N = \frac{1}{N}\sum_{i=1}^N \mathbf{X}_i$ is approximately normal, provided N is large enough and Cov[$\mathbf{X}_i$] exists. More precisely, if $E[\mathbf{X}_i] = \boldsymbol{\mu}$ and Cov[$\mathbf{X}_i$] $= \psi$, $(i = 1, \ldots, N)$, then Theorem 3.3.3 says that the distribution of $\sqrt{N}\,(\bar{\mathbf{X}}_N - \boldsymbol{\mu})$ is approximately $\mathcal{N}_p(\mathbf{0}, \psi)$, which implies that the distribution of $\bar{\mathbf{X}}_N = \boldsymbol{\mu} + \frac{1}{\sqrt{N}}\sqrt{N}(\bar{\mathbf{X}}_N - \boldsymbol{\mu})$ is approximately $\mathcal{N}_p(\boldsymbol{\mu}, \psi/N)$. Exactly what constitutes a "large enough sample size N" is a difficult problem. As a general rule, convergence will be better for samples from symmetric (see Section 3.4) than for skewed distributions.

The Multivariate Central Limit Theorem has far-reaching applications to "simple" statistics like $\bar{\mathbf{X}}_N$, to more complicated statistics like the sample covariance matrix, and even to statistics that cannot be written in explicit formulas, but arise from systems of equations that have to be solved numerically. Maximum likelihood estimators in logistic regression (Sections 7.5 and 7.6) are an example.

Our next result is actually useful in univariate statistics, but the technique used is somewhat characteristic for multivariate calculations. The multivariate generalization will be given in Section 4.2.

Theorem 3.3.4 *Suppose that $X_1, \ldots, X_N$, $N \ge 2$, are independent, identically distributed, $\mathcal{N}(\mu, \sigma^2)$, random variables. Then the sample mean $\bar{X} = \frac{1}{N}\sum_{i=1}^N X_i$ and the sample variance $S^2 = \frac{1}{N}\sum_{i=1}^N (X_i - \bar{X})^2$ [or $\frac{1}{N-1}\sum_{i=1}^N (X_i - \bar{X})^2$] are stochastically independent.*

Proof We will show that all random variables $D_i = X_i - \bar{X}$ are independent of $\bar{X}$, $i = 1, \ldots, N$. Since S^2 is a function of the D_i alone, this implies independence of $\bar{X}$ and S^2 (see Exercise 8). Because the X_i have identical distributions, it suffices to show independence of just *one* D_i from $\bar{X}$, say D_1.

Let $\mathbf{X} = (X_1, \ldots, X_N)'$, let $\mathbf{1} \in \mathbb{R}^N$ denote a vector of length N with 1 in each position, and let $\mathbf{e}_1 = (1, 0, \ldots, 0)' \in \mathbb{R}^N$ denote the first unit basis vector. Then $X_1 = \mathbf{e}_1'\mathbf{X}$, and $\sum_{i=1}^N X_i = \mathbf{1}'\mathbf{X}$. Written jointly,

$$\begin{pmatrix} X_1 \\ \sum_{i=1}^N X_i \end{pmatrix} = \begin{pmatrix} \mathbf{e}_1' \\ \mathbf{1}' \end{pmatrix} \mathbf{X}. \tag{15}$$

Thus, X_1 and $\sum_{i=1}^N X_i$ are jointly (bivariate) normal, with covariance matrix

$$\begin{aligned} \text{Cov}\begin{pmatrix} X_1 \\ \sum_{i=1}^N X_i \end{pmatrix} &= \begin{pmatrix} \mathbf{e}_1' \\ \mathbf{1}' \end{pmatrix} \sigma^2 \mathbf{I}_p \begin{pmatrix} \mathbf{e}_1 & \mathbf{1} \end{pmatrix} \\ &= \sigma^2 \begin{pmatrix} \mathbf{e}_1'\mathbf{e}_1 & \mathbf{e}_1'\mathbf{1} \\ \mathbf{1}'\mathbf{e}_1 & \mathbf{1}'\mathbf{1} \end{pmatrix} \\ &= \sigma^2 \begin{pmatrix} 1 & 1 \\ 1 & N \end{pmatrix}. \end{aligned} \tag{16}$$

Next, $X_1 - \bar{X} = X_1 - \frac{1}{N}\sum_{i=1}^N X_i$, $\bar{X} = \frac{1}{N}\sum_{i=1}^N X_i$, and so,

$$\begin{pmatrix} X_1 - \bar{X} \\ \bar{X} \end{pmatrix} = \begin{pmatrix} 1 & -\frac{1}{N} \\ 0 & \frac{1}{N} \end{pmatrix} \begin{pmatrix} X_1 \\ \sum_{i=1}^N X_i \end{pmatrix}. \tag{17}$$

Thus, $X_1 - \bar{X}$ and $\bar{X}$ are jointly normal with covariance matrix

$$\begin{aligned} \text{Cov}\begin{pmatrix} X_1 - \bar{X} \\ \bar{X} \end{pmatrix} &= \begin{pmatrix} 1 & -\frac{1}{N} \\ 0 & \frac{1}{N} \end{pmatrix} \sigma^2 \begin{pmatrix} 1 & 1 \\ 1 & N \end{pmatrix} \begin{pmatrix} 1 & 0 \\ -\frac{1}{N} & \frac{1}{N} \end{pmatrix} \\ &= \sigma^2 \begin{pmatrix} 1 - \frac{1}{N} & 0 \\ 0 & \frac{1}{N} \end{pmatrix}. \end{aligned} \tag{18}$$

Thus, $X_1 - \bar{X}$ and $\bar{X}$ are uncorrelated and, by joint normality, also independent. An alternative proof that $X_1 - \bar{X}$ and $\bar{X}$ are uncorrelated is given in Exercise 19. ∎

Here are two remarks on Theorem 3.3.4. First, notice that in the calculations (15) to (18), multivariate normality was never used, except to argue that uncorrelatedness implies independence. Thus, if the X_i follow *any* distributions with the same variance σ^2 and are uncorrelated, then the $D_i = X_i - \bar{X}$ are still uncorrelated with $\bar{X}$. This points out why the technique is characteristic for multivariate normal theory: Moment calculations often suffice to prove a result. Second, using the same methods, it can be shown (Exercise 9) that the joint distribution of all D_i and $\bar{X}$ is $(N+1)$-variate normal. Let $\mathbf{D} = (D_1, \ldots, D_N)'$. Then there exists an $(N+1) \times N$ matrix $\mathbf{A}$ such that

$$\begin{pmatrix} \mathbf{D} \\ \bar{X} \end{pmatrix} = \mathbf{AX}, \tag{19}$$

that is, the matrix $\mathbf{A}$ transforms the N-variate normal vector $\mathbf{X}$ into an $(N+1)$-dimensional normal vector. This says that the covariance matrix of (19) has rank at most, N, i.e., the joint $(N+1)$-dimensional normal distribution of $\mathbf{D}$ and $\bar{X}$ is singular. See Exercise 9 for more detail.

The last theorem in this section gives a generalization of Theorem 3.3.4 to the classical linear model and is of interest mostly to the reader familiar with multiple linear regression. Its main purpose here is to illustrate, again, some typical calculations involving the multivariate normal distribution. The setup is as follows. Suppose that

$$\mathbf{Y} \sim \mathcal{N}_N(\mathbf{T}\boldsymbol{\beta}, \sigma^2 \mathbf{I}_N), \tag{20}$$

where $\mathbf{T}$ is a known matrix of dimension $N \times k$, of rank $k < N$, $\boldsymbol{\beta} \in \mathbb{R}^k$ is a vector of known parameters, and $\sigma^2 > 0$ is an unknown parameter. The matrix $\mathbf{T}$ is called the *design* matrix. The least squares estimator of $\boldsymbol{\beta}$ is

$$\mathbf{b} = (\mathbf{T}'\mathbf{T})^{-1}\mathbf{T}'\mathbf{Y}, \tag{21}$$

and the predicted (fitted) values are defined as

$$\begin{aligned}\hat{\mathbf{Y}} &= \mathbf{Tb} = \mathbf{T}(\mathbf{T}'\mathbf{T})^{-1}\mathbf{T}'\mathbf{Y} \\ &= \mathbf{HY},\end{aligned} \tag{22}$$

where $\mathbf{H} = \mathbf{T}(\mathbf{T}'\mathbf{T})^{-1}\mathbf{T}'$. The $N \times N$ matrix $\mathbf{H}$ is often referred to as the *hat* matrix. Finally, residuals are defined as

$$\mathbf{e} = \mathbf{Y} - \hat{\mathbf{Y}}. \tag{23}$$

Then we have the following result.

Theorem 3.3.5 *In the setup of equations* (20) *to* (23),

(i) *the* $(N + p)$*-variate vector* $\begin{pmatrix}\mathbf{e}\\ \mathbf{b}\end{pmatrix}$ *is multivariate normal,*

(ii) $\mathbf{e}$ *and* $\mathbf{b}$ *are independent,*

(iii) $E[\mathbf{e}] = \mathbf{0}$, $\quad E[\mathbf{b}] = \boldsymbol{\beta}$, *and*

(iv) $\text{Cov}[\mathbf{b}] = \sigma^2(\mathbf{T}'\mathbf{T})^{-1}$, $\quad \text{Cov}[\mathbf{e}] = \sigma^2(\mathbf{I}_N - \mathbf{H})$.

Proof First, notice that

$$\begin{aligned}\mathbf{e} &= \mathbf{Y} - \mathbf{Tb} = \mathbf{Y} - \mathbf{T}(\mathbf{T}'\mathbf{T})^{-1}\mathbf{T}'\mathbf{Y} \\ &= \mathbf{Y} - \mathbf{HY} \\ &= (\mathbf{I}_N - \mathbf{H})\mathbf{Y}.\end{aligned} \tag{24}$$

Thus,

$$\begin{pmatrix}\mathbf{e}\\ \mathbf{b}\end{pmatrix} = \begin{bmatrix}(\mathbf{I}_N - \mathbf{H})\mathbf{Y}\\ (\mathbf{T}'\mathbf{T})^{-1}\mathbf{T}'\mathbf{Y}\end{bmatrix} = \begin{bmatrix}\mathbf{I}_N - \mathbf{H}\\ (\mathbf{T}'\mathbf{T})^{-1}\mathbf{T}'\end{bmatrix}\mathbf{Y}, \tag{25}$$

and joint multivariate normality (i) follows from the assumption that $\mathbf{Y}$ is N-variate normal.

Next, find the covariance matrix of $\begin{pmatrix}\mathbf{e}\\ \mathbf{b}\end{pmatrix}$. Using the fact that $\mathbf{H}$ is indempotent, i.e., $\mathbf{H}^2 = \mathbf{H}$, and symmetric, we find that

$$\begin{aligned}\text{Cov}\begin{bmatrix}\mathbf{e}\\ \mathbf{b}\end{bmatrix} &= \begin{bmatrix}\mathbf{I}_N - \mathbf{H}\\ (\mathbf{T}'\mathbf{T})^{-1}\mathbf{T}'\end{bmatrix}\sigma^2\mathbf{I}_N\begin{bmatrix}\mathbf{I}_N - \mathbf{H}, & \mathbf{T}(\mathbf{T}'\mathbf{T})^{-1}\end{bmatrix} \\ &= \sigma^2\begin{bmatrix}\mathbf{I}_N - \mathbf{H} & \mathbf{0}\\ \mathbf{0} & (\mathbf{T}'\mathbf{T})^{-1}\end{bmatrix},\end{aligned} \tag{26}$$

see Exercise 10. This proves (iv). Independence, as claimed in (ii), follows from Theorem 3.2.8. Finally, part (iii) is left to the reader as Exercise 11. ■

As in Theorem 3.3.4, we are using a linear transformation of an N-variate normal vector into a random vector of higher dimension (see equation (25)), meaning that the joint distribution of $\mathbf{e}$ and $\mathbf{b}$ is singular. Since the matrix $(\mathbf{T}'\mathbf{T})^{-1}$ is nonsingular and the rank of (26) can be, at most, N, the rank of $\mathbf{I}_N - \mathbf{H}$ can be, at most, $N - k$. It turns out that the rank of $\mathbf{H}$ is k and the rank of $\mathbf{I}_N - \mathbf{H}$ is $N - k$; see Exercise 12. This means that the joint distribution of the residuals is singular in a linear subspace of dimension $N - k$. Note also that the assumption of multivariate normality of $\mathbf{Y}$ is not used in calculating (26), that is, the random vectors $\mathbf{e}$ and $\mathbf{b}$ will be uncorrelated under more general conditions.

We leave it to the student (Exercise 13) to verify that Theorem 3.3.5 implies Theorem 3.3.4 as a special case.

At this point, we leave the area of multivariate normal theory, well aware that we have really had only a glimpse of this fascinating area. Hopefully the student has found this section motivating for attempting a more advanced book, e.g., Anderson (1984); Mardia, Kent, and Bibby (1979); Muirhead (1982); or Seber (1984).

Exercises for Section 3.3

1. Prove Theorem 3.3.1 using densities, assuming that the covariance matrix ψ is positive definite.

   Hint: Let $f_1(\mathbf{x}_1)$ be the marginal *pdf* of $\mathbf{X}_1$, and denote by $f_{2.1}(\mathbf{x}_2|\mathbf{x}_1)$ the *pdf* of a $(p-q)$-variate normal random variable with mean vector $\mu_{2.1}$ and covariance matrix $\psi_{22.1}$. Then show that $f_{2.1}(\mathbf{x}_2|\mathbf{x}_1)f_1(\mathbf{x}_1)$ is the joint *pdf* of $\mathbf{X}_1$ and $\mathbf{X}_2$.

2. Suppose that $\mathbf{X}$ is an $(r+s+t)$-variate normal random vector, partitioned in vectors $\mathbf{X}_1$, $\mathbf{X}_2$, and $\mathbf{X}_3$ of dimensions r, s, and t, respectively, with mean vector μ and covariance matrix ψ partitioned analogously as

$$\mu = \begin{pmatrix} \mu_1 \\ \mu_1 \\ \mu_3 \end{pmatrix},$$

   and

$$\psi = \begin{pmatrix} \psi_{11} & \psi_{12} & \psi_{13} \\ \psi_{21} & \psi_{22} & \psi_{23} \\ \psi_{31} & \psi_{32} & \psi_{33} \end{pmatrix}.$$

   (a) Find the joint conditional distribution of $\mathbf{X}_2$ and $\mathbf{X}_3$, given $\mathbf{X}_1 = \mathbf{x}_1$.

   (b) Using the result in (a), find the marginal distribution of $\mathbf{X}_2$, given $\mathbf{X}_1 = \mathbf{x}_1$.

   (c) Verify that you get the same result for the conditional distribution of $\mathbf{X}_2$, given $\mathbf{X}_1 = \mathbf{x}_1$, by first computing the joint marginal distribution of $\mathbf{X}_1$ and $\mathbf{X}_2$ and then conditioning on $\mathbf{X}_1 = \mathbf{x}_1$.

3. Verify Exercise 2 numerically with the following three-dimensional example ($r = s = t = 1$):

$$\mu = \begin{pmatrix} 1 \\ 2 \\ 3 \end{pmatrix},$$

and

$$\psi = \begin{pmatrix} 1 & 1 & 1 \\ 1 & 2 & 2 \\ 1 & 2 & 3 \end{pmatrix}.$$

4. Prove equation (12).

5. This exercise highlights an aspect of the multivariate normal distribution which, at first, may seem counterintuitive, but actually just reflects the fact that high-dimensional distributions are, after all, not that easy to understand. Suppose that $\mathbf{X} \sim \mathcal{N}_k(\mathbf{0}, \mathbf{I}_k)$, for $k \geq 1$. Then the density of $\mathbf{X}$ has its peak at $\mathbf{0}$, and one might intuitively expect that "much of the probability mass" is centered at the origin. The exercise investigates to what extent this is true by computing $\Pr[\mathbf{X}'\mathbf{X} \leq 1]$, i.e., the probability of a k-variate standard normal vector falling inside the k-dimensional unit ball.

(a) Compute $\Pr[\mathbf{X}'\mathbf{X} \leq 1]$ using a table of the chi-square distribution with k degrees of freedom (or a numerical approximation as provided by some software packages), for $k = 1, 2, 3, 4, 5$. Conjecture what happens for large k.

(b) For $\mathbf{X} \sim \mathcal{N}_k(\mathbf{0}, \mathbf{I}_k)$, find

$$\lim_{k \to \infty} \Pr[\mathbf{X}'\mathbf{X} \leq 1]$$

(The *pdf* of the chi-square distribution is given in Exercise 9 of Section 3.1).

6. Let $\mathbf{X} = \begin{pmatrix} X_1 \\ \vdots \\ X_k \end{pmatrix} \sim \mathcal{N}_k(\mathbf{0}, \mathbf{I}_k)$, as in the preceding exercise, and repeat parts (a) and (b) for $\Pr[\text{all } |X_i| \leq 1]$, i.e., the probability of $\mathbf{X}$ falling in a cube of side length two centered at the origin. (This cube contains the unit ball as a subset).

7. Suppose that (X, Y, and Z) are three random variables with joint *pdf*

$$f(x, y, z) = \begin{cases} \dfrac{1}{\pi} \exp[x(y + z - x - 2) - \dfrac{1}{2}(y^2 + z^2)] & \text{if } x \geq 0,\ y \in \mathbb{R},\ z \in \mathbb{R}, \\ 0 & \text{otherwise.} \end{cases}$$

(a) Find the joint conditional distribution of Y and Z, given $X = x$.

(b) Find $E[Y|X = x]$ and $E[Z|X = x]$.

(c) Show that, conditionally on $X = x$, Y and Z are independent for all $x > 0$, but, unconditionally, they are not independent.

8. Let $\mathbf{X}$ denote a p-variate random vector and $\mathbf{Y}$ a q-variate random vector, $\mathbf{X}$ and $\mathbf{Y}$ independent. Let f and g denote real-valued functions of p and q variables, respectively. Show that the random variables $f(\mathbf{X})$ and $g(\mathbf{Y})$ are independent. (This result is used in the proof of Theorem 3.3.4).

9. Find the matrix $\mathbf{A}$ in equation (19), explicitly, to prove joint normality of $\bar{\mathbf{X}}$ and $\mathbf{D}$.

10. This exercise fills in some details of the proof of Theorem 3.3.5.
   (a) Show that the matrix $\mathbf{H} = \mathbf{T}(\mathbf{T}'\mathbf{T})^{-1}\mathbf{T}'$ is symmetric and idempotent (i.e., $\mathbf{H}^2 = \mathbf{H} = \mathbf{H}'$).
   (b) Show that $\mathbf{I}_N - \mathbf{H}$ is symmetric and idempotent.
   (c) Using (a) and (b), verify equation (26).

11. Prove part (iii) of Theorem 3.3.5.

12. Let $\mathbf{H} = \mathbf{T}(\mathbf{T}'\mathbf{T})^{-1}\mathbf{T}'$, as in equation (22). Show that $\text{rank}(\mathbf{H}) = k$ and $\text{rank}(\mathbf{I}_N - \mathbf{H}) = N - k$. *Hint*: Use properties of the trace of a symmetric matrix and the fact that all eigenvalues of an idempotent matrix are 0 or 1; see appendix A.5.

13. Show that Theorem 3.3.5 implies Theorem 3.3.4. *Hint*: In equation (20), use $\boldsymbol{\beta} = \boldsymbol{\mu}$ and $\mathbf{T} = \mathbf{1_N}$ (a vector of length N with 1 in each position).

14. In the same setup as Exercise 13, find the matrices $\mathbf{H}$ and $\mathbf{I}_N - \mathbf{H}$, explicitly. How is this related to Exercise 9?

15. Prove the following result. Suppose that $\mathbf{Y} \sim \mathcal{N}_k(\mathbf{0}, \mathbf{I}_k)$, and let $\mathbf{M}$ denote a symmetric, idempotent matrix of dimension $k \times k$, of rank m $(1 \le m \le k-1)$. Then the random variables $\mathbf{Y}'\mathbf{M}\mathbf{Y}$ and $\mathbf{Y}'(\mathbf{I}_k - \mathbf{M})\mathbf{Y}$ are independent and distributed as chi-square on m and $k-m$ degrees of freedom, respectively. *Hint*: Use the fact that $\mathbf{M}$ has m eigenvalues equal to 1 and $k-m$ eigenvalues equal to 0. Then use the spectral decomposition of $\mathbf{M}$.

16. Let $f_k(x; \boldsymbol{\mu}, \boldsymbol{\psi})$ denote the density function of the k-variate normal distribution with mean vector $\boldsymbol{\mu}$ and covariance matrix $\boldsymbol{\psi}$. For $0 < p < 1$, define a multivariate normal mixture density by

$$f_{\mathbf{X}}(x) = p\, f_k(x; \boldsymbol{\mu}_1, \boldsymbol{\psi}_1) + (1-p)\, f_k(x; \boldsymbol{\mu}_2, \boldsymbol{\psi}_2). \tag{27}$$

   (a) Show that each of the components of a k-variate $\mathbf{X}$ with density (27) follows a univariate normal mixture density.
   (b) If $\mathbf{X}$ has density (27) and $\mathbf{A}$ is a nonsingular $k \times k$ matrix, find the distribution of $\mathbf{AX}$.
   Note: Both (a) and (b) can be done technically, using integration in (a) and the transformation theorem in (b), or more easily by using the first principles of finite mixtures; see Section 2.8.

17. This exercise is concerned with a result that demonstrates the importance of the inverse of the covariance matrix of the multivariate normal distribution. Suppose that $\mathbf{X}$ is an $(r+s+t)$-variate normal random vector, partitioned as in Exercise 2, where $\boldsymbol{\psi}$ is

positive definite. Let

$$\psi^{-1} = \begin{pmatrix} \psi^{11} & \psi^{12} & \psi^{13} \\ \psi^{21} & \psi^{22} & \psi^{23} \\ \psi^{31} & \psi^{32} & \psi^{33} \end{pmatrix}$$

denote the inverse of ψ, partitioned analogously.

(a) Prove the following theorem: Conditionally on $\mathbf{X}_3$, the random vectors $\mathbf{X}_1$ and $\mathbf{X}_2$ are independent exactly if $\psi^{12} = \mathbf{0}$. *Hint*: Use the formula for the inverse of a partitioned matrix; see appendix A.1.

(b) Use part (a) to show the following result. Suppose that $\mathbf{X} = (X_1, \ldots, X_p)'$ is a p-variate normal, with positive definite covariance matrix ψ. If the (i, j)th entry of ψ^{-1} is zero, then X_i and X_j are conditionally independent, given the other $(p-2)$ variables.

18. The moment-generating function of a p-variate random vector $\mathbf{X}$ is defined as

$$M_{\mathbf{X}}(\mathbf{t}) = E\left[e^{\mathbf{t}'\mathbf{X}}\right], \qquad \mathbf{t} = (t_1, \ldots, t_p)' \in \mathbb{R}^p, \tag{28}$$

provided this expectation exists. Show that the moment-generating function of $\mathbf{X}$, distributed as $\mathcal{N}_p(\boldsymbol{\mu}, \psi)$, is given by

$$M_{\mathbf{X}}(\mathbf{t}) = \exp\left(\mathbf{t}'\boldsymbol{\mu} + \frac{1}{2}\mathbf{t}'\psi\mathbf{t}\right). \tag{29}$$

Hint: Let $U_Y(s)$ denote the moment-generating function of a univariate normal variable; see Exercise 10 in Section 3.1. Then use the fact that $M_{\mathbf{X}}(\mathbf{t}) = U_{\mathbf{t}'\mathbf{X}}(1)$.

19. Let $X_1, \ldots, X_N$ be independent, identically distributed, random variables. Show that $\text{cov}[X_1 - \bar{X}, \bar{X}] = 0$ using that, for any two random variables U and V, $\text{var}[U + V] = \text{var}[U] + \text{var}[V]$ exactly if $\text{cov}[U, V] = 0$. Set $U = X_1 - \bar{X}$ and $V = \bar{X}$.

3.4 Spherical and Elliptical Distributions

The theory of spherically and elliptically symmetric distributions is a relatively modern topic in multivariate statistics. In this section we give a brief and rather informal introduction to this interesting class of distributions, illustrating it mostly for the two-dimensional case. Mostly for simplicity, we also restrict our discussion to random vectors that have a density function. Good introductions to this topic can be found in Muirhead (1982), and most comprehensively, in Fang, Kotz, and Ng (1990).

A random vector $\mathbf{X}$ of dimension p is called spherical (or spherically symmetric) if its distribution does not change under rotations of the coordinate system, i.e., if the distribution of $\mathbf{BX}$ is the same as the distribution of $\mathbf{X}$ for any orthogonal $p \times p$ matrix $\mathbf{B}$. If $\mathbf{X}$ has a density function $f_{\mathbf{X}}$, then this is equivalent to the following definition (see Exercise 1).

Definition 3.4.1 The p-variate random vector $\mathbf{X}$ with density function $f_{\mathbf{X}}(\mathbf{x})$ is spherical, or spherically symmetric, if $f_{\mathbf{X}}$ depends on $\mathbf{x}$ only through $\mathbf{x}'\mathbf{x} = \sum_{i=1}^{p} x_i^2$, i.e., through the squared Euclidean distance from the origin. □

This definition says that contours of constant density of a spherical random vector $\mathbf{X}$ are circles, or spheres for $p > 2$, centered at the origin. The $\mathcal{N}_p(\mathbf{0}, \sigma^2 \mathbf{I}_p)$ distribution is an example, with *pdf*

$$f_{\mathbf{X}}(\mathbf{x}) = (2\pi)^{-p/2} (\sigma^2)^{-p/2} \exp\left(-\frac{1}{2\sigma^2}\mathbf{x}'\mathbf{x}\right). \tag{1}$$

Other examples of spherical distributions already encountered in this text are given by the density function

$$f_{\mathbf{X}}(x_1, x_2) = \begin{cases} \frac{2}{\pi}[1 - (x_1^2 + x_2^2)] & \text{if } x_1^2 + x_2^2 \le 1, \\ 0 & \text{otherwise,} \end{cases} \tag{2}$$

which has been graphed in Figure 2.4.2, and by the p-variate uniform distribution inside the unit sphere, with density

$$f_{\mathbf{X}}(\mathbf{x}) = \begin{cases} c & \text{if } \mathbf{x}'\mathbf{x} \le 1, \\ 0 & \text{otherwise,} \end{cases} \tag{3}$$

for some constant c; see Example 2.9.6.

Example 3.4.1 Let $\mathbf{X} = \begin{pmatrix} X_1 \\ X_2 \end{pmatrix}$ denote a bivariate random vector with *pdf*

$$f_{\mathbf{X}}(x) = \frac{1}{2\pi} \exp[-(\mathbf{x}'\mathbf{x})^{1/2}], \qquad \mathbf{x} \in \mathbb{R}^2. \tag{4}$$

Sphericity of $\mathbf{X}$ is clear from equation (4), but despite the superficial similarity of $f_{\mathbf{X}}$ to the density function of the bivariate standard normal distribution, the two distributions are quite different, as we will see later in Example 3.4.7. □

Example 3.4.1 may be considered as a bivariate generalization of the univariate double-exponential distribution, which has *pdf*

$$f_X(x) = \frac{1}{2}e^{-|x|}, \qquad x \in \mathbb{R}. \tag{5}$$

Another bivariate generalization of the double-exponential distribution is given by the *pdf*

$$g_{\mathbf{X}}(\mathbf{x}) = \frac{1}{4} \exp\Big[-\big(|x_1| + |x_2|\big)\Big], \qquad \mathbf{x} \in \mathbb{R}^2. \tag{6}$$

In Example 3.4.1, contours of constant density are circles, and X_1 and X_2 are not independent. In (6), contours of equal density are squares, and the two variables are independent; see Exercise 3.

Example 3.4.2 If $\mathbf{X} = \begin{pmatrix} X_1 \\ X_2 \end{pmatrix}$ has density

$$f_{\mathbf{X}}(\mathbf{x}) = \frac{1}{2\pi}\left(1 + \mathbf{x}'\mathbf{x}\right)^{-3/2}, \qquad \mathbf{x} \in \mathbb{R}^2, \tag{7}$$

then $\mathbf{X}$ is said to follow a bivariate Cauchy distribution, which, again, is spherical. Equation (7), in fact, is a special case of the so-called multivariate t-distribution. As in Example 3.4.1, the two components of $\mathbf{X}$ are not independent, and the density function $f_{\mathbf{X}}$ in (7) looks quite different from the joint *pdf* of two independent univariate Cauchy random variables; see Exercise 4. □

Example 3.4.3 Let V denote a discrete random variable, taking two different positive values v_1 and v_2 with probabilities p_1 and p_2, $p_1 + p_2 = 1$. Define a k-variate random vector $\mathbf{X}$ as follows: Conditionally on $V = v_i$, $\mathbf{X}$ has a $\mathcal{N}_k(\mathbf{0},\ v_i\,\mathbf{I}_k)$ distribution, that is, the conditional density of $\mathbf{X}$, given V, is

$$f_{\mathbf{X}|V}(\mathbf{x}|v_i) = (2\pi)^{-k/2}\, v_i^{-k/2} \exp\left(-\frac{1}{2v_i}\mathbf{x}'\mathbf{x}\right), \qquad \mathbf{x} \in \mathbb{R}^k. \tag{8}$$

Then (see Section 2.8), the unconditional distribution of $\mathbf{X}$ has the mixture *pdf*

$$f_{\mathbf{X}}(\mathbf{x}) = (2\pi)^{-k/2}\left[p_1 v_1^{-k/2} \exp\left(-\frac{1}{2v_1}\mathbf{x}'\mathbf{x}\right) + p_2 v_2^{-k/2} \exp\left(-\frac{1}{2v_2}\mathbf{x}'\mathbf{x}\right)\right]. \tag{9}$$

This distribution is called a scale mixture of multivariate normals; again, sphericity is clear from equation (9). By varying the mixing weights p_i and the values v_i, one can generate a rather flexible class of spherical distributions. Also, equation (9) can be generalized to mixtures with more than two components. A particular case of such a mixture results if we set $p_1 = 1 - \epsilon$ and $p_2 = \epsilon$ for some small ϵ, e.g., $\epsilon = 0.05$, $v_1 = 1$, and $v_2 = \sigma^2$ (large). Then the *pdf*

$$f_{\mathbf{X}}(\mathbf{x}) = (2\pi)^{-k/2}\left[(1-\epsilon)\exp\left(-\frac{1}{2}\mathbf{x}'\mathbf{x}\right) + \epsilon\,\sigma^{-k} \exp\left(-\frac{1}{2\sigma^2}\mathbf{x}'\mathbf{x}\right)\right], \quad \mathbf{x} \in \mathbb{R}^k, \tag{10}$$

is called an ϵ-contaminated multivariate normal distribution. This generalizes the scale mixture of univariate normals in equation (17) of Section 2.8. It turns out that all marginal distributions of a scale mixture like (9) are themselves scale mixtures of normals of proper dimension; see Exercise 5. □

A convenient way of generating bivariate, spherical distributions is to start with polar coordinates. Let U denote a uniform random variable in the interval $[0, 2\pi)$, and let R denote a random variable with density $f_R(r)$, independent of U, where

$f_R(r) = 0$ for $r < 0$, that is, R is a nonnegative, continuous random variable. The joint *pdf* of U and R then is, by independence,

$$f_{UR}(u,r) = \begin{cases} \frac{1}{2\pi} f_R(r) & \text{if } 0 \le u < 2\pi, \quad r \ge 0, \\ 0 & \text{otherwise.} \end{cases} \tag{11}$$

Now set

$$\begin{aligned} X_1 &= R\cos U\ , \\ X_2 &= R\sin U\ , \end{aligned} \tag{12}$$

and find the joint *pdf* of X_1 and X_2.

The Jacobian of the transformation from (U, R) to (X_1, X_2) is given by $|J| = 1/\sqrt{x_1^2 + x_2^2}$ (Exercise 7). Moreover, since $\sqrt{X_1^2 + X_2^2} = R$, we can use Theorem 2.11.3 to obtain

$$f_{\mathbf{X}}(x_1, x_2) = \frac{1}{2\pi\sqrt{x_1^2 + x_2^2}}\, f_R\left(\sqrt{x_1^2 + x_2^2}\right), \qquad \mathbf{x} \in \mathbb{R}^2. \tag{13}$$

Recall that to obtain (13), we started out with polar coordinates U (angle) and R (distance), assumed independent. Then $\begin{pmatrix} \cos U \\ \sin U \end{pmatrix}$ is a randomly (uniformly) chosen point on the unit circle, i.e., U gives the direction, and R gives the distance of (X_1, X_2) from the origin. As seen from (13), $f_{\mathbf{X}}$ depends on $\mathbf{x}$ only through $\mathbf{x}'\mathbf{x} = x_1^2 + x_2^2$, and therefore any distribution generated this way is spherically symmetric.

Example 3.4.4 Suppose that

$$f_R(r) = re^{-r^2/2}, \qquad r > 0. \tag{14}$$

Then (13) gives the bivariate standard normal density function. This is not surprising in view of the fact that R^2 has a chi-square distribution on two degrees of freedom (Exercise 8). □

Example 3.4.5 Suppose that

$$f_R(r) = \begin{cases} 2r & 0 \le r \le 1, \\ 0 & \text{otherwise.} \end{cases} \tag{15}$$

Then equation (13) gives the following joint *pdf* of x_1 and x_2:

$$f_{\mathbf{X}}(x_1, x_2) = \begin{cases} \frac{1}{2\pi\sqrt{x_1^2 + x_2^2}}\, 2\sqrt{x_1^2 + x_2^2} = \frac{1}{\pi} & \text{if } x_1^2 + x_2^2 \le 1, \\ 0 & \text{otherwise.} \end{cases} \tag{16}$$

Thus, we can generate the uniform distribution inside the unit circle using this technique. Actually, this result has already been derived in Example 2.4.4. □

Bivariate scale mixtures of normals can also be generated using this technique, as the next example shows.

Example 3.4.6 Let

$$f_R(r) = r\left[(1-\epsilon)\,e^{-r^2/2} + \frac{\epsilon}{\sigma}\,e^{-r^2/(2\sigma^2)}\right], \qquad r > 0, \tag{17}$$

which is a mixture of two densities of the form given in (14). Then the joint density of $X_1 = R\cos U$ and $X_2 = R\sin U$ is given by

$$f_{\mathbf{X}}(x_1, x_2) = \frac{1}{2\pi}\left\{(1-\epsilon)\exp\left[-\frac{1}{2}(x_1^2 + x_2^2)\right] + \frac{\epsilon}{\sigma^2}\exp\left[-\frac{1}{2\sigma}(x_1^2 + x_2^2)\right]\right\}; \tag{18}$$

see Exercise 11. □

The main appeal of spherical distributions perhaps is that many results for the multivariate spherical normal distribution can be shown for the general class of spherical distributions. Let us explain this with a two-dimensional example. Suppose $\mathbf{X} \sim \mathcal{N}_2(\mathbf{0},\ \sigma^2\mathbf{I}_2)$, i.e., X_1 and X_2 are two independent normal random variables, both with mean 0 and variance $\sigma^2 > 0$. Then (see Exercise 22), $T = X_1/X_2$ follows a Cauchy distribution, with *pdf*

$$f_T(t) = \frac{1}{\pi}\cdot\frac{1}{1+t^2}, \qquad t \in \mathbb{R}. \tag{19}$$

From Example 3.4.4, we know that we can write $X_1 = R\cos U$ and $X_2 = R\sin U$, where R and U are independent, R has *pdf* (14), and U is uniform in $[0, 2\pi)$. However,

$$T = \frac{X_1}{X_2} = \frac{\cos U}{\sin U}, \tag{20}$$

i.e., the distribution of T does not depend on R, only on U. This means that, for *any* spherically distributed $\mathbf{X} = \begin{pmatrix} X_1 \\ X_2 \end{pmatrix}$, the ratio $T = X_1/X_2$ follows a Cauchy distribution, as long as $\Pr[X_2 = 0] = 0$. In polar coordinates, the latter condition means $\Pr[R = 0] = 0$, so only spherical distributions that put nonzero probability mass at the origin must be excluded.

This remarkable result is a special case of a more general result that we will not prove here, namely: if $\mathbf{Z} = (Z_1, \ldots, Z_p)'$ is p-variate spherical, $p \geq 2$, with $\Pr[\mathbf{Z} = \mathbf{0}] = 0$, then the random variable

$$T = \frac{Z_1}{\sqrt{(Z_2^2 + \cdots + Z_p^2)/(p-1)}} \tag{21}$$

has a t-distribution on $(p-1)$ degrees of freedom. See Muirhead (1982, Chapter 1).

Elliptical distributions generalize spherical distributions in the same way the multivariate normal distribution with mean μ and covariance matrix ψ generalizes the multivariate, standard normal distribution.

Definition 3.4.2 Suppose that $\mathbf{Z}$ follows a p-variate spherical distribution, $\mathbf{A}$ is a matrix of dimension $p \times p$, and $\mathbf{m} \in \mathbb{R}^p$ is a fixed vector. Then the random vector

$$\mathbf{X} = \mathbf{m} + \mathbf{A}\mathbf{Z} \tag{22}$$

is said to have an elliptical, or elliptically symmetric distribution. □

An immediate consequence of Definition 3.4.2 is that

$$E[\mathbf{X}] = \mathbf{m} \tag{23}$$

and

$$\text{Cov}[\mathbf{X}] = c\mathbf{A}\mathbf{A}' \tag{24}$$

for some constant $c \geq 0$, provided the mean vector and the covariance matrix exist. This follows from the fact that, for a spherical p-variate $\mathbf{Z}$, $E[\mathbf{Z}] = \mathbf{0}$ and $\text{Cov}[\mathbf{Z}] = c\,\mathbf{I}_p$; see Exercise 14.

Again, we will be mostly interested in the case where $\mathbf{X}$ has a density function. Suppose the spherical random vector $\mathbf{Z}$ in (22) has *pdf* $f_{\mathbf{Z}}(\mathbf{z})$, then $f_{\mathbf{Z}}$ depends on $\mathbf{z}$ only through $\mathbf{z}'\mathbf{z}$, i.e., we can write $f_{\mathbf{Z}}(\mathbf{z}) = g(\mathbf{z}'\mathbf{z})$ for some function g. Furthermore, suppose that the matrix $\mathbf{A}$ in (22) is nonsingular. Then the transformation (22) from $\mathbf{Z}$ to $\mathbf{X}$ has Jacobian $J = \det(\mathbf{A}^{-1})$, and Theorem 2.11.3 can be used to obtain the density function of $\mathbf{X}$ as

$$\begin{aligned} f_{\mathbf{X}}(\mathbf{x}) &= |\det(\mathbf{A}^{-1})| f_{\mathbf{Z}}\left[\mathbf{A}^{-1}(\mathbf{x}-\mathbf{m})\right] \\ &= \left[\det(\mathbf{A})^{-1} \det(\mathbf{A}^{-1})\right]^{1/2} g\left[(\mathbf{x}-\mathbf{m})'(\mathbf{A}')^{-1}\mathbf{A}^{-1}(\mathbf{x}-\mathbf{m})\right] \\ &= \det\left[(\mathbf{A}\mathbf{A}')^{-1}\right]^{1/2} g\left[(\mathbf{x}-\mathbf{m})'(\mathbf{A}\mathbf{A}')^{-1}(\mathbf{x}-\mathbf{m})\right] \\ &= (\det \mathbf{V})^{-1/2} g\left[(\mathbf{x}-\mathbf{m})'\mathbf{V}^{-1}(\mathbf{x}-\mathbf{m})\right], \qquad \mathbf{x} \in \mathbb{R}^p, \end{aligned} \tag{25}$$

where $\mathbf{V} = \mathbf{A}\mathbf{A}'$. In fact, this is the same calculation used in the derivation of the multivariate normal density function in Section 3.2.

From (25) we see that the contours of constant density of an elliptical distribution are ellipses, or ellipsoids in higher dimension, centered at $\mathbf{m}$. From (24) it also follows that $\text{Cov}[\mathbf{X}] = c\mathbf{V}$ for some constant $c > 0$ (Exercise 14). Thus, the matrix $\mathbf{V}$ is not necessarily the covariance matrix, but rather, proportional to the covariance matrix. Therefore, it is often called the scale matrix. The correlation matrix of $\mathbf{X}$, however, is completely determined by $\mathbf{V}$; see Exercise 15.

The multivariate normal distribution is certainly the most prominent member of the class of elliptical distributions, but there are many other interesting members, as we show with some final examples.

Example 3.4.7 continuation of Example 3.4.1

Suppose $\mathbf{Z} = \begin{pmatrix} Z_1 \\ Z_2 \end{pmatrix}$ has *pdf*

$$f_{\mathbf{Z}}(\mathbf{z}) = \frac{1}{2\pi} \exp\left[-(\mathbf{z}'\mathbf{z})^{1/2}\right], \qquad \mathbf{z} \in \mathbb{R}^2. \tag{26}$$

Let $\mathbf{A}$ denote a nonsingular 2×2 matrix, and $\mathbf{m} \in \mathbb{R}^2$ a fixed vector. Then the *pdf* of $\mathbf{X} = \mathbf{m} + \mathbf{AZ}$ is given by

$$f_{\mathbf{X}}(\mathbf{x}) = \frac{1}{2\pi(\det \mathbf{V})^{1/2}} \exp\left\{-\left[(\mathbf{x}-\mathbf{m})'\mathbf{V}^{-1}(\mathbf{x}-\mathbf{m})\right]^{1/2}\right\}, \quad \mathbf{x} \in \mathbb{R}^2, \tag{27}$$

where $\mathbf{V} = \mathbf{AA}'$; see Exercise 16. Figure 3.4.1 shows this density function as a contour plot and in a three-dimensional view, for $\mathbf{m} = \begin{pmatrix} 4 \\ 2 \end{pmatrix}$ and $\mathbf{V} = \begin{pmatrix} 4 & 2 \\ 2 & 2 \end{pmatrix}$. Compared with the bivariate normal, this distribution is much more peaked at the center and has wider tails. The correlation between X_1 and X_2 is

$$\text{Corr}[X_1, X_2] = \frac{v_{12}}{(v_{11}v_{22})^{1/2}} = \frac{1}{\sqrt{2}},$$

as seen from Exercise 15. □

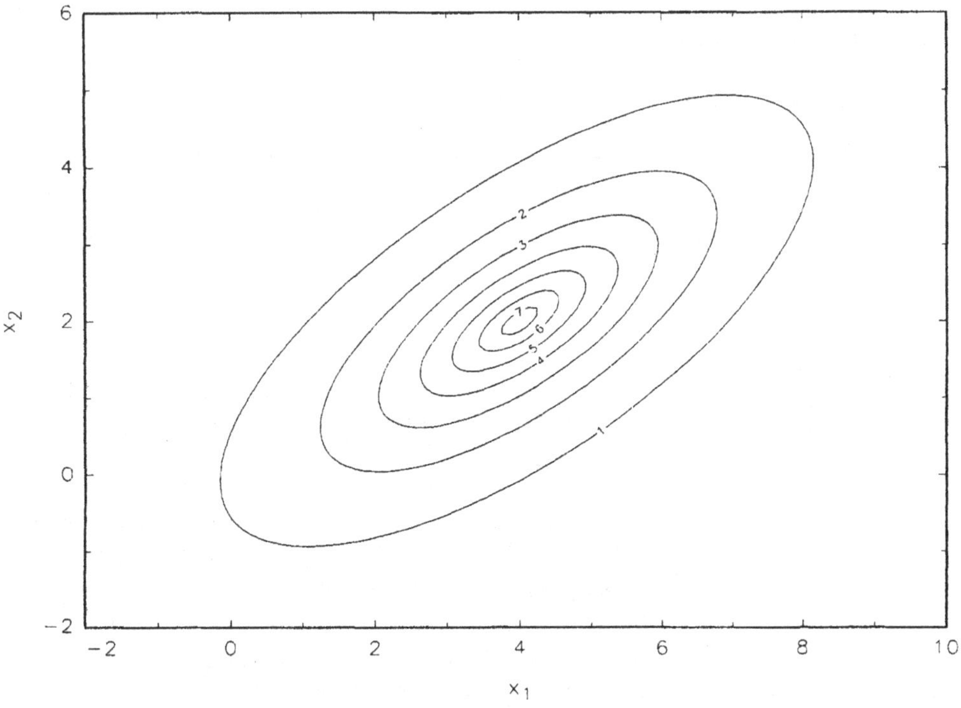

Figure 3-4-1A

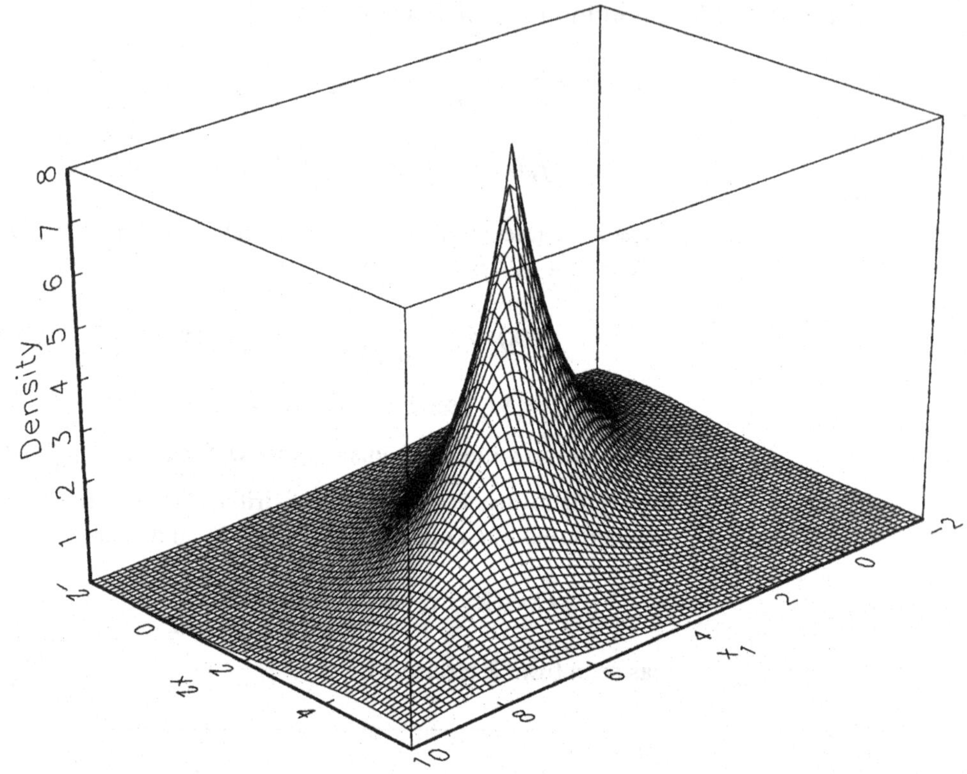

Figure 3-4-1B Density function of the bivariate double-exponential distribution in Example 3.4.7: (a) contour plot; (b) three-dimensional view.

Example 3.4.8 continuation of Example 3.4.5

Suppose that $\mathbf{Z}$ is a bivariate random variable with uniform distribution inside the unit circle, and set $\mathbf{X} = \mathbf{m} + \mathbf{AZ}$, where $\mathbf{A}$ is nonsingular. Then the density function of $\mathbf{X}$ is given by

$$f_{\mathbf{X}}(\mathbf{x}) = \begin{cases} \dfrac{1}{\pi(\det \mathbf{V})^{1/2}} & \text{if } (\mathbf{x}-\mathbf{m})'\mathbf{V}^{-1}(\mathbf{x}-\mathbf{m}) \leq 1, \\ 0 & \text{otherwise,} \end{cases} \tag{28}$$

where $\mathbf{V} = \mathbf{AA}'$; see Exercise 17. This is a uniform distribution inside the ellipse $(\mathbf{x}-\mathbf{m})'\mathbf{V}^{-1}(\mathbf{x}-\mathbf{m}) = 1$. Incidentally, equation (28) shows that the area of this ellipse is $\pi(\det \mathbf{V})^{1/2}$. □

More examples of elliptical distributions are given in Exercises 18 and 19. Elliptical distributions share many properties with the multivariate normal. For instance, marginal and conditional distributions are themselves elliptical, and conditional means are linear. An example is given in Exercise 20. However, we abandon further developments here, and, once again, refer the interested student to the books of Muirhead (1982) and Fang, Kotz, and Ng (1990).

There is one last astonishing result, however, which we present despite the fact that its proof is beyond the level of this book. The theorem, already been alluded to in Section 2.4, characterizes the multivariate normal distribution within the class of elliptical distributions.

Theorem 3.4.1 *Suppose that* $\mathbf{X}$ *has a* p*-variate elliptical distribution with covariance matrix* ψ. *If diagonality of* ψ *implies independence of all components of* $\mathbf{X}$, *then* $\mathbf{X}$ *is multivariate normal.*

For a proof, see Muirhead (1982, Theorem 1.3.5).

Theorem 3.4.1 implies that the components of an elliptically symmetric random vector $\mathbf{X}$ are never independent, except when $\mathbf{X}$ is normal with diagonal covariance matrix. Thus, if we ask for ellipticity and independence at the same time, we automatically specifiy the normal distribution. This remarkable fact limits the practical usefulness of spherical and elliptical distributions, to some extent, without making them less interesting.

Exercises for Section 3.4

1. Let $\mathbf{X}$ denote a p-variate random vector, with density function $f_{\mathbf{X}}(\mathbf{x})$. Show that the following two conditions are equivalent:
   (i) The distribution of $\mathbf{BX}$ is the same as the distribution of $\mathbf{X}$ for any orthogonal $p \times p$ matrix $\mathbf{B}$.
   (ii) $f_{\mathbf{X}}$ depends on $\mathbf{x}$ only through $\mathbf{x}'\mathbf{x}$.
   Note: For dimension $p = 2$, it may be instructive to give a proof using polar coordinates.

2. Show that for dimension $p = 1$, spherical symmetry of a random variable X means symmetry of its *pdf* to 0.

3. This Exercise refers to Example 3.4.1.
   (a) Show that the two variables in (4) are dependent.
   (b) Graph the *pdf* $g_{\mathbf{X}}$ in equation (6) as a contour plot.

4. Let X_1 and X_2 denote independent Cauchy variables, each with *pdf* (19). Graph contours of equal density of the joint *pdf* of X_1 and X_2, and compare them to contours of equal density of the bivariate Cauchy distribution in equation (7).

5. Show that all marginal distributions of dimension d, $1 \le d < k$, of a k-variate, random vector $\mathbf{X}$ with mixture *pdf* (9), are themselves scale mixtures of normals. *Hint*: See Exercise 16 in Section 3.3.

6. Let

$$f_X(x) = (1-\epsilon)\frac{1}{\sqrt{2\pi}}e^{-x^2/2} + \epsilon\frac{1}{\sqrt{2\pi}\sigma}e^{-x^2/(2\sigma^2)}, \qquad x \in \mathbb{R}, \tag{29}$$

denote a univariate normal scale mixture, where $0 < \epsilon < 1$ and $\sigma^2 > 0$, $\sigma^2 \neq 1$. Suppose that X_1 and X_2 are two independent, random variables, both with *pdf* (29).

(a) Show that the joint distribution of X_1 and X_2 is not spherical.

(b) Graph contours of constant density of the joint distribution of X_1 and X_2 for $\epsilon = 0.1$ and $\sigma^2 = 25$, and compare them to contours of equal density of (10), with the same ϵ and σ^2, and $k = 2$.

7. Find the Jacobian of the transformation (12) from polar coordinates to coordinates (x_1, x_2) in the Euclidean plane. *Hint*: Find the Jacobian of the inverse transformation.

8. Suppose that the random variable R has density function $f_R(r) = r\,e^{-r^2/2}$, $r > 0$. Show that R^2 follows a chi-square distribution on two degrees of freedom.

9. Suppose that U and Y are independent random variables, U distributed as uniform in $[0, 2\pi)$, and Y as chi-square on two degrees of freedom. Find the joint *pdf* of $X_1 = \sqrt{c^2 + Y}\cos U$ and $X_2 = \sqrt{c^2 + Y}\sin U$, where $c^2 \geq 0$ is a constant.

10. Suppose that U and R are independent random variables, U uniform in $[0, 2\pi)$ and R with density

$$f_R(r) = \begin{cases} c\,r^{c-1} & \text{if } 0 < r \leq 1, \\ 0 & \text{otherwise,} \end{cases}$$

where $c > 0$ is a fixed constant. Find the joint *pdf* of $X_1 = R\cos U$ and $X_2 = R\sin U$.

11. Prove equation (18).

12. If U and R are independent, U uniform in $[0, 2\pi)$, and R with density

$$f_R(r) = \begin{cases} 6r(1-r) & \text{if } 0 \leq r \leq 1 \\ 0 & \text{otherwise,} \end{cases}$$

find the joint *pdf* of $X_1 = R\cos U$ and $X_2 = R\sin U$.

13. If $\mathbf{Z}$ is p-variate spherical, show that all marginal distributions of dimension d, $1 \leq d \leq p - 1$, are themselves spherical.

14. Suppose that $\mathbf{Z}$ is p-variate spherical, and assume that $E[\mathbf{Z}]$ and $\text{Cov}[\mathbf{Z}]$ exist. Show that $E[\mathbf{Z}] = \mathbf{0}$ and $\text{Cov}[\mathbf{Z}] = c \cdot \mathbf{I}_p$ for some constant $c \geq 0$,

(a) for $p = 2$ and

(b) for general $p \geq 2$. *Hint*: Use Exercise 13.

15. Suppose that $\mathbf{X}$ is p-variate elliptical with scale matrix $\mathbf{V}$, and assume that the covariance matrix of $\mathbf{X}$ exists. Show that the correlation matrix of $\mathbf{X}$ is $(\text{diag}\mathbf{V})^{-1/2}\mathbf{V}(\text{diag}\mathbf{V})^{-1/2}$.

16. Prove equation (27).

17. Prove equation (28). *Hint*: Let

$$g(s) = \begin{cases} \dfrac{1}{\pi} & \text{if } 0 \leq s \leq 1, \\ 0 & \text{otherwise,} \end{cases}$$

then the *pdf* of $\mathbf{Z}$ is $f_{\mathbf{Z}}(\mathbf{z}) = g(\mathbf{z}'\mathbf{z})$.

18. If $\mathbf{Z}$ has a bivariate Cauchy distribution, as in Example 3.4.2, find the *pdf* of $\mathbf{X} = +\mathbf{AZ}$.

19. If the distribution of $\mathbf{Z}$ is a scale mixture of multivariate normals, as in Example 3.4.3, find the *pdf* of $\mathbf{X} = +\mathbf{AZ}$, and the covariance matrix of $\mathbf{X}$. *Hint*: Use results from Section 2.8 to find the covariance matrix of $\mathbf{Z}$.

20. Suppose $\mathbf{X} = \begin{pmatrix} X_1 \\ X_2 \end{pmatrix}$ has a uniform distribution inside the ellipse $\mathbf{x}'\mathbf{V}^{-1}\mathbf{x} = 1$, where $\mathbf{V} = \begin{pmatrix} 2 & 1 \\ 1 & 1 \end{pmatrix}$; see Example 3.4.8.

(a) Find the marginal distribution of X_1.

(b) Find the conditional distribution of X_2, given X_1.

(c) Find $E[X_2|X_1]$ and $\text{var}[X_2|X_1]$. Show that the conditional mean is a linear function of X_1 and that the conditional variance depends on X_1. (The latter property distinguishes this example from the bivariate normal).

21. Prove the following result. If $\mathbf{X} = (X_1, X_2)'$ has a spherical distribution, with *pdf* $f_{\mathbf{X}}$, then the polar coordinates R and U (i.e., $X_1 = R\cos U$, $X_2 = R\sin U$) are independent, and U is uniform in $[0, 2\pi)$.

22. If X_1 and X_2 are independent, $\mathcal{N}(0, \sigma^2)$, random variables, show that $T = X_1/X_2$ follows a Cauchy distribution with *pdf* given by equation (19).

Suggested Further Reading

General: Muirhead (1982), Chapter 1.
Section 3.4: Efron (1969), Fang et al. (1990), Kelker (1970).

4 Parameter Estimation

4.1 Introduction

Estimation of parameters is a central topic in statistics. In probability theory we study the distribution of random variables, assuming they follow certain distributions, and try to find out what is likely to happen and what is unlikely. Conversely, in statistics we observe data and try to find out which distribution generated the data. In the words of my colleague R.B. Fisher: "In probability, God gives us the parameters and we figure out what is going to happen. In statistics, things have already happened, and we are trying to figure out how God set the parameters."

Parameter estimation goes beyond description and summarization of data by making statements that refer to a probabilistic model from which the data is assumed to come. Such an underlying model is often called a *population*, but the notion of a model is more appropriate because, in many cases, there is no such thing as an actual physical population from which a sample is drawn. For instance, if we measure electrodes produced by two machines (Example 5.3.4), then any statements we make about the difference between the two machines refer to the production process, not to an existing large set of items already produced.

In this chapter we will consider approaches to parameter estimation that are particularly useful in the multivariate setup, ignoring much of the classical theory such as minimum variance unbiased estimation, sufficiency, invariance, and Bayesian estimation, which many books treat extensively (e.g., Casella and Berger, 1990). It is assumed that the student is at least vaguely familiar with elementary ideas of parameter estimation. In Section 4.2 we use the heuristic plug-in principle to define simple estimators of mean vectors and covariance matrices, and study their properties. Section 4.3 deals with the powerful machinery of maximum likelihood, which is the most useful method of estimation in multivariate statistics. Finally, in Section 4.4, we discuss the *EM*-algorithm, a highly valuable tool for finding maximum likelihood estimates numerically in situations with missing data. Most of our treatment of finite mixtures in Chapter 9 will be based on the *EM*-algorithm.

We will need some terminology and notation, as given by the following definitions.

Definition 4.1.1 Suppose that $\mathbf{X}_1, \ldots, \mathbf{X}_N$ are N independent random vectors, all with the same distribution function F. Then $\mathcal{X} = (\mathbf{X}_1, \ldots, \mathbf{X}_N)$ is called a *random sample* from the distribution F. Alternatively, we will call $\mathbf{X}_1, \ldots, \mathbf{X}_N$ *iid* (independent, identically distributed) with distribution F. □

For simplicity, most often, we will omit the word "random" when talking about random samples and simply refer to a sample from a distribution F.

Now, suppose that the distribution F depends on an unknown parameter θ, which may be a real number, a vector, or even a more complicated object, as we shall see shortly. The distribution is specified except for the value of θ, and the purpose of parameter estimation is to "guess" this unknown value, using the information provided by the sample. Usually we will specify a set Ω, called the *parameter space*, in which θ is allowed to vary. For instance, if we are sampling from a Bernoulli distribution with unknown success probability θ, then it is natural to assume that $\theta \in \Omega = [0, 1]$. If we are sampling from a p-variate normal distribution with unknown mean vector μ and unknown covariance matrix ψ, then we can set $\theta = (\mu, \psi) \in \Omega$, where $\Omega = \{(\mu, \psi) : \mu \in \mathbb{R}^p,\ \psi \text{ positive definite symmetric of dimension } p \times p\}$.

Definition 4.1.2 Let $\mathcal{X} = (\mathbf{X}_1, \ldots, \mathbf{X}_N)$ be a random sample of size N from a distribution F, and let $\mathbf{t}(\mathbf{x}_1, \ldots, \mathbf{x}_N)$ denote a function defined on the sample space of $\mathcal{X}$, such that the function $\mathbf{t}$ does not depend on any unknown parameter. Then the random variable, or random vector, $\mathbf{T} = \mathbf{t}(\mathbf{X}_1, \ldots, \mathbf{X}_N)$ is called a *statistic*. If the purpose of $\mathbf{T}$ is to estimate an unknown parameter θ on which F depends, then we will refer to $\mathbf{T}$ as an *estimator* of θ. □

Somewhat more formally, we might define an estimator as a function defined on the sample space of $\mathcal{X}$, taking values in the parameter space Ω in which θ is allowed to vary. However, this would exclude certain "simple" estimators. See Exercises 3 and 4 for illustrative examples.

Example 4.1.1 Estimation of the parameters of a binomial distribution

Let $X_1, \ldots, X_N$ denote a sample from the binomial distribution with parameters K (number of trials) and π (success probability). Suppose that K is known, but π is unknown. Then the family of distributions under consideration consists of all binomial distributions with K trials, and it is natural to define the parameter space Ω as the unit interval. Using the function $t(x_1, \ldots, x_N) = \sum_{i=1}^{N} x_i/(NK)$, the random variable

$$T = \frac{1}{NK} \sum_{i=1}^{N} X_i \tag{1}$$

is a statistic and appears to be a natural candidate for estimating π. However, if K is also unknown, the unknown parameter is a vector $\boldsymbol{\theta} = (K, \pi)$, where K may vary in the set of nonnegative integers and $\pi \in [0, 1]$. In this case, equation (1) is no longer a statistic because the function t depends on the unknown parameter K. □

According to Definition 4.1.2, an estimator is itself random: it may be a random variable, vector, matrix, or a combination thereof. For instance, we may estimate the parameters $\boldsymbol{\theta} = (\boldsymbol{\mu}, \boldsymbol{\psi})$ of a multivariate, normal distribution by the sample quantities $\hat{\boldsymbol{\theta}} = (\bar{\mathbf{x}}, \mathbf{S})$, as suggested by the plug-in principle of Section 4.2. The distribution of the estimator $\mathbf{T}$ as a function of the N random variables $\mathbf{X}_1, \ldots, \mathbf{X}_N$ is called the *sampling distribution* of $\mathbf{T}$. We will follow the usual convention of distinguishing between estimators and estimates: An estimate is the numerical value of the estimator obtained by substituting observed values (i.e., the observed sample $\mathbf{x}_1, \ldots, \mathbf{x}_N$) in the definition of the estimator. Also, we will often denote estimators by the "^" symbol, that is, write $\hat{\boldsymbol{\theta}}$ for a statistic that is supposed to estimate $\boldsymbol{\theta}$, and it will usually be clear from the context whether we mean the random variable or its observed value, the estimate.

In some sense, a good estimator should, be "close" to the true but unknown parameter value. The following definition introduces two frequently used properties of estimators.

Definition 4.1.3 Suppose that $F_{\boldsymbol{\theta}}$ denotes a family of distributions indexed by a parameter $\boldsymbol{\theta} \in \Omega \subset \mathbb{R}^k$, $\mathbf{X}_1, \ldots, \mathbf{X}_N$ is a sample of size N from $F_{\boldsymbol{\theta}}$, and $\mathbf{T} = \mathbf{T}(\mathbf{X}_1, \ldots, \mathbf{X}_N)$ is an estimator of $\boldsymbol{\theta}$, taking values in $\mathbb{R}^k$. Then

(a) $\mathbf{T}$ is *unbiased* for $\boldsymbol{\theta}$ if $E[\mathbf{T}] = \boldsymbol{\theta}$ for all $\boldsymbol{\theta} \in \Omega$, and

(b) $\mathbf{T}$ is *consistent* for $\boldsymbol{\theta}$ if, as the sample size N tends to infinity, the sampling distribution of $\mathbf{T}$ converges to a distribution that puts probability 1 on $\boldsymbol{\theta}$. □

Unbiasedness states that, on the average, if we use $\mathbf{T}$ to estimate $\boldsymbol{\theta}$, we "estimate the right thing." It does not mean that, in any given case, the estimate is guaranteed to be near the parameter value. Rather, it states that the sampling distribution of $\mathbf{T}$ is centered at the right place, thus, justifying the use of $\mathbf{T}$. Consistency is an asymptotic concept. In a somewhat more precise formulation than in Definition 4.1.3, set $\mathbf{T}_N = \mathbf{T}(\mathbf{X}_1, \ldots, \mathbf{X}_N)$. Then it is required that, as the sample size N tends to infinity, the sequence of random variables $\mathbf{T}_N$ converges in distribution to a degenerated random variable $\mathbf{T}_\infty$ which takes only one value — the parameter $\boldsymbol{\theta}$. Again, this must hold for all $\boldsymbol{\theta} \in \boldsymbol{\Omega}$. In the spirit of this introduction to multivariate statistics, we do not expand much on asymptotic distributions, and the reader who would like to understand the concept of convergence in distribution better is referred to Casella and Berger (1990).

The *bias* of an estimator $\mathbf{T}$ of $\boldsymbol{\theta}$ is

$$\text{Bias}[\mathbf{T}; \boldsymbol{\theta}] = E[\mathbf{T} - \boldsymbol{\theta}], \tag{2}$$

where the expectation is calculated using the sampling distribution of $\mathbf{T}$ if the sample $\mathbf{X}_1, \ldots, \mathbf{X}_N$ is from $F_{\boldsymbol{\theta}}$. It follows that $\mathbf{T}$ is unbiased for $\boldsymbol{\theta}$ exactly if $\text{Bias}[\mathbf{T}; \boldsymbol{\theta}] = \mathbf{0}$. An estimator is consistent if its asymptotic variance and its asymptotic bias (as the sample size tends to infinity) are both zero.

It is often desirable to compare different estimators of the same parameter. The notions of consistency and bias alone do not necessarily lead to a reasonable criterion for such a comparison. *Mean squared error* (MSE) is a frequently used concept for comparing competing estimators. For a real-valued parameter θ and a univariate estimator T, the mean squared error of T for θ is defined as

$$\text{MSE}[T; \theta] = E\left[(T - \theta)^2\right]. \tag{3}$$

In Exercise 5, it is shown that

$$\text{MSE}[T; \theta] = \text{var}[T] + (\text{Bias}[T; \theta])^2. \tag{4}$$

In the multivariate case, suppose that $\boldsymbol{\theta}$ is a k-dimensional vector of parameters and $\mathbf{T}$ a k-variate estimator of $\boldsymbol{\theta}$. Then a multivariate generalization of (3) is

$$\text{MSE}[\mathbf{T}; \boldsymbol{\theta}] = \sum_{j=1}^{k} E\left[(T_j - \theta_j)^2\right] = E\left[(\mathbf{T} - \boldsymbol{\theta})'(\mathbf{T} - \boldsymbol{\theta})\right], \tag{5}$$

that is, in this multivariate generalization the mean squared errors of the k components are added to obtain a simple criterion for assessing the performance of an estimator. Similarly to equation (4), it is shown in Exercise 5 that

$$\text{MSE}[\mathbf{T}; \boldsymbol{\theta}] = \text{tr}\left(\text{Cov}[\mathbf{T}]\right) + (\text{Bias}[\mathbf{T}; \boldsymbol{\theta}])'\left(\text{Bias}[\mathbf{T}; \boldsymbol{\theta}]\right). \tag{6}$$

Here, "tr" is the trace of a matrix. Mean squared error is a meaningful concept mostly for comparing estimators of location, but it is rather questionable in the case of scale parameters: a variance is always nonnegative, and thus, one may argue that negative and positive deviations of the estimator from the parameter value should not be penalized equally. We will use the concept of mean squared error extensively in

Chapter 8, in the somewhat different context of approximating multivariate random vectors.

For proving consistency of an estimator, it is often simpler to work with mean squared error rather than with a sequence of distribution functions. If the mean squared error of an estimator tends to zero as the sample size goes to infinity, the estimator is consistent. A proof of this statement is left to the technically inclined student; see Exercise 6.

Usually, we will also want to estimate the variability of a parameter estimate. A well-known example is the so-called *standard error of the mean*: Let $X_1, \ldots, X_N$ be a sample from some distribution with mean μ and variance σ^2, and suppose that we estimate μ by $\bar{X} = \frac{1}{N}\sum_{i=1}^{N} X_i$. Then

$$\operatorname{var}\left[\bar{X}\right] = \frac{1}{N}\sigma^2, \tag{7}$$

but, if the variance σ^2 is unknown, this formula is only of theoretical value. If an estimate $\hat{\sigma}^2$ of the variance is available at the same time, then we might replace σ^2 in equation (7) by $\hat{\sigma}^2$, and thus, define

$$\widehat{\operatorname{var}}\left[\bar{X}\right] = \frac{1}{N}\hat{\sigma}^2. \tag{8}$$

More commonly in numerical applications, we will use the square root of (8) and refer to it as the standard error of $\bar{x}$, or $\operatorname{se}(\bar{x})$, for short. A multivariate generalization is given in the following definition.

Definition 4.1.4 Let $\mathbf{X}_1, \ldots, \mathbf{X}_N$ be a sample from a distribution that depends on an unknown parameter $\boldsymbol{\theta} = (\theta_1, \ldots, \theta_k)' \in \mathbb{R}^k$. Let $\hat{\boldsymbol{\theta}}$ denote an estimator of $\boldsymbol{\theta}$ based on the sample, and $\boldsymbol{\Sigma} = \operatorname{Cov}[\hat{\boldsymbol{\theta}}]$ its covariance matrix. Let $\hat{\boldsymbol{\Sigma}}$ be an estimate of $\boldsymbol{\Sigma}$ obtained by replacing parameters in $\boldsymbol{\Sigma}$ with estimates obtained from the same sample. Then the *standard error* of $\hat{\theta}_j$, written $\operatorname{se}(\hat{\theta}_j)$, is defined as the square root of the jth diagonal entry of $\hat{\boldsymbol{\Sigma}}$. □

Perhaps, this is best illustrated with a classical example.

Example 4.1.2 This example anticipates material from Section 4.2, but it is appropriate here. For a sample $\mathbf{X}_1, \ldots, \mathbf{X}_N$ from some p-variate distribution with mean vector $\boldsymbol{\mu}$ and covariance matrix $\boldsymbol{\psi}$, let $\bar{\mathbf{X}}$ and $\mathbf{S}$ denote the usual sample mean vector and sample covariance matrix. It is shown in Theorem 4.2.1 that $\operatorname{Cov}[\bar{\mathbf{X}}] = \frac{1}{N}\boldsymbol{\psi}$. Since the diagonal entries of $\mathbf{S}$ are the sample variances s_j^2, we obtain the standard error of the jth average as $\operatorname{se}(\bar{x}_j) = s_j/\sqrt{N}$. □

Standard errors are a useful tool in many multivariate methods; we will use them extensively in Chapters 6 to 9. Basically, standard errors provide us with an assessment of the stability of an estimate: if a standard error is relatively large, we know that the parameter estimate may be far off the true value of the parameter, whereas a small standard error would make us feel confident about the estimate. Calculating standard errors is rarely as straightforward as in Example 4.1.2, but, for maximum likelihood estimators, it is often possible to use the powerful large sample theory to obtain at least an approximation to the variability of an estimate. A good example is logistic regression, the topic of Sections 7.5 and 7.6. However, sometimes large sample theory may not be available, or we may not be willing to trust some of the assumptions made in the asymptotic approximation. In such cases, we will refer to an alternative method of computing standard errors based on heavy computation, the *bootstrap*. A brief introduction is given at the end of Section 4.2.

Exercises for Section 4.1

1. Let $X_1, \ldots, X_N$ denote a sample from a distribution F with mean μ. Show that the sample average $\bar{X} = \sum_{i=1}^{N} X_i/N$ is an unbiased estimator of μ.

2. Let $X_1, \ldots, X_N$, $N \geq 2$, denote a sample from a distribution F with mean μ and variance σ^2. Show that the sample variance $s^2 = \sum_{i=1}^{N}(X_i - \bar{X})^2/(N-1)$ is an unbiased estimator of σ^2. *Hint*: Compute $E\left[(X_1 - \bar{X})^2\right]$, using $E[X_1 - \bar{X}] = 0$.

3. Three envelopes contain $\theta, \theta + 1$, and $\theta + 2$ rubles each, θ being an unknown integer. N people independently play the game of choosing one of the three envelopes at random; they report amounts $X_1, \ldots, X_N$ found in the envelope chosen. (The chosen envelope is replaced after each game; thus, the X_i are independent).
   (a) Show that $\hat{\theta} = \bar{X} - 1$ is an unbiased and consistent estimator of θ, but $\hat{\theta}$ is not necessarily in the parameter space.
   (b) Show that, if the range of the sample (i.e., $\max_{1 \leq i \leq N} X_i - \min_{1 \leq i \leq N} X_i$) is 2, then the parameter θ is determined by the sample.
   (c) Find the probability that the parameter is determined by the sample as a function of the sample size N. How should you choose N if you want to be at least 99.99 percent sure that the parameter is determined?
   (d) Show that $\hat{\theta} = \min_{1 \leq i \leq N} X_i$ is a biased but consistent estimator of θ. *Hint*: it is not necessary to compute the bias of $\hat{\theta}$ explicitly. Instead, argue that the mean of a discrete random variable must be between the smallest and the largest value which the random variable takes with positive probability.

4. Let $X_1, \ldots, X_N$ denote a sample of size N from some distribution with unknown mean and unknown variance σ^2, $\sigma^2 > 0$. Let $Y_1, \ldots, Y_M$ denote a sample of size M, independent of the first sample, from some distribution with unknown mean and unknown variance $\sigma^2 + \tau^2$, $\tau^2 \geq 0$. For instance, the X_i might be measurements of blood pressure of N healthy people made by a careful physician, and the Y_i are blood pressures measured on M different healthy individuals by a physician who reads the numbers off less carefully and, thus, adds some error variability τ^2 to the biological

variability σ^2. Let s_X^2 and s_Y^2 denote the respective sample variances, defined as in Exercise 2.

(a) Show that $\hat{\tau}^2 = s_Y^2 - s_X^2$ is an unbiased estimator of τ^2.

(b) Show that $\hat{\tau}^2$ can take values outside the parameter space.

(c) Consider the modified estimator

$$\hat{\tau}_m^2 = \begin{cases} \hat{\tau}^2 & \text{if } \hat{\tau}^2 > 0, \\ 0 & \text{otherwise.} \end{cases}$$

Show that $\hat{\tau}_m^2$ is a biased estimator of τ. Also show that $\text{MSE}[\hat{\tau}_m^2; \tau^2] < \text{MSE}[\hat{\tau}^2; \tau^2]$. *Hint*: Consider a random variable U that takes both negative and positive values, and define a new random variable V by $V = U$ if $U \geq 0$, and $V = 0$ otherwise. Show that $E[V] > E[U]$. Similarly, with $\mu = E[U]$, show that $E[(V - \mu)^2] < E[(U - \mu)^2]$.

5. Prove equations (4) and (6).

6. Let $X_1, \ldots, X_N$ be a sample from some distribution that depends on a parameter $\boldsymbol{\theta}$, and denote by $\hat{\boldsymbol{\theta}}$ an estimator of $\boldsymbol{\theta}$ based on the sample. If $\lim_{N\to\infty} \text{MSE}[\hat{\boldsymbol{\theta}}; \boldsymbol{\theta}] = 0$, show that $\hat{\boldsymbol{\theta}}$ is a consistent estimator of $\boldsymbol{\theta}$.

7. Let $\mathbf{X}_1, \ldots, \mathbf{X}_N$ be a sample from some p-variate distribution with mean $\boldsymbol{\mu}$ and covariance matrix $\boldsymbol{\psi}$. Find the mean squared error of $\bar{\mathbf{X}}$ for $\boldsymbol{\mu}$, and show that the mean squared error goes to zero as the sample size N tends to infinity.

8. Let $X_1, \ldots, X_N$ be a sample from the uniform distribution in the interval $[0, \theta]$, where $\theta > 0$ is an unknown parameter. Show that $2\bar{X}$ is an unbiased estimator of θ, with $\text{MSE}[2\bar{X}; \theta] = \theta^2/3N$.

9. In the same setup as problem 8, let $Y = \max_{1 \leq i \leq N} X_i$.

(a) Show that, for any finite N, Y is a biased estimator of θ. *Hint*: Use an argument similar to part (d) in problem 3.

(b) If $U_1, \ldots, U_N$ denote *independent, identically distributed* random variables with a uniform distribution in the interval $[0, 1]$, and $V = \max_{1 \leq i \leq N} U_i$, show that the *pdf* of V is given by

$$f_V(v) = \begin{cases} Nv^{N-1} & \text{if } 0 \leq v \leq 1, \\ 0 & \text{otherwise.} \end{cases}$$

Hint: First, find the *cdf* of V, $F_V(a) = \Pr[V \leq a]$, and notice that $V \leq a$ exactly if all $U_i \leq a$.

(c) Use part (b) to show that $\text{Bias}[Y; \theta] = -\theta/(N+1)$ and $\text{MSE}[Y; \theta] = 2\theta^2/[(N+1)(N+2)]$. *Hint*: Use $Y = \theta V$, where V is the maximum of N *iid* random variables with uniform distribution in the interval $[0, 1]$.

(d) Find a constant c such that the statistic $Y^* = cY$ is unbiased for θ. Compute the mean squared error of Y^* for θ.

(e) Show that both Y and Y^* are consistent estimators of θ.

10. Compare the three estimators $2\bar{X}$ (from problem 8), Y, and Y^* (from problem 9) in terms of their mean squared errors. For a fixed value of θ (e.g. $\theta = 1$), graph the three mean squared errors as functions of N, for $N = 1$ to 5. Which estimator would you prefer?

4.2 Plug-in Estimators

The plug-in principle, perhaps, gives the simplest of all approaches to estimation. It states that parameters should be estimated by their corresponding sample counterparts. Formally, let F be a distribution function and $\theta = h(F)$ a parameter which is uniquely determined as a function of F. Simple examples are mean, variance, and median — but notice that the median is not necessarily unique. For observed data $\mathbf{x}_1, \ldots, \mathbf{x}_N$, let F_N denote the distribution that puts probability $1/N$ on each of the N data points. Then the plug-in estimate is defined as the corresponding parameter computed from F_N, i.e., as $\hat{\theta} = h(F_N)$. Since F_N is defined in terms of the N data points, we may also consider $\hat{\theta}$ as a function of $\mathbf{x}_1, \ldots, \mathbf{x}_N$, say $\hat{\theta} = h^*(\mathbf{x}_1, \ldots, \mathbf{x}_N)$. Using a random sample $\mathbf{X}_1, \ldots, \mathbf{X}_N$ in place of observed data, thus, we get a *plug-in estimator*

$$\hat{\theta} = h^*(\mathbf{X}_1, \ldots, \mathbf{X}_N). \tag{1}$$

Note that we do not distinguish in notation between the random variable $\hat{\theta}$ and its observed value for a particular set of observations, as it will be clear from the context whether we are talking about an estimator or an estimate.

The most important application of the plug-in principle is estimation of mean vectors and covariance matrices. For vectors $\mathbf{x}_1, \ldots, \mathbf{x}_N$ in $\mathbb{R}^p$, define a p-variate, random vector $\mathbf{X}_{\text{plug-in}}$ by

$$\Pr\left[\mathbf{X}_{\text{plug-in}} = \mathbf{x}_i\right] = \frac{1}{N} \quad \text{for } i = 1, \ldots, N. \tag{2}$$

Some care is required if some of the vectors $\mathbf{x}_i$ are identical, a case which is not excluded in equation (2). Suppose, for instance, that m of the vectors $\mathbf{x}_i$ are identical and equal to (say) $\mathbf{x}^*$. Then we would set

$$\Pr\left[\mathbf{X}_{\text{plug-in}} = \mathbf{x}^*\right] = \frac{m}{N}.$$

This difficulty can be avoided if we think of $\mathbf{x}_1, \ldots, \mathbf{x}_N$ as a list of N distinct items. When we sample from the random variable $\mathbf{X}_{\text{plug-in}}$, each of the items is selected with probability $1/N$, but some of the items may have identical numerical values. In this way, we can apply equation (2) to any list of N vectors, treating them formally as different. For the random vector $\mathbf{X}_{\text{plug-in}}$,

$$\begin{aligned} E\left[\mathbf{X}_{\text{plug-in}}\right] &= \sum_{i=1}^{N} \mathbf{x}_i \Pr\left[\mathbf{X}_{\text{plug-in}} = \mathbf{x}_i\right] \\ &= \frac{1}{N}\sum_{i=1}^{N} \mathbf{x}_i \\ &= \bar{\mathbf{x}}, \end{aligned} \tag{3}$$

and

$$\begin{aligned} \text{Cov}\left[\mathbf{X}_{\text{plug-in}}\right] &= E\left[\left(\mathbf{X}_{\text{plug-in}} - \bar{\mathbf{x}}\right)\left(\mathbf{X}_{\text{plug-in}} - \bar{\mathbf{x}}\right)'\right] \\ &= \frac{1}{N}\sum_{i=1}^{N} (\mathbf{x}_i - \bar{\mathbf{x}})(\mathbf{x}_i - \bar{\mathbf{x}})'. \end{aligned} \tag{4}$$

For a sample $\mathbf{X}_1, \ldots, \mathbf{X}_N$ from *any* distribution F, the plug-in estimators of mean vector and covariance matrix therefore are

$$\bar{\mathbf{X}} = \frac{1}{N}\sum_{i=1}^{N} \mathbf{X}_i \tag{5}$$

and

$$\mathbf{S}_P = \frac{1}{N}\sum_{i=1}^{N} (\mathbf{X}_i - \bar{\mathbf{X}})(\mathbf{X}_i - \bar{\mathbf{X}})' = \frac{1}{N}\sum_{i=1}^{N} \mathbf{X}_i\mathbf{X}_i' - \bar{\mathbf{X}}\bar{\mathbf{X}}'. \tag{6}$$

Note that equations (5) and (6) define a random vector and a random matrix, respectively, whose distribution we may study if $\mathbf{X}_1, \ldots, \mathbf{X}_N$ is a sample from a particular *cdf* F. In (3) and (4), the expectation was taken with respect to the distribution of the random vector $\mathbf{X}_{\text{plug-in}}$. This served the purpose of finding a function which can, then be applied to a random sample. Later, when we study $E[\bar{\mathbf{X}}]$ or $\text{Cov}[\bar{\mathbf{X}}]$, the expectation will be taken with respect to the distribution F from which $\mathbf{X}_1, \ldots, \mathbf{X}_N$ form a sample.

In Theorem 4.2.2, we will see that $\mathbf{S}_P$ is a biased estimator of the covariance matrix. Therefore, it is common to use

$$\mathbf{S} = \frac{1}{N-1}\sum_{i=1}^{N} (\mathbf{X}_i - \bar{\mathbf{X}})(\mathbf{X}_i - \bar{\mathbf{X}})' = \frac{N}{N-1}\mathbf{S}_P \tag{7}$$

instead. We will refer to $\mathbf{S}$ as the *sample covariance matrix*. In view of the plug-in principle, this is somewhat unnatural; however, the use of $\mathbf{S}$ is so common that we will not attempt to change this convention.

At times we may refer to selected entries of a sample mean vector or covariance matrix. Let $\mathbf{x}_1, \ldots, \mathbf{x}_N$ denote observed vectors, and write

$$\mathbf{x}_i = \begin{pmatrix} x_{i1} \\ \vdots \\ x_{ip} \end{pmatrix}, \qquad i = 1, \ldots, N, \tag{8}$$

for the ith vector. Then the jth component of the sample mean vector is given by

$$\bar{x}_j = \frac{1}{N} \sum_{i=1}^{N} x_{ij}, \qquad j = 1, \ldots, p, \tag{9}$$

and the (j, h)th entry of the sample covariance matrix is given by

$$s_{jh} = \frac{1}{N-1} \sum_{i=1}^{N} (x_{ij} - \bar{x}_j)(x_{ih} - \bar{x}_h), \qquad j, h = 1, \ldots, p. \tag{10}$$

The diagonal and off-diagonal entries of $\mathbf{S}$ are the *sample variances* and *sample covariances*, respectively. For the sample variances, we may write

$$s_j^2 := s_{jj} = \frac{1}{N-1} \sum_{i=1}^{N} (x_{ij} - \bar{x}_j)^2, \qquad j = 1, \ldots, p. \tag{11}$$

In analogy to the theoretical counterparts, the $s_j = \sqrt{s_j^2}$ are called the (sample) *standard deviations*. The *sample correlations* are

$$r_{jh} = \frac{s_{jh}}{s_j s_h} \qquad j, h = 1, \ldots, p. \tag{12}$$

The sample correlations (12) are plug-in estimates of the correlations in the model, since they can be computed from $\mathbf{S}_P$ and from $\mathbf{S}$; see Exercise 1. They can be arranged conveniently as a *correlation matrix*,

$$\mathbf{R} = (r_{jh}) = \begin{pmatrix} 1 & r_{12} & \cdots & r_{1p} \\ r_{21} & 1 & \cdots & r_{2p} \\ \vdots & \vdots & \ddots & \vdots \\ r_{p1} & r_{p2} & \cdots & 1 \end{pmatrix}. \tag{13}$$

In matrix notation, the correlation matrix can be computed analogously to equation (9) in Section 2.10 as

$$\mathbf{R} = (\text{diag}\mathbf{S})^{-1/2}\mathbf{S}(\text{diag}\mathbf{S})^{-1/2}. \tag{14}$$

Again, the latter equation can be written in terms of the plug-in estimate $\mathbf{S}_P$.

Some properties of plug-in estimates and estimators follow directly from their definition because they can be viewed as parameters of the distribution of the random vector $\mathbf{X}_{\text{plug-in}}$. For instance, in Section 2.11 we saw that the variance of a linear combination $\mathbf{a}'\mathbf{X}$ of a random vector $\mathbf{X}$ with covariance matrix ψ is $\mathbf{a}'\psi\mathbf{a}$. Since variances are by definition nonnegative, it follows that $\mathbf{a}'\psi\mathbf{a} \geq 0$ for all vectors $\mathbf{a} \in \mathbb{R}^p$, i.e., ψ is positive semidefinite. Because $\mathbf{S}_P$ is the covariance matrix of a random vector $\mathbf{X}_{\text{plug-in}}$, the same property holds for $\mathbf{S}_P$ and, therefore, for $\mathbf{S}$. Similarly, it follows that the sample correlation matrix $\mathbf{R}$ is positive semidefinite. In particular, any sample correlation is between -1 and 1, as follows from Theorem 2.5.4. Further properties are examined in Exercise 6.

Next, we will study some sampling properties of $\bar{\mathbf{X}}$ and $\mathbf{S}$.

Theorem 4.2.1 *Let $\mathbf{X}_1, \ldots, \mathbf{X}_N$ be a sample from any distribution with mean vector μ and covariance matrix ψ. Then $\bar{\mathbf{X}}$ is an unbiased and consistent estimator of μ, with Cov$[\mathbf{X}] = \psi/N$ and mean squared error* $\text{MSE}[\bar{\mathbf{X}}; \mu] = \text{tr}(\psi)/N$.

Proof Unbiasedness follows from $E[\bar{\mathbf{X}}] = \frac{1}{N}\sum_{i=1}^{N} E[\mathbf{X}_i] = \mu$. Since the $\mathbf{X}_i$ are independent, Exercise 21 in Section 2.11 implies $\text{Cov}\left[\sum_{i=1}^{N} \mathbf{X}_i\right] = N\psi$ and, therefore, $\text{Cov}[\bar{\mathbf{X}}] = \psi/N$. Since $\text{Bias}[\bar{\mathbf{X}}; \mu] = \mathbf{0}$, equation (6) in Section 4.1 gives $\text{MSE}[\bar{\mathbf{X}}; \mu] = \text{tr}(\psi)/N$. As N tends to infinity, the mean squared error goes to zero, which implies consistency. ■

Theorem 4.2.2 *Under the same assumptions as in Theorem 4.2.1, the asymptotic distribution of $\sqrt{N}(\bar{\mathbf{X}} - \mu)$, as N tends to infinity, is multivariate normal with mean vector* $\mathbf{0}$ *and covariance matrix ψ.*

Proof This follows directly from the Multivariate Central Limit Theorem 3.3.3. ■

Approximate multivariate normality of $\bar{\mathbf{X}}$ is a very important property that we will use extensively in Chapter 6. Even in cases where the multivariate normal distribution is questionable as a model for actual data, Theorem 4.2.2 often allows us to treat a mean vector as if it were obtained from multivariate normal data. This ultimately justifies the emphasis put on multivariate normal theory.

Theorem 4.2.3 *In the same setup as Theorem 4.2.1, the sample covariance matrix* $\mathbf{S}$ *is an unbiased estimator of ψ.*

Proof By linearity of the expectation,

$$E[\mathbf{S}] = \frac{1}{N-1}\sum_{i=1}^{N} E\left[(\mathbf{X}_i - \bar{\mathbf{X}})(\mathbf{X}_i - \bar{\mathbf{X}})'\right].$$

Setting $\mathbf{Y}_i = \mathbf{X}_i - \bar{\mathbf{X}}$ for $i = 1, \ldots, N$, we need to show that $\sum_{i=1}^{N} E[\mathbf{Y}_i\mathbf{Y}_i'] = (N-1)\psi$. Because the $\mathbf{X}_i$ are *iid*, all $\mathbf{Y}_i$ have the same distribution (although they are not independent), and it suffices to find, e.g., $E[\mathbf{Y}_1\mathbf{Y}_1']$. But $E[\mathbf{Y}_1] = \mathbf{0}$, and therefore,

$E[\mathbf{Y}_1\mathbf{Y}_1'] = \text{Cov}[\mathbf{Y}_1]$. Writing

$$\mathbf{Y}_1 = \mathbf{X}_1 - \bar{\mathbf{X}} = \frac{N-1}{N}\mathbf{X}_1 - \frac{1}{N}\mathbf{X}_2 - \cdots - \frac{1}{N}\mathbf{X}_N$$

and using Exercise 21 in Section 2.11, we obtain

$$\text{Cov}[\mathbf{Y}_1] = \left(\frac{N-1}{N}\right)^2 \text{Cov}[\mathbf{X}_1] + \sum_{i=2}^{N} \frac{1}{N^2}\text{Cov}[\mathbf{X}_i] = \frac{N-1}{N}\psi,$$

and unbiasedness follows. A different proof is given in Exercise 4. ■

Although unbiasedness is a nice property, its importance has probably been overrated in the statistical literature, leading to the almost universally accepted convention of defining sample variances and covariances in terms of the denominator $N-1$ instead of the more natural denominator N. As mentioned before, we will follow this convention, albeit half-heartedly, to keep in line with most textbooks.

Under the additional assumption that all fourth moments exist for the distribution from which the sample $\mathbf{X}_1, \ldots, \mathbf{X}_N$ is taken, it can also be shown that the sample covariance matrix $\mathbf{S}$, as well as $\mathbf{S}_P$, is a consistent estimator of the model covariance matrix ψ. A proof for the univariate case is given in Exercise 7. The distribution theory for $\mathbf{S}$ is considerably more difficult, even asymptotically, than the distribution theory for $\bar{\mathbf{X}}$ and is not treated in this text. If the sample is obtained from a multivariate normal distribution, then $\mathbf{S}$ follows a so-called *Wishart distribution*, which generalizes the familiar chi-square distribution to the multivariate case. We will use properties of the Wishart distribution in Chapter 6, but its formal treatment is beyond the level of this book. Using the Multivariate Central Limit Theorem, it can also be shown that the joint distribution of the p^2 elements of $\mathbf{S}$ is approximately normal for large sample size N. The interested reader is referred to the book by Muirhead (1982).

In Section 3.3, we saw that, for samples from a univariate normal distribution, the sample mean and sample variance are stochastically independent. The proof was based on the observation that the covariance between $\bar{X}$ and $X_i - \bar{X}$ is zero, which in the normal case implies independence. However, uncorrelatedness did not require any specific distributional assumptions. Now, we are going to prove a similar result for the multivariate case.

Theorem 4.2.4 *Let $\mathbf{X}_1, \ldots, \mathbf{X}_N$ be a sample from the $\mathcal{N}_p(\mu, \psi)$ distribution, with sample mean vector $\bar{\mathbf{X}}$ and sample covariance matrix $\mathbf{S}$. Then $\bar{\mathbf{X}}$ and $\mathbf{S}$ are stochastically independent.*

Proof We will show that all random vectors $\mathbf{Y}_i = \mathbf{X}_i - \bar{\mathbf{X}}$ are independent of $\bar{\mathbf{X}}$. Since $\mathbf{S}$ is a function of the $\mathbf{Y}_i$ alone, this implies independence of $\bar{\mathbf{X}}$ and $\mathbf{S}$. Because all $\mathbf{Y}_i$ have identical distributions, it suffices to show independence of $\mathbf{X}_1 - \bar{\mathbf{X}}$ and $\bar{\mathbf{X}}$. Let

$\mathbf{V} = \begin{pmatrix} \mathbf{X}_1 \\ \vdots \\ \mathbf{X}_N \end{pmatrix}$ denote the (Np)-dimensional, random vector obtained by stacking the $\mathbf{X}_i$ on top of each other; then $\mathbf{V}$ is (Np)-variate normal with mean vector $\begin{pmatrix} \boldsymbol{\mu} \\ \vdots \\ \boldsymbol{\mu} \end{pmatrix}$ and block-diagonal covariance matrix

$$\text{Cov}[\mathbf{V}] = \begin{pmatrix} \psi & \mathbf{O} & \cdots & \mathbf{O} \\ \mathbf{O} & \psi & \cdots & \mathbf{O} \\ \vdots & \vdots & \ddots & \vdots \\ \mathbf{O} & \mathbf{O} & \cdots & \psi \end{pmatrix}.$$

Writing

$$\begin{pmatrix} \mathbf{X}_1 \\ \sum_{i=1}^N \mathbf{X}_i \end{pmatrix} = \begin{pmatrix} \mathbf{I}_p & \mathbf{O} & \cdots & \mathbf{O} \\ \mathbf{I}_p & \mathbf{I}_p & \cdots & \mathbf{I}_p \end{pmatrix} \begin{pmatrix} \mathbf{X}_1 \\ \vdots \\ \mathbf{X}_N \end{pmatrix},$$

$$\text{Cov}\left[\begin{pmatrix} \mathbf{X}_1 \\ \sum_{i=1}^N \mathbf{X}_i \end{pmatrix}\right] = \begin{pmatrix} \mathbf{I}_p & \mathbf{O} & \cdots & \mathbf{O} \\ \mathbf{I}_p & \mathbf{I}_p & \cdots & \mathbf{I}_p \end{pmatrix} \begin{pmatrix} \psi & \mathbf{O} & \cdots & \mathbf{O} \\ \mathbf{O} & \psi & \cdots & \mathbf{O} \\ \vdots & \vdots & \ddots & \vdots \\ \mathbf{O} & \mathbf{O} & \cdots & \psi \end{pmatrix} \begin{pmatrix} \mathbf{I}_p & \mathbf{I}_p \\ \mathbf{O} & \mathbf{I}_p \\ \vdots & \vdots \\ \mathbf{O} & \mathbf{I}_p \end{pmatrix}$$

$$= \begin{pmatrix} \psi & \psi \\ \psi & N\psi \end{pmatrix}.$$

But

$$\begin{pmatrix} \mathbf{X}_1 - \bar{\mathbf{X}} \\ \bar{\mathbf{X}} \end{pmatrix} = \begin{pmatrix} \mathbf{I}_p & -\frac{1}{N}\mathbf{I}_p \\ \mathbf{O} & \frac{1}{N}\mathbf{I}_p \end{pmatrix} \begin{pmatrix} \mathbf{X}_1 \\ \sum_{i=1}^N \mathbf{X}_i \end{pmatrix},$$

and therefore,

$$\text{Cov}\left[\begin{pmatrix} \mathbf{X}_1 - \bar{\mathbf{X}} \\ \bar{\mathbf{X}} \end{pmatrix}\right] = \begin{pmatrix} \mathbf{I}_p & -\frac{1}{N}\mathbf{I}_p \\ \mathbf{O} & \frac{1}{N}\mathbf{I}_p \end{pmatrix} \begin{pmatrix} \psi & \psi \\ \psi & N\psi \end{pmatrix} \begin{pmatrix} \mathbf{I}_p & \mathbf{O} \\ -\frac{1}{N}\mathbf{I}_p & \frac{1}{N}\mathbf{I}_p \end{pmatrix}$$

$$= \begin{bmatrix} \left(1 - \frac{1}{N}\right)\psi & \mathbf{O} \\ \mathbf{O} & \frac{1}{N}\psi \end{bmatrix}.$$

Thus, all covariances between $\bar{\mathbf{X}}$ and the deviation vectors $\mathbf{X}_i - \bar{\mathbf{X}}$ are zero. By Theorem 3.2.1, the joint distribution of the $\mathbf{X}_i - \bar{\mathbf{X}}$ and $\bar{\mathbf{X}}$ is $(N+1)p$-variate normal, and therefore, by Theorem 3.2.8, all $\mathbf{X}_i - \bar{\mathbf{X}}$ are independent of $\bar{\mathbf{X}}$. This concludes the proof. ∎

As in the proof of Theorem 3.3.4, multivariate normality is not used except to argue that uncorrelatedness implies independence. Thus, the deviations $\mathbf{X}_i - \bar{\mathbf{X}}$ are uncorrelated with $\bar{\mathbf{X}}$ for samples from *any* distribution. Also see Exercise 5.

In many multivariate techniques, we will encounter the inverse of the sample covariance matrix. As we have seen from the definition of the plug-in estimates, a sample covariance matrix is always positive semidefinite. However, for $\mathbf{S}$ to have an inverse, we need positive definiteness, i.e., $\mathbf{a}'\mathbf{S}\mathbf{a} > 0$ for all $\mathbf{a} \in \mathbb{R}^p$, $\mathbf{a} \neq \mathbf{0}$. Can this be taken for granted? Later, we will see that the sample covariance matrix of N p-variate observations $\mathbf{x}_1, \ldots, \mathbf{x}_N$ is positive definite exactly if the N centered observations $\mathbf{x}_1 - \bar{\mathbf{x}}, \ldots, \mathbf{x}_N - \bar{\mathbf{x}}$ contain a set of p linearly independent vectors. In particular, this means that if $N \leq p$, then $\mathbf{S}$ is *not* positive definite. On the other hand, $N \geq p + 1$ is a necessary condition, but not a sufficient one for $\mathbf{S}$ to be invertible. What then are the chances that, for a fixed sample size $N \geq p + 1$, we will be able to invert the sample covariance matrix and, therefore, perform the analyses that depend on $\mathbf{S}^{-1}$?

An answer is given in Theorem 4.2.5, the proof of which is left to the technically inclined student as an exercise. The theorem generalizes the fact that, for a sample X_1, X_2 from a univariate continuous distribution, $\Pr[X_1 = X_2] = 0$. Therefore, the sample variance is positive with probability 1 for samples of size $N \geq 2$ from a continuous random variable. However, see the remarks following the theorem.

Theorem 4.2.5 *Let $\mathbf{X}_1, \ldots, \mathbf{X}_N$ be a sample from a p-variate, continuous, random variable with density function $f(\mathbf{x})$. If $N \geq p + 1$, then $\mathbf{S}$ is positive definite with probability 1.*

Proof See Exercise 8. ■

While Theorem 4.2.5 appears to suggest that we should be "safe" if we collect at least $p + 1$ observations from a p-variate, continuous distribution, in reality, this is not always the case because, strictly speaking, we do not sample from a continuous distribution. Due to rounding of measurements, in fact, we are always sampling from a discrete distribution, which means that the probability of $\mathbf{S}$ being positive definite is not quite 1, although it may be very close to 1. See Exercise 11.

The main appeal of the plug-in principle of estimation is its simplicity. At the same time, this leads to serious limitations. For instance, the plug-in principle can suggest estimates that are in contradiction with the data, as shown in Exercise 12. Another limitation appears in a situation that we will encounter extensively in Chapter 5 and later. Suppose $\mathbf{X}_1, \ldots, \mathbf{X}_{N_1}$ is a sample from a p-variate distribution with mean vector $\boldsymbol{\mu}_1$ and covariance matrix $\boldsymbol{\psi}$, and $\mathbf{Y}_1, \ldots, \mathbf{Y}_{N_2}$ is a sample from some other p-variate distribution with mean vector $\boldsymbol{\mu}_2$ and the same covariance matrix $\boldsymbol{\psi}$. While the plug-in principle suggests using $\bar{\mathbf{X}}$ and $\bar{\mathbf{Y}}$ as estimators of $\boldsymbol{\mu}_1$ and $\boldsymbol{\mu}_2$, it is not clear whether we should use the $\mathbf{X}$-data or the $\mathbf{Y}$-data to estimate the common $\boldsymbol{\psi}$. It is intuitively clear that one should use all the data to estimate $\boldsymbol{\psi}$, but the plug-in principle fails to suggest a combined estimator. The most popular solution to the problem is to take a weighted average: If $\mathbf{S}_1$ denotes the sample covariance matrix of the $\mathbf{X}$-data, and $\mathbf{S}_2$ the sample covariance matrix of the $\mathbf{Y}$-data, then the *pooled sample covariance matrix* is defined as

$$\mathbf{S}_{\text{pooled}} = \frac{(N_1 - 1)\mathbf{S}_1 + (N_2 - 1)\mathbf{S}_2}{N_1 + N_2 - 2} . \tag{15}$$

This combined estimator is neither a plug-in estimator nor a maximum likelihood estimator (see Section 4.3), but may be justified again by unbiasedness. See Exercises 9, 10, and 13 for more details on the pooled-sample covariance matrix. Later, we will usually write just $\mathbf{S}$ instead of $\mathbf{S}_{\text{pooled}}$ for pooled covariance matrices, as long as there is no danger of confusion.

For computational and for theoretical reasons, it will be useful to develop some algebra for sample mean vectors and covariance matrices. Let $\mathbf{x}_1, \ldots, \mathbf{x}_N$ denote N observed data vectors, and write

$$\mathbf{x}_i = \begin{pmatrix} x_{i1} \\ x_{i2} \\ \vdots \\ x_{ip} \end{pmatrix}$$

for the ith vector. The N data points are usually arranged in the form of a *data matrix* $\mathbf{D}$ of dimension $N \times p$,

$$\mathbf{D} = \begin{pmatrix} \mathbf{x}_1' \\ \mathbf{x}_2' \\ \vdots \\ \mathbf{x}_N' \end{pmatrix} = \begin{pmatrix} x_{11} & x_{12} & \cdots & x_{1p} \\ x_{21} & x_{22} & \cdots & x_{2p} \\ \vdots & \vdots & & \vdots \\ x_{N1} & x_{N2} & \cdots & x_{Np} \end{pmatrix} . \tag{16}$$

In the data matrix, the rows correspond to observations, and the columns to variables, that is, in the sampling setup, the rows correspond to the *iid* random vectors. Let $\mathbf{1}_N$ denote a vector in $\mathbb{R}^N$ whose entries are all 1. Then we can express the sample mean vector $\bar{\mathbf{x}}$ as

$$\bar{\mathbf{x}} = \begin{pmatrix} \bar{x}_1 \\ \vdots \\ \bar{x}_p \end{pmatrix} = \frac{1}{N}\mathbf{D}'\mathbf{1}_N . \tag{17}$$

The *mean-corrected* data matrix, or *centered* data matrix $\mathbf{D}_c$ is obtained by subtracting $\bar{\mathbf{x}}'$ from each row of $\mathbf{D}$, i.e.,

$$\mathbf{D}_c = \begin{pmatrix} (\mathbf{x}_1 - \bar{\mathbf{x}})' \\ \vdots \\ (\mathbf{x}_N - \bar{\mathbf{x}})' \end{pmatrix} = \mathbf{D} - \begin{pmatrix} \bar{\mathbf{x}}' \\ \vdots \\ \bar{\mathbf{x}}' \end{pmatrix} = \mathbf{D} - \mathbf{1}_N\bar{\mathbf{x}}' . \tag{18}$$

Now, the plug-in estimate $\mathbf{S}_P$ can be written as

$$\mathbf{S}_P = \frac{1}{N}\sum_{i=1}^{N}(\mathbf{x}_i - \bar{\mathbf{x}})(\mathbf{x}_i - \bar{\mathbf{x}})' = \frac{1}{N}\mathbf{D}_c'\mathbf{D}_c \,. \tag{19}$$

For matrix-oriented software, equations (17) and (19) provide a convenient way to compute mean vectors and covariance matrices. For the unbiased version $\mathbf{S}$, one needs to replace the denominator N in (19) by $N-1$. Equation (19) also gives some theoretical insight, since the rank of $\mathbf{S}_P$ is the same as the rank of $\mathbf{D}_c$. Notice that

$$\mathbf{D}_c'\mathbf{1}_N = \sum_{i=1}^{N}(\mathbf{x}_i - \bar{\mathbf{x}}) = \mathbf{0},$$

that is, the N rows of the centered data matrix $\mathbf{D}_c$ are linearly dependent. Therefore, the rank of $\mathbf{S}$ is at most $N-1$, and we need at least $p+1$ observations if we want the sample covariance matrix to be invertible. Also see Exercise 10 and the remarks following Theorem 4.2.5. For some further algebra of the centered data matrix and the sample covariance matrix, see Exercise 14.

For an intuitive understanding of why at least $p+1$ observations are needed, one may visualize the case $p = 2$ graphically. For any two points $\mathbf{x}_1$ and $\mathbf{x}_2 \in \mathbb{R}^2$, there is always a line that contains the two points and their average, that is, the two points span a one-dimensional linear manifold, and there is no variability in the direction orthogonal to the line. Similarly, for up to p points in $\mathbb{R}^p$, there is always a plane or hyperplane of dimension at most $p-1$ that contains all points and their average. Then we may define a coordinate system with origin at $\bar{\mathbf{x}}$. At most, $p-1$ basis vectors are needed to span the manifold that contains all data points, and there is no variability in directions orthogonal to the manifold. This theme will be treated in some detail in Chapter 8.

At the beginning of this section, we introduced plug-in estimators as parameters of the empirical distribution F_N which puts weight $1/N$ on each of N observed data points. Now, we will use the same concept to introduce a computer-intensive method for estimating the variability of sample statistics, the nonparametric *bootstrap*. Actually, the bootstrap methodology goes far beyond computation of standard errors of parameter estimates, but standard errors will be its main application in this book, justifying the relatively narrow focus. For a general introduction, see the book by Efron and Tibshirani (1993).

Suppose that we wish to study the distribution of a statistic T defined for samples of size N from some distribution F, that is, we would like to find the distribution of $T = T(\mathbf{X}_1, \ldots, \mathbf{X}_N)$, where $\mathbf{X}_1, \ldots, \mathbf{X}_N$ are *iid* random variables with distribution F. Since F is unknown (or at least some parameter determining F is unknown), finding the sampling distribution of T is not possible. In classical theory, we assume (temporarily) that F is known, so we can calculate the sampling distribution of T. Then we declare the distribution again as unknown and substitute parameter estimates for parameters to calculate standard errors. This approach has been illustrated in Example 4.1.2. The bootstrap chooses a different way: It studies the *exact* distribution of T when sampling from the empirical distribution F_N, assuming that this will give a reasonable approximation to the distribution of T when sampling from the unknown F. This may be viewed again as an application of the plug-in principle: The empirical distribution F_N is substituted for the unknown distribution F.

In the "bootstrap world," F_N is the model — a discrete distribution that puts probability $1/N$ on each of the observed data points $\mathbf{x}_1, \ldots, \mathbf{x}_N$. The distribution of the statistic T for samples of size N from F_N is called the *bootstrap distribution* of T. Since F_N is a discrete distribution, taking only N values, the bootstrap distribution of T can be computed exactly (at least in theory) by exhaustive calculation. We illustrate this with an example.

Example 4.2.1 For reasons that will be clear quite soon, this example uses a very small data set: the midge data from Table 1.1 and only the $N = 6$ observations from group *Apf*. For numerical convenience, the data were multiplied by 100, i.e., results are for variables measured in mm/100. First, consider a univariate case, using variable "antenna length." The empirical distribution is given by a random variable Y which takes values $x_1 = 114$, $x_2 = 120$, $x_3 = 118$, $x_4 = 130$, $x_5 = 126$, and $x_6 = 128$ with probability $1/6$ each. Suppose that we are interested in the sampling distribution of $\bar{\mathbf{x}}$. Let T denote the function that assigns the average to a list of six numbers. Then the distribution of the statistic T for random samples of size 6 taken from Y is the bootstrap distribution of $\bar{x}$. But Y has a discrete distribution, with $\Pr[Y = y] = 1/6$ for $y \in \{x_1, \ldots, x_6\}$. If $Y_1, \ldots, Y_6$ is a sample of size 6 from Y, then

$$\Pr[Y_1 = y_1, \ldots, Y_6 = y_6] = \begin{cases} 1/6^6 & \text{if all } y_i \in \{x_1, \ldots, x_6\}, \\ 0 & \text{otherwise.} \end{cases}$$

There are $6^6 = 46,656$ samples of size 6 from this distribution, called the bootstrap samples, which we can enumerate systematically. One of the possible samples will have all Y_j equal to x_1. Another possible sample will be $Y_1 = x_1$, $Y_2 = x_2, \ldots,$ $Y_6 = x_6$, i.e., the original sample. Many other samples will contain all six data points but in different order. Most of the samples, however, will contain at least one of the data points more than once, and at least one other data point will not occur in the sample. Each of the 6^6 samples has the same probability $1/6^6$, and we can compute the value of T for each of them. This is illustrated in Table 4.2.1. Let t_b denote the value of T calculated from the bth bootstrap sample, $b = 1, \ldots, 6^6$. Then the distribution, which puts probability $1/6^6$ on each of the t_b, is the bootstrap distribution of T (in our case, of the sample mean of antenna length of *Apf* midges). This distribution is shown as a histogram in Figure 4.2.1(a). Many of the t_b in the list of 6^6 values are identical, and we could count exact frequencies for all the different values, but there is no need here to introduce more formalism.

Table 4.2.1 Bootstrap computations in Example 4.2.1. The table displays details of the first four, the last four, and nine arbitrarily chosen intermediate bootstrap samples in a systematic generation of all 6^6 bootstrap samples. For each bootstrap sample, the indexes of the data points included in the sample are listed. For instance, in bootstrap sample # 17171, $Y_1 = x_3$, $Y_2 = x_2$, $Y_3 = x_2$, $Y_4 = x_3$, $Y_5 = x_6$, $Y_6 = x_5$. "Average" is the mean of the first variable (antenna length). "Corr" is the correlation between the two variables. If at least one of the standard deviations is zero, the correlation is set equal to 0.

Bootstrap Sample #	Indexes of Items in Sample						Average	Corr.
1	1	1	1	1	1	1	114.000	0.000
2	1	1	1	1	1	2	115.000	1.000
3	1	1	1	1	1	3	114.667	1.000
4	1	1	1	1	1	4	116.667	1.000
⋮								
17171	3	2	2	3	6	5	121.667	0.551
17172	3	2	2	3	6	6	122.000	0.577
17173	3	2	2	4	1	1	119.333	0.752
17174	3	2	2	4	1	2	120.333	0.678
17175	3	2	2	4	1	3	120.000	0.535
17176	3	2	2	4	1	4	122.000	0.744
17177	3	2	2	4	1	5	121.333	0.740
17178	3	2	2	4	1	6	121.667	0.771
17179	3	2	2	4	2	1	120.333	0.678
⋮								
46653	6	6	6	6	6	3	126.333	1.000
46654	6	6	6	6	6	4	128.333	−1.000
46655	6	6	6	6	6	5	127.667	0.000
46656	6	6	6	6	6	6	128.000	0.000

Similarly we can generate a bootstrap distribution for more complicated statistics, such as a correlation coefficient. We illustrate this again using the data of the *Apf* midges. Let $\mathbf{Y} = (Y_1, Y_2)'$ denote a bivariate random variable that takes values $\mathbf{x}_1 = (114,\ 178)', \ldots, \mathbf{x}_6 = (128,\ 200)'$ with probability 1/6 each. Again, there are 6^6 different bootstrap samples, each of them now consisting of six data pairs. For each of the bootstrap samples, we can calculate the value of the correlation coefficient, and thus generate a distribution of the sample correlation when sampling from $\mathbf{Y}$. There is a little difficulty in this case: For some bootstrap samples, at least one of the standard deviations is zero (see Exercise 18), and therefore, the correlation coefficient, strictly speaking, is not defined. Since the proportion of such samples is relatively small, we simply set the correlation coefficient equal to 0 in these cases. For some other bootstrap samples, the correlation coefficient is ± 1; see Exercise 19.

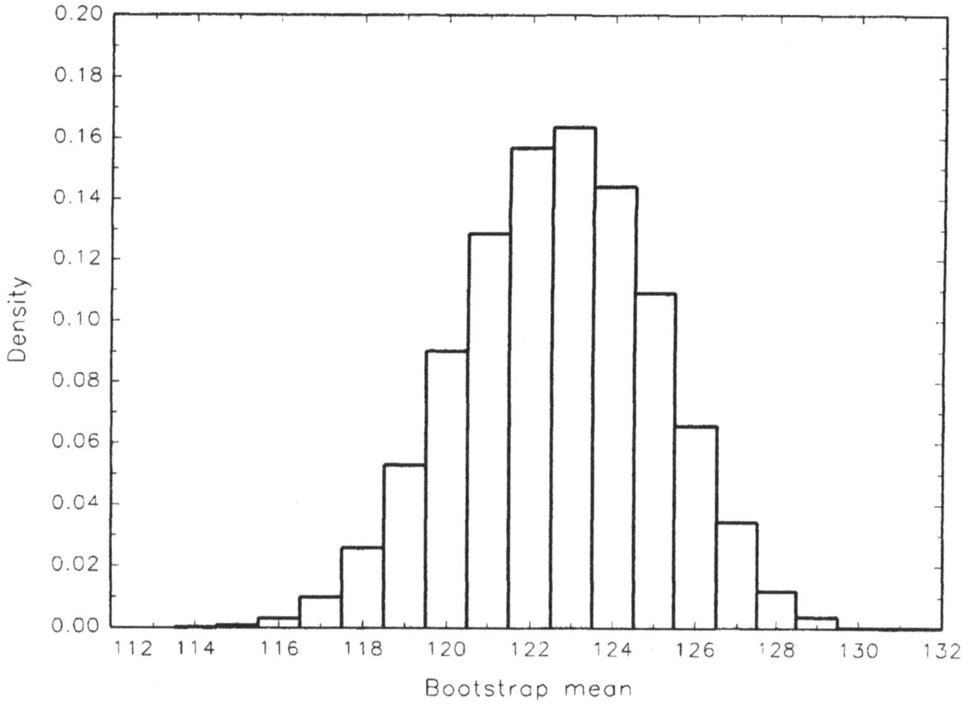

Figure 4-2-1A

The bootstrap distribution of the correlation coefficient is shown as a histogram in Figure 4.2.1(b). □

Example 4.2.1 illustrates the computation of an exact bootstrap distribution by exhaustive enumeration. For an analysis based on a sample with N observations, there will be N^N different bootstrap samples, a number which is typically far larger than we would be able to handle with even the fastest computers. The good news is that there is really no need to consider all N^N possible bootstrap samples. Instead, we can use a computer to randomly generate a large number B of bootstrap samples. Recommendations vary on how to choose B. We will usually choose $B = 10^3$ or $B = 10^4$, which is larger than the minimal number of bootstrap samples recommended by Efron and Tibshirani (1993). A convenient way of generating a bootstrap sample is to choose, with replacement, N "items" from the list $\mathbf{x}_1, \ldots, \mathbf{x}_N$. The fact that sampling is *with* replacement is sometimes confusing to the novice; however, this is exactly what sampling from the empirical distribution F_N means and is relatively easy to implement on a computer. For a large number B of bootstrap samples, thus, we get a good approximation to the exact bootstrap distribution.

While all of the bootstrap methodology explained so far sounds like doing lots of computation for no apparent benefit, the theory of the bootstrap shows that, in

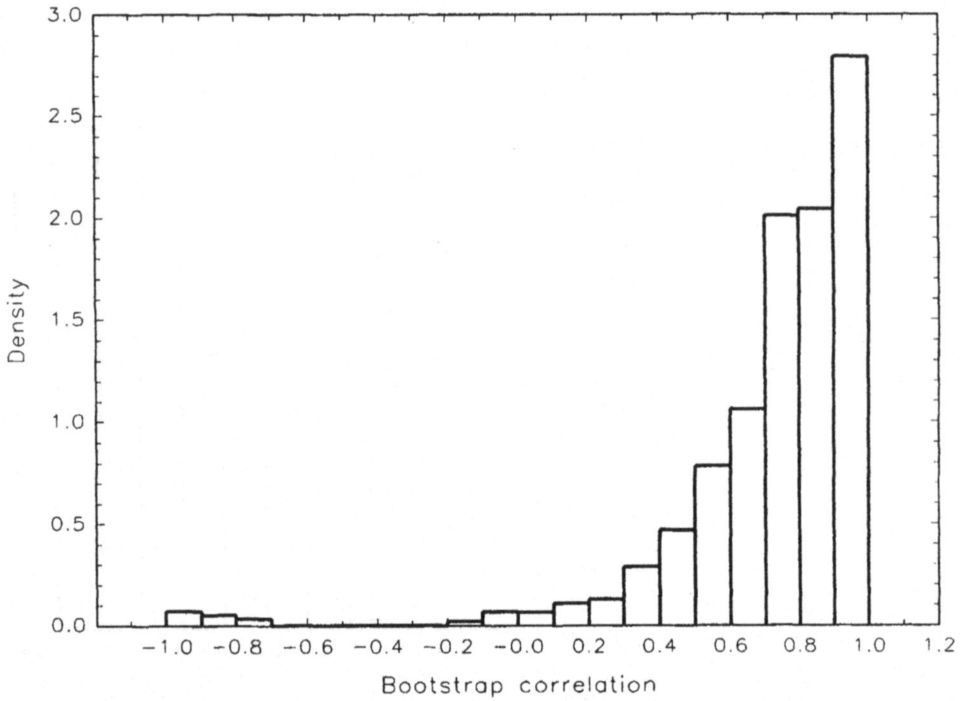

Figure 4-2-1B Exact bootstrap distributions in Example 4.2.1, antenna length and wing length of six *Apf* midges: (a) bootstrap distribution of average antenna length; (b) bootstrap distribution of the correlation between antenna length and wing length.

many situations, the bootstrap distribution of a statistic T is, indeed, a good "guess" at the sampling distribution of T when sampling from the unknown F. In particular, the standard deviation of the bootstrap distribution may serve as a standard error of T, obtained without making specific parametric assumptions (like normality) on the distribution F, that is, if we define a random variable $\mathbf{Y}$ by the plug-in principle, as illustrated in Example 4.2.1 and compute the distribution of a statistic T either exactly or approximately using B random samples from $\mathbf{Y}$, then the standard deviation of T under sampling from $\mathbf{Y}$ can be used as an estimate of the standard deviation of T under sampling from the unknown distribution F. That is, we can use the standard deviation of T calculated from the bootstrap distribution as a standard error of the corresponding parameter estimate. If we wish to construct confidence intervals for an unknown parameter, then we can also use quantiles of the bootstrap distribution directly, to get nonparametric bootstrap confidence intervals. We will use this approach repeatedly in Chapters 8 and 9; for now, we conclude with a simple example. The reader who is interested in more details is, once again, referred to the book by Efron and Tibshirani (1993). In "standard" cases where classical theory is applicable, the classical approach and the bootstrap approach will give similar answers. However, if some distributional assumptions made in the classical approach are violated or if no appropriate theory is available, the bootstrap may be the only solution to an otherwise intractable problem.

Example 4.2.2 continuation of Example 4.2.1

Antenna length and wing length of *Apf* midges. First, consider the bootstrap distribution of the mean of antenna length, shown in Figure 4.2.1(a). The standard deviation of this distribution is approximately 2.34, and therefore, the bootstrap standard error of $\bar{x}$ is $\text{se}(\bar{x}) = 2.34$. A nonparametric 95 percent bootstrap confidence interval for μ, the (theoretical) mean antenna length of *Apf* midges, is obtained by finding the 2.5 percent quantile and the 97.5 percent quantile of the bootstrap distribution; these are 118.0 and 127.0, respectively. This confidence interval is quite close to the one obtained from the classical theory based on the t-distribution; see Exercise 20. Actually, there is little need to use the bootstrap machinery for inference on means because (due to the central limit theorem) classical normal theory works quite well. This is also reflected by the fact that the bootstrap distribution in Figure 4.2.1(a) looks quite normal. The bootstrap distribution of the correlation between antenna length and wing length shown in Figure 4.2.1(b) has a mean of 0.7282 and a standard deviation of 0.2905; the latter quantity can, again, serve as a standard error of the observed correlation between antenna length and wing length. See Exercise 21 for more details. The bootstrap distribution of the correlation coefficient is highly asymmetric. As in the case of the mean, we may construct a nonparametric bootstrap confidence interval for the correlation between antenna length and wing length by taking appropriate quantiles from the bootstrap distribution; for instance, the 2.5 percent quantile and the 97.5 percent quantile of the bootstrap distribution are 0.034 and 0.997, respectively. Thus, the 95 percent bootstrap confidence interval just barely excludes a correlation of zero between the two variables. □

Exercises for Section 4.2

1. This exercise deals with sample correlations.
   (a) Prove that the sample correlation coefficient is less or equal to 1 in absolute value. (If this takes you more than 30 seconds, you have not read this section carefully enough).
   (b) Show that the sample covariance matrix is the plug-in estimate of the model correlation matrix. (Assume that all sample variances are positive).

2. For a univariate random variable X with distribution function F and for some fixed $\alpha \in (0, 1)$, let θ be the α-percentile, i.e., $F(\theta) = \alpha$. Find the plug-in estimator of θ based on a sample of size N. Is it necessarily unique?

3. Find the bias of $\mathbf{S}_P$ as an estimator of the model covariance matrix ψ. *Note*: You need to expand the definition of bias to random matrices.

4. Show that the plug-in estimator $\mathbf{S}_P$ can be written as $\mathbf{S}_P = \frac{1}{N}\sum_{i=1}^{N} \mathbf{X}_i\mathbf{X}_i' - \bar{\mathbf{X}}\bar{\mathbf{X}}'$. Use this formula to prove unbiasedness of $\mathbf{S}$.

5. Let $\mathbf{X}_1, \ldots, \mathbf{X}_N$ denote a sample from $\mathcal{N}_p(\boldsymbol{\mu}, \psi)$. Show that the joint distribution of $\bar{\mathbf{X}}$ and the N deviation vectors $\mathbf{X}_i - \bar{\mathbf{X}}$ is $(N+1)p$-dimensional normal. *Hint*: Show that

$$\begin{pmatrix} \bar{\mathbf{X}} \\ \mathbf{X}_1 - \bar{\mathbf{X}} \\ \vdots \\ \mathbf{X}_N - \bar{\mathbf{X}} \end{pmatrix} = \mathbf{A} \begin{pmatrix} \mathbf{X}_1 \\ \vdots \\ \mathbf{X}_N \end{pmatrix}$$

for some matrix $\mathbf{A}$. Find $\mathbf{A}$.

6. Let $\mathbf{x}_1, \ldots, \mathbf{x}_N$ denote observed data vectors in $\mathbb{R}^p$, with sample mean $\bar{\mathbf{x}}$ and sample covariance matrix $\mathbf{S}_\mathbf{x}$, respectively. For a fixed matrix $\mathbf{A}$ of dimension $q \times p$ and a vector $\mathbf{b} \in \mathbb{R}^q$, transform the data by

$$\mathbf{y}_i = \mathbf{A}\mathbf{x}_i + \mathbf{b} \qquad i = 1, \ldots, N.$$

Show that the sample mean vector $\bar{\mathbf{y}}$ and sample covariance matrix $\mathbf{S}_\mathbf{y}$ of the $\mathbf{y}$-data are given by

$$\bar{\mathbf{y}} = \mathbf{A}\bar{\mathbf{x}} + \mathbf{b}$$

and

$$\mathbf{S}_\mathbf{y} = \mathbf{A}\mathbf{S}_\mathbf{x}\mathbf{A}',$$

respectively. *Hint*: If $\mathbf{D}_\mathbf{x}$ and $\mathbf{D}_\mathbf{y}$ are the data matrices of the $\mathbf{x}$-data and the $\mathbf{y}$-data, respectively. Show that $\mathbf{D}_\mathbf{y} = \mathbf{D}_\mathbf{x}\mathbf{A}' + \mathbf{1}_N'\mathbf{b}$. Alternatively, use the principle of plug-in estimation.

7. Let $X_1, \ldots, X_N$ denote a sample from a univariate random variable with $\text{var}[X] = \sigma^2$, and assume $E[X^4] < \infty$. Show that the sample variance is a consistent estimator of σ^2.

8. Let $\mathbf{X}_1, \ldots, \mathbf{X}_N$ be a sample of size $N > p$ from a continuous p-variate random vector $\mathbf{X}$ with density function $f(\mathbf{x})$, and let $\mathbf{S}$ be the sample covariance matrix. Show that $\mathbf{S}$ is positive definite with probability 1. *Hint*: Show that the probability of the $(p+1)$ *iid* random vectors $\mathbf{X}_1, \ldots, \mathbf{X}_{p+1}$ spanning a linear manifold of dimension less than p is zero.

9. Let $\mathbf{X}_{h1}, \ldots, \mathbf{X}_{hN_h}$, $h = 1, \ldots, k$, denote independent samples of size $N_1, \ldots, N_k$ from k p-variate distributions with mean vectors $\boldsymbol{\mu}_1, \ldots, \boldsymbol{\mu}_k$, respectively, and with common covariance matrix ψ. Let $\bar{\mathbf{X}}_h$ and $\mathbf{S}_h$ denote the sample mean vector and covariance matrix from the hth sample. Show that

$$\mathbf{S}_{\text{pooled}} = \frac{1}{\sum_{h=1}^{k} N_h - k} \sum_{h=1}^{k} (N_h - 1)\mathbf{S}_h$$

is an unbiased estimator of ψ.

10. This exercise refers to the pooled sample covariance matrix of two groups as defined in equation (15). Let $\mathbf{D}_1$ denote the *centered* data matrix of the first sample, and $\mathbf{D}_2$ the centered data matrix of the second sample. Show that $\mathbf{S}_{\text{pooled}}$ is positive definite exactly if the combined matrix $\mathbf{D} = \begin{pmatrix} \mathbf{D}_1 \\ \mathbf{D}_2 \end{pmatrix}$ of dimension $(N_1 + N_2) \times p$ contains at least p linearly independent rows. Argue that this requires $N_1 + N_2 \geq p + 2$. Then generalize this exercise to an arbitrary number k of samples. *Hint*: write $\mathbf{S}_{\text{pooled}}$ in terms of the combined matrix $\mathbf{D}$. This allows you, at the same time, to define the pooled sample covariance matrix even in cases where one of the sample sizes is 1.

11. Let $\mathbf{X}_1, \ldots, \mathbf{X}_N$ be a sample from a discrete p-variate distribution, and $\mathbf{S}$ the sample covariance matrix. Show that, for any finite N, the probability of $\mathbf{S}$ being positive definite is less than 1.

12. Let $\mathbf{X}$ be a bivariate random vector with a uniform distribution inside a circle of radius 1 centered at some unknown $\boldsymbol{\theta} \in \mathbb{R}^2$. Show that $E[\mathbf{X}] = \boldsymbol{\theta}$. Suppose that $\mathbf{x}_1 = \begin{pmatrix} 0.9 \\ 0 \end{pmatrix}$, $\mathbf{x}_2 = \begin{pmatrix} 0.7 \\ 0.7 \end{pmatrix}$, $\mathbf{x}_3 = \begin{pmatrix} 0.7 \\ -0.7 \end{pmatrix}$ and $\mathbf{x}_4 = \begin{pmatrix} -0.9 \\ 0 \end{pmatrix}$ is an observed sample from this distribution. Compute the plug-in estimate $\bar{\mathbf{x}}$ for this sample, and argue that this is a bad estimate of $\boldsymbol{\theta}$.

13. Let $\mathbf{S}_1$ and $\mathbf{S}_2$ denote the sample covariance matrices obtained from independent samples of size N_1 and N_2 from two continuous p-variate distributions. Assume that $N_1 \geq 2$ and $N_2 \geq 2$. Show that the pooled-sample covariance matrix is positive definite exactly if $N_1 + N_2 \geq p + 2$. *Hint*: Use Exercises 8 and 10.

14. Let $\mathbf{1}_N$ denote a vector in $\mathbb{R}^N$ with 1 in each position, and define an $N \times N$ matrix $\mathbf{H} = \mathbf{I}_N - \frac{1}{N}\mathbf{1}_N\mathbf{1}_N'$.

(a) Show that $\mathbf{H}$ is symmetric and idempotent (i.e., $\mathbf{H}^2 = \mathbf{H}$). Also show that $\text{rank}(\mathbf{H}) = N - 1$.

(b) Show that the centered data matrix $\mathbf{D}_c$ is related to the data matrix $\mathbf{D}$ by $\mathbf{D}_c = \mathbf{HD}$. (This explains why $\mathbf{H}$ is sometimes called the *centering matrix*).

(c) Show that the sample covariance matrix can be computed as $\mathbf{S} = \frac{1}{N-1}\mathbf{D}'\mathbf{HD}$.

(d) Using parts (a) to (c), show that $\text{rank}(\mathbf{S}) \leq \min(p, N - 1)$.

15. This example shows that the same parameter may be estimated by different plug-in estimators, depending on how the parameter is defined as a function of the *cdf* F. Let X be a random variable that takes values $\theta, \theta + 1$, and $\theta + 2$ with probability 1/3 each, where θ is an unknown integer. Let $X_1, \ldots, X_N$ be a sample from this distribution.

(a) If θ is defined as $E[X] - 1$, what is the plug-in estimator of θ?

(b) If θ is defined as $\inf\{x \in \mathbb{R} : F(x) > 0\}$, what is the plug-in estimator of θ?

16. This exercise concerns the midge data of Example 1.1; see Table 1.1 (p. 1-1).

(a) Compute mean vectors, covariance matrices, and correlation matrices separately for the two species. Then compute a pooled-sample covariance matrix. (These results will be used in Chapter 5).

(b) With X_1 denoting antenna length and X_2 denoting wing length, define two new variables $Y_1 = X_1 + X_2$ (sum) and $Y_2 = X_2 - X_1$ (difference). Repeat part (a) of this exercise by transforming the raw data to sum and difference. Then confirm the numerical results using Exercise 6.

17. This exercise is based on the head dimension data; see Example 1.2 and Table 1.2 (p. 8). Repeat Exercise 3 of Chapter 1, using the matrix transformation of Exercise 6 above to find the covariance matrix and correlation matrix of the two variables U and V. Also compute the mean vector, covariance matrix, and correlation matrix of all six variables in Table 1.2 (p. 8); these results will be used in Chapter 8.

18. In Example 4.2.1 show that, in exactly $2^7 + 2 = 130$ of the 6^6 bootstrap samples, the correlation coefficient is not defined because at least one of the standard deviations is zero.

19. In how many of the 6^6 bootstrap samples in Example 4.2.1 is the correlation $+1$? In how many samples is it -1?

20. In Example 4.2.1, compute a standard error of the mean of *antenna length* using the familiar formula $\text{se}(\bar{x}) = s/\sqrt{N}$, where s is the sample standard deviation. Compare this to the bootstrap standard error of $\bar{x}$. Compute a parametric 95 percent confidence interval, using the formula $\bar{x} \pm t_{0.975}\,\text{se}(\bar{x})$, where t_α is the α-quantile of the t-distribution with $N-1$ degrees of freedom. Compare this to the bootstrap confidence interval reported in Example 4.2.1, and comment.

21. Based on the assumption of bivariate normality, the standard error of a correlation coefficient r based on a (bivariate) sample of size N can be calculated as

$$\text{se}(r) = \frac{1 - r^2}{\sqrt{N - 3}};$$

see Efron and Tibshirani (1993, p. 54). Apply this formula to the midge data of Example 4.2.2 (p. 237), and compare the result to the bootstrap standard error. Would it be reasonable to give a symmetric confidence interval of the form $r \pm c \cdot \text{se}(r)$ for the correlation coefficient? Explain.

22. Let $\mathbf{x}_1, \ldots, \mathbf{x}_N$ denote observed data, and let $\mathbf{Y}_1, \ldots, \mathbf{Y}_N$ be a random sample from the empirical distribution of the $\mathbf{x}_i$ (i.e., a bootstrap sample). Set $\mathbf{T} = (\mathbf{Y}_1 + \ldots + \mathbf{Y}_N)/N$. Show that the average of $\mathbf{T}$ over all N^N bootstrap samples is $\bar{\mathbf{x}}$.

23. The following is an algorithm for generating a bootstrap sample on a computer, using observed data $\mathbf{x}_1, \ldots, \mathbf{x}_N$.

- Generate N independent random numbers $u_1, \ldots, u_N$, with a uniform distribution in the interval $[0, 1)$.
- Set $v_i = \lfloor Nu_i + 1 \rfloor$ for $i = 1, \ldots, N$. Here, $\lfloor z \rfloor$ is the floor function of z, i.e., the largest integer less or equal to z.
- Set $\mathbf{Y}_1 = \mathbf{x}_{v_1}, \ldots, \mathbf{Y}_N = \mathbf{x}_{v_N}$, that is, choose the "items" with labels $v_1, \ldots, v_N$ from the list $\mathbf{x}_1, \ldots, \mathbf{x}_N$.

Show that this algorithm produces a random sample from the empirical distribution function F_N. Implement it on a computer, using a programming language of your choice.

24. Repeat the analysis of Example 4.2.1 using the data of the *Af* midges. In particular, graph the bootstrap distributions of the average of one of the variables and the correlation between the two variables. Find a 95 percent bootstrap confidence interval for the correlation between *antenna length* and *wing length* of *Af* midges, following the technique used in Example 4.2.1. *Note*: If you are not patient enough to wait until all 9^9 bootstrap samples have been generated by the computer, use the algorithm of Exercise 23 to generate $B = 10,000$ random bootstrap samples.

25. Let $X_1, \ldots, X_N$ be a sample from some distribution F with mean μ and variance σ^2. If F is a normal distribution, then Theorem 4.2.4 says that the sample mean and the sample variance are independent. In particular, this means that $E[S^2|\bar{X}] = \sigma^2$. Now, suppose that F is a Poisson distribution with parameter λ. Show that $E[S^2|\bar{X}] = \bar{X}$. *Hint*: Use Exercise 14 in Section 2.9.

4.3 Maximum Likelihood Estimation

In this section we explore the most popular and most useful approach to parametric estimation, called maximum likelihood. Its popularity is due to various reasons. First, in many situations the method of maximum likelihood allows us to find estimators where other approaches (such as the plug-in principle) fail. In fact, in areas such as generalized linear models, of which the logistic regression model treated in Sections 7.5 and 7.6 is a special case, it is essentially the only method of estimation used. Second, in many situations where different estimators exist, the maximum likelihood estimator has better properties than its competitors. An illustration is given in Exercise 1. Third, in many standard situations, the theory of maximum likelihood estimation gives an approximation to the distribution of the estimator, thus, providing us with tools for the construction of confidence regions and tests. Last but not least, the principle of maximum likelihood has considerable intuitive appeal, as we will see shortly: among all models considered in an estimation problem, it singles out the "most plausible" one, or the one for which the observed data gives most support.

The introduction to maximum likelihood given in this section focuses on multivariate applications. For more background the interested student is referred to the books by Edwards (1984) and Kalbfleisch (1985).

Let $\mathcal{X}$ denote a random variable, vector, or matrix, with *pdf* $g(\mathbf{x}; \boldsymbol{\theta})$ depending on an unknown parameter $\boldsymbol{\theta}$. As before, $\boldsymbol{\theta}$ may be a scalar, a vector, a matrix, or a combination thereof, and $\boldsymbol{\theta}$ is assumed to be in a set Ω called the parameter space. In the current setup, $\mathcal{X}$ is not necessarily a random sample from a given distribution, although in many applications this will be the case. All we assume for now is that $\mathcal{X}$ has a *pdf*, which may be a probability function, a density function, or even a combination of the two, as introduced in Section 2.8.

Definition 4.3.1 Let $g(\mathbf{x}; \boldsymbol{\theta})$ denote the *pdf* of $\mathcal{X}$, depending on an unknown parameter $\boldsymbol{\theta} \in \boldsymbol{\Omega}$. Then for every $\mathbf{x}$ in the sample space of $\mathcal{X}$, the real-valued function

$$L(\boldsymbol{\theta}; \mathbf{x}) = g(\mathbf{x}; \boldsymbol{\theta}) \qquad \text{for } \boldsymbol{\theta} \in \boldsymbol{\Omega} \tag{1}$$

is called the *likelihood function* of $\boldsymbol{\theta}$. □

The likelihood function marks the transition from probability to statistics mentioned in the introductory paragraph of Section 4.1. In probability theory, we study the chances of events $\mathbf{x}$ in the sample space, considering the parameter $\boldsymbol{\theta}$ as fixed. In statistics, we assess how likely or how plausible various values of the parameter $\boldsymbol{\theta}$ are in view of the observed data $\mathbf{x}$. Since probability functions and density functions take only nonnegative values, the same holds true for likelihood functions; however, it is, in general, not correct to view a likelihood function as the *pdf* of a random variable $\boldsymbol{\theta}$ with sample space $\boldsymbol{\Omega}$. This means, in particular, that values of the likelihood function should not be interpreted as probabilities, even in cases where $\boldsymbol{\theta}$ can take only finitely many values.

Example 4.3.1 Let $\mathcal{X} = (X_1, \ldots, X_k)$ have a joint multinomial distribution with *pdf*

$$g(x_1, \ldots, x_k) = \Pr(X_1 = x_1, \ldots, X_k = x_k) = \binom{N}{x_1\ x_2\ \ldots\ x_k} \theta_1^{x_1} \theta_2^{x_2} \cdots \theta_k^{x_k}$$

for $x_1, x_2, \ldots, x_k = 0, 1, \ldots, N$, $\sum_{i=1}^k x_i = N$. Here, $\boldsymbol{\theta} = (\theta_1, \ldots, \theta_k)' \in \mathbb{R}^k$ is a probability vector, i.e., $0 \le \theta_i \le 1$ for all i, and $\sum_{j=1}^k \theta_i = 1$. This model has been studied before in Section 2.10.

For the sake of illustration, we consider the case $k = 3$. Substituting $\theta_3 = 1 - (\theta_1 + \theta_2)$, the parameter space is $\boldsymbol{\Omega} = \{(\theta_1, \theta_2) : \theta_1 \ge 0, \theta_2 \ge 0, \theta_1 + \theta_2 \le 1\}$. For observed counts x_1, x_2, and x_3, the likelihood function is given by

$$L(\theta_1, \theta_2; x_1, x_2, x_3) = \binom{N}{x_1\ x_2\ x_3} \theta_1^{x_1} \theta_2^{x_2} (1 - \theta_1 - \theta_2)^{x_3}. \tag{2}$$

Figure 4.3.1 shows a contour plot of this likelihood function for $N = 10$ and $(x_1, x_2, x_3) = (3, 4, 3)$. The parameter space corresponds to the lower left triangle of the graph, in which the likelihood function rises as a mountain with a peak at $(\theta_1, \theta_2) = (0.3, 0.4)$. □

Example 4.3.2 Suppose that $\mathcal{X} = (\mathbf{X}_1, \ldots, \mathbf{X}_N)$ is a random sample from the uniform distribution inside a circle of radius 1, centered at an unknown location $\boldsymbol{\theta} \in \mathbb{R}^2$. Thus, each of the N random variables has *pdf*

$$f(\mathbf{x}_i; \boldsymbol{\theta}) = \begin{cases} \dfrac{1}{\pi} & \text{if } \|\mathbf{x}_i - \boldsymbol{\theta}\| \le 1, \\ 0 & \text{otherwise.} \end{cases} \tag{3}$$

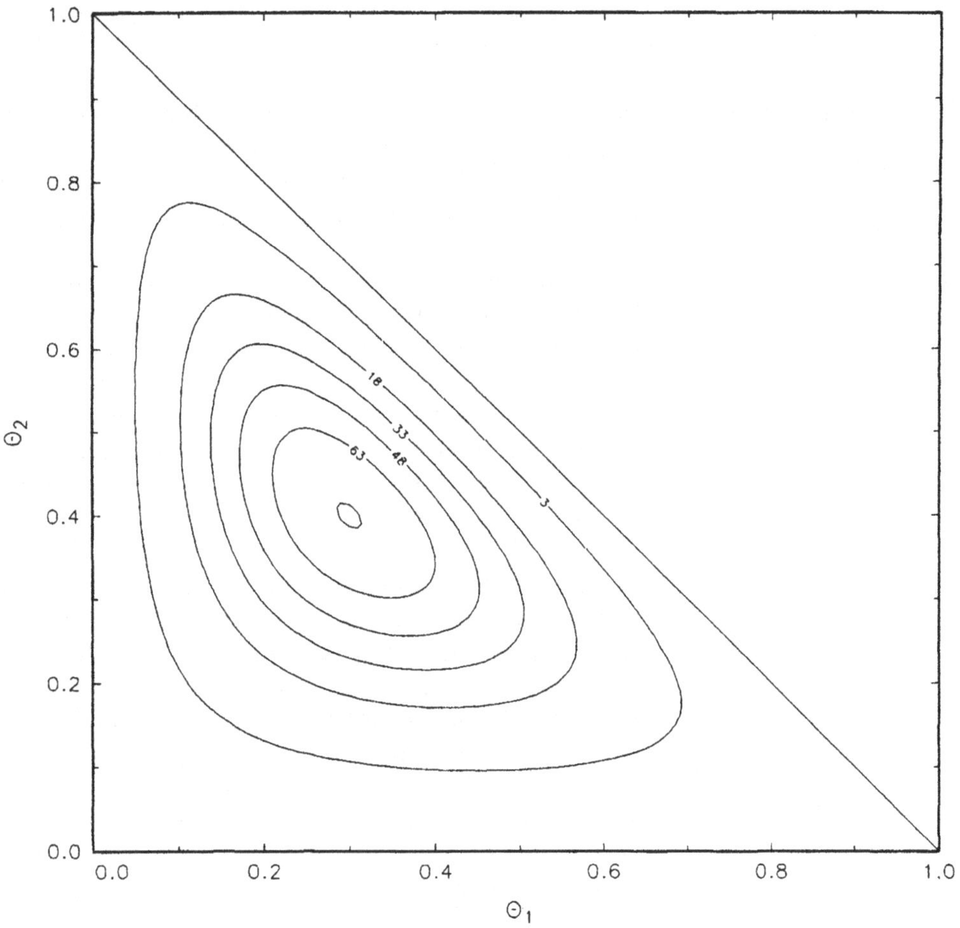

Figure 4-3-1
The multinomial likelihood function of Example 4.3.1.

The likelihood function of $\boldsymbol{\theta}$, given fixed data vectors $\mathbf{x}_1, \ldots, \mathbf{x}_N$ is the product of the N density functions (3), taking the value π^{-N} if all $\|\mathbf{x}_i - \boldsymbol{\theta}\| \leq 1$, and zero otherwise. The condition $\|\mathbf{x}_i - \boldsymbol{\theta}\| \leq 1$ for all i is satisfied exactly if

$$\max_{1 \leq i \leq N} \|\mathbf{x}_i - \boldsymbol{\theta}\| \leq 1.$$

Thus, the likelihood function $L(\boldsymbol{\theta}; \mathbf{x}_1, \ldots, \mathbf{x}_N)$ is rather strange. It takes the constant value π^{-N} for all $\boldsymbol{\theta}$ such that a circle of unit radius centered at $\boldsymbol{\theta}$ covers all N points and is zero outside this region. This means that, in view of the observed sample, any circle that covers all N points is equally plausible. See Exercise 3 for a graphical illustration. □

Example 4.3.2 illustrates, rather impressively, the idea of looking at the same function either as a *pdf* or as a likelihood function. Students who see this type of example for the first time tend to have some difficulty accepting the conclusions, but the logic is impeccable.

Definition 4.3.2 In the setup of Definition 4.3.1, for each $\mathbf{x}$ in the sample space of $\mathcal{X}$, let $\hat{\boldsymbol{\theta}}(\mathbf{x})$ denote a parameter value for which the likelihood function $L(\boldsymbol{\theta}; \mathbf{x})$ takes its maximum in Ω as a function of $\boldsymbol{\theta}$. Then $\hat{\boldsymbol{\theta}}(\mathbf{x})$ is called a maximum likelihood estimate of $\hat{\boldsymbol{\theta}}$ for the given data, and $\hat{\boldsymbol{\theta}}(\mathcal{X})$ is called a maximum likelihood estimator of $\boldsymbol{\theta}$. □

For discrete distributions, the principle of maximum likelihood estimation may be formulated in terms of probabilities: The observed data vectors $\mathbf{x}_1, \ldots, \mathbf{x}_N$ form an event $\mathcal{E}$ in the sample space, and, for each value of $\boldsymbol{\theta} \in \Omega$, we can compute $L(\boldsymbol{\theta}; \mathbf{x}_1, \ldots, \mathbf{x}_N) = \Pr[\mathcal{E}; \boldsymbol{\theta}]$. Maximum likelihood then tells us to "guess" the parameter value as the $\boldsymbol{\theta} \in \Omega$ for which the event $\mathcal{E}$ has the highest probability. In the continuous case this interpretation fails because the probability of each point $(\mathbf{x}_1, \ldots, \mathbf{x}_N)$ in the sample space is zero; however, a similar interpretation holds, as is shown in Exercise 4.

Example 4.3.3 In the numerical setup of Example 4.3.1, we have already seen (although not proven) that the maximum likelihood estimate of $\boldsymbol{\theta} = (\theta_1, \theta_2, \theta_3)'$ for observed data $\mathbf{x} = (x_1, x_2, x_3) = (4, 3, 4)$ is $\hat{\boldsymbol{\theta}} = (0.4, 0.3, 0.4)'$. Now, we are going to study maximum likelihood estimation for the multinomial distribution in general. In the setup given at the beginning of Example 4.3.1, the likelihood function for observed counts $(x_1, \ldots, x_k)$ with $x_1 + \cdots + x_k = N$ is given by

$$L(\boldsymbol{\theta}; x_1, \ldots, x_k) = \binom{N}{x_1 \;\cdots\; x_k} \theta_1^{x_1} \theta_2^{x_2} \cdots \theta_k^{x_k}, \tag{4}$$

for $\boldsymbol{\theta}$ in a parameter space

$$\Omega = \left\{ \boldsymbol{\theta} = (\theta_1, \ldots, \theta_k)' \in \mathbb{R}^k : \text{ all } \theta_i \geq 0,\ \theta_1 + \cdots + \theta_k = 1 \right\}.$$

Thus we need to maximize (4) over $\boldsymbol{\theta}$ in Ω. Since the likelihood function takes only nonnegative values and since the logarithm is a monotonically increasing function, we can equivalently find the value of $\boldsymbol{\theta}$ for which

$$\begin{aligned} \ell\, \boldsymbol{\theta}) &:= \log L(\boldsymbol{\theta}; x_1, \ldots, x_k) \\ &= C + x_1 \log \theta_1 + \cdots + x_k \log \theta_k \end{aligned} \tag{5}$$

takes its maximum in Ω. In (5), we have written C for $\log \binom{N}{x_1 \cdots x_k}$, because this term does not depend on the parameter $\boldsymbol{\theta}$ and therefore is a constant for the purpose of the maximization problem.

Now, we can use the standard tools for maximization: Take partial derivatives with respect to the θ_i, and set them equal to zero to find candidates for the value at which the maximum is taken. There is an additional difficulty: the constraint $\theta_1 + \cdots + \theta_N = 1$. This constraint can be accomodated by using one of the θ_i, for instance, θ_k, as a function of the other ones, i.e., $\theta_k = 1 - (\theta_1 + \cdots + \theta_{k-1})$. This solution is the topic of Exercise 5. Alternatively, we may introduce a Lagrange multiplier λ for the constraint $\sum_{i=1}^k \theta_i - 1 = 0$ and, thus, find stationary points of the function

$$\begin{aligned} \ell^*(\boldsymbol{\theta}) &= \ell(\boldsymbol{\theta}) - \lambda \left(\sum_{i=1}^k \theta_i - 1 \right) \\ &= C + \sum_{i=1}^k x_i \log \theta_i - \lambda \left(\sum_{i=1}^k \theta_i - 1 \right). \end{aligned} \tag{6}$$

Assuming for the moment that all $x_i > 0$ and taking partial derivatives, we obtain

$$\frac{\partial \ell^*}{\partial \theta_h} = \frac{x_h}{\theta_h} - \lambda, \qquad h = 1, \ldots, k. \tag{7}$$

Setting these partial derivatives equal to zero gives

$$\theta_h \lambda = x_h, \qquad h = 1, \ldots, N. \tag{8}$$

Adding equations (8) for $h = 1, \ldots, k$ and using the constraint $\sum_{h=1}^k \theta_h = 1$, then gives

$$\lambda = \sum_{h=1}^k x_h = N,$$

and therefore,

$$\theta_h = \frac{x_h}{N}, \qquad h = 1, \ldots, k, \tag{9}$$

are our candidate values for the maximum likelihood estimates. Since (9) is the only candidate and is inside the parameter space, and since the likelihood function is zero on the boundary of Ω and positive inside, we conclude that the maximum likelihood estimate for the given data $(x_1, \ldots, x_N)$ is given by

$$\hat{\boldsymbol{\theta}} = \begin{pmatrix} \hat{\theta}_1 \\ \vdots \\ \hat{\theta}_k \end{pmatrix} = \begin{pmatrix} x_1/N \\ \vdots \\ x_k/N \end{pmatrix}. \tag{10}$$

Notice that we have implicitly assumed $x_i > 0$ for all $i = 1, \ldots, k$. Exercise 6 shows that the same solution (10) is obtained even if some of the x_i are zero. Thus, the maximum likelihood *estimator* is given by

$$\hat{\boldsymbol{\theta}} = \begin{pmatrix} \hat{\theta}_1 \\ \vdots \\ \hat{\theta}_k \end{pmatrix} = \begin{pmatrix} X_1/N \\ \vdots \\ X_k/N \end{pmatrix}. \tag{11}$$

As usual, we do not distinguish in notation between estimate and estimator, i.e., we write $\hat{\boldsymbol{\theta}}$ for both. □

As illustrated in Example 4.3.3, it is often more convenient to work with the logarithm of the likelihood function rather than with the likelihood function, because the logarithm turns the product into a sum, which is easier to handle for differentiation. There are also deeper reasons for using the logarithm, as we shall see later. In any case, the logarithmically transformed likelihood function is so common that we give it a special name, as in the following definition.

Definition 4.3.3 Let $L(\boldsymbol{\theta}; \mathbf{x})$ denote the likelihood function of $\boldsymbol{\theta}$ in an estimation problem, given observed data $\mathbf{x}$. Then

$$\ell(\boldsymbol{\theta}; \mathbf{x}) = \log L(\boldsymbol{\theta}; \mathbf{x}) \tag{12}$$

is called the *log-likelihood function* for the same estimation problem. □

See Exercises 8 and 9 for some applications. Whenever there is no danger of confusion, for simplicity, we will write $\ell(\boldsymbol{\theta})$ instead of $\ell(\boldsymbol{\theta}; \mathbf{x}_1, \ldots, \mathbf{x}_N)$. In standard applications of maximum likelihood where calculus is used to find the maximum, it is most often preferable to use the log-likelihood function. In other cases though, like the following example, it would not be helpful at all.

Example 4.3.4 continuation of Example 4.3.2

Again, suppose that $\mathbf{x}_1, \ldots, \mathbf{x}_N$ is an observed sample from the uniform distribution inside a circle of radius 1 centered at an unknown location $\boldsymbol{\theta} \in \mathbb{R}^2$. We have seen that the likelihood function is given by

$$L(\boldsymbol{\theta}; \mathbf{x}_1, \ldots, \mathbf{x}_N) = \begin{cases} \pi^{-N} & \text{if } \max\limits_{1 \le i \le N} \|\boldsymbol{\theta} - \mathbf{x}_i\| \le 1, \\ 0 & \text{otherwise,} \end{cases} \tag{13}$$

that is, every $\boldsymbol{\theta}$ which has a distance of 1, at most, from all observed data points has equal likelihood, and any $\boldsymbol{\theta}$ inside the set $\{\mathbf{x} \in \mathbb{R}^2 : \|\mathbf{x} - \mathbf{x}_i\| \le 1 \text{ for all } i = 1, \ldots, N\}$ is a maximum likelihood estimate of $\boldsymbol{\theta}$. For a sample $\mathbf{X}_1, \ldots, \mathbf{X}_N$, the centers of all circles of unit radius that cover all $\mathbf{X}_i$ are maximum likelihood estimators. Thus, the maximum likelihood estimate, typically, is not unique in this situation. Variations on Example 4.3.4 are given in Exercise 7. □

Now, we turn to a central topic of this section: maximum likelihood estimation for the multivariate normal distribution. Suppose that $\mathbf{X}_1, \ldots, \mathbf{X}_N$ is a sample from $\mathcal{N}_p(\boldsymbol{\mu}, \boldsymbol{\psi})$, where both $\boldsymbol{\mu}$ and $\boldsymbol{\psi}$ are unknown. The parameter space is

$$\boldsymbol{\Omega} = \{(\boldsymbol{\mu}, \boldsymbol{\psi}) : \boldsymbol{\mu} \in \mathbb{R}^p, \boldsymbol{\psi} \text{ positive definite symmetric}\}. \tag{14}$$

By Theorem 3.2.7, the likelihood function of $(\boldsymbol{\mu}, \boldsymbol{\psi})$, given N fixed vectors $\mathbf{x}_1, \ldots, \mathbf{x}_N \in \mathbb{R}^p$, is expressed by

$$
\begin{aligned}
L(\boldsymbol{\mu}, \boldsymbol{\psi}; \mathbf{x}_1, \ldots, \mathbf{x}_N) &= \prod_{i=1}^{N} (2\pi)^{-p/2} (\det \boldsymbol{\psi})^{-1/2} \exp\left[-\frac{1}{2}(\mathbf{x}_i - \boldsymbol{\mu})' \boldsymbol{\psi}^{-1} (\mathbf{x}_i - \boldsymbol{\mu})\right] \\
&= (2\pi)^{-Np/2} (\det \boldsymbol{\psi})^{-N/2} \exp\left[-\frac{1}{2} \sum_{i=1}^{N} (\mathbf{x}_i - \boldsymbol{\mu})' \boldsymbol{\psi}^{-1} (\mathbf{x}_i - \boldsymbol{\mu})\right].
\end{aligned}
\tag{15}
$$

The log-likelihood function then is given by

$$
\begin{aligned}
\ell(\boldsymbol{\mu}, \boldsymbol{\psi}; \mathbf{x}_1, \ldots, \mathbf{x}_N) &= \log L(\boldsymbol{\mu}, \boldsymbol{\psi}; \mathbf{x}_1, \ldots, \mathbf{x}_N) \\
&= -\frac{Np}{2} \log(2\pi) - \frac{N}{2} \log(\det \boldsymbol{\psi}) - \frac{1}{2} \sum_{i=1}^{N} (\mathbf{x}_i - \boldsymbol{\mu})' \boldsymbol{\psi}^{-1} (\mathbf{x}_i - \boldsymbol{\mu}).
\end{aligned}
\tag{16}
$$

Now, we are going to find a different expression for the last term in equation (16). Using the rules for the trace of a matrix from Exercise 1 of the Appendix and noticing that $\text{tr}(a) = a$ for any $a \in \mathbb{R}$, we obtain

$$
\begin{aligned}
\sum_{i=1}^{N} (\mathbf{x}_i - \boldsymbol{\mu})' \boldsymbol{\psi}^{-1} (\mathbf{x}_i - \boldsymbol{\mu}) &= \text{tr}\left[\sum_{i=1}^{N} (\mathbf{x}_i - \boldsymbol{\mu})' \boldsymbol{\psi}^{-1} (\mathbf{x}_i - \boldsymbol{\mu})\right] \\
&= \sum_{i=1}^{N} \text{tr}\left[(\mathbf{x}_i - \boldsymbol{\mu})' \boldsymbol{\psi}^{-1} (\mathbf{x}_i - \boldsymbol{\mu})\right] \\
&= \sum_{i=1}^{N} \text{tr}\left[\boldsymbol{\psi}^{-1} (\mathbf{x}_i - \boldsymbol{\mu})(\mathbf{x}_i - \boldsymbol{\mu})'\right] \\
&= \text{tr}\left[\sum_{i=1}^{N} \boldsymbol{\psi}^{-1} (\mathbf{x}_i - \boldsymbol{\mu})(\mathbf{x}_i - \boldsymbol{\mu})'\right] \\
&= \text{tr}\left[\boldsymbol{\psi}^{-1} \sum_{i=1}^{N} (\mathbf{x}_i - \boldsymbol{\mu})(\mathbf{x}_i - \boldsymbol{\mu})'\right].
\end{aligned}
\tag{17}
$$

But with $\bar{\mathbf{x}} = \frac{1}{N} \sum_{i=1}^{N} \mathbf{x}_i$,

$$
\begin{aligned}
\sum_{i=1}^{N} (\mathbf{x}_i - \boldsymbol{\mu})(\mathbf{x}_i - \boldsymbol{\mu})' &= \sum_{i=1}^{N} \left[(\mathbf{x}_i - \bar{\mathbf{x}}) + (\bar{\mathbf{x}} - \boldsymbol{\mu})\right]\left[(\mathbf{x}_i - \bar{\mathbf{x}}) + (\bar{\mathbf{x}} - \boldsymbol{\mu})\right]' \\
&= \sum_{i=1}^{N} (\mathbf{x}_i - \bar{\mathbf{x}})(\mathbf{x}_i - \bar{\mathbf{x}})' + N(\bar{\mathbf{x}} - \boldsymbol{\mu})(\bar{\mathbf{x}} - \boldsymbol{\mu})' \\
&= N\left[\mathbf{S}_P + (\bar{\mathbf{x}} - \boldsymbol{\mu})(\bar{\mathbf{x}} - \boldsymbol{\mu})'\right],
\end{aligned}
$$

where $\mathbf{S}_P = \frac{1}{N}\sum_{i=1}^{N}(\mathbf{x}_i - \bar{\mathbf{x}})(\mathbf{x}_i - \bar{\mathbf{x}})'$ is the plug-in estimate of the covariance matrix defined in equation (6) of Section 4.2. Thus equation (17) gives

$$\begin{aligned}\sum_{i=1}^{N}(\mathbf{x}_i - \mu)\psi^{-1}(\mathbf{x}_i - \mu)' &= N \operatorname{tr}\left\{\psi^{-1}\left[\mathbf{S}_P + (\bar{\mathbf{x}} - \mu)(\bar{\mathbf{x}} - \mu)'\right]\right\} \\ &= N\left\{\operatorname{tr}\left(\psi^{-1}\mathbf{S}_P\right) + \operatorname{tr}\left[\psi^{-1}(\bar{\mathbf{x}} - \mu)(\bar{\mathbf{x}} - \mu)'\right]\right\} \\ &= N\left[\operatorname{tr}\left(\psi^{-1}\mathbf{S}_P\right) + (\bar{\mathbf{x}} - \mu)'\psi^{-1}(\bar{\mathbf{x}} - \mu)\right].\end{aligned}$$

The log-likelihood function, therefore, is

$$\ell(\mu, \psi) = -\frac{N}{2}\left[p\log(2\pi) + \log(\det\psi) + (\bar{\mathbf{x}} - \mu)'\psi^{-1}(\bar{\mathbf{x}} - \mu) + \operatorname{tr}(\psi^{-1}\mathbf{S}_P)\right]. \quad (18)$$

This shows that the likelihood function depends on the data only through the values of $\bar{\mathbf{x}}$ and $\mathbf{S}_P$. In fact the pair $(\bar{\mathbf{x}}, \mathbf{S}_P)$ is a *sufficient statistic* for (μ, ψ), but we do not elaborate on this because sufficiency is not otherwise treated in this text.

Example 4.3.5 For $p \geq 2$, it is difficult to graph the likelihood function because there are too many parameters. For $p = 1$, however, there are only two parameters $\mu \in \mathbb{R}$ and $\sigma^2 > 0$. For observed data $x_1, \ldots, x_N$, the log-likelihood function (18) is given by

$$\ell(\mu, \sigma^2) = -\frac{N}{2}\left[\log(2\pi) + \log(\sigma^2) + \frac{(\bar{x} - \mu)^2}{\sigma^2} + \frac{s_P^2}{\sigma^2}\right], \quad (19)$$

where $\bar{x} = \frac{1}{N}\sum_{i=1}^{N} x_i$ and $s_P^2 = \frac{1}{N}\sum_{i=1}^{N}(x_i - \bar{x})^2$. Figure 4.3.2 shows this log-likelihood function for a numerical example with $N = 10$, $\bar{x} = 0.35$, and $s_P^2 = 4.0$. Again, note that the likelihood depends on the data only through $\bar{x}$ and s_P^2. For every fixed value of $\sigma^2 > 0$, the log-likelihood is a quadratic function of μ, reaching its maximum at $\mu = \bar{x}$. The overall maximum is attained at $(\mu, \sigma^2) = (\bar{x}, s_P^2)$. A proof of this result, although it is a special case of the more general result given in Theorem 4.3.1 below, is left to the reader; see Exercise 10. □

Theorem 4.3.1 *Let $\mathbf{X}_1, \ldots, \mathbf{X}_N$, $N > p$, be a sample from $\mathcal{N}(\mu, \psi)$, where μ and ψ are unknown, $\mu \in \mathbb{R}^p$, and ψ is positive definite. Then the maximum likelihood estimator of (μ, ψ) is given by*

$$\hat{\mu} = \bar{\mathbf{x}} = \frac{1}{N}\sum_{i=1}^{N}\mathbf{X}_i$$

and

$$\hat{\psi} = \mathbf{S}_P = \frac{1}{N}\sum_{i=1}^{N}(\mathbf{X}_i - \bar{\mathbf{X}})(\mathbf{X}_i - \bar{\mathbf{X}})'. \quad (20)$$

Proof The proof given here avoids differentiation but requires some clever matrix techniques. It is commonly ascribed to Watson (1964) but is rapidly becoming part of

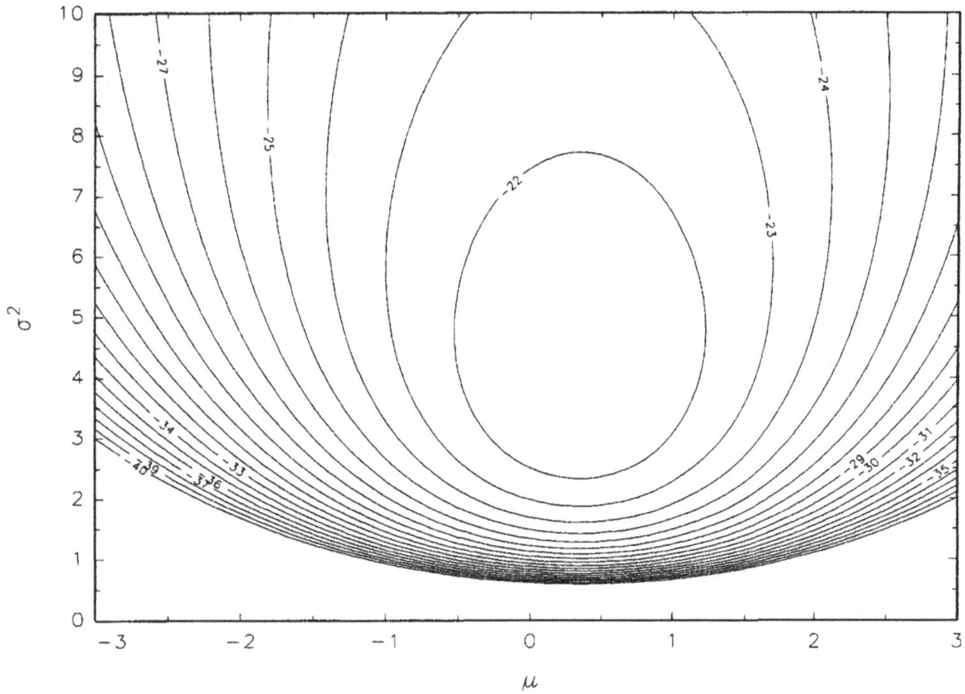

Figure 4-3-2
Graph of a normal log-likelihood function, as in Example 4.3.5.

multivariate statistical folklore. For N fixed vectors $\mathbf{x}_1, \ldots, \mathbf{x}_N \in \mathbb{R}^p$, let $\bar{\mathbf{x}}$ and $\mathbf{S}_P$ denote the plug-in estimates of mean vector and covariance matrix, and assume, for the moment, that $\mathbf{S}_P$ is positive definite. By equation (18), we need to minimize the function

$$\ell^*(\boldsymbol{\mu}, \boldsymbol{\psi}) = (\bar{\mathbf{x}} - \boldsymbol{\mu})'\boldsymbol{\psi}^{-1}(\bar{\mathbf{x}} - \boldsymbol{\mu}) + \log(\det \boldsymbol{\psi}) + \operatorname{tr}(\boldsymbol{\psi}^{-1}\mathbf{S}_P) \tag{21}$$

over all $\boldsymbol{\mu} \in \mathbb{R}^p$ and over all positive definite symmetric matrices $\boldsymbol{\psi}$. Since $\boldsymbol{\psi}$ is positive definite, the same holds for $\boldsymbol{\psi}^{-1}$, and therefore,

$$(\bar{\mathbf{x}} - \boldsymbol{\mu})'\boldsymbol{\psi}^{-1}(\bar{\mathbf{x}} - \boldsymbol{\mu}) \geq 0, \tag{22}$$

with equality exactly if $\boldsymbol{\mu} = \bar{\mathbf{x}}$. That is, whatever the estimate of $\boldsymbol{\psi}$ will be (as long as it is positive definite) maximum likelihood tells us that we should estimate $\boldsymbol{\mu}$ by $\bar{\mathbf{x}}$. It remains to minimize

$$\ell^*(\bar{\mathbf{x}}, \boldsymbol{\psi}) = \log(\det \boldsymbol{\psi}) + \operatorname{tr}(\boldsymbol{\psi}^{-1}\mathbf{S}_P) \tag{23}$$

with respect to $\boldsymbol{\psi}$. Equivalently, we minimize

$$\begin{aligned} \ell^*(\bar{\mathbf{x}}, \boldsymbol{\psi}) - \log(\det \mathbf{S}_P) &= \operatorname{tr}(\boldsymbol{\psi}^{-1}\mathbf{S}_P) - \log \frac{\det \mathbf{S}_P}{\det \boldsymbol{\psi}} \\ &= \operatorname{tr}(\boldsymbol{\psi}^{-1}\mathbf{S}_P) - \log \det(\boldsymbol{\psi}^{-1}\mathbf{S}_P). \end{aligned} \tag{24}$$

Writing $\mathbf{S}_P^{1/2}$ for the positive definite symmetric square root of $\mathbf{S}_P$ (see Appendix A.7), $\text{tr}(\psi^{-1}\mathbf{S}_P) = \text{tr}(\psi^{-1}\mathbf{S}_P^{1/2}\mathbf{S}_P^{1/2}) = \text{tr}(\mathbf{S}_P^{1/2}\psi^{-1}\mathbf{S}_P^{1/2})$, and $\det(\psi^{-1}\mathbf{S}_P) = \det(\mathbf{S}_P^{1/2}\psi^{-1}\mathbf{S}_P^{1/2})$. But $\mathbf{A} = \mathbf{S}_P^{1/2}\psi^{-1}\mathbf{S}_P^{1/2}$ is a symmetric matrix, and therefore,

$$\text{tr}(\mathbf{A}) = \sum_{i=1}^{p} \lambda_i ,$$

and

$$\det(\mathbf{A}) = \prod_{i=1}^{p} \lambda_i ,$$

where $\lambda_1, \ldots, \lambda_p$ are the eigenvalues of $\mathbf{A}$. By positive definiteness, all λ_i must be positive. Thus, equation (24) can be written as

$$\begin{aligned} \ell^*(\bar{\mathbf{x}}, \psi) - \log(\det \mathbf{S}_P) &= \sum_{i=1}^{p} \lambda_i - \log \prod_{i=1}^{p} \lambda_i \\ &= \sum_{i=1}^{p} (\lambda_i - \log \lambda_i). \end{aligned} \tag{25}$$

We have to choose the λ_i so as to minimize (25). But the function $h(u) = u - \log(u)$ takes a unique minimum at $u = 1$, and (25) is the sum of p such functions. Therefore, we should choose $\lambda_1 = \cdots = \lambda_p = 1$. This means that, at the minimum value of (24), $\mathbf{A} = \mathbf{S}_P^{1/2}\psi^{-1}\mathbf{S}_P^{1/2}$ is a matrix with p eigenvalues all equal to 1. The only such matrix is $\mathbf{I}_p$, and therefore,

$$\mathbf{S}_P^{1/2}\psi^{-1}\mathbf{S}_P^{1/2} = \mathbf{I}_p . \tag{26}$$

Multiplying (26) on the left and the right by $\mathbf{S}_P^{1/2}$, then gives $\psi^{-1} = \mathbf{S}_P^{-1}$, and inversion provides the final result.

This proof is valid under the assumption that the sample covariance matrix of the data vectors $\mathbf{x}_1, \ldots, \mathbf{x}_N$ is positive definite. When sampling from a p-variate normal distribution with positive definite covariance matrix ψ, this assumption is satisfied with probability 1 if $N > p$; see Exercise 8 in Section 4.2. Thus, the assumption $N > p$ is crucial. ∎

By Theorem 4.3.1, the normal theory maximum likelihood estimators of the mean vector and covariance matrix of a multivariate normal distribution are the same as the plug-in estimators. As seen before, $\mathbf{S}_P$ is a biased estimator of ψ, and we will most often follow the standard convention of using the unbiased sample covariance matrix $\mathbf{S}$ instead.

Theorem 4.3.1 also allows us to find the maximum likelihood estimators in somewhat more general situations. A typical and important case is given by the following example.

Example 4.3.6 Suppose that $\mathbf{X}_{11}, \ldots, \mathbf{X}_{1N_1}$ and $\mathbf{X}_{21}, \ldots, \mathbf{X}_{2N_2}$ are independent samples from $\mathcal{N}_p(\boldsymbol{\mu}_1, \boldsymbol{\psi})$ and $\mathcal{N}_p(\boldsymbol{\mu}_2, \boldsymbol{\psi})$, respectively, that is, the covariance matrices in the two distributions are assumed to be identical, whereas the mean vectors are allowed to differ. Thus, the parameter space is given by

$$\boldsymbol{\Omega} = \{(\boldsymbol{\mu}_1, \boldsymbol{\mu}_2, \boldsymbol{\psi}) : \boldsymbol{\mu}_1 \in \mathbb{R}^p,\ \ \boldsymbol{\mu}_2 \in \mathbb{R}^p,\ \ \boldsymbol{\psi} \text{ positive definite}\}.$$

For $N_1 + N_2$ fixed vectors $\mathbf{x}_{11}, \ldots, \mathbf{x}_{1N_1}$ and $\mathbf{x}_{21}, \ldots, \mathbf{x}_{2N_2}$, the log-likelihood function by (18) is

$$\begin{aligned}\ell\,(\boldsymbol{\mu}_1, \boldsymbol{\mu}_2, \boldsymbol{\psi}) = &-\frac{N_1}{2}\left[p\log(2\pi) + \log(\det\boldsymbol{\psi}) + (\bar{\mathbf{x}}_1 - \boldsymbol{\mu}_1)'\boldsymbol{\psi}^{-1}(\bar{\mathbf{x}}_1 - \boldsymbol{\mu}_1) + \operatorname{tr}(\boldsymbol{\psi}^{-1}\mathbf{T}_1)\right] \\ &-\frac{N_2}{2}\left[p\log(2\pi) + \log(\det\boldsymbol{\psi}) + (\bar{\mathbf{x}}_2 - \boldsymbol{\mu}_2)'\boldsymbol{\psi}^{-1}(\bar{\mathbf{x}}_2 - \boldsymbol{\mu}_2) + \operatorname{tr}(\boldsymbol{\psi}^{-1}\mathbf{T}_2)\right],\end{aligned} \tag{27}$$

where $\bar{\mathbf{x}}_1$ and $\bar{\mathbf{x}}_2$ are the two sample mean vectors and $\mathbf{T}_j = \frac{1}{N_j}\sum_{i=1}^{N_j}(\mathbf{x}_{ji} - \bar{\mathbf{x}}_j)(\mathbf{x}_{ji} - \bar{\mathbf{x}}_j)'$, $j = 1, 2$. But

$$\begin{aligned}N_1\operatorname{tr}(\boldsymbol{\psi}^{-1}\mathbf{T}_1) + N_2\operatorname{tr}(\boldsymbol{\psi}^{-1}\mathbf{T}_2) &= \operatorname{tr}(\boldsymbol{\psi}^{-1}N_1\mathbf{T}_1) + \operatorname{tr}(\boldsymbol{\psi}^{-1}N_2\mathbf{T}_2) \\ &= \operatorname{tr}\left[\boldsymbol{\psi}^{-1}(N_1\mathbf{T}_1 + N_2\mathbf{T}_2)\right] \\ &= (N_1 + N_2)\operatorname{tr}(\boldsymbol{\psi}^{-1}\mathbf{T}),\end{aligned} \tag{28}$$

where we set $\mathbf{T} = (N_1\mathbf{T}_1 + N_2\mathbf{T}_2)/(N_1 + N_2)$. Therefore,

$$\begin{aligned}\ell\,(\boldsymbol{\mu}_1, \boldsymbol{\mu}_2, \boldsymbol{\psi}) = &-\frac{N_1 + N_2}{2}\left[p\log(2\pi) + \log(\det\boldsymbol{\psi}) + \operatorname{tr}(\boldsymbol{\psi}^{-1}\mathbf{T})\right] \\ &-\frac{N_1}{2}(\bar{\mathbf{x}}_1 - \boldsymbol{\mu}_1)'\boldsymbol{\psi}^{-1}(\bar{\mathbf{x}}_1 - \boldsymbol{\mu}_1) - \frac{N_2}{2}(\bar{\mathbf{x}}_2 - \boldsymbol{\mu}_2)'\boldsymbol{\psi}^{-1}(\bar{\mathbf{x}}_2 - \boldsymbol{\mu}_2).\end{aligned} \tag{29}$$

Following the arguments in the proof of Theorem 4.3.1, the maximum likelihood estimates of $\boldsymbol{\mu}_1$ and $\boldsymbol{\mu}_2$ are $\bar{\mathbf{x}}_1$ and $\bar{\mathbf{x}}_2$, respectively, as each of the two last terms in (29) is zero exactly if $\boldsymbol{\mu}_j = \bar{\mathbf{x}}_j$. It remains to maximize

$$\ell\,(\bar{\mathbf{x}}_1, \bar{\mathbf{x}}_2, \boldsymbol{\psi}) = -\frac{N_1 + N_2}{2}\left[p\log(2\pi) + \log(\det\boldsymbol{\psi}) + \operatorname{tr}(\boldsymbol{\psi}^{-1}\mathbf{T})\right].$$

This is the same problem as minimization of (23), with $\mathbf{S}_P$ replaced by $\mathbf{T}$. Assuming that $\mathbf{T}$ is positive definite (see Exercises 10 and 13 in Section 4.2), it follows that the maximum likelihood estimate of $\boldsymbol{\psi}$ is

$$\hat{\boldsymbol{\psi}} = \mathbf{T} = \frac{1}{N_1 + N_2}\sum_{j=1}^{2}\sum_{i=1}^{N_j}(\mathbf{x}_{ji} - \bar{\mathbf{x}}_j)(\mathbf{x}_{ji} - \bar{\mathbf{x}}_j)'. \tag{30}$$

By Exercise 13 of Section 4.2, $N_1 + N_2 \geq p + 2$ is required to make sure that $\mathbf{T}$ is positive definite. This estimator is a weighted sum of the groupwise plug-in estimators and is biased, as we have seen in Section 4.2. Again, we will follow the convention of using an unbiased version,

$$\mathbf{S}_{\text{pooled}} = \frac{(N_1 - 1)\mathbf{S}_1 + (N_2 - 1)\mathbf{S}_2}{N_1 + N_2 - 2} \tag{31}$$

in most applications, where $\mathbf{S}_1$ and $\mathbf{S}_2$ are the two sample covariance matrices. This has already been discussed in Section 4.2. A generalization of the case of data from k independent distributions with identical covariance matrices is given in Exercise 12. □

Much of the popularity of the method of maximum likelihood is because of the fact that, in many situations, it provides an approximation to the sampling distribution of the estimator. Before elaborating on this, we need some more notation and terminology. Calculus of several variables will be used extensively from now on.

Now, we proceed to some important large-sample results, without proving them because rigorous proofs are beyond the level of this text. The results will be used extensively in Chapters 6 and 7.

Definition 4.3.4 Let $\mathcal{X} = (\mathbf{X}_1, \ldots, \mathbf{X}_N)$ denote a sample from a random variable with *pdf* $f(\cdot\,; \boldsymbol{\theta})$, where $\boldsymbol{\theta} = (\theta_1, \ldots, \theta_k)' \in \mathbb{R}^k$ is an unknown parameter. Let $L(\boldsymbol{\theta}; \mathbf{x}) = \prod_{i=1}^{N} f(\mathbf{x}_i; \boldsymbol{\theta})$ be the likelihood function of $\boldsymbol{\theta}$ for observed values $\mathbf{x} = (\mathbf{x}_1, \ldots, \mathbf{x}_N)$ in the sample space, and $\ell(\boldsymbol{\theta}; \mathbf{x}) = \log L(\boldsymbol{\theta}; \mathbf{x})$ the log-likelihood function. Then

(a) the vector of first derivatives of $\ell(\boldsymbol{\theta}; \mathbf{x})$ with respect to the θ_j is called the *score function* of $\boldsymbol{\theta}$, written as $\mathcal{S}(\boldsymbol{\theta}; \mathbf{x})$. That is, the score function consists of k functions

$$S_j(\boldsymbol{\theta}; \mathbf{x}) = \frac{\partial}{\partial \theta_j} \ell(\boldsymbol{\theta}; \mathbf{x}), \qquad j = 1, \ldots, k, \tag{32}$$

and

(b) the matrix of second derivatives of the log-likelihood function with respect to the θ_j, multiplied by -1, is called the *information function* of $\boldsymbol{\theta}$, written as $\mathcal{I}(\boldsymbol{\theta}; \mathbf{x})$. That is, the (j, k)th entry of the information function is expressed by

$$\begin{aligned} I_{jk}(\boldsymbol{\theta}; \mathbf{x}) &= -\frac{\partial^2}{\partial \theta_j \partial \theta_h} \ell(\boldsymbol{\theta}; \mathbf{x}) \\ &= -\frac{\partial}{\partial \theta_h} S_j(\boldsymbol{\theta}; \mathbf{x}), \qquad j, h = 1, \ldots, k. \end{aligned} \tag{33}$$

□

The notions of score and information function are appropriate only in cases where calculus can be used to find maximum likelihood estimates. In cases like Example 4.3.4, they are not useful.

In vector and matrix notation, we will usually write

$$\mathcal{S}(\boldsymbol{\theta};\mathbf{x}) = \begin{bmatrix} S_1(\boldsymbol{\theta};\mathbf{x}) \\ \vdots \\ S_k(\boldsymbol{\theta};\mathbf{x}) \end{bmatrix} = \begin{bmatrix} \frac{\partial}{\partial\theta_1}\ell(\boldsymbol{\theta};\mathbf{x}) \\ \vdots \\ \frac{\partial}{\partial\theta_k}\ell(\boldsymbol{\theta};\mathbf{x}) \end{bmatrix}, \tag{34}$$

and

$$\begin{aligned}
\mathcal{I}(\boldsymbol{\theta};\mathbf{x}) &= \begin{bmatrix} I_{11}(\boldsymbol{\theta};\mathbf{x}) & \cdots & I_{1k}(\boldsymbol{\theta};\mathbf{x}) \\ \vdots & \ddots & \vdots \\ I_{k1}(\boldsymbol{\theta};\mathbf{x}) & \cdots & I_{kk}(\boldsymbol{\theta};\mathbf{x}) \end{bmatrix} \\
&= -\begin{bmatrix} \frac{\partial}{\partial\theta_1}S_1(\boldsymbol{\theta};\mathbf{x}) & \cdots & \frac{\partial}{\partial\theta_k}S_1(\boldsymbol{\theta};\mathbf{x}) \\ \vdots & \ddots & \vdots \\ \frac{\partial}{\partial\theta_1}S_k(\boldsymbol{\theta};\mathbf{x}) & \cdots & \frac{\partial}{\partial\theta_k}S_k(\boldsymbol{\theta};\mathbf{x}) \end{bmatrix} \\
&= -\begin{bmatrix} \frac{\partial^2}{\partial\theta_1^2}\ell(\boldsymbol{\theta};\mathbf{x}) & \cdots & \frac{\partial^2}{\partial\theta_1\,\partial\theta_k}\ell(\boldsymbol{\theta};\mathbf{x}) \\ \vdots & \ddots & \vdots \\ \frac{\partial^2}{\partial\theta_k\,\partial\theta_1}\ell(\boldsymbol{\theta};\mathbf{x}) & \cdots & \frac{\partial^2}{\partial\theta_k^2}\ell(\boldsymbol{\theta};\mathbf{x}) \end{bmatrix}.
\end{aligned} \tag{35}$$

The obvious application of the notions of score and information functions is in finding maximum likelihood estimates using calculus. In many cases, we can do the maximization by solving the equation system $\mathcal{S}(\boldsymbol{\theta};\mathbf{x}) = \mathbf{0}$. If $\boldsymbol{\theta}^*$ is a solution and if $\mathcal{I}(\boldsymbol{\theta}^*;\mathbf{x})$ is a postive-definite matrix, then the log-likelihood function takes a local maximum at $\boldsymbol{\theta}^*$. Unless the maximum is reached at the boundary of the parameter space, evaluation of the log-likelihood function for all candidates $\boldsymbol{\theta}^*$ then will allow us to identify the maximum likelihood estimate $\hat{\boldsymbol{\theta}}$, or estimates in the case of identical local maxima. Often there will be a single solution to the equation system $\mathcal{S}(\boldsymbol{\theta};\mathbf{x}) = \mathbf{0}$. Also, in the case of a single parameter $\theta \in \mathbb{R}$, the score and information functions are real-valued functions of a single variable, and postive definiteness means simply $\mathcal{I}(\theta^*) > 0$.

In some cases, for instance in the logistic regression model of Sections 7.5 and 7.6, it may not be possible to obtain explicit solutions to the equation system $\mathcal{S}(\boldsymbol{\theta};\mathbf{x}) = \mathbf{0}$. In such cases, the score and information functions are still useful tools in a numerical maximization procedure such as the Newton–Raphson algorithm. The ultimate justification for introducing new terminology however, is the use of these derivatives to find an approximation to the sampling distribution of the maximum likelihood estimator. Before stating the main result, we illustrate this with an example.

Example 4.3.7 Let $\mathcal{X} = (X_1, \ldots, X_N)$ be a sample from the exponential distribution with unknown mean $\theta > 0$, i.e., from the distribution with *pdf*

$$f(x;\theta) = \begin{cases} \dfrac{1}{\theta}e^{-x/\theta} & \text{if } x \geq 0, \\ 0 & \text{otherwise.} \end{cases}$$

For observed data $\mathbf{x} = (x_1, \ldots, x_N)$, we obtain

$$L(\theta; \mathbf{x}) = \prod_{i=1}^{N} f(x_i; \theta) = \theta^{-N} \exp\left(-\frac{1}{\theta}\sum_{i=1}^{N} x_i\right),$$

$$\ell(\theta; \mathbf{x}) = -N\left(\log\theta + \frac{1}{N\theta}\sum_{i=1}^{N} x_i\right),$$

$$\mathcal{S}(\theta; \mathbf{x}) = -N\left(\frac{1}{\theta} - \frac{1}{N\theta^2}\sum_{i=1}^{N} x_i\right),$$

and

$$\mathcal{I}(\theta; \mathbf{x}) = N\left(-\frac{1}{\theta^2} + \frac{2}{N\theta^3}\sum_{i=1}^{N} x_i\right).$$

The only solution to the equation $\mathcal{S}(\theta; \mathbf{x}) = 0$ is $\theta = \bar{x} = \frac{1}{N}\sum_{i=1}^{N} x_i$. But $\mathcal{I}(\bar{x}; \mathbf{x}) = N/\bar{x}^2 > 0$ (unless all $x_i = 0$, in which case the likelihood function has an infinite peak at $\theta = 0$). Thus,

$$\hat{\theta} = \bar{X} = \frac{1}{N}\sum_{i=1}^{N} X_i$$

is the maximum likelihood estimator.

What is the sampling distribution of $\hat{\theta}$? The exact distribution of $\hat{\theta}$ is a gamma distribution; see Exercise 13, but, for large N, we can apply the central limit theorem. Since $E[X_i] = \theta$ and $\text{var}[X_i] = \theta^2$, we can consider $\hat{\theta}$ as approximately normal with mean θ and variance θ^2/N. No particular properties of maximum likelihood need to be used in this approximation.

The more interesting (and more difficult) part follows. So far, we have considered the score and information functions as functions of the parameter θ, for a fixed vector $\mathbf{x} = (x_1, \ldots, x_N)$ in the sample space of $\mathcal{X}$. Now, we will turn around and look at them as functions of a random sample, considering the parameter as fixed, that is, we will write

$$\mathcal{S}(\theta; \mathcal{X}) = N\left(-\frac{1}{\theta} + \frac{1}{\theta^2}\cdot\frac{1}{N}\sum_{i=1}^{N} X_i\right) = N\left(\frac{1}{\theta^2}\bar{X} - \frac{1}{\theta}\right)$$

and

$$\mathcal{I}(\theta; \mathcal{X}) = N\left(-\frac{1}{\theta^2} + \frac{2}{\theta^3}\cdot\frac{1}{N}\sum_{i=1}^{N} X_i\right) = N\left(\frac{2}{\theta^3}\bar{X} - \frac{1}{\theta^2}\right),$$

that is, $\mathcal{S}(\theta; \mathcal{X})$ and $\mathcal{I}(\theta; \mathcal{X})$ are now random variables, but they are not statistics because they depend on the unknown parameter. We can study their properties; for instance (see Exercise 14)

$$E[\mathcal{S}(\theta;\mathcal{X})]=0,$$
$$\text{var}[\mathcal{S}(\theta;\mathcal{X})]=\frac{N}{\theta^2}, \tag{36}$$

and

$$E\left[\mathcal{I}(\theta;\mathcal{X})\right]=\frac{N}{\theta^2}=\text{var}\left[\mathcal{S}(\theta;\mathcal{X})\right]. \tag{37}$$

Next, consider the random variable

$$\begin{aligned}\sqrt{N}\cdot\frac{\mathcal{S}(\theta;\mathcal{X})}{\mathcal{I}(\theta;\mathcal{X})}&=\frac{\frac{1}{\sqrt{N}}\mathcal{S}(\theta;\mathcal{X})}{\frac{1}{N}\mathcal{I}(\theta;\mathcal{X})}\\&=\frac{\sqrt{N}(\bar{X}-\theta)}{\frac{2}{\theta}\bar{X}-1}.\end{aligned} \tag{38}$$

For large N, the denominator $\frac{2}{\theta}\bar{X}-1$ is approximately 1 (more precisely, converges to 1 in probability), and thus, the distribution of (38) is approximately the same as the distribution of $\sqrt{N}(\bar{X}-\theta)=\sqrt{N}(\hat{\theta}-\theta)$, which, by the central limit theorem, is approximately normal.

Thus, we have found the following rather interesting results:

- $\sqrt{N}\,\frac{\mathcal{S}(\theta;\mathcal{X})}{\mathcal{I}(\theta;\mathcal{X})}$ is approximately normal and has approximately the same distribution as $\sqrt{N}(\hat{\theta}-\theta)$.
- $E\left[\mathcal{S}(\theta;\mathcal{X})\right]=0$, and $\text{var}\left[\mathcal{S}(\theta;\mathcal{X})\right]=E\left[\mathcal{I}(\theta;\mathcal{X})\right]$.

Finally, we notice that

- $\text{var}[\hat{\theta}]=\frac{1}{E[\mathcal{I}(\theta;\mathcal{X})]}$. □

The results pointed out at the end of Example 4.3.7, in fact, are not specific for this example but hold under relatively general conditions. In a formal proof of the one-parameter version of Theorem 4.3.2, we would show that, under certain technical conditions, the following results hold for the maximum likelihood estimator $\hat{\theta}$ of a parameter $\theta\in\mathbb{R}$ based on a sample $\mathcal{X}=(X_1,\ldots,X_N)$:

- $E\left[\mathcal{S}(\theta;\mathcal{X})\right]=0$, and $\text{var}\left[\mathcal{S}(\theta;\mathcal{X})\right]=E\left[\mathcal{I}(\theta;\mathcal{X})\right]$,
- the asymptotic distribution (as $N\to\infty$) of

$$\sqrt{N}\frac{\mathcal{S}(\theta;\mathcal{X})}{\mathcal{I}(\theta;\mathcal{X})}$$

is normal, and the same as the asymptotic distribution of $\sqrt{N}(\hat{\theta}-\theta)$, and

- $\text{var}[\hat{\theta}]$ is approximately the same as $1/E\,[\mathcal{I}(\theta;\mathcal{X})]$.

The formal proof is riddled with regularity conditions that concern existence and continuity of derivatives and interchange of the order of differentiation and integration (or summation). At the heart of the proof however, is the central limit theorem, which is used to show that $\sqrt{N}(\theta;\mathcal{X})$ is approximately normal. This is actually fairly straightforward since

$$\begin{aligned}\mathcal{S}(\theta;\mathcal{X}) &= \frac{\partial}{\partial\theta}\ell(\theta;\mathcal{X}) \\ &= \frac{\partial}{\partial\theta}\left[\sum_{i=1}^{N}\log f(X_i;\theta)\right] \\ &= \sum_{i=1}^{N}\frac{\partial}{\partial\theta}\log f(X_i;\theta),\end{aligned}$$

that is, the score function, considered as a random variable, is the sum of N *iid* random variables $Y_i = \frac{\partial}{\partial\theta}\log f(X_i;\theta)$, which, by the central limit theorem, is approximately normal.

Now, we are ready for the main result. For a proof and a precise statement of the conditions under which it holds, the reader is referred to Silvey (1975) or to Cox and Hinkley (1974).

Theorem 4.3.2 *Let $\mathcal{X} = (\mathbf{X}_1,\ldots,\mathbf{X}_N)$ be a sample from some distribution that depends on an unknown parameter $\boldsymbol{\theta}\in\mathbb{R}^k$. Under appropriate regularity conditions, the distribution of the maximum likelihood estimator $\hat{\boldsymbol{\theta}}$ for sufficiently large N is approximately normal with mean vector $\boldsymbol{\theta}$ and covariance matrix $\boldsymbol{\Sigma} = \{E[\mathcal{I}(\boldsymbol{\theta};\mathcal{X})]\}^{-1}$.*

Here are some remarks on the regularity conditions assumed in Theorem 4.3.2. The most important assumption is that the log-likelihood function can be expanded into a second-order Taylor series about $\hat{\boldsymbol{\theta}}$, at which point the first derivatives are zero. This means, in particular, that cases are excluded where the support of the probability function or density function $f(\cdot\,;\boldsymbol{\theta})$ depends on the unknown $\boldsymbol{\theta}$. See, e.g., Example 4.3.4.

It would be difficult to overestimate the power of Theorem 4.3.2. Even in cases where it is possible to find the exact or approximate distribution of a maximum likelihood estimator directly, the theorem may give a surprisingly easy solution, as we are going to show with two examples.

Example 4.3.8 continuation of Example 4.3.5

Let $\mathcal{X} = (X_1,\ldots,X_N)$ be a sample from $\mathcal{N}(\mu,\sigma^2)$, where $\mu\in\mathbb{R}$ and $\sigma^2 > 0$ are unknown. Leaving details to the reader (see Exercise 10), the information function with μ as the first parameter and σ^2 as the second parameter is given by

$$\mathcal{I}(\mu, \sigma^2; \mathcal{X}) = \frac{N}{\sigma^2} \begin{bmatrix} 1 & \frac{1}{\sigma^2} \cdot \frac{1}{N} \sum_{i=1}^N (X_i - \mu) \\ \frac{1}{\sigma^2} \cdot \frac{1}{N} \sum_{i=1}^N (X_i - \mu) & \frac{1}{\sigma^4} \cdot \frac{1}{N} \sum_{i=1}^N (X_i - \mu)^2 - \frac{1}{2\sigma^2} \end{bmatrix}.$$

Taking the expectation gives

$$E\left[\mathcal{I}(\mu, \sigma^2; \mathcal{X})\right] = \begin{pmatrix} N/\sigma^2 & 0 \\ 0 & N/2\sigma^4 \end{pmatrix},$$

with inverse

$$\boldsymbol{\Sigma} = \begin{pmatrix} \sigma^2/N & 0 \\ 0 & 2\sigma^4/N \end{pmatrix}.$$

Thus, we conclude that the maximum likelihood estimators $\hat{\mu} = \bar{X}$ and $\hat{\sigma}^2 = \frac{1}{N} \sum_{i=1}^N (X_i - \bar{X})^2$ have the following properties:

- They are approximately jointly normal (for large N), with expected values μ and σ^2 and variances σ^2/N and $2\sigma^4/N$, respectively.
- They are asymptotically independent, as implied by joint normality and zero covariance.

Remarkably, in this case, Theorem 4.3.2 gives the *exact* distribution of $\hat{\mu} = \bar{X}$. Also, independence between $\bar{X}$ and $\hat{\sigma}^2$ holds for any finite $N \geq 2$, as we have seen in Section 3.3. In contrast to this, the exact distribution of $\hat{\sigma}^2$ is not normal for finite N; thus, the theorem gives only an approximation. □

Example 4.3.9 continuation of Example 4.3.3

Suppose that $\mathbf{X} = (X_1, X_2, X_3)'$ is multinomial with parameter $(\theta_1, \theta_2, \theta_3)' \in \mathbb{R}^3$, and $X_1 + X_2 + X_3 = N$. Note that, with observed counts $\mathbf{x} = (x_1, x_2, x_3)$, we have only *one* observation from the multivariate random vector $\mathbf{X}$, a point to be discussed later. Because of the constraint $\theta_1 + \theta_2 + \theta_3 = 1$, write $\theta_3 = 1 - \theta_1 - \theta_2$, and, thus, consider $\boldsymbol{\theta} = (\theta_1, \theta_2)'$ as the parameter, with parameter space $\boldsymbol{\Omega} = \{\boldsymbol{\theta} \in \mathbb{R}^2 : \theta_1 \geq 0, \theta_2 \geq 0, \theta_1 + \theta_2 \leq 1\}$. From (5), the log-likelihood function is given by

$$\begin{aligned} \ell(\boldsymbol{\theta}; \mathbf{x}) &= C + x_1 \log\theta_1 + x_2 \log\theta_2 + x_3 \log\theta_3 \\ &= C + x_1 \log\theta_1 + x_2 \log\theta_2 + x_3 \log(1 - \theta_1 - \theta_2), \end{aligned} \tag{39}$$

where C is a constant that does not depend on the unknown parameter. We will continue using θ_3, but have to remember, when taking derivatives, that it is a function of θ_1 and θ_2. Leaving details to the reader (see Exercise 15), we obtain

$$\mathcal{S}(\boldsymbol{\theta}; \mathbf{x}) = \begin{bmatrix} S_1(\boldsymbol{\theta}; \mathbf{x}) \\ S_2(\boldsymbol{\theta}; \mathbf{x}) \end{bmatrix} = \begin{pmatrix} \frac{x_1}{\theta_1} - \frac{x_3}{\theta_3} \\ \frac{x_2}{\theta_2} - \frac{x_3}{\theta_3} \end{pmatrix} \tag{40}$$

and

$$\mathcal{I}(\boldsymbol{\theta};\mathbf{x}) = \begin{bmatrix} I_{11}(\boldsymbol{\theta};\mathbf{x}) & I_{12}(\boldsymbol{\theta};\mathbf{x}) \\ I_{21}(\boldsymbol{\theta};\mathbf{x}) & I_{22}(\boldsymbol{\theta};\mathbf{x}) \end{bmatrix} = \begin{pmatrix} \frac{x_1}{\theta_1^2} + \frac{x_3}{\theta_3^2} & \frac{x_3}{\theta_3^2} \\ \frac{x_3}{\theta_3^2} & \frac{x_2}{\theta_2^2} + \frac{x_3}{\theta_3^2} \end{pmatrix}. \tag{41}$$

Since $E[X_i] = N\theta_i$ for $i = 1, 2, 3$ (see Example 2.10.9), the expected value of the information function is given by

$$E\left[\mathcal{I}(\boldsymbol{\theta};\mathbf{X})\right] = N \begin{pmatrix} \frac{1}{\theta_1} + \frac{1}{\theta_3} & \frac{1}{\theta_3} \\ \frac{1}{\theta_3} & \frac{1}{\theta_2} + \frac{1}{\theta_3} \end{pmatrix}, \tag{42}$$

and its inverse is given by

$$\boldsymbol{\Sigma} = \{E\left[\mathcal{I}(\boldsymbol{\theta};\mathbf{X})\right]\}^{-1} = \frac{1}{N}\begin{bmatrix} \theta_1(1-\theta_1) & -\theta_1\theta_2 \\ -\theta_1\theta_2 & \theta_2(1-\theta_2) \end{bmatrix}. \tag{43}$$

Thus, we conclude that the distribution of the maximum likelihood estimator $(\hat{\theta}_1, \hat{\theta}_2) = (\frac{1}{N}X_1, \frac{1}{N}X_2)$ is approximately bivariate normal with mean (θ_1, θ_2) and covariance matrix $\boldsymbol{\Sigma}$ given by (43). In particular, this means that $\hat{\theta}_i = \frac{1}{N}X_i$ is approximately normal with mean θ_i and variance $\theta_i(1-\theta_i)/N$, a result well known from the theory of estimation of a single binomial proportion. In fact, this is the exact variance. We can also conclude (Exercise 16) that the joint distribution of $(\hat{\theta}_1, \hat{\theta}_2, \hat{\theta}_3)$ is approximately three-dimensional normal with mean vector $(\theta_1, \theta_2, \theta_3)$ and covariance matrix

$$\frac{1}{N}\begin{bmatrix} \theta_1(1-\theta_1) & -\theta_1\theta_2 & -\theta_1\theta_3 \\ -\theta_2\theta_1 & \theta_2(1-\theta_2) & -\theta_2\theta_3 \\ -\theta_3\theta_1 & -\theta_3\theta_2 & \theta_3(1-\theta_3) \end{bmatrix}. \tag{44}$$

This joint normal distribution is singular because $\hat{\theta}_1 + \hat{\theta}_2 + \hat{\theta}_3 = 1$. The extension to a general multinomial random vector with k probabilities, again, is left to the reader; see Exercise 17. The result may well be anticipated though.

But how can we justify the use of asymptotic theory when we have only *one* observation from the distribution of $\mathbf{X}$? The answer is that we can write $\mathbf{X} = \mathbf{Y}_1 + \cdots + \mathbf{Y}_N$, where all $\mathbf{Y}_i$ are independent and identically distributed, with the same parameters $\theta_1, \theta_2, \theta_3$, and exactly one of the counts in each $\mathbf{Y}_i$ is 1. That is, $\mathbf{Y}_i$ has three components,

$$\mathbf{Y}_i = \begin{pmatrix} Y_{1i} \\ Y_{2i} \\ Y_{3i} \end{pmatrix},$$

and the *pdf* of $\mathbf{Y}_i$ is given by

$$f(y_{1i}, y_{2i}, y_{3i}; \boldsymbol{\theta}) = \begin{cases} \theta_1^{y_{1i}}\theta_2^{y_{2i}}\theta_3^{y_{3i}} & \text{if all } y_{ji} = 0 \text{ or } 1,\ y_{1i} + y_{2i} + y_{3i} = 1, \\ 0 & \text{otherwise.} \end{cases}$$

The likelihood function based on $\mathcal{Y} = (\mathbf{Y}_1, \ldots, \mathbf{Y}_N)$ then is given by

$$\begin{aligned} L(\boldsymbol{\theta}; \mathcal{Y}) &= \prod_{i=1}^{N} \theta_1^{y_{1i}} \theta_2^{y_{2i}} \theta_3^{y_{3i}} \\ &= \theta_1^{x_1} \theta_2^{x_2} \theta_3^{x_3}, \end{aligned}$$

where $x_j = \sum_{i=1}^{N} y_{ji}$. This is the same log-likelihood function as (39), except for the constant that disappears in the score and information functions. Thus, the use of the large-sample theory is justified. □

We conclude the section with a review of another important application of likelihood theory, likelihood ratio testing. This will be used extensively in Sections 6.6 and in parts of Chapter 7. It is assumed that the reader is familiar with the basic ideas of hypothesis testing.

Suppose that the parameter in a testing problem can vary in a parameter space $\Omega \in \mathbb{R}^k$, and we wish to test the hypothesis

$$H : \boldsymbol{\theta} \in \omega,$$

where ω is a subset of Ω. As before, let $\mathcal{X}$ denote the random variable, vector, or matrix whose distribution depends on the unknown $\boldsymbol{\theta}$, and let $\mathbf{x}$ denote any element in the sample space of $\mathcal{X}$. Let $L(\boldsymbol{\theta}; \mathbf{x})$ and $\ell(\boldsymbol{\theta}; \mathbf{x})$ be the likelihood function and the log-likelihood function, respectively, based on observed data $\mathbf{x}$. For simplicity, we will now often write $L(\boldsymbol{\theta})$ and $\ell(\boldsymbol{\theta})$, oppressing the dependency on $\mathbf{x}$.

The likelihood ratio test compares the maxima of the likelihood function reached in Ω and in ω, respectively. Since ω is a subset of Ω, the maximum in Ω will always be larger or equal to the maximum in ω, and

$$0 \le \frac{\max_{\boldsymbol{\theta} \in \omega} L(\boldsymbol{\theta})}{\max_{\boldsymbol{\theta} \in \Omega} L(\boldsymbol{\theta})} \le 1.$$

The upper limit 1 is reached exactly if the two maxima are identical. This setup is used in the following definition.

Definition 4.3.5 The *likelihood ratio statistic* for the hypothesis $\boldsymbol{\theta} \in \omega$ is

$$\text{LRS} = \frac{\max_{\boldsymbol{\theta} \in \omega} L(\boldsymbol{\theta})}{\max_{\boldsymbol{\theta} \in \Omega} L(\boldsymbol{\theta})}, \tag{45}$$

and the *log-likelihood ratio statistic* is given by

$$\begin{aligned} \text{LLRS} &= -2 \log(\text{LRS}) \\ &= 2 \left[\max_{\boldsymbol{\theta} \in \Omega} \ell(\boldsymbol{\theta}) - \max_{\boldsymbol{\theta} \in \omega} \ell(\boldsymbol{\theta}) \right]. \end{aligned} \tag{46}$$

□

We will refer to Ω and ω as the *unconstrained parameter space* and the *constrained parameter space*, respectively. If $\hat{\theta}$ denotes the maximum likelihood estimate in Ω and $\tilde{\theta}$ the maximum likelihood estimate in ω, then equations (45) and (46) can also be written as

$$\text{LRS} = \frac{L(\tilde{\theta})}{L(\hat{\theta})}$$

and

$$\text{LLRS} = 2\left[\ell(\hat{\theta}) - \ell(\tilde{\theta})\right],$$

respectively. Note that we do not state a formal "alternative hypothesis" to the hypothesis $\theta \in \omega$. In all testing problems to be considered, the constrained parameter space ω will be a subset of Ω, and the test may be viewed as an assessment of the plausibility of the hypothesis $\theta \in \omega$ inside the full parameter space Ω. Most authors call the statement $\theta \in \Omega \backslash \omega$ (the complement of ω in Ω) the alternative hypothesis.

It also follows from Definition 4.3.5 that the log-likelihood ratio statistic takes only nonnegative values. Large values of LLRS constitute "strong evidence" against the hypothesis, that is, a testing rule will be of the form

$$\begin{aligned} &\text{accept } H: \theta \in \omega \quad &&\text{if LLRS} \le c, \\ &\text{reject } H: \theta \in \omega \quad &&\text{if LLRS} > c, \end{aligned} \tag{47}$$

where c is a suitably chosen constant.

Example 4.3.10 Testing for equality of two exponential distributions.

Let $X_1, \ldots, X_N$ be a sample from an exponential distribution with mean $\lambda_1 > 0$, and $Y_1, \ldots, Y_M$ an independent sample from an exponential distribution with mean $\lambda_2 > 0$. The unconstrained parameter space is

$$\Omega = \left\{(\lambda_1, \lambda_2) \in \mathbb{R}^2 : \lambda_1 > 0,\ \lambda_2 > 0\right\}.$$

Suppose we wish to test the hypothesis $\lambda_1 = \lambda_2$, i.e., the hypothesis that (λ_1, λ_2) is in the constrained set

$$\omega = \left\{(\lambda_1, \lambda_2) \in \mathbb{R}^2 : \lambda_1 = \lambda_2 > 0\right\}.$$

By independence of the two samples, the joint likelihood function of λ_1 and λ_2 is the product of the likelihood function of λ_1 and the likelihood function of λ_2. Using the same technique as in Example 4.3.7, the log likelihood function for observed data $x_1, \ldots, x_N$ and $y_1, \ldots, y_M$ is given by

$$\ell(\lambda_1, \lambda_2) = -N\left(\log\lambda_1 + \frac{1}{\lambda_1} \cdot \frac{1}{N}\sum_{i=1}^{N} x_i\right) - M\left(\log\lambda_2 + \frac{1}{\lambda_2} \cdot \frac{1}{M}\sum_{i=1}^{M} y_i\right).$$

For maximization in Ω, each of the two parts in the log-likelihood function can be maximized independently, giving maximum likelihood estimates

$$\hat{\lambda}_1 = \bar{x} = \frac{1}{N}\sum_{i=1}^{N} x_i$$

and

$$\hat{\lambda}_2 = \bar{y} = \frac{1}{M}\sum_{i=1}^{M} y_i \, .$$

The function $2\ell(\lambda_1, \lambda_2)$ is shown as a contour plot in Figure 4.3.3, for a numerical example with $N = 10$, $\sum_{i=1}^{N} x_i = 35$, $M = 12$, and $\sum_{i=1}^{M} y_i = 66$. The maximum in Ω is taken at $(\lambda_1, \lambda_2) = (5, 5.5)$.

For maximization in ω, we can set $\lambda_1 = \lambda_2 := \lambda$. The function

$$\ell(\lambda, \lambda) = -(N + M)\left[\log\lambda + \frac{1}{\lambda}\cdot\frac{1}{N+M}\left(\sum_{i=1}^{N} x_i + \sum_{i=1}^{M} y_i\right)\right],$$

takes its maximum at $\tilde{\lambda} = (N\bar{x} + M\bar{y})/(N + M) = 5.191$. This is also illustrated in Figure 4.3.3. Maximization of $\ell(\lambda_1, \lambda_2)$ in ω corresponds to maximization on the straight line $\lambda_1 = \lambda_2$. The log-likelihood ratio statistic is

$$\begin{aligned}
\text{LLRS} &= 2\left[\ell(\hat{\lambda}_1, \hat{\lambda}_2) - \ell(\tilde{\lambda}_1, \tilde{\lambda}_2)\right] \\
&= 2\left[-N(\log\hat{\lambda}_1 + 1) - M(\log\hat{\lambda}_2 + 1) + (N + M)(\log\tilde{\lambda} + 1)\right] \qquad (48) \\
&= 2\left(N\log\frac{\tilde{\lambda}}{\hat{\lambda}_1} + M\log\frac{\tilde{\lambda}}{\hat{\lambda}_2}\right).
\end{aligned}$$

By construction, LLRS can take only nonnegative values, and is zero exactly if $\hat{\lambda}_1 = \hat{\lambda}_2$. See Exercise 18. □

As illustrated by Example 4.3.10 and by Figure 4.3.3, the log-likelihood ratio statistic measures how far we have to "walk down" from the top of the log-likelihood mountain to reach the constrained maximum. A test for the hypothesis $\theta \in \omega$ would reject the hypothesis if the constrained maximum $\ell(\tilde{\theta})$ is too small compared to the unconstrained maximum $\ell(\hat{\theta})$. This procedure has considerable intuitive appeal.

Much of the popularity of likelihood ratio testing is because of the fact that, in many situations, the likelihood theory provides an approximation to the distribution of LLRS. As in the case of Theorem 4.3.2, regularity conditions need to be satisfied. In particular, the second-order Taylor series expansions of the log-likelihood function about the maximum likelihood estimates of θ in Ω and in ω, again, is a central technical tool, excluding cases where the maximum is reached at some point on the boundary of the parameter space. Once again, we will give a theorem without proof,

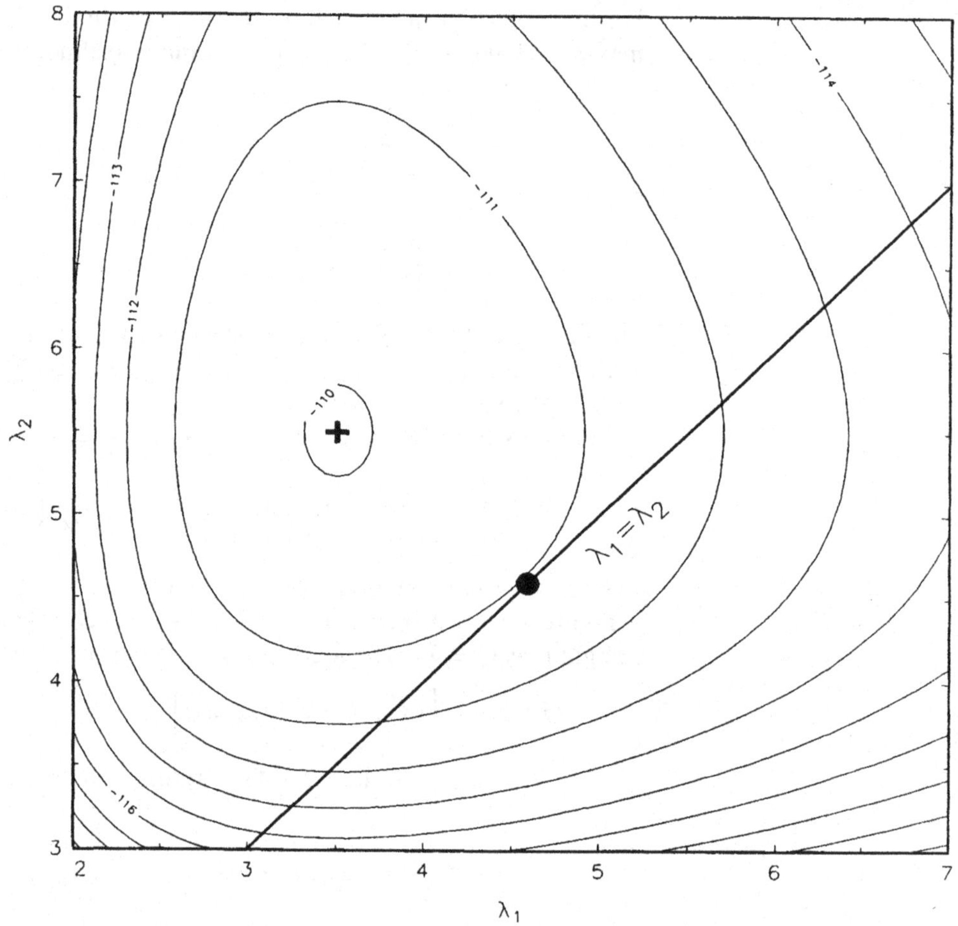

Figure 4-3-3 Contour plot of the log-likelihood function (multiplied by 2) in Example 4.3.10. The solid line corresponds to the constrained parameter space $\lambda_1 = \lambda_2$. The unconstrained and the constrained maxima are marked by a solid plus sign and a solid circle, respectively.

referring the interested reader to the books by Silvey (1975) and by Cox and Hinkley (1974).

Theorem 4.3.3 *In the setup of Definition 4.3.5, suppose that Ω is a k-dimensional subset of $\mathbb{R}^k$ and ω is an r-dimensional subset of Ω. Under appropriate regularity conditions, the asymptotic distribution of the log-likelihood ratio statistic is chi-square with $(k - r)$ degrees of freedom for any $\boldsymbol{\theta} \in \omega$.*

It is important to interpret the word "asymptotic" properly in Theorem 4.3.3. In the case of a single sample, it means that the sample size N tends to infinity. In two-sample cases like Example 4.3.10, it means that both sample sizes have to go to infinity such that their ratio converges to some positive constant. However, a rigorous treatment of this topic is beyond the level of this book.

Example 4.3.11 continuation of Example 4.3.10

In the problem of testing for equality of two exponential distributions, the dimensions of the parameter spaces Ω and ω are 2 and 1, respectively. Thus the distribution of the log-likelihood ratio statistic (48) under the hypothesis $\lambda_1 = \lambda_2$, for large N and M, is approximately chi-square on one degree of freedom. That is, with c denoting the $(1-\alpha)$-quantile of the chi-square distribution with one degree of freedom, we would reject the hypothesis if LLRS $> c$, and accept it otherwise. Such a test has level approximately α. □

No further examples are given here because numerous applications of Theorem 4.3.3 will appear in Chapters 6 and 7. See also Exercises 21 to 23.

Exercises for Section 4.3

1. Let $X_1, \ldots, X_N$ denote a sample from the uniform distribution in the interval $[0, 2\theta]$, where $\theta > 0$ is an unknown parameter.
   (a) Show that $\bar{x}$ is an unbiased estimator of θ.
   (b) Show that the maximum likelihood estimator of θ is $\hat{\theta} = \frac{1}{2} \max_{1 \le i \le N} X_i$.
   (c) Suppose that $x_1 = x_2 = x_3 = 3,\ x_4 = 11$ is an observed sample. Why is $\bar{x}$ a bad estimate of θ?

2. This exercise expands on Example 4.3.1. Let (X_1, X_2, X_3) be a multinomial random vector with parameters θ_i. Graph the likelihood function, similar to Figure 4.3.1, if the observed counts are
   (a) $x_1 = 5,\ x_2 = 0,\ x_3 = 5$.
   (b) $x_1 = 10,\ x_2 = 0,\ x_3 = 0$.

3. This exercise refers to Examples 4.3.2 and 4.3.4. Suppose that $\mathbf{x}_1 = (1.5, 1.8)'$, $\mathbf{x}_2 = (1.8, 1.5)'$ is an observed sample from the uniform distribution inside a circle of radius 1 centered at an unknown location $\boldsymbol{\theta} \in \mathbb{R}^2$. Graph the set of all $\boldsymbol{\theta} \in \mathbb{R}^2$ on which the likelihood function is positive. How does this set change if you add a third observation $\mathbf{x}_3 = (0.3, 0.3)'$?

4. This exercise gives some justification for the use of maximum likelihood estimation when the joint distribution of $X_1, \ldots, X_N$ is continuous. Suppose, for simplicity, that the X_i are *iid*, univariate, each with density function $f(x; \boldsymbol{\theta})$, and cumulative distribution function $F(x; \boldsymbol{\theta})$. Assume that f is a continuous function of x for x in the support of X_i.

(a) Argue that, for small $\epsilon > 0$,

$$\Pr[x_i - \epsilon \le X_i \le x_i + \epsilon] \approx 2\epsilon \cdot f(x_i; \theta).$$

(b) Show that the likelihood function for observed data $\mathbf{x} = (x_1, \ldots, x_N)$ can be written as

$$L(\theta; \mathbf{x}) = \lim_{\epsilon \to 0} \prod_{i=1}^{N} \frac{F(x_i + \epsilon; \theta) - F(x_i - \epsilon_i; \theta)}{2\epsilon}.$$

Thus, argue that, for any two parameter values θ_1 and θ_2, the ratio $L(\theta_1; \mathbf{x})/L(\theta_2; \mathbf{x})$ may be viewed approximately as a ratio of two probabilities.

5. This exercise refers to maximum likelihood estimation in the multinomial model; see Example 4.3.3. Find the maximum likelihood estimates without using a Lagrange multiplier. Instead, substitute $\theta_k = 1 - \theta_1 - \cdots - \theta_{k-1}$, and take partial derivatives of the log-likelihood function with respect to $\theta_1, \ldots, \theta_{k-1}$. Assume that all $x_i > 0$.

6. Show that, in maximum likelihood estimation of the parameters of a multinomial distribution, the same estimator (11) is obtained if some counts x_i are zero. *Hint*: Assume that $x_1 = 0$, write the likelihood function, and argue that $\theta_1 = 0$ at the maximum of the likelihood function in the parameter space.

7. Suppose that $\mathbf{X}_1, \ldots, \mathbf{X}_N$ is a sample from the uniform distribution inside a circle of radius $R > 0$, centered at $\theta \in \mathbb{R}^2$.
   (a) Find the maximum likelihood estimator of R if θ is known. Is it unique?
   (b) Find the maximum likelihood estimator of (R, θ) if both parameters are unknown. Is it unique?

8. A geometric random variable X has probability function

$$f(x; p) = \Pr[X = x; p] = p(1-p)^{x-1} \qquad \text{for } x = 1, 2, \ldots$$

Here, $p \in [0, 1]$ is an unknown parameter.
   (a) Find the likelihood function and the log-likelihood function for a sample of size $N \ge 1$. What happens if all x_i are 1?
   (b) Find the maximum likelihood estimator.
   (c) Verify that the expected value of the score function is zero and that the variance of the score function is the same as the expected value of the information function.
   (d) Find the approximate distribution of the maximum likelihood estimator $\hat{p}$, using Theorem 4.3.2.

9. Repeat Exercise 8 for a sample from a continuous random variable X with *pdf*

$$f(x; \alpha) = \begin{cases} (\alpha + 1)(1 - x)^{\alpha} & \text{for } 0 < x < 1, \\ 0 & \text{otherwise.} \end{cases}$$

Here, $\alpha > -1$ is an unknown parameter.

10. Let $X_1, \ldots, X_N$ be a sample from the univariate normal distribution with unkown mean μ and unknown variance σ^2.

(a) Find the likelihood function and the log-likelihood function.

(b) Use derivatives of the log-likelihood function to show that the maximum likelihood estimators are $\hat{\mu} = \frac{1}{N}\sum_{i=1}^{N} X_i$ and $\hat{\sigma}^2 = \frac{1}{N}\sum_{i=1}^{N}(X_i - \bar{X})^2$. Assume $N \geq 2$ (why?). Take second derivatives to show that this solution corresponds to a maximum rather than just a stationary point.

(c) Let $\mathcal{S}(\mu, \sigma^2; \mathcal{X})$ and $\mathcal{I}(\mu, \sigma^2; \mathcal{X})$ denote the score and information functions, respectively. Show that $E[\mathcal{S}(\mu, \sigma^2; \mathcal{X})] = \mathbf{0}$. Find $\text{Cov}[\mathcal{S}(\mu, \sigma^2; \mathcal{X})]$, and show that it is the same as $E[\mathcal{I}(\mu, \sigma^2; \mathcal{X})]$.

11. Let $\mathbf{X}_1, \ldots, \mathbf{X}_N$ be a sample of size N from $\mathcal{N}_p(\boldsymbol{\mu}, \boldsymbol{\psi})$.

(a) Find the maximum likelihood estimator of $\boldsymbol{\mu}$ if $\boldsymbol{\psi}$ is known.

(b) Find the maximum likelihood estimator of $\boldsymbol{\psi}$ if $\boldsymbol{\mu}$ is known.

(c) What assumptions do you need to make on the sample size N in parts (a) and (b)?

12. This exercise generalizes Example 4.3.6. Suppose that $\mathbf{X}_{j1}, \ldots, \mathbf{X}_{jN_j}$, $j = 1, \ldots, k$, are k independent samples from multivariate normal distributions with mean vectors $\boldsymbol{\mu}_j$ and common covariance matrix $\boldsymbol{\psi}$. Show that the maximum likelihood estimators are $\hat{\boldsymbol{\mu}}_j = \bar{\mathbf{x}}_j = \sum_{i=1}^{N_j} \mathbf{x}_{ji}/N_j$ $(j = 1, \ldots, k)$, and

$$\hat{\boldsymbol{\psi}} = \frac{1}{N_1 + \ldots + N_k} \sum_{j=1}^{k}\sum_{i=1}^{N_j}(\mathbf{x}_{ji} - \bar{\mathbf{x}}_j)(\mathbf{x}_{ji} - \bar{\mathbf{x}}_j)'.$$

What assumptions do you need to make on the sample sizes N_j?

13. This exercise expands on Example 4.3.7. Show that the maximum likelihood estimator $\hat{\theta}$ has *pdf*

$$h(u) = \frac{1}{(N-1)!\,\lambda^N}\, u^{N-1} e^{-u/\lambda}, \quad u \geq 0.$$

(This is a member of the *gamma* family of distributions; see Exercise 12 in Section 2.9).

14. Prove equations (36) and (37).

15. Prove equations (40) to (43). *Hint*: Remember that θ_3 is a function of θ_1 and θ_2.

16. This exercise expands on Example 4.3.9.

(a) Instead of writing θ_3 as a function of θ_1 and θ_2, write $\theta_2 = 1 - \theta_1 - \theta_3$, and find the maximum likelihood estimators (and their joint asymptotic distribution) of θ_1 and θ_3. In particular, this will give you the variance of $\hat{\theta}_3$ and the covariance between $\hat{\theta}_1$ and $\hat{\theta}_3$. Similarly, find the covariance between θ_2 and θ_3.

(c) Prove (44) using the fact that $\hat{\theta}_3 = 1 - \hat{\theta}_1 - \hat{\theta}_2$. Argue that this proves the asymptotic joint normality of $(\hat{\theta}_1, \hat{\theta}_2, \hat{\theta}_3)$.

17. Generalize Example 4.3.9 to the case of a multinomial distribution with $k \geq 2$ categories. In particular, show that

$$\mathrm{cov}(\hat{\theta}_j, \hat{\theta}_h) = \begin{cases} \theta_j(1-\theta_j)/N & \text{if } j = h, \\ -\theta_j\theta_h/N & \text{if } j \neq h. \end{cases}$$

18. This exercise expands on Example 4.3.11.

(a) Show that, for positive numbers a and b, $(a+b)/2 \geq \sqrt{ab}$, with equality exactly if $a = b$. *Hint*: Maximize the function $h(a, b) = ab$ subject to the constraint $a + b = c$, for a positive constant c, and show that the maximum is attained if $a = b = c/2$.

(b) Use part (a) to show that the log-likelihood ratio statistic (48) in the special case $N = M = 1$ can take only nonnegative values, and is zero exactly if $\hat{\lambda}_1 = \hat{\lambda}_2$.

(c) Show that, for positive numbers a and b and positive integers N and M,

$$\frac{Na + Mb}{N + M} \geq \left(a^N b^M\right)^{1/(N+M)},$$

with equality exactly if $a = b$. Use this to show that the log-likelihood ratio statistic (48) can take only nonnegative values.

(d) Argue that this exercise is superfluous because the inequality in part (c) has implicitly been proven by the derivation of the log-likelihood ratio statistic (48).

19. Let $\mathcal{X} = (X_1, \ldots, X_N)$ be a sample from a random variable with *pdf* $f(x; p) = p\, f_1(x) + (1-p)\, f_2(x)$, where $p \in [0, 1]$ is an unknown parameter. Here, $f_1(x)$ is the uniform density in the interval $(0, 2)$, and $f_2(x)$ is the uniform density in the interval $(1, 3)$.

(a) Show that the likelihood function is given by

$$L(p; \mathcal{X}) = 2^{-N} p^b (1-p)^c,$$

where b is the number of observations in the interval $(0, 1)$ and c is the number of observations in the interval $(2, 3)$.

(b) The likelihood function given in part (a) is the same (up to a multiplicative constant) as a binomial likelihood with b successes out of $b+c$ trials, with success probability p. Explain why this makes sense. *Hint*: Graph the *pdf* $f(x; p)$ for one or two selected values of p.

(c) Find the maximum likelihood estimator of p. Does it always exist?

(d) Argue that maximum likelihood estimation (including the application of Theorem 4.3.2) in this example is the same as maximum likelihood estimation of the success probability p based on the value of a binomial random variable Y with N' trials, where N' is itself distributed as binomial$(N, 1/2)$.

20. Let $\theta \in \Omega$ be the unknown parameter in an estimation problem, and let $g(\theta)$ be a one-to-one function from Ω into $\Omega^* = g(\Omega)$. Suppose that $\hat{\theta}$ is the maximum likelihood estimate of θ based on a sample of size N. Show that the maximum likelihood estimate of $\beta = g(\theta)$ is $g(\hat{\theta})$. (This is called the *invariance* property of maximum likelihood estimates).

21. Let $X_1, \ldots, X_N$ denote a sample of size $N \geq 2$ from $\mathcal{N}(\mu, \sigma^2)$, where both parameters are unknown.

(a) Find the maximum likelihood estimate of σ^2 under the hypothesis $\mu = \mu_0$, where μ_0 is a specified hypothetical mean.

(b) Show that the log-likelihood ratio statistic for the hypothesis $\mu = \mu_0$ is given by

$$\text{LLRS} = N \log \frac{\sum_{i=1}^N (X_i - \mu_0)^2}{\sum_{i=1}^N (X_i - \bar{X})^2} .$$

(c) Show that

$$\text{LLRS} = N \log \left(1 + \frac{1}{N-1} t^2\right),$$

where $t = \sqrt{N}(\bar{X} - \mu_0)/s$ is the well-known one-sample t-statistic for the hypothesis $\mu = \mu_0$. (Here, s is the sample standard deviation, defined in terms of the denominator $N - 1$).

(d) Find the large-sample distribution of the log-likelihood ratio statistic under the hypothesis, using Theorem 4.3.3.

22. Let $X_{11}, \ldots, X_{1N_1}$ and $X_{21}, \ldots, X_{2N_2}$ denote independent samples of size $N_j \geq 2$ from $\mathcal{N}(\mu_1, \sigma_1^2)$ and $\mathcal{N}(\mu_2, \sigma_2^2)$, respectively.

(a) Let L_1 denote the log-likelihood ratio statistic for the hypothesis $\sigma_1 = \sigma_2$. Show that L_1 depends on the data only through the two sample variances. Find the asymptotic distribution of L_1, using Theorem 4.3.3.

(b) Let L_2 denote the log-likelihood ratio statistic for the hypothesis $\mu_1 = \mu_2$, assuming $\sigma_1 = \sigma 2$. Find L_2 and show (similarly to Exercise 21) that

$$L_2 = (N_1 + N_2) \log \left(1 + \frac{1}{N_1 + N_2 - 2} t^2\right),$$

where t is the well known t-statistic for equality of the means of two normal distributions. Find the large-sample distribution of L_2, using Theorem 4.3.3.

(c) Let L denote the log-likelihood ratio statistic for the hypothesis ($\mu_1 = \mu_2$, $\sigma_1 = \sigma_2$), i.e., the hypothesis of equality of the two distributions. Find L, and show that $L = L_1 + L_2$. Find the large-sample distribution of L, using Theorem 4.3.3.

23. Let $\begin{pmatrix} X_1 \\ Y_1 \end{pmatrix}, \ldots, \begin{pmatrix} X_1 \\ Y_1 \end{pmatrix}$ be a sample of size $N > 2$ from a bivariate normal distribution with mean vector μ and covariance matrix ψ, assumed positive definite. Consider the hypothesis $\text{cov}[X, Y] = 0$.

(a) Find the unconstrained and the constrained parameter spaces.

(b) Find the maximum likelihood estimators of μ and ψ under the hypothesis. *Hint*: If $\text{cov}[X, Y] = 0$, then X and Y are independent by their joint normality.

(c) Show that the log-likelihood ratio statistic for the hypothesis $\text{cov}[X, Y] = 0$ is given by

$$\text{LLRS} = N \log \frac{1}{1 - r^2},$$

where r is the sample correlation coefficient between the X_i and the Y_i. Show that the large sample distribution of LLRS under the hypothesis is chi-square with one degree of freedom.

24. This exercise gives a generalization of Exercise 23 to a p-variate normal distribution. Let $\mathbf{X}_1, \ldots, \mathbf{X}_N$ $(N > p)$ be a sample from $\mathcal{N}_p(\boldsymbol{\mu}, \boldsymbol{\psi})$, and consider testing the hypothesis that the covariance matrix $\boldsymbol{\psi}$ is diagonal, i.e., $\boldsymbol{\psi} = \text{diag}(\boldsymbol{\psi})$.

(a) Find the maximum likelihood estimate of $\boldsymbol{\mu}$ and $\boldsymbol{\psi}$ under the hypothesis $\boldsymbol{\psi} = \text{diag}(\boldsymbol{\psi})$. *Hint*: By multivariate normality, diagonality of the covariance matrix is equivalent to mutual independence of the p components.

(b) Show that the log-likelihood ratio statistic for hypothesis of diagonality of $\boldsymbol{\psi}$ is given by

$$\text{LLRS} = N \log \frac{\det(\text{diag}\, \mathbf{S})}{\det(\mathbf{S})},$$

where $\mathbf{S}$ is the sample covariance matrix. Use Theorem 4.3.3 to find the large-sample distribution of the test statistic under the hypothesis. What are the degrees of freedom in the chi-square approximation? *Hint*: For $p = 2$, this problem reduces to Exercise 23.

(c) Show that the test statistic from part (b) depends on the data only through the sample correlation matrix.

4.4 Maximum Likelihood Estimation with Incomplete Data

This section presents a rather ingenious method to compute maximum likelihood estimates for incomplete data, commonly called the *EM* -algorithm. The main reason for including the *EM*-algorithm in this book is its application to estimation of the parameters of finite mixture distributions; see Chapter 9. Accordingly, the presentation is relatively short, and proofs are omitted altogether. Comprehensive treatments of the *EM*-algorithm can be found in the fundamental article by Dempster, Laird, and Rubin (1977), and in the books by Little and Rubin (1987) and by McLachlan and Krishnan (1997). Much of the appeal of the *EM*-algorithm is due to the fact that it can be applied to situations that one would typically not associate with missing data. Finite mixtures are such a case. Another interesting application of this type is given in Example 4.4.8.

The *EM*-algorithm is most useful in cases where the computation of maximum likelihood estimates would be simple if complete data had been observed. Suppose for instance that we are sampling from a bivariate normal distribution, but for some of the observations one of the two variables has not been recorded. We will discuss this particular case in Example 4.4.6. However, to introduce the main ideas, we start with an example in which the *EM*-algorithm would not be necessary because an explicit solution is available. This will allow us to verify that both the *EM*-algorithm and the explicit solution give the same results.

Example 4.4.1 Suppose we wish to estimate the probabilities θ_1, θ_2, and θ_3 $(\theta_1 + \theta_2 + \theta_3 = 1)$ in a multinomial random vector as in Example 4.3.9. However, this time we have data from two independent experiments available. In the first experiment there are N_1 observations, with counts Z_1, Z_2, and Z_3 $(Z_1 + Z_2 + Z_3 = N_1)$. In the second experiment with N_2 observations, we register only whether they fall inside or outside category 1. Let Z_1^* be the number of observations in the second experiment that fall into category 1, and $Z_{23}^* = N_2 - Z_1^*$.

The log-likelihood functions, omitting constants that do not depend on any parameter, are $Z_1 \log\theta_1 + Z_2 \log\theta_2 + Z_3 \log\theta_3$ for experiment 1, and $Z_1^* \log\theta_1 + Z_{23}^* \log(1 - \theta_1)$ for experiment 2. By independence, the combined log-likelihood function of the two experiments is

$$\ell^*(\boldsymbol{\theta}) = (Z_1 + Z_1^*)\log\theta_1 + Z_{23}^* \log(1 - \theta_1) + Z_2 \log\theta_2 + Z_3 \log\theta_3\,, \qquad (1)$$

for $\boldsymbol{\theta}$ in the parameter space $\boldsymbol{\Omega} = \left\{\boldsymbol{\theta} \in \mathbb{R}^3 :\ \text{all } \theta_i \geq 0,\ \theta_1 + \theta_2 + \theta_3 = 1\right\}$. In Exercise 1 it is shown that the maximum likelihood estimator of $\boldsymbol{\theta}$ (provided $Z_2 + Z_3 > 0$) is

$$\begin{aligned}\hat{\theta}_1 &= \frac{Z_1 + Z_1^*}{N_1 + N_2}\,,\\ \hat{\theta}_2 &= (1 - \hat{\theta}_1)\,\frac{Z_2}{Z_2 + Z_3}\,,\\ \hat{\theta}_3 &= (1 - \hat{\theta}_1)\,\frac{Z_3}{Z_2 + Z_3}\,. \end{aligned} \qquad (2)$$

We will now outline how a missing data approach would handle the same estimation problem. Suppose that (hypothetically) all observations from experiment 2 had been fully classified as well, with Z_2^* observations falling into category 2, and Z_3^* observations into category 3, where $Z_2^* + Z_3^* = Z_{23}^*$. Then the log-likelihood function, ignoring again constants that do not depend on $\boldsymbol{\theta}$, would be

$$\ell(\boldsymbol{\theta}) = (Z_1 + Z_1^*)\log\theta_1 + (Z_2 + Z_2^*)\log\theta_2 + (Z_3 + Z_3^*)\log\theta_3\,, \qquad (3)$$

and its maximum in $\boldsymbol{\Omega}$ would be taken at $\theta_i = (Z_i + Z_i^*)/(N_1 + N_2)$, $i = 1, 2, 3$. Equation (3) is called the *complete data log-likelihood*, which we could use if no information were missing. Since maximization of the complete data log-likelihood is simple, it would be convenient to substitute some proper values for Z_2^* and Z_3^*. (Again, in view of the explicit solution (2) this is not really necessary for the current example, but it illustrates the point well).

Now, it will be convenient to change to the notation used later in this section. We will use symbols Y for observed data and X for missing data. Thus, let

$$Y_1 = Z_1 + Z_1^* = \# \text{ of observations in category } 1,$$
$$Y_2 = Z_2 = \# \text{ of observations in category } 2,$$
$$Y_3 = Z_3 = \# \text{ of observations in category } 3,$$

and

$$Y_{23} = Z_{23}^* = \# \text{ of observations in category 2 or 3, but not fully classified.}$$

Similarly, let X_2 and X_3 ($X_2 + X_3 = Y_{23}$) denote the missing counts in categories 2 and 3. Write $\mathcal{X} = (X_2, X_3)$, and $\mathcal{Y} = (Y_1, Y_2, Y_3, Y_{23})$. With this notation, the (hypothetical) complete data log-likelihood is given by

$$\ell(\theta; \mathcal{X}, \mathcal{Y}) = Y_1 \log \theta_1 + (Y_2 + X_2) \log \theta_2 + (Y_3 + X_3) \log \theta_3 \,. \tag{4}$$

In contrast, the log-likelihood for the observed data $\mathcal{Y}$ is given by

$$\ell^*(\theta; \mathcal{Y}) = Y_1 \log \theta_1 + Y_{23} \log(1 - \theta_1) + Y_2 \log \theta_2 + Y_3 \log \theta_3 \,. \tag{5}$$

As we will see, the *EM*-algorithm finds the maximum of ℓ^* by maximizing the hypothetical complete data log-likelihood ℓ of equation (4), with "expected values of X_i" substituted for X_i. The quotation marks indicate that this will need some further explanation. □

In the general estimation problem with missing data, we will denote the observed part of the data by $\mathcal{Y}$ and the missing part of the data by $\mathcal{X}$. Together, $(\mathcal{X}, \mathcal{Y})$ constitute the hypothetical complete data. For the methodology of this section to be applicable, we need to know how the complete data likelihood and the observed data likelihood functions are connected. More precisely, we need to be able to specify the distribution of the missing data, given the observed data. Let $f_\theta(\mathbf{y})$ denote the joint *pdf* of the observed data $\mathcal{Y}$, and $g_\theta(\mathbf{x}|\mathbf{y})$ the conditional *pdf* of the missing data $\mathcal{X}$, given $\mathcal{Y} = \mathbf{y}$. Then the joint *pdf* of $\mathcal{Y}$ and $\mathcal{X}$ is $g_\theta(\mathbf{x}|\mathbf{y}) \cdot f_\theta(\mathbf{y})$, which (considered as a function of θ) is the complete data likelihood. At the same time $f_\theta(\mathbf{y})$, considered as a function of θ, is the observed data likelihood. Taking logarithms gives

$$\log [g_\theta(\mathbf{x}|\mathbf{y}) \cdot f_\theta(\mathbf{y})] = \log [f_\theta(\mathbf{y})] + \log [g_\theta(\mathbf{x}|\mathbf{y})] \,,$$

or in the notation of Example 4.4.1,

$$\ell(\theta; \mathbf{x}, \mathbf{y}) = \ell^*(\theta; \mathbf{y}) + \log [g_\theta(\mathbf{x}|\mathbf{y})] \,, \tag{6}$$

where ℓ is the complete data log-likelihood, and ℓ^* is the observed data log-likelihood. This illustrates why the specification of $g_\theta(\mathbf{x}|\mathbf{y})$ is crucial.

Example 4.4.2 continuation of Example 4.4.1

Suppose that in the second experiment the investigator decided that she is mostly interested in θ_1, the probability of the first category, and therefore, did not bother to distinguish between categories 2 and 3. Then conditionally on $Y_{23} = y_{23}$,

$$X_2 \sim \text{binomial}\left(y_{23}, \frac{\theta_2}{\theta_2 + \theta_3}\right),$$

and

$$X_3 \sim \text{binomial}\left(y_{23}, \frac{\theta_3}{\theta_2 + \theta_3}\right) \tag{7}$$

(see Exercise 2), that is, we can specify the distribution of the missing data, given the observed data. In fact, (7) gives the conditional distribution of X_2 and X_3, given $\mathcal{Y} = \mathbf{y}$, because X_2 and X_3 depend on $\mathcal{Y}$ only through Y_{23}.

On the other hand, suppose that in experiment 2 the investigator first observed all three categories, but found that there were no observations in category 2, and therefore, collapsed categories 2 and 3. Then the specification of the conditional distribution (7) would not be appropriate because conditionally on the observed data, $\Pr[X_3 = y_{23}] = 1$. □

The *EM*-algorithm is a method for maximizing the observed data log-likelihood $\ell^*(\boldsymbol{\theta}; \mathcal{Y})$ iteratively, using maximization of the complete data log-likelihood $\ell(\boldsymbol{\theta}; \mathcal{X}, \mathcal{Y})$ with the missing part $\mathcal{X}$ "properly replaced." The algorithm starts with some initial parameter value $\boldsymbol{\theta}^{(0)}$ in the parameter space Ω and computes the expectation of $\ell(\boldsymbol{\theta}; \mathcal{X}, \mathcal{Y})$, conditionally on the observed data $\mathcal{Y}$, using $\boldsymbol{\theta}^{(0)}$ as the parameter in the conditional distribution of $\mathcal{X}$. That is, we compute

$$\ell_1(\boldsymbol{\theta}; \mathcal{X}, \mathcal{Y}) = E\left[\ell(\boldsymbol{\theta}) | \mathcal{Y}, \boldsymbol{\theta}^{(0)}\right]. \tag{8}$$

This is called the E-step (expectation step) of the algorithm. Next, we maximize the function $\ell_1(\boldsymbol{\theta})$ with respect to $\boldsymbol{\theta}$ in the parameter space Ω and call the parameter value, for which the maximum is reached, $\boldsymbol{\theta}^{(1)}$. This is called the M-step (maximization step) of the algorithm. Next, follows another E-step, giving

$$\ell_2(\boldsymbol{\theta}; \mathcal{X}, \mathcal{Y}) = E\left[\ell(\boldsymbol{\theta}) | \mathcal{Y}, \boldsymbol{\theta}^{(1)}\right], \tag{9}$$

and another M-step, yielding $\boldsymbol{\theta}^{(2)}$ as the value of $\boldsymbol{\theta}$ that maximizes ℓ_2. Iteration between the two steps is continued until the values $\boldsymbol{\theta}^{(t)}$ and $\boldsymbol{\theta}^{(t+1)}$ are identical in two consecutive iterations. At this point, the algorithm is stopped because we would always obtain the same parameter value in any further iterations. In practice, we will stop iterating when two consecutive parameter values $\boldsymbol{\theta}^{(t)}$ and $\boldsymbol{\theta}^{(t+1)}$ differ only by some very small amount. We will return to this point later.

Example 4.4.3 continuation of Example 4.4.1

Suppose that we have observed the following counts: $Y_1 = 5$, $Y_2 = 3$, $Y_3 = 2$, and $Y_{23} = 10$. Then the observed data log-likelihood, ignoring terms that do not depend on the parameter, is given by

$$\ell^*(\boldsymbol{\theta}) = 5\log\theta_1 + 10\log(1-\theta_1) + 3\log\theta_2 + 2\log\theta_3. \tag{10}$$

This function is shown as a contour plot in Figure 4.4.1. By (2) and Exercise 3, it takes its maximum at $\hat{\theta}_1 = 0.25$, $\hat{\theta}_2 = 0.45$, and $\hat{\theta}_3 = 0.3$. Let X_2 and X_3 be the unobserved counts in categories 2 and 3. Then $X_2 + X_3 = Y_{23} = 10$. The complete data log-likelihood, again ignoring constants, is given by

$$\ell(\boldsymbol{\theta}) = 5\log\theta_1 + (3+X_2)\log\theta_2 + (2+X_3)\log\theta_3. \tag{11}$$

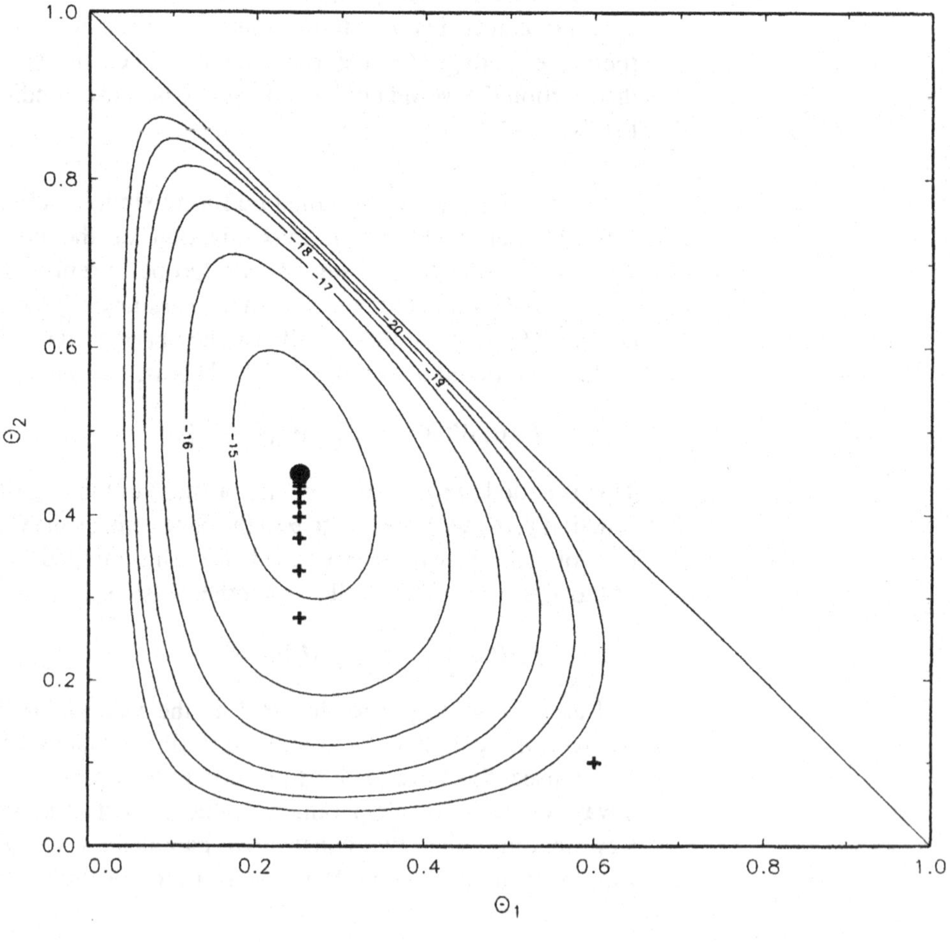

Figure 4-4-1 Contour plot of the observed data log-likelihood function in Example 4.4.3. The "+" signs represent parameter values in 20 successive iterations of the *EM*-algorithm. See Table 4.4.1.

Initializing the algorithm arbitrarily at $\boldsymbol{\theta}^{(0)} = \left[\theta_1^{(0)}, \theta_2^{(0)}, \theta_3^{(0)}\right]' = (0.6, 0.1, 0.3)'$, the value of ℓ^* at $\boldsymbol{\theta}^{(0)}$ is $\ell^*\left[\boldsymbol{\theta}^{(0)}\right] = -21.0327$. The first iteration of the *EM*-algorithm takes the expectation of $\ell(\boldsymbol{\theta})$ with respect to the variables X_2 and X_3, given the observed counts $\mathcal{Y}$ and using the current parameter value $\boldsymbol{\theta}^{(0)}$ to compute the expectation. Because $\ell(\boldsymbol{\theta})$ is linear in X_2 and X_3, this amounts to computing $E\left[X_2|\mathcal{Y}; \boldsymbol{\theta}^{(0)}\right]$ and $E\left[X_3|\mathcal{Y}; \boldsymbol{\theta}^{(0)}\right]$. By (7), we consider X_2 binomial with $y_{23} = 10$ trials and success probability $\theta_2^{(0)}/(\theta_2^{(0)}+\theta_3^{(0)}) = 1/4$. Similarly, X_3 is considered binomial with 10 trials and success probability 3/4. Hence, $E\left[X_2|\mathcal{Y}; \boldsymbol{\theta}^{(0)}\right] = 2.5$ and $E\left[X_3|\mathcal{Y}; \boldsymbol{\theta}^{(0)}\right] = 7.5$. These two values are substituted for X_2 and X_3 in (11), giving

$$\ell_1(\boldsymbol{\theta}) = 5\log\theta_1 + 5.5\log\theta_2 + 9.5\log\theta_3.$$

This completes the first E-step. The M-step finds the maximum of $\ell_1(\boldsymbol{\theta})$ for $\boldsymbol{\theta} \in \boldsymbol{\Omega}$. By Example 4.3.3, the maximum is taken for

$$\boldsymbol{\theta}^{(1)} = \begin{bmatrix} \theta_1^{(1)} \\ \theta_2^{(1)} \\ \theta_3^{(1)} \end{bmatrix} = \begin{pmatrix} 5/20 \\ 5.5/20 \\ 9.5/20 \end{pmatrix} = \begin{pmatrix} 0.25 \\ 0.275 \\ 0.475 \end{pmatrix}.$$

At this point, the value of the observed data log-likelihood is $\ell^*\left(\boldsymbol{\theta}^{(1)}\right) = -15.1701$, which is an improvement over the initial value. This completes the first iteration of the *EM*-algorithm.

In the second iteration, again, we take the expectation of $\ell(\boldsymbol{\theta})$ with respect to the missing data, given the observed data, and using $\boldsymbol{\theta}^{(1)}$ as the parameter in the conditional distribution. Following the same steps as above, this amounts to considering the conditional distributions of X_2 and X_3, given $\mathcal{Y}$, as

$$X_2 \sim \text{binomial}\left[y_{23}, \frac{\theta_2^{(1)}}{\theta_2^{(1)} + \theta_3^{(1)}}\right],$$

and

$$X_3 \sim \text{binomial}\left[y_{23}, \frac{\theta_3^{(1)}}{\theta_2^{(1)} + \theta_3^{(1)}}\right],$$

with expectations $E\left[X_2|\mathcal{Y}; \boldsymbol{\theta}^{(1)}\right] = 11/3$ and $E\left[X_3|\mathcal{Y}; \boldsymbol{\theta}^{(1)}\right] = 19/3$. Thus, the second E-step updates the complete data log-likelihood to

$$\ell_2(\boldsymbol{\theta}) = 5\log\theta_1 + \left(3 + \frac{11}{3}\right)\log\theta_2 + \left(2 + \frac{19}{3}\right)\log\theta_3,$$

and the second M-step finds the maximum of $\ell_2(\boldsymbol{\theta})$ for $\boldsymbol{\theta} \in \boldsymbol{\Omega}$, which is taken at

$$\boldsymbol{\theta}^{(2)} = \begin{bmatrix} \theta_1^{(2)} \\ \theta_2^{(2)} \\ \theta_3^{(2)} \end{bmatrix} = \begin{pmatrix} \frac{5}{20} \\ \frac{20/3}{20} \\ \frac{25/3}{20} \end{pmatrix} \approx \begin{pmatrix} 0.25 \\ 0.3333 \\ 0.4167 \end{pmatrix}.$$

At this point, the value of the observed data log-likelihood is $\ell^*\left(\boldsymbol{\theta}^{(2)}\right) = -14.8551$, which, again, is an improvement.

Continuing the iterations, the jth execution of the E-step consists of finding

$$E\left[X_2|\mathcal{Y};\boldsymbol{\theta}^{(j-1)}\right] = y_{23}\cdot\frac{\theta_2^{(j-1)}}{\theta_2^{(j-1)}+\theta_3^{(j-1)}}$$

and

$$E\left[X_3|\mathcal{Y};\boldsymbol{\theta}^{(j-1)}\right] = y_{23}\cdot\frac{\theta_3^{(j-1)}}{\theta_2^{(j-1)}+\theta_3^{(j-1)}},$$

replacing X_2 and X_3 by these conditional expectations in the complete data log-likelihood, and therefore obtaining an updated log-likelihood function $\ell_j(\boldsymbol{\theta})$. The jth execution of the M-step finds $\boldsymbol{\theta}^{(j)}$ which maximizes $\ell_j(\boldsymbol{\theta})$ over $\boldsymbol{\theta}\in\boldsymbol{\Omega}$. This is summarized in Table 4.4.1 for 20 iterations. The table shows that the sequence $\boldsymbol{\theta}^{(j)}$ approaches the maximum likelihood estimates $\hat{\theta}_1 = 0.25$, $\hat{\theta}_2 = 0.45$, and $\hat{\theta}_3 = 0.3$, and the observed data log-likelihood $\ell^*(\boldsymbol{\theta}^{(j)})$ increases in each step. The sequence $\boldsymbol{\theta}^{(0)}, \boldsymbol{\theta}^{(1)}, \ldots$ is also illustrated in Figure 4.4.1, where "+" signs indicate the pairs $\left[(\theta_1^{(j)},\theta_2^{(j)}\right]$ for $j = 0, 1, \ldots, 20$. Because $\theta_1^{(j)}$ does not depend on the missing data for $j\geq 1$, the final value $\hat{\theta}_1$ is reached in the first iteration, and all subsequent iterations change only the values of θ_2 and θ_3.

A generalization of this example is given in Exercise 4. □

Summarizing the material presented so far and using the same notation, we can formally define the *EM*-algorithm as a computational procedure, which iterates (until convergence) between the following two steps:

- *E-step*: compute $\ell_j(\boldsymbol{\theta}) = E\left[\ell(\boldsymbol{\theta})|\mathcal{Y};\boldsymbol{\theta}^{(j-1)}\right]$, where $\ell(\boldsymbol{\theta})$ is the complete data log-likelihood and the expectation is taken with respect to the conditional distribution of the missing data $\mathcal{X}$, given the observed data $\mathcal{Y}$, using the parametric value $\boldsymbol{\theta}^{(j-1)}$ to calculate the expectation.
- *M-step*: Find the value $\boldsymbol{\theta}^{(j)}$ in the parameteric space which maximizes $\ell_j(\boldsymbol{\theta})$.

As illustrated in Example 4.4.3, iterations between the two steps increase the log-likelihood function $\ell^*(\boldsymbol{\theta};\mathcal{Y})$ of the observed data in each step. This is a central result, so we state it as a theorem.

Theorem 4.4.1 *Under general conditions, each iteration of the EM-algorithm increases the observed data log-likelihood, unless a stationary point has been reached. That is,* $\ell^*(\boldsymbol{\theta}^{(j+1)})\geq\ell^*(\boldsymbol{\theta}^{(j)})$. *The maximum likelihood estimate* $\hat{\boldsymbol{\theta}}$ *based on the observed data is a stationary point of the sequence* $\left\{\boldsymbol{\theta}^{(j)}\right\}_{j\geq 0}$.

Table 4.4.1 The first 20 iterations of the *EM*-algorithm in Example 4.4.3

Iteration (j)	$\theta_1^{(j)}$	$\theta_2^{(j)}$	$\theta_3^{(j)}$	$\ell^*(\theta^{(j)})$
0	0.6000	0.1000	0.3000	−21.0327
1	0.2500	0.2750	0.4750	−15.1701
2	0.2500	0.3333	0.4167	−14.8551
3	0.2500	0.3722	0.3778	−14.7200
4	0.2500	0.3981	0.3519	−14.6602
5	0.2500	0.4154	0.3346	−14.6334
6	0.2500	0.4270	0.3230	−14.6215
7	0.2500	0.4346	0.3154	−14.6161
8	0.2500	0.4398	0.3102	−14.6137
9	0.2500	0.4432	0.3068	−14.6126
10	0.2500	0.4454	0.3046	−14.6121
11	0.2500	0.4470	0.3030	−14.6119
12	0.2500	0.4480	0.3020	−14.6118
13	0.2500	0.4487	0.3013	−14.6118
14	0.2500	0.4491	0.3009	−14.6118
15	0.2500	0.4494	0.3006	−14.6118
16	0.2500	0.4496	0.3004	−14.6118
17	0.2500	0.4497	0.3003	−14.6118
18	0.2500	0.4498	0.3002	−14.6118
19	0.2500	0.4499	0.3001	−14.6118
20	0.2500	0.4499	0.3001	−14.6118

The first line displays the initial parameter values. Also see Figure 4.4.1.

Proof See Dempster, Laird, and Rubin (1977), Little and Rubin (1987, Chapter 7), or McLachlan and Krishnan (1997, Chapter 3) for a proof as well as technical conditions. ∎

Theorem 4.4.1 tells us that the value of the observed data log-likelihood will increase or stay the same in each iteration of the algorithm (i.e., execution of an E-step followed by an M-step). In the latter case, a stationary point has been reached, which is a candidate for the maximum likelihood estimate $\hat{\boldsymbol{\theta}}$. If the observed data log-likelihood has a unique maximum (as in Example 4.4.3) and no other stationary point, this means that the sequence $\{\boldsymbol{\theta}^{(j)}\}_{j\geq 0}$, produced by the EM-algorithm, converges to $\hat{\boldsymbol{\theta}}$.

How many iterations are needed to find the maximum likelihood estimate? In theory we would stop when $\ell^*(\boldsymbol{\theta}^{(j+1)}) = \ell^*(\boldsymbol{\theta}^{(j)})$ or when $\boldsymbol{\theta}^{(j+1)} = \boldsymbol{\theta}^{(j)}$ at some point in the sequence. In practice, we have to stop when the results produced in two consecutive iterations are almost identical. For instance, we may declare convergence when $\ell^*(\boldsymbol{\theta}^{(j+1)}) - \ell^*(\boldsymbol{\theta}^{(j)}) < \epsilon$ for some small $\epsilon > 0$. Alternatively, we may stop when the maximum absolute value in the vector of differences $\boldsymbol{\theta}^{(j+1)} - \boldsymbol{\theta}^{(j)}$ is less than ϵ. We will usually follow the latter criterion.

The theoretical convergence criterion can sometimes also be used to find an explicit formula for the maximum likelihood estimator $\hat{\boldsymbol{\theta}}$. This method works when it is possible to write $\boldsymbol{\theta}^{(j+1)}$ explicitly as a function of $\boldsymbol{\theta}^{(j)}$, say $\boldsymbol{\theta}^{(j+1)} = h(\boldsymbol{\theta}^{(j)})$, and solve the equation $\hat{\boldsymbol{\theta}} = h(\hat{\boldsymbol{\theta}})$. This is illustrated by the continuation of our now familiar trinomial example.

Example 4.4.4 continuation of Example 4.4.3

As seen before, the first parameter (θ_1) reaches its final value $\hat{\theta}_1$ in the first iteration, because $\theta_1^{(1)}$ does not depend on the missing data X_2 and X_3. For θ_2, the jth E-step and M-step can be combined to give

$$\theta_2^{(j)} = \frac{Y_2 + E[X_2|\mathcal{Y}; \theta^{(j-1)}]}{N}, \tag{12}$$

where $N = Y_1 + Y_2 + Y_3 + Y_{23}$. But $E[X_2|\mathcal{Y}, \theta^{(j-1)}] = Y_{23}\theta_2^{(j-1)}/[1 - \theta_1^{(j-1)}]$, and therefore, we obtain the equation

$$N\theta_2^{(j)} = Y_2 + Y_{23}\frac{\theta_2^{(j-1)}}{1 - \theta_1^{(j-1)}}. \tag{13}$$

For $j > 1$, we have seen that $\theta_1^{(j-1)} = Y_1/N = \hat{\theta}_1$. Therefore, setting $\theta_2^{(j)} = \theta_2^{(j-1)} = \hat{\theta}_2$ in equation (13) and solving for $\hat{\theta}_2$ gives

$$\hat{\theta}_2 = (1 - \hat{\theta}_1)\frac{Y_2}{N - Y_{23} - Y_1} = (1 - \hat{\theta}_1)\frac{Y_2}{Y_2 + Y_3}. \tag{14}$$

This is the same as the solution found in Example 4.4.1; see also Exercise 3. The equation $\hat{\theta}_3 = (1 - \hat{\theta}_1)Y_3/(Y_2 + Y_3)$ is obtained similarly. □

Example 4.4.5 The *univariate normal distribution* with randomly missing data

This is an artificial example because the missing data approach is not needed at all. However, it serves two purposes: First, it gives some more theoretical insight; and second, it prepares us for the more relevant case of a multivariate normal distribution with randomly missing measurements. Suppose that N observations have been made from the normal distribution with mean μ and variance τ, but, for some reason unrelated to the outcome of the experiment, only the first N_1 measurements were recorded, and the last $N_2 = N - N_1$ values were lost. Let $Y_1, \ldots, Y_{N_1}$ denote the observed data, and $X_1, \ldots, X_{N_2}$ the lost data. Because the missing part of the data $\mathcal{X}$ is independent of the observed part $\mathcal{Y}$, the observed data likelihood is just a normal likelihood function based on N_1 *iid* random variables, and the maximum likelihood estimates are $\hat{\mu} = \bar{Y}$ and $\hat{\tau} = \frac{1}{N_1}\sum_{i=1}^{N_1}(Y_i - \hat{\mu})^2 = \frac{1}{N_1}\sum_{i=1}^{N_1} Y_i^2 - \bar{Y}^2$. But let us see how the *EM*-algorithm approaches this problem. The complete data log-likelihood, up to a constant that does not depend on the parameters, is given by

$$\begin{aligned}\ell(\mu, \tau; \mathcal{X}, \mathcal{Y}) &= -\frac{N_1}{2}\log\tau - \frac{1}{2\tau}\sum_{i=1}^{N_1}(Y_i-\mu)^2 - \frac{N_2}{2}\log\tau - \frac{1}{2\tau}\sum_{i=1}^{N_2}(X_i-\mu)^2 \\ &= -\frac{N}{2}\log\tau - \frac{1}{2\tau}\left[N\mu^2 - 2\mu\left(\sum_{i=1}^{N_1}Y_i + \sum_{i=1}^{N_2}X_i\right) + \sum_{i=1}^{N_1}Y_i^2 + \sum_{i=1}^{N_2}X_i^2\right].\end{aligned} \tag{15}$$

The X_i are missing at random, and therefore, the conditional distribution of X_i, given $\mathcal{Y}$, is just the same $\mathcal{N}(\mu, \tau)$ distribution. Suppose that the current parameter estimates in the *EM*-algorithm are $\mu^{(j)}$ and $\tau^{(j)}$. Because the complete data log-likelihood function (15) is linear in the X_i and X_i^2, computing the conditional expectation of $\ell(\mu, \tau)$ with respect to the missing data amounts to finding $E[X_i]$ and $E[X_i^2]$, using the current $\mu^{(j)}$ and $\tau^{(j)}$ as parameters. That is, we will need $E[X_i|\mathcal{Y}; \mu^{(j)}, \tau^{(j)}] = \mu^{(j)}$ and $E[X_i^2|\mathcal{Y}; \mu^{(j)}, \tau^{(j)}] = \left(\mu^{(j)}\right)^2 + \tau^{(j)}$. Therefore, the jth E-step of the algorithm updates the complete data log-likelihood as

$$\ell_j(\mu, \tau) = -\frac{N}{2}\log\tau - \frac{1}{2\tau}\left[N\mu^2 - 2\mu A^{(j)} + B^{(j)}\right], \tag{16}$$

where

$$A^{(j)} = \sum_{i=1}^{N_1} Y_i + N_2\mu^{(j-1)} \tag{17}$$

and

$$B^{(j)} = \sum_{i=1}^{N_1} Y_i^2 + N_2\left[\left(\mu^{(j-1)}\right)^2 + \tau^{(j-1)}\right]. \tag{18}$$

The maximum of ℓ_j is taken at

$$\mu^{(j)} = \frac{1}{N}A^{(j)}$$

and

$$\tau^{(j)} = \frac{1}{N}B^{(j)} - \left(\mu^{(j)}\right)^2 . \tag{19}$$

Thus, the E-step consists of equations (17) and (18), and the M-step is represented by (19).

Although it is not obvious from equations (17) to (19), the algorithm, indeed, converges to the solution $\hat{\mu} = \bar{Y}$ and $\hat{\tau} = \frac{1}{N_1}\sum_{i=1}^{N_1}(Y_i - \hat{\mu})^2 = \frac{1}{N_1}\sum_{i=1}^{N_1} Y_i^2 - \bar{Y}^2$. This can be seen again by setting $\mu^{(j+1)} = \mu^{(j)} = \hat{\mu}$, which, by (17), gives

$$\hat{\mu} = \frac{1}{N}\left(\sum_{i=1}^{N_1} Y_i + N_2\hat{\mu}\right). \tag{20}$$

Similarly, setting $\tau^{(j+1)} = \tau^{(j)} = \hat{\tau}$ gives, by (17) to (19),

$$\hat{\tau} = \frac{1}{N}\left[\sum_{i=1}^{N_1} Y_i^2 + N_2(\hat{\mu}^2 + \hat{\tau})\right] - \hat{\mu}^2 . \tag{21}$$

Solving (20) and (21) for $\hat{\mu}$ and $\hat{\tau}$ gives the expected solution; see Exercise 5. A multivariate generalization is given in Exercise 7. □

The main lesson from Example 4.4.5 is that the E-step in the EM-algorithm, in general, is not the same as replacing the missing data by their expectation. Rather, the E-step consists of finding the expectation of the complete data log-likelihood function $\ell(\boldsymbol{\theta}; \mathcal{Y}, \mathcal{X})$ with respect to the conditional distribution of the missing data $\mathcal{X}$, given $\mathcal{Y}$. In Example 4.4.5, this means finding both $E[X_i]$ and $E[X_i^2]$.

A more relevant situation of incomplete univariate data occurs when observations are censored. Suppose, for instance, that the missing observations are not missing at random in Example 4.4.5. Instead, they are missing because they exceed the largest value that can be registered by the measuring instrument. Suppose this largest value is some constant c; then the observed data Y_i are a sample from the conditional distribution of a random variable Y, given $Y \le c$. Similarly, the missing data are a sample from the same random variable, conditionally on $Y > c$. The EM-algorithm is applicable to this situation, but not in the form of Example 4.4.5. In fact, censored data are an important topic in survival analysis; see, e.g., the book by Cox and Oakes (1984). In Example 4.4.1, we might consider the second experiment as a case of censoring because, for observations outside category 1, the exact category is not recorded. Also see Exercise 16.

Example 4.4.6 Bivariate normal data with randomly missing measurements

First, we will discuss the general p-variate case and then study details for $p = 2$. Suppose that $\mathbf{Z}_1, \ldots, \mathbf{Z}_N$ $(N > p)$ is a sample from $\mathcal{N}_p(\boldsymbol{\mu}, \boldsymbol{\psi})$, and the data is complete. Then the log-likelihood function, up to an additive constant that does not depend on the parameters, is given by

$$\ell(\boldsymbol{\mu}, \boldsymbol{\psi}) = -\frac{1}{2}\left[N \log(\det \boldsymbol{\psi}) + N\boldsymbol{\mu}'\boldsymbol{\psi}^{-1}\boldsymbol{\mu} + \operatorname{tr}\left(\boldsymbol{\psi}^{-1}\sum_{i=1}^{N}\mathbf{Z}_i\mathbf{Z}_i'\right) - 2\boldsymbol{\mu}'\boldsymbol{\psi}^{-1}\sum_{i=1}^{N}\mathbf{Z}_i\right]; \tag{22}$$

see Exercise 8. For the complete data, we know that the maximum likelihood estimates are

$$\hat{\boldsymbol{\mu}} = \bar{\mathbf{Z}} = \frac{1}{N}\sum_{i=1}^{N}\mathbf{Z}_i$$

and

$$\hat{\psi} = \frac{1}{N}\sum_{i=1}^{N} \mathbf{Z}_i \mathbf{Z}_j' - \hat{\mu}\hat{\mu}'. \tag{23}$$

Thus, the M-step will be simple. Now, suppose that some of the observations have one or several components missing at random. (By Exercise 7, any $\mathbf{Z}_i$ missing completely can be omitted from the analysis.) Let $\mathcal{X}$ denote all the missing data, and $\mathcal{Y}$ the observed data. Because

$$E\left[\operatorname{tr}\left(\psi^{-1}\sum_{i=1}^{N}\mathbf{Z}_i\mathbf{Z}_i'\right)\right] = \operatorname{tr}\left(\psi^{-1}\sum_{i=1}^{N}E[\mathbf{Z}_i\mathbf{Z}_i']\right)$$

and

$$E[\mu'\psi^{-1}\sum_{i=1}^{N}\mathbf{Z}_i] = \mu'\psi^{-1}\sum_{i=1}^{N}E[\mathbf{Z}_i],$$

equation (22) shows that the E-step requires finding $E[\mathbf{Z}_i\mathbf{Z}_i']$ and $E[\mathbf{Z}_i]$ for each observation with at least one missing component. Also, because the $\mathbf{Z}_i$ are independent, conditioning on $\mathcal{Y}$ when finding these expectations means just conditioning on the observed part of $\mathbf{Z}_i$. Writing $\theta^{(j)} = \left[\mu^{(j)}, \psi^{(j)}\right]$ for the optimal parameter values obtained after j iterations of the EM-algorithm, the jth E-step finds

$$\mathbf{A}^{(j)} = E\left[\sum_{i=1}^{N}\mathbf{Z}_i|\mathcal{Y};\theta^{(j-1)}\right] = \sum_{i=1}^{N}E\left[\mathbf{Z}_i|\mathcal{Y};\theta^{(j-1)}\right] \tag{24}$$

and

$$\mathbf{B}^{(j)} = E\left[\sum_{i=1}^{N}\mathbf{Z}_i\mathbf{Z}_i'|\mathcal{Y};\theta^{(j-1)}\right] = \sum_{i=1}^{N}E\left[\mathbf{Z}_i\mathbf{Z}_i'|\mathcal{Y};\theta^{(j-1)}\right], \tag{25}$$

thereby updating the complete data log-likelihood to

$$\ell(\mu,\psi) = -\frac{1}{2}\left[N\log(\det\psi) + N\mu'\psi^{-1}\mu + 2\mu'\psi^{-1}\mathbf{A}^{(j)} + \operatorname{tr}\left(\psi^{-1}\mathbf{B}^{(j)}\right)\right]. \tag{26}$$

The jth M-step updates the parameters to

$$\mu^{(j)} = \frac{1}{N}\mathbf{A}^{(j)},$$

and

$$\psi^{(j)} = \frac{1}{N}\mathbf{B}^{(j)} - \mu^{(j)}\mu^{(j)\prime}. \tag{27}$$

Now, we will work out the details for the bivariate case ($p = 2$), but refer the reader to Little and Rubin (1987, Chapter 8) for the general multivariate case. Let $\mathbf{Z}_i = \begin{pmatrix} Z_{i1} \\ Z_{i2} \end{pmatrix}$ denote the jth observation, where, at most, one of the Z_{i1} and Z_{i2} is missing. Write

$\mu = \begin{pmatrix} \mu_1 \\ \mu_2 \end{pmatrix}$ and $\psi = \begin{pmatrix} \sigma_{11} & \sigma_{12} \\ \sigma_{21} & \sigma_{22} \end{pmatrix}$. There are three possible cases: (i) both Z_{i1} and Z_{i2} are observed; (ii) Z_{i1} is observed but Z_{i2} is missing; and (iii) Z_{i2} is observed but Z_{i1} is missing. Equations (24) and (25) mean that the statistics

$$a_1 = \sum_{i=1}^{N} Z_{i1},$$

$$a_2 = \sum_{i=1}^{N} Z_{i2},$$

$$b_{11} = \sum_{i=1}^{N} Z_{i1}^2,$$

$$b_{22} = \sum_{i=1}^{N} Z_{i2}^2,$$

and

$$b_{12} = \sum_{i=1}^{N} Z_{i1} Z_{i2} \tag{28}$$

must be updated according to the missing value pattern. In case (i), the observed Z_{i1} and Z_{i2} are used. In case (ii), the conditional distribution of Z_{i2}, given Z_{i1}, is normal with mean $\alpha + \beta Z_{i1}$ and variance $\sigma_{22.1}$, where $\beta = \sigma_{12}/\sigma_{11}$, $\alpha = \mu_2 - \beta\mu_1$, and $\sigma_{22.1} = \sigma_{22} - \sigma_{12}^2/\sigma_{11}$. Denoting by $\alpha^{(j)}$, $\beta^{(j)}$, and $\sigma_{22.1}^{(j)}$ the analogous quantities based on the current $\theta^{(j)}$, the jth E-step replaces the missing Z_{i2} in the update of a_2 in equation (28) by

$$E\left[Z_{i2}|\mathcal{Y};\theta^{(j-1)}\right] = E\left[Z_{i2}|Z_{i1};\theta^{(j-1)}\right] = \alpha^{(j-1)} + \beta^{(j-1)}Z_{i1}. \tag{29}$$

Similarly, when updating b_{12}, we replace $Z_{i1}Z_{i2}$ by

$$E\left[Z_{i1}Z_{i2}|\mathcal{Y};\theta^{(j-1)}\right] = Z_{i1}E\left[Z_{i2}|Z_{i1};\theta^{(j-1)}\right] = Z_{i1}\left[\alpha^{(j-1)} + \beta^{(j-1)}Z_{i1}\right], \tag{30}$$

and, finally, in the update of b_{22}, Z_{i2}^2 is replaced by

$$\begin{aligned} E\left[Z_{i2}^2|\mathcal{Y};\theta^{(j-1)}\right] &= \operatorname{var}\left[Z_{i2}|Z_{i1};\theta^{(j-1)}\right] + \left\{E\left[Z_{i2}|Z_{i1};\theta^{(j-1)}\right]\right\}^2 \\ &= \sigma_{22.1}^{(j-1)} + \left\{\alpha^{(j-1)} + \beta^{(j-1)}Z_{i1}\right\}^2. \end{aligned} \tag{31}$$

Similarly, the regression of the first variable on the second variable is used for the missing Z_{i1} in case (iii); see Exercise 9.

As already seen in Example 4.4.5, the E-step does not just replace the missing data points by their conditional expectation. Rather, it is the conditional expectation of the complete data log-likelihood that needs to be calculated. Notice also that we assumed that the missing data points are missing at random, which enables us to use the theory of conditional distributions in a multivariate normal. If data points are missing because they are too small or too large, then we would be in a case of censoring, and the method given in this example would not be correct. □

Numerical applications illustrating the theory of Example 4.4.6 are given in Exercises 10 and 11.

Example 4.4.7 This example anticipates some material from Chapter 9. Suppose X is a binary variable taking value 1 with probability π and value 0 with probability $1-\pi$, $(0 \le \pi \le 1)$. Let Y be a (discrete or continuous) random variable such that, conditionally on $X = j$, Y has *pdf* $f_j(y)$, $j = 0, 1$. Then the joint *pdf* of X and Y is given by

$$f_{XY}(x, y) = [\pi\, f_1(y)]^x\, [(1-\pi)\, f_0(y)]^{1-x} \tag{32}$$

for $x = 0, 1$. As seen in Section 2.8, the marginal *pdf* of Y is a mixture density,

$$f_Y(y) = \pi\, f_1(y) + (1-\pi)\, f_0(y). \tag{33}$$

Suppose that f_1 and f_2 contain no unknown parameters and we wish to estimate the unknown proportion π from a sample of size N. As we will see in Chapter 9, the observed data log-likelihood function based on the mixture density (33) is quite tricky to handle. On the other hand, if we assume a complete data sample $\mathbf{Z}_i = (X_i, Y_i)$, $i = 1, \ldots, N$, then the complete data likelihood function is

$$\begin{aligned} L(\pi; \mathcal{X}, \mathcal{Y}) &= \prod_{i=1}^{N} [\pi\, f_1(Y_i)]^{X_i}\, [(1-\pi)\, f_0(Y_i)]^{1-X_i} \\ &= \left\{ \prod_{i=1}^{N} [f_1(Y_i)]^{X_i}\, [f_0(Y_i)]^{1-X_i} \right\} \cdot \pi^S (1-\pi)^{N-S}, \end{aligned} \tag{34}$$

where $S = \sum_{i=1}^N X_i$. The term in braces of (34) does not depend on any parameters, and, therefore, is a constant for the purpose of maximization. The complete data log-likelihood, up to an additive constant, is given by

$$\ell(\pi) = \left(\sum_{i=1}^{N} X_i \right) \log \pi + \left(N - \sum_{i=1}^{N} X_i \right) \log(1-\pi). \tag{35}$$

This is just an ordinary Bernoulli log-likelihood function, taking its maximum at $\hat{\pi} = \frac{1}{N} \sum_{i=1}^N X_i$.

With $X_1, \ldots, X_N$ missing, we can apply the *EM*-algorithm as follows. For a binary variable X, $E[X] = \Pr[X = 1]$. Moreover, since the $\mathbf{Z}_i = (X_i, Y_i)$ are assumed mutually independent, $E[X_i|\mathcal{Y}] = E[X_i|Y_i]$. But from Section 2.8, we know that

$$E\,[X_i|Y_i = y_i] = \frac{\pi\, f_1(y_i)}{\pi\, f_1(y_i) + (1-\pi)\, f_0(y_i)}, \tag{36}$$

which is the posterior probability of membership in group 1 for the ith observation.

This already gives us the *EM*-algorithm. Since the complete data log-likelihood is linear in the X_i, the E-step consists, once again, of replacing the X_i by their conditional expectation, that is, the jth E-step updates the complete data log-likelihood as

$$\ell_j(\pi) = (\log \pi) \cdot \sum_{i=1}^{N} e_i + \log(1-\pi) \cdot \left(N - \sum_{i=1}^{N} e_i\right), \tag{37}$$

where

$$e_i = \frac{\pi^{(j-1)} f_1(Y_i)}{\pi^{(j-1)} f_1(Y_i) + [1 - \pi^{(j-1)}] f_0(Y_i)}. \tag{38}$$

Then the jth M-step updates the parameter value

$$\pi^{(j)} = \frac{1}{N} \sum_{i=1}^{N} e_i. \tag{39}$$

Thus, the *EM*-algorithm consists of iterations until convergence, starting with some initial value $\pi^{(0)}$, between equations (38) and (39). A numerical illustration is given in Exercise 13. □

Example 4.4.8 Another multinomial application

This is the introductory example of Dempster, Laird, and Rubin (1977), illustrating a rather unexpected way of using incomplete data. Suppose $\mathbf{Y} = (Y_1, Y_2, Y_3, Y_4)$ is multinomial with $\sum_{i=1}^{4} Y_i = N$ and probabilities $\theta_1 = \frac{1}{2} + \frac{1}{4}\pi$, $\theta_2 = \theta_3 = \frac{1}{4}(1-\pi)$, and $\theta_4 = \frac{1}{4}\pi$, for some unknown $\pi \in [0, 1]$. The observed data log-likelihood, up to an additive constant, is given by

$$\ell^*(\pi) = Y_1 \log\left(\frac{1}{2} + \frac{1}{4}\pi\right) + (Y_2 + Y_3)\log\left(\frac{1-\pi}{4}\right) + Y_4 \log\left(\frac{1}{4}\pi\right). \tag{40}$$

The maximum of (40), for given $\mathcal{Y} = (Y_1, Y_2, Y_3, Y_4)$, would not be difficult to find with standard numerical methods, but the *EM*-algorithm offers a surprisingly simple alternative. Suppose that the first category is further subdivided into a category 1A with probability $\frac{1}{2}$ and a category 1B with probability $\frac{1}{4}\pi$. Let X_A and X_B denote the (unobserved) counts in these two categories. The complete data log-likelihood is given by

$$\begin{aligned} \ell(\pi; \mathcal{X}, \mathcal{Y}) &= X_A \log\left(\frac{1}{2}\right) + (X_B + Y_4)\log\left(\frac{1}{4}\pi\right) + (Y_2 + Y_3)\log\left(\frac{1-\pi}{4}\right) \\ &= C + (X_B + Y_4)\log(\pi) + (Y_2 + Y_3)\log(1-\pi). \end{aligned} \tag{41}$$

The maximum of ℓ is attained for $\pi = (X_B + Y_4)/(X_B + Y_2 + Y_3 + Y_4)$. Since the complete data log-likelihood is linear in X_A and X_B, the E-step consists simply of replacing X_A and X_B in (41) by their conditional expectation, given $\mathcal{Y}$, and using the current parameter value $\pi^{(j)}$. As in Example 4.4.3, the conditional distribution of X_A and X_B is binomial, giving

$$X_B^{(j)} = E\left[X_B|\mathcal{Y};\pi^{(j-1)}\right] = Y_1 \cdot \frac{\frac{1}{4}\pi^{(j-1)}}{\frac{1}{2}+\frac{1}{4}\pi^{(j-1)}} \tag{42}$$

and a similar expression for X_A. Thus the *EM*-algorithm iterates until convergence between equation (42) and

$$\pi^{(j)} = \frac{X_B^{(j)}+Y_4}{X_B^{(j)}+Y_2+Y_3+Y_4}. \tag{43}$$

The approach used in this example was inspired by a genetic model discussed by Rao (1965, pp. 368–369) and illustrated with a numerical example based on $N = 197$ animals. The individual cell counts are $Y_1 = 125$, $Y_2 = 18$, $Y_3 = 20$, and $Y_4 = 34$. Table 4.4.2 reports the results of using the *EM*-algorithm, starting with an initial value of $\pi^{(0)} = 0.0001$. Using the criterion that iterations stop when two successive values of $\pi^{(j)}$ differ by at most 10^{-10} in absolute value, the algorithm converged after 12 iterations, giving a parameter estimate of $\hat{\pi} \approx 0.6268$. Details are given in Table 4.4.2. This value can also be found by solving a quadratic equation; see Exercise 14. □

Exercises for Section 4.4

1. Prove equation (2). *Hint*: Define $\tau = \theta_2/(1-\theta_1)$, and write the log-likelihood function in terms of the parameters θ_1 and τ.
2. Prove equation (7).

Table 4.4.2 The *EM*-algorithm in Example 4.4.8.

Iteration (j)	$\pi^{(j)}$	$\ell^*(\pi^{(j)})$
0	0.0001000000	−499.6057145641
1	0.4722680300	−209.7534230947
2	0.6036641383	−205.8148352471
3	0.6236933415	−205.7177281859
4	0.6264051861	−205.7159197466
5	0.6267662036	−205.7158876230
6	0.6268141557	−205.7158870561
7	0.6268205230	−205.7158870461
8	0.6268213684	−205.7158870459
9	0.6268214807	−205.7158870459
10	0.6268214956	−205.7158870459
11	0.6268214976	−205.7158870459
12	0.6268214978	−205.7158870459

3. This exercise refers to Example 4.4.3. Use (2) to show that the maximum likelihood estimates are $\hat{\theta}_1 = Y_1/N, \hat{\theta}_2 = (1-\hat{\theta}_1)Y_2/(Y_2+Y_3)$, and $\hat{\theta}_3 = (1-\hat{\theta}_1)Y_3/(Y_2+Y_3)$.

4. This exercise generalizes Example 4.4.1. Suppose that the observed counts in categories 1, 2, and 3 are Y_1, Y_2, and Y_3. Moreover, suppose that there are Y_{12} observations that fall into either category 1 or category 2, but are not fully classified. Define Y_{13} and Y_{23} similarly.

   (a) Using a setup as in Example 4.4.1, show that the likelihood function is proportional to

   $$\theta_1^{Y_1}(1-\theta_1)^{Y_{23}}\theta_2^{Y_2}(1-\theta_2)^{Y_{13}}\theta_3^{Y_3}(1-\theta_3)^{Y_{12}}.$$

   (b) Find an *EM*-algorithm for this situation by introducing two missing variables for each of Y_{12}, Y_{13}, and Y_{23}.

   (c) Verify that your algorithm in part (b) is the same as the algorithm presented in Example 4.4.3 if $Y_{12} = Y_{13} = 0$. Also verify that the algorithm finds the maximum likelihood estimate in one iteration if $Y_{12} = Y_{13} = Y_{23} = 0$.

5. In Example 4.4.6, solve equations (20) and (21) for $\hat{\mu}$ and $\hat{\tau}$ to show that the stationary point of the *EM*-algorithm corresponds to the maximum of the observed data likelihood function.

6. In Example 4.4.5 (univariate normal with randomly missing observations), show that, if $N_2 = 0$, the *EM*-algorithm finds the maximum likelihood estimate in one iteration.

7. Generalize Example 4.4.5 to the multivariate normal distribution, that is, suppose that a sample of size N has been obtained from a multivariate normal distribution with unknown parameters, but only $N_1 < N$ observations are reported, and $N_2 = N - N_1$ observations are missing at random. (In contrast to Example 4.4.6, only entire observations are missing in the current setup.) Show that the *EM*-algorithm converges to the sample mean and the plug-in estimate of the covariance matrix of the N_1 observed data vectors.

8. Prove equation (22).

9. In Example 4.4.6, find equations similar to (29) to (31), to be used for observations where Z_{i2} has been observed, but Z_{i1} is missing.

10. This exercise refers to Example 4.4.6.

    (a) Write a computer program for the *EM*-algorithm with randomly missing data in the components of a bivariate normal distribution.

    (b) Test your program with an example of your choice, using complete data. The algorithm should find the solution in one iteration, and the solution should just be the ordinary sample mean and the plug-in estimate of the covariance matrix.

    (c) Test your program using the midge data (Table 1.1) of species *Af*. Assume that, for observations 4 to 6, variable 1 (antenna length) is missing, and, for observations 7 to 9, variable 2 (wing length) is missing. With both variables measured in mm/100, the solution should be $\hat{\mu} = \begin{pmatrix} 143.47 \\ 174.82 \end{pmatrix}$ and $\hat{\psi} = \begin{pmatrix} 117.66 & 33.09 \\ 33.09 & 41.28 \end{pmatrix}$.

(d) Explain why it is not a good idea to run this algorithm if all observations have one component missing. (Exercise 12 might help).

11. Consider the following bivariate data from Murray (1977), where "M" indicates missing data points.

x	1	−1	−1	1	2	2	−2	−2	M	M	M	M
y	1	1	−1	−1	M	M	M	M	2	2	−2	−2

(a) Run your program from Exercise 10 using $\mu = \mathbf{0}$ and $\psi = \mathbf{I}_2$ as initial values of the parameters.

(b) Run your program again, this time using $\mu = \mathbf{0}$ and $\psi = \begin{pmatrix} 1 & \rho \\ \rho & 1 \end{pmatrix}$ as inital parameter values, where $0 < |\rho| < 1$. Try it for both positive and negative values of ρ.

(c) What conclusions can you draw from a comparison of the solutions obtained in (a) and (b)?

12. In the situation of Example 4.4.6 and for dimension $p = 2$, find the observed data log-likelihood function. *Hint*: Partition the observations in three groups, corresponding to cases (*i*) to (*iii*) introduced in the text following equation (27).

13. This exercise refers to Example 4.4.7.

(a) Write a computer program for estimating the parameter π of a mixture of two known distributions. Use subprograms to evaluate the component densities f_1 and f_0, so you can adapt your program to component densities of your choice.

(b) Specify normal distributions with means −1 (for f_1) and 1 (for f_0) and common variance 1. Run your program with the data $Y_1 = -3$, $Y_2 = -2$, $Y_3 = -1$, $Y_4 = 1$, $Y_5 = 2$, $Y_6 = 3$. Your program should converge to $\hat{\pi} = 0.5$.

(c) Let f_1 be the uniform density in $[0, 2]$, and f_0 the uniform density in $[1, 3]$. Run your program with a sample of your choice, all Y_i between 0 and 3, but not all Y_i between 1 and 2. If there are no Y_i in the interval $[1, 2]$, the algorithm should find the solution in just one iteration. Explain why this happens, and find an explicit solution for $\hat{\pi}$.

14. This exercise expands on Example 4.4.8. Use equations (42) and (43) to show that the maximum likelihood estimator is a root of a quadratic equation. Evaluate the formula found in part (a) for the numerical data given in the example.

15. Let $Z_1, \ldots, Z_N$ denote a sample from a Bernoulli random variable with parameter θ $(0 \le \theta \le 1)$. Suppose that out of the N observations only $N_1 < N$ were reported, while $N_2 = N - N_1$ observations were (randomly) lost during transmission of the data from Jupiter to Earth. In the usual notation, write $Y_1, \ldots, Y_{N_1}$ for the observed data and $X_1, \ldots, X_{N_2}$ for the missing data.

(a) Find the *EM*-algorithm for this situation, and show that it converges to $\hat{\theta} = \sum_{i=1}^{N_1} Y_i / N_1$.

(b) Implement the algorithm from part (a) on a computer, and run the program for the following two cases: (*i*) $N_1 = 2$, $\sum_{i=1}^{N_1} Y_i = 1$, and $N_2 = 1$; (*ii*) $N_1 = 2$,

$\sum_{i=1}^{N_1} Y_i = 1$, and $N_2 = 1000$. Starting with the same inital value θ_0 (excluding $\theta_0 = 0.5$), in which case is convergence faster?

16. An exponential distribution with censored data.

(a) If Y has an exponential distribution with mean $\theta > 0$, find the conditional *pdf* of Y, given $Y > t$, and show that $E[Y|Y > t] = t + \theta$ for all $t > 0$.

(b) Suppose that N_1 independent observations $Y_1, \ldots, Y_{N_1}$ were made from the exponential distribution with mean $\theta > 0$. For an additional N_2 observations $X_1, \ldots, X_{N_2}$ from the same distribution, all we know is that $X_i > t$, i.e, t is the common censoring time. Find the *EM*-algorithm to estimate θ.

(c) Show that the single stationary point of the algorithm (provided $N_1 \geq 1$) is given by

$$\hat{\theta} = \frac{\left(\sum_{i=1}^{N_1} Y_i + N_2 t\right)}{N_1}.$$

Suggested Further Reading

General: Casella and Berger (1990), Silvey (1975).

Section 4.2: Diaconis and Efron (1983), Efron (1982), Efron and Gong (1983), Efron and Tibshirani (1993).

Section 4.3: Edwards (1984), Kalbfleisch (1985), Watson (1964).

Section 4.4: Dempster et al. (1977), Little and Rubin (1987), McLachlan and Krishnan (1997), Rubin and Szatrowski (1982).

5 Discrimination and Classification, Round 1

5.1 Introduction

Discriminant analysis and related methods, treated in Chapters 5 to 7, are the central topics of this course. Chapter 5 gives an introduction on a mostly descriptive level, ignoring questions of statistical inference. The mathematical level of Chapter 5 is moderate, and all concepts are explained at great length, hoping that even students without a strong mathematical background will be able to master most of the material. Chapter 6 gives an introduction to problems of statistical inference that arise naturally from the setup of discriminant analysis: testing for equality of mean vectors, confidence regions for mean vectors, and related problems for discriminant functions. Then Chapter 7 resumes the classification theory on a more abstract level and gives brief introductions to related topics, such as logistic regression and multivariate analysis of variance.

The setup and purpose of discriminant analysis has already been given informally in Chapter 1, by the midge example. Some important concepts of classification theory, mostly the notion of posterior probability, have been introduced more formally in connection with finite mixtures in Section 2.8. In the current chapter, we will focus on the problem of p-variate observations measured in two groups, as illustrated in the midge example, and the formulation of rules for classification of future observations. We will start out by defining a measure of distance between two multivariate distributions, called the multivariate *standard distance*. Standard distance is closely related to the *linear discriminant function*, which we may think of as a linear combination of the variables that separates the two distributions as well as possible. Section 5.2 develops and illustrates these notions in detail. In Section 5.3, we apply the concepts of standard distance and discriminant function to various data sets to illustrate the power of this methodology. While Sections 5.2 and 5.3 are based on heuristic concepts, without making any specific distributional assumptions, Section 5.4 rediscovers the linear discriminant function in a normal theory setup, using the principle of classification based on maximum posterior probability. We will see that the assumption of multivariate normal distributions in two groups with identical covariance matrices leads to the same method of reducing the p-variate data to a single variable as the heuristic principle approach in Section 5.2. In Section 5.5, we study various types of error rates to measure (or estimate) the success of a classification rule. Finally, Section 5.6 explores a relatively little known but highly interesting connection between linear discriminant functions and conditional mean differences, thus, connecting the areas of regression and discrimination. At the same time, the theory of Section 5.6 provides a basis for inference on linear discriminant function coefficients, a topic to be developed in Chapter 6.

This is the first chapter to use real data examples extensively. It is essential for the student to use appropriate computer software for the practical applications. Although discriminant analysis can be performed with many popular statistics packages, such as SAS or SPSS, it is better at the learning stage to use more mathematical software, such as GAUSS, S-PLUS or MATLAB, which allow one to do matrix calculations interactively. In the author's experience, it is helpful for the student to actually "see" the matrices and vectors involved in the calculations, and intermediate results, rather than having a black-box program produce just the final results.

5.2 Standard Distance and the Linear Discriminant Function

In regression analysis, linear regression functions are introduced as linear combinations of p regressors $X_1, \ldots, X_p$ such that the dependent variable Y is optimally approximated using the method of least squares. A similar approach will lead to the definition of the linear discriminant function. Instead of minimization of residual variance, however, we shall use another criterion, maximization of the distance between two groups. Thus, we need to talk about the notion of distance.

Definition 5.2.1 Let X denote a random variable with mean μ and variance $\sigma^2 > 0$. Then the *standard distance* between two numbers x_1 and x_2, with respect to the random variable X, is given by

$$\Delta_X(x_1, x_2) = \frac{|x_1 - x_2|}{\sigma}, \tag{1}$$

that is, the standard distance between x_1 and x_2 is the absolute difference between x_1 and x_2 in units of the standard deviation of X. □

Standard distance is going to play such a crucial role in this chapter that we discuss it immediately with some remarks.

1. If $\sigma = 1$, then standard distance is the same as Euclidean distance.

2. Standard distance is invariant under nondegenerate linear transformations: Let $Y = aX + b$, where $a \neq 0$ and b are fixed constants. Similarly, transform x_1 and x_2 to $y_i = ax_i + b$, $i = 1, 2$. Then

$$\begin{aligned}\Delta_Y(y_1, y_2) &= \frac{|y_1 - y_2|}{\sqrt{\text{var}[Y]}} \\ &= \frac{|a(x_1 - x_2)|}{\sqrt{a^2\sigma^2}} \\ &= \Delta_X(x_1, x_2).\end{aligned} \tag{2}$$

3. Standard distance is most useful for symmetric distributions. If the distribution of X is symmetric with mean μ, then the *pdf* of X depends on the argument x only through $\Delta_X(x, \mu)$.

Let us illustrate this with examples. For simplicity, we will write $\Delta(x)$ instead of $\Delta_X(x, \mu)$ because there is no danger of confusion.

Example 5.2.1 Suppose that X is normal with mean μ and variance σ^2. Then the density function of X is given by

$$f_X(x) = \frac{1}{\sqrt{2\pi}\,\sigma} \exp\left[-\frac{1}{2}\Delta^2(x)\right], \quad x \in \mathbb{R}.$$

□

Example 5.2.2 If X is uniform in the interval $[a, b]$, then $E[X] = (a+b)/2$, and $\text{var}[X] = (b-a)^2/12$. The density function of X may be written as

$$f_X(x) = \begin{cases} \dfrac{1}{b-a} & \text{if } \Delta(x) \leq \sqrt{3}, \\ 0 & \text{otherwise.} \end{cases}$$

Again, the value of the density function depends on x only through $\Delta(x)$. Further examples are given in Exercise 1. □

Now let us turn to a situation more closely related to discriminant analysis. Suppose a variable X is measured on items from two populations, with different means μ_j but common variance σ^2. When taking expectations, we will sometimes attach an index to the expectation operator to indicate the distribution on which the calculation is done, that is, we write

$$E_1[X] = \mu_1, \quad \text{var}[X] = \sigma^2 \quad \text{in population 1,}$$

and

$$E_2[X] = \mu_2, \ \text{var}[X] = \sigma^2 \quad \text{in population 2.} \tag{3}$$

In close analogy to Definition 5.2.1, we can measure the distance between the two means by

$$\Delta_X(\mu_1, \mu_2) = \frac{|\mu_1 - \mu_2|}{\sigma}. \tag{4}$$

Invariance of standard distance between two means under linear transformations follows as in Remark 2 above; see Exercise 2.

If the distribution of X is symmetric in both groups, then any given value of $\Delta = \Delta(\mu_1, \mu_2)$ is uniquely related to a certain amount of overlap. It is important, mostly for the purpose of classification analysis, to develop some intuition for the meaning of standard distance. The following example serves this purpose.

Example 5.2.3 This example leads us straight into classical normal distribution theory of classification. Suppose that $X \sim \mathcal{N}(\mu_j, \sigma^2)$ in population j, $j = 1, 2$. Let $\Delta = \Delta(\mu_1, \mu_2)$ denote the standard distance between the two means, and write $\gamma = (\mu_1 + \mu_2)/2$ for the midpoint between the two means. Figure 5.2.1 shows density curves for $\Delta = 1, 2, 3, 4, 5$. The overlap between the density curves is considerable for $\Delta = 1$ and almost zero for $\Delta = 5$. But how should we measure overlap? In anticipation of error rates, to be discussed later, we consider areas under the density curve to the right and to the left of the midpoint γ. Assume for simplicity, that $\mu_1 > \mu_2$. Then we may ask for the probability that X falls below γ, given that X is from population 1, that is, we shall evaluate the integral given by the shaded area in Figure 5.2.1. For $Z \sim \mathcal{N}(0, 1)$, let

$$\Phi(z) = \Pr[Z \le z], \tag{5}$$

and write $\Pr[X \le \gamma | P_1]$ for the desired integral, where "P_1" is to indicate that the calculation is based on the distribution in population 1. Then

$$\begin{aligned} \Pr[X \le \gamma | P_1] &= \Pr\left[\frac{X - \mu_1}{\sigma} \le \frac{\gamma - \mu_1}{\sigma} | P_1\right] \\ &= \Pr\left[Z \le \frac{\gamma - \mu_1}{\sigma}\right] \\ &= \Phi\left[\frac{\frac{1}{2}(\mu_1 + \mu_2) - \mu_1}{\sigma}\right] \\ &= \Phi(-\Delta/2). \end{aligned} \tag{6}$$

By the same argument, $\Pr[X \ge \gamma | P_2] = \Phi(-\Delta/2)$. So, we can declare $\Phi(-\Delta/2)$ as a measure of overlap between the two groups. The overlap decreases monotonically as Δ goes from 0 to infinity. In Figure 5.2.1, the numerical values are as follows:

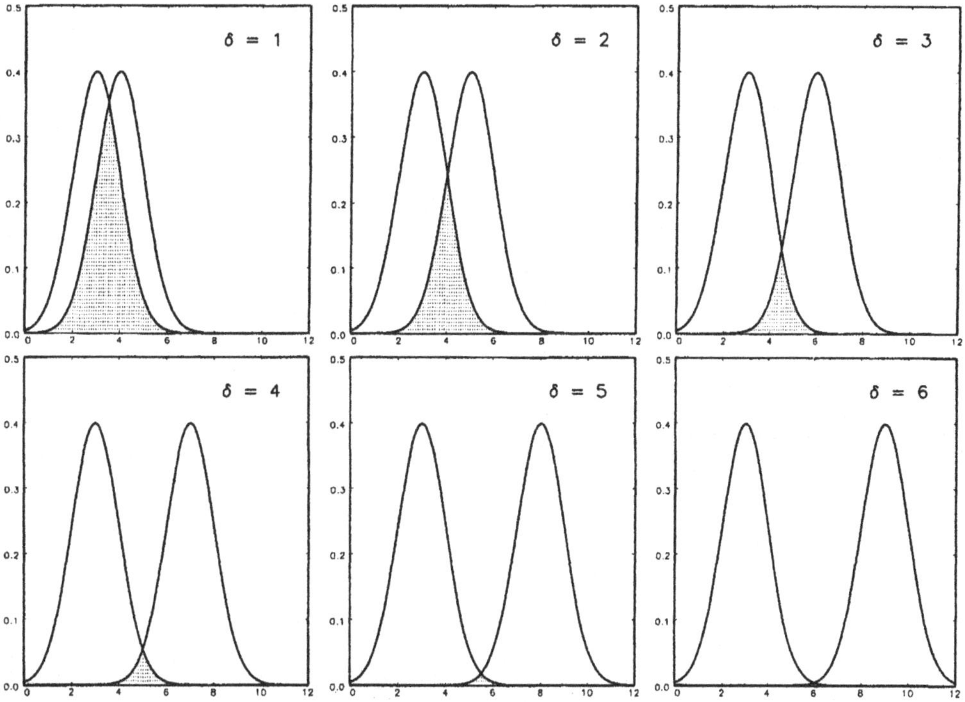

Figure 5-2-1 Normal density curves with standard distances from $\Delta = 1$ to $\Delta = 5$ between the two means.

Δ	1	2	3	4	5	6
$\Phi(-\Delta/2)$	0.308	0.159	0.067	0.023	0.006	0.001

Figure 5.2.2 shows a graph of the overlap function $o(\Delta) = \Phi(-\Delta/2)$ for Δ in the range from 0 to 6. For Δ larger than about 5, the separation of the two distributions is almost perfect. Similar calculations can also be done for other symmetric distributions; see Exercise 3. □

Standard distance between two means occurs in elementary statistics, although textbooks usually do not use this concept. It occurs most predominantly in the one and two-sample t-tests. Let us explain this briefly.

- *One-Sample situation.* Suppose that we take a sample $x_1, \ldots, x_N$ from a distribution with mean μ and variance σ^2, where both parameters are unknown. (In the classical setup, the distribution is assumed normal, but this is irrelevant for the current purpose). The t-test for the hypothesis $\mu = \mu_0$, where μ_0 is a fixed, hypothetical constant, is based on the statistic

$$t = \sqrt{N}\,\frac{\bar{x} - \mu_0}{s}\,. \tag{7}$$

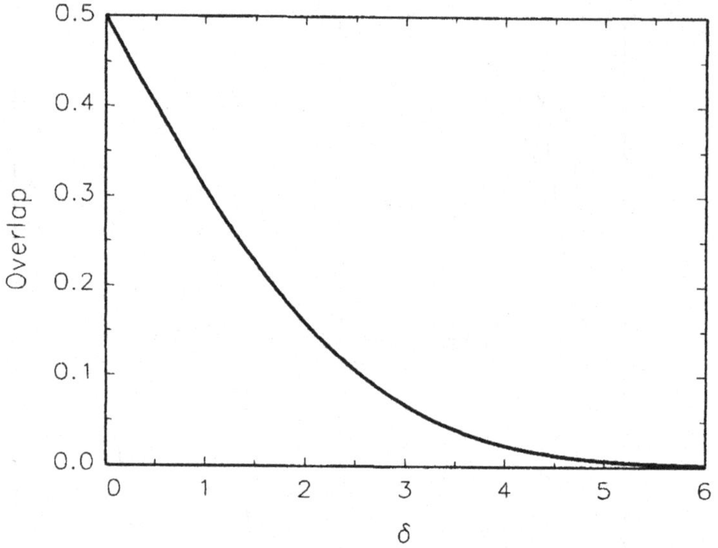

Figure 5-2-2
Graph of the overlap function $o(\Delta) = \Phi(-\frac{1}{2}\Delta)$, for $0 \leq \Delta \leq 6$.

Here, $\bar{x} = \frac{1}{N}\sum_{i=1}^{N} x_i$ and $s = \left[\frac{1}{N-1}\sum_{i=1}^{N}(x_i - \bar{x})^2\right]^{1/2}$ denote the sample mean and standard deviation, respectively. Defining *sample standard distance* between two numbers u_1 and u_2 with respect to the given sample as

$$D(u_1, u_2) = |u_1 - u_2|/s\,, \tag{8}$$

we obtain $D(\bar{x}, \mu_0) = |\bar{x} - \mu_0|/s$, and therefore,

$$|t| = \sqrt{N} \cdot D(\bar{x}, \mu_0). \tag{9}$$

Since there is such a close relationship, why use standard distance at all? The answer is that the standard distance and the t-statistic serve different purposes. The standard distance

$$D(\bar{x}, \mu_0) = \frac{|\mu_0 - \bar{x}|}{s} \tag{10}$$

is a descriptive measure, telling us how far apart are the sample mean $\bar{x}$ and the hypothetical mean μ_0 in terms of the standard deviation. It can also be considered as an estimate of the theoretical standard distance

$$\Delta(\mu_0, \mu) = \frac{|\mu_0 - \mu|}{\sigma}\,. \tag{11}$$

The t-statistic, on the other hand, is better suited for testing the hypothesis $\mu = \mu_0$. By itself, it does not tell much about how distant μ_0 and $\bar{x}$ are. Indeed, a large value of $|t|$ may result from a large standard distance or from a large sample size N.

- *Two-Sample situation.* Consider independent samples $x_{11}, \ldots, x_{1N_1}$, and $x_{21}, \ldots, x_{2N_2}$ from distributions with means μ_j ($j = 1, 2$), and common variance σ^2. Let

$$\bar{x}_j = \frac{1}{N_j} \sum_{i=1}^{n_j} x_{ji} , \qquad j = 1, 2,$$

and

$$s_j = \left[\frac{1}{N_j - 1} \sum_{i=1}^{n_j} (x_{ji} - \bar{x}_j)^2 \right]^{1/2}, \quad j = 1, 2,$$

be the usual sample means and standard deviations, respectively, and denote by

$$s^2 = \left[(N_1 - 1)s_1^2 + (N_2 - 1)s_2^2 \right] / (N_1 + N_2 - 2)$$

the estimate of the common variance σ^2, also called the pooled variance. The two-sample t-test for the hypothesis $\mu_1 = \mu_2$ is based on the statistic

$$t = \frac{\bar{x}_1 - \bar{x}_2}{s} \cdot \sqrt{\frac{N_1 N_2}{N_1 + N_2}} . \tag{12}$$

As in the one-sample situation, we define a sample version of $\Delta(\mu_1, \mu_2)$ as

$$D(\bar{x}_1, \bar{x}_2) = \frac{|\bar{x}_1 - \bar{x}_2|}{s} , \tag{13}$$

which can be considered an estimate of $\Delta(\mu_1, \mu_2)$. Then

$$|t| = \sqrt{\frac{N_1 N_2}{N_1 + N_2}} \cdot D(\bar{x}_1, \bar{x}_2) . \tag{14}$$

As in the one-sample situation, the t-statistic is suitable for testing the hypothesis of equal means. The standard distance $D(\bar{x}_1, \bar{x}_2)$, on the other hand, serves the purpose of a descriptive measure of separation between the two groups. Intuitive guidelines, such as those provided by the normal theory example in Figure 5.2.1, are helpful for understanding the meaning of $D(\bar{x}_1, \bar{x}_2)$.

Example 5.2.4 continuation of Example 1.1

Back to the midges. Table 5.2.1 gives means and standard deviations for both variables and both groups, as well as the pooled standard deviations and the standard distances for both variables. The standard distances indicate that there is considerable overlap between the two groups. Yet Figure 1.1 indicates that the two groups are very well separated, if the data are viewed in both dimensions simultaneously. Hence, we need a generalization of standard distance to the situation where two or more variables are measured simultaneously.

The key idea is that we do not have to stick to the original variables. For instance, we could consider variables such as $U = X + Y$ or $V = Y - X$. The transformed

Table 5.2.1 Sample statistics, including standard distances, for two variables in the midge example.

Variable	Group	Mean	Standard Deviation	Pooled Standard Deviation	Standard Distance
X (antenna length)	*Af*	141.33	9.90	8.69	2.15
	Apf	122.67	6.28		
Y (wing length)	*Af*	180.44	12.99	11.57	1.06
	Apf	192.67	8.82		

Note: The unit of measurement for both variables is mm/100.

data are given in Table 5.2.2. Treating U and V like measured variables, we can use them to compute means, standard deviations, and standard distances as we did for X and Y. This is illustrated in Table 5.2.3. Variable U seems useless for discrimination purposes, but the difference V separates the two groups much better than either X or Y, with a standard distance of 3.53. So the question naturally arises: Could we do even better than this? In other words, are there any linear combinations that would lead to an even better separation in terms of their standard distance? □

Table 5.2.2 Raw data and new variables $U = X + Y$ and $V = Y - Y$ in the midge example.

	X Antenna	Y Wing	$X + Y$ Sum	$Y - X$ Diff.
Af	138	164	302	26
	140	170	310	30
	124	172	296	48
	136	174	310	38
	138	182	320	44
	148	182	330	34
	154	182	336	28
	138	190	328	52
	156	208	364	52
Apf	114	178	292	64
	120	186	306	66
	118	196	314	78
	130	196	326	66
	126	200	326	74
	128	200	328	72

Note: The unit of measurement for X and Y is mm/100.

Table 5.2.3 Sample statistics, including standard distances, for variables $U = X + Y$ and $V = Y - X$ in the midge example.

Variable	Group	Mean	Standard Deviation	Pooled Standard Deviation	Standard Distance
U (sum)	*Af*	321.78	20.70	18.50	0.35
	Apf	315.33	14.29		
V (difference)	*Af*	39.11	10.25	8.74	3.53
	Apf	70.00	5.51		

Example 5.2.4 leads us straight into the generalization of the notion of standard distance to the general p-variate situation. We will discuss it, first, on the theoretical level. Suppose that $\mathbf{X} = (X_1, \ldots, X_p)'$ is a p-variate random vector with

$$E[\mathbf{X}] = \boldsymbol{\mu}, \quad \text{Cov}[\mathbf{X}] = \boldsymbol{\psi}.$$

Now take two fixed points $\mathbf{x}_1 \in \mathbb{R}^p$ and $\mathbf{x}_2 \in \mathbb{R}^p$. How can we measure their distance with respect to $\mathbf{X}$? The key idea is to reduce the p-dimensional problem to a univariate problem by constructing linear combinations. Then we can apply the univariate notion of standard distance to each given linear combination, just as we did in Example 5.2.4. Finally, among all possible linear combinations of $\mathbf{X}$, we shall choose the "most interesting" one—"interesting" being defined as having a large standard distance.

More formally, let $\mathbf{a} = (a_1, \ldots, a_p)' \in \mathbb{R}^p$ denote the vector of coefficients of a linear combination, and set $Y = \mathbf{a}'\mathbf{X}$. From Theorem 2.11.1, we obtain $E[Y] = \mu_Y = \mathbf{a}'\boldsymbol{\mu}$, and $\text{var}[Y] = \sigma_Y^2 = \mathbf{a}'\boldsymbol{\psi}\mathbf{a}$. Applying the same linear transformation to the points $\mathbf{x}_1$ and $\mathbf{x}_2$, we get $y_1 = \mathbf{a}'\mathbf{x}_1$ and $y_2 = \mathbf{a}'\mathbf{x}_2$, which are fixed numbers in $\mathbb{R}$. Hence we can apply our one-dimensional notion of standard distance to the linear combination $Y = \mathbf{a}'\mathbf{X}$, and get

$$\begin{aligned} \Delta_Y(y_1, y_2) &= \Delta_{\mathbf{a}'\mathbf{X}}(\mathbf{a}'\mathbf{x}_1, \mathbf{a}'\mathbf{x}_2) \\ &= \frac{|\mathbf{a}'(\mathbf{x}_1 - \mathbf{x}_2)|}{(\mathbf{a}'\boldsymbol{\psi}\mathbf{a})^{1/2}} . \end{aligned} \tag{15}$$

At this point, we need to make sure that the preceding expression is well defined, i.e., that the denominator is not zero. Since $\mathbf{a}'\boldsymbol{\psi}\mathbf{a} = \text{var}[\mathbf{a}'\mathbf{X}]$ is the variance of a random variable, it is clear that $\mathbf{a}'\boldsymbol{\psi}\mathbf{a} \geq 0$. Obviously, we will have equality if $\mathbf{a} = \mathbf{0}$, so let us exclude the degenerate linear combination given by the vector with all coefficients zero. Hence, we require that

$$\mathbf{a}'\boldsymbol{\psi}\mathbf{a} > 0 \quad \text{for all } \mathbf{a} \in \mathbb{R}^p \quad (\mathbf{a} \neq \mathbf{0}). \tag{16}$$

This condition is commonly referred to as *positive definiteness* of the covariance matrix $\boldsymbol{\psi}$, see Appendix A.2. It means that no nontrivial linear combination of $\mathbf{X}$

is allowed to have variance zero. Positive definiteness of ψ is a generalization of positive variance in the univariate case.

As mentioned before, "interesting" linear combinations are those which yield a large standard distance. This leads to the definition of multivariate standard distance, one of the key definitions for understanding multivariate concepts.

Definition 5.2.2 Let $\mathbf{X}$ denote a p-variate random vector with mean $\mu = E[\mathbf{X}]$ and covariance matrix $\text{Cov}[\mathbf{X}] = \psi$, assumed positive definite. The *p-variate standard distance* between two vectors $\mathbf{x}_1 \in \mathbb{R}^p$ and $\mathbf{x}_2 \in \mathbb{R}^p$, with respect to $\mathbf{X}$, is the maximum univariate standard distance between $\mathbf{a}'\mathbf{x}_1$ and $\mathbf{a}'\mathbf{x}_2$ with respect to the random variable $\mathbf{a}'\mathbf{X}$, the maximum being taken over all linear combinations, that is,

$$\begin{aligned}\Delta_{\mathbf{X}}(\mathbf{x}_1, \mathbf{x}_2) &= \max_{\substack{\mathbf{a}\in R^p\\ \mathbf{a}\neq \mathbf{0}}} \Delta_{\mathbf{a}'\mathbf{X}}(\mathbf{a}'\mathbf{x}_1, \mathbf{a}'\mathbf{x}_2) \\ &= \max_{\substack{\mathbf{a}\in R^p\\ \mathbf{a}\neq \mathbf{0}}} \frac{|\mathbf{a}'(\mathbf{x}_1 - \mathbf{x}_2)|}{(\mathbf{a}'\psi\mathbf{a})^{1/2}}.\end{aligned} \tag{17}$$

□

To clarify this definition, let us explain it one more time. The definition says that we should look at all possible linear combinations $\mathbf{a}'\mathbf{X}$ of $\mathbf{X}$, except the trivial one given by $\mathbf{a} = \mathbf{0}$, and find the one or ones for which $\mathbf{a}'\mathbf{x}_1$ and $\mathbf{a}'\mathbf{x}_2$ are as distant from each other as possible—"as distant as possible" in terms of the standard deviation of $\mathbf{a}'\mathbf{X}$. Then the maximum univariate standard distance, provided it exists (a point still to be shown), will be called the multivariate standard distance between $\mathbf{x}_1$ and $\mathbf{x}_2$, with respect to the random vector $\mathbf{X}$.

Before turning to the mathematics of multivariate standard distance, we will investigate some elementary properties.

Definition 5.2.3 Two linear combinations $\mathbf{a}'\mathbf{X}$ and $\mathbf{b}'\mathbf{X}$ of a random vector $\mathbf{X}$ are called *equivalent* if they are proportional to each other, i.e., if $\mathbf{b} = c \cdot \mathbf{a}$ for some nonzero $c \in \mathbb{R}$. □

The importance of the notion of equivalent linear combinations is apparent from the following result.

Lemma 5.2.1 *Equivalent linear combinations lead to the same value of the (univariate) standard distance between two points.*

Proof Let $Y = \mathbf{a}'\mathbf{X}$ denote a linear combination of $\mathbf{X}$, and let Y^* denote an equivalent linear combination, i.e., $Y^* = c\,\mathbf{a}'\mathbf{X}$ for some $c \neq 0$. For two points $\mathbf{x}_1$ and $\mathbf{x}_2 \in \mathbb{R}^p$,

$$\Delta_{Y^*}(c\,\mathbf{a}'\mathbf{x}_1, c\,\mathbf{a}'\mathbf{x}_2) = \frac{|c\,\mathbf{a}'(\mathbf{x}_1 - \mathbf{x}_2)|}{(c^2\mathbf{a}'\psi\mathbf{a})^{1/2}} = \frac{|\mathbf{a}'(\mathbf{x}_1 - \mathbf{x}_2)|}{(\mathbf{a}'\psi\mathbf{a})^{1/2}} = \Delta_Y(\mathbf{a}'\mathbf{x}_1, \mathbf{a}'\mathbf{x}_2).$$

■

Lemma 5.2.1 tells us that, in searching for the maximum standard distance, we can restrict ourselves to just one member in each class of equivalent linear combinations. For instance, we can require that all linear combinations considered to be *normalized*, i.e., we can impose the constraint $\mathbf{a}'\mathbf{a} = 1$ on the vector of coefficients. For any given linear combination $Y = \mathbf{a}'\mathbf{X}$, the equivalent normalized linear combination is given by

$$Y^* = \frac{1}{(\mathbf{a}'\mathbf{a})^{1/2}}\, Y = \frac{1}{(\mathbf{a}'\mathbf{a})^{1/2}}\, \mathbf{a}'\mathbf{X};$$

see Exercise 9. Actually, later we are going to adopt a different convention, one that is easier to handle algebraically, but, for the purpose of developing some geometric intuition, the idea of looking only at normalized linear combinations is helpful. As usual, the case $p = 2$ is easier to understand. Let $\mathbf{X} = \begin{pmatrix} X_1 \\ X_2 \end{pmatrix}$ denote a bivariate random vector and $Y = \mathbf{a}'\mathbf{X} = a_1X_1 + a_2X_2$ a linear combination of $\mathbf{X}$. An equivalent normalized linear combination is $Y^* = \mathbf{b}'\mathbf{X}$, where $\mathbf{b} = \mathbf{a}/(\mathbf{a}'\mathbf{a})^{1/2}$, i.e., $b_1 = a_1/\sqrt{a_1^2 + a_2^2}$ and $b_2 = a_2/\sqrt{a_1^2 + a_2^2}$. Since $\mathbf{b}'\mathbf{b} = b_1^2 + b_2^2 = 1$, we can write $b_1 = \cos(\phi)$ and $b_2 = \sin(\phi)$ for some angle $\phi \in [0, 2\pi)$. The normalized linear combination Y^* is then given by

$$Y^* = (\cos\phi)X_1 + (\sin\phi)X_2 \,.$$

Thus, looking at all normalized linear combinations of $\mathbf{X}$ amounts to varying the angle ϕ from 0 to 2π (actually from 0 to π is sufficient) and determining the distribution of Y^* for each ϕ. For instance, choosing $\phi = \pi/4$ gives $\cos(\phi) = \sin(\phi) = 1/\sqrt{2}$, and we could find the distribution of $Y^* = (X_1 + X_2)/\sqrt{2}$ using the method of Section 2.9. Heuristically speaking, we can visualize the distribution of the normalized linear combination Y^* for any ϕ by projecting the bivariate distribution on a line corresponding to the angle ϕ. For instance, for $\phi = 0$, this would be a line parallel to the x_1-axis, for $\phi = \pi/2$, a line parallel to the x_2-axis, and for $\phi = \pi/4$, a line parallel to the line $x_1 = -x_2$. How this can be done technically is not explained here, and we are not going to worry about angles of projection in this text, but visualization helps understanding, as illustrated in the following example.

Example 5.2.5 Let $\mathbf{X} = \begin{pmatrix} X_1 \\ X_2 \end{pmatrix}$ follow a bivariate normal distribution with mean vector $\boldsymbol{\mu} = \begin{pmatrix} 0 \\ 0 \end{pmatrix}$ and covariance matrix $\psi = \begin{pmatrix} 2 & 0 \\ 0 & 1 \end{pmatrix}$, and suppose we are interested in the bivariate standard distance of the point $\mathbf{x} = \begin{bmatrix} 2 \\ 1 \end{bmatrix}$ from $\boldsymbol{\mu}$. Figure 5.2.3 shows marginal distributions of the three normalized linear combinations

$$Y_1 = \frac{1}{\sqrt{2}}X_1 + \frac{1}{\sqrt{2}}X_2,$$

$$Y_2 = -0.966X_1 + 0.259X_2,$$

and

$$Y_3 = 0.259X_1 - 0.966X_2,$$

corresponding to angles of $\frac{1}{4}\pi$, $\frac{11}{12}\pi$, and $\frac{19}{12}\pi$, respectively. Among the three linear combinations, Y_1 yields the largest standard distance, with a value of $\sqrt{3}$. This is also larger than the standard distances for both variables X_1 and X_2. It turns out (without proof at this point) that $\sqrt{3}$ is the maximum that can be achieved, and hence, $\sqrt{3}$ is the multivariate standard distance between $\mathbf{x} = \begin{pmatrix} 2 \\ 1 \end{pmatrix}$ and $\boldsymbol{\mu} = \mathbf{0}$. Note that this result depends in no way on the normality of $\mathbf{X}$. Any other distribution with the same mean vector and the same covariance matrix as $\mathbf{X}$ would yield the same result. □

Now we turn to the mathematics of standard distance. To find an explicit expression for the multivariate standard distance, we will use a result known as the *extended Cauchy-Schwartz inequality*, which is itself a generalization of the (ordinary) Cauchy-Schwartz inequality given in the following lemma.

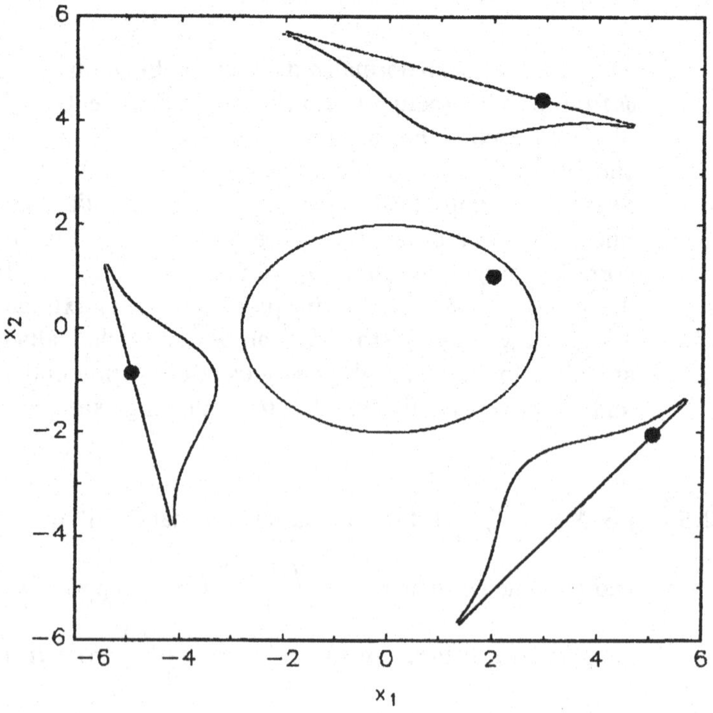

Figure 5-2-3 Distribution of three normalized linear combinations of X_1 and X_2 as in Example 5.2.5.

Lemma 5.2.2 (Cauchy-Schwartz inequality) *For any two nonzero vectors* $\mathbf{x} \in \mathbb{R}^p$ *and* $\mathbf{y} \in \mathbb{R}^p$,

$$(\mathbf{x}'\mathbf{y})^2 \le (\mathbf{x}'\mathbf{x})(\mathbf{y}'\mathbf{y}), \tag{18}$$

with equality exactly if $\mathbf{y} = c\,\mathbf{x}$ *for some* $c \in \mathbb{R}$.

Proof See Exercise 10. ∎

Lemma 5.2.3 (Extended Cauchy-Schwartz inequality) *For any two nonzero vectors* $\mathbf{u} \in \mathbb{R}^p$ *and* $\mathbf{v} \in \mathbb{R}^p$ *and any positive definite symmetric* $p \times p$ *matrix* $\mathbf{M}$,

$$(\mathbf{u}'\mathbf{v})^2 \le (\mathbf{u}'\mathbf{M}\mathbf{u})(\mathbf{v}'\mathbf{M}^{-1}\mathbf{v}), \tag{19}$$

with equality exactly if $\mathbf{v} = c\mathbf{M}\mathbf{u}$ *for some* $c \in \mathbb{R}$ *(or, equivalently,* $\mathbf{u} = c^*\mathbf{M}^{-1}\mathbf{v}$ *for some* $c^* \in \mathbb{R}$*).*

Proof Since $\mathbf{M}$ is positive definite, symmetric, there exists a symmetric, nonsingular matrix $\mathbf{M}^{1/2}$ such that $\left(\mathbf{M}^{1/2}\right)^2 = \mathbf{M}$, with inverse $\mathbf{M}^{-1/2} = \left(\mathbf{M}^{1/2}\right)^{-1}$. Let $\mathbf{x} = \mathbf{M}^{1/2}\mathbf{u}$ and $\mathbf{y} = \mathbf{M}^{-1/2}\mathbf{v}$. Then Lemma 5.2.2 implies

$$\begin{aligned}(\mathbf{u}'\mathbf{v})^2 &= \left(\mathbf{x}'\mathbf{M}^{-1/2}\mathbf{M}^{1/2}\mathbf{y}\right)^2 \\ &= (\mathbf{x}'\mathbf{y})^2 \\ &\le (\mathbf{x}'\mathbf{x})(\mathbf{y}'\mathbf{y}) \\ &= (\mathbf{u}'\mathbf{M}\mathbf{u})(\mathbf{v}'\mathbf{M}^{-1}\mathbf{v}).\end{aligned} \tag{20}$$

Equality in (20) holds exactly if $\mathbf{y} = c\,\mathbf{x}$ for some $c \in \mathbb{R}$, i.e., if $\mathbf{M}^{-1/2}\mathbf{v} = c\mathbf{M}^{1/2}\mathbf{u}$ or $\mathbf{v} = c\mathbf{M}\mathbf{u}$ for some $c \in \mathbb{R}$. ∎

As a consequence of Lemma 5.2.3, for a given vector $\mathbf{v} \in \mathbb{R}^p$ and a given positive definite, symmetric matrix $\mathbf{M}$,

$$\max_{\mathbf{u}\in\mathbb{R}^p} \frac{(\mathbf{u}'\mathbf{v})^2}{\mathbf{u}'\mathbf{M}\mathbf{u}} = \mathbf{v}'\mathbf{M}^{-1}\mathbf{v}, \tag{21}$$

and the maximum is attained for any vector $\mathbf{u}$ proportional to $\mathbf{M}^{-1}\mathbf{v}$.

Now we are ready for the main result.

Theorem 5.2.4 *Let* $\mathbf{X}$ *denote a* p*-variate random vector with positive definite covariance matrix* ψ. *Then the multivariate standard distance between two points* $\mathbf{x}_1 \in \mathbb{R}^p$ *and* $\mathbf{x}_2 \in \mathbb{R}^p$ *is given by*

$$\Delta_{\mathbf{X}}(\mathbf{x}_1, \mathbf{x}_2) = \left[(\mathbf{x}_1 - \mathbf{x}_2)'\psi^{-1}(\mathbf{x}_1 - \mathbf{x}_2)\right]^{1/2}. \tag{22}$$

Proof Instead of maximizing $\Delta_{\mathbf{a}'\mathbf{X}}(\mathbf{a}'\mathbf{x}_1, \mathbf{a}'\mathbf{x}_2)$, we can equivalently maximize its square $\Delta^2_{\mathbf{a}'\mathbf{X}}(\mathbf{a}'\mathbf{x}_1, \mathbf{a}'\mathbf{x}_2) = \left[\mathbf{a}'(\mathbf{x}_1 - \mathbf{x}_2)\right]^2/(\mathbf{a}'\psi\mathbf{a})$ over $\mathbf{a} \in \mathbb{R}^p$, $\mathbf{a} \neq \mathbf{0}$. Using equation (21) with $\mathbf{u} = \mathbf{a}$, $\mathbf{v} = \mathbf{x}_1 - \mathbf{x}_2$, and $\mathbf{M} = \psi$, the maximum is attained if we choose $\mathbf{a}$ proportional to $\psi^{-1}(\mathbf{x}_1 - \mathbf{x}_2)$, and

$$\max_{\substack{\mathbf{a} \in \mathbb{R}^p \\ \mathbf{a} \neq \mathbf{0}}} \frac{\left[\mathbf{a}'(\mathbf{x}_1 - \mathbf{x}_2)\right]^2}{\mathbf{a}'\psi\mathbf{a}} = (\mathbf{x}_1 - \mathbf{x}_2)'\psi^{-1}(\mathbf{x}_1 - \mathbf{x}_2). \tag{23}$$

Then the result follows by taking the square root on both sides of (23). ■

Because of the importance of Theorem 5.2.4, we will add some remarks.

(1) If the random vector $\mathbf{X}$ consists of pairwise uncorrelated, random variables, all with the same variance $\sigma^2 > 0$, then the standard distance between two points $\mathbf{x}_1$ and $\mathbf{x}_2$ is given by

$$\begin{aligned} \Delta_{\mathbf{X}}(\mathbf{x}_1, \mathbf{x}_2) &= \left[(\mathbf{x}_1 - \mathbf{x}_2)'\left(\sigma^2 \mathbf{I}_p\right)^{-1}(\mathbf{x}_1 - \mathbf{x}_2)\right]^{1/2} \\ &= \frac{1}{\sigma}\sqrt{(\mathbf{x}_1 - \mathbf{x}_2)'(\mathbf{x}_1 - \mathbf{x}_2)}. \end{aligned} \tag{24}$$

that is, standard distance is the same as Euclidean distance in units of the common standard deviation.

(2) Most textbooks on multivariate statistics give standard distance a different name. *Statistical distance, elliptical distance*, and *Mahalanobis distance* are commonly used terms. To make things worse, Δ^2 is sometimes referred to as Mahalanobis distance, although Δ^2 does not satisfy the usual axioms of a measure of distance; see Exercise 4. Related notions are *effect size* in meta-analysis, and *leverage* in linear regression. See Exercise 5 for the relationship between leverage and standard distance.

(3) The term "elliptical distance" refers to the fact that, for a given positive definite, symmetric matrix $\mathbf{A}$ and a given vector $\mathbf{m} \in \mathbb{R}^p$, the set of points

$$\mathcal{E}_c = \{\mathbf{x} \in \mathbb{R}^p : (\mathbf{x} - \mathbf{m})'\mathbf{A}^{-1}(\mathbf{x} - \mathbf{m}) = c^2\} \tag{25}$$

is an ellipse ($p = 2$), ellipsoid ($p = 3$), or hyperellipsoid ($p > 3$) for any $c > 0$. The multivariate normal distribution and multivariate elliptical distributions, in general, have the property that the density is constant on sets $\mathcal{E}_c$, with $\mathbf{m}$ and $\mathbf{A}$ corresponding to the mean vector and to a multiple of the covariance matrix, respectively; see Chapter 3.

Now we are ready to approach the two-group situation, which will lead us straight into linear discriminant analysis. Suppose that a random vector $\mathbf{X}$ of dimension p is measured in two populations, and

$$\begin{aligned} E_1[\mathbf{X}] &= \boldsymbol{\mu}_1, \quad \text{Cov}[\mathbf{X}] = \psi \quad \text{in population 1,} \\ E_2[\mathbf{X}] &= \boldsymbol{\mu}_2, \quad \text{Cov}[\mathbf{X}] = \psi \quad \text{in population 2,} \end{aligned} \tag{26}$$

where ψ is positive definite. This setup parallels the univariate situation given after Example 5.2.2. Again, no distributional assumptions are made at this point. The following definition is similar to the one just discussed and needs no further explanation.

Definition 5.2.4 The multivariate standard distance between μ_1 and μ_2 for a random vector $\mathbf{X}$ as in (26) is given by

$$\begin{aligned}\Delta_{\mathbf{X}}(\mu_1, \mu_2) &= \max_{\substack{\mathbf{a}\in\mathbf{R}^p \\ \mathbf{a}\neq\mathbf{0}}} \Delta_{\mathbf{a}'\mathbf{X}}(\mathbf{a}'\mu_1, \mathbf{a}'\mu_2) \\ &= \max_{\substack{\mathbf{a}\in\mathbf{R}^p \\ \mathbf{a}\neq\mathbf{0}}} \frac{|\mathbf{a}'(\mu_1 - \mu_2)|}{(\mathbf{a}'\psi\mathbf{a})^{1/2}}.\end{aligned} \tag{27}$$

□

Now we can define the notion of a discriminant function.

Definition 5.2.5 Let $\mathbf{X}$ denote a $p-$variate random vector with $E_1[\mathbf{X}] = \mu_1$ in population 1, $E_2[\mathbf{X}] = \mu_2$ in population 2, and $\text{Cov}[\mathbf{X}] = \psi$ in both populations, with ψ positive definite. Let $Y = \beta'\mathbf{X}$ denote a linear combination of $\mathbf{X}$. If

$$\Delta_Y(\beta'\mu_1\,,\ \beta'\mu_2) = \Delta_{\mathbf{X}}(\mu_1, \mu_2),$$

then $Y = \beta'\mathbf{X}$ is called a *linear discriminant function* for the two populations. □

Definitions 5.2.4 and 5.2.5 are crucial for understanding linear discriminant analysis. The multivariate standard distance between two populations with the same covariance matrix is the maximum univariate standard distance over all linear combinations. Any linear combination for which the maximum is attained is a linear discriminant function for the two populations.

The next theorem is a direct consequence of Theorem 5.2.4 and its proof.

Theorem 5.2.5 *The multivariate standard distance between two populations with mean vectors μ_1, μ_2 and common covariance matrix ψ is given by*

$$\Delta(\mu_1, \mu_2) = [(\mu_1 - \mu_2)'\psi^{-1}(\mu_1 - \mu_2)]^{1/2}. \tag{28}$$

Any linear combination $Y = \beta'\mathbf{X}$, with

$$\beta = c \cdot \psi^{-1}(\mu_1 - \mu_2),$$

where $c \neq 0$ is an arbitrary constant and a linear discriminant function for the two populations.

Example 5.2.6 Suppose that $\mathbf{X} = \begin{pmatrix} X_1 \\ X_2 \end{pmatrix}$ is a bivariate random vector with

$$E_1[\mathbf{X}] = \mu_1 = \begin{pmatrix} 2 \\ 5 \end{pmatrix} \text{ in population 1,}$$

$$E_2[\mathbf{X}] = \mu_2 = \begin{pmatrix} 0 \\ 0 \end{pmatrix} \text{ in population 2,}$$

and

$$\text{Cov}[\mathbf{X}] = \psi = \begin{pmatrix} 1 & 0 \\ 0 & 5 \end{pmatrix} \text{ in both populations.}$$

The univariate standard distances are $|2-0|/1 = 2$ for variable X_1, and $|5-0|/\sqrt{5} = \sqrt{5} \approx 2.24$ for variable X_2. The bivariate standard distance, from equation (28), is given by

$$\Delta(\mu_1, \mu_2) = \left[(2,\ 5)\begin{pmatrix} 1 & 0 \\ 0 & \frac{1}{5} \end{pmatrix}\begin{pmatrix} 2 \\ 5 \end{pmatrix}\right]^{1/2} = 3,$$

and the coefficient vector of the discriminant function is given by

$$\beta = c \cdot \psi^{-1}(\mu_1 - \mu_2) = c \cdot \begin{pmatrix} 1 & 0 \\ 0 & \frac{1}{5} \end{pmatrix}\begin{pmatrix} 2 \\ 5 \end{pmatrix} = c \cdot \begin{pmatrix} 2 \\ 1 \end{pmatrix},$$

where $c \neq 0$ is an arbitrary constant. If we choose $c = 1$, then the discriminant function is given by $Y = 2X_1 + X_2$. This example is continued in Exercise 6. □

From now , we shall usually talk about *the* linear discriminant function as if it were uniquely defined, and we will set $c = 1$, unless stated otherwise. That is, the coefficient vector will be

$$\beta = \psi^{-1}(\mu_1 - \mu_2). \tag{29}$$

This choice of the constant c typically yields a nonnormalized discriminant function, but it has the simple property that

$$\begin{aligned} \text{var}[Y] &= \text{var}[\beta'\mathbf{X}] = \beta'\psi\beta \\ &= (\mu_1 - \mu_2)'\psi^{-1}(\mu_1 - \mu_2) \\ &= \Delta^2_{\mathbf{X}}(\mu_1, \mu_2). \end{aligned} \tag{30}$$

Moreover, for $c = 1$,

$$\Delta^2_{\mathbf{X}}(\mu_1, \mu_2) = \beta'(\mu_1 - \mu_2). \tag{31}$$

But $\beta'(\mu_1 - \mu_2)$ is the mean difference between the two groups for the discriminant function. Thus, for $c = 1$, both the variance and the mean difference of the discriminant function are identical with the squared multivariate standard distance. Note that this holds only for $c = 1$, but not for an arbitrary choice of c.

Depending on the particular convention used, on the investigator's taste, or on standards used in software for the numerical computation of a discriminant function, one may get a different solution β^* for the coefficients of the linear discriminant function. To make things worse, sometimes a constant (or intercept) is added, resulting in a discriminant function

$$Y^* = \beta_0 + c \cdot Y = \beta_0 + c \cdot \beta'\mathbf{X}. \tag{32}$$

For the purpose of discrimination, any Y^* of this type is as good as $Y = (\mu_1 - \mu_2)'\psi^{-1}\mathbf{X}$ because it leads to the same standard distance, provided that $c \neq 0$. Although this is somewhat confusing to the novice, one may use the freedom provided by the choice of two parameters β_0 and c to transform the discriminant function to a form convenient for classification purposes. For instance, if we write $\Delta = \Delta_{\mathbf{X}}(\mu_1, \mu_2)$ and choose

$$c = 1/\Delta,$$

and

$$\beta_0 = -\frac{1}{2\Delta}(\mu_1 - \mu_2)'\psi^{-1}(\mu_1 + \mu_2), \tag{33}$$

then the transformed discriminant function is given by

$$\begin{aligned} Y^* &= \beta_0 + c \cdot \beta'\mathbf{X} \\ &= -\frac{1}{2\Delta} \cdot (\mu_1 - \mu_2)'\psi^{-1}(\mu_1 + \mu_2) + \frac{1}{\Delta}(\mu_1 - \mu_2)'\psi^{-1}\mathbf{X}. \end{aligned} \tag{34}$$

The variable Y^* has the property that

$$\text{var}[Y^*] = \frac{1}{\Delta^2}(\mu_1 - \mu_2)'\psi^{-1}\psi\psi^{-1}(\mu_1 - \mu_2) = 1 \tag{35}$$

in both populations. Moreover, with $\gamma_1 = E[Y^*|P_1]$ and $\gamma_2 = E[Y|P_2]$ denoting the means of Y^* in the two populations, we obtain

$$\begin{aligned} \gamma_1 &= -\frac{1}{2\Delta}(\mu_1 - \mu_2)'\psi^{-1}(\mu_1 + \mu_2) + \frac{1}{\Delta}(\mu_1 - \mu_2)'\psi^{-1}\mu_1 \\ &= \frac{1}{\Delta}(\mu_1 - \mu_2)'\psi^{-1}\Big[-\frac{1}{2}(\mu_1 + \mu_2) + \mu_1\Big] \\ &= \frac{1}{\Delta}(\mu_1 - \mu_2)'\psi^{-1}(\mu_1 - \mu_2)/2 \\ &= \frac{1}{2}\Delta, \end{aligned} \tag{36}$$

and similarly,

$$\gamma_2 = -\frac{1}{2}\Delta. \tag{37}$$

Hence, the midpoint between the means of Y^* is zero, and the variance of Y^* is one. Standard distance then is identical with Euclidean distance on the Y^*-scale.

Example 5.2.7 Suppose $\mathbf{X} = \begin{pmatrix} X_1 \\ X_2 \end{pmatrix}$ has a normal distribution with mean vector $\mu_1 = \begin{pmatrix} 5 \\ 1 \end{pmatrix}$ in population 1, $\mu_2 = \begin{pmatrix} 2 \\ 1 \end{pmatrix}$ in population 2, and covariance matrix $\psi = \begin{pmatrix} 2 & 1 \\ 1 & 1 \end{pmatrix}$ in both populations. The univariate standard distances are $|5 - 2|/\sqrt{2} \approx 2.121$ for X_1 and 0 for X_2. The coefficients of the linear discriminant function are

$$\beta = \begin{pmatrix} 2 & 1 \\ 1 & 1 \end{pmatrix}^{-1} \begin{pmatrix} 3 \\ 0 \end{pmatrix} = \begin{pmatrix} 1 & -1 \\ -1 & 2 \end{pmatrix} \begin{pmatrix} 3 \\ 0 \end{pmatrix} = \begin{pmatrix} 3 \\ -3 \end{pmatrix},$$

and therefore, the linear discriminant function is $Y = 3X_1 - 3X_2$ or any linear combination equivalent to $X_1 - X_2$. The bivariate standard distance is calculated as

$$\Delta^2 = (\mu_1 - \mu_2)'\psi^{-1}(\mu_1 - \mu_2) = \beta'(\mu_1 - \mu_2) = (3, -3) \begin{pmatrix} 3 \\ 0 \end{pmatrix} = 9,$$

and hence, $\Delta = 3$. This is larger than the univariate standard distance of X_1. Notice that the linear discriminant function depends on X_2, although the mean difference in X_2 is zero.

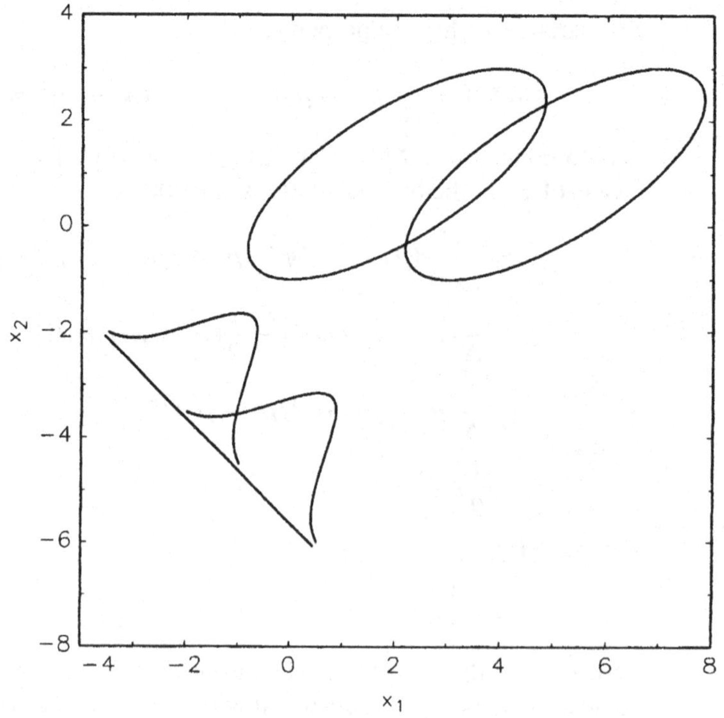

Figure 5-2-4 The distribution of the linear discriminant function in Example 5.2.7.

Figure 5.2.4 shows a contour plot for this example, with the distribution of the normalized discriminant function

$$Y' = \frac{1}{\sqrt{2}}(X_1 - X_2) = \frac{1}{3\sqrt{2}}\, Y$$

shown as the projection of the distribution on a line. On the line corresponding to Y', the two means are $\Delta = 3$ standard deviations apart.
Finally, let us look at the transformation suggested in (34). Note that

$$E[Y] = E[3X_1 - 3X_2] = (3,\ -3)\mu_i$$
$$= \begin{cases} 12 & \text{in group 1} \\ 3 & \text{in group 2,} \end{cases}$$

and $\text{var}[Y] = (3,\ -3)\begin{pmatrix} 2 & 1 \\ 1 & 1 \end{pmatrix}\begin{pmatrix} 3 \\ -3 \end{pmatrix} = 9$ in both groups. Thus, we choose a transformation $Y^* = \beta_0 + c \cdot Y$, with $c = 1/3$ and $\beta_0 = -2.5$. Then

$$Y^* = -2.5 + \frac{1}{3}Y = X_1 - X_2 - 2.5\,,$$

and $E_1[Y^*] = 1.5$ in group 1, $E_2[Y^*] = -1.5$ in group 2, and $\text{var}[Y^*] = 1$ in both groups. Hence, in a graph of the densities of Y^* in both groups, they intersect exactly at 0, and the means differ by $\Delta = 3$.

As in Example 5.2.5, the numerical results depend in no way on the normality of $\mathbf{X}$. The normality assumptions were introduced only to allow graphing contours of constant density. □

Exercises for Section 5.2

1. Show that the probability density functions of the following distributions depend on the argument x only through the standard distance $\Delta(x) = \Delta_X(x, \mu)$:
   (a) The distribution of the sum of three fair and independent dice; see Exercise 1 in Section 2.9.
   (b) Any binomial distribution with parameter $p = \frac{1}{2}$.
   (c) All beta distributions with identical exponents, i.e., with density functions

   $$f_X(x) = c \cdot [x(1-x)]^\gamma \text{ for } x \in (0, 1)\,,$$

   where $\gamma > -1$ is a parameter and c is a proper constant depending on γ.
   (d) The distribution of the x-coordinate of the first raindrop in Example 2.3.4.
   (e) The *logistic distribution* with density function

   $$f_X(x) = \beta \cdot \exp(\alpha + \beta x)/[1 + \exp(\alpha + \beta x)]^2\,,\ x \in \mathbb{R},$$

where $\beta > 0$ and α are parameters.

(f) Are you sure about part (e)?

2. Let X denote a random variable with

$$E_1[X] = \mu_1 \; , \; \text{var}[X] = \sigma^2 \text{ in population 1,}$$

$$E_2[X] = \mu_2 \; , \; \text{var}[X] = \sigma^2 \text{ in population 2,}$$

and let $Y = \alpha + \beta X$ for some fixed $\alpha, \beta \in \mathbb{R}, \beta \neq 0$. Show that the standard distance between the two means is the same for Y as for X; see equation (4).

3. This exercise is to illustrate the computation of error rates, or overlap, as in Example 5.2.3, for selected nonnormal distributions. Four examples of continuous distributions are given, with densities symmetric to their mean μ. Suppose X follows the respective distribution with mean $\mu_1 = \delta \geq 0$ in population 1, and mean $\mu_2 = 0$ in population 2. Hence, δ is the (absolute) mean difference. Let

$$h(\delta) := \Pr\left[X \leq \frac{\delta}{2} \mid P_1\right]$$

denote the "overlap", as in Example 5.2.3, written as a function of the mean difference δ. To make this comparable to the case of the normal distribution, we have to express overlap in terms of standard distance. Thus, do the following:

(a) Compute the function $h(\delta)$.

(b) Find $\sigma^2 = \text{Var}[X]$, which is identical in both groups.

(c) Since $\Delta(\mu_1, \mu_2) = \delta/\sigma =: \Delta$, write the mean difference as $\delta = \sigma\Delta$. Hence, find the "overlap function"

$$o(\Delta) = h(\sigma\Delta),$$

and graph $o(\Delta)$ for $0 \leq \Delta \leq 6$ as in Figure 5.2.2

(d) Compare the graph to Figure 5.2.2, and comment on differences. The four distributions to be considered are given here in the form of density functions; they are also shown graphically in Figure 5.2.5.

(i) The *double-exponential distribution*, with density

$$f(x; \mu) = \frac{1}{2} \exp\left[-|x - \mu|\right] \; , \; x \in \mathbb{R} .$$

(ii) The *triangle distribution*, with density

$$f(x; \mu) = \begin{cases} 1 - |x - \mu| & \text{if } \mu - 1 \leq x \leq \mu + 1, \\ 0 & \text{otherwise.} \end{cases}$$

(iii) The *uniform distribution*, with density

$$f(x; \mu) = \begin{cases} \frac{1}{2} & \text{if } \mu - 1 \leq x \leq \mu + 1, \\ 0 & \text{otherwise.} \end{cases}$$

(iv) The *inverse triangle distribution*, with density

$$f(x;\mu) = \begin{cases} |x-\mu| & \text{if } \mu - 1 \le x \le \mu + 1 \\ 0 & \text{otherwise.} \end{cases}$$

4. This exercise is to show that standard distance satisfies the axioms of a measure of distance. Let $\mathbf{M}$ denote a positive definite symmetric matrix of dimension $p \times p$. Then $\mathbf{M}^{-1}$ is positive definite; see Appendix A.2. With the notation

$$d(\mathbf{x}, \mathbf{y}) = [(\mathbf{x}-\mathbf{y})'\mathbf{M}^{-1}(\mathbf{x}-\mathbf{y})]^{1/2},$$

show the following:

(a) $d(\mathbf{x}, \mathbf{y}) \ge 0$ for all $\mathbf{x}, \mathbf{y} \in \mathbb{R}^p$, with equality exactly if $\mathbf{x} = \mathbf{y}$.

(b) $d(\mathbf{x}, \mathbf{y}) = d(\mathbf{y}, \mathbf{x})$ for all $\mathbf{x}, \mathbf{y} \in \mathbb{R}^p$.

(c) $d(\mathbf{x}, \mathbf{y}) + d(\mathbf{y}, \mathbf{z}) \ge d(\mathbf{x}, \mathbf{z})$ for all $\mathbf{x}, \mathbf{y}, \mathbf{z} \in \mathbb{R}^p$. *Hint*: Square the right-hand side, write $\mathbf{x} - \mathbf{z} = (\mathbf{x} - \mathbf{y}) + (\mathbf{y} - \mathbf{z})$, use the fact that $u \le |u|$ for all $u \in \mathbb{R}$, and apply the extended Cauchy-Schwartz inequality.

(d) Show that $d^2(\mathbf{x}, \mathbf{y})$ is not a measure of distance. Which of (a), (b), or (c) does it violate?

5. This is an exercise for students who have taken a course in multiple regression and who are familiar with the notion of leverage. In a least squares regression setup, let $\mathbf{T}$ denote the $N \times (p+1)$ design matrix with a column of 1's. Then the leverage of the

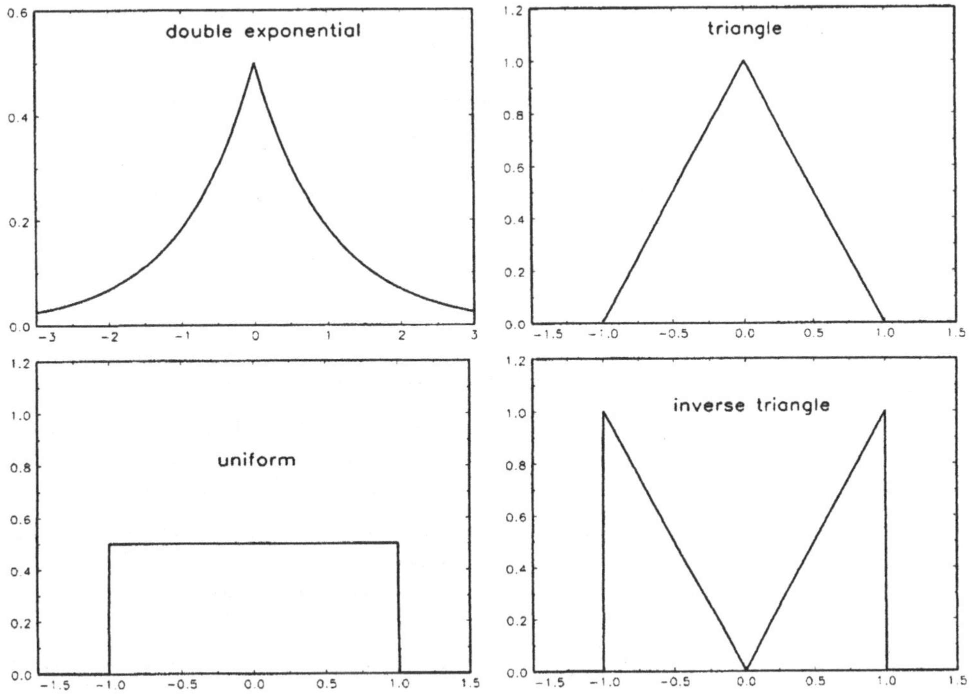

Figure 5-2-5
Graphs of the density functions used in Exercise 3 of Section 5.2: (i) double-exponential distribution; (ii) triangle distribution; (iii) uniform distribution; (iv) V-distribution.

ith observation is the ith diagonal entry of the hat matrix $\mathbf{H} = \mathbf{T}(\mathbf{T}'\mathbf{T})^{-1}\mathbf{T}'$. Denote this entry by h_{ii}, and show that

$$\begin{aligned} h_{ii} &= \frac{1}{N} + \frac{1}{N-1}(\mathbf{x}_i - \bar{\mathbf{x}})'\mathbf{S}^{-1}(x_i - \bar{x}) \\ &= \frac{1}{N} + \frac{1}{N-1} D^2(\mathbf{x}_i, \bar{\mathbf{x}}) \quad i = 1, \ldots, N, \end{aligned}$$

where $\mathbf{x}_i \in \mathbb{R}^p$ is the column vector of values of the p regressors for the ith observation, $\bar{\mathbf{x}} \in \mathbb{R}^p$ is the mean vector of the regressors, and $\mathbf{S}$ is the covariance matrix of the regressors. *Hint*: If you find this exercise somewhat messy, assume that all regressors are centered, i.e., $\bar{\mathbf{x}} = \mathbf{0}$. Use this to show that the lower right $p \times p$ part of $(\mathbf{T}'\mathbf{T})^{-1}$ is $\mathbf{S}^{-1}/(N-1)$. Note also that the ith row of $\mathbf{T}$ is $\mathbf{x}_i'$, with an additional entry "1" in the first position.

6. This exercise refers to Example 5.2.6.

   (a) With $Y = 2X_1 + X_2$ denoting the linear discriminant function, find the mean of Y in both groups and the variance of Y.

   (b) Verify that the univariate standard distance between the means of Y in the two groups is 3.

   (c) Find the normalized version of the discriminant function.

   (d) Assuming that $\mathbf{X}$ is bivariate normal with parameters as indicated, sketch ellipses of constant density according to

   $$[(\mathbf{x} - \boldsymbol{\mu}_i)'\boldsymbol{\psi}^{-1}(\mathbf{x} - \boldsymbol{\mu}_i)]^{1/2} = 2\,, \; i = 1, 2.$$

   (e) In the same graph, sketch the distribution of the discriminant function as in Example 5.2.7 and Figure 5.2.4.

7. Suppose that $\mathbf{X}$ follows a bivariate distribution with $E_1[\mathbf{X}] = \boldsymbol{\mu}_1 = \begin{pmatrix} 1 \\ 1 \end{pmatrix}$ in group 1, $E_2[\mathbf{X}] = \boldsymbol{\mu}_2 = \begin{pmatrix} 0 \\ 0 \end{pmatrix}$ in group 2, and common covariance matrix $\boldsymbol{\psi} = \begin{pmatrix} 1 & \rho \\ \rho & 1 \end{pmatrix}$, $-1 < \rho < 1$.

   (a) Compute the vector of discriminant function coefficients. How does the discriminant function depend on ρ?

   (b) Compute the bivariate standard distance Δ as a function of ρ, and plot this function in the interval $[-1, 1]$. What are the minimum and the maximum of Δ, and for which value(s) of ρ are they attained?

   (c) Assuming a bivariate normal distribution for $\mathbf{X}$, sketch ellipses of constant density as in Exercise 6, part (d), for $\rho = \pm 0.8$, and $\rho = 0$. These graphs will help you to understand the results in parts (a) and (b).

8. Same as Exercise 7, with (i) $\boldsymbol{\mu}_1 = \begin{pmatrix} 1 \\ 0 \end{pmatrix}$, and (ii) $\boldsymbol{\mu}_1 = \begin{pmatrix} 2 \\ 1 \end{pmatrix}$.

9. Let $Y = \mathbf{a}'\mathbf{X}$ denote a linear combination of $\mathbf{X}$. Show that the linear combinations

$$Y^* = \pm \frac{\mathbf{a}}{(\mathbf{a}'\mathbf{a})^{1/2}} \mathbf{X}$$

are normalized and equivalent to Y in the sense of Definition 7.2.2.

10. Prove the Cauchy-Schwartz inequality. *Hint*: Set $\mathbf{a} = \mathbf{x}/(\mathbf{x}'\mathbf{x})^{1/2}$ and $\mathbf{b} = \mathbf{y}/(\mathbf{y}'\mathbf{y})^{1/2}$. Use $(\mathbf{a}+\mathbf{b})'(\mathbf{a}+\mathbf{b}) \geq 0$ and $(\mathbf{a}-\mathbf{b})'(\mathbf{a}-\mathbf{b}) \geq 0$ to show that $(\mathbf{a}'\mathbf{b})^2 \leq 1$.

11. Prove Theorem 5.2.4 along the following lines. First show that, for any linear combination $Y = \mathbf{a}'\mathbf{X}$, there is an equivalent one such that $\text{var}[Y] = 1$. Then introduce a Lagrange multiplier λ for the constraint $\mathbf{a}'\psi\mathbf{a} = 1$, and find stationary point of the function

$$h(\mathbf{a}) = \left[\mathbf{a}'(\mathbf{x}_1 - \mathbf{x}_2)\right]^2 - \lambda(\mathbf{a}'\psi\mathbf{a} - 1),$$

using partial derivatives with respect to $\mathbf{a}$.

12. Suppose that $\mathbf{X}$ is a p-variate random vector with means μ_1 and μ_2 in groups 1 and 2 and with common covariance matrix ψ. Let Y denote the linear discriminant function for the two groups, and let $Z = \gamma'\mathbf{X}$ denote an arbitrary linear combination of $\mathbf{X}$ which is uncorrelated with Y. Show that the mean of Z is the same in both groups, i.e., $\gamma'\mu_1 = \gamma'\mu_2$.

13. Let X denote a univariate random variable with mean μ and variance σ^2. Define $k \geq 1$ random variables $Y_1, \ldots, Y_k$ as $Y_h = X + e_h$, $h = 1, \ldots, k$, where all e_h are mutally independent and independent of X, and identically distributed, with mean zero and variance $\tau^2 > 0$. Define $\mathbf{Y} = (Y_1, \cdots, Y_k)'$. Let $\mathbf{1}_k \in \mathbb{R}^k$ denote a vector with 1 in all entries.

   (a) Show that $E[\mathbf{Y}] = \mu\, \mathbf{1}_k$ and

$$\psi(\sigma,\tau) := \text{Cov}[\mathbf{Y}] = \sigma^2 \mathbf{1}_k \mathbf{1}_k' + \text{diag}(\tau^2, \ldots, \tau^2)$$
$$= \begin{pmatrix} \sigma^2+\tau^2 & \sigma^2 & \cdots & \sigma^2 \\ \sigma^2 & \sigma^2+\tau^2 & \cdots & \sigma^2 \\ \vdots & \vdots & \ddots & \vdots \\ \sigma^2 & \sigma^2 & \cdots & \sigma^2+\tau^2 \end{pmatrix}. \tag{38}$$

   (b) Suppose that $\mathbf{Y}$ is measured in two groups, with mean vectors $E_1[\mathbf{Y}] = \mu_1 = \mu_1 \mathbf{1}_k$ in group 1, $E_2[\mathbf{Y}] = \mu_2 = \mu_2 \mathbf{1}_k$ in group 2, and common covariance matrix $\psi(\sigma,\tau)$ in both groups. Show that the vector of coefficients of the linear discriminant function is proportional to the vector $\mathbf{1}_k$. *Hint*: This can be done by finding the inverse of $\psi(\sigma,\tau)$, but there is a much simpler proof by arguing that interchanging any two variables Y_h and Y_ℓ in $\mathbf{Y}$ does not change the mean vectors and the covariance matrix.

   (c) Show that the squared standard distance between μ_1 and μ_2 is given by $\Delta_{\mathbf{Y}}^2 = (\mu_1 - \mu_2)^2/(\sigma^2 + \tau^2/k)$. *Hint*: Avoid inversion of the covariance matrix by recalling that the multivariate standard distance is the same as the univariate standard distance based on the discriminant function.

Note: The model used in this exercise has some practical relevance for the following reason. One may think of the variable X as the variable the investigator would really like to measure; however, X can only be measured with error. The variables Y_h represent repeated measurements, all affected by independent errors. The standard distance based on X would be $\Delta_X(\mu_1, \mu_2) = |\mu_1 - \mu_2|/\sigma$. The same standard distance is achieved by the k variables $\mathbf{Y}$ only asymptotically as $k \to \infty$.

5.3 Using the Linear Discriminant Function

In this section, we apply the theory developed so far to practical examples, by substituting sample statistics for model parameters, as justified by the plug-in principle of estimation from Section 4.2. This is a heuristic and mostly descriptive approach, in which we ignore problems of statistical inference. No distributional assumptions are made at this point.

Let $\mathbf{x}_{11}, \mathbf{x}_{12}, \ldots, \mathbf{x}_{1N_1}$ denote the observed data vectors from group 1, and $\mathbf{x}_{21}, \mathbf{x}_{22}, \ldots, \mathbf{x}_{2N_2}$ the data vectors from group 2. These $N_1 + N_2$ observations constitute the *training samples*. Let

$$\bar{\mathbf{x}}_j = \frac{1}{N_j} \sum_{i=1}^{N_j} \mathbf{x}_{ji}, \quad j = 1, 2, \tag{1}$$

denote the sample mean vectors, and

$$\mathbf{S}_j = \frac{1}{N_j - 1} \sum_{i=1}^{N_j} (\mathbf{x}_{ji} - \bar{\mathbf{x}}_j)(\mathbf{x}_{ji} - \bar{\mathbf{x}}_j)', \quad j = 1, 2, \tag{2}$$

the usual sample covariance matrices, with denominators $N_j - 1$. By using denominators $N_j - 1$ instead of N_j we actually violate the plug-in principle, but follow the convention accepted by most statisticians.

The only difficulty in applying the notions of multivariate standard distance and linear discriminant function to the observed data is that we have two covariance matrices $\mathbf{S}_1$ and $\mathbf{S}_2$, rather than a single common covariance matrix. This difficulty is overcome by using a weighted average of $\mathbf{S}_1$ and $\mathbf{S}_2$, the *pooled sample covariance matrix*

$$\mathbf{S} = \frac{1}{N_1 + N_2 - 2} \left[(N_1 - 1)\mathbf{S}_1 + (N_2 - 1)\mathbf{S}_2 \right]. \tag{3}$$

From the plug-in principle of estimation alone, it would be difficult to justify the use of (3), but we may recall from Theorem 4.2.3 that the pooled sample covariance matrix is an unbiased estimator of the common covariance matrix of two populations, irrespective of the exact distribution. This may serve, for the moment, as justification.

In analogy to the definitions of Section 5.2, we now define sample multivariate standard distance and the sample linear discriminant function.

Definition 5.3.1 In the setup of equations (1) to (3), the *multivariate standard distance* between $\bar{\mathbf{x}}_1$ and $\bar{\mathbf{x}}_2$ is given by

$$D(\bar{\mathbf{x}}_1, \bar{\mathbf{x}}_2) = \max_{\substack{\mathbf{a} \in R^p \\ \mathbf{a} \neq \mathbf{0}}} \frac{|\mathbf{a}'(\bar{\mathbf{x}}_1 - \bar{\mathbf{x}}_2)|}{(\mathbf{a}'\mathbf{S}\mathbf{a})^{1/2}}, \tag{4}$$

i.e., the maximum univariate standard distance over all linear combinations, provided the maximum exists. Any linear combination for which the maximum is attained is called a *linear discriminant function* for the given samples. □

This maximization problem is mathematically identical to the one we dealt with in Definition 5.2.4 and Theorem 5.2.5. Hence, we immediately obtain the following theorem.

Theorem 5.3.1 *In the setup of equations (1) to (3), assume that the pooled sample covariance matrix* $\mathbf{S}$ *is nonsingular. Then the multivariate standard distance between* $\bar{\mathbf{x}}_1$ *and* $\bar{\mathbf{x}}_2$ *is given by*

$$D(\bar{\mathbf{x}}_1, \bar{\mathbf{x}}_2) = \left[(\bar{\mathbf{x}}_1 - \bar{\mathbf{x}}_2)'\, \mathbf{S}^{-1}\, (\bar{\mathbf{x}}_1 - \bar{\mathbf{x}}_2)\right]^{1/2}, \tag{5}$$

and the vector of coefficients of the linear discriminant function is given by

$$\mathbf{b} = \mathbf{S}^{-1}(\bar{\mathbf{x}}_1 - \bar{\mathbf{x}}_2) \tag{6}$$

or any vector proportional to (6).

The assumption of nonsingularity of $\mathbf{S}$ is explored in more detail in Exercise 8. Note, however, that if $N_1 + N_2 < p + 2$, then $\mathbf{S}$ is always singular. This restricts the use of linear discriminant analysis to samples large enough for the pooled covariance matrix to be positive definite. Note also that, in equation (4), the term $\mathbf{a}'\mathbf{S}\mathbf{a}$ is the pooled-sample variance of the linear combination $Y = \mathbf{a}'\mathbf{X}$; see Exercise 13. This means that sample standard distance automatically has the same properties as its theoretical analog; for instance, it satisfies the axioms of a measure of distance. As in Section 5.2, we will use equation (6) as the definition of the vector of coefficients of the linear discriminant function, i.e., choose the proportionality constant to be $c = 1$.

In the current descriptive approach, we will not make any assumptions about the data beyond positive definiteness of $\mathbf{S}$. In particular, we do not assume that the data constitute random samples from a particular family of distributions or that the covariance matrices in the populations are identical. However, we will have to discuss the use of the pooled-sample covariance matrix to understand under what circumstances it is reasonable.

Example 5.3.1 This is the continuation of the midge example. Univariate statistics have been given in Table 5.2.1. The covariance matrices (with X_1 = antenna length, X_2 = wing length) are given by

$$\mathbf{S}_1 = \begin{pmatrix} 98.00 & 80.83 \\ 80.83 & 168.78 \end{pmatrix} \quad \text{(for } Af\text{)}$$

and

$$\mathbf{S}_2 = \begin{pmatrix} 39.47 & 43.47 \\ 43.47 & 77.87 \end{pmatrix} \quad \text{(for } Apf\text{)}.$$

With sample sizes $N_1 = 9,\ N_2 = 6$, the pooled covariance matrix is

$$\mathbf{S} = \frac{1}{13}[8\mathbf{S}_1 + 5\mathbf{S}_2] = \begin{pmatrix} 75.49 & 66.46 \\ 66.46 & 133.81 \end{pmatrix},$$

and its inverse as

$$\mathbf{S}^{-1} = \begin{pmatrix} 23.54 & -11.69 \\ -11.69 & 13.28 \end{pmatrix} \cdot 10^{-3}.$$

The vector of mean differences is given by

$$\mathbf{d} = \bar{\mathbf{x}}_1 - \bar{\mathbf{x}}_2 = \begin{pmatrix} 18.67 \\ -12.22 \end{pmatrix}.$$

Finally, we get the coefficients of the linear discriminant function as

$$\mathbf{b} = \mathbf{S}^{-1}\mathbf{d} = \begin{pmatrix} b_1 \\ b_2 \end{pmatrix} = \begin{pmatrix} 0.582 \\ -0.381 \end{pmatrix},$$

and the bivariate standard distance as

$$D = [\mathbf{d}'\mathbf{S}^{-1}\mathbf{d}]^{1/2} = [\mathbf{d}'\mathbf{b}]^{1/2} = 3.94.$$

Thus, in the optimal linear combination of antenna length and wing length, the two means are $D = 3.94$ standard deviations apart. Now, we can compute the value of

$$V = b_1 X_1 + b_2 X_2 = 0.582 X_1 - 0.381 X_2$$

for all 15 observations. Table 5.3.1 gives the enlarged data matrix thus obtained. We leave it to the student to verify by direct calculation that the univariate standard distance of variable V is $D = 3.94$, see Exercise 1.

Let us now attempt classification, again, using a heuristic approach. Either from the data in Table 5.3.1 or by matrix calculations, we obtain the sample means of the discriminant function in both groups: $\bar{v}_1 = 13.633$ for group 1 (Af), and $\bar{v}_2 = -1.890$ for group 2 (Apf). The midpoint between $\bar{v}_1$ and $\bar{v}_2$ is $m = (\bar{v}_1 + \bar{v}_2)/2 = 5.872$. Hence, a simple classification rule could be

$$\text{assign to} \begin{cases} Af & \text{if } V > 5.87 \\ Apf & \text{if } V < 5.87. \end{cases}$$

Table 5.3.1 Raw data (in mm/100) and two versions of the discriminant function in the midge example

	Antenna (X_1)	Wing (X_2)	V	V^*
Af	138	164	17.95	3.07
	140	170	16.83	2.78
	124	172	6.75	0.22
	136	174	12.98	1.80
	138	182	11.10	1.33
	148	182	16.92	2.81
	154	182	20.42	3.69
	138	190	8.05	0.55
	156	208	11.69	1.48
Apf	114	178	−1.36	−1.83
	120	186	−0.91	−1.72
	118	196	−5.88	−2.98
	130	196	1.11	−1.21
	126	200	−2.74	−2.19
	128	200	−1.58	−1.89

Thus, to classify a midge with unknown group membership but known values x_1 and x_2, one would compute $v = 0.582x_1 - 0.381x_2$ and check if $v < m$ or $v > m$. See also Exercise 2.

Viewed in the space of the original variables, the cutoff point for classification corresponds to a straight line. To see this, notice that $v = m$ corresponds to the equation

$$b_1x_1 + b_2x_2 = m,$$

or (assuming $b_2 \neq 0$)

$$x_2 = \frac{m}{b_2} - \frac{b_1}{b_2}x_1.$$

In the midge example, this equation is $x_2 = -15.428 + 1.53x_1$ and is shown in Figure 5.3.1. Hence, we have found a preliminary solution to the problem stated in Chapter 1. Principles of optimal classification will be discussed in more detail in Section 7.1.

Finally, let us look at an equivalent version of the discriminant function, suggested by the transformation (34) in Section 5.2. Since the pooled variance of V is $D^2 = 15.52$ and the midpoint between $\bar{v}_1$ and $\bar{v}_2$ is $m = 5.87$, we may prefer to use

$$\begin{aligned} V^* &= \frac{1}{D}(V - m) = (V - 5.87)/3.94 \\ &= 1.49 + 0.1478X_1 - 0.0966X_2. \end{aligned}$$

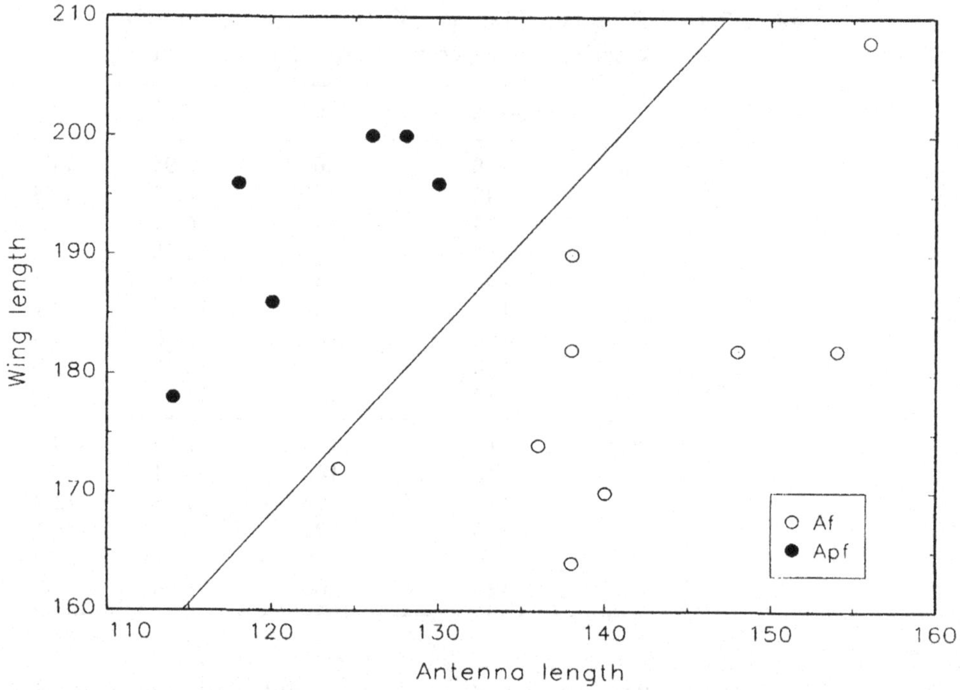

Figure 5-3-1 Classification of midges by the linear discriminant function.

The data for V^* is also given in Table 5.3.1. We leave it to the student to verify by direct calculation that V^* has pooled variance 1 and means $\pm D/2$. Hence the cutoff point for classification based on V^* would be zero. This cutoff corresponds exactly to the same straight line as the one shown in Figure 5.3.1. □

Example 5.3.2 This example is similar in nature to the midge example, but this time it consists of two species of flea beetles and four variables, namely,

X_1 = distance of the transverse groove to the posterior border of the prothorax,

X_2 = length of the elytra,

X_3 = length of the second antennal joint,

and

X_4 = length of the third antennal joint.

Variable X_2 is in 0.01mm; all others are in microns. The data of $N_1 = 19$ specimens of *Haltica oleracea* and $N_2 = 20$ specimens of *H. carduorum* are reproduced in Table 5.3.2. Univariate sample statistics for all four variables are given in Table 5.3.3. The

Table 5.3.2 Flea beetle data[a]

Species	X_1	X_2	X_3	X_4
1	189	245	137	163
1	192	260	132	217
1	217	276	141	192
1	221	299	142	213
1	171	239	128	158
1	192	262	147	173
1	213	278	136	201
1	192	255	128	185
1	170	244	128	192
1	201	276	146	186
1	195	242	128	192
1	205	263	147	192
1	180	252	121	167
1	192	283	138	183
1	200	294	138	188
1	192	277	150	177
1	200	287	136	173
1	181	255	146	183
1	192	287	141	198
2	181	305	184	209
2	158	237	133	188
2	184	300	166	231
2	171	273	162	213
2	181	297	163	224
2	181	308	160	223
2	177	301	166	221
2	198	308	141	197
2	180	286	146	214
2	177	299	171	192
2	176	317	166	213
2	192	312	166	209
2	176	285	141	200
2	169	287	162	214
2	164	265	147	192
2	181	308	157	204
2	192	276	154	209
2	181	278	149	235
2	175	271	140	192
2	197	303	170	205

Source: Lubischew (1962)

[a] Variables are explained in Example 5.3.2.

Table 5.3.3 Univariate summary statistics in the flea beetle example

Variable	Group	Mean	Std. Dev.	Standard Distance
X_1	*H. oleracea*	194.47	13.70	1.25
	H. carduorum	179.55	10.09	
X_2	*H. oleracea*	267.05	18.58	1.24
	H. carduorum	290.80	19.72	
X_3	*H. oleracea*	137.37	8.15	1.82
	H. carduorum	157.20	12.94	
X_4	*H. oleracea*	185.95	15.49	1.62
	H. carduorum	209.25	13.34	

univariate standard distances range from 1.24 for variable X_2 to 1.82 for variable X_3; thus, none of the four variables allows good discrimination by itself.

The groupwise sample covariance matrices are

$$\mathbf{S}_1 = \begin{pmatrix} 187.60 & 176.86 & 48.37 & 113.58 \\ 176.86 & 345.39 & 75.98 & 118.78 \\ 48.37 & 75.98 & 66.36 & 16.24 \\ 113.58 & 118.78 & 16.24 & 239.94 \end{pmatrix}$$

and

$$\mathbf{S}_2 = \begin{pmatrix} 101.84 & 128.06 & 36.99 & 32.59 \\ 128.06 & 389.01 & 165.36 & 94.37 \\ 36.99 & 165.36 & 167.54 & 66.53 \\ 32.59 & 94.37 & 66.53 & 177.88 \end{pmatrix},$$

respectively, and the pooled covariance matrix is

$$\mathbf{S} = \frac{1}{37}\left[18\mathbf{S}_1 + 19\mathbf{S}_2\right] = \begin{pmatrix} 143.56 & 151.80 & 42.53 & 71.99 \\ 151.80 & 367.79 & 121.88 & 106.24 \\ 42.53 & 121.88 & 118.31 & 42.06 \\ 71.99 & 106.24 & 42.06 & 208.07 \end{pmatrix}.$$

With $\bar{\mathbf{x}}_1 = (194.47, 267.05, 137.37, 185.95)'$ and $\bar{\mathbf{x}}_2 = (179.55, 290.80, 157.20, 209.25)'$ from Table 5.3.3, we obtain the coefficients of the linear discriminant function as

$$\mathbf{b} = \mathbf{S}^{-1}(\bar{\mathbf{x}}_1 - \bar{\mathbf{x}}_2) = \begin{pmatrix} 0.345 \\ -0.130 \\ -0.106 \\ -0.143 \end{pmatrix},$$

and the multivariate standard distance as

$$D(\bar{\mathbf{x}}_1, \bar{\mathbf{x}}_2) = \left[(\bar{\mathbf{x}}_1 - \bar{\mathbf{x}}_2)'\mathbf{S}^{-1}(\bar{\mathbf{x}}_1 - \bar{\mathbf{x}}_2)\right]^{1/2} = 3.70.$$

Thus, the linear discriminant function is

$$V = 0.345X_1 - 0.130X_2 - 0.106X_3 - 0.143X_4.$$

We leave it to the student to graph the distribution of V and to study classification aspects of this example; see Exercise 3. We note, however, that discrimination based on all four variables is considerably better than discrimination based on any single one of the four variables alone. □

Example 5.3.2 raises the question under what conditions a variable should be included in an analysis or excluded from it. A naive answer to this question is that including a variable never hurts because the multivariate standard distance cannot decrease by adding variables to the analysis; see Exercises 5 and 6. In the model setup, this answer might be acceptable. For sample data, however, including additional variables may actually impair the ability of the discriminant function to classify future observations correctly. We will return to this topic in Section 5.5. On the current descriptive level, we can discuss the problem informally by looking at subsets of variables to be used for discrimination and deciding, for each variable, whether or not the increase in standard distance justifies its inclusion in the analysis.

Example 5.3.3 continuation of Example 5.3.2

Table 5.3.4 gives a summary of an "all subsets linear discriminant analysis." The student who knows multiple regression will recognize a familiar phenomenon: using all four variables doesn't seem to be much better, in terms of multivariate standard distance, than using the optimal subset of three. Standard distance plays a role here similar to the role of the coefficient of determination (R^2) in multiple regression. It gives an overall assessment of the success of the analysis.

Table 5.3.4 reveals some interesting details. For instance, eliminating variable X_3 from the analysis affects the multivariate standard distance only moderately: it decreases from 3.70 to 3.58. This is somewhat surprising because X_3 is the best variable in the univariate approach. On the other hand, elimination of X_1 reduces the standard distance from 3.70 to 2.17, although X_1 by itself does not seem to be a very good discriminator.

We are dealing here with the same type of problem as in multiple regression, when it comes to assessing the importance of regressors or subsets of regressors. We shall approach this problem on two levels: in Section 5.6, we will introduce the notion of redundancy of variables and discuss under what circumstances a seemingly "worthless" variable may contribute to multivariate discrimination. In Section 6.5, we will present the associated testing theory.

Suppose we decide to omit variable X_3. The discriminant function based on variables X_1, X_2, and X_4, then is given by

$$V^* = 0.358X_1 - 0.169X_2 - 0.149X_4,$$

Table 5.3.4 Multivariate standard distances for all subsets of variables in the flea beetle example

Variables Included				Standard Distance
X_1	X_2	X_3	X_4	3.70
X_1	X_2	X_3		3.21
X_1	X_2		X_4	3.58
X_1		X_3	X_4	3.34
	X_2	X_3	X_4	2.17
X_1	X_2			3.01
X_1		X_3		2.67
X_1			X_4	2.66
	X_2	X_3		1.84
	X_2		X_4	1.75
		X_3	X_4	2.17
X_1				1.25
	X_2			1.24
		X_3		1.82
			X_4	1.62

Note: The standard distances for all models with only one variable are the same as those in Table 5.3.3.

see Exercise 7. To illustrate the seemingly paradoxical result that X_3 does not contribute to the multivariate standard distance despite its relatively large univariate standard distance, Figure 5.3.2 shows a scatterplot of V^* vs. X_3. Both V^* and X_3 show differences in location, but, in the bivariate setup, X_3 does not contribute much to the discrimination provided by V^*, the discriminant function based on X_1, X_2, and X_4. This is a typical case of redundancy of a variable, to which we will return in Section 5.6. □

In none of the examples discussed so far have we questioned the pooling of sample covariance matrices. In both examples, $\mathbf{S}_1$ and $\mathbf{S}_2$ looked sufficiently "similar" to justify the pooling. The next example will provide a situation where pooling is more questionable.

Example 5.3.4 The data used in this example are from the area of quality control. Two machines producing the same type of electrodes are compared. Samples of size $N_1 = N_2 = 50$ are available from the two machines. There are five variables, labeled X_1 to X_5, that measure the dimensions of electrodes illustrated in Figure 5.3.3. The raw data (which, for reasons of confidentiality, are linear transformations of the original data) are listed in Table 5.3.5. Univariate summary statistics are given in Table 5.3.6. There are quite substantial differences between the two machines, not only in location, but also in variability. For instance, the standard deviations for variable X_5 differ by a factor of

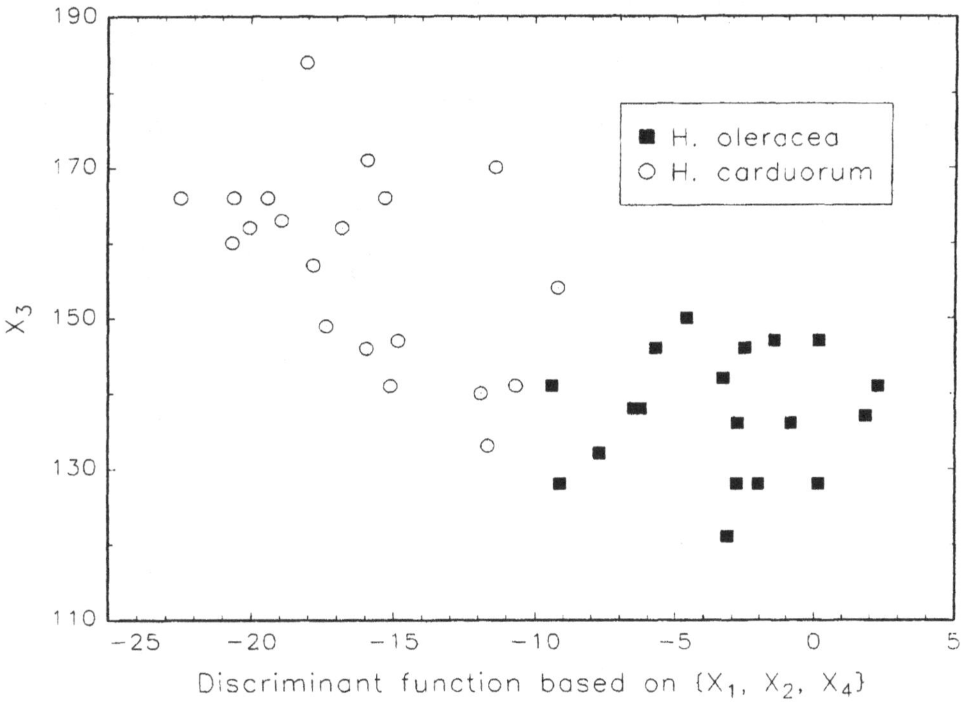

Figure 5-3-2
Plot of the discriminant function based on $\{X_1, X_2, X_4\}$ vs X_3 in the flea beetle example.

roughly 2. Hence the question arises: how does this affect linear discrimination? Is the linear discriminant function still a useful tool under these circumstances?

For the time being, let us be pragmatic and simply apply the linear discriminant function technique. Good separation of the two groups is certainly to be expected, in view of the fact that the best variable (X_4) alone has a standard distance larger

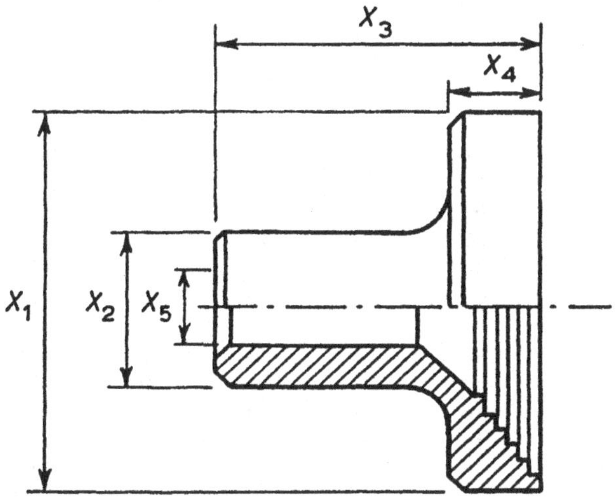

Figure 5-3-3
Definition of variables in the electrode example. Reproduced from Flury and Riedwyl (1988), with permission from Chapman & Hall. Figure 7.20, p. 128.

Table 5.3.5 Electrode data[a]

Machine	X_1	X_2	X_3	X_4	X_5
1	40	58	31	44	64
1	39	59	33	40	60
1	40	58	35	46	59
1	39	59	31	47	58
1	40	60	36	41	56
1	45	60	45	45	58
1	42	64	39	38	63
1	44	59	41	40	60
1	42	66	48	20	61
1	40	60	35	40	58
1	40	61	40	41	58
1	40	58	38	45	60
1	38	59	39	46	58
1	42	59	32	36	61
1	40	61	45	45	59
1	40	59	45	52	59
1	42	58	38	51	59
1	40	59	37	44	60
1	39	60	35	49	59
1	39	60	37	46	56
1	40	58	35	39	58
1	39	59	34	41	60
1	39	60	37	39	59
1	40	59	42	43	57
1	40	59	37	46	60
1	43	60	35	38	62
1	40	59	29	41	60
1	40	59	37	41	59
1	40	60	37	46	60
1	40	58	42	45	61
1	42	63	48	47	64
1	41	59	37	49	60
1	39	58	31	47	60
1	42	60	43	49	61
1	42	59	37	53	62
1	40	58	35	40	59
1	40	59	35	48	58
1	39	60	35	46	59
1	38	59	30	47	57
1	40	60	38	48	62
1	44	60	36	44	60
1	40	58	34	41	58
1	38	60	31	49	60
1	38	58	29	46	60
1	39	59	35	43	56
1	40	60	37	45	59

Table 5.3.5 *Continued*

Machine	X_1	X_2	X_3	X_4	X_5
1	40	60	37	44	61
1	42	62	37	35	60
1	40	59	35	44	58
1	42	58	35	43	61
2	44	58	32	25	57
2	43	58	25	19	60
2	44	57	30	24	59
2	42	59	36	20	59
2	42	60	38	29	59
2	43	56	38	32	58
2	43	57	26	18	59
2	45	60	27	27	59
2	45	59	33	18	60
2	43	58	29	26	59
2	43	59	39	22	58
2	43	59	35	29	59
2	44	57	37	19	58
2	43	58	29	20	58
2	43	58	27	8	58
2	44	60	39	15	60
2	43	58	35	13	58
2	44	58	38	19	58
2	43	58	36	19	58
2	43	58	29	19	60
2	43	58	29	21	58
2	42	59	43	26	58
2	43	58	26	20	58
2	44	59	22	17	59
2	43	59	36	25	59
2	44	57	33	11	59
2	44	60	25	10	59
2	44	58	22	16	59
2	44	60	36	18	57
2	46	61	39	14	59
2	42	58	36	27	57
2	43	60	20	19	60
2	42	59	27	23	59
2	43	58	28	12	58
2	42	57	41	24	58
2	44	60	28	20	60
2	43	58	45	25	59
2	43	59	35	21	59
2	43	60	29	2	60
2	44	59	22	11	59
2	44	58	46	25	58
2	43	60	28	9	60

Table 5.3.5 *Continued*

Machine	X_1	X_2	X_3	X_4	X_5
2	43	59	38	29	59
2	43	58	47	24	57
2	42	58	24	19	59
2	43	60	35	22	58
2	45	60	28	18	60
2	43	57	38	23	60
2	44	60	31	22	58
2	43	58	22	20	57

Source: Flury and Riedwyl (1988)

[a] See Figure 5.3.3 for definitions of the variables.

than 4. Details of the calculations are left to the student as an exercise. The linear discriminant function is given by

$$V = -1.900X_1 + 2.224X_2 - 0.071X_3 + 0.879X_4 + 0.833X_5,$$

with a multivariate standard distance of $D = 5.39$. The means and standard deviations of V are as follows:

Group	Mean	Standard Deviation
Machine 1	141.08	5.21
Machine 2	112.06	5.56

Hence, the differences in variability observed in Table 5.3.6 do not seem to affect the discriminant function. Indeed, the standard deviation of V is almost identical in

Table 5.3.6 Univariate summary statistics in the electrode example

Variable		Mean	Standard Deviation	Standard Distance
X_1	Machine 1	40.36	1.575	2.33
	Machine 2	43.32	0.868	
X_2	Machine 1	59.54	1.555	0.70
	Machine 2	58.60	1.107	
X_3	Machine 1	36.80	4.472	0.78
	Machine 2	32.34	6.736	
X_4	Machine 1	43.66	5.255	4.19
	Machine 2	19.88	6.077	
X_5	Machine 1	59.54	1.809	0.60
	Machine 2	58.68	0.913	

both groups. Figure 5.3.4 shows a frequency polygon of the discriminant function, illustrating the excellent separation. We leave it to the student to compute the values of the discriminant function for all 100 observations and to verify that the two groups are completely separated. Another question of potential interest is, since variable X_4 alone is such a good discriminator, how well could we do if this variable is omitted? See Exercise 10. □

In anticipation of quadratic normal theory discrimination to be discussed in Chapter 7, it is not always a good idea to simply ignore differences between covariance matrices, because such differences may affect the optimal classification procedure. However, for the time being, we will study properties of the linear discriminant function, and it is interesting to know that there are situations under which even considerable differences between the covariance matrices seem irrelevant in the following sense. Suppose that the random vector $\mathbf{X}$ has mean vectors $\boldsymbol{\mu}_1$ and $\boldsymbol{\mu}_2$ in two populations and covariance matrices $\boldsymbol{\psi}_1 \neq \boldsymbol{\psi}_2$. Since the covariance matrices are not equal, we might use some convex combination of them, say $\boldsymbol{\psi}_\alpha = \alpha\boldsymbol{\psi}_1 + (1-\alpha)\boldsymbol{\psi}_2$, where $\alpha \in [0, 1]$, and define the vector of linear discriminant function coefficients as $\boldsymbol{\beta} = \boldsymbol{\psi}_\alpha^{-1}(\boldsymbol{\mu}_1 - \boldsymbol{\mu}_2)$, which depends on the particular choice of α. (Note that the pooling of covariance matrices in the sample situation is analogous to this). Or one might argue that we should choose either $\boldsymbol{\psi}_1$ or $\boldsymbol{\psi}_2$ in the formula for the coefficients

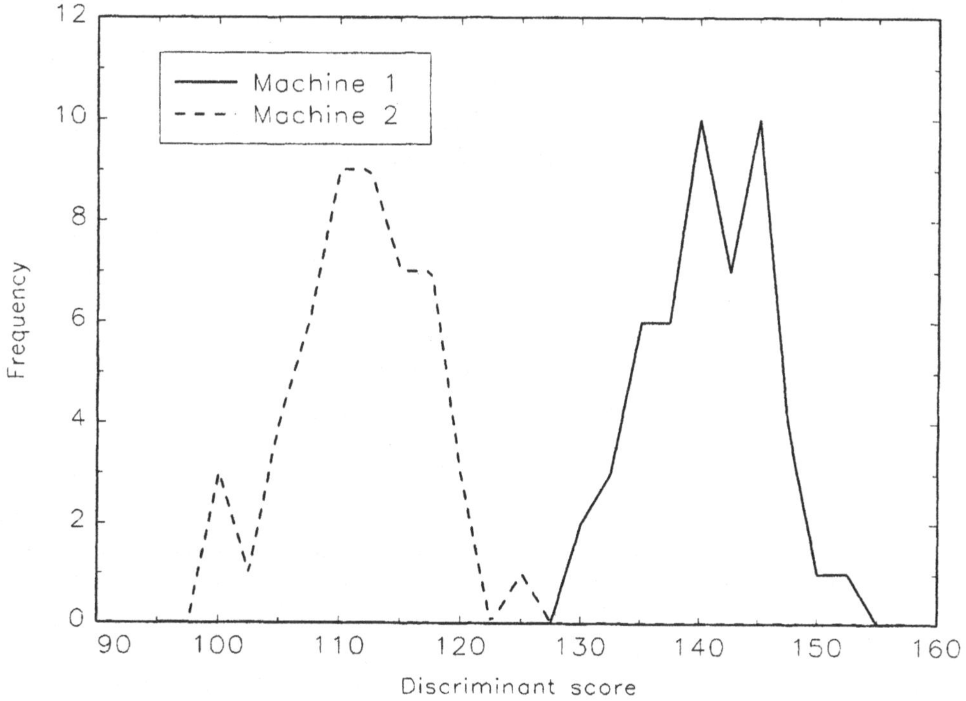

Figure 5-3-4
Frequency plot of the linear discriminant function in the electrode example.

of the linear discriminant function, that is, we should actually compute *two* linear discriminant functions $Y_1 = \beta_1'\mathbf{X}$ and $Y_2 = \beta_2'\mathbf{X}$, where $\beta_1 = \psi_1^{-1}(\mu_1 - \mu_2)$ and $\beta_2 = \psi_2^{-1}(\mu_1 - \mu_2)$, and then decide which one is better. Surprisingly, such a decision may sometimes not be necessary, as the following example shows.

Example 5.3.5 Suppose $\mathbf{X}$ is a bivariate random vector with $E[\mathbf{X}] = \mu_1 = \begin{pmatrix} 7 \\ 7 \end{pmatrix}$, $\text{Cov}[\mathbf{X}] = \psi_1 = \begin{pmatrix} 2 & 0 \\ 0 & 2 \end{pmatrix}$ in population 1 and $E[\mathbf{X}] = \mu_2 = \begin{pmatrix} 4 \\ 4 \end{pmatrix}$, $\text{Cov}[\mathbf{X}] = \psi_2 = \begin{pmatrix} 5 & -3 \\ -3 & 5 \end{pmatrix}$ in population 2. If we use ψ_1 for computing the discriminant function coefficients, we obtain

$$\beta_1 = \psi_1^{-1}(\mu_1 - \mu_2) = \frac{1}{2} \cdot \begin{pmatrix} 1 & 0 \\ 0 & 1 \end{pmatrix} \cdot \begin{pmatrix} 3 \\ 3 \end{pmatrix} = \frac{3}{2} \cdot \begin{pmatrix} 1 \\ 1 \end{pmatrix}.$$

If we choose ψ_2, we get coefficients

$$\beta_2 = \psi_2^{-1}(\mu_1 - \mu_2) = \frac{1}{16} \cdot \begin{pmatrix} 5 & 3 \\ 3 & 5 \end{pmatrix} \cdot \begin{pmatrix} 3 \\ 3 \end{pmatrix} = \frac{3}{2} \cdot \begin{pmatrix} 1 \\ 1 \end{pmatrix}.$$

In other words, whether we use ψ_1 or ψ_2, in both cases we get

$$V = \frac{3}{2}(X_1 + X_2)$$

as the discriminant function, or any linear combination equivalent to V. Even the bivariate standard distances are identical, whether we use ψ_1 or ψ_2—and this despite the fact that the univariate standard distances *do* depend on the choice of ψ_1 or ψ_2. See problem 11 for more details of this example. □

In general, the two vectors $\beta_1 = \psi_1^{-1}(\mu_1 - \mu_2)$ and $\beta_2 = \psi_2^{-1}(\mu_1 - \mu_2)$ are neither identical nor proportional, but, nevertheless, the preceding example shows that a linear discriminant function may be uniquely defined, up to multiplication by a constant, even in cases where the covariance matrices are distinctly different.

In practical applications, one might compute two vectors of discriminant function coefficients

$$\mathbf{b}_1 = \mathbf{S}_1^{-1}(\bar{\mathbf{x}}_1 - \bar{\mathbf{x}}_2) \qquad \text{and} \qquad \mathbf{b}_2 = \mathbf{S}_2^{-1}(\bar{\mathbf{x}}_1 - \bar{\mathbf{x}}_2) \tag{7}$$

to assess the effect of differences between the covariance matrices on the linear discriminant function. Then we can compute the values of $V_1 = \mathbf{b}_1'\mathbf{X}$ and $V_2 = \mathbf{b}_2'\mathbf{X}$ for all observations and study the joint distribution of V_1 and V_2 in a scatterplot. Ideally, if the differences between $\mathbf{S}_1$ and $\mathbf{S}_2$ do not affect the linear discriminant function at all, the correlation between V_1 and V_2 would be 1. This is not a commonly used procedure in discriminant analysis but is instructive, nonetheless.

Example 5.3.6 continuation of Example 5.3.4

Computing the vectors $\mathbf{b}_1$ and $\mathbf{b}_2$ of equation (7) for the electrode data, the linear combinations

$$V_1 = -1.326X_1 + 2.485X_2 + 0.120X_3 + 1.160X_4 + 0.308X_5$$

and

$$V_2 = -3.900X_1 + 2.006X_2 - 0.110X_3 + 0.714X_4 + 1.960X_5$$

were obtained. Although these two linear combinations appear to be quite different, they are highly correlated and yield about the same group separation, as illustrated in Figure 5.3.5. The relationship between V_1 and V_2 is far from "perfect," but the plot suggests that a single linear combination would be adequate. In other words, this gives support to the analysis in Example 5.3.4. □

Some further aspects of the idea of computing two linear discriminant functions are explored in Exercise 12.

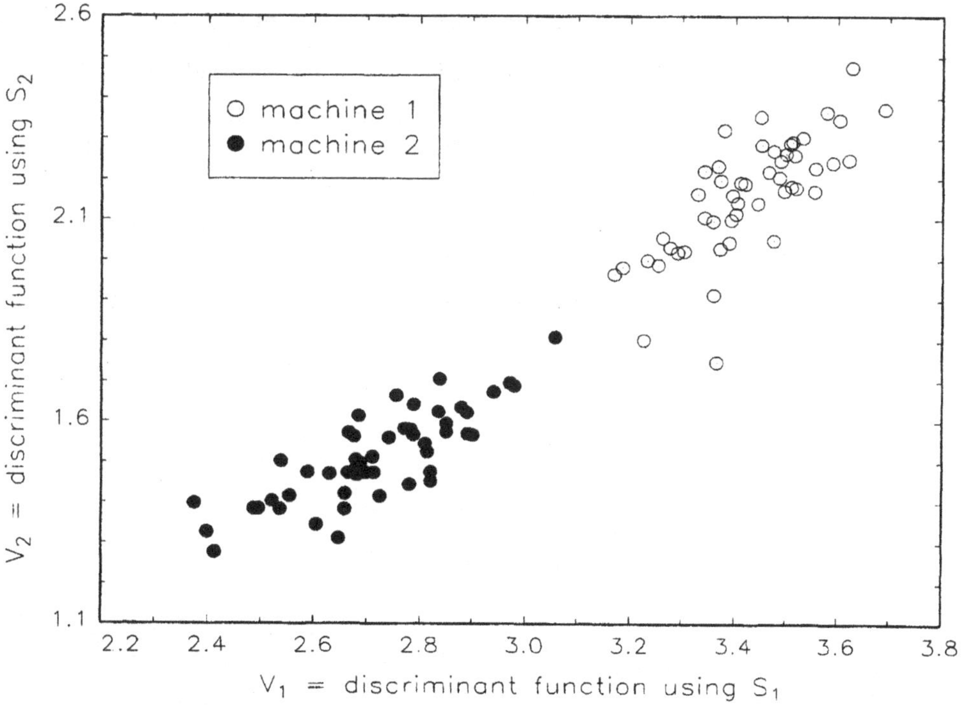

Figure 5-3-5 Plot of the joint distribution of two linear discriminant functions in the electrode example.

Exercises for Section 5.3

1. In Example 5.3.1, apply the linear discriminant function $V = 0.582X_1 - 0.381X_2$ to the data, as in Table 5.3.1. Compute univariate summary statistics (i.e., groupwise means and standard deviations) and the univariate standard distance between the two groups for variable V. Thus, verify that the univariate standard distance of V is the same as the multivariate standard distance based on X_1 and X_2.

2. In the midge example, use the linear discriminant function and the midpoint $m = 5.87$ to classify the following four specimens on which antenna length (X_1) and wing length (X_2) were measured:

Specimen	X_1	X_2
1	124	174
2	138	196
3	124	190
4	138	186

In each case, describe verbally how confident you would feel with the classification. *Note*: In the case of two variables, you may mark the four specimens in a graph like Figure 5.3.1, where regions of classification are given by a boundary line.

3. This exercise is based on the flea beetle data of Example 5.3.2.

(a) Apply the discriminant function $V = 0.345X_1 - 0.130X_2 - 0.106X_3 - 0.143X_4$ to the data of all 39 observations, and plot histograms or frequency polygons of V (see, e.g., Figure 5.3.4).

(b) Compute $\bar{v}_1$, $\bar{v}_2$, the standard deviation of V in both groups, and the standard distance based on V. Thus, verify that the multivariate standard distance is 3.70.

(c) Formulate a classification rule based on the midpoint m between $\bar{v}_1$ and $\bar{v}_2$.

(d) Apply the classification rule to four new specimens with the following measurements:

X_1	X_2	X_3	X_4
195	290	145	195
180	265	145	195
180	250	145	195
195	310	145	195

In each of the four cases, how certain would you be about the classification? (Only a verbal answer is expected at this point.)

(e) Find the "transformed" version of the discriminant function as suggested in equations (32) to (35) of Section 5.2, similar to variable V^* in Example 5.3.1, and formulate the corresponding classification rule.

4. Repeat Exercise 3, but use only variables X_1 and X_2. In part (b), verify that the bivariate standard distance is 3.01. In part (d) use a scatterplot of X_1 vs. X_2 with classification boundary as in Figure 5.3.1.

5. Let $\begin{pmatrix} X \\ Y \end{pmatrix}$ denote a bivariate random vector with mean vector μ_1 in group 1, mean vector μ_2 in group 2, and common covariance matrix ψ. Let Δ_X denote the standard distance between the two groups based on X, Δ_Y the standard distance based on Y, and Δ_{XY} the bivariate standard distance. Show that

$$\Delta_{XY} \geq \max(\Delta_X, \Delta_Y).$$

Hint: Argue that $\Delta_X \leq \Delta_{XY}$ and $\Delta_Y \leq \Delta_{XY}$. A graph like Figure 5.2.3 should help.

6. Let $\mathbf{X}$ denote a p-variate random vector with mean μ_i in population $i = 1, 2$ and common covariance matrix ψ. Let $1 \leq q \leq p$ denote a fixed integer. Show that the standard distance between the two groups based on all p variables is always larger or equal to the standard distance based on a subset of q variables. *Hint*: Let Δ_p denote the standard distance based on all p variables and Δ_q the standard distance based on, say, the first q variables. This exercise asks you to show that $\Delta_q \leq \Delta_p$, but you do not need to establish the exact relationship between Δ_q and Δ_p.

7. This exercise, again, is based on the flea beetle data of Example 5.3.2. Part (a) is straightforward, but part (b) may be somewhat confusing for the student who is just starting to understand the concepts of linear discrimination.

   (a) Compute the discriminant function based on (X_1, X_2, X_4), and confirm the numerical results given in the text. Call this discriminant function V_1', and compute its value for all 39 observations in the data set.

   (b) Now, treat V' like a measured variable, and compute a linear discriminant function based on the two variables V' and X_3. If X_3 is redundant (the word "redundant" has yet to be given a precise meaning), what would you expect the coefficient of X_3 to be? In Figure 5.3.2, how should the optimal classification boundary look if X_3 does not contribute to discrimination in addition to the discrimination provided by (X_1, X_2, X_4)?

8. This exercise refers to the assumption of nonsingularity of the pooled-sample covariance matrix $\mathbf{S}$; see equation (3).

   (a) Show that, for $N_1 \geq 1$, $N_2 \geq 1$, $N_1 + N_2 < p + 2$, the pooled-sample covariance matrix is always singular. (If $N_j = 1$, set $\mathbf{S}_j = \mathbf{O}$ for equation (3) to be applicable).

   (b) Show that, for any $N_1 \geq 1$ and $N_2 \geq 1$, the pooled-sample covariance matrix is positive semidefinite, irrespective of the distributions from which the samples are taken.

   (c) If $N_1 + N_2 \geq p + 2$ and if the samples are from multivariate normal distributions with positive definite covariance matrices, show that Pr[$\mathbf{S}$ is positive definite]$= 1$.

   (d) If the samples are from multivariate discrete distributions, show that Pr[$\mathbf{S}$ is positive definite]< 1 for all finite sample sizes.

9. This exercise is based on the female vole data reproduced in Table 5.3.7.
   (a) Compute univariate summary statistics for all variables and for both groups.
   (b) Compute a linear discriminant function between the two groups based on variables L_7 and B_4. Also compute the bivariate standard distance based on these two variables, draw a scatterplot, and draw a classification boundary line corresponding to the midpoint between the means of the discriminant function.
   (c) From part (a), it appears that the two groups differ in age, so there is some suspicion that the discriminant function might express mostly age differences rather than morphometric ones. Plot the discriminant function found in part (b) vs.variable AGE and comment.

10. This exercise is based on the electrode data used in Example 5.3.4.
   (a) Compute the covariance matrices for all five variables in both groups. Also compute correlation matrices. Without doing a formal test of equality of covariance matrices, comment on differences.
   (b) Compute the linear discriminant function, and thus, verify the numerical results given in the text.
   (c) Compute the values of the discriminant function for all 100 observations, and verify that the two groups are completely separated.
   (d) Compute a discriminant function with all variables except X_4. Is complete separation of the two groups still possible?

11. This exercise illustrates how to construct examples like 5.3.5, where the two covariance matrices ψ_1 and ψ_2 are unequal, yet the linear discriminant function based on ψ_1 is identical with the linear discriminant function based on ψ_2.

   Let $\mathbf{X} = \begin{pmatrix} X_1 \\ X_2 \end{pmatrix}$ denote a bivariate random vector with

$$E[\mathbf{X}] = \mathbf{0} = \begin{pmatrix} 0 \\ 0 \end{pmatrix}, \quad \text{Cov}[\mathbf{X}] = \mathbf{I}_2 = \begin{pmatrix} 1 & 0 \\ 0 & 1 \end{pmatrix} \quad \text{in group 1}$$

   and

$$E[\mathbf{X}] = \begin{pmatrix} \delta \\ 0 \end{pmatrix}, \quad \text{Cov}[\mathbf{X}] = \text{diag}(\sigma^2, \tau^2) = \begin{pmatrix} \sigma^2 & 0 \\ 0 & \tau^2 \end{pmatrix} \quad \text{in group 2.}$$

   Let $\mathbf{A} = \begin{pmatrix} a_{11} & a_{12} \\ a_{21} & a_{22} \end{pmatrix}$ denote a fixed, nonsingular matrix, $\mathbf{b} = \begin{pmatrix} b_1 \\ b_2 \end{pmatrix} \in \mathbb{R}^2$ a fixed vector, and set $\mathbf{Y} = \mathbf{AX} + \mathbf{b}$. Let $\mu_1 = E[\mathbf{Y}]$ in group 1, $\mu_2 = E[\mathbf{Y}]$ in group 2, and denote by ψ_i $(i = 1, 2)$ the covariance matrices of $\mathbf{Y}$ in groups 1 and 2, respectively.
   (a) Find μ_1, μ_2, ψ_1, and ψ_2.
   (b) How were δ, σ^2, τ^2, $\mathbf{A}$, and $\mathbf{b}$ chosen in Example 5.3.5?
   (c) Let β_i denote the vector of discriminant function coefficients based on ψ_i, i.e.,

$$\beta_1 = \psi_1^{-1}(\mu_1 - \mu_2), \quad \beta_2 = \psi_2^{-1}(\mu_1 - \mu_2).$$

   Show that β_2 is proportional to β_1, i.e., $\beta_2 = c \cdot \beta_1$ for some $c \in \mathbb{R}$. What is the value of c?

Table 5.3.7 Female vole data

Species	Age	L_2	L_9	L_7	B_3	B_4	H_1
1	345	304	55	75	168	34	110
1	127	273	47	69	155	33	108
1	132	272	42	71	157	38	110
1	218	289	48	74	165	38	114
1	189	291	47	75	164	36	114
1	288	298	52	73	161	32	104
1	121	280	49	72	156	35	106
1	188	271	44	67	154	34	107
1	231	283	52	73	168	35	108
1	139	285	45	73	158	38	111
1	126	292	50	80	174	35	117
1	196	288	50	71	161	33	110
1	164	285	48	75	163	37	114
1	181	277	51	73	156	33	106
1	128	256	44	66	148	32	102
1	129	268	42	71	159	37	112
1	166	281	50	68	155	33	107
1	178	267	48	70	150	36	108
1	242	288	51	72	165	34	108
1	189	297	52	73	161	33	109
1	161	283	49	72	156	34	112
1	164	288	49	73	156	33	108
1	163	281	46	70	151	32	105
1	185	250	39	66	148	33	100
1	160	284	50	76	162	34	108
1	165	281	52	71	160	37	108
1	151	273	56	75	159	35	108
1	149	277	53	70	161	35	112
1	249	282	53	72	165	33	110
1	168	277	49	74	164	35	113
1	129	274	51	72	159	36	109
1	129	284	55	74	164	36	110
1	148	299	56	75	166	34	110
1	130	303	62	79	177	35	118
1	296	297	58	77	168	35	115
1	140	264	50	71	151	37	105
1	129	262	52	67	152	32	104
1	129	269	52	73	165	33	107
1	129	281	56	73	161	33	106
1	121	293	55	78	168	34	118
1	129	267	53	68	159	35	103
2	152	265	44	64	144	41	104
2	183	261	47	64	143	42	103
2	183	249	43	58	141	42	100
2	183	245	42	61	145	40	102
2	183	263	42	63	157	39	103

Table 5.3.7 *Continued*

Species	Age	L_2	L_9	L_7	B_3	B_4	H_1
2	183	255	42	63	142	41	101
2	213	273	47	65	155	41	107
2	213	250	41	60	142	37	98
2	213	252	39	60	142	37	99
2	213	265	41	62	147	40	105
2	213	268	45	63	152	40	103
2	213	259	44	63	149	39	103
2	213	268	47	62	151	39	103
2	213	270	45	66	149	39	105
2	213	260	39	63	147	40	101
2	213	269	46	65	148	41	106
2	244	261	43	63	146	42	103
2	244	272	43	63	158	40	106
2	244	277	49	64	158	39	111
2	244	257	42	67	152	41	104
2	244	261	44	64	146	40	106
2	244	268	48	64	157	40	106
2	274	288	51	67	153	41	109
2	305	274	46	68	158	39	108
2	305	262	46	65	151	39	108
2	305	269	47	63	147	42	106
2	305	275	43	66	149	39	103
2	305	277	44	66	145	40	102
2	305	271	44	64	152	41	102
2	305	278	43	68	155	38	107
2	305	274	49	63	157	39	107
2	335	264	40	64	158	40	102
2	335	273	47	66	154	42	107
2	365	274	48	64	159	38	106
2	365	271	39	65	160	40	108
2	365	268	46	61	147	40	100
2	365	274	50	66	159	39	105
2	365	271	47	67	155	39	103
2	274	289	51	71	167	41	108
2	122	254	42	66	143	37	99
2	122	269	49	68	146	38	107
2	152	264	46	67	147	40	105
2	124	263	45	64	138	39	102
2	138	267	45	66	149	42	107
2	126	262	43	63	144	41	102

Courtesy of Dr. J.P. Airoldi, Department of Biology, University of Berne, Switzerland. The two species are (1) *Microtus Californicus*, and (2) *M. ochrogaster*. Age is in days. The remaining variables, with variable names as in the original analysis of this data set by Airoldi and Hoffmann (1984), are: L_2 = condylo-incisive length; L_9 = length of incisive foramen; L_7 = alveolar length of upper molar tooth row; B_3 = zygomatic width; B_4 = interorbital width; H_1 = skull height. All variables are measured in units of 0.1mm.

(d) Find the standard distance between μ_1 and μ_2 relative to ψ_1 and ψ_2, i.e., find

$$\Delta_1 = [(\mu_1 - \mu_2)'\psi_1^{-1}(\mu_1 - \mu_2)]^{1/2},$$
$$\Delta_2 = [(\mu_1 - \mu_2)'\psi_2^{-1}(\mu_1 - \mu_2)]^{1/2},$$

and show that $\Delta_1 = \Delta_2$ exactly if $\sigma = 1$.

12. This exercise expands on the idea of computing two linear discriminant functions as in Example 5.3.6.

(a) Verify the computation of Example 5.3.6, i.e., compute $\mathbf{b}_1 = \mathbf{S}_1^{-1}(\bar{\mathbf{x}}_1 - \bar{\mathbf{x}}_2)$ and $\mathbf{b}_2 = \mathbf{S}_2^{-1}(\bar{\mathbf{x}}_1 - \bar{\mathbf{x}}_2)$.

(b) Compute the standard distances between the two groups with respect to the two covariance matrices, i.e., compute $D_1 = \left[(\bar{\mathbf{x}}_1 - \bar{\mathbf{x}}_2)' \mathbf{S}_1^{-1} (\bar{\mathbf{x}}_1 - \bar{\mathbf{x}}_2)\right]^{1/2}$ and $D_2 = \left[(\bar{\mathbf{x}}_1 - \bar{\mathbf{x}}_2)' \mathbf{S}_2^{-1} (\bar{\mathbf{x}}_1 - \bar{\mathbf{x}}_2)\right]^{1/2}$.

(c) Compute the covariance matrix of $\mathbf{V} = \begin{pmatrix} V_1 \\ V_2 \end{pmatrix}$, where $V_1 = \mathbf{b}_1'\mathbf{X}$ and $V_2 = \mathbf{b}_2'\mathbf{X}$ are the two discriminant functions, and compare to part (b).

(d) In a theoretical setup, let $\beta_j = \psi_j^{-1}(\mu_1 - \mu_2)$ and $\Delta_j = \left[(\mu_1 - \mu_2)' \psi_j^{-1} (\mu_1 - \mu_2)\right]^{1/2}$, $j = 1, 2$. Set $Y_j = \beta_j'\mathbf{X}$, $j = 1, 2$, and $\mathbf{Y} = \begin{pmatrix} Y_1 \\ Y_2 \end{pmatrix}$. Find $\text{Cov}[\mathbf{Y}]$ in both groups, and relate this result to parts (b) and (c).

(e) Show that the correlation between Y_1 and Y_2 is always nonnegative in both groups, assuming $\mu_1 \neq \mu_2$.

13. Let $\mathbf{x}_{11}, \ldots, \mathbf{x}_{1,N_1}$ and $\mathbf{x}_{21}, \ldots, \mathbf{x}_{2,N_2}$ denote observed p-dimensional vectors in two groups. For some vector $\mathbf{a} \in \mathbb{R}^p$, let $\mathbf{y}_{ji} = \mathbf{a}'\mathbf{x}_{ji}$, $j = 1, 2$; $i = 1, \ldots, N_j$. Show that the pooled sample variance of the $\mathbf{y}_{ji}$ is $\mathbf{a}'\mathbf{S}\mathbf{a}$, where $\mathbf{S}$ is the pooled sample covariance matrix of the $\mathbf{x}_{ji}$.

5.4 Normal Theory Linear Discrimination

In this section we will rediscover the linear discriminant function using the multivariate normal distribution. More specifically, we will specify multivariate normal distributions for a random vector in two groups, formulate a classification principle in terms of posterior probabilities, and, then show in Section 5.2 that classification leads to the use of the linear discriminant function derived from heuristic principles. This anticipates, to some extent, the theory of optimal classification to be presented in Chapter 7.

Although we are not dealing explicitly with mixtures at this point, the theoretical setup from Section 2.8 will be useful here. Changing the notation used in Sections 5.2 and 5.3, X will now denote a discrete random variable indicating group membership. Suppose that X takes values 1 and 2 with probabilities π_1 and π_2, $\pi_1 + \pi_2 = 1$. Assume furthermore that, conditionally on $X = 1$, $\mathbf{Y}$ follows some p-variate distribution with *pdf* $f_1(\mathbf{y})$. Similarly, assume that conditionally on $X = 2$, $\mathbf{Y}$ has *pdf* $f_2(\mathbf{y})$.

Repeating the terminology from Section 2.8, the π_j are the *prior probabilities* of membership in group j, $j = 1, 2$. Denoting by $f_j(\mathbf{y})$ the conditional *pdf* of $\mathbf{Y}$, given $X = j$, the unconditional (marginal) *pdf* of $\mathbf{Y}$ is the mixture density

$$f_{\mathbf{Y}}(\mathbf{y}) = \pi_1 f_1(\mathbf{y}) + \pi_2 f_2(\mathbf{y}). \tag{1}$$

Finally, the conditional probability of membership in group j, given $\mathbf{Y} = \mathbf{y}$, is expressed by

$$\begin{aligned} \pi_{j\mathbf{y}} = \Pr[X = j | \mathbf{Y} = \mathbf{y}] &= \frac{\pi_j f_j(\mathbf{y})}{\pi_1 f_1(\mathbf{y}) + \pi_2 f_2(\mathbf{y})} \\ &= \frac{\pi_j f_j(\mathbf{y})}{f_{\mathbf{Y}}(\mathbf{y})}, \qquad j = 1, 2. \end{aligned} \tag{2}$$

The $\pi_{j\mathbf{y}}$ are usually called *posterior probabilities.*

With this notation, we can formulate an intuitively reasonable principle of classification as follows. Classify an observation $\mathbf{y} \in \mathbb{R}^p$ into group 1 if $\pi_{1\mathbf{y}} > \pi_{2\mathbf{y}}$, and classify it into group 2 if $\pi_{2\mathbf{y}} > \pi_{1\mathbf{y}}$. For the moment, we do not worry about the possibility that $\pi_{1\mathbf{y}} = \pi_{2\mathbf{y}}$. Since $\pi_{1\mathbf{y}} + \pi_{2\mathbf{y}} = 1$ for all $\mathbf{y} \in \mathbb{R}^p$, this rule is equivalent to the following. Classify into group 1 if $\pi_{1\mathbf{y}} > \frac{1}{2}$, and classify into group 2 if $\pi_{2\mathbf{y}} > \frac{1}{2}$. This defines two classification regions $C_j = \{\mathbf{y} \in \mathbb{R}^p : \pi_{j\mathbf{y}} > \frac{1}{2}\}$ $j = 1, 2$, with a classification boundary given by the set $\{\mathbf{y} \in \mathbb{R}^p : \pi_{1\mathbf{y}} = \frac{1}{2}\}$. The main purpose of this section is to show that, if the conditional distributions of $\mathbf{Y}$ are multivariate normal with equal covariance matrices, then this rule implies that classification should be based on the linear discriminant function. But first let us study a more general example.

Example 5.4.1 Suppose the conditional distributions of $\mathbf{Y}$, given $X = j$, are bivariate normal with mean vectors $\boldsymbol{\mu}_j$ and covariance matrices $\boldsymbol{\psi}_j$ given by

$$\boldsymbol{\mu}_1 = \begin{pmatrix} -1.0 \\ 0.5 \end{pmatrix},$$

$$\boldsymbol{\psi}_1 = \begin{pmatrix} 1 & 0.6 \\ 0.6 & 1 \end{pmatrix},$$

$$\boldsymbol{\mu}_2 = \begin{pmatrix} 1.5 \\ -0.5 \end{pmatrix},$$

$$\boldsymbol{\psi}_2 = \begin{pmatrix} 1 & -0.8 \\ -0.8 & 1 \end{pmatrix}.$$

Suppose, furthermore, that the prior probabilities are $\pi_1 = 0.75$ and $\pi_2 = 0.25$. A contour plot of $\pi_1 f_1(\mathbf{y})$ and $\pi_2 f_2(\mathbf{y})$ is shown in Figure 5.4.1(a). The marginal *pdf*

$f_{\mathbf{Y}}(\mathbf{y})$, as given in equation (1), is a mixture of two bivariate normal distributions. Actually, it is the mixture illustrated earlier in Figures 2.6.7 and 2.8.10. Using equation (2), we can compute the posterior probabilities $\pi_{1\mathbf{y}}$ and $\pi_{2\mathbf{y}}$ for $\mathbf{y} \in \mathbb{R}^2$. Figure 5.4.1(b) shows a contour plot of $\pi_{1\mathbf{y}}$. The contour $\pi_{1\mathbf{y}} = 0.5$ is the boundary between the two classification regions. It consists of two separate curves, yielding classification regions that have quite complicated shapes. We will study similar examples in Section 7.2. □

Now we turn to the situation of two normal distributions with equal covariance matrices, i.e.,

$$\mathbf{Y} \sim \mathcal{N}_p(\boldsymbol{\mu}_1, \boldsymbol{\psi}) \qquad \text{in group 1,}$$

and

$$\mathbf{Y} \sim \mathcal{N}_p(\boldsymbol{\mu}_2, \boldsymbol{\psi}) \qquad \text{in group 2,}$$

where $\boldsymbol{\psi}$ is positive definite. Then the *pdf* of $\mathbf{Y}$ in group j is given by

$$f_j(\mathbf{y}) = (2\pi)^{-p/2}(\det \boldsymbol{\psi})^{-1/2} \exp\left[-\frac{1}{2}(\mathbf{y} - \boldsymbol{\mu}_j)'\boldsymbol{\psi}^{-1}(\mathbf{y} - \boldsymbol{\mu}_j)\right], \quad \mathbf{y} \in \mathbb{R}^p,\ j = 1, 2. \qquad (3)$$

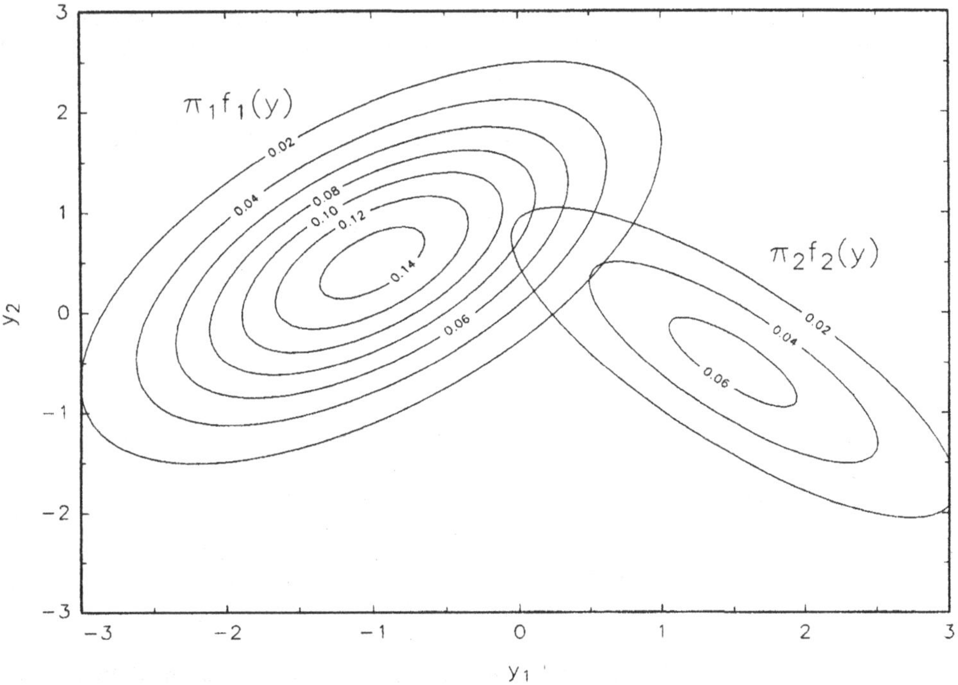

Figure 5-4-1A

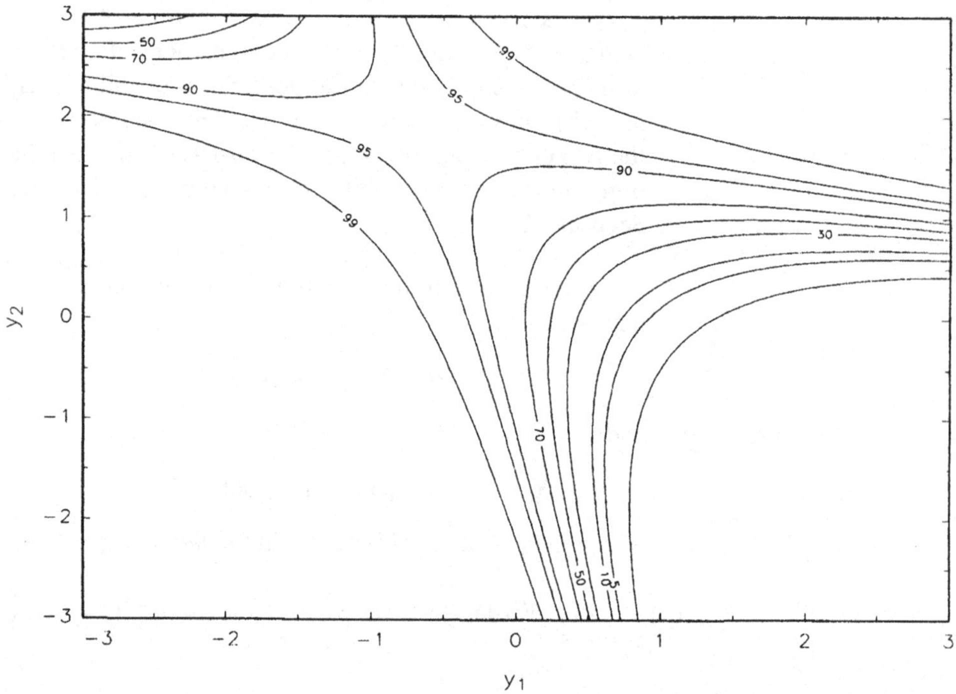

Figure 5-4-1B Classification regions in a bivariate normal example: (a) contour plots of $\pi_1 f_1(\mathbf{y})$ and $\pi_2 f_2(\mathbf{y})$; (b) contour plot of the posterior probabilities $\pi_{1\mathbf{y}}$. Posterior probabilities are in percent.

The condition $\pi_{1\mathbf{y}} > \pi_{2\mathbf{y}}$ is equivalent to $\pi_1 f_1(\mathbf{y}) > \pi_2 f_2(\mathbf{y})$, which in turn is equivalent to

$$\log \frac{\pi_1 f_1(\mathbf{y})}{\pi_2 f_2(\mathbf{y})} > 0. \tag{4}$$

But

$$\begin{aligned} \log \frac{f_1(\mathbf{y})}{f_2(\mathbf{y})} &= \frac{1}{2}\left[(\mathbf{y}-\boldsymbol{\mu}_2)'\boldsymbol{\psi}^{-1}(\mathbf{y}-\boldsymbol{\mu}_2) - (\mathbf{y}-\boldsymbol{\mu}_1)'\boldsymbol{\psi}^{-1}(\mathbf{y}-\boldsymbol{\mu}_1)\right] \\ &= (\boldsymbol{\mu}_1-\boldsymbol{\mu}_2)'\boldsymbol{\psi}^{-1}\mathbf{y} + \frac{1}{2}(\boldsymbol{\mu}_1+\boldsymbol{\mu}_2)'\boldsymbol{\psi}^{-1}(\boldsymbol{\mu}_2-\boldsymbol{\mu}_1) \qquad (5) \\ &= \boldsymbol{\beta}'\left[\mathbf{y} - \frac{1}{2}(\boldsymbol{\mu}_1+\boldsymbol{\mu}_2)\right], \end{aligned}$$

where

$$\boldsymbol{\beta} = \boldsymbol{\psi}^{-1}(\boldsymbol{\mu}_1-\boldsymbol{\mu}_2) \tag{6}$$

is the vector of linear discriminant function coefficients encountered in the preceding sections. Thus the principle of assigning $\mathbf{y} \in \mathbb{R}^p$ to the group with the higher posterior probability leads to a classification rule based on the value of $\beta'\mathbf{y}$:

$$\text{Classify into group} \begin{cases} 1 & \text{if } \beta'\mathbf{y} - \dfrac{1}{2}\beta'(\mu_1 + \mu_2) > \log(\pi_2/\pi_1) \\ 2 & \text{if } \beta'\mathbf{y} - \dfrac{1}{2}\beta'(\mu_1 + \mu_2) < \log(\pi_2/\pi_1)\,. \end{cases} \tag{7}$$

If $\pi_1 = \pi_2 = \frac{1}{2}$, then this rule simplifies further to a cutoff at 0. However, even for prior probabilities different from $\frac{1}{2}$, the rule says that classification should depend on the observed data vector $\mathbf{y}$ only through the value $\beta'\mathbf{y}$ of the linear discriminant function. This is a remarkable and highly desirable result because it implies that the multivariate data can be reduced without loss of information (at least under the ideal setup of two multivariate, normal distributions with identical covariance matrices) to values of the linear discriminant function, irrespective of the prior probabilities. In other words, finding the classification boundary is equivalent to finding a single cutoff point for the univariate distribution of the discriminant function.

Example 5.4.2 This is the continuation of the midge example in a normal theory setup. Let $\mathbf{Y} = \begin{pmatrix} Y_1 \\ Y_2 \end{pmatrix}$ denote a bivariate normal random vector with means $\hat{\mu}_1 = \begin{pmatrix} 141.33 \\ 180.44 \end{pmatrix}$ in group 1, $\hat{\mu}_2 = \begin{pmatrix} 122.67 \\ 192.67 \end{pmatrix}$ in group 2, and common covariance matrix $\hat{\psi} = \begin{pmatrix} 75.49 & 66.46 \\ 66.46 & 133.81 \end{pmatrix}$. That is, we are substituting the sample quantities for the model parameters. Suppose that the prior probabilities are $\pi_1 = \pi_2 = \frac{1}{2}$. Figure 5.4.2 shows a contour plot of $\pi_1 f_1(\mathbf{y})$ and $\pi_2 f_2(\mathbf{y})$, where f_1 and f_2 are the normal densities. The boundary between the two regions of classification is the set of all points where the curves of equal altitude of the two "normal mountains" intersect. According to the theory just developed, the boundary is defined by the equation $\hat{\beta}\left[\mathbf{y} - \frac{1}{2}(\hat{\mu}_1 + \hat{\mu}_2)\right] = \log(\pi_2/\pi_1)$, where $\hat{\beta} = \hat{\psi}^{-1}(\hat{\mu}_1 - \hat{\mu}_2)$. Because we have chosen $\pi_1 = \pi_2 = \frac{1}{2}$, this is exactly the same as the boundary line seen earlier in Figure 5.3.1.

As in Figure 5.4.1, we might draw curves of equal posterior probability . All such curves are straight lines parallel to the boundary line in Figure 5.4.2; see Exercises 1 and 2. Also, any other choice of the prior probabilities π_1 and π_2 would move the boundary line to a parallel line. □

In the normal theory setup, the linear discriminant function $V = \beta'\mathbf{Y}$ has itself a univariate normal distribution with means $\nu_1 = \beta'\mu_1 = (\mu_1 - \mu_2)'\psi^{-1}\mu_1$ and $\nu_2 = \beta'\mu_2 = (\mu_1 - \mu_2)'\psi^{-1}\mu_2$, and with common variance $\sigma^2 = (\mu_1 - \mu_2)'\psi^{-1}(\mu_1 - \mu_2)$, which is identical to the squared standard distance Δ^2. This implies that the calculation of posterior probabilities can be based on the discriminant variable V; see Exercise 3.

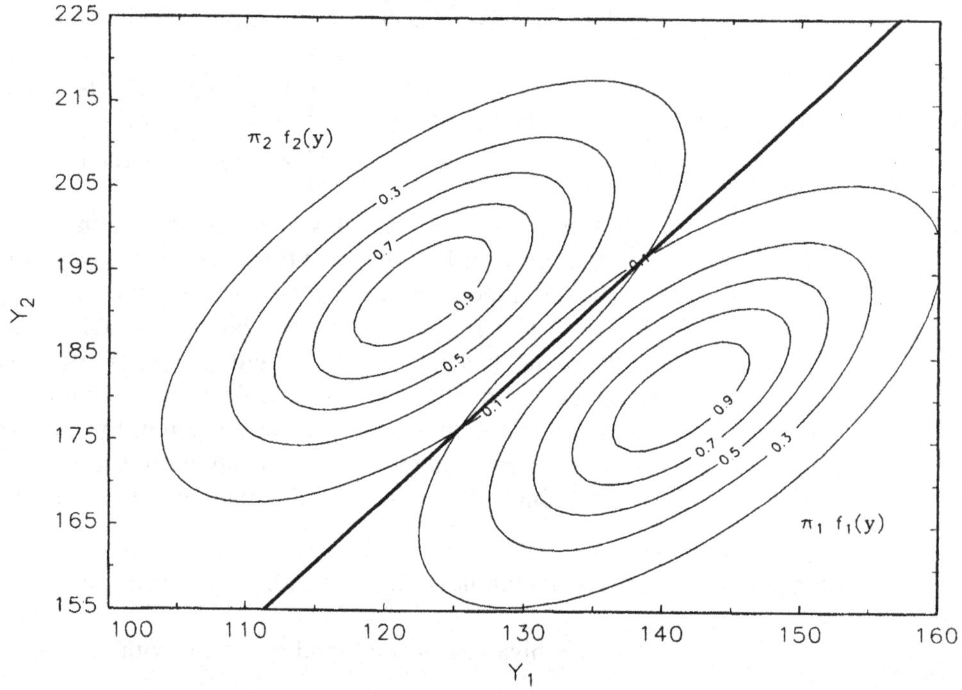

Figure 5-4-2 Normal theory classification in the midge example. Ellipses are contours of constant value of $\pi_1 f_1(\mathbf{y})$ and $\pi_2 f_2(\mathbf{y})$. The unit of measurement of both variables is mm/100. The classification boundary corresponds to the set of points where contours of equal altitudes of the two mountains intersect. Contour levels are scaled by 1000.

The posterior probability for membership in group j can be computed as

$$\begin{aligned}\pi_{jv} &= \Pr[X = j | V = v] \\ &= \Pr[X = j | \beta' \mathbf{Y} = v] \\ &= \frac{\pi_j h_j(v)}{\pi_1 h_1(v) + \pi_2 h_2(v)}, \qquad j = 1, 2,\end{aligned} \tag{8}$$

where $h_j(v)$ is the *pdf* of a normal random variable with mean $\nu_j = \beta' \mu_j$ and variance Δ^2. The boundary line for classification corresponds to the single value of z such that $\pi_{1v} = \pi_{2v} = \frac{1}{2}$.

Example 5.4.3 continuation of Example 5.4.2

We use estimated parameters from the midge example to illustrate the calculation of posterior probabilities based on equation (8). From Example 5.3.1, $\hat{\nu}_1 = 13.632$, $\hat{\nu}_2 = -1.890$, and $\hat{\Delta} = 3.94$. Taking the prior probabilities, again, as $\pi_1 = \pi_2 = \frac{1}{2}$, Figure 5.4.3(a) shows the graphs of $\pi_1 h_1(v)$ and $\pi_2 h_2(v)$, where h_1 and h_2 are the normal densities with parameters $(\hat{\nu}_1, \hat{\Delta}^2)$ and $(\hat{\nu}_2, \hat{\Delta}^2)$, respectively. Part (b) of Figure

5.4.3 shows the corresponding curves for the posterior probabilities. The point of intersection between the two curves corresponds to the boundary line between the two regions of classification in Figure 5.4.2.

The values of the linear discriminant function are also marked in Figure 5.4.3(b). The estimated posterior probabilities for all 15 data points can be read off this graph; it appears that we would be fairly certain in all cases how to classify a midge. Notice, however, that these posterior probabilities are explicitly based on normality assumptions and that we are currently ignoring sampling properties altogether. □

An important question in practical applications is how to choose proper values for π_1 and π_2 in determining posterior probabilities, and hence, the regions of classification. Often in situations where discriminant analysis is used, the training samples are obtained from the conditional distributions of $\mathbf{Y}$, given $X = j$, and the sample sizes N_1 and N_2 are fixed. The electrode example is a typical case. The data contain no information about the prior probabilities, and reasonable values have to be chosen according to the investigator's experience. For instance, if both machines producing electrodes have about the same output, then one might argue for $\pi_1 = \pi_2 = \frac{1}{2}$. Sometimes, equal prior probabilities are used to express the investigator's ignorance about the likelihood that future observations will be from groups 1 and 2, respectively.

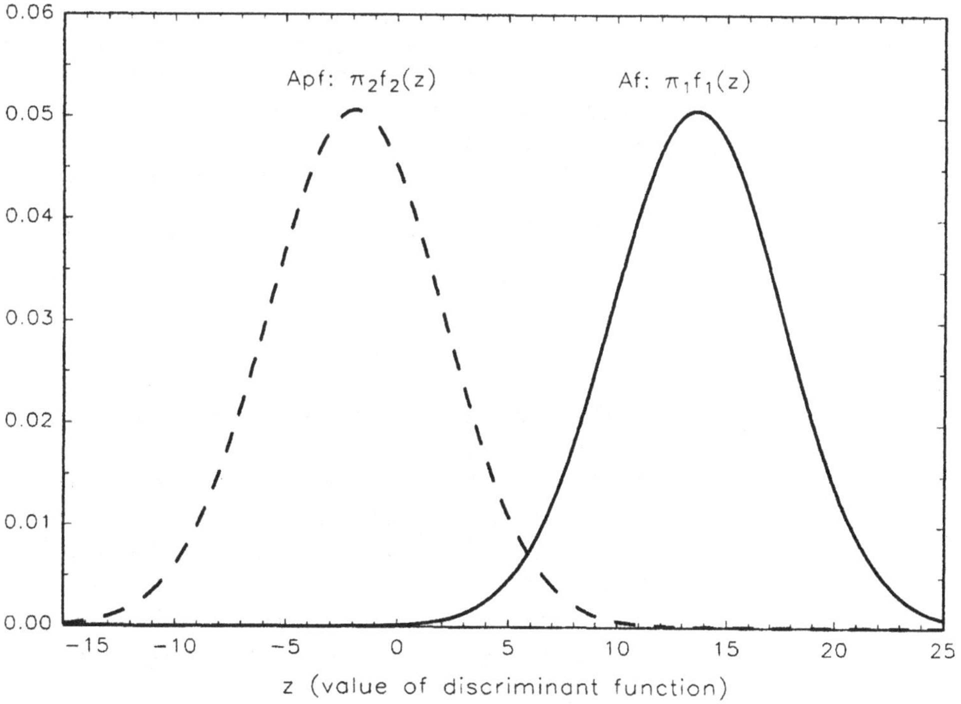

Figure 5-4-3A

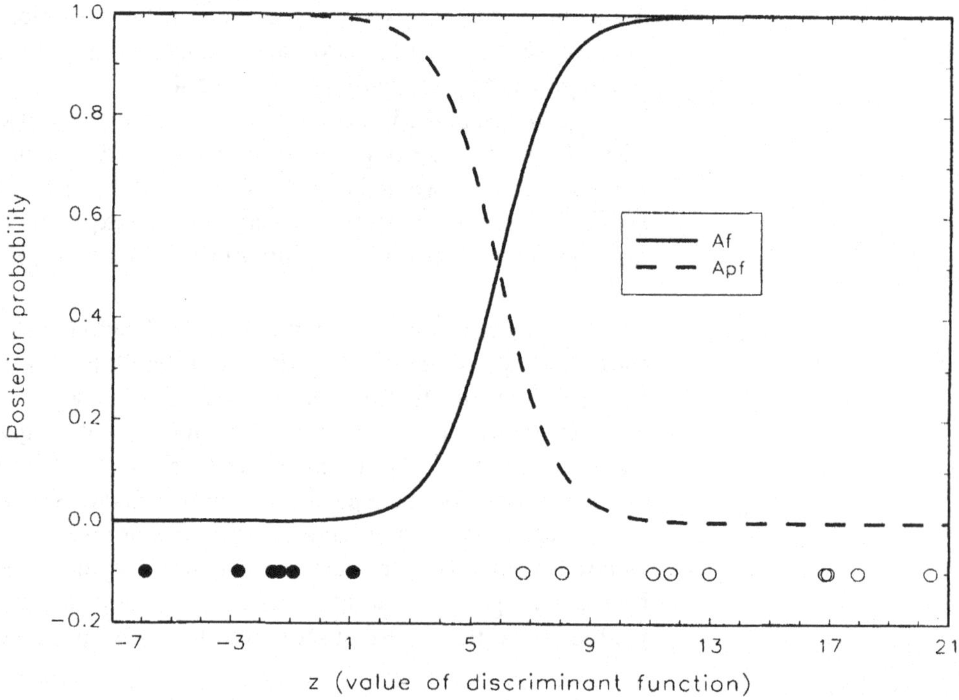

Figure 5-4-3B Normal theory classification based on the distribution of the discriminant function $Z = 0.582Y_1 - 0.381Y_2$: (a) graph of the *pdf* in two groups; (b) posterior probabilities. Parameters correspond to sample estimates in the midge example.

In other cases, sampling may be from the joint distribution of X and $\mathbf{Y}$, that is, both the group code X and the measured variables $\mathbf{Y}$ are random. The sample sizes N_j, then are random also, and it makes intuitive sense to estimate π_j by $\hat{\pi}_j = N_j/(N_1 + N_2)$, $j = 1, 2$. Actually these $\hat{\pi}_j$ are also maximum likelihood estimates, as is shown in Section 9.3. Although the type of sampling is not known to this author for the midge and the flea beetle examples, it seems reasonable to assume, in both examples, that group membership was random, and estimating π_j by $\hat{\pi}_j = N_j/(N_1 + N_2)$ would therefore be our "best guess" at the probabilities that future observations will come from the respective groups.

The question of randomness of X brings us to the more general question under which circumstances is discriminant analysis an appropriate method. In the general setup introduced first in Section 2.8, both X and $\mathbf{Y}$ are random. Discriminant analysis looks at the conditional distribution of $\mathbf{Y}$, given X. In particular, in this section we have discussed the case where the conditional distribution of $\mathbf{Y}$, given X, is multivariate normal for both $X = 1$ and $X = 2$. By classifying a new data vector $\mathbf{y}$ into one of two groups, we changed perspective and studied the conditional distribution of X, given $\mathbf{Y} = \mathbf{y}$, which yielded the posterior probabilities $\pi_{j\mathbf{y}}$. However, in some cases it may be more reasonable to model the conditional distribution of X, given $\mathbf{Y} = \mathbf{y}$, directly, without making any distributional assumptions on $\mathbf{Y}$. For instance, suppose that a basketball player throws the ball at various distances from the basket (Y) and records success or failure (X). In this case, it would make no sense to model the conditional distribution of Y, given X. Rather, one would want to model the probability of success as a function of Y. In a more general setup, one would like to treat X as the dependent variable and $\mathbf{Y}$ as a vector of p "independent," or better, regressor variables. The most popular statistical technique based on this approach is *logistic regression*, the topic of Sections 7.5 and 7.6. Logistic regression assumes that

$$\pi_{1\mathbf{y}} = \Pr[X = 1|\mathbf{Y} = \mathbf{y}] = \frac{\exp\left[\alpha + \boldsymbol{\beta}'\mathbf{y}\right]}{1 + \exp\left[\alpha + \boldsymbol{\beta}'\mathbf{y}\right]}, \qquad \mathbf{y} \in \mathbb{R}^p, \tag{9}$$

or, equivalently, that

$$\log \frac{\pi_{1\mathbf{y}}}{1 - \pi_{1\mathbf{y}}} = \alpha + \boldsymbol{\beta}'\mathbf{y}, \qquad \mathbf{y} \in \mathbb{R}^p, \tag{10}$$

for some parameters $\alpha \in \mathbb{R}$ and $\boldsymbol{\beta} \in \mathbb{R}^p$.

Interestingly, the logistic regression model (10) holds for the case where the conditional distribution of $\mathbf{Y}$, given X, is multivariate normal with covariance matrix ψ in both groups. To see this, recall from equation (2) that

$$\pi_{1\mathbf{y}} = \frac{\pi_1 f_1(\mathbf{y})}{\pi_1 f_1(\mathbf{y}) + \pi_2 f_2(\mathbf{y})}, \qquad \mathbf{y} \in \mathbb{R}^p,$$

and $\pi_{2\mathbf{y}} = 1 - \pi_{1\mathbf{y}}$. Hence,

$$\log \frac{\pi_{1\mathbf{y}}}{1 - \pi_{1\mathbf{y}}} = \log \frac{\pi_1 f_1(\mathbf{y})}{\pi_2 f_2(\mathbf{y})}.$$

Using multivariate normal densities for f_1 and f_2, equation (5) gives

$$\log \frac{\pi_{1\mathbf{y}}}{1 - \pi_{1\mathbf{y}}} = \alpha + \boldsymbol{\beta}'\mathbf{y}, \tag{11}$$

where $\boldsymbol{\beta} = \psi^{-1}(\boldsymbol{\mu}_1 - \boldsymbol{\mu}_2)$ and $\alpha = \log(\pi_1/\pi_2) - \frac{1}{2}\boldsymbol{\beta}'(\boldsymbol{\mu}_1 + \boldsymbol{\mu}_2)$, That is, $\log[\pi_{1\mathbf{y}}/(1 - \pi_{1\mathbf{y}})]$ has the form of equation (10). Moreover, now, the vector of discriminant function coefficients $\boldsymbol{\beta}$ has another interpretation as a vector of logistic regression coefficients.

Equation (9) allows for simple computation of normal theory posterior probabilities. Setting

$$\begin{aligned} z = \alpha + \boldsymbol{\beta}'\mathbf{y} &= \log \frac{\pi_1}{\pi_2} - \frac{1}{2}(\boldsymbol{\mu}_1 - \boldsymbol{\mu}_2)'\psi^{-1}(\boldsymbol{\mu}_1 + \boldsymbol{\mu}_2) + \boldsymbol{\beta}'\mathbf{y} \\ &= \log \frac{\pi_1}{\pi_2} + \boldsymbol{\beta}'\left[\mathbf{y} - \frac{1}{2}(\boldsymbol{\mu}_1 + \boldsymbol{\mu}_2)\right], \end{aligned} \tag{12}$$

we obtain

$$\pi_{1\mathbf{y}} = \frac{e^z}{1 + e^z}. \tag{13}$$

In the sample situation, parameter estimates replace parameters in equation (12) as usual, whereas (13) remains unchanged.

Example 5.4.4 Classification of two species of voles. The two species *Microtus multiplex* and *M. subterraneus* are morphometrically difficult to distinguish. Table 5.4.1 lists data for eight variables measured on the skulls of 288 specimens found at various places in central Europe. For 89 of the skulls, the chromosomes were analyzed to identify their species; $N_1 = 43$ specimens were from *M. multiplex*, and $N_2 = 46$ from *M. subterraneus*. Species was not determined for the remaining 199 specimens. Airoldi, Flury, and Salvioni (1996) report a discriminant analysis and finite mixture analysis of this data set. For the time being, we will concentrate on the discriminant analysis, but we will continue the example in Section 9.5.

The analysis based on all eight variables is, at most, marginally better than an analysis based on selected subsets of two variables; see Exercise 13 in the current section and Exercise 10 in Section 5.5. Here we present the results using variables Y_1 = M1LEFT and Y_4 = FORAMEN. Details of the numerical calculations are left to the reader; see Exercise 12. The vector of coefficients of the linear discriminant function, based on the training samples of size $N_1 = 43$ and $N_2 = 46$, is $\mathbf{b} = 10^{-3}(32.165, -4.780)$. Thus, the linear discriminant function is given by

$$V = 10^{-3}(32.165Y_1 - 4.780Y_4).$$

The associated standard distance $D = 2.954$. Since the training samples were themselves drawn from the mixture and not from the conditional distributions, it is reasonable to estimate the prior probabilities by the relative frequencies $N_j/(N_1+N_2)$, yielding $\hat{\pi}_1 = 0.483$ and $\hat{\pi}_2 = 0.517$. The normal theory classification rule of equation (7) for an observation $\mathbf{y}$, with parameters substituted by estimates, is as follows:

$$\text{Classify } \mathbf{y} \text{ in group } \begin{cases} 1 & \text{if } \mathbf{b}'\mathbf{y} - \frac{1}{2}\mathbf{b}'(\bar{\mathbf{y}}_1 + \bar{\mathbf{y}}_2) + \log(\hat{\pi}_1/\hat{\pi}_2) > 0 \\ 2 & \text{if } \mathbf{b}'\mathbf{y} - \frac{1}{2}\mathbf{b}'(\bar{\mathbf{y}}_1 + \bar{\mathbf{y}}_2) + \log(\hat{\pi}_1/\hat{\pi}_2) < 0 . \end{cases}$$

Numerically, $v = 10^{-3}(32.165y_1 - 4.780y_4)$ gives the following:

$$\text{Classify in group } \begin{cases} 1 & \text{if } v - 42.831 > 0 \\ 2 & \text{if } v - 42.831 < 0. \end{cases}$$

Equivalently, we can classify into group 1 if $\mathbf{b}'\mathbf{y} > 42.696$, and classify into group 2 if $\mathbf{b}'\mathbf{y} < 42.696$. Numerical values of $z = v - 42.831$ are displayed in Table 5.4.2 for the first 15 observations in each of groups 1 and 2, and for the first 15 unclassified observations; the remaining observations are omitted. In the training samples, a total of 5 observations would be misclassified; one of them is the third observation in group 1. Of the 199 observations with unknown group membership, 113 would be classified in group 1 (*M. multiplex*), and the remaining 86 in group 2 (*M. subterraneus*). The last column of Table 5.4.2 shows estimated posterior probabilities calculated from

Table 5.4.1 *Microtus* data

Group	Y_1	Y_2	Y_3	Y_4	Y_5	Y_6	Y_7	Y_8
1	2078	1649	1708	3868	5463	2355	805	475
1	1929	1551	1550	3825	4741	2305	760	450
1	1888	1613	1674	4440	4807	2388	775	460
1	2020	1670	1829	3800	4974	2370	766	460
1	2223	1814	1933	4222	5460	2470	815	475
1	2190	1800	2066	4662	4860	2535	838	521
1	2136	1640	1767	4070	5372	2385	815	480
1	2150	1761	1859	4053	5231	2445	840	480
1	2040	1694	1958	3977	5579	2435	835	440
1	2052	1551	1712	3877	5401	2330	830	475
1	2282	1706	1896	3976	5560	2500	855	500
1	1892	1626	1763	3538	5149	2270	810	446
1	1977	1556	1935	3576	5346	2330	785	462
1	2220	1680	2054	4226	5130	2465	880	490
1	2070	1604	1616	3633	5037	2345	845	475
1	2000	1602	1818	3997	5304	2410	790	460
1	2140	1612	1719	3490	5254	2305	790	450
1	2084	1565	1793	3834	5078	2345	760	450
1	2072	1651	1772	3970	5402	2396	804	462
1	2132	1784	1875	4150	5422	2390	845	460
1	1826	1548	1815	3519	5230	2250	800	425
1	2073	1588	1919	4239	5203	2385	790	475
1	2187	1801	2145	4464	5874	2600	910	524
1	1802	1363	1458	3631	4842	2145	760	416
1	2054	1569	1745	3678	5445	2305	791	462
1	2479	1880	2065	4195	6104	2590	860	535
1	2102	1506	1660	3871	5212	2300	772	437
1	2158	1612	1869	4015	5652	2500	828	480
1	1907	1549	1672	4050	5307	2350	770	456
1	2084	1660	1906	4000	5061	2355	805	465
1	1987	1592	1720	3741	5245	2475	810	470
1	1933	1486	1742	4007	5032	2345	810	465
1	1914	1583	1722	3677	4871	2237	805	437
1	2015	1695	1997	4404	5453	2525	815	495
1	1930	1688	1883	3941	5004	2370	795	469
1	2155	1656	2150	4070	5473	2457	796	477
1	1988	1599	1779	3856	5165	2352	770	475
1	2027	1645	1966	4334	5293	2452	775	470
1	2023	1612	1781	4148	4940	2340	796	455
1	1885	1549	1628	3718	5286	2300	810	455
1	1945	1580	1739	3801	5567	2370	800	475
1	2186	1847	1896	4160	5587	2470	845	500
1	2110	1631	1703	3856	4773	2350	850	465
2	1888	1548	1763	4112	4814	2350	735	450
2	1898	1568	1734	4169	4919	2285	750	420
2	1735	1534	1566	3947	4773	2170	738	415

Table 5.4.1 *Continued*

Group	Y_1	Y_2	Y_3	Y_4	Y_5	Y_6	Y_7	Y_8
2	1746	1394	1397	3657	4771	2060	720	415
2	1734	1495	1561	3859	5229	2275	785	417
2	1741	1530	1683	3999	4745	2330	790	450
2	1746	1562	1456	3807	5108	2260	760	426
2	1722	1558	1757	4097	4379	2290	750	432
2	1873	1524	1885	3921	5007	2340	795	450
2	1738	1419	1634	4039	4228	2270	771	420
2	1731	1546	1560	3764	4866	2175	755	424
2	1815	1436	1361	3728	4911	2150	750	412
2	1790	1524	1606	3890	4700	2189	770	427
2	1814	1454	1672	3890	5282	2275	795	425
2	1819	1506	1809	3564	5062	2290	790	435
2	1814	1550	1552	4265	4801	2298	776	440
2	1773	1355	1447	3717	4649	2135	745	415
2	1783	1465	1487	4141	4459	2240	760	415
2	1762	1657	1717	4262	4982	2270	750	435
2	1766	1585	1557	3805	4474	2235	731	430
2	1823	1504	1591	3928	4611	2275	725	425
2	1795	1487	1478	3762	5000	2200	775	415
2	1702	1522	1725	4155	5065	2350	779	488
2	1755	1517	1536	4098	4634	2265	746	420
2	1811	1611	1537	4081	4998	2365	740	461
2	1776	1582	1715	3989	5118	2342	788	460
2	1674	1491	1433	3521	4724	2042	770	425
2	1770	1490	1586	3762	4971	2250	740	425
2	1902	1499	1680	4056	5178	2300	755	450
2	1814	1510	1677	3856	4689	2245	805	430
2	1728	1505	1544	3726	4746	2120	750	420
2	1714	1525	1590	3973	4957	2230	725	425
2	1895	1480	1561	3991	4816	2210	772	450
2	1758	1507	1631	3852	4979	2221	765	430
2	1640	1416	1542	3687	4601	2095	740	410
2	1770	1621	1567	4156	4773	2286	745	436
2	1746	1419	1700	4021	4368	2182	735	400
2	1784	1502	1417	3959	4815	2168	750	424
2	1781	1504	1731	3649	5104	2260	748	427
2	1770	1396	1509	3864	3980	2061	715	400
2	1702	1443	1500	3451	4977	2060	745	395
2	1779	1572	1771	4016	5199	2355	792	425
2	1747	1411	1566	3803	4537	2180	750	417
2	1878	1549	1844	4078	4747	2295	795	430
2	1619	1458	1402	3492	4439	1965	740	395
2	1749	1482	1462	3797	4855	2218	765	415
0	1841	1562	1585	3750	5024	2232	821	430
0	1770	1459	1542	3856	4542	2140	755	405
0	1785	1573	1616	4165	3928	2295	767	425

Table 5.4.1 *Continued*

Group	Y_1	Y_2	Y_3	Y_4	Y_5	Y_6	Y_7	Y_8
0	2095	1660	1870	3937	5218	2355	842	490
0	1976	1666	1704	4058	5235	2335	814	481
0	1980	1643	1950	3569	6020	2355	815	460
0	2183	1726	1939	3825	5049	2416	850	500
0	1750	1564	1656	3155	4714	2423	760	445
0	2227	1798	1923	4194	5521	2504	835	487
0	1682	1480	1531	3543	4680	2100	720	400
0	2205	1729	1924	4337	5480	2555	850	515
0	2054	1663	2002	4171	5273	2490	800	469
0	2128	1772	1933	4425	5952	2570	875	530
0	1720	1539	1758	4041	4637	2306	755	462
0	1985	1575	1887	4018	5195	2595	775	480
0	2235	1806	1829	4286	5464	2470	858	490
0	1943	1407	1571	3527	4715	2110	750	445
0	1630	1521	1417	3241	4540	1908	700	380
0	1795	1525	1758	3974	4940	2250	800	410
0	1775	1458	1617	3733	4706	2205	755	412
0	1786	1476	1659	4022	5045	2235	790	420
0	1877	1480	1569	3821	4820	2235	770	435
0	2134	1805	1856	4160	5789	2525	866	505
0	1956	1538	1602	3679	5040	2265	755	430
0	2102	1783	1912	3862	5777	2470	880	525
0	2355	1809	1986	4310	5719	2545	856	500
0	2007	1569	1746	3972	4850	2312	820	490
0	2100	1670	1890	3935	5798	2450	796	470
0	2196	1773	1980	4047	5206	2395	850	477
0	1786	1559	1809	3566	5148	2390	770	465
0	2245	1850	2043	4460	5448	2570	870	480
0	1875	1444	1699	3625	5099	2200	775	440
0	2226	1764	1873	4086	5153	2405	857	478
0	2137	1670	1908	4074	4865	2353	780	474
0	1705	1491	1434	3694	4815	2210	755	400
0	2129	1697	2037	4369	5443	2530	865	530
0	1867	1567	1580	3685	5068	2224	810	415
0	1703	1576	1553	3835	5026	2240	768	428
0	1696	1447	1417	3384	4469	1975	730	375
0	2147	1793	2057	4360	5313	2495	872	520
0	1787	1471	1477	4015	4584	2171	766	415
0	2104	1699	1876	3832	5561	2400	795	466
0	2104	1702	1967	4185	5405	2434	820	476
0	2161	1710	1792	4096	5491	2415	805	495
0	2067	1636	1892	3728	5404	2355	820	440
0	2098	1606	1833	3959	5695	2480	825	480
0	2088	1705	1929	3864	5583	2420	828	485
0	1964	1564	1781	3984	4875	2450	786	440
0	1837	1504	1571	3967	4654	2275	751	448

Table 5.4.1 *Continued*

Group	Y_1	Y_2	Y_3	Y_4	Y_5	Y_6	Y_7	Y_8
0	1726	1531	1432	3385	4924	2160	750	400
0	1855	1511	1555	3840	4605	2195	760	415
0	1670	1426	1518	3684	4581	2100	727	385
0	2018	1638	1782	3701	5634	2326	812	477
0	1914	1574	1744	3795	5093	2178	820	424
0	2269	1847	1947	4292	5823	2567	871	521
0	2151	1796	2010	4253	5294	2440	830	488
0	2177	1775	1839	3473	5743	2425	830	474
0	1840	1517	1493	3467	4642	2048	805	407
0	1852	1592	1657	3736	4768	2215	790	425
0	2140	1865	2030	4228	5283	2483	825	458
0	1986	1424	1610	3705	4938	2190	800	452
0	2138	1668	1919	4138	5264	2425	835	450
0	2245	1835	1984	4156	5614	2455	825	498
0	2029	1558	1824	4075	5315	2440	810	475
0	2132	1585	1849	3862	5223	2360	820	451
0	1724	1587	1571	4102	4834	2270	765	465
0	2065	1756	2047	3986	5781	2519	865	516
0	2055	1777	1887	3643	5604	2350	840	466
0	1698	1531	1465	3981	4737	2255	778	420
0	2053	1702	1765	4077	5483	2363	878	489
0	1841	1589	1806	4065	5081	2355	790	457
0	1944	1577	1595	3908	5371	2294	815	457
0	2095	1742	1959	3876	5886	2425	815	480
0	1737	1463	1532	3765	4849	2202	754	432
0	1807	1501	1646	3903	4544	2109	760	420
0	1846	1579	1743	3668	4862	2206	735	420
0	2145	1670	1857	4091	5460	2436	835	475
0	1931	1410	1667	3726	4355	2145	735	425
0	1697	1364	1637	3817	4799	2092	760	395
0	2065	1626	1675	3823	5240	2340	866	455
0	1975	1599	1707	4011	5449	2407	895	516
0	1950	1658	1627	3820	5250	2315	770	470
0	2002	1521	1649	3852	5174	2326	825	475
0	1973	1481	1773	3922	5384	2360	770	470
0	1969	1649	1743	3537	5393	2240	811	438
0	1956	1672	1870	3773	5316	2325	811	475
0	1813	1422	1608	3199	5142	2125	775	435
0	1883	1532	1577	4069	4803	2315	757	445
0	1837	1542	1618	4145	5037	2298	775	442
0	1740	1659	1678	4152	4735	2320	760	445
0	2220	1749	1874	3927	5908	2461	833	510
0	1922	1497	1609	4028	5335	2300	755	430
0	1767	1464	1515	3742	4785	2072	750	400
0	1845	1441	1571	3453	5008	2080	772	412
0	1693	1423	1629	3400	4992	2065	727	406

Table 5.4.1 *Continued*

Group	Y_1	Y_2	Y_3	Y_4	Y_5	Y_6	Y_7	Y_8
0	2434	1707	1766	4028	5277	2388	815	460
0	1741	1550	1647	3870	4854	2245	730	420
0	1995	1745	1776	4076	5281	2386	806	455
0	1834	1547	1696	4050	4755	2220	762	416
0	1870	1604	1762	4087	4688	2325	760	435
0	2036	1533	1672	3875	5725	2445	808	483
0	2145	1617	1685	4177	5111	2285	846	471
0	2070	1539	1876	4162	5138	2320	800	453
0	1829	1535	1639	3793	4950	2230	815	440
0	1993	1565	1668	3960	5152	2350	780	433
0	1534	1433	1416	3360	4583	1945	728	390
0	1974	1710	1935	3500	5201	2270	790	460
0	2137	1768	1990	4253	5692	2566	904	545
0	1808	1521	1643	4258	4678	2305	775	452
0	1933	1541	1718	3875	5175	2245	765	435
0	1919	1613	1772	3753	5476	2370	790	466
0	1898	1522	1775	4224	5018	2384	825	435
0	1829	1597	1815	4350	5026	2370	790	440
0	2247	1689	1974	3981	5466	2485	850	478
0	1933	1557	1762	4055	5261	2360	780	450
0	2022	1784	1794	4169	5131	2385	815	468
0	1975	1566	1702	4143	5417	2465	798	472
0	2068	1734	1879	3989	5344	2355	796	475
0	1783	1480	1434	3385	5056	2105	785	430
0	1817	1508	1498	3882	4973	2270	770	450
0	2353	1765	2187	4500	5703	2605	883	501
0	1978	1655	1946	3683	5283	2369	810	461
0	2247	1765	1909	3991	5325	2335	811	475
0	2001	1649	1895	4091	5459	2470	840	485
0	1859	1662	1783	4137	4878	2350	750	440
0	1645	1362	1476	3436	4457	2038	714	400
0	1695	1411	1451	3823	4693	2155	774	420
0	1999	1633	1780	4045	5091	2397	795	453
0	1824	1566	1735	3948	5156	2298	760	435
0	2197	1847	1845	4122	5280	2410	810	465
0	2050	1625	1828	3623	5942	2380	830	470
0	1735	1454	1702	4045	4866	2210	745	428
0	1815	1623	1527	3963	4883	2280	764	446
0	2015	1664	1722	4098	5163	2345	815	485
0	1879	1608	1796	4005	4985	2362	760	455
0	1852	1493	1501	3796	5140	2275	770	420
0	2077	1632	1777	3982	5384	2380	820	470
0	1979	1699	1730	3608	5258	2289	871	431
0	1994	1706	1854	3857	5245	2335	771	460
0	2318	1837	1970	4277	5738	2545	831	485
0	2189	1625	1739	4242	4874	2330	817	476

Table 5.4.1 *Continued*

Group	Y_1	Y_2	Y_3	Y_4	Y_5	Y_6	Y_7	Y_8
0	2139	1762	1896	4341	5342	2554	868	492
0	1667	1412	1469	3620	4495	2070	740	403
0	2227	1830	1983	4195	5586	2441	840	485
0	1981	1475	1882	3784	4973	2283	775	475
0	2002	1647	1737	4075	5212	2300	810	450
0	2334	1801	1907	3993	5747	2462	867	520
0	1768	1454	1682	3855	4623	2195	750	435
0	2100	1863	1862	3995	5689	2450	870	490
0	1766	1500	1552	3941	4659	2225	740	420
0	2245	1684	1968	4010	5561	2416	821	495
0	1968	1502	1564	3418	4958	2190	780	445
0	1829	1594	1721	3741	5089	2265	805	434
0	2088	1637	1670	3842	5100	2300	851	462
0	2202	1607	2005	4205	5317	2335	800	471
0	1820	1501	1596	3864	4864	2175	780	425
0	1639	1460	1536	3787	4481	2220	740	435
0	1869	1465	1469	3844	4856	2245	752	425
0	1688	1402	1560	3524	4318	1980	720	376
0	1881	1465	1775	3940	4864	2295	745	426
0	1855	1431	1611	4043	4978	2335	875	490
0	1693	1503	1573	3477	4612	2100	730	400
0	1657	1502	1764	3686	4429	2170	712	420
0	1678	1509	1521	3647	4879	2115	710	420
0	2159	1689	1915	4175	5510	2482	865	500
0	1902	1645	1832	3837	5051	2430	755	460
0	1739	1436	1471	3481	4426	2025	755	410
0	1730	1442	1648	3750	4417	2200	770	435
0	1702	1495	1685	3925	4649	2210	765	420
0	2103	1655	1713	4042	4840	2295	795	445
0	2197	1800	2131	4120	5711	2448	835	498
0	1826	1564	1649	4028	5035	2350	790	456
0	1746	1461	1650	3780	4380	2200	755	410
0	1752	1573	1732	4087	4737	2255	746	416
0	1969	1635	1825	3759	5558	2368	835	470
0	2014	1671	1877	4079	5312	2336	830	461
0	1868	1494	1753	4091	5171	2370	755	465
0	1959	1644	1674	3801	5436	2352	782	456
0	2033	1466	1594	3751	4355	2152	825	427
0	1967	1646	1766	3930	5492	2365	785	455
0	1940	1464	1443	3532	5061	2135	752	438
0	2083	1635	1715	3605	5125	2311	802	451
0	1726	1506	1617	3981	4289	2165	755	415
0	1946	1570	1758	3916	5482	2335	790	440
0	1800	1598	1769	3921	4977	2320	772	458
0	2014	1551	1727	3597	5119	2166	780	452
0	1723	1505	1625	4127	4502	2284	770	453

Table 5.4.1 *Continued*

Group	Y_1	Y_2	Y_3	Y_4	Y_5	Y_6	Y_7	Y_8
0	1776	1489	1498	4273	4753	2240	745	415
0	1624	1438	1623	3793	4592	2135	738	404
0	1840	1598	1672	3922	5267	2300	760	455
0	1924	1531	1655	4131	4932	2312	762	445
0	1786	1540	1757	3731	5096	2295	750	455
0	1991	1557	1649	3568	5293	2260	765	450
0	1646	1541	1532	3598	5000	2185	756	410
0	1869	1574	1777	3784	5080	2310	795	459
0	2282	1784	1836	4049	5059	2525	815	503
0	2236	1761	2103	4159	5815	2596	912	522
0	1718	1417	1566	3996	4652	2228	730	436
0	1907	1518	1760	3683	5177	2260	775	457

Courtesy of Dr. M. Salvioni, Ufficio Caccia e Pesca, Bellinzona, Switzerland. Variables are Y_1 = width of upper left molar 1 (M1LEFT); Y_2 = width of upper left molar 2 (M2LEFT); Y_3 = width of upper left molar 3 (M3LEFT); Y_4 = length of incisive foramen (FORAMEN); Y_5 = length of palatal bone (PBONE); Y_6 = condylo incisive length or skull length (LENGTH); Y_7 = skull height above bullae (HEIGHT); Y_8 = skull width across rostrum (ROSTRUM). Variables Y_1 to Y_5 are in mm/1000; variables Y_6 to Y_8 are in mm/100. Variable "Group" indicates species: 1 = *Microtus multiplex*, 2 = *M. subterraneus*, 0 = species not determined.

equations (11) and (12), which in the notation of this example reads

$$\hat{\pi}_{1y} = \frac{e^{v-42.831}}{1 + e^{v-42.831}}.$$

The classification criterion given above is equivalent to classifying in group 1 if $\hat{\pi}_{1y} > \frac{1}{2}$, and in group 2 otherwise.

A graph helps to understand the performance of the linear discriminant function for classification of observations. Figure 5.4.4 shows the frequency distribution of $V = 10^{-3}(32.165Y_1 - 4.780Y_4)$ for the data in the training samples and for the 199 specimens with unknown group membership. The cutoff for classification at 42.831 is also marked in the graph. For many of the 199 specimens with unknown group membership, there would be high uncertainty regarding their classification because the posterior probabilities are close to 50 percent. □

Example 5.4.4 suggests some critical questions about the use of the linear discriminant function. As seen in the first part of this section, the linear discriminant function is a good tool if the two distributions are multivariate normal with equal covariance matrices. However, it would be too restrictive to apply the technique only in cases where one is certain about multivariate normality. A more pragmatic approach would be to ask for univariate normality of the discriminant function; this is a much weaker assumption that seems to be met, at least approximately, in many applied

Table 5.4.2 Classification of voles using the linear discriminant function

i	x	y_1	y_4	z	$\hat{x}$	$\hat{\pi}_{1y}$
1	1	2078	3868	5.52	1	99.6
2	1	1929	3825	0.93	1	71.8
3	1	1888	4440	−3.32	2	3.5
4	1	2020	3800	3.98	1	98.2
5	1	2223	4222	8.49	1	100.0
6	1	2190	4662	5.33	1	99.5
7	1	2136	4070	6.42	1	99.8
8	1	2150	4053	6.95	1	99.9
9	1	2040	3977	3.78	1	97.8
10	1	2052	3877	4.64	1	99.0
11	1	2282	3976	11.57	1	100.0
12	1	1892	3538	1.12	1	75.3
13	1	1977	3576	3.67	1	97.5
14	1	2220	4226	8.38	1	100.0
15	1	2070	3633	6.39	1	99.8
44	2	1888	4112	−1.76	2	14.7
45	2	1898	4169	−1.71	2	15.3
46	2	1735	3947	−5.89	2	0.3
47	2	1746	3657	−4.15	2	1.6
48	2	1734	3859	−5.50	2	0.4
49	2	1741	3999	−5.95	2	0.3
50	2	1746	3807	−4.87	2	0.8
51	2	1722	4097	−7.02	2	0.1
52	2	1873	3921	−1.33	2	21.0
53	2	1738	4039	−6.23	2	0.2
54	2	1731	3764	−5.14	2	0.6
55	2	1815	3728	−2.27	2	9.4
56	2	1790	3890	−3.85	2	2.1
57	2	1814	3890	−3.08	2	4.4
58	2	1819	3564	−1.36	2	20.5
90	0	1841	3750	−1.54	2	17.7
91	0	1770	3856	−4.33	2	1.3
92	0	1785	4165	−5.32	2	0.5
93	0	2095	3937	5.74	1	99.7
94	0	1976	4058	1.33	1	79.1
95	0	1980	3569	3.80	1	97.8
96	0	2183	3825	9.10	1	100.0
97	0	1750	3155	−1.62	2	16.5
98	0	2227	4194	8.75	1	100.0
99	0	1682	3543	−5.66	2	0.3
100	0	2205	4337	7.36	1	99.9
101	0	2054	4171	3.30	1	96.4
102	0	2128	4425	4.47	1	98.9
103	0	1720	4041	−6.82	2	0.1

Table 5.4.2 *Continued*

i	x	y_1	y_4	z	$\hat{x}$	$\hat{\pi}_{1y}$
104	0	1985	4018	1.81	1	86.0

The first 15 observations are shown for each of the two groups (*Microtus multiplex, M. subterraneus*), and for the specimens with unknown group membership. i = sequential number of the observation in the data file; y_1 = value of M1LEFT; y_4 = value of FORAMEN; z = value of $v - 42.831$; $\hat{x}$ = group membership according to the classification rule; $\hat{\pi}_{1y}$ = estimated posterior probability (in percent) for membership in group 1.

real-world problems. Without doing a formal test, Figure 5.4.4 seems to support, or at least not contradict, normality of the discriminant function in both groups. Recall from Section 5.2 that the linear discriminant function can be justified from a heuristic approach, without making any distributional assumptions. The use of normal theory posterior probabilities, on the other hand, is justified only if we believe in approximate normality of the distributions.

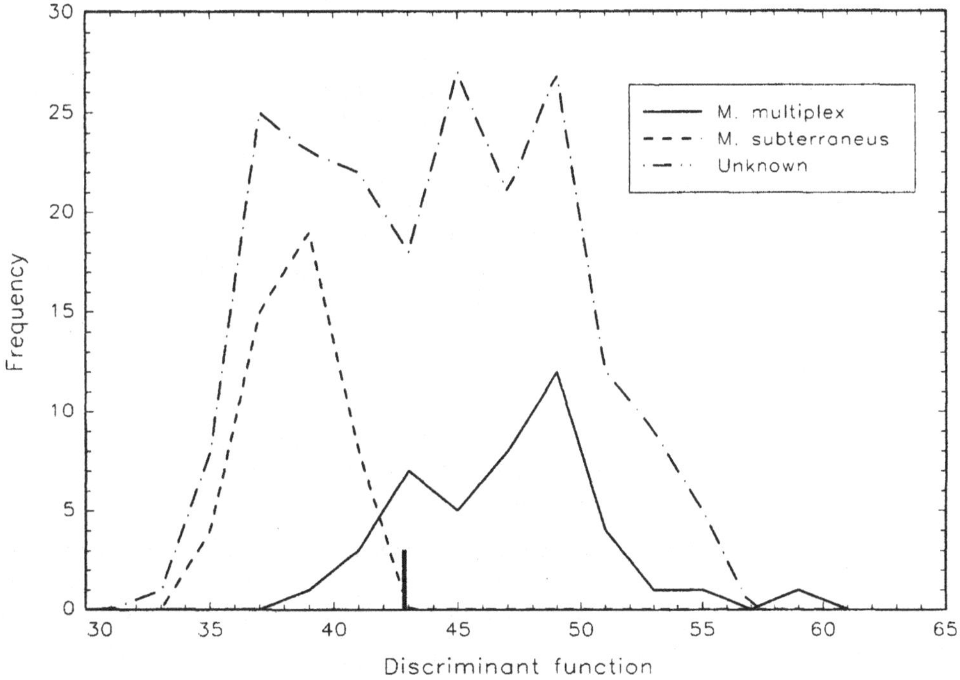

Figure 5-4-4 Distribution of the discriminant function $V = 10^{-3}(32.165Y_1 - 4.780Y_4)$ in the *Microtus* example. The cutoff point for classification is marked as a short vertical line. Reproduced from Airoldi et al (1995) with permission by Academic Press, Inc.

What about equality of variances? As seen in Figure 5.4.4, the data in group *M. multiplex* have somewhat more variability than the data in group *M. subterraneus*. This might mean that classification based on the linear discriminant function is not optimal, and we should use, instead, a more general method, such as quadratic discrimination. We delay the discussion of this topic until Section 7.2.

Finally, a remark on the use of observations with unknown group membership. As seen in Example 5.4.4 and from parts (b) and (c) of Exercise 12, there are many observations that we can assign, almost with certainty, to one of the two groups. Why not include these observations in the training samples? Are we not wasting information by treating them as if we knew nothing about their group membership, when, in fact, we are fairly certain? The answer is that there are, indeed, ways in which observations of unknown group membership can be used to estimate the parameters of a discrimination rule, but, first, we will have to go through quite a sophisticated treatment of the finite mixture model. The particular situation where some observations have known group membership and some are unclassified a priori is treated in Section 9.5.

Exercises for Section 5.4

1. Suppose that $\mathbf{Y} \sim \mathcal{N}_2(\boldsymbol{\mu}_j, \boldsymbol{\psi})$ in groups $j = 1, 2$, where $\boldsymbol{\psi}$ is positive definite. Suppose that the prior probabilities are π_1 and $\pi_2 = 1 - \pi_1$. Show that curves of constant posterior probability $\pi_{1\mathbf{y}}$ = constant or $\pi_{2\mathbf{y}}$ = constant are parallel straight lines in the (y_1, y_2)-plane. Also show that any change in the prior probabilities moves the boundary line $\pi_{1\mathbf{y}} = 0.5$ to a parallel line.

2. In the midge example, construct a scatterplot of Y_1 (antenna length) vs .Y_2 (wing length), and draw lines of constant estimated posterior probabilities $\hat{\pi}_{1\mathbf{y}} = c$ for $c =$ 0.01, 0.05, 0.1, 0.3, 0.5, 0.7, 0.9, 0.95, and 0.99. As in Example 5.4.2, substitute the sample quantities $\bar{\mathbf{y}}_1$, $\bar{\mathbf{y}}_2$, and $\mathbf{S}$ for the parameters, and use prior probabilities $\pi_1 = \pi_2 = \frac{1}{2}$.

3. Suppose that $\mathbf{Y} \sim \mathcal{N}_p(\boldsymbol{\mu}_j, \boldsymbol{\psi})$ in groups $j = 1, 2$, with prior probabilities π_1 and π_2, where $\boldsymbol{\psi}$ is positive definite. Let $\pi_{j\mathbf{y}} = \Pr[X = j | \mathbf{Y} = \mathbf{y}]$ denote the posterior probability of membership in group j. Show that $\pi_{j\mathbf{y}}$ can be computed as in equation (8), where $v = \boldsymbol{\beta}'\mathbf{y}$.

4. This exercise continues Exercise 2 from Section 5.3. For each of the four specimens, calculate the value of the estimated posterior probability of membership in the two groups, using equal prior probabilities and equation (8).

5. This exercise is based on the flea beetle data of Example 5.3.2.
   (a) Use the normal theory approach with equal covariance matrices, parameter estimates replaced by sample quantities from Example 5.3.2, and prior probabilities $\hat{\pi}_j = N_j/(N_1 + N_2)$ to graph the estimated posterior probabilities as functions

of the discriminant function, similar to Figure 5.4.3. Use the same scale as in Exercise 3 of Section 5.3, so that you can compare the two graphs.

(b) Calculate estimated posterior probabilities for the four new specimens from Exercise 3(d) of Section 5.3, and thus, quantify your earlier answer.

6. Suppose that X is a discrete random variable taking values 1 and 2 with probability $\frac{1}{2}$ each. Let $\mathbf{Y}$ be a bivariate, continuous, random variable with the following conditional densities:

$$f_1(\mathbf{y}) = f_{\mathbf{Y}|X}(\mathbf{y}|X=1) = \begin{cases} y_1 + y_2 & \text{if } 0 \le y_1 \le 1,\ 0 \le y_2 \le 1, \\ 0 & \text{otherwise,} \end{cases}$$

and

$$f_2(\mathbf{y}) = f_{\mathbf{Y}|X}(\mathbf{y}|X=2) = \begin{cases} 2-(y_1 + y_2) & \text{if } 0 \le y_1 \le 1,\ 0 \le y_2 \le 1, \\ 0 & \text{otherwise.} \end{cases}$$

(a) Find the unconditional (marginal) *pdf* of $\mathbf{Y}$.

(b) Find $\pi_{j\mathbf{y}} = \Pr[X=j|\mathbf{Y}=\mathbf{y}]$ for $j=1,2$. Thus, show that the classification rule based on posterior probabilities depends on (Y_1, Y_2) through a linear function of Y_1 and Y_2.

(c) Find the classification boundary corresponding to $\pi_{1\mathbf{y}} = \pi_{2\mathbf{y}} = \frac{1}{2}$.

(d) Find $\boldsymbol{\mu}_j = E[\mathbf{Y}|X=j]$ and $\boldsymbol{\psi}_j = \text{Cov}[\mathbf{Y}|X=j]$ for $j=1,2$, and verify that $\boldsymbol{\psi}_1 = \boldsymbol{\psi}_2 = \boldsymbol{\psi}$, say.

(e) Let $\boldsymbol{\beta} = \boldsymbol{\psi}^{-1}(\boldsymbol{\mu}_1 - \boldsymbol{\mu}_2)$, and show that the classification rule found in part (b) depends on the data vector $\mathbf{y}$ through $\boldsymbol{\beta}'\mathbf{y}$.

(f) Show that $\log\left[\pi_{1\mathbf{y}}/(1-\pi_{1\mathbf{y}})\right]$ is not a linear function of $\mathbf{y}$.

7. In the normal theory setup with equal covariance matrices, show that the probability of $\mathbf{Y}$ falling on the boundary line is zero conditionally on both $X=1$ and $X=2$.

8. Suppose that $\mathbf{Y}$ is a bivariate continuous random variable with a uniform distribution in $[0,2] \times [0,2]$ in group 1 and a uniform distribution in $[1,3] \times [1,3]$ in group 2.

(a) Find the linear discriminant function for the two groups, i.e., compute $\boldsymbol{\beta} = \boldsymbol{\psi}^{-1}(\boldsymbol{\mu}_1 - \boldsymbol{\mu}_2)$.

(b) Suppose that the prior probabilities are $\pi_1 = \pi_2 = \frac{1}{2}$. Show that the set $\{\mathbf{y} \in \mathbb{R}^p : \pi_{1\mathbf{y}} = \pi_{2\mathbf{y}} = \frac{1}{2}\}$ has positive probability in both groups.

(c) Suppose that $\pi_1 \neq \pi_2$. Show that classification according to the rule of assigning $\mathbf{y} \in \mathbb{R}^2$ to the group with higher posterior probability is not the same as classification based on the linear discriminant function. That is, show that the classification rule based on posterior probabilities depends on $\mathbf{y}$ through a function of $\mathbf{y}$ different from $\boldsymbol{\beta}'\mathbf{y}$.

9. Suppose that $\mathbf{Y}$ is a p-variate random variable such that in group $j=1,2$, the distribution of $\mathbf{Y}$ is elliptical with mean vector $\boldsymbol{\mu}_j$ and covariance matrix $\boldsymbol{\psi}$. Assume furthermore that the functional form of the two elliptical distributions is the same, that is, if $f_j(\mathbf{y})$ is the *pdf* of $\mathbf{Y}$ in group j, then $f_2(\mathbf{y}) = f_1(\mathbf{y}+\boldsymbol{\delta})$ for some $\boldsymbol{\delta} \in \mathbb{R}^p$. Is it true for all elliptical distributions that classification according to highest posterior probability depends on $\mathbf{y}$ only through $\boldsymbol{\beta}'\mathbf{y}$, where $\boldsymbol{\beta} = \boldsymbol{\psi}^{-1}(\boldsymbol{\mu}_1 - \boldsymbol{\mu}_2)$? If so, prove the result. If not, give a counterexample. *Note*: The multivariate normal distribution is a special case of the elliptical family, so the result holds true in at least one example.

10. Let $\mathbf{Y} = (Y_1, \ldots, Y_p)'$ denote a p-variate discrete random vector such that all Y_h, in group 1, are independent binary random variables, with success probabilities q_{1h}, $h = 1, \ldots, p$. Similarly, assume that all Y_h in group 2, are independent binary random variables with success probabilities q_{2h}, $h = 1, \ldots, p$. Show that

$$\log \frac{\pi_{1\mathbf{y}}}{\pi_{2\mathbf{y}}} = \alpha + \boldsymbol{\beta}'\mathbf{y}$$

for some $\alpha \in \mathbb{R}$ and $\boldsymbol{\beta} \in \mathbb{R}^p$. Find α and $\boldsymbol{\beta}$ explicitly in terms of the prior probabilities π_1 and π_2 and the success probabilities q_{jh}. Show that $\boldsymbol{\beta}$ does not depend on the prior probabilities.

11. Show that equations (9) and (10) are equivalent.

12. This exercise expands on Example 5.4.4.

(a) Verify the calculations done in Example 5.4.4, using variables M1LEFT and FORAMEN. In particular, compute mean vectors, sample covariance matrices, the pooled covariance matrix, and the vector of discriminant function coefficients. Complete Table 5.4.2 by computing values of the linear discriminant function, predicted group membership, and estimated posterior probabilities for all 288 observations.

(b) Draw a scatterplot of M1LEFT vs. FORAMEN for (i) the 89 specimens in the training samples, marking group membership with some symbol, and (ii) the 199 specimens with unknown group membership. Draw a boundary line according to the cutoff used in Example 5.4.4.

(c) Graph the estimated posterior probabilities as curves similar to Figure 5.4.3 (b), and comment on the ability of the discriminant function to classify observations with unknown group membership. Are there any specimens of which you are practically sure which group they belong to? Are there any for which group membership is highly uncertain?

13. Compute univariate standard distances between the two groups for all eight variables in the *Microtus* data set. Which variables seem good for discrimination? Repeat the calculations of Example 5.4.4 using all eight variables and decide if it is better to use eight variables, or better to restrict the analysis to M1LEFT and FORAMEN. Only an informal assessment is expected here; also see Exercise 10 in Section 5.5.

5.5 Error Rates

In this section we discuss various types of error rates commonly used to assess the performance of classification procedures. The setup is as in Sections 2.8 and 5.4. A discrete variable X takes values 1 and 2, indicating group membership, with probabilities π_1 and π_2, and the p-variate random vector $\mathbf{Y}$ has *pdf* $f_j(\mathbf{y})$ conditionally on $X = j$, $j = 1, 2$. Sometimes, we will refer to the two conditional distributions as populations, or simply groups. A classification procedure serves the purpose of identifying the group membership of a new observed vector $\mathbf{y} \in \mathbb{R}^p$, without knowing the value of X. Typically, errors cannot be completely avoided, and, therefore, one would like to find a procedure that makes, in some sense, as few errors as possible.

Now, we present a formal setup. A method of classification into two groups partitions the sample space of $\mathbf{Y}$ into two regions of classification C_1 and C_2. For simplicity, we assume here that the sample space is $\mathbb{R}^p$, so $C_1 \cup C_2 = \mathbb{R}^p$. Now, define a function $x^* : \mathbb{R}^p \to \{1, 2\}$ such that

$$x^*(\mathbf{y}) = \begin{cases} 1 & \text{if } \mathbf{y} \in C_1, \\ 2 & \text{if } \mathbf{y} \in C_2, \end{cases} \tag{1}$$

that is, $x^*(\mathbf{y})$ is the function that assigns to $\mathbf{y}$ a "predicted group membership" or a guess of the group membership, according to the partition of the sample space given by the method of classification.

Now, we can give a formal definition of a classification rule.

Definition 5.5.1 A *classification rule* is a random variable

$$X^* = x^*(\mathbf{Y}), \tag{2}$$

taking value 1 if $\mathbf{Y} \in C_1$, and value 2 if $\mathbf{Y} \in C_2$. □

According to this definition, X^* may be regarded as predicted group membership, whereas X represents actual group membership. Ideally, one would have $X^* = X$ with probability one, i.e., the correct classification in all cases.

The performance of a classification rule can be evaluated by a table of the joint distribution of X and X^* as follows:

	$X = 1$	$X = 2$	Marginal
$X^* = 1$	$q_{11}\pi_1$	$q_{12}\pi_2$	$q_{11}\pi_1 + q_{12}\pi_2$
$X^* = 2$	$q_{21}\pi_1$	$q_{22}\pi_2$	$q_{21}\pi_1 + q_{22}\pi_2$
Marginal	π_1	π_2	

In this table, $\pi_j = \Pr[X = j]$, and the q_{ij} represent the conditional probabilities of X^*, given X. That is, $q_{ij} = \Pr[X^* = i | X = j]$, with $q_{11} + q_{21} = q_{12} + q_{22} = 1$. The two entries, such that $X^* = X$ correspond to correct classification, whereas the entries ($X^* = 1$, $X = 2$) and ($X^* = 2$, $X = 1$) represent incorrect assignments. A good classification rule will attempt to make the probabilities of incorrect assignments small.

In some cases, one may want to focus on the conditional probabilities of misclassification q_{12} and q_{21}, making them simultaneously as small as possible or fixing one of them, irrespective of the prior probabilities π_j. In other cases, one may want to find a classification rule that minimizes an overall assessment of error, as given in the following definition.

Definition 5.5.2 The *overall probability of misclassification*, or *error rate*, of a classification rule X^* is given by

$$\gamma(X^*) := \Pr[X^* \neq X] = q_{21}\pi_1 + q_{12}\pi_2, \tag{3}$$

where $\pi_j = \Pr[x = j]$ and $q_{ij} = \Pr[X^* = i | X = j]$, for $i, j = 1, 2$. □

Example 5.5.1 This is a simple univariate example to illustrate the notions introduced so far. Suppose that the conditional distribution of $\mathbf{Y}$, given $X = j$, is $\mathcal{N}(\mu_j, \sigma^2)$, $j = 1, 2$. Assume that $\mu_1 > \mu_2$, and define a classification rule as

$$X^* = \begin{cases} 1 & \text{if } Y > c, \\ 2 & \text{if } Y \leq c. \end{cases}$$

Here, c is a real constant that is not yet specified. Then

$$q_{21} = \Pr[X^* = 2 | X = 1] = \Phi\left(\frac{c - \mu_1}{\sigma}\right),$$

and

$$q_{12} = \Pr[X^* = 1 | X = 2] = 1 - \Phi\left(\frac{c - \mu_2}{\sigma}\right),$$

where Φ is the distribution function of a standard, normal, random variable. For any $c \in \mathbb{R}$, the error rate $\gamma(X^*)$ can be computed now from equation (3), provided prior probabilities are known. Figure 5.5.1 illustrates this situation for parameter values $\pi_1 = 0.6$, $\pi_2 = 0.4$, $\mu_1 = 2$, $\mu_2 = 0$, and $\sigma = 1$. Suppose we choose a cutoff point $c = 1.5$. Then the conditional probabilities of misclassification are $q_{21} = 0.309$ and $q_{12} = 0.067$. The error rate is $\gamma(X^*) = 0.212$, shown in Figure 5.5.1 as the shaded area. □

The setup of Example 5.5.1 might represent a situation where the two groups correspond to people who are immune (group 1) or susceptible (group 2) to a certain disease, and Y is the result of a screening test for immunity. In such a case, one might want to keep the probability q_{12} = Pr[a person is declared immune | the person is suspectible] small, irrespective of the error probability q_{21} or the error rate. Exercise 1 expands on this idea in the context of Example 5.5.1.

If prior probabilities are either known or estimated, one may want to find a classification rule X^* that minimizes $\gamma(X^*)$. The following theorem is a special case of a result shown in Section 7.1, and, therefore, is given here without proof.

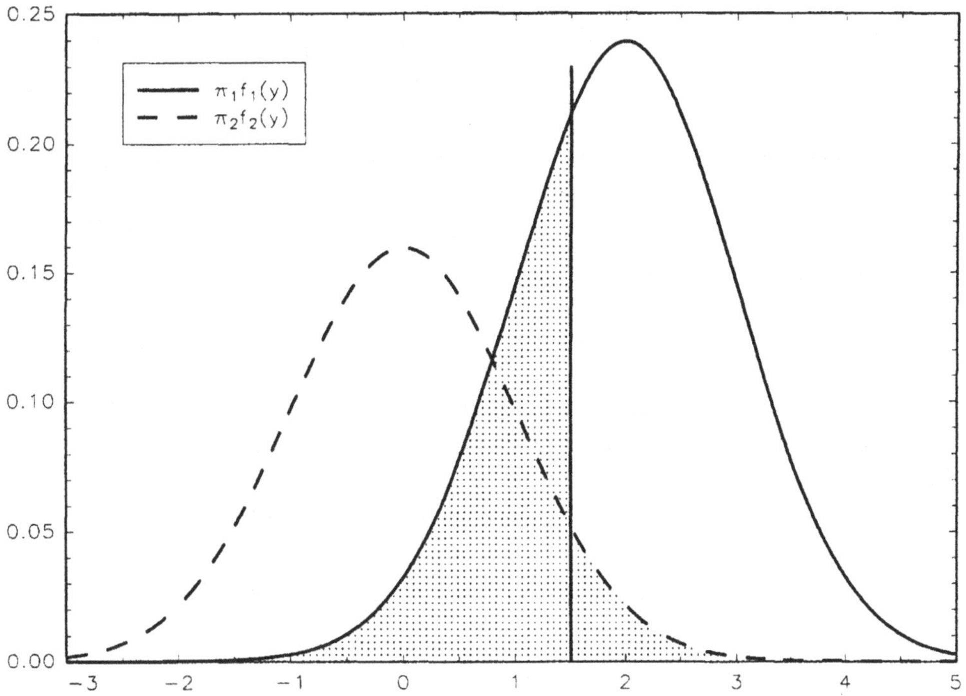

Figure 5-5-1 Illustration of the error rate in Example 5.5.1. The two bell-shaped curves represent $\pi_1 f_1(y)$ and $\pi_2 f_2(y)$, respectively. The error rate corresponds to the shaded area.

Theorem 5.5.1 *A classification rule X^* that minimizes the error rate $\gamma(X^*)$ is given by*

$$X^*_{\text{opt}} = \begin{cases} 1 & \text{if } \mathbf{y} \in C_1, \\ 2 & \text{if } \mathbf{y} \in C_2, \end{cases} \tag{4}$$

where

$$C_1 = \{\mathbf{y} \in \mathbb{R}^p : \pi_1 f_1(\mathbf{y}) \geq \pi_2 f_2(\mathbf{y})\}$$

and

$$C_2 = \{\mathbf{y} \in \mathbb{R}^p : \pi_1 f_1(\mathbf{y}) < \pi_2 f_2(\mathbf{y})\}. \tag{5}$$

Theorem 5.5.1 says that the best classification procedure, in terms of the error rate, is the same as the classification rule based on maximum posterior probability introduced in Section 5.4. It is often referred to as the *Bayes rule*. For the case of normal distributions with equal covariance matrices, this means, in turn, that optimal classification is based on the linear discriminant function. Note that, in equation (5), the boundary set $\{\mathbf{y} \in \mathbb{R}^p : \pi_1 f_1(\mathbf{y}) = \pi_2 f_2(\mathbf{y})\}$ is associated arbitrarily with region C_1; one might as well associate it with C_2 or split it between C_1 and C_2. Thus X^*_{opt} is not uniquely defined, but if the boundary set has probability zero in both groups, then this non-uniqueness is irrelevant. From now on, we will refer to a classification rule X^*_{opt} that minimizes the error rate as an *optimal classification rule* and to the associated error rate $\gamma_{\text{opt}} = \gamma(X^*_{\text{opt}})$ as the *optimal error rate* for a given classification problem.

Example 5.5.2 In the normal theory setup with identical covariance matrices, the optimal classification rule is

$$X^*_{\text{opt}} = \begin{cases} 1 & \text{if } \beta'\left[\mathbf{Y} - \frac{1}{2}(\mu_1 + \mu_2)\right] \geq \log(\pi_2/\pi_1), \\ 2 & \text{otherwise,} \end{cases} \tag{6}$$

where $\beta' = \psi^{-1}(\mu_1 - \mu_2)$. This follows from equation (4) and (5) in Section 5.4 and Theorem 5.5.1. With $\Delta = \left[(\mu_1 - \mu_2)'\psi^{-1}(\mu_1 - \mu_2)\right]^{1/2}$ denoting the standard distance, the optimal error rate is given by

$$\gamma_{\text{opt}} = \pi_1 \Phi\left(-\frac{1}{2}\Delta + \frac{1}{\Delta}\log\frac{\pi_2}{\pi_1}\right) + \pi_2 \Phi\left(-\frac{1}{2}\Delta - \frac{1}{\Delta}\log\frac{\pi_2}{\pi_1}\right); \tag{7}$$

see Exercise 2. For $\pi_1 = \pi_2 = \frac{1}{2}$, this reduces to

$$\gamma_{\text{opt}} = \Phi\left(-\frac{1}{2}\Delta\right), \tag{8}$$

which we have defined as "overlap function" in Section 5.2. □

A classification rule obtained from training samples, rather than from known models, will be denoted by $\hat{X}^*$. For instance, the classification rule based on the sample linear discriminant function may be described as

$$\hat{X}^* = \begin{cases} 1 & \text{if } \mathbf{b}'\left[\mathbf{y} - \frac{1}{2}(\bar{\mathbf{y}}_1 - \bar{\mathbf{y}}_2)\right] \geq c \\ 2 & \text{otherwise,} \end{cases}$$

where $\mathbf{b} = \mathbf{S}^{-1}(\bar{\mathbf{y}}_1 - \bar{\mathbf{y}}_2)$. In accordance with the optimality principle, we would choose $c = \log(\pi_2/\pi_1)$, if the prior probabilities are assumed known and fixed, and $c = \log(\hat{\pi}_2/\hat{\pi}_1) = \log(N_2/N_1)$, if the prior probabilities are estimated from the sample sizes. Other choices of c, as outlined in the discussion following Example 5.5.1, may also be considered.

Now, we are going to study various ways to assess the performance of classification rules estimated from samples. The simplest and most popular measure, called the *plug-in error rate*, is the proportion of observations misclassified when the classification rule $\hat{X}^*$ is applied to the data in the training samples. Formally, for each of the $N_1 + N_2$ observations in the training samples we have an observed group membership x_i and a predicted membership $\hat{x}_i^*$. Let $e_i = 0$, if $x_i = x_i^*$, and $e_i = 1$ if $x_i \neq x_i^*$. Then the plug-in error rate is given by

$$\hat{\gamma}_{\text{plug-in}} = \frac{1}{N_1 + N_2} \sum_{i=1}^{N} e_i \,. \tag{9}$$

If classification is based on the linear discriminant function and if the investigator is willing to assume that the linear discriminant function is approximately normal with equal variance in both groups, then one may also use a *normal theory estimate of error rate* inspired by equation (7):

$$\hat{\gamma}_{\text{normal}} = \pi_1 \, \Phi\left(-\frac{1}{2}D + \frac{1}{D}\log\frac{\pi_2}{\pi_1}\right) + \pi_2 \, \Phi\left(-\frac{1}{2}D - \frac{1}{D}\log\frac{\pi_2}{\pi_1}\right). \tag{10}$$

Here, D is the sample standard distance as usual, and the prior probabilities π_j may be replaced by estimates, depending on the situation. In particular, for $\pi_1 = \pi_2 = \frac{1}{2}$, we get $\hat{\gamma}_{\text{normal}} = \Phi\left(-\frac{1}{2}D\right)$.

Finally, the *leave-one-out error rate* is obtained as follows. Omit the ith observation from the training sample, and calculate the classification rule $\hat{X}^*_{-i}$ using the remaining $N_1 + N_2 - 1$ observations. Apply the classification rule $\hat{X}^*_{-i}$ to the omitted observation, and check if it would be classified correctly. This process is repeated $N_1 + N_2$ times, once for each observation. Define $e_{i,-i} = 0$ if group membership of the ith observation was predicted correctly by $\hat{X}^*_{-i}$, and $e_{i,-i} = 1$ otherwise. The leave-one-out error rate ,then is given by

$$\hat{\gamma}_{\text{leave-one-out}} = \frac{1}{N_1 + N_2} \sum_{i=1}^{N_1+N_2} e_{i,-i} \,. \tag{11}$$

Which one of the error rates should be used in a practical situation? We will try to answer this question with some remarks.

The plug-in error rate is usually easy to compute and can be applied to any discrimination rule. Its main disadvantage is that $\hat{\gamma}_{\text{plug-in}}$ tends to be overly optimistic, that is, it tends to underestimate the probability of misclassifying future observations. The following example serves as an illustration.

Example 5.5.3 Suppose that Y is univariate and we sample a single observation from each population. Label these observations y_1 and y_2, and assume $y_1 > y_2$. Define a classification rule for a new observation y as

$$\hat{X}^* = \begin{cases} 1 & \text{if } y \geq \frac{1}{2}(y_1 + y_2), \\ 2 & \text{otherwise.} \end{cases}$$

Then $\hat{\gamma}_{\text{plug-in}} = 0$. This is clearly overly optimistic for the classification of future observations, except in some simple cases where the optimal error rate for the two populations is itself 0. Also see Exercise 4. □

As Example 5.5.3 illustrates, the plug-in error rate is too optimistic because the validation of the procedure (i.e., computation of $\hat{\gamma}_{\text{plug-in}}$) is done on the same data as the estimation of the classification rule itself. In situations with small samples and complicated classification rules that involve estimation of many parameters, the amount of underestimation is particularly bad, as we will see in Section 7.2.

The normal theory error rate is appropriate only in cases where the linear discriminant function is approximately normal with identical variances in both groups. (This does not necessarily mean that the p-dimensional distributions have to be normal; see Exercise 6.) The normal theory error rate can take any value between 0 and $\max(\pi_1, \pi_2)$, as opposed to $\hat{\gamma}_{\text{plug-in}}$ and $\hat{\gamma}_{\text{leave-one-out}}$ which can take only the discrete values $j/(N_1 + N_2)$, $j = 0, 1, \ldots, N_1 + N_2$. However, like the plug-in error rate, $\hat{\gamma}_{\text{normal}}$ tends to be overly optimistic, as seen from the following result. If the distribution of $\mathbf{Y}$ is p-variate normal in two groups, with identical covariance matrices and standard distance Δ between the means, and if D denotes the sample standard distance based on training samples of size N_1 and N_2, then

$$E\left[D^2\right] = \frac{N_1 + N_2 - 2}{N_1 + N_2 - p - 3}\left[\Delta^2 + p\left(\frac{1}{N_1} + \frac{1}{N_2}\right)\right]. \tag{12}$$

The proof of equation (12) for general $p \geq 1$ is beyond the level of this text; see Anderson (1984, Section 6.6). A proof for dimension $p = 1$ is left to the student; see Exercise 7. From equation (12), it follows that $E\left[D^2\right] > \Delta^2$ for all finite sample sizes. The bias of D^2 is more pronounced for small sample sizes. It is straightforward though to remove the bias from D^2; see Exercise 8.

Finally, the leave-one-out error rate is based on the same idea as the plug-in error rate but avoids the pitfalls of $\hat{\gamma}_{\text{plug-in}}$ by giving an "honest" assessment of the performance of a classification rule. "Honesty" is achieved by not letting the ith observation influence the determination of the classification rule $\hat{X}^*_{-i}$. Obviously, the amount of computing necessary to evaluate $\hat{\gamma}_{\text{leave-one-out}}$ is considerable, but, nevertheless, it should be used, whenever possible, instead of $\hat{\gamma}_{\text{plug-in}}$. The student who knows multiple regression may have found similar ideas in the form of "predicted residuals."

Example 5.5.4 This is the continuation of the midge example, with equal prior probabilities. As seen in Figure 5.4.3, $\hat{\gamma}_{\text{plug-in}} = 0$, as all 15 observations are correctly classified. The normal theory error rate is $\hat{\gamma}_{\text{normal}} = \Phi\left(-\frac{1}{2}D\right) = 2.44\%$, which may not be too reliable in view of the remarks preceding this example. Finally, in a leave-one-out procedure the third observation, with antenna length 124 and wing length 172, is wrongly classified in group 2, and therefore, $\hat{\gamma}_{\text{leave-one-out}} = 1/15$. The leave-one-out error rate is a more realistic assessment of the performance of the classification procedure than the plug-in error rate. The misclassification of the third observation is somewhat surprising but can be explained by the fact that it is relatively extreme for group *Af*. See Exercise 9. □

Example 5.5.5 continuation of Example 5.3.2

In the flea beetle example, we used the linear discriminant function and estimated prior probabilities $\hat{\pi}_j = N_j/(N_1 + N_2)$ to get the following error rates, based on all four variables:

$$\hat{\gamma}_{\text{plug-in}} = 1/39 = 2.56\%,$$

$$\hat{\gamma}_{\text{leave-one-out}} = 3/39 = 7.69\%,$$

and

$$\hat{\gamma}_{\text{normal}} = 3.21\%.$$

Thus, the leave-one-out method gives a distinctly more pessimistic assessment. Table 5.5.1 gives more details of the analysis, including posterior probabilities of group membership according to both the analysis using all 39 observations and the analysis omitting the ith observation.

If variable Y_3 is omitted from the analysis, as suggested in Example 5.3.2, the following error rates are obtained:

$$\hat{\gamma}_{\text{plug-in}} = 1/39 = 2.56\%,$$

$$\hat{\gamma}_{\text{leave-one-out}} = 2/39 = 5.13\%,$$

and

$$\hat{\gamma}_{\text{normal}} = 3.66\%.$$

Thus, the leave-one-out assessment of error would support the choice of the reduced model with three variables. □

As Example 5.5.5 shows, the leave-one-out assessment of error, at times, may indicate that a simpler model (i.e., a model based on a subset of the available variables) has better predictive power for future classification than a more complicated one. This reflects an important aspect of multivariate modelling: models with few parameters are often preferable over models with many parameters because estimating a large number of parameters leads to increased variability of parameter estimates, and thereby, to lower predictive power. We will return to this theme in Section 7.2.

We conclude this section with yet another type of error rate which is useful for assessing theoretical properties of classification rules. In the usual setup, let X^*_{opt} denote the optimal classification rule according to Theorem 5.5.1, and write $\gamma_{\text{opt}} = \gamma(X^*_{\text{opt}})$ for the optimal error rate. Let $\hat{X}^*$ denote a classification rule obtained from training samples of size N_1 and N_2, with associated classification regions $\hat{C}_1$ and $\hat{C}_2$.

Table 5.5.1 Details of the computation of error rates in the flea beetle example

i	x_i	$\hat{z}_i$	$\hat{\pi}_{1i}$	$\hat{x}_i^*$	$\hat{\pi}_{1i,-i}$	$\hat{x}_{i,-i}^*$
1	1	11.16	100.00	1	100.00	1
2	1	3.04	95.19	1	84.65	1
3	1	12.21	100.00	1	100.00	1
4	1	7.47	99.94	1	99.90	1
5	1	7.41	99.94	1	99.89	1
6	1	7.49	99.94	1	99.92	1
7	1	9.81	99.99	1	99.99	1
8	1	8.70	99.98	1	99.98	1
9	1	1.54	81.52	1	57.23	1
10	1	7.01	99.91	1	99.87	1
11	1	10.43	100.00	1	100.00	1
12	1	9.12	99.99	1	99.99	1
13	1	8.27	99.97	1	99.96	1
14	1	4.27	98.55	1	97.92	1
15	1	4.88	99.21	1	98.65	1
16	1	4.64	98.99	1	98.48	1
17	1	8.16	99.97	1	99.96	1
18	1	3.27	96.17	1	93.96	1
19	1	1.28	77.37	1	69.71	1
20	2	−11.02	0.00	2	0.00	2
21	2	−1.65	15.39	2	52.91	1*
22	2	−10.57	0.00	2	0.00	2
23	2	−8.53	0.02	2	0.02	2
24	2	−9.89	0.00	2	0.01	2
25	2	−10.86	0.00	2	0.00	2
26	2	−11.68	0.00	2	0.00	2
27	2	0.76	66.96	1*	96.56	1*
28	2	−5.56	0.37	2	0.50	2
29	2	−7.80	0.04	2	0.06	2
30	2	−12.97	0.00	2	0.00	2
31	2	−6.22	0.19	2	0.27	2
32	2	−4.27	1.31	2	2.08	2
33	2	−11.19	0.00	2	0.00	2
34	2	−5.30	0.47	2	0.72	2
35	2	−7.82	0.04	2	0.05	2
36	2	−0.25	42.63	2	61.00	1*
37	2	−7.50	0.05	2	0.09	2
38	2	−1.54	16.98	2	23.58	2
39	2	−3.17	3.84	2	7.80	2

The symbols used are x_i, actual group of observation i; $\hat{z}_i$ = value of the linear discriminant function using all observations; $\hat{\pi}_{1i}$ = estimated posterior probability (percent) in group 1 using the discriminant function based on all observations; $\hat{x}_i^*$ = predicted group membership using the discriminant function based on all observations; $\hat{\pi}_{1i,-i}$ and $\hat{x}_{i,-i}^*$ = analogous quantities using the discriminant function based on all but the ith observation. Wrongly predicted group membership is marked with an asterisk (*).

The error rate of $\hat{X}^*$ is given by

$$\begin{aligned}\gamma(\hat{X}^*) &= \Pr[X \neq \hat{X}^*] \\ &= \Pr[X = 1]\Pr[\hat{X}^* = 2|X = 1] + \Pr[X = 2]\Pr[\hat{X}^* = 1|X = 2] \\ &= \pi_1 \int_{\hat{C}_2} f_1(\mathbf{y})\,\mathbf{dy} + \pi_2 \int_{\hat{C}_1} f_2(\mathbf{y})\,\mathbf{dy} \\ &\geq \gamma_{\text{opt}}\,.\end{aligned} \tag{13}$$

The inequality in (13) follows from the fact that γ_{opt} is the smallest error rate achievable. It is important to interpret $\gamma(\hat{X}^*)$ correctly. It is the probability that a future observation will be misclassified if the given classification rule $\hat{X}^*$ is applied. In the literature, $\gamma(\hat{X}^*)$ is usually called the *actual error rate* of the classification rule, or AER, for short. The actual error rate measures how well the given classification rule will do in the future, but, in any given sample situation, it cannot be computed because the distribution of $\mathbf{Y}$ is not known and the integrals in (13) cannot be evaluated. The plug-in and leave-one-out error rates may be regarded as attempts to estimate the actual error rate of a classification rule $\hat{X}^*$.

The concept of actual error rate refers to the performance of a given classification rule when applied to future observations. However, we might also ask about properties of a classification rule in general, not restricted to a particular pair of training samples. For instance, we might ask how well classification based on the linear discriminant function works, in general. This is analogous to investigating properties of the average of N observations as an estimator of the mean of a distribution. Thus, we seek to find the distribution of the actual error rate $\gamma(\hat{X}^*)$ under repeated generation of training samples.

To clarify this, take, for instance, a classification rule $\hat{X}^*$ based on the linear discriminant function, with known prior probabilities and training samples of fixed size N_1 and N_2. Then we can consider the actual error rate as a function of $N_1 + N_2$ random variables,

$$\gamma(\hat{X}^*) = h(\mathbf{Y}_{11}, \ldots, \mathbf{Y}_{1N_1}, \mathbf{Y}_{21}, \ldots, \mathbf{Y}_{2N_2}), \tag{14}$$

that is, $\gamma(\hat{X}^*)$ is now itself a random variable, typically with a rather complicated distribution. In some cases, one may be able to compute its distribution or at least its expected value. This leads to yet another notion. The *expected actual error rate* of a classification rule $\hat{X}^*$ is defined as

$$E\left[\gamma(\hat{X}^*)\right] = E\left[h(\mathbf{Y}_{11}, \ldots, \mathbf{Y}_{1N_1}, \mathbf{Y}_{21}, \ldots, \mathbf{Y}_{2N_2})\right]. \tag{15}$$

Here, the expectation is taken with respect to the joint distribution of all $\mathbf{Y}_{ij}$ from their respective distributions. The expected actual error rate is a concept that refers to the long-term average performance of a classification method. Having a small expected actual error rate is a desirable property for a classification procedure, just like small mean squared error is a good property of an estimator.

Computing the distribution of AER and the expected actual error rate is typically difficult, so we illustrate these concepts with an example that is artificially simple but allows some explicit calculations.

Example 5.5.6 Suppose that Y follows a uniform distribution in the interval $[0, 1]$ in group 1 and a uniform distribution in $[1, 2]$ in group 2, with equal prior probabilities. The optimal classification rule for this model has error probability zero. Now, suppose that we have training samples of size 1 from each group and denote the two observations as Y_1 and Y_2. Define a classification rule for a future observation Y as

$$\hat{X}^* = \begin{cases} 1 & \text{if } Y \leq C, \\ 2 & \text{otherwise}, \end{cases} \tag{16}$$

where $C = (Y_1 + Y_2)/2$. For the computation of the actual error rate, consider C as fixed, that is,

$$\begin{aligned} \text{AER} = \gamma(\hat{X}^*) &= \Pr[X \neq \hat{X}^*] \\ &= \frac{1}{2}\Pr[Y > C|X = 1] + \frac{1}{2}\Pr[Y \leq C|X = 2] \\ &= \begin{cases} (1 - C)/2 & \text{if } C \leq 1 \\ (C - 1)/2 & \text{if } C > 1 \end{cases} \\ &= \frac{1}{2}|C - 1| . \end{aligned} \tag{17}$$

This is the actual error rate of $\hat{X}^*$, which depends on the two observations in the training samples through their average.

To find the expected actual error rate, we consider $\gamma(\hat{X}^*)$ as a function of two independent random variables Y_1 and Y_2, where Y_1 is uniform in $[0, 1]$ and Y_2 is uniform in $[1, 2]$. That is, we define a random variable

$$G = \gamma(\hat{X}^*) = \frac{1}{2}\left|\frac{1}{2}(Y_1 + Y_2) - 1\right| . \tag{18}$$

In Exercise 12, it is shown that G has *pdf*

$$f_G(g) = \begin{cases} 8(1 - 4g) & \text{if } 0 \leq g \leq \frac{1}{4}, \\ 0 & \text{otherwise}. \end{cases} \tag{19}$$

The expected actual error rate of the classification rule $\hat{X}^*$, therefore, is $E[G] = 1/12$.

Now, suppose that, instead of a single observation, we have $N \geq 2$ observations in each training sample, with averages $\bar{Y}_j = \sum_{i=1}^{N} Y_{ji}$, $j = 1, 2$. Define a classification rule $\hat{X}_N^*$ as in equation (16), where $C = (\bar{Y}_1 + \bar{Y}_2)/2$. Then the actual error rate, again, is given by equation (17). For $N = 2$, an exact calculation of the expected actual error rate is still possible, but complicated. Instead, notice that, for any $N \geq 1$, the random variable $C = (\bar{Y}_1 + \bar{Y}_2)/2 - 1$ has mean and variance

$$E[C] = 0,$$

and

$$\text{var}[C] = \frac{1}{24N}; \tag{20}$$

see Exercise 12. Moreover, for sufficiently large N, the central limit theorem implies that C is approximately normal. Now, the actual error rate G is a function of C, $G = |C|/2$. By the approximate normality of C,

$$E\left[|C|\right] \approx \frac{1}{\sqrt{24N}} \cdot \frac{\sqrt{2}}{\sqrt{\pi}}, \tag{21}$$

and therefore, we obtain the expected actual error rate of $\hat{X}_N^*$ as

$$E[G] \approx \frac{1}{4\sqrt{3\pi N}}. \tag{22}$$

As N tends to infinity, the expected actual error rate tends to $\gamma_{\text{opt}} = 0$, that is, asymptotically the optimal error rate is reached. □

For more realistic classification procedures, such as the classification rule based on the linear discriminant function, it is difficult to find even an approximation to the distribution of the actual error rate. Therefore, problems of comparing expected actual error rates for different methods of classification are often treated by simulation rather than analytical calculation.

Exercises for Section 5.5

1. This exercise refers to Example 5.5.1.
   (a) If the two conditional error probabilities are to be equal ($q_{12} = q_{21}$), show that the cutoff c should be chosen as $c = \frac{1}{2}(\mu_1 + \mu_2)$.
   (b) Suppose we want to find the cutoff c such that $q_{12} = 0.01$. How should c be chosen and what is the error probability q_{21}? Use the parameter values from Figure 5.5.1.
   (c) For the numerical values of π_1, π_2, μ_1, μ_2, and σ used in Example 5.5.1, evaluate the error rate $\gamma(X^*)$ as a function of c for $-3 \leq c \leq 6$, and graph this function. Where does it take its minimum? What happens if $c \to \pm\infty$?

2. Prove equation (7).

3. In the normal theory setup with equal covariance matrices and equal prior probabilities, show that optimal classification is equivalent to: classify $\mathbf{y}$ into group 1 if $\Delta_{\mathbf{Y}}(\mathbf{y}, \boldsymbol{\mu}_1) \le \Delta_{\mathbf{Y}}(\mathbf{y}, \boldsymbol{\mu}_2)$; otherwise, classify into group 2.

4. Consider the following classification procedure for a univariate random variable Y: N_j observations are sampled from population j, $j = 1, 2$, and a classification rule is defined by dividing the real line into two half-lines at a cutoff point c such that the plug-in error rate is minimal. Suppose that the two populations are identical, in fact, and argue that $\hat{\gamma}_{\text{plug-in}}$ is overly optimistic. (Assume that none of the $N_1 + N_2$ observations are identical).

5. Show that, if the two populations are identical, the optimal error rate is $\min(\pi_1, \pi_2)$, irrespective of the form of the distribution.

6. Find an example of a p-variate random vector $\mathbf{Y}$ such that the distribution of $\mathbf{Y}$ in groups 1 and 2 is not multivariate normal, yet the linear discriminant function is univariate normal.

7. Prove equation (12) for $p = 1$. *Hint*: Write the squared standard distance as
$$D^2 = \frac{(\bar{y}_1 - \bar{y}_2)^2/\sigma^2}{s^2/\sigma^2},$$
and use the fact that $\bar{y}_1$, $\bar{y}_2$, and s^2 are independent. Recall that $(N_1 + N_2 - 2)s^2/\sigma^2$ has a chi-square distribution, and find $E\left[\sigma^2/s^2\right]$ assuming $N_1 + N_2 > 4$ (why?).

8. Use equation (12) to find a constant c such that $D_u^2 = cD^2$ is an unbiased estimator of Δ^2.

9. Compute a linear discriminant function in the midge example, omitting the third observation. Draw the boundary line for equal prior probabilities in a graph of Y_1 vs. Y_2, and verify that the omitted observation is misclassified.

10. Compute the plug-in, normal theory, and leave-one-out error rates for the *Microtus* data of Example 5.4.4, using estimated prior probabilities, and
    (a) all eight variables,
    (b) only variables M1LEFT and FORAMEN.

    Do you think it is better to use all eight variables or only M1LEFT and FORAMEN, if the classification rule based on this training sample is to be applied to future observations?

11. Show that the normal theory error rate cannot decrease if additional variables are entered in an analysis. *Hint*: See Exercise 5 in Section 5.3.

12. This exercise refers to Example 5.5.6.
    (a) Show that the random variable G of equation (18) has *pdf* (19), and compute $E[G]$. *Hint*: First, find the *pdf* of $Y_1 + Y_2$, using the technique of Section 2.9 and observing that Y_2 can be represented as $Y_2^* + 1$, where Y_2^* is uniform in the interval [0, 1].

(b) Prove equation (20).

(c) Prove equation (21). *Hint*: Find $E[|Z|]$, where $Z \sim \mathcal{N}(0, 1)$.

13. In Exercises 7 and 8 of Section 5.2, find the error rate of the optimal classification rule as a function of ρ, assuming bivariate normality and equal prior probabilities.

5.6 Linear Discriminant Functions and Conditional Means

In this section we will study properties of the linear discriminant function in some detail. In particular we will investigate the problem of redundancy, i.e. the question, under what circumstances a variable or a set of variables can be excluded from the analysis without loss of information. The section is mostly theoretical, but on a moderate mathematical level. The results presented here are of considerable value for a thorough understanding of linear discriminant analysis. Moreover, they will serve as a basis for tests on discriminant function coefficients to be presented in Section 6.5.

Throughout this section it is assumed that $\mathbf{Y} = (Y_1, \ldots, Y_p)'$ is a p-variate random vector with $E_1[\mathbf{Y}] = \boldsymbol{\mu}_1$ in group 1, $E_2[\mathbf{Y}] = \boldsymbol{\mu}_2$ in group 2, and $\text{Cov}[\mathbf{Y}] = \boldsymbol{\psi}$ in both groups, $\boldsymbol{\psi}$ being positive definite. We will also write $\boldsymbol{\delta} = \boldsymbol{\mu}_1 - \boldsymbol{\mu}_2$ for the vector of mean differences and $\boldsymbol{\beta} = \boldsymbol{\psi}^{-1}\boldsymbol{\delta}$ for the vector of discriminant function coefficients.

The first result tells how the coefficients of the linear discriminant function change if the variables are transformed linearly.

Lemma 5.6.1 *Let* $\mathbf{A}$ *denote a nonsingular matrix of dimension* $p \times p$, *and* $\mathbf{b} \in \mathbb{R}^p$ *a fixed vector. Define* $\mathbf{Z} = \mathbf{A}\mathbf{Y} + \mathbf{b}$, *and denote by* $\boldsymbol{\beta}_\mathbf{Y}$ *and* $\boldsymbol{\beta}_\mathbf{Z}$ *the vectors of discriminant function coefficients based on* $\mathbf{Y}$ *and* $\mathbf{Z}$, *respectively. Then*

$$\boldsymbol{\beta}_\mathbf{Y} = \mathbf{A}'\boldsymbol{\beta}_\mathbf{Z}, \tag{1}$$

or, equivalently,

$$\boldsymbol{\beta}_\mathbf{Z} = (\mathbf{A}')^{-1}\boldsymbol{\beta}_\mathbf{Y}. \tag{2}$$

Proof For all parameters involved, we will use subscripts 'Y' and 'Z' to indicate that they refer to the random vectors $\mathbf{Y}$ and $\mathbf{Z}$, respectively. Since $\mathbf{Z} = \mathbf{A}\mathbf{Y} + \mathbf{b}$,

$$E_1[\mathbf{Z}] = \mathbf{A}\boldsymbol{\mu}_1 + \mathbf{b} \quad \text{in group 1,}$$

$$E_2[\mathbf{Z}] = \mathbf{A}\boldsymbol{\mu}_2 + \mathbf{b} \quad \text{in group 2,}$$

and

$$\text{Cov}[\mathbf{Z}] = \boldsymbol{\psi}_\mathbf{Z} = \mathbf{A}\boldsymbol{\psi}_\mathbf{Y}\mathbf{A}' \quad \text{in both groups.}$$

Hence, the vector of mean differences in $\mathbf{Z}$ is given by $\delta_{\mathbf{Z}} = \mathbf{A}(\mu_1 - \mu_2) = \mathbf{A}\delta_{\mathbf{Y}}$, and the vector of discriminant function coefficients in $\mathbf{Z}$ is given by

$$\begin{aligned}\beta_{\mathbf{Z}} &= \psi_{\mathbf{Z}}^{-1}\delta_{\mathbf{Z}} \\ &= (\mathbf{A}\psi_{\mathbf{Y}}\mathbf{A}')^{-1}\mathbf{A}\delta_{\mathbf{Y}} \\ &= (\mathbf{A}')^{-1}\psi_{\mathbf{Y}}^{-1}\delta_{\mathbf{Y}} \\ &= (\mathbf{A}')^{-1}\beta_{\mathbf{Y}},\end{aligned}$$

as was to be shown. ■

In Section 5.2 we found that the multivariate standard distance is invariant under nonsingular linear transformations and that the squared standard distance between the two mean vectors can be written as

$$\Delta_{\mathbf{Y}}^2 = \beta_{\mathbf{Y}}'\delta_{\mathbf{Y}};$$

see equation (31) in Section 5.2. For a random vector $\mathbf{Z}$ as in Lemma 5.6.1,

$$\begin{aligned}\Delta_{\mathbf{Z}}^2 &= \beta_{\mathbf{Z}}'\delta_{\mathbf{Z}} \\ &= (\beta_{\mathbf{Y}}'\mathbf{A}^{-1})(\mathbf{A}\delta_{\mathbf{Y}}), \\ &= \Delta_{\mathbf{Y}}^2\end{aligned}$$

which confirms the invariance of standard distance.

Next, we partition the random vector $\mathbf{Y}$ into subvectors of dimension q and $p-q$, respectively:

$$\mathbf{Y} = \begin{pmatrix}\mathbf{Y}_1 \\ \mathbf{Y}_2\end{pmatrix}, \tag{3}$$

where

$$\mathbf{Y}_1 = \begin{pmatrix}Y_1 \\ \vdots \\ Y_q\end{pmatrix},$$

and

$$\mathbf{Y}_2 = \begin{pmatrix}Y_{q+1} \\ \vdots \\ Y_p\end{pmatrix}$$

contain the first q and the last $(p-q)$ variables, respectively. Partition the vector of mean differences δ and the covariance matrix ψ analogously, i.e., write

$$\delta = \begin{pmatrix}\delta_1 \\ \delta_2\end{pmatrix},$$

and

$$\psi = \begin{pmatrix} \psi_{11} & \psi_{12} \\ \psi_{21} & \psi_{22} \end{pmatrix}. \tag{4}$$

Furthermore, let

$$\Delta^2_{\mathbf{Y}_1} = \delta_1' \psi_{11}^{-1} \delta_1$$

and

$$\Delta^2_{\mathbf{Y}_2} = \delta_2' \psi_{22}^{-1} \delta_2 \tag{5}$$

denote the squared multivariate standard distances based on $\mathbf{Y}_1$ and $\mathbf{Y}_2$, respectively. Then we can establish the following result, the proof of which is left to the reader (see Exercise 1).

Lemma 5.6.2 *In the setup of equations (3) to (5), if $\psi_{12} = \mathbf{0}$, i.e., if all covariances between $\mathbf{Y}_1$ and $\mathbf{Y}_2$ are zero, then*

(a) *the vector of coefficients of the linear discriminant function based on all p variables is*

$$\beta = \begin{pmatrix} \beta_1 \\ \beta_2 \end{pmatrix} = \begin{pmatrix} \psi_{11}^{-1}\delta_1 \\ \psi_{22}^{-1}\delta_2 \end{pmatrix}; \tag{6}$$

(b) *the p-variate squared standard distance is the sum of the squared standard distances based on $\mathbf{Y}_1$ and $\mathbf{Y}_2$, respectively, i.e.,*

$$\Delta^2(\mu_1, \mu_2) = \delta' \psi^{-1} \delta = \Delta^2_{\mathbf{Y}_1} + \Delta^2_{\mathbf{Y}_2}. \tag{7}$$

Note that equation (7) is not true in general, that is, if $\psi_{12} \neq \mathbf{0}$, then usually,

$$\Delta^2(\mu_1, \mu_2) \neq \Delta^2_{\mathbf{Y}_1} + \Delta^2_{\mathbf{Y}_2},$$

see Exercise 2.

We are now approaching the main result of this section, a characterization of conditions under which a variable or a set of variables can be omitted from the analysis without loss of information. First, we define the notion of redundancy.

Definition 5.6.1 Let $\mathbf{Y} = \begin{pmatrix} \mathbf{Y}_1 \\ \mathbf{Y}_2 \end{pmatrix}$ be partitioned into q and $(p-q)$ components as in equation (3), and let $\beta = \begin{pmatrix} \beta_1 \\ \beta_2 \end{pmatrix}$ denote the vector of discriminant function coefficients, partitioned analogously to $\mathbf{Y}$. Then the variables $\mathbf{Y}_2$ are *redundant* if $\beta_2 = \mathbf{0}$. Equivalently, we say that the subset of variables contained in $\mathbf{Y}_1$ is *sufficient*. □

The definition of redundancy is quite natural. A variable is redundant if it does not appear in the discriminant function, or, in other words, if its omission does not affect the discriminant function. For simplicity in notation, the definition is in terms of redundancy of the last $(p-q)$ variables, but it applies to any subset of variables by rearranging the variables properly.

As in regression analysis, the notion of redundancy of variables is not as trivial as it might seem at first. In particular, it is important to realize that the redundancy of a subset of variables depends crucially on the complete set of variables under consideration. This is best explained by an example.

Example 5.6.1 Suppose that $\mathbf{Y} = \begin{pmatrix} Y_1 \\ Y_2 \end{pmatrix}$ is a bivariate random vector with covariance matrix $\psi = \begin{pmatrix} 2 & 2 \\ 2 & 3 \end{pmatrix}$ and mean vectors $\mu_1 = \begin{pmatrix} 2 \\ 2 \end{pmatrix}$ and $\mu_2 = \begin{pmatrix} 0 \\ 0 \end{pmatrix}$. Then $\beta = \begin{pmatrix} 1 \\ 0 \end{pmatrix}$, i.e., the discriminant function is given by $Z = Y_1$. Therefore, variable Y_2 is redundant in the set of variables $\{Y_1, Y_2\}$. However, if the set of variables consists only of Y_2, then Y_2 is not redundant because its mean difference is not zero.

The opposite phenomenon can also happen, as illustrated by the case $\psi = \begin{pmatrix} 2 & 2 \\ 2 & 3 \end{pmatrix}$, $\delta = \begin{pmatrix} 2 \\ 0 \end{pmatrix}$. Then $\beta = \begin{pmatrix} 3 \\ -2 \end{pmatrix}$, i.e., the discriminant function is given by $Z = 3Y_1 - 2Y_2$. In the set of variables $\{Y_1, Y_2\}$, none of the variables is redundant, although variable Y_2 by itself does not provide any information about differences in location. □

There are also situations where variables are trivially redundant. For instance, if ψ is a diagonal matrix, then any variable with a zero mean difference is redundant in any subset of variables in which it is contained. See Exercise 3.

For the main theorem in this section, we need additional notation. Let $\mathbf{Y} = \begin{pmatrix} \mathbf{Y}_1 \\ \mathbf{Y}_2 \end{pmatrix}$ be partitioned into q and $(p-q)$ variables, and denote by

$$\Delta_q^2 = \delta_1' \psi_{11}^{-1} \delta_1$$

and

$$\Delta_p^2 = \delta' \psi^{-1} \delta \tag{8}$$

the squared standard distances based on the first q variables (Δ_q^2) and based on all p variables (Δ_p^2), respectively. Partition the vector of discriminant function coefficients as $\beta = \begin{pmatrix} \beta_1 \\ \beta_2 \end{pmatrix}$, the vector of mean differences as $\delta = \begin{pmatrix} \delta_1 \\ \delta_2 \end{pmatrix}$, and let

$$\delta_{2.1} = \delta_2 - \psi_{21} \psi_{11}^{-1} \delta_1 \, . \tag{9}$$

Theorem 5.6.3 *The following three conditions are equivalent:*

(a) $\beta_2 = \mathbf{0}$, *i.e., the last $(p-q)$ variables are redundant.*

(b) $\Delta_p^2 = \Delta_q^2$.

(c) $\delta_{2.1} = \mathbf{0}$.

Proof Consider the linear transformation

$$\mathbf{Z} = \begin{pmatrix} \mathbf{Z}_1 \\ \mathbf{Z}_2 \end{pmatrix} = \begin{pmatrix} \mathbf{I}_q & \mathbf{0} \\ -\psi_{21}\psi_{11}^{-1} & \mathbf{I}_{p-q} \end{pmatrix} \begin{pmatrix} \mathbf{Y}_1 \\ \mathbf{Y}_2 \end{pmatrix}, \tag{10}$$

corresponding to $\mathbf{Z}_1 = \mathbf{Y}_1$ and $\mathbf{Z}_2 = \mathbf{Y}_2 - \psi_{21}\psi_{11}^{-1}\mathbf{Y}_1$. Then the vector of mean differences of $\mathbf{Z}$, written in the usual partitioned form, is given by (see Exercise 4)

$$\delta_{\mathbf{Z}} = \begin{pmatrix} \delta_1 \\ \delta_{2.1} \end{pmatrix}, \tag{11}$$

and the covariance matrix of $\mathbf{Z}$ is

$$\psi_{\mathbf{Z}} = \mathrm{Cov}[\mathbf{Z}] = \begin{pmatrix} \psi_{11} & \mathbf{0} \\ \mathbf{0} & \psi_{22.1} \end{pmatrix}, \tag{12}$$

where

$$\psi_{22.1} = \psi_{22} - \psi_{21}\psi_{11}^{-1}\psi_{12}\,. \tag{13}$$

Because the transformation matrix in equation (10) is nonsingular, we can apply Lemmas 5.6.1 and 5.6.2. Denoting the vector of discriminant function coefficients based on $\mathbf{Z}$ as $\beta^* = \begin{pmatrix} \beta_1^* \\ \beta_2^* \end{pmatrix}$,

$$\begin{aligned} \begin{pmatrix} \beta_1^* \\ \beta_2^* \end{pmatrix} &= \begin{pmatrix} \mathbf{I}_q & \psi_{11}^{-1}\psi_{12} \\ \mathbf{0} & \mathbf{I}_{p-q} \end{pmatrix} \begin{pmatrix} \beta_1 \\ \beta_2 \end{pmatrix} \\ &= \begin{pmatrix} \beta_1 + \psi_{11}^{-1}\psi_{12}\beta_2 \\ \beta_2 \end{pmatrix} \\ &= \begin{pmatrix} \psi_{11}^{-1}\delta_1 \\ \psi_{22.1}^{-1}\delta_{2.1} \end{pmatrix}, \end{aligned} \tag{14}$$

and

$$\begin{aligned} \Delta_p^2 &= \delta_1'\psi_{11}^{-1}\delta_1 + \delta_{2.1}'\psi_{22.1}^{-1}\delta_{2.1} \\ &= \Delta_q^2 + \delta_{2.1}'\psi_{22.1}^{-1}\delta_{2.1}\,. \end{aligned} \tag{15}$$

From equation (15) and because $\psi_{22.1}$ is positive definite (see Exercise 6), $\Delta_p^2 = \Delta_q^2$ implies $\delta_{2.1} = \mathbf{0}$. Next, from equation (14), $\delta_{2.1} = \mathbf{0}$ implies $\beta_2 = \mathbf{0}$. Finally, $\beta_2 = \mathbf{0}$ implies

$$\Delta_p^2 = \beta'\delta = (\beta_1' \quad \mathbf{0}') \begin{pmatrix} \delta_1 \\ \delta_2 \end{pmatrix} = \beta_1'\delta_1 = \Delta_q^2\,. \qquad \blacksquare$$

Note that equation (14) implies

$$\beta_2 = \psi_{22.1}^{-1}\delta_{2.1} \tag{16}$$

and, by analogy,

$$\beta_1 = \psi_{11.2}^{-1}\delta_{1.2}\,, \tag{17}$$

where $\delta_{1.2}$ and $\psi_{11.2}$ are defined analogously to (9) and (13). Also note that equation (15) allows for simplified computation of Δ_q^2; see Exercise 18.

Example 5.6.2 For $\mathbf{Y} = \begin{pmatrix} Y_1 \\ Y_2 \end{pmatrix}$ with covariance matrix $\psi = \begin{pmatrix} 2 & 2 \\ 2 & 3 \end{pmatrix}$ and mean difference $\delta = \begin{pmatrix} 2 \\ 2 \end{pmatrix}$, as in Example 5.6.1, $\delta_{2.1} = 0$, and $\Delta_2 = \Delta_1 = \sqrt{2}$. □

In Section 6.5, we will present a test for the hypothesis $\beta_2 = \mathbf{0}$, i.e., the hypothesis of redundancy of $(p-q)$ variables, and we will see that it is based on a comparison of D_p and D_q, the sample analogs of the quantities Δ_p and Δ_q just discussed. For now, we will relate discriminant function coefficients to conditional means, and thus, link linear discriminant analysis with regression analysis. In Theorem 2.7.1, we saw that, for two jointly distributed, random variables Y and Z, if $E[Z|Y=y] = \alpha + \beta y$ for some $\alpha \in \mathbb{R}$ and $\beta \in \mathbb{R}$, then

$$\beta = \frac{\text{cov}[Y,Z]}{\text{var}[Y]}\,, \qquad \alpha = E[Z] - \beta\, E[Y]\,. \tag{18}$$

Now, we will state a multivariate generalization of this result, leaving the proof as an exercise for the mathematically inclined reader; see Exercise 7. A particular case where all assumptions of Theorem 5.6.4 hold for all subsets of variables is given by the multivariate normal distribution. However, the multivariate normal distribution is by no means the only one where the theorem can be used; see Exercise 8.

Theorem 5.6.4 *Let* $\mathbf{Y} = \begin{pmatrix} \mathbf{Y}_1 \\ \mathbf{Y}_2 \end{pmatrix}$ *denote a p-variate random vector with mean* $\boldsymbol{\mu} = \begin{pmatrix} \boldsymbol{\mu}_1 \\ \boldsymbol{\mu}_2 \end{pmatrix}$ *and covariance matrix* $\psi = \begin{pmatrix} \psi_{11} & \psi_{12} \\ \psi_{21} & \psi_{22} \end{pmatrix}$, *partitioned into q and* $(p-q)$ *components. Assume that the conditional mean of* $\mathbf{Y}_2$, *given* $\mathbf{Y}_1 = \mathbf{y}_1$, *is a linear function of* $\mathbf{y}_1$, *i.e.,* $E[\mathbf{Y}_2|\mathbf{Y}_1 = \mathbf{y}_1] = \mathbf{c} + \mathbf{D}\mathbf{y}_1$, *where* $\mathbf{c} \in \mathbb{R}^{p-q}$ *and* $\mathbf{D}$ *is a matrix of dimension* $(p-q)\times q$. *Then*

$$\mathbf{D} = \psi_{21}\psi_{11}^{-1},$$

and

$$\mathbf{c} = \boldsymbol{\mu}_2 - \mathbf{D}\boldsymbol{\mu}_1\,. \tag{19}$$

Now, we are going to apply Theorem 5.6.4 to the setup of two groups that differ in location but not in their covariance matrix ψ. Writing

$$\mu^{(j)} = \begin{pmatrix} \mu_1^{(j)} \\ \mu_2^{(j)} \end{pmatrix}, \quad j = 1, 2, \tag{20}$$

for the mean vector in the jth group, and assuming linearity of the conditional means, Theorem 5.6.4 says that

$$E_1[\mathbf{Y}_2|\mathbf{Y}_1 = \mathbf{y}_1] = \mu_2^{(1)} + \psi_{21}\psi_{11}^{-1}(\mathbf{y}_1 - \mu_1^{(1)}) \quad \text{in group 1}$$

and

$$E_2[\mathbf{Y}_2|\mathbf{Y}_1 = \mathbf{y}_1] = \mu_2^{(2)} + \psi_{21}\psi_{11}^{-1}(\mathbf{y}_1 - \mu_1^{(2)}) \quad \text{in group 2.} \tag{21}$$

Conditionally on $\mathbf{Y}_1 = \mathbf{y}_1$, the mean difference between the two groups for variable $\mathbf{Y}_2$ is given by

$$\begin{aligned} \delta_{\mathbf{Y}_2}(\mathbf{y}_1) &= E_2[\mathbf{Y}_2|\mathbf{Y}_1 = \mathbf{y}_1] - E_1[\mathbf{Y}_2|\mathbf{Y}_1 = \mathbf{y}_1] \\ &= \left(\mu_2^{(1)} - \mu_2^{(2)}\right) - \psi_{21}\psi_{11}^{-1}\left(\mu_1^{(1)} - \mu_1^{(2)}\right) \\ &= \delta_2 - \psi_{21}\psi_{11}^{-1}\delta_1 \\ &= \delta_{2.1}, \end{aligned} \tag{22}$$

as in equation (9). Note that $\delta_{2.1}$ is a vector of constants that does not depend on $\mathbf{y}_1$, showing that the regression lines or planes are parallel. Equation (22) provides us with an interesting interpretation of the vector $\delta_{2.1}$, which we first encountered in Theorem 5.6.3. It is the vector of conditional mean differences of $\mathbf{Y}_2$, given $\mathbf{Y}_1$, under the assumption of linearity of the conditional expectation.

A related interpretation can be attached to the matrix $\psi_{22.1}$. We have already seen this type of matrix in conditional normal distributions; see Section 3.3. The following corollary to Theorem 5.6.4 shows that the interpretation of $\psi_{22.1}$ as the covariance matrix of a conditional distribution is not restricted to the case of multivariate normality. The proof, again, is left to the advanced student as an exercise; see Exercise 7.

Corollary 5.6.5 *Under the assumptions of Theorem 5.6.4, suppose that* $\mathrm{Cov}[\mathbf{Y}_2|\mathbf{Y}_1 = \mathbf{y}_1]$ *does not depend on* $\mathbf{y}_1$. *Then*

$$\mathrm{Cov}[\mathbf{Y}_2|\mathbf{Y}_1 = \mathbf{y}_1] = \psi_{22.1} = \psi_{22} - \psi_{21}\psi_{11}^{-1}\psi_{12}\,. \tag{23}$$

Thus, we have found a rather interesting relationship between regression and discriminant analysis, valid under the assumption of Theorem 5.6.4 and Corollary 5.6.5: For any subset of variables, the discriminant function coefficients of the variables in the subset are functions of the mean and covariance matrix of the variables in the subset, given all other variables, as expressed in equations (16) and (17). In particular, for any single variable Y_h, its coefficient in the linear discriminant function is the same as the mean difference divided by the variance of the conditional distribution of Y_h, given all other variables. This demonstrates, again, that discriminant function coefficients depend always on the full set of variables considered. If the set of variables changes, then the coefficients of the discriminant function may also change. A variable is redundant exactly if its conditional mean difference, given all other variables, is zero. If variables are added to or removed from the set of variables to be used, then formerly important variables may become redundant, or vice versa. See also Exercise 10.

Example 5.6.3 continuation of Example 5.6.2

Assume that the distribution of $\mathbf{Y} = \begin{pmatrix} Y_1 \\ Y_2 \end{pmatrix}$ is such that Theorem 5.6.4 can be applied to the conditional distributions in both groups. Then $E[Y_2|Y_1 = y_1] = y_1$ in both groups, i.e., the two parallel regression lines of Y_2 on Y_1 coincide, and Y_2 is redundant. Conversely, the conditional expectation of Y_1, given $Y_2 = y_2$, is given by

$$E_1[Y_1|Y_2 = y_2] = \tfrac{2}{3} + \tfrac{2}{3}y_2 \quad \text{in group 1}$$

and

$$E_2[Y_1|Y_2 = y_2] = 0 + \tfrac{2}{3}y_2 \quad \text{in group 2,}$$

and the conditional mean difference is $\delta_{1.2} = \frac{2}{3}$. Figure 5.6.1 shows a plot of both the regressions of Y_2 on Y_1, and Y_1 on Y_2, where Y_1 and Y_2 are jointly normal. Exercise 17 gives a generalization of this example. □

Example 5.6.4 In Example 5.3.2, we studied the two species of flea beetles *Haltica oleracea* and *H. carduorum*, and discriminated between them using four variables. In the current example, we use only the two variables Z_1 = length of the second antennal joint and Z_2 = length of the elytra (these variables were labelled X_3 and X_2 in Example 5.3.2). Substituting the pooled sample covariance matrix $\mathbf{S}$ for ψ and the vector of sample mean differences $\mathbf{d}$ for $\boldsymbol{\delta}$,

$$\mathbf{S} = \begin{pmatrix} 118.31 & 121.88 \\ 121.88 & 367.79 \end{pmatrix},$$

$$\mathbf{d} = \begin{pmatrix} d_1 \\ d_2 \end{pmatrix} = \begin{pmatrix} -19.83 \\ -23.75 \end{pmatrix},$$

and the vector of discriminant function coefficients is given by $\mathbf{b} = \mathbf{S}^{-1}\mathbf{d} = \begin{pmatrix} 0.153 \\ 0.014 \end{pmatrix}$. Variable Z_2 has a much smaller coefficient than Z_1 and seems "almost" redundant. Calculating the sample analog of equation (9), we get the conditional mean difference $d_{2.1} = d_2 - (s_{21}/s_{11})d_1 = -3.32$. Without an associated measure of variability, the

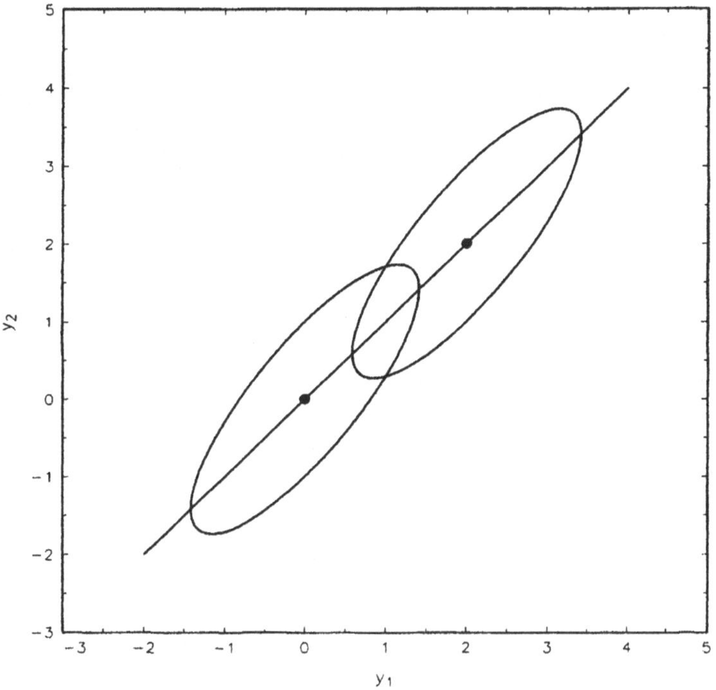

Figure 5-6-1A

value of the conditional mean difference does not really tell whether the two regression lines should be assumed to coincide; this is left as an exercise (see problem 11). For the time being, a better informal assessment of redundancy may be based on a comparison of standard distances. Writing D_1 for the standard distance based on variable Z_1, and D_2 for the standard distance based on both variables, $D_1 = |d_1|/s_{11} = 1.82$, and $D_2 = (\mathbf{d}'\mathbf{S}^{-1}\mathbf{d})^{1/2} = (\mathbf{b}'\mathbf{d})^{1/2} = 1.84$. Hence, the standard distance increases only by a very small amount if variable Z_2 is included in addition to variable Z_1. A scatterplot of the two variables (see Exercise 11) helps to understand this better. □

Example 5.6.5 This is an abstract version of a seeming paradox that has puzzled lawyers and statisticians alike, the so-called discrimination paradox. To add to the confusion, the discrimination paradox has nothing to do with discriminant analysis, but rather with regression. However, the connection between the coefficients of the linear discriminant function and conditional mean differences can help in understanding the paradox better, as we shall outline. Another rather puzzling aspect (at least to the novice) is that the phenomenon is much easier to understand in the abstract form than using data.

The paradox can be described as follows. A company is accused of discriminating against its female employees, based on a study of salary (S) and qualification (Q). It turns out that, for each level of qualification, the average male income is higher than the average female income, so the ladies' claim has a solid base. Interestingly, as the

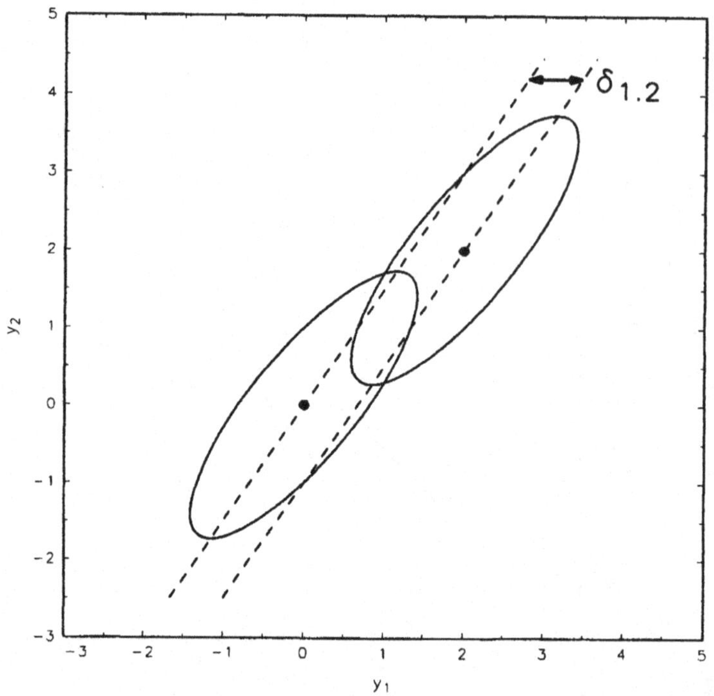

Figure 5-6-1B
(a) Regression of Y_2 on Y_1 and (b) regression of Y_1 on Y_2 in two groups. The conditional mean difference $\delta_{1.2}$ is the horizontal distance between the two lines corresponding to the groupwise regressions of Y_1 on Y_2.

men look at the same data, they discover that, for each level of salary, the average qualification of female employees is lower than the average qualification of their male colleagues. Hence, men are discriminated against, not women.

To understand how this is possible, consider the following simplified setup. Let $\begin{pmatrix} S \\ Q \end{pmatrix}$ denote a bivariate random vector with mean $\begin{pmatrix} \mu_1 \\ \mu_1 \end{pmatrix}$ in group 1 (males), mean $\begin{pmatrix} \mu_2 \\ \mu_2 \end{pmatrix}$ in group 2 (females), and common covariance matrix $\begin{pmatrix} 1 & \rho \\ \rho & 1 \end{pmatrix}$. Assume that $\mu_1 > \mu_2$ and $\rho > 0$, since qualification and salary are supposed to be positively related. Assume, furthermore, that the joint distribution of S and Q is such that, in each group, the regression of S on Q and the regression of Q on S are linear. By equation (18),

- Group 1 (male): $f_1(q) := E_1[S|Q = q] = \mu_1 + \rho(q - \mu_1) = (1 - \rho)\mu_1 + \rho\, q\,.$
- Group 2 (female): $f_2(q) := E_2[S|Q = q] = (1 - \rho)\mu_2 + \rho\, q\,.$

The conditional mean difference in salary, given qualification, is then is given by

$$\delta_{S|q} = f_1(q) - f_2(q) = (1 - \rho)(\mu_1 - \mu_2)\,.$$

As $\mu_1 > \mu_2$, the conditional mean difference is positive, and hence, women are discriminated against. Note that $\delta_{S|q}$ is strictly greater than zero, except in the degenerate case $\rho = 1$.

Let us turn around and look at average qualification, given salary. By the same calculations as above, we get the following:

- Group 1 (male): $g_1(s) := E_1[Q|S = s] = (1 - \rho)\mu_1 + \rho s$.
- Group 2 (female): $g_2(s) := E_2[Q|S = s] = (1 - \rho)\mu_2 + \rho s$.

The conditional mean difference is given by

$$\delta_{Q|s} = g_1(s) - g_2(s) = (1 - \rho)(\mu_1 - \mu_2).$$

This conditional mean difference, again, is positive, and therefore, men are discriminated against.

All four regression lines are shown in Figure 5.6.2. The paradox occurs for all positive values of $\rho \neq 1$, i.e., as long as the relationship between S and Q is not deterministic. Actually, in the abstract setup of Figure 5.6.2, there is no paradox at all — it is really just our tendency to interpret a relationship that holds "on the average" (i.e., a regression function) in a deterministic way.

Now we turn to the discriminant function. The vector of discriminant function coefficients is proportional to $(1, 1)'$, so we can take $V = S + Q$ as the discriminant

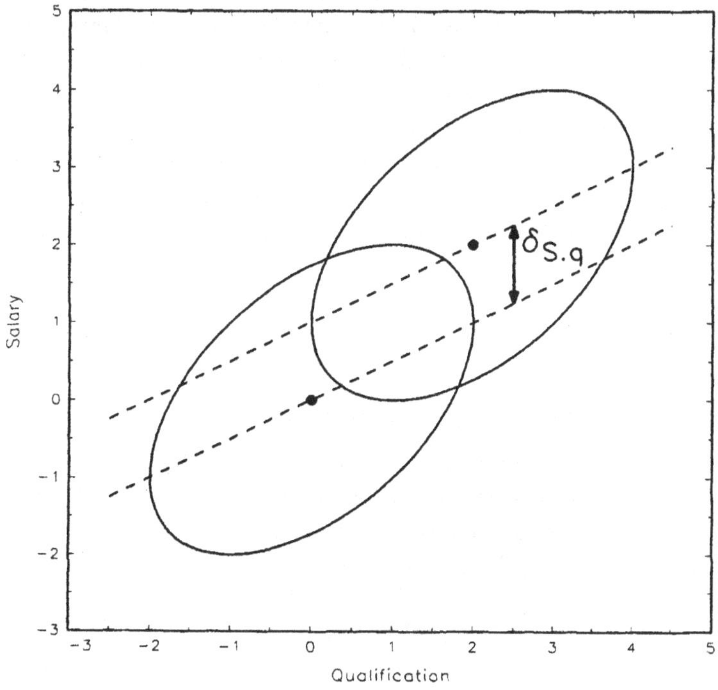

Figure 5-6-2A

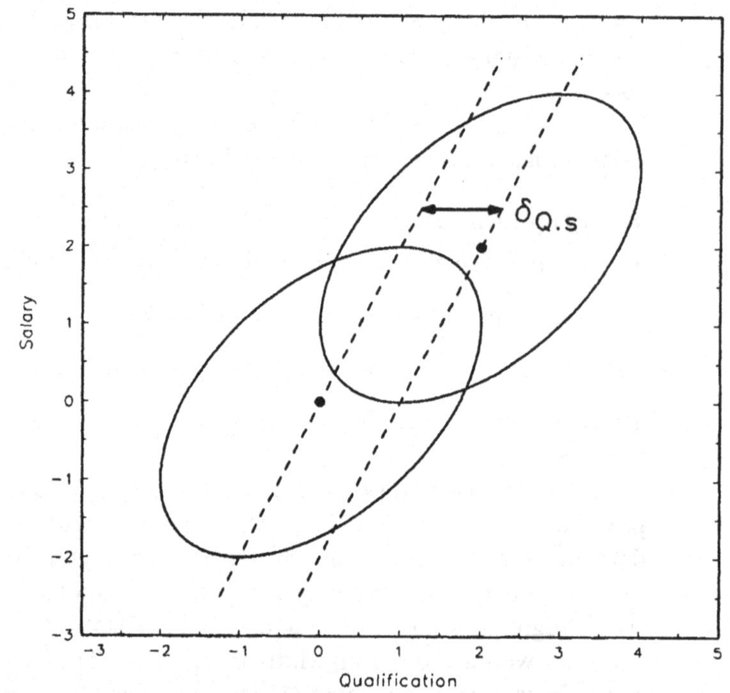

Figure 5-6-2B (a) Regression of S on Q and (b) regression of Q on S in the discrimination paradox example, for two bivariate normal distributions with common covariance matrix and correlation $\rho = 0.5$.

function. Since $\mu_1 > \mu_2$ was assumed, men are expected to have larger values of V than women. The discriminant function says that the higher S and the higher Q, the more likely it is that the person is male. There is nothing paradoxical about this, but the connection between linear discriminant functions and conditional mean differences tells us that if both discriminant function coefficients have the same sign, then both conditional mean differences have the same sign, and hence, the paradox must occur.

In this example we used a particularly simple setup for the parameters. In Exercise 14, this is generalized. Further aspects are discussed in Exercises 15 and 16. □

Exercises for Section 5.6

1. Prove equations (6) and (7)

2. In the notation of Lemma 5.6.2, find examples of covariance matrices ψ and mean difference vectors $\boldsymbol{\delta}$ such that
   (a) $\Delta^2(\mu_1, \mu_2) < \Delta^2_{Y_1} + \Delta^2_{Y_2}$,
   (b) $\Delta^2(\mu_1, \mu_2) > \Delta^2_{Y_1} + \Delta^2_{Y_2}$, and
   (c) $\Delta^2(\mu_1, \mu_2) = \Delta^2_{Y_1} + \Delta^2_{Y_2}$ but $\psi_{12} \neq 0$. *Hint*: Take $\psi = \begin{pmatrix} 1 & \rho \\ \rho & 1 \end{pmatrix}$, and use equation (15).

3. Prove the following results for a random vector $\mathbf{Y}$ with mean difference vector $\delta = (\delta_1, \ldots, \delta_p)'$ and covariance matrix ψ.
   (a) If ψ is diagonal, then the squared p-variate standard distance is the sum of the univariate squared standard distances.
   (b) If ψ is diagonal, then variable Y_j is redundant exactly if $\delta_j = 0$.
   (c) If variable Y_j is uncorrelated with all other variables and $\delta_j = 0$, then Y_j is redundant in *any* subset of variables. *Note*: Part (c) confirms that adding "pure noise" to the list of variables does not improve discrimination.

4. Prove equation (11).

5. Suppose that the p-variate random vector $\mathbf{Y}$ is partitioned into q variables $\mathbf{Y}_1$ and $(p-q)$ variables $\mathbf{Y}_2$, and $\mathbf{Y}_2$ is redundant for discrimination. Let $\mathbf{A}_1$ and $\mathbf{A}_2$ denote nonsingular matrices of dimension $q \times q$ and $(p-q) \times (p-q)$, and set $\mathbf{Z} = \begin{pmatrix} \mathbf{A}_1 & \mathbf{O} \\ \mathbf{O} & \mathbf{A}_2 \end{pmatrix} \begin{pmatrix} \mathbf{Y}_1 \\ \mathbf{Y}_2 \end{pmatrix}$. Show that if discrimination is based on $\mathbf{Z}$, then $\mathbf{Z}_2$ is redundant.

6. Show that positive definiteness of ψ implies that $\psi_{11.2}$ and $\psi_{22.1}$ are positive definite.

7. Prove (a) Theorem 5.6.4 and (b) Corollary 5.6.5, using the following techniques.
   (a) Theorem 5.6.4: Construct a proof for the continuous case along the lines of the proof of Theorem 2.7.1.
   (b) Corollary 5.6.5: First, without making assumptions of linearity of conditional means, show that,

$$\text{Cov}[Y_2] = \text{Cov}\big[E[Y_2|Y_1]\big] + E\big[\text{Cov}[Y_2|Y_1]\big],$$

   which is a generalization of Theorem 2.7.3. Then use Theorem 5.6.4 to evaluate $\text{Cov}\,[E[Y_2|Y_1]]$.

8. Consider a bivariate continuous random variable (Y, Z) with density function

$$f_{YZ}(y, z) = \begin{cases} \frac{1}{20}\exp(-|z-y|) & \text{if } 0 \le y \le 10, \quad z \in \mathbb{R}, \\ 0 & \text{otherwise.} \end{cases}$$

   (a) Show that the regression of Z on Y is linear.
   (b) Show that $\text{var}[Z|Y = y]$ does not depend on y.

9. Show that Theorem 5.6.4 implies equation (18) in the case $p = 2$ and $q = 1$.

10. Suppose $(X, Y, Z)'$ is multivariate normal in two groups, with vector of mean differences $(\delta_X, \delta_Y, \delta_Z) = (0, 2, 3)$ and common covariance matrix

$$\psi = \begin{pmatrix} 2 & 0 & 1 \\ 0 & 2 & 1 \\ 1 & 1 & 2 \end{pmatrix}.$$

   Show that
   (a) X is redundant by itself and in the set $\{X, Y\}$, but not redundant in the set $\{X, Z\}$ and in the full set $\{X, Y, Z\}$, and

(b) Y is redundant in the full set $\{X, Y, Z\}$, but not redundant by itself and in the subsets $\{X, Y\}$ and $\{Y, Z\}$.

11. This exercise refers to the flea beetle data of Example 5.6.4, using variables Y_1 = length of the second antennal joint and Y_2 = length of the elytra.

(a) Fit two parallel regression lines of Y_2 on Y_1, i.e., find the least squares estimates of the parameters in the model

$$E[Y_2|Y_1 = y_1] = \alpha_1 + \beta y_1 \qquad \text{in group 1}$$

and

$$E[Y_2|Y_1 = y_1] = \alpha_2 + \beta y_1 \qquad \text{in group 2.}$$

Hint: Write the parallel regression model as $E[Y_2] = \gamma_0 + \gamma_1 Y_1 + \gamma_2 B$, where B is a binary variable (indicator variable) taking value 1 for the specimens in group 1 and value 0 for those in group 2.

(b) Verify that the least squares estimator of γ_2 is $\hat{\gamma}_2 = d_{2.1}$. *Note*: This will be considered again in Section 6.5.

(c) Plot Y_1 vs Y_2, marking group membership with two different symbols, and draw the two parallel regression lines.

(d) Test the hypothesis $\gamma_2 = 0$ (and hence, the hypothesis of redundancy of Y_2 in the linear discriminant function), using a partial t-test from multiple regression and a significance level of five percent.

(e) In the same graph as in part (c), plot the classification boundary associated with the linear discriminant function $0.153Y_1 + 0.014Y_2$, assuming equal prior probabilities and equal cost of misclassification.

12. Figure 5.6.3 gives six typical patterns of linear discrimination. In each case, the two ellipses represent contours of equal density of two bivariate normal distributions centered at means $\boldsymbol{\mu}_1$ and $\boldsymbol{\mu}_2$, with common covariance matrix $\boldsymbol{\psi}$. For simplicity, $\boldsymbol{\mu}_2 = \mathbf{0}$ (and hence, $\boldsymbol{\delta} = \boldsymbol{\mu}_1 - \boldsymbol{\mu}_2 = \boldsymbol{\mu}_1$) is assumed in all six cases. The remaining parameters are as follows:

(i) $\boldsymbol{\mu}_1 = \begin{pmatrix} 2 \\ 2 \end{pmatrix}, \quad \boldsymbol{\psi} = \begin{pmatrix} 1 & 0 \\ 0 & 1 \end{pmatrix}.$

(ii) $\boldsymbol{\mu}_1 = \begin{pmatrix} 2 \\ 2 \end{pmatrix}, \quad \boldsymbol{\psi} = \begin{pmatrix} 2 & 2 \\ 2 & 3 \end{pmatrix}.$

(iii) $\boldsymbol{\mu}_1 = \begin{pmatrix} 2 \\ 2 \end{pmatrix}, \quad \boldsymbol{\psi} = \begin{pmatrix} 2 & 1.9 \\ 1.9 & 2 \end{pmatrix}.$

(iv) $\boldsymbol{\mu}_1 = \begin{pmatrix} 2 \\ 0 \end{pmatrix}, \quad \boldsymbol{\psi} = \begin{pmatrix} 2 & -1.6 \\ -1.6 & 2 \end{pmatrix}.$

(v) $\boldsymbol{\mu}_1 = \begin{pmatrix} 2 \\ 0 \end{pmatrix}, \quad \boldsymbol{\psi} = \begin{pmatrix} 2 & 0 \\ 0 & 4 \end{pmatrix}.$

(vi) $\boldsymbol{\mu}_1 = \begin{pmatrix} 1 \\ 1 \end{pmatrix}, \quad \boldsymbol{\psi} = \begin{pmatrix} 2 & -1.8 \\ -1.8 & 2 \end{pmatrix}.$

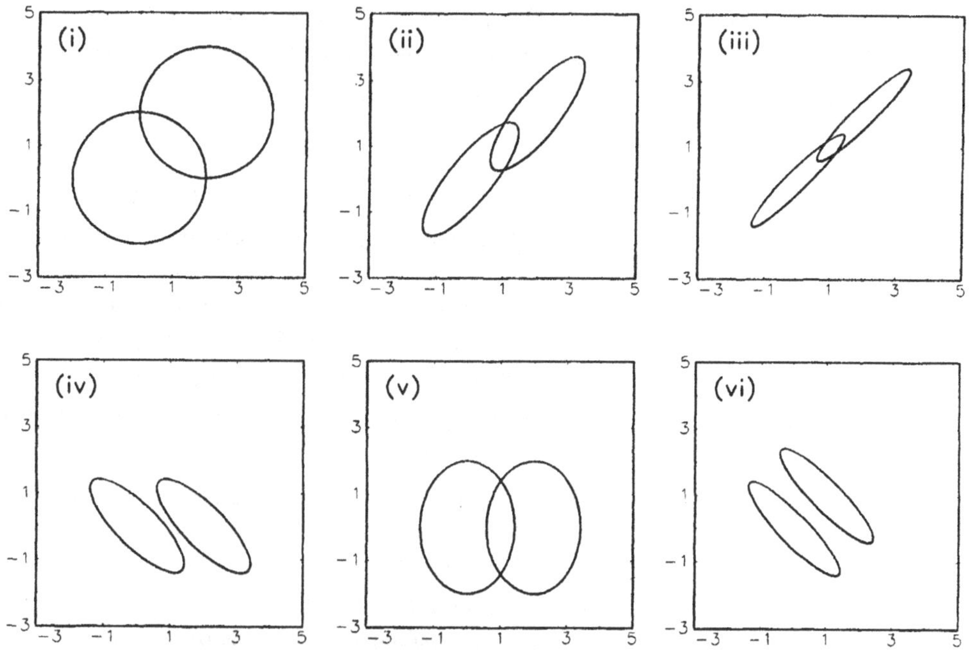

Figure 5-6-3
Six typical patterns in linear discrimination.

For each pattern, do the following:

(a) Compute the two univariate standard distances.

(b) Compute the bivariate standard distance and the linear discriminant function.

(c) Decide if any of the two variables is redundant, either by itself or in combination with the other variable.

(d) In the associated part of Figure 5.6.3, draw the classification boundary for equal prior probabilities.

13. Let

$$f(y;\alpha) = \begin{cases} \frac{1}{10} & \text{if } \alpha \le y \le \alpha + 10, \\ 0 & \text{otherwise} \end{cases}$$

and

$$g(z;\beta) = \frac{1}{\sqrt{2\pi}} \exp[-\tfrac{1}{2}(z-\beta)^2], \quad z \in \mathbb{R},$$

denote two univariate density functions.

(a) Show that

$$h(y, z; \alpha, \beta) := f(y;\alpha) g(z; \beta + y/2)$$

is a density function of a bivariate random variable (Y, Z).

(b) Find the regression of Z on Y for (Y, Z) with density function $h(y, z; \alpha, \beta)$.

(c) Suppose that (Y, Z) has density h as in part (a) but is measured in two groups with equal prior probabilities and with parameters

$$\alpha = 0, \quad \beta = 0 \qquad \text{in group 1,}$$

and

$$\alpha = 0, \quad \beta = 2 \qquad \text{in group 2.}$$

Show that the regression lines of Z on Y are parallel and the covariance matrix of (Y, Z) is the same in both groups.

(d) For the parameter values in part (c), show that the optimal classification boundary according to Theorem 5.5.1 is a straight line.

(e) Same as part (c), but with parameter values $\alpha = 5$, $\beta = 2$ in group 2.

(f) For the parameter values in part (e), show that the optimal classification boundary is not a straight line.

14. In the setup of Example 5.6.5, transform the variables as $Q^* = a + bQ$ and $S^* = a + cS$, where a, b, and c are fixed constants, $b > 0$, $c > 0$. Show that the "paradox" occurs for variables S^* and Q^*.

15. In the setup of Example 5.6.5, suppose you can add an amount of α to every lady's salary, thus, making the mean vector for females

$$E_2\left[\begin{pmatrix} S \\ Q \end{pmatrix}\right] = \begin{pmatrix} \mu_2 + \alpha \\ \mu_2 \end{pmatrix}.$$

(a) How should you choose α such that the two parallel regression lines coincide?

(b) With α chosen as in part (a), find the conditional mean difference of Q, given S.

(c) With α chosen as in part (a), show that the discriminant function has coefficient 0 for variable S.

16. Let $\mathbf{Y} = \begin{pmatrix} Y_1 \\ Y_2 \end{pmatrix}$ denote a bivariate random vector with

$$E_j[\mathbf{Y}] = \mu_j\,, \quad \text{Cov}[\mathbf{Y}] = \psi_j\,,$$

for groups $j = 1, 2$.

(a) Assume that the following is known:

- The conditional mean of Y_1, given $Y_2 = y_2$, is a linear function of y_2 in both groups, with identical slopes.
- The conditional mean of Y_2, given $Y_1 = y_1$, is a linear function of y_1 in both groups, with identical slopes.

Show that ψ_1 and ψ_2 are proportional, i.e., $\psi_2 = c\,\psi_1$ for some $c > 0$. *Hint*: Read Section 2.7.

(b) If ψ_1 and ψ_2 are proportional and it is known that all regression functions are linear, show that they are parallel.

17. This exercise is similar to Exercise 13 in Section 5.2 and generalizes Examples 5.6.1 to 5.6.3. Let Y_0 denote a univariate random variable with mean μ and variance σ^2. Define $k \geq 1$ random variables $Y_1, \ldots, Y_k$ as $Y_h = Y_0 + e_h$, $h = 1, \ldots, k$, where all e_h are mutually independent, independent also of Y_0, and identically distributed, with mean zero and variance $\tau^2 > 0$. Define $\mathbf{Y} = (Y_0, Y_1, \cdots, Y_k)'$. Let $\mathbf{1}_m \in \mathbb{R}^m$ denote a vector with 1 in all entries.

(a) Show that $E[\mathbf{Y}] = \mu\, \mathbf{1}_k$ and

$$\text{Cov}[\mathbf{Y}] = \begin{pmatrix} \sigma^2 & \sigma^2 & \sigma^2 & \cdots & \sigma^2 \\ \sigma^2 & \sigma^2+\tau^2 & \sigma^2 & \cdots & \sigma^2 \\ \sigma^2 & \sigma^2 & \sigma^2+\tau^2 & \cdots & \sigma^2 \\ \vdots & \vdots & \vdots & \ddots & \vdots \\ \sigma^2 & \sigma^2 & \sigma^2 & \cdots & \sigma^2+\tau^2 \end{pmatrix} = \begin{pmatrix} \sigma^2 & \sigma^2 \mathbf{1}_k' \\ \sigma^2 \mathbf{1}_k & \psi(\sigma, \tau) \end{pmatrix}, \qquad (24)$$

where $\psi(\sigma, \tau)$ is defined as in equation (38) of Section 5.2.

(b) Suppose that $\mathbf{Y}$ is measured in two groups, with mean vectors $E_1[\mathbf{Y}] = \boldsymbol{\mu}_1 = \mu_1 \mathbf{1}_{k+1}$ in group 1, $E_2[\mathbf{Y}] = \boldsymbol{\mu}_2 = \mu_2 \mathbf{1}_{k+1}$ in group 2, and common covariance matrix of the form (24). Show that variables Y_1 to Y_k are all redundant, i.e., linear discrimination depends on $\mathbf{Y}$ only through Y_0. *Hint*: Although it is possible to prove this result by inversion of the covariance matrix and explicit calculation of the vector of discriminant function coefficients, it is easier to use the equivalent condition given by part (c) of Theorem 5.6.3, that is, show that the conditional mean difference of Y_h, given Y_0, is zero for $h = 1, \ldots, k$.

(c) Compare this example to Exercise 13 in Section 5.2. Compute the multivariate standard distance between $\boldsymbol{\mu}_1$ and $\boldsymbol{\mu}_2$, and comment.

18. Let $\mathbf{Y} = \begin{pmatrix} \mathbf{Y}_1 \\ \mathbf{Y}_2 \end{pmatrix}$ be partitioned into q and $(p-q)$ variables, let $\boldsymbol{\beta} = \begin{pmatrix} \boldsymbol{\beta}_1 \\ \boldsymbol{\beta}_2 \end{pmatrix} = \psi^{-1}\boldsymbol{\delta}$ denote the the vector of coefficients of the linear discriminant function, and use the notation and setup of equations (8) and (9).

(a) If $\begin{pmatrix} \psi^{11} & \psi^{12} \\ \psi^{21} & \psi^{22} \end{pmatrix}$ denotes the inverse of ψ, show that

$$\Delta_q^2 = \Delta_p^2 - \boldsymbol{\beta}_2' \left(\psi^{22}\right)^{-1} \boldsymbol{\beta}_2 \,. \qquad (25)$$

Hint: Use equations (15) and (16), and recall that $\psi^{22} = \psi_{22.1}^{-1}$.

(b) Let Δ_{-j} denote the standard distance between the two mean vectors based on all variables except Y_j. Write $\boldsymbol{\beta} = (\beta_1, \ldots, \beta_p)'$ for the vector of discriminant function coefficients. Show that

$$\Delta_p^2 - \Delta_{-j}^2 = \beta_j^2 / a_{jj}\,, \qquad (26)$$

where a_{jj} is the jth diagonal entry of $\mathbf{A} = \psi^{-1}$.

Suggested Further Reading

General: Hand (1981), Mardia et al. (1979), McLachlan (1992), Seber (1984).
Sections 5.2 and 5.3: Fisher (1936).
Section 5.4: Anderson (1984, Chapter 6).
Section 5.5: Lachenbruch (1975), Lachenbruch and Mickey (1968).
Section 5.6: Kaye and Aickin (1986), Rao (1970).

6 Statistical Inference for Means

6.1 Introduction

In this chapter we study selected problems of hypothesis testing and confidence regions in multivariate statistics. We will focus mostly on T^2-tests, or *Hotelling's* T^2, after the statistician Harold Hotelling (1895–1973). In the spirit of this book, which emphasizes parameter estimation more than testing, we will give rather less attention to aspects of hypotheses testing than traditional textbooks on multivariate statistics. In particular, we will largely ignore problems like optimality criteria or power of tests. Instead, we will focus on a heuristic foundation to the T^2-test methodology, for which we are well prepared from Chapter 5.

Traditional parametric statistics, as taught in many academic institutions, is often perverted into a mechanism of calling certain numbers (e.g., means, mean differences, or coefficients of a regression function) "significant" or "not significant," without reflecting on the meaning of these words. But tests of significance by themselves are really only a small part of a statistical analysis. Imagine, for instance, a biologist who spends two years catching and measuring animals of two species. Certainly the biologist would hope to get more information out of the data than just a statement like "the mean difference between the two species is significant." The descriptive methods of Chapter 5 are likely to give more valuable insights than tests of hypotheses.

This is not meant to imply that statistical tests should be abandoned altogether. They are useful tools for deciding which one of two or several statistical models should be used. The point is that a statistical analysis, particularly in multivariate situations, should not consist of a test alone. For instance, in the discriminant analysis examples of Chapter 5, statistical tests will help to make sure that an empirical result is not just due to "white noise." Standard errors and confidence intervals will help to assess the stability of coefficients of a discriminant function and to decide whether or not a variable is important for discrimination.

We will derive the T^2-tests from heuristic principles, using the notion of multivariate standard distance developed in Chapter 5. In the current chapter, we will need to do a substantial amount of hand waving, that is, quote results from multivariate statistical theory without proving them. Although the one-sample case is less important than the two-sample case, we will start out with the former in Section 6.2, because the one-sample situation is mathematically somewhat simpler and provides us with all the ideas and techniques needed to handle the two-sample case appropriately. Section 6.3 applies the one-sample T^2 methodology to the construction of confidence regions for mean vectors. In Section 6.4 we discuss the most important test, the two-sample T^2-test for equality of mean vectors, which we may also regard as an overall test of significance in linear discriminant analysis. Sections 6.2 to 6.4 rely heavily on heuristic arguments and should be readily understood.

In Section 6.5 we present a test for redundancy of variables in a linear discriminant function and relate it to the problem of standard errors of discriminant function coefficients. Section 6.5 is also interesting from a theoretical point of view because it connects the redundancy problem in discriminant analysis to a problem of parallel regression lines or planes, thus, continuing the theory developed in Section 5.6. Section 6.6 is more technical and may be skipped by the mathematically less inclined reader; however, from a theoretical perspective it is important and rather attractive because it rederives the tests studied before using the powerful principles of union-intersection testing and likelihood ratio testing. Actually, Section 6.6 goes somewhat beyond statistical inference for means, by studying a model of exchangeability of random vectors, illustrating the power and versatility of the likelihood approach. Finally, Section 6.7 presents resampling procedures for computing the distribution of test statistics under the null hypothesis; this section should, again, be of considerable interest to both theoretically and practically inclined students. More precisely, Section 6.7 presents a randomization procedure in the two-sample case and a bootstrap procedure in the one-sample case, both meant as alternatives to the normal theory computation of the significance of a statistical test. To fully appreciate the material of Section 6.7, the student will need to use appropriate software for randomization and bootstrap sampling.

Methods for comparing more than two mean vectors, commonly called MANOVA (multivariate analysis of variance), are not treated in the current chapter because we have not yet studied the problem of classification into more than two groups. The reader who is mostly interested in this topic may skip Sections 6.5 to 7.2 and proceed directly to the canonical discriminant functions of Section 7.3, on which the MANOVA tests of Section 7.4 are based.

6.2 The One-Sample T^2-Test

Let us first recall the univariate t–test in the one-sample case. Given a sample $x_1, \ldots, x_N$ of N observations from a random variable X with unknown mean μ, we wish to test the null hypothesis

$$H_0 : \mu = \mu_0 ,$$

where μ_0 is a fixed number, usually called the hypothetical mean. The t–statistic is given by

$$t = \sqrt{N} \cdot \frac{\bar{x} - \mu_0}{s} , \tag{1}$$

where $\bar{x}$ and s denote sample mean and standard deviation, as usual. In a two-sided test, the null hypothesis is rejected if $|t| > c$, where the constant c is to be chosen so as to fix the probability of an erroneous rejection of the null hypothesis at a given value.

From Section 5.2, recall that, for a given sample with statistics $\bar{x}$ and s, the standard distance between μ_0 and $\bar{x}$ is defined as

$$D(\bar{x}, \mu_0) = \frac{|\mu_0 - \bar{x}|}{s} . \tag{2}$$

Hence, the two-sided t-test rejects the null hypothesis exactly if the standard distance $D(\bar{x}, \mu_0)$ exceeds a critical value. This fact is the heuristic justification of a similar procedure in the multivariate case. Note that no distributional assumptions have been made yet.

Now, we turn to the multivariate case. Suppose that $\mathbf{x}_1, \ldots, \mathbf{x}_N$ is a sample from a p–variate distribution with (unknown) mean vector $\boldsymbol{\mu} = (\mu_1, \ldots, \mu_p)'$, and we wish to test the hypothesis

$$H_0 : \boldsymbol{\mu} = \boldsymbol{\mu}_0 , \tag{3}$$

where $\boldsymbol{\mu}_0 = (\mu_{10}, \ldots, \mu_{p0})' \in \mathbb{R}^p$ is a given vector, called the hypothetical mean vector. Let $\bar{\mathbf{x}}$ and $\mathbf{S}$ denote the sample mean vector and the sample covariance matrix, respectively. Analogous to the univariate test based on the univariate sample standard distance, we can base a test of the hypothesis (3) on the multivariate standard distance between $\bar{\mathbf{x}}$ and $\boldsymbol{\mu}_0$. Much of the necessary theory has already been developed. From Theorem 5.2.4, we know that the multivariate, standard distance of a point $\mathbf{x} \in \mathbb{R}^p$ from the mean $\boldsymbol{\mu}$ of a p-variate random vector $\mathbf{X}$ with covariance matrix ψ is given by

$$\Delta_{\mathbf{X}}(\mathbf{x}, \boldsymbol{\mu}) = \left[(\mathbf{x} - \boldsymbol{\mu})'\psi^{-1}(\mathbf{x} - \boldsymbol{\mu})\right]^{1/2} . \tag{4}$$

The sample analog of (4) is

$$D(\mathbf{x}, \bar{\mathbf{x}}) = \left[(\mathbf{x} - \bar{\mathbf{x}})'\mathbf{S}^{-1}(\mathbf{x} - \bar{\mathbf{x}})\right]^{1/2} , \tag{5}$$

which is well defined provided the sample covariance matrix $\mathbf{S}$ is positive definite. The multivariate standard distance of the hypothetical mean vector $\boldsymbol{\mu}_0$ from the observed mean vector $\bar{\mathbf{x}}$, then is given by

$$D(\boldsymbol{\mu}_0, \bar{\mathbf{x}}) = \left[(\boldsymbol{\mu}_0 - \bar{\mathbf{x}})'\mathbf{S}^{-1}(\boldsymbol{\mu}_0 - \bar{\mathbf{x}})\right]^{1/2} . \tag{6}$$

Now, we can formulate a rule for a test, based entirely on the heuristic principle of multivariate standard distance.

Heuristic Test. *For a sample* $\mathbf{x}_1, \ldots, \mathbf{x}_N$ *from a p-variate distribution with mean vector* $\boldsymbol{\mu}$,

$$\text{accept } H_0 : \boldsymbol{\mu} = \boldsymbol{\mu}_0 \qquad \text{if } D(\boldsymbol{\mu}_0, \bar{\mathbf{x}}) \le c ,$$

and

$$\text{reject } H_0 : \boldsymbol{\mu} = \boldsymbol{\mu}_0 \qquad \text{if } D(\boldsymbol{\mu}_0, \bar{\mathbf{x}}) > c ,$$

where the critical value c is to be determined such that Pr[reject $H_0 | H_0$ is true] *takes a fixed value.*

We call this test "heuristic" because it is based entirely on the assumption that the multivariate standard distance (6) is a reasonable summary of the observed mean differences. More support for this test will be given in Section 6.6. The constant c depends in general on the sample size N, and at least for small samples, on the exact form of the distribution from which the sample is obtained. In general, the critical value c might depend on the unknown value of the parameter $\boldsymbol{\mu}$, and hence, finding a proper constant c might not be possible. Fortunately, at least in the normal theory setup, this is not the case, as we will see.

Instead of using the standard distance as a test statistic, the so-called T^2*-statistic* is commonly calculated:

$$\begin{aligned} T^2 &= N \cdot D^2(\boldsymbol{\mu}_0, \bar{\mathbf{x}}) \\ &= N \cdot (\boldsymbol{\mu}_0 - \bar{\mathbf{x}})'\mathbf{S}^{-1}(\boldsymbol{\mu}_0 - \bar{\mathbf{x}}) . \end{aligned} \tag{7}$$

As in the univariate situation, $D(\boldsymbol{\mu}_0, \bar{\mathbf{x}})$ serves as a descriptive measure of distance, whereas T^2 is aimed at hypothesis testing. The notation T^2, by the way, reflects the fact that equation (7) reduces to the square of a t-statistic in the case of a single variable.

Some properties of the T^2-statistic are straightforward to prove. For instance, T^2 is invariant under nonsingular linear transformations; see Exercise 1. Finding the distribution of T^2 is more involved and beyond the level of this introduction, so, for once, we will quote results without proving them. In the classical setup, it is assumed that the sample $\mathbf{x}_1, \ldots, \mathbf{x}_N$ is from a p-variate normal distribution with mean vector $\boldsymbol{\mu}$ and covariance matrix $\boldsymbol{\psi}$, both $\boldsymbol{\mu}$ and $\boldsymbol{\psi}$ unknown. The distribution of $N(\bar{\mathbf{x}} - \boldsymbol{\mu})'\mathbf{S}^{-1}(\bar{\mathbf{x}} - \boldsymbol{\mu})$ under these assumptions is called the T^2*-distribution*, but it is usually not tabulated because there is a simple one-to-one correspondence between

T^2-distributions and F-distributions, the latter ones tabulated in many books. The relevant result is given in the following theorem.

Theorem 6.2.1 *Let $\mathbf{x}_1 \dots, \mathbf{x}_N$ denote a sample of size N from a p-variate, normal distribution with mean vector $\boldsymbol{\mu}$ and covariance matrix $\boldsymbol{\psi}$. Assume that $\boldsymbol{\psi}$ is positive definite and $N > p$. Let $\bar{\mathbf{x}}$ and $\mathbf{S}$ denote the sample mean and sample covariance matrix, respectively. Then the random variable*

$$\frac{N}{N-1} \cdot \frac{N-p}{p} (\bar{\mathbf{x}} - \boldsymbol{\mu})' \mathbf{S}^{-1} (\bar{\mathbf{x}} - \boldsymbol{\mu})$$

follows an F-distribution with p and $(N-p)$ degrees of freedom.

Proof See Anderson (1984), Muirhead(1982), or Seber (1984). Note that the assumption $N > p$ is needed to make sure that $\mathbf{S}$ is nonsingular; see Exercise 8 in Section 4.2 and Exercise 8 in Section 5.3. ∎

Theorem 6.2.1 gives us the exact distribution of T^2 or, equivalently, of $D^2(\boldsymbol{\mu}_0, \bar{\mathbf{x}})$, under the null hypothesis $\boldsymbol{\mu} = \boldsymbol{\mu}_0$, assuming multivariate normality of the population. We will discuss the relevance of the normality assumptions at the end of this section. A particularly appealing aspect of Theorem 6.2.1 is that the distribution of $(\bar{\mathbf{x}} - \boldsymbol{\mu})' \mathbf{S}^{-1} (\bar{\mathbf{x}} - \boldsymbol{\mu})$ depends neither on $\boldsymbol{\mu}$ nor on $\boldsymbol{\psi}$. This is a highly desirable property, paralleling the analogous result for the univariate t-statistic. Proving that the distribution does not depend on $\boldsymbol{\mu}$ is left to the student; see Exercise 4.

Now, we can formulate a rule for testing the hypothesis $\boldsymbol{\mu} = \boldsymbol{\mu}_0$. For simplicity, from now on, we will write D for $D(\boldsymbol{\mu}_0, \bar{\mathbf{x}})$. The test, then is based on any of the three equivalent statistics

(i) $D = \left[(\boldsymbol{\mu}_0 - \bar{\mathbf{x}})' \mathbf{S}^{-1} (\boldsymbol{\mu}_0 - \bar{\mathbf{x}})\right]^{1/2}$,

(ii) $T^2 = N D^2$, or

(iii) $F = \frac{N}{N-1} \cdot \frac{N-p}{p} D^2 = \frac{N-p}{(N-1)p} T^2$.

Rule for testing the hypothesis $\boldsymbol{\mu} = \boldsymbol{\mu}_0$: Accept *the hypothesis H_0 if any of the following three equivalent conditions is satisfied:*

(i) $D \leq \left[\frac{N-1}{N} \cdot \frac{p}{N-p} \cdot f\right]^{1/2}$,

(ii) $T^2 \leq \frac{(N-1)p}{N-p} \cdot f$,

(iii) $F \leq f$,

where f is the $(1-\alpha)$ quantile of the F-distribution with p and $(N-p)$ degrees of freedom. Otherwise, reject H_0.

It follows from Theorem 6.2.1 that a test based on this rule has level α under the normal theory setup, that is,

$$\Pr[\text{reject } H_0 | H_0 \text{ is true}] = \alpha.$$

Statistical software will most often give an F-statistic, according to version (iii) of the decision rule, which makes the comparison easy with an appropriate quantile or the computation of a p-value. However, formulating the rule in terms of the standard distance $D(\mu_0, \bar{\mathbf{x}})$, as in version (i), is somewhat more intuitive.

Example 6.2.1 Head length and breadth of siblings. Table 6.2.1 reproduces a well-known data set first published by Frets (1921). The four variables are X_1 = head length of first son, X_2 = head breadth of first son, X_3 = head length of second son, and X_4 = head breadth of second son, obtained from 25 families. Do first and second sons differ in their average head dimensions? The mean vector for the sample of 25 observations is $\bar{\mathbf{x}} = (187.40,\ 151.12,\ 183.32,\ 149.36)'$, that is, the observed means are somewhat larger for first sons than for second sons. In terms of a statistical hypothesis, the assumption of equal average head dimensions of first and second sons can be written as

$$H_0: \quad E[X_1] = E[X_3]$$

and

$$E[X_2] = E[X_4].$$

Defining two new variables

$$\mathbf{Y} = \begin{pmatrix} Y_1 \\ Y_2 \end{pmatrix} = \begin{pmatrix} X_1 - X_3 \\ X_2 - X_4 \end{pmatrix},$$

and setting $\mu = E[\mathbf{Y}]$, the above hypothesis is equivalent to $\mu = \mathbf{0}$. Now, we can apply the T^2-test to the $N = 25$ observations and the $p = 2$ variables in $\mathbf{Y}$, with $\mu_0 = \mathbf{0}$. After transforming the data from Table 6.2.1 to the differences Y_1 and Y_2, numerical calculations give

$$\bar{\mathbf{y}} = \begin{pmatrix} 4.08 \\ 1.76 \end{pmatrix},$$

and

$$\mathbf{S} = \begin{pmatrix} 60.160 & 14.395 \\ 14.395 & 17.690 \end{pmatrix}$$

for the sample mean vector and the sample covariance matrix of $\mathbf{Y}$. With $\mu_0 = \mathbf{0}$, we obtain $D(\mu_0, \bar{\mathbf{y}}) = [\bar{\mathbf{y}}'\mathbf{S}^{-1}\bar{\mathbf{y}}]^{1/2} = 0.566$, $T^2 = 7.996$, and $F = 3.831$. This value of F is between the 95th percentile ($f_{0.95} = 3.42$) and the 99th percentile ($f_{0.99} = 5.66$) of the F-distribution with $p = 2$ and $(N - p) = 23$ degrees of freedom. The equivalent

Table 6.2.1 Head dimension data from Frets (1921), measured on siblings in 25 families.

X_1	X_2	X_3	X_4
191	155	179	145
195	149	201	152
181	148	185	149
183	153	188	149
176	144	171	142
208	157	192	152
189	150	190	149
197	159	189	152
188	152	197	159
192	150	187	151
186	161	179	158
179	147	183	147
195	153	174	150
202	160	190	159
194	154	188	151
163	137	161	130
195	155	183	158
186	153	173	148
181	145	182	146
175	140	165	137
192	154	185	152
174	143	178	147
176	139	176	143
197	167	200	158
190	153	187	150

Variables are: $(X_1, X_2) =$ head length and head breadth of first son; $(X_3, X_4) =$ head length and head breadth of second son.

critical values for the test based on D are $c_{0.95} = 0.534$ and $c_{0.99} = 0.688$. The result is, therefore, somewhat inconclusive. The same "nonconclusion" is obtained from a scatterplot of Y_1 vs. Y_2; see Figure 6.2.1. Although the observed mean $\bar{\mathbf{y}}$ and the hypothetical mean $\mathbf{0}$ are visibly apart, the evidence against the null hypothesis is not overwhelming.

Suppose we decide to accept the null hypothesis. Then the logical next step would be to estimate the parameters of the distribution of the original random vector $\mathbf{X}$ accordingly, that is, with $\nu_i = E[X_i]$ denoting the expected values of the X_i, one would like to estimate the ν_i subject to the constraints $\nu_1 = \nu_3$ and $\nu_2 = \nu_4$. An intuitively reasonable solution is to average over sons, i.e., to set

$$\hat{\nu}_1 = \hat{\nu}_3 = \frac{1}{2}(\bar{x}_1 + \bar{x}_3) = 185.36$$

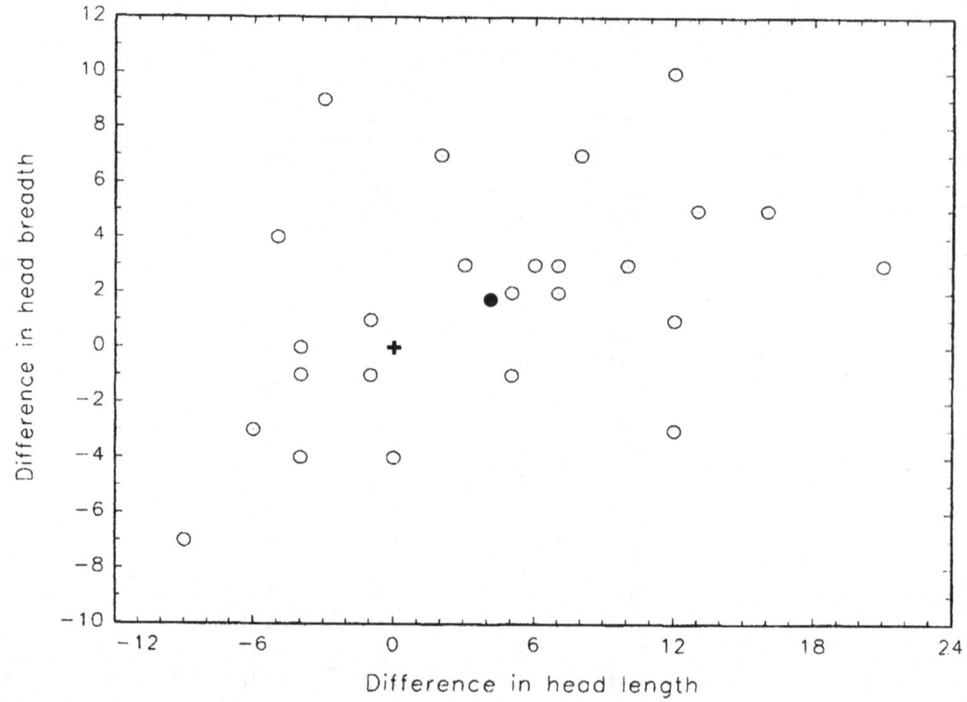

Figure 6-2-1 Scatterplot of difference in head length between first and second sons vs difference in head breadth. The sample mean is marked as a solid circle, and the hypothetical mean **0** as a plus-sign.

and

$$\hat{\nu}_1 = \hat{\nu}_4 = \frac{1}{2}(\bar{x}_2 + \bar{x}_4) = 150.24\,.$$

This solution can also be justified from a theoretical perspective; see Exercise 12 and Example 6.6.5. Finally, notice that for the T^2 test to be valid we do not need to assume multivariate normality of $\mathbf{X} = (X_1, X_2, X_3, X_4)'$. Rather, the relevant assumption is that $\mathbf{Y} = (Y_1, Y_2)'$ is bivariate normal. This assumption is weaker than multivariate normality of $\mathbf{X}$. See also Exercises 5 and 13. □

Since the multivariate hypothesis $\boldsymbol{\mu} = \boldsymbol{\mu}_0$ is equivalent to p univariate hypotheses $\mu_i = \mu_{i0}$, one hypothesis for each variable, the question arises why it would be better to perform a multivariate test rather than p univariate tests. There are two reasons. First, if the variables considered are not independent, then the p univariate tests are not independent, and it would be difficult to adjust the levels of the univariate tests to maintain a specified significance level for the combined tests. Second, in some situations, the variable by variable approach will fail to detect that a hypothesis is wrong, but a multivariate T^2-test will reveal it. This situation occurs typically when the variables are highly correlated. The phenomenon is easier to understand in terms of confidence regions, to which we will turn in Section 6.3.

Testing for a specified hypothetical mean vector μ_0 is a procedure of rather limited practical value because it hardly ever occurs that a theoretical mean vector is known, except perhaps for situations in quality control where deviations of produced goods from their specifications may be analyzed. The one-sample T^2-test is useful though, to test for equality of the means of correlated variables, a problem that can be treated by taking differences between variables and testing if the mean of the difference is zero. Example 6.2.1 was actually of this type. Now, we are going to discuss, in some detail, a problem of testing for equality of the means of k correlated variables.

Suppose that $\mathbf{X} = (X_1, \ldots, X_k)'$ is a p-variate random vector, with $E[X_i] = \mu_i$ $(i = 1, \ldots, k)$, and we wish to test the hypothesis $\mu_1 = \mu_2 = \cdots = \mu_k$. For instance, the X_i might represent measurements taken on the same objects at k different times. Or for $k = 2$, X_1 might represent a measurement taken on the left eye, and X_2 the same variable measured on the right eye of a subject. Note that the common value of the k means is not specified, and hence, we cannot just apply the one-sample T^2 statistic to the k variables. Formally, consider a k-variate random vector

$$\mathbf{X} = \begin{pmatrix} X_1 \\ \vdots \\ X_k \end{pmatrix}, \qquad \text{with } E[\mathbf{X}] = \mu_{\mathbf{X}} = \begin{pmatrix} \mu_1 \\ \vdots \\ \mu_k \end{pmatrix},$$

and the hypothesis

$$H_0 : \mu_1 = \cdots = \mu_k \,, \tag{8}$$

where the common value of the μ_i is unspecified. This hypothesis will be true exactly if $E[X_i - X_{i+1}] = 0$ for $i = 1, \ldots, k-1$. Thus, consider the transformations

$$\begin{aligned} Y_1 &= X_1 - X_2 \,, \\ Y_2 &= X_2 - X_3 \,, \\ &\vdots \\ Y_{k-1} &= X_{k-1} - X_k \,. \end{aligned}$$

In matrix notation, this reads

$$\mathbf{Y} = \mathbf{A}\mathbf{X} \,, \tag{9}$$

where

$$\mathbf{A} = \begin{pmatrix} 1 & -1 & 0 & \cdots & 0 & 0 \\ 0 & 1 & -1 & \cdots & 0 & 0 \\ \vdots & \vdots & \vdots & & \vdots & \vdots \\ 0 & 0 & 0 & \cdots & 1 & -1 \end{pmatrix} \tag{10}$$

is a matrix of dimension $(k-1) \times k$. The mean vector of $\mathbf{Y}$ is $E[\mathbf{Y}] = \boldsymbol{\mu}_{\mathbf{Y}} = \mathbf{A}\boldsymbol{\mu}_{\mathbf{X}}$, and the hypothesis H_0 of equation (8) is equivalent to

$$H_0^* : \boldsymbol{\mu}_{\mathbf{Y}} = \mathbf{0}. \tag{11}$$

Thus we can test H_0 by transforming the k original variables into $p = (k-1)$ differences (or contrasts) $\mathbf{Y}$, and apply the one-sample T^2-test to $\mathbf{Y}$ with hypothetical mean vector $\mathbf{0}$. Note that, in the case $k = 2$, this procedure reduces the bivariate problem to a univariate problem of testing $E[Y_1] = 0$, a method often referred to as paired comparison.

Example 6.2.2 Cork deposits on trees. Table 6.2.2 reproduces another classical data set. The thickness of cork deposits in four directions (North, East, South, West) was measured by cork borings on $N = 28$ trees. If the average cork deposits are not equal, this might indicate that thickness of cork depends on ecological circumstances, such as dominant wind direction. Define three contrasts

$$\mathbf{Y} = \begin{pmatrix} Y_1 \\ Y_2 \\ Y_3 \end{pmatrix} = \begin{pmatrix} \text{North} - \text{East} \\ \text{East} - \text{South} \\ \text{South} - \text{West} \end{pmatrix}.$$

Then the hypothesis to be tested is $E[\mathbf{Y}] = \mathbf{0}$. Numerical calculations give the following sample mean vector and covariance matrix:

$$\bar{\mathbf{y}} = \begin{pmatrix} 4.357 \\ -3.500 \\ 4.500 \end{pmatrix}, \qquad \mathbf{S}_{\mathbf{Y}} = \begin{pmatrix} 62.831 & -55.556 & 4.481 \\ -55.556 & 111.815 & -32.778 \\ 4.481 & -32.778 & 56.926 \end{pmatrix}.$$

The standard distance of $\mathbf{0}$ from $\bar{\mathbf{y}}$ is $D = \left(\bar{\mathbf{y}}'\mathbf{S}_{\mathbf{Y}}^{-1}\bar{\mathbf{y}}\right)^{1/2} = 0.861$. With $N = 28$ and $p = 3$ and choosing $\alpha = 0.05$, we find the 99th percentile of the F-distribution with $p = 3$ and $(N-p) = 25$ degrees of freedom: $f = 2.99$. The equivalent cutoff for the statistic D is $c = 0.588$, and the hypothesis of equal average growth is, therefore, rejected. Again, notice as in Example 6.2.1, that multivariate normality of $\mathbf{X}$ is not required for the T^2 test to be valid; rather, it is normality of the vector of differences $\mathbf{Y}$ that matters. □

In Example 6.2.2 it might be interesting to ask which of the contrasts are mostly responsible for the rejection. More important, we have to ask whether we would have obtained the same decision if we had chosen the contrasts differently, for instance, as $Y_1 = \text{North} - \text{South}$, $Y_2 = \text{East} - \text{West}$, and $Y_3 = \text{South} - \text{East}$. The answer is that the result would be exactly the same, as we will show now.

Let us return to the transformation $\mathbf{Y} = \mathbf{A}\mathbf{X}$ of equation (9). With $\mathbf{1}_k$ denoting the vector $(1, 1, \ldots, 1)' \in \mathbb{R}^k$, a characteristic property of $\mathbf{A}$ is that

$$\mathbf{A}\mathbf{1}_k = \mathbf{0}, \tag{12}$$

Table 6.2.2 Cork deposit data from Rao (1948).

North	East	South	West
72	66	76	77
60	53	66	63
56	57	64	58
41	29	36	38
32	32	35	36
30	35	34	26
39	39	31	27
42	43	31	25
37	40	31	25
33	29	27	36
32	30	34	28
63	45	74	63
54	46	60	52
47	51	52	43
91	79	100	75
56	68	47	50
79	65	70	61
81	80	68	58
78	55	67	60
46	38	37	38
39	35	34	37
32	30	30	32
60	50	67	54
35	37	48	39
39	36	39	31
50	34	37	40
43	37	39	50
48	54	57	43

Variable names North, East, South, and West indicate thickness in mm of cork deposits measured by borings from four directions on 28 cork trees.

that is, the sum of the entries in each row of $\mathbf{A}$ is zero. Let $\mathbf{A}^*$ denote any matrix of dimension $(k-1) \times k$ such that $\mathbf{A}^*\mathbf{1}_k = \mathbf{0}$, and assume that the $k-1$ rows of $\mathbf{A}^*$ are linearly independent, i.e., rank $(\mathbf{A}^*) = k-1$. Then there exists a nonsingular matrix $\mathbf{C}$ of dimension $(k-1) \times (k-1)$ such that

$$\mathbf{A}^* = \mathbf{C}\mathbf{A}\,, \quad \mathbf{A} = \mathbf{C}^{-1}\mathbf{A}^*\,, \tag{13}$$

see Exercise 9. Suppose that instead of the transformation $\mathbf{Y} = \mathbf{A}\mathbf{X}$ we wish to use $\mathbf{Y}^* = \mathbf{A}^*\mathbf{X}$. Because

$$\mathbf{Y}^* = \mathbf{A}^*\mathbf{X} = \mathbf{C}\mathbf{A}\mathbf{X} = \mathbf{C}\mathbf{Y} \tag{14}$$

and because $\mathbf{C}$ is nonsingular, it follows that $E[\mathbf{Y}] = \mathbf{0}$ is equivalent to $E[\mathbf{Y}^*] = \mathbf{0}$. In other words, the original hypothesis $\mu_1 = \cdots = \mu_k$ can be formulated equivalently in terms of any $(k-1) \times k$ matrix $\mathbf{A}^*$ of rank $k-1$ that satisfies $\mathbf{A}^*\mathbf{1}_k = \mathbf{0}$, i.e., whose row sums are all zero.

But do different choices of the contrasts lead to the same value of the test statistic? For $\mathbf{Y} = \mathbf{AX}$ as in equation (9), let $\bar{\mathbf{y}}$ and $\mathbf{S_Y}$ denote the sample mean vector and the sample covariance matrix, respectively. Then the multivariate standard distance based on $\mathbf{Y}$ is given by

$$D_{\mathbf{Y}}(\mathbf{0}, \bar{\mathbf{y}}) = \left(\bar{\mathbf{y}}'\mathbf{S}_{\mathbf{Y}}^{-1}\bar{\mathbf{y}}\right)^{1/2}. \tag{15}$$

For variables $\mathbf{Y}^*$ we have from equation (14), $\bar{\mathbf{y}}^* = \mathbf{C}\bar{\mathbf{y}}$ and $\mathbf{S}_{\mathbf{Y}^*} = \mathbf{CS_YC}'$. Because $\mathbf{C}$ is nonsingular, the squared multivariate standard distance based on $\mathbf{Y}^*$ is given by

$$\begin{aligned} D^2_{\mathbf{Y}^*}(\mathbf{0}, \bar{\mathbf{y}}^*) &= (\bar{\mathbf{y}}^*)'(\mathbf{S}_{\mathbf{Y}^*})^{-1}\bar{\mathbf{y}}^* \\ &= \bar{\mathbf{y}}'\mathbf{C}'\left(\mathbf{CS_YC}'\right)^{-1}\mathbf{C}\bar{\mathbf{y}} \\ &= \bar{\mathbf{y}}'\mathbf{C}'(\mathbf{C}')^{-1}\mathbf{S}_{\mathbf{Y}}^{-1}\mathbf{C}^{-1}\mathbf{C}\bar{\mathbf{y}} \\ &= D^2_{\mathbf{Y}}(\mathbf{0}, \bar{\mathbf{y}}). \end{aligned} \tag{16}$$

Thus, the value of the multivariate standard distance and, hence, the value of the T^2-statistic are the same for all choices of $\mathbf{A}^*$, as long as we make sure that the conditions $\mathbf{A}^*\mathbf{1}_k = \mathbf{0}$ and rank $(\mathbf{A}^*) = k-1$ are satisfied.

Example 6.2.3 continuation of Example 6.2.2

Suppose that we would like to use the contrasts

$$V_1 = \text{North} - \text{South},$$

$$V_2 = \text{East} - \text{West},$$

and

$$V_3 = (\text{East} + \text{West}) - (\text{North} + \text{South}),$$

instead of the variables Y_1, Y_2, Y_3 from Example 6.2.2. We can write

$$\mathbf{V} = \begin{pmatrix} V_1 \\ V_2 \\ V_3 \end{pmatrix} = \begin{pmatrix} 1 & 0 & -1 & 0 \\ 0 & 1 & 0 & -1 \\ -1 & 1 & -1 & 1 \end{pmatrix} \begin{pmatrix} \text{North} \\ \text{East} \\ \text{South} \\ \text{West} \end{pmatrix}$$

and verify that the rank of the transformation matrix is 3, as required. Thus, a test for $E[\mathbf{V}] = \mathbf{0}$ will give exactly the same value of the T^2-statistic as obtained in Example 6.2.2.

It is interesting to go beyond a mere test of a hypothesis and to ask which of the contrasts is most informative. Table 6.2.3 gives a list of the multivariate standard distances obtained for all seven nonempty subsets of variables from $\{V_1, V_2, V_3\}$. It appears that variable V_3 alone accounts for most of the multivariate standard distance, and therefore, that the average cork deposits along the north-south axis differ from those along the east-west axis. Numerical calculations give $\bar{v}_3 = -8.86$, that is, the average cork deposit along the north-south axis is about 8.86 units larger than the average deposit along the east-west axis. Further details are discussed in Exercise 7. □

We conclude this section with some remarks about the significance of the normality assumptions and some large sample considerations. To fully understand where and why the normality assumptions are important, one needs to know a little more about the background to Theorem 6.2.1. Since $D^2(\boldsymbol{\mu}, \bar{\mathbf{x}})$ depends on the data only through the sample mean vector $\bar{\mathbf{x}}$ and the sample covariance matrix $\mathbf{S}$, all that matters is the joint distribution of these two statistics. If the sample $\mathbf{x}_1, \ldots, \mathbf{x}_N$ is from $\mathcal{N}_p(\boldsymbol{\mu}, \boldsymbol{\psi})$, then by Exercise 7 in Section 3.2, $\sqrt{N}(\bar{\mathbf{x}} - \boldsymbol{\mu})$ is p-variate normal with mean vector $\mathbf{0}$ and covariance matrix $\boldsymbol{\psi}$. Moreover, from Theorem 4.2.4 in Section 4.2, we know that $\bar{\mathbf{x}}$ and $\mathbf{S}$ are independent in the normal theory setup. The normality assumptions also imply that the distribution of $\mathbf{S}$ is of a specific kind, called the *Wishart distribution* of dimension p, with $m = (N-1)$ degrees of freedom and parametric matrix $\boldsymbol{\psi}/m$, written symbolically as $\mathbf{S} \sim \mathcal{W}_p(m, \ \boldsymbol{\psi}/m)$. In our context, this is just a name, but, for the current purposes, it is sufficient to think of the Wishart distribution as the distribution of the sample covariance matrix from a normal sample. Thus, we can summarize the conditions for Theorem 6.2.1 to hold as follows:

(i) $\sqrt{N}(\bar{\mathbf{x}} - \boldsymbol{\mu}) \sim \mathcal{N}_p(\mathbf{0}, \boldsymbol{\psi})$ for some positive definite symmetric matrix $\boldsymbol{\psi}$,

(ii) $\mathbf{S} \sim \mathcal{W}_p[N-1, \ \boldsymbol{\psi}/(N-1)]$, and

Table 6.2.3 Standard distance from $\mathbf{0}$ for subsets of contrasts in the cork example

Variables Included[a]			Standard Distance
V_1			0.108
	V_2		0.098
		V_3	0.781
V_1	V_2		0.126
V_1		V_3	0.835
	V_2	V_3	0.832
V_1	V_2	V_3	0.861

Contrasts are V_1 = North − South, V_2 = East − West, and V_3 = (East + West) − (North + South).

(iii) $\bar{\mathbf{x}}$ and $\mathbf{S}$ are stochastically independent.

That means it is not really normality of the population that matters; rather it is normality of $\bar{\mathbf{x}}$ along with conditions (ii) and (iii). Although this does not help to decide whether or not the T^2-test is valid in any given case, it allows some relatively straightforward asymptotic calculations. The following is mostly heuristic reasoning because formal asymptotic theory would lead us too far into technical calculations and is beyond the level of this book.

Suppose that $\bar{\mathbf{x}}_N$ is the average of N independent observations from some arbitrary, p-variate distribution with mean vector $\boldsymbol{\mu}$ and positive definite covariance matrix $\boldsymbol{\psi}$, and $\mathbf{S}_N$ is the sample covariance matrix. By the Multivariate Central Limit Theorem 3.3.3, as N tends to infinity, the limiting distribution of $\mathbf{Z}_N = \sqrt{N}(\bar{\mathbf{x}}_N - \boldsymbol{\mu})$ is $\mathcal{N}(\mathbf{0}, \boldsymbol{\psi})$. Moreover, by the law of large numbers, $\mathbf{S}_N$ converges to $\boldsymbol{\psi}$. Therefore, for large N, we would expect the distribution of $\mathbf{Z}_N'\mathbf{S}_N^{-1}\mathbf{Z}_N = N(\bar{\mathbf{x}}_N - \boldsymbol{\mu})'\mathbf{S}_N^{-1}(\bar{\mathbf{x}}_N - \boldsymbol{\mu})$ to be approximately the same as the distribution of $Y = \mathbf{Z}'\boldsymbol{\psi}^{-1}\mathbf{Z}$, where $\mathbf{Z} \sim \mathcal{N}_p(\mathbf{0}, \boldsymbol{\psi})$. By Theorem 3.3.2, the random variable Y follows a chi-square distribution with p degrees of freedom, that is, for large N, the distribution of the test statistic $T^2 = N D^2(\boldsymbol{\mu}_0, \bar{\mathbf{x}})$ under the hypothesis $\boldsymbol{\mu} = \boldsymbol{\mu}_0$ is approximately chi-square with p degrees of freedom, irrespective of the exact form of the distribution from which the sample was taken.

These heuristic arguments are confirmed by the fact that a random variable F that follows an F-distribution with m_1 and m_2 degrees of freedom can be represented as

$$F = \frac{Y_1/m_1}{Y_2/m_2} ,$$

where Y_1 and Y_2 are independent chi-square random variables with m_1 and m_2 degrees of freedom, respectively. As m_2 tends to infinity, the denominator Y_2/m_2 tends to 1; see Exercise 11. Hence, for large degrees of freedom m_2, the random variable $m_1 F$ is approximately chi-square on m_1 degrees of freedom. In the situation of the one-sample T^2-test, m_1 corresponds to p, and m_2 corresponds to $N - p$. The F-statistic and the T^2-statistic from our test rule are related by

$$T^2 = \frac{N-1}{N-p} \cdot pF .$$

As N tends to infinity while p remains constant, the ratio $(N-1)/(N-p)$ approaches 1, and the asymptotic distribution of the random variable pF under the hypothesis is approximately chi-square with p degrees of freedom, thus confirming the earlier result.

While this is comforting for large samples, it is exactly for small samples that testing is often important, and asymptotic theory may not be of great value in such cases. A good alternative to parametric tests in such a situation is provided by *bootstrap* tests. We will discuss this approach in Section 6.7.

Exercises for Section 6.2

1. Consider the T^2-statistic $N(\bar{\mathbf{x}} - \mu_0)'\mathbf{S}_{\mathbf{X}}^{-1}(\bar{\mathbf{x}} - \mu_0)$ for $H_0 : E[\mathbf{X}] = \mu_0$ based on a sample of size N with sample mean vector $\bar{\mathbf{x}}$ and covariance matrix $\mathbf{S}_{\mathbf{X}}$. Suppose that the data are transformed according to $\mathbf{Y} = \mathbf{AX} + \mathbf{b}$, where $\mathbf{A}$ is a nonsingular $p \times p$ matrix and $\mathbf{b} \in \mathbb{R}^p$. Show that the T^2-statistic for $H_0^* : E[\mathbf{Y}] = \mathbf{A}\mu_0 + \mathbf{b}$ based on $\bar{\mathbf{y}}$ and $\mathbf{S}_{\mathbf{Y}}$ is identical to $N(\bar{\mathbf{x}} - \mu_0)'\mathbf{S}_{\mathbf{x}}^{-1}(\bar{\mathbf{x}} - \mu_0)$.

2. For dimension $p = 1$, show that the T^2-test for $H_0 : \mu = \mu_0$ rejects the hypothesis exactly if a two-sided t-test rejects the hypothesis. *Hint*: If U has a t-distribution with m degrees of freedom, then U^2 follows an F-distribution with 1 and m degrees of freedom.

3. This exercise is based on the vole data of Example 5.4.4. Based on the 43 specimens in group *Microtus multiplex*, test the hypothesis of equal average width of the first three molars, i.e., the hypothesis $E[Y_1] = E[Y_2] = E[Y_3]$, at a significance level of 0.01.

4. Suppose $\mathbf{x}_1, \ldots, \mathbf{x}_N$ is a sample from $\mathcal{N}_p(\mu, \psi)$, with sample mean vector $\bar{\mathbf{x}}$ and sample covariance matrix $\mathbf{S}$. Without attempting to find the exact distributional form, show that the distribution of $(\bar{\mathbf{x}} - \mu)'\mathbf{S}^{-1}(\bar{\mathbf{x}} - \mu)$ does not depend on μ. *Hint*: Argue that neither the distribution of $\bar{\mathbf{x}} - \mu$ nor the distribution of $\mathbf{S}$ depends on μ.

5. Suppose that in Example 6.2.1 there are three sons per family, and you would like to test the hypothesis of equal average head length and head breadth of first, second, and third sons. How would you test this hypothesis? Can you treat this problem in the framework of a one-sample T^2-test?

6. Consider testing the hypothesis H as in equation (8). Show that, for $k = 2$, the T^2-statistic is the square of a univariate t-statistic based on the contrast $X_1 - X_2$.

7. This exercise expands on Examples 6.2.2 and 6.2.3.

   (a) Construct a table similar to Table 6.2.3 using the three contrasts

   $$W_1 = \text{East} - \text{North}\,,$$

   $$W_2 = \text{North} - \text{South}\,,$$

   and

   $$W_3 = \text{East} + \text{West} - (\text{North} + \text{South}) \quad (= V_3)\,.$$

   (b) Draw a scatterplot of W_1 vs. W_3 similar to Figure 6.2.1, and mark the hypothetical mean.

   (c) In part (a) you may have noticed that the standard distance based on $\{W_1, W_3\}$ is not much larger than the standard distance of W_3. Use the figure in part (b) to explain why this happens.

8. In the setup of the one-sample T^2-test, consider the multivariate standard distance $D(\mu_0, \bar{\mathbf{x}}) = \left[(\mu_0 - \bar{\mathbf{x}})'\mathbf{S}^{-1}(\mu_0 - \bar{\mathbf{x}})\right]^{1/2}$. According to the results of Section 5.2,

$D(\boldsymbol{\mu}_0, \bar{\mathbf{x}})$ is the univariate, standard distance obtained from a linear combination $Y = \mathbf{b}'\mathbf{X}$, where $\mathbf{b}$ is the vector

$$\mathbf{b} = \mathbf{S}^{-1}(\boldsymbol{\mu}_0 - \bar{\mathbf{x}}) \in \mathbb{R}^p .$$

Note that the transformation $Y = \mathbf{b}'\mathbf{X}$ is similar to a linear discriminant function, with the hypothetical mean vector $\boldsymbol{\mu}_0$ in place of the sample mean vector $\bar{\mathbf{x}}_2$ of a second group.

(a) In Example 6.2.3, compute $\mathbf{b} = \mathbf{S}_v^{-1}(\bar{\mathbf{v}} - \mathbf{0})$ for $\mathbf{V} = \begin{pmatrix} V_1 \\ V_2 \\ V_3 \end{pmatrix}$.

(b) Apply the transformation $Y = \mathbf{b}'\mathbf{V} = b_1 V_1 + b_2 V_2 + b_3 V_3$ to the data, and graph the distribution of Y.

(c) Compute $\bar{y}$ (sample mean of Y) and $s(Y)$ (standard deviation of Y), and verify that $D(\mathbf{0}) = |\bar{y}|/s(Y)$.

9. Prove equation (13), that is, show the following. If $\mathbf{A}$ and $\mathbf{A}^*$ are $(k-1) \times k$ matrices such that $\mathbf{A1}_k = \mathbf{A}^*\mathbf{1}_k = \mathbf{0}$, then there exists a nonsingular matrix $\mathbf{C}$ of dimension $(k-1) \times (k-1)$ such that $\mathbf{A}^* = \mathbf{CA}$. *Hint*: The $k \times k$ matrices $\begin{pmatrix} \mathbf{A} \\ \mathbf{1}_k' \end{pmatrix}$ and $\begin{pmatrix} \mathbf{A}^* \\ \mathbf{1}_k' \end{pmatrix}$ are nonsingular. Hence, there exists a nonsingular matrix $\mathbf{D} = \begin{pmatrix} \mathbf{D}_{11} & \mathbf{d}_{12} \\ \mathbf{d}_{21} & d_{22} \end{pmatrix}$, where $\mathbf{D}_{11}$ has dimension $(k-1) \times (k-1)$, such that $\begin{pmatrix} \mathbf{A}^* \\ \mathbf{1}_k' \end{pmatrix} = \mathbf{D} \begin{pmatrix} \mathbf{A} \\ \mathbf{1}_k' \end{pmatrix}$.

10. Testing for linear trend in a growth curve. Let $X_1, \ldots, X_k$ denote measurements of a growing organism obtained at times $1, \ldots, k$; $k > 2$. Suppose that you wish to test the hypothesis of a linear growth trend, i.e., the hypothesis $E[X_{i+1}] = E[X_i] + \alpha$, $i = 1, \ldots, k-1$, where α is an unspecifed constant. This hypothesis states that the average amount of growth between two successive measurements is α.

(a) Show that the hypothesis of linear growth is equivalent to the hypothesis $E[Z_1] = \cdots = E[Z_{k-2}] = 0$, where

$$Z_i = X_i - 2X_{i+1} + X_{i+2}, \quad i = 1, \ldots, k-2 .$$

Hint: $Z_i = (X_i - X_{i+1}) - (X_{i+1} - X_{i+2})$.

(b) Apply this technique to test for linear growth in the water strider data of Table 8.5.3 (p. 612), using $k = 3$ stages of growth and variables F_i = femur length at growth stage i, $i = 1, 2, 3$. Use a significance level of 0.01.

11. Suppose that the random variable Y_m follows a chi-square distribution on m degrees of freedom. Show that the limiting distribution of $Z = Y_m/m$, as m tends to infinity, is degenerate, such that Z takes the value 1 with probability 1. *Note*: If you are not familiar with limit theorems, find $\mu_m = E[Y_m/m]$ and $\sigma_m^2 = \text{var}[Y_m/m]$. Then show that $\lim_{m\to\infty} \sigma_m^2 = 0$.

12. This exercise gives some more theoretical background to Example 6.2.1. Let $\mathbf{X} = \begin{pmatrix} \mathbf{X}_1 \\ \mathbf{X}_2 \end{pmatrix}$ denote a $2p$-variate random vector, where both $\mathbf{X}_1$ and $\mathbf{X}_2$ have dimension p.

Suppose that $\mathbf{X}_1$ and $\mathbf{X}_2$ are exchangeable, i.e., the distribution of $\mathbf{X}^* = \begin{pmatrix} \mathbf{X}_2 \\ \mathbf{X}_1 \end{pmatrix}$ is the same as the distribution of $\mathbf{X}$.

(a) Show that $E[\mathbf{X}] = \begin{pmatrix} \boldsymbol{\mu} \\ \boldsymbol{\mu} \end{pmatrix}$ for some $\boldsymbol{\mu} \in \mathbb{R}^p$ and $\text{Cov}[\mathbf{X}] = \begin{pmatrix} \boldsymbol{\psi} & \boldsymbol{\Sigma} \\ \boldsymbol{\Sigma} & \boldsymbol{\psi} \end{pmatrix}$ for some symmetric $p \times p$ matrices $\boldsymbol{\psi}$ and $\boldsymbol{\Sigma}$.

(b) Let $\mathbf{Y} = \begin{pmatrix} \mathbf{Y}_1 \\ \mathbf{Y}_2 \end{pmatrix} = \begin{pmatrix} \mathbf{X}_1 - \mathbf{X}_2 \\ \mathbf{X}_1 + \mathbf{X}_2 \end{pmatrix}$. Find $E[\mathbf{Y}]$ and $\text{Cov}[\mathbf{Y}]$. How does this justify estimation of $E[\mathbf{X}]$ as proposed at the end of Example 6.2.1?

13. Find an example of a joint distribution of k variables $X_1, \ldots, X_k$ that is not k-variate normal, yet the $k-1$ differences $Y_i = X_i - X_{i+1}, i = 1, \ldots, k-1$, are jointly $(k-1)$-variate normal. *Hint*: The following is an example for $k = 2$. Let $\mathbf{Z} = (Z_1, Z_2)'$ denote an arbitrary, bivariate normal random vector, and let U denote a random variable which is independent of $\mathbf{Z}$. Set $X_1 = Z_1 + U$ and $X_2 = Z_2 + U$. For instance, if U takes values 0 and 1 with probabilities α and $1 - \alpha$, respectively, then $\mathbf{X} = (X_1, X_2)'$ follows the distribution of a mixture of two bivariate normals, and $X_1 - X_2$ is univariate normal.

6.3 Confidence Regions for Mean Vectors

In Theorem 6.2.1, we gave the distribution of the multivariate standard distance

$$D(\boldsymbol{\mu}, \bar{\mathbf{x}}) = \left[(\boldsymbol{\mu} - \bar{\mathbf{x}})'\mathbf{S}^{-1}(\boldsymbol{\mu} - \bar{\mathbf{x}})\right]^{1/2} \tag{1}$$

under normality assumptions. Then the theorem was used to find a rule for testing the hypothesis $\boldsymbol{\mu} = \boldsymbol{\mu}_0$. Now, we turn to a different application of the theorem, the construction of confidence regions for mean vectors.

Let f denote the $(1-\alpha)$ quantile of the F-distribution with p and $(N-p)$ degrees of freedom, and set

$$c = \left[\frac{N-1}{N} \cdot \frac{p}{N-p} f\right]^{1/2}. \tag{2}$$

Then we know that

$$\Pr[D(\boldsymbol{\mu}, \bar{\mathbf{x}}) \le c] = 1 - \alpha, \tag{3}$$

or, equivalently,

$$\Pr[(\bar{\mathbf{x}} - \boldsymbol{\mu})'\mathbf{S}^{-1}(\bar{\mathbf{x}} - \boldsymbol{\mu}) \le c^2] = 1 - \alpha. \tag{4}$$

The inequality

$$(\bar{\mathbf{x}} - \boldsymbol{\mu})'\mathbf{S}^{-1}(\bar{\mathbf{x}} - \boldsymbol{\mu}) \le c^2 \tag{5}$$

has two interpretations: (i) If $\boldsymbol{\mu}$ is known, it gives an elliptically shaped area, centered at $\boldsymbol{\mu}$, in which $\bar{\mathbf{x}}$ falls with probability $1 - \alpha$. (ii) If $\boldsymbol{\mu}$ is unknown, it gives a random area of elliptical shape, centered at $\bar{\mathbf{x}}$, that covers the mean vector $\boldsymbol{\mu}$ with probability

$1 - \alpha$. The latter interpretation parallels the interpretation of a confidence interval in univariate statistics. Thus, we see that the natural generalization of the notion of a confidence interval leads to elliptically shaped confidence regions in dimension $p \geq 2$. In fact, a univariate confidence interval symmetric to the sample mean is a special case of (5), the interior of a "one-dimensional ellipse" being an interval. A multivariate confidence region of the form given by equation (5) consists of all points $\mathbf{x} \in \mathbb{R}^p$ that have a standard distance of at most c from the sample mean vector $\bar{\mathbf{x}}$. It is difficult to visualize such a confidence region in general dimension p, but, for any given $\mathbf{x} \in \mathbb{R}^p$, one can decide whether or not it is inside the confidence region by computing $D(\mathbf{x}, \bar{\mathbf{x}})$. In the case $p = 2$, it is often instructive to graph the elliptical confidence region, using the technique of the software instructions to Chapter 3. Given the constant c from (2), we have to find points $\mathbf{x} = (x_1, x_2)' \in \mathbb{R}^2$ that satisfy the equation

$$(\mathbf{x} - \bar{\mathbf{x}})'\mathbf{S}^{-1}(\mathbf{x} - \bar{\mathbf{x}}) = c^2 . \tag{6}$$

These points form an ellipse centered at $\bar{\mathbf{x}}$. The set of all points inside the ellipse is the confidence region, i.e., all such points are "plausible" or "acceptable" values for the mean vector.

Example 6.3.1 continuation of Example 6.2.1

Head length and breadth of siblings. With variables $Y_1 = X_1 - X_3$ and $Y_2 = X_2 - X_4$ defined as in Example 6.2.1, let us find a confidence region for $\boldsymbol{\mu} = E[\mathbf{Y}]$. Figure 6.3.1 shows the same scatterplot as Figure 6.2.1, now with 99 percent and 95 percent confidence ellipses for $\boldsymbol{\mu}$, corresponding to the equation $(\mathbf{y} - \bar{\mathbf{y}})'\mathbf{S}^{-1}(\mathbf{y} - \bar{\mathbf{y}}) = c^2$. Here, $c = 0.5345$ for $\alpha = 0.05$, and $c = 0.6876$ for $\alpha = 0.01$. In agreement with the conclusions of Example 6.2.1, the hypothetical mean $(0, 0)$ is outside the 95 percent confidence region, but inside the 99 percent ellipse. Note that the shape and orientation of the ellipses follow the shape and orientation of the scatterplot because the covariance matrix of the average is proportional to the covariance matrix of the sample. □

What advantage, if any, does the bivariate confidence ellipse have over a variable by variable approach, with a confidence interval for each mean? Or in terms of testing a hypothesis, why is it better to do a single multivariate test than p univariate tests? If the variables considered are stochastically independent, then there is indeed little reason for preferring the multivariate approach. Suppose that variables $X_1, \ldots, X_p$ are all mutually independent. Then independent confidence intervals $[A_i, B_i]$ for $\mu_i = E[X_i]$ can be constructed, each with the same coverage probability, say $1 - \gamma$, that is,

$$\Pr[A_i \leq \mu_i \leq B_i] = 1 - \gamma, \quad i = 1, \ldots, p.$$

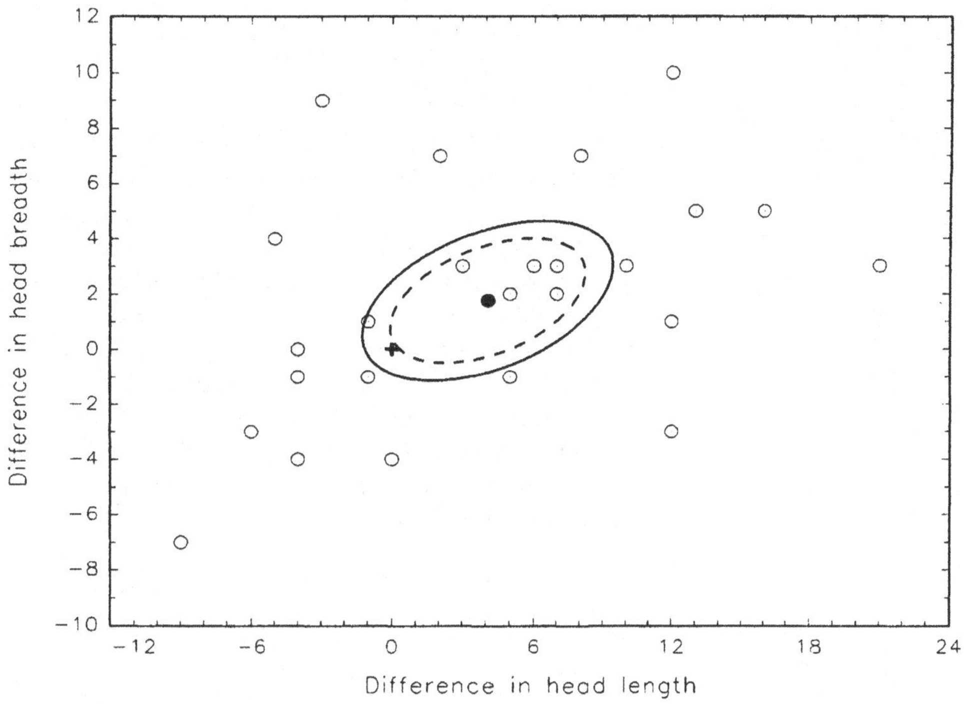

Figure 6-3-1 Scatterplot of Y_1 vs. Y_2 in Example 6.3.1, along with 95 percent and 99 percent confidence ellipses for the mean vector.

The rectangular region $\{(x_1, \ldots, x_p) : A_i \le x_i \le B_i\}$ covers the mean vector μ with probability $(1-\gamma)^p$. Thus, if the total coverage probability is $1-\alpha$ for some fixed α, then we would need to choose

$$\gamma = 1 - (1-\alpha)^{1/p} . \tag{7}$$

All this fails, however, if the variables are not independent. Let us see with another example what happens if we construct such a rectangular region as a confidence set.

Example 6.3.2 Shell dimensions of turtles. Table 1.4 (p. 21) reproduces data measured on 24 male and 24 female turtles of the species *Chrysemys picta marginata*. Shell length, width, and height are very strongly correlated, which makes the example interesting for the current purpose. Figure 6.3.2 shows a scatterplot of X_1 = carapace length vs. X_2 = carapace width for the $N = 24$ male turtles. Sample statistics are

$$\bar{\mathbf{x}} = \begin{pmatrix} 113.375 \\ 88.292 \end{pmatrix}$$

and

$$\mathbf{S} = \begin{pmatrix} 138.766 & 79.147 \\ 79.147 & 50.042 \end{pmatrix}.$$

The correlation between X_1 and X_2 is about 0.95. Assuming that the normality conditions hold, individual 99 percent confidence intervals for the means $\mu_i = E[X_i]$ are $106.62 \le \mu_1 \le 120.13$ and $84.24 \le \mu_2 \le 92.35$; see Exercise 2. If X_1 and X_2 were independent, then the rectangle $\{(x_1, x_2)' \in \mathbb{R}^2 : \ 106.62 \le x_1 \le 120.13, \ 84.24 \le x_2 \le 92.35\}$ would be a correct confidence region for $\boldsymbol{\mu}$, with coverage probability $0.99^2 \approx 0.98$. Hence, let us construct a 98 percent confidence ellipse for $\boldsymbol{\mu}$. From the F-distribution with $p = 2$ and $(N - p) = 22$ degrees of freedom, we find the 98th percentile $f = 4.705$, corresponding to a value $c = 0.640$ of the standard distance; see equation (2). Thus, a 98 percent confidence ellipse for $\boldsymbol{\mu}$ consists of all points in $\mathbb{R}^2$ with a standard distance of at most 0.640 from $\bar{\mathbf{x}}$. Both the elliptical confidence region and the rectangular region obtained from the variable-by-variable approach are illustrated in Figure 6.3.2. □

As Example 6.3.2 illustrates, the rectangular area formed by univariate confidence intervals may be quite inappropriate, especially when there is a large correlation be-

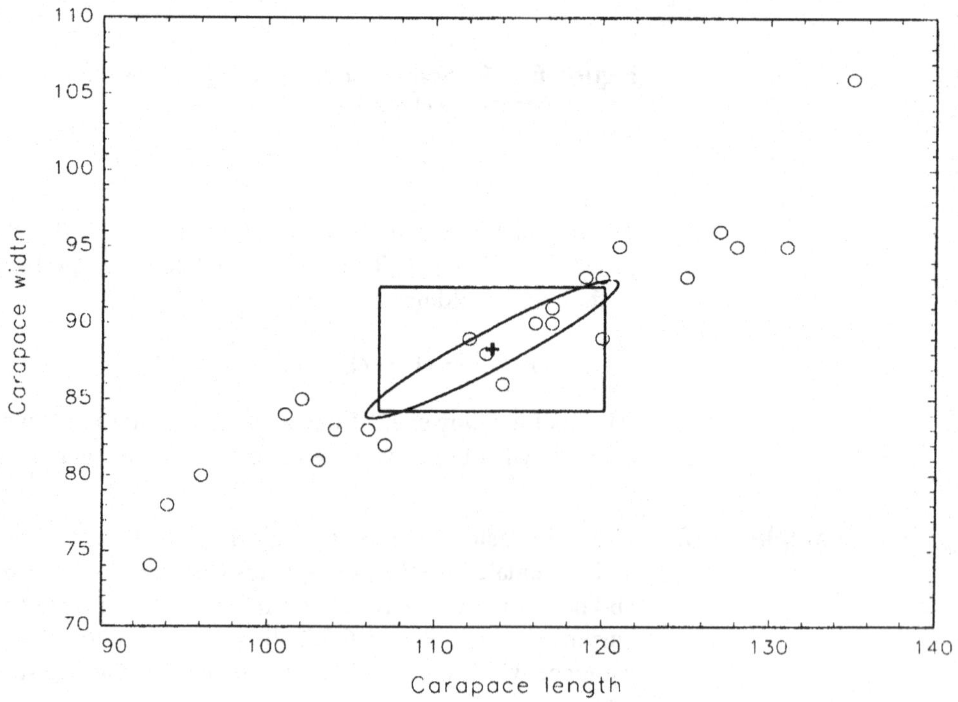

Figure 6-3-2 Elliptical confidence region and univariate confidence intervals for the means of two variables in Example 6.3.2.

tween the variables. If the correlation coefficient approaches ± 1, then the confidence ellipse is highly eccentric, and the rectangular area formed by the univariate confidence intervals will cover mostly points outside the ellipse. This is important in view of the interpretation of a confidence region as the set of all points μ_0 for which the hypothesis $\mu = \mu_0$ would be accepted. In Figure 6.3.2, the point $(107, 92)$, which is an almost ridiculous candidate for the mean vector, would be accepted in univariate t-tests, each at the one percent level. The same point is far outside the 98 percent confidence ellipse for μ and is therefore rejected in the bivariate test.

As outlined in Section 3.2, exact normality is not required for the T^2-test to be valid, provided the sample size N is large enough. This is due to the Multivariate Central Limit Theorem. The same remark holds for elliptical confidence regions. Ellipticity of the confidence set results from the approximate normality of $\bar{\mathbf{x}}$ for sufficiently large N. Also, for large N, the distribution of $N D^2(\mu, \bar{\mathbf{x}})$ is approximately chi-square with p degrees of freedom, irrespective of the exact distribution from which the sample is taken. Hence, the constant c from equation (2) is approximately $c \approx \sqrt{g/N}$, where g is the $(1 - \alpha)$ quantile of the chi-square distribution with p degrees of freedom.

Yet, in some situations, one may not feel comfortable with the assumptions that justify the use of normal theory. A possible alternative, based on the bootstrap method, is outlined in Section 6.6.

We conclude this section by connecting confidence ellipses to tests and confidence regions in linear regression for the student who is familiar with linear model theory. Recall the linear model

$$\mathbf{Y} = \mathbf{X}\beta + \varepsilon, \tag{8}$$

where $\mathbf{Y}$ is the vector of N responses, $\mathbf{X}$ is the design matrix, $\beta = (\beta_0, \beta_1, \ldots, \beta_k)'$ is the vector of unknown parameters, and ε is the vector of "errors." Under the assumptions

$$E[\varepsilon] = \mathbf{0}$$

and

$$\text{Cov}[\varepsilon] = \sigma^2 \cdot \mathbf{I}_N, \tag{9}$$

the least squares estimator $\hat{\beta} = (\mathbf{X}'\mathbf{X})^{-1}\mathbf{X}'\mathbf{Y}$ has mean β and covariance matrix

$$\psi = \text{Cov}[\hat{\beta}] = \sigma^2 \cdot (\mathbf{X}'\mathbf{X})^{-1}. \tag{10}$$

Substituting the mean squared error $s^2 = (\mathbf{y}'\mathbf{y} - \mathbf{y}'\mathbf{X}\hat{\beta})/(N-k-1)$ for σ^2, an unbiased estimator $\hat{\psi} = s^2(\mathbf{X}'\mathbf{X})^{-1}$ of $\text{Cov}[\hat{\beta}]$ is obtained. Under normality assumptions, a confidence region for β with coverage probability $1-\alpha$ is given by all vectors $\mathbf{b}$ such that

$$(\mathbf{b} - \hat{\beta})'\hat{\psi}^{-1}(\mathbf{b} - \hat{\beta}) \leq (k+1)f, \tag{11}$$

or, equivalently,

$$(\mathbf{b} - \hat{\beta})'\mathbf{X}'\mathbf{X}(\mathbf{b} - \hat{\beta}) \le s^2(k+1)f \,, \tag{12}$$

where f is the $(1-\alpha)$ quantile of the F-distribution with $(k+1)$ and $(N-k-1)$ degrees of freedom. This confidence region has elliptical shape and is determined by the standard distance between $\mathbf{b}$ and $\hat{\beta}$, the metric being given by $\hat{\psi}$.

Confidence regions for arbitrary subsets of parameters are obtained by choosing from ψ those entries that correspond to the selected variables. For instance, if we wish to construct a confidence region for $(\beta_1, \beta_2, \ldots, \beta_k)$, that is, for all parameters except the intercept, then we have to partition ψ as

$$\psi = \begin{pmatrix} \psi_{00} & \psi_{01} \\ \psi_{10} & \psi_{11} \end{pmatrix}$$

and use $\text{Cov}\begin{pmatrix} \hat{\beta}_1 \\ \vdots \\ \hat{\beta}_k \end{pmatrix} = \psi_{11}$. The $(1-\alpha)$ confidence region has the form

$$\begin{pmatrix} b_1 - \hat{\beta}_1 \\ \vdots \\ b_k - \hat{\beta}_k \end{pmatrix}' \hat{\psi}_{11}^{-1} \begin{pmatrix} b_1 - \hat{\beta}_1 \\ \vdots \\ b_k - \hat{\beta}_k \end{pmatrix} \le k \cdot f^* \,, \tag{13}$$

where f^* is the $(1-\alpha)$ quantile of the F-distribution with k and $(N-k-1)$ degrees of freedom. If this confidence region covers the vector $\mathbf{0} \in \mathbb{R}^k$, then the overall hypothesis $\beta_1 = \cdots = \beta_k = 0$ is accepted in a test with level α. Although it is not evident from equation (13), this test is equivalent to the usual analysis-of-variance F-test based on the sums-of-squares decomposition. A proof is left as an exercise to the student who knows linear models; see Exercise 6.

Example 6.3.3 Volume of trees. Table 6.3.1 reproduces data measured on $N = 31$ black cherry trees, to find a simple formula for estimating the volume of usable timber in the trunk of a tree, based on easily obtained measurements of diameter and length. The example follows, in part, an analysis of the same data set by Fairley (1986). The variables are V = *volume of usable timber*, D = *diameter*, and H = *height*. Variable V is in cubic feet and H and D are in feet. (In Table 6.3.1, the values of *diameter* are given in inches, and, therefore, must be divided by 12 before the analysis). To fully understand and appreciate this example, the reader should be familiar with multiple linear regression.

If trunks of cherry trees are approximately cone-shaped, then the formula for the volume of a cone suggests that the three variables should be approximately related by the equation

$$V = \frac{\pi}{12} D^2 H \,.$$

Table 6.3.1 Tree data from Ryan, Joiner, and Ryan (1976, p. 278)

Diameter	Height	Volume
8.3	70	10.3
8.6	65	10.3
8.8	63	10.2
10.5	72	16.4
10.7	81	18.8
10.8	83	19.7
11.0	66	15.6
11.0	75	18.2
11.1	80	22.6
11.2	75	19.9
11.3	79	24.2
11.4	76	21.0
11.4	76	21.4
11.7	69	21.3
12.0	75	19.1
12.9	74	22.2
12.9	85	33.8
13.3	86	27.4
13.7	71	25.7
13.8	64	24.9
14.0	78	34.5
14.2	80	31.7
14.5	74	36.3
16.0	72	38.3
16.3	77	42.6
17.3	81	55.4
17.5	82	55.7
17.9	80	58.3
18.0	80	51.5
18.0	80	51.0
20.6	87	77.0

Variables are diameter (one foot above ground), height, and volume of 31 black cherry trees. Diameter is in inches and should be transformed to feet before the analysis. Height is in feet, and volume is in cubic feet.

Taking logarithms, this equation reads

$$\log V = \log\left(\frac{\pi}{12}\right) + 2\log D + \log H\,. \tag{14}$$

Let $Y = \log V$, $X_1 = \log D$, and $X_2 = \log H$. We could use either common logarithms or natural logarithms, the latter being chosen here. If we regress Y on X_1 and X_2, i.e., fit a linear model

$$E[Y] = \beta_0 + \beta_1 X_1 + \beta_2 X_2 , \tag{15}$$

and if the cone is a good approximation to the shape of the trunk, then we would expect the least squares estimates of β_0, β_1, and β_2 to be approximately $\log(\pi/12) \approx -1.340, 2$, and 1, respectively. We will refer to the hypothetical values $(\beta_0 , \beta_1 , \beta_2) = (-1.340,\ 2,\ 1)$ as the *cone model*, or the cone hypothesis.

Numerical calculations give the least squares estimates

$$\hat{\beta} = \begin{pmatrix} \hat{\beta}_0 \\ \hat{\beta}_1 \\ \hat{\beta}_2 \end{pmatrix} = \begin{pmatrix} -1.7049 \\ 1.9826 \\ 1.1171 \end{pmatrix}$$

and the associated estimate of the covariance matrix of $\hat{\beta}$,

$$\widehat{\psi} = 10^{-2} \begin{pmatrix} 77.7746 & 3.4775 & -18.0266 \\ 3.4775 & 0.5627 & -0.1831 \\ -18.0266 & -0.8131 & 4.1795 \end{pmatrix}.$$

It is important here to do all calculations with high precision since the matrix $\widehat{\psi}$ is almost singular. The square roots of the diagonal entries of $\widehat{\psi}$ give standard errors of the regression coefficients; these are $\text{se}(\hat{\beta}_0) = 0.882$, $\text{se}(\hat{\beta}_1) = 0.075$, and $\text{se}(\hat{\beta}_2) = 0.204$. For each of the three regression coefficients, then we can construct a confidence interval with coverage probability $1 - \alpha$ as

$$\hat{\beta}_j \pm t \cdot \text{se}(\hat{\beta}_j),$$

where t is the $(1 - \alpha/2)$-quantile of the t-distribution with $(N - 3) = 28$ degrees of freedom. With $\alpha = 0.05$, and $t = 2.048$, the three confidence intervals are

$$-3.511 \le \beta_0 \le 0.102,$$

$$1.829 \le \beta_1 \le 2.136,$$

and

$$0.698 \le \beta_2 \le 1.536.$$

Each of the three hypothetical values suggested by the cone model is well inside the respective confidence interval, and, therefore, the cone hypothesis seems acceptable. However, a joint confidence region for all three parameters tells a different story. By equation (11), an elliptical confidence region for $\beta = (\beta_0 , \beta_1 , \beta_2)'$ is given by all $\mathbf{b} = (b_0 , b_1 , b_2)' \in \mathbb{R}^3$ that satisfy the inequality

$$(\mathbf{b} - \hat{\beta})'\widehat{\psi}^{-1}(\mathbf{b} - \hat{\beta}) \le 3f , \tag{16}$$

where f is the $(1-\alpha)$-quantile of the F distribution with 3 and 28 degrees of freedom. For $\alpha = 0.05$, $f = 2.947$. Although it would be difficult to graph this confidence

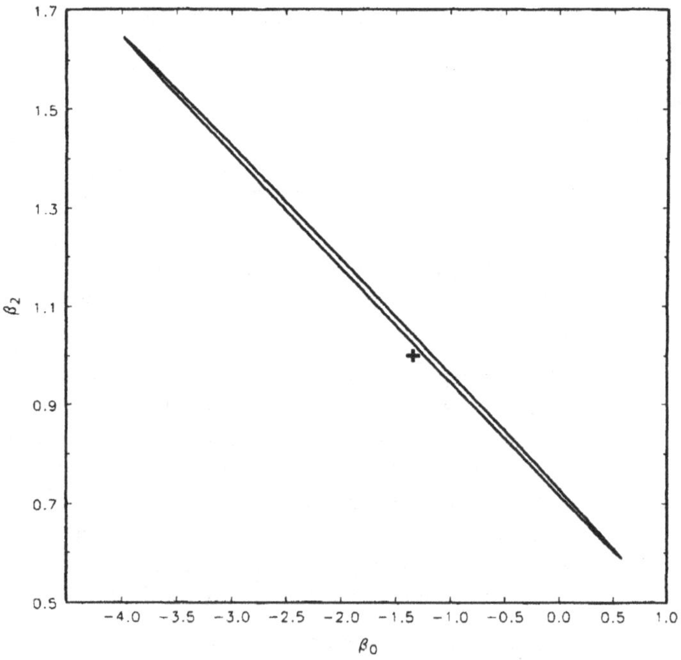

Figure 6-3-3 Elliptical 95 percent confidence region for the parameters β_0 and β_2 in Example 6.3.3. The confidence ellipse is extremely narrow due to the high correlation between the two parameter estimates. The plus sign marks the hypothetical parameter values.

region, we can still decide, for any given vector $\mathbf{b} \in \mathbb{R}^3$, if it falls inside or outside the confidence region. For the vector $\mathbf{b} = (-1.340, 2, 1)'$ specified by the cone hypothesis, the left-hand side of equation (16) is 47121.6, which is much larger than $3f = 8.84$. Therefore, we have clear evidence against the cone hypothesis.

How can this be? The answer is that this is a case of the type alluded to in Figure 6.3.2, but it is even more extreme because of the high correlations between the estimated regression coefficient; see Exercise 7. We can visualize this by taking only two of the coefficients. Using coefficients $\hat{\beta}_0$ and $\hat{\beta}_2$ and choosing from $\widehat{\psi}$ the proper four entries to form the estimated covariance matrix $\widehat{\psi}^*$ of $\hat{\beta}_0$ and $\hat{\beta}_2$, a confidence ellipse with coverage probability $1 - \alpha$ is given by all pairs (b_0, b_2) such that

$$\begin{pmatrix} b_0 - \hat{\beta}_0 \\ b_2 - \hat{\beta}_2 \end{pmatrix}' \left(\hat{\psi}^*\right)^{-1} \begin{pmatrix} b_0 - \hat{\beta}_0 \\ b_2 - \hat{\beta}_2 \end{pmatrix} \leq 2f^*,$$

where f^* is the $(1-\alpha)$-quantile of the F-distribution with 2 and 28 degrees of freedom. For $\alpha = 0.05$, this confidence region is shown graphically in Figure 6.3.3. The confidence ellipse is so narrow that it appears practically as a a line segment, due to the high correlation between $\hat{\beta}_0$ and $\hat{\beta}_2$. The hypothetical point $(\beta_0, \beta_2) = (-1.340, 1)$ is marked in Figure 6.3.3 as well; it is outside the confidence ellipse, indicating that the cone model is not appropriate. This illustrates quite drastically the difference between univariate confidence intervals and multivariate confidence regions. The message is that individual confidence intervals may be badly misleading for simultaneous inference on several variables. □

A simpler example is given in Exercise 5.

Exercises for Section 6.3

1. Verify the computations leading to the construction of the 95 percent confidence ellipse in Example 6.3.1.

2. Find 99 percent confidence intervals for the means μ_1 and μ_2 in Example 6.3.2, using percentiles of the t-distribution. Thus, verify the numerical results presented in Example 6.3.2.

3. The Bonferroni method of constructing simultaneous confidence intervals (see, e.g., Morrison 1976, Chapter 4.3) works as follows. If $m \geq 2$ confidence intervals are to be computed, each one of them should be constructed so as to have a coverage probability of $1 - \alpha/m$. This will guarantee that the m simultaneous confidence statements hold with an associated confidence probability of *at least* $1 - \alpha$. Apply this procedure to Example 6.3.2, using the same two variables and $\alpha = 0.05$. Show that this leads to a graph similar to Figure 6.3.2 with a slightly larger rectangle. Hence, argue that the Bonferroni method is not a remedy for the problem outlined in Example 6.3.2.

4. In Example 6.3.1, head dimensions of brothers, find a 95 percent confidence interval of the mean of $Y_1 = X_1 - X_3$. Draw a scatterplot of X_1 vs. X_3, and find the area in $\mathbb{R}^2$ corresponding to the confidence interval for $E[Y_1]$. *Hint*: This area should be an infinite strip bounded by pairs (x_1, x_3) corresponding to the lower and upper end of the confidence interval.

5. This exercise is to illustrate elliptical confidence regions for regression coefficients, as outlined in equations (8) to (13). Table 6.3.2 gives $N = 14$ observations on repair times of 14 computers. With X = *# of units to repair*, Y = *repair time in minutes*, consider the regression model $E[Y] = \beta_0 + \beta_1 X$.

   (a) Let $\mathbf{X} = \begin{pmatrix} 1 & 1 \\ \vdots & \vdots \\ 1 & 10 \end{pmatrix}$ denote the design matrix, and $\mathbf{y} = \begin{pmatrix} 23 \\ \vdots \\ 166 \end{pmatrix}$ the vector of responses. Compute $\mathbf{X}'\mathbf{X}$, $(\mathbf{X}'\mathbf{X})^{-1}$, the least squares estimates $\hat{\beta} = \begin{bmatrix} \hat{\beta}_0 \\ \hat{\beta}_1 \end{bmatrix} = (\mathbf{X}'\mathbf{X})^{-1}\mathbf{X}'\mathbf{Y}$, and the mean squared error s^2 = (sum of squared residuals)/$(N - 2)$.

   (b) Compute the estimated covariance matrix of $\hat{\beta}$, i.e.,
   $$\hat{\psi} = s^2(\mathbf{X}'\mathbf{X})^{-1},$$
   and the associated correlation matrix. Why do you think the correlation between $\hat{\beta}_0$ and $\hat{\beta}_1$ is so high?

   (c) Compute and graph a 95% confidence ellipse for (β_0, β_1) and univariate 95% confidence intervals for β_0 and β_1.

Table 6.3.2 Computer repair data from Chatterjee and Price (1977, p. 10)

Units	Minutes
1	23
2	29
3	49
4	64
4	74
5	87
6	96
6	97
7	109
8	119
9	149
9	145
10	154
10	166

Units is the number of parts that need to be replaced. *Time* is the repair time in minutes.

(d) Suppose that hypothetical values of the two regression parameters are $\beta_0 = 0$ (no setup time) and $\beta_1 = 15$ (average time to repair one unit is 15 minutes). For confidence level $1 - \alpha = 0.95$, show that

- the hypothesis $\beta_0 = 0$ would be accepted, based on the confidence interval for β_0,
- the hypothesis $\beta_1 = 15$ would be accepted, based on the confidence interval for β_1, but
- the hypothesis $(\beta_0, \beta_1) = (0, 15)$ is rejected, based on the bivariate confidence region.

6. This exercise is for the student who is familiar with the theory of linear models. In a regression model $E[Y] = \beta_0 + \beta_1 X_1 + \cdots + \beta_k X_k$, consider the hypothesis $\beta_1 = \cdots = \beta_k = 0$, and the associated overall F-statistic for significance of the regression,

$$F = \frac{\left(\text{SS}_{\text{total}} - \text{SS}_{\text{residual}}\right)/k}{\text{SS}_{\text{residual}}/(N-k-1)},$$

where SS_{total} and $\text{SS}_{\text{residual}}$ are the (corrected) total and the residual sums of squares, respectively. The hypothesis is rejected if F is larger than the $(1-\alpha)$-quantile of the F-distribution with k and $N-k-1$ degrees of freedom. Show that this test rule is the same as rejecting the hypothesis if the vector $\mathbf{0}$ is outside the elliptical confidence region (14).

7. This exercise is based on the tree data of Example 6.3.3.
   (a) Verify the numerical calculations of Example 6.3.3.
   (b) Compute the estimated correlation matrix for the regression coefficients $\hat{\beta}_0$, $\hat{\beta}_1$, and $\hat{\beta}_2$. Can you explain the high correlation between $\hat{\beta}_0$ and $\hat{\beta}_2$?
   (c) Find $\widehat{\psi}^*$ (the estimated covariance matrix of $\hat{\beta}_0$ and $\hat{\beta}_2$), and verify that the vector $(-1.340,\ 1)$ is outside the 95 percent confidence region for $(\beta_0\,,\ \beta_2)$.
   (d) A possible remedy to the problem of instability of regression coefficients due to high correlations between the parameter estimates is provided by fixing one parameter at some reasonable value and estimating only the remaining parameters. For instance, we could fix the parameter β_2 in equation (15) at $\beta_2 = 1$ and estimate β_0 and β_1 in the model

$$E[Y] = \beta_0 + \beta_1 X_1 + X_2\,,$$

   which is equivalent to the model

$$E[Y - X_2] = \beta_0 + \beta_1 X_1\,.$$

   Find the least squares estimates of β_0 and β_1 by regressing $Y^* = Y - X_2$ on X_1.
   (e) Using the results from part (d), find individual 95 percent confidence intervals for β_0 and β_1.
   (f) Graph a 95 percent confidence ellipse for $(\beta_0\,,\ \beta_1)$ based on the results from part (d).

6.4 The Two-Sample T^2-test

In this section we will discuss the most important multivariate test procedure, Hotelling's T^2-test, for equality of two mean vectors, or the *two-sample T^2-test* for short. The test closely parallels the univariate t-test for equality of the means μ_1 and μ_2 of two normal distributions. In the multivariate setup, we have a p-variate random vector $\mathbf{X}$ with

$$\boldsymbol{\mu}_1 = E_1[\mathbf{X}] = \begin{pmatrix} \mu_{11} \\ \vdots \\ \mu_{p1} \end{pmatrix} \qquad \text{in population 1,}$$

and

$$\boldsymbol{\mu}_2 = E_2[\mathbf{X}] = \begin{pmatrix} \mu_{12} \\ \vdots \\ \mu_{p2} \end{pmatrix} \qquad \text{in population 2,} \tag{1}$$

and we wish to test the hypothesis

$$H_0: \ \boldsymbol{\mu}_1 = \boldsymbol{\mu}_2 \tag{2}$$

without specifying the common value of the two mean vectors. The hypothesis H_0 means that all p pairwise equalities $\mu_{11} = \mu_{21}, \ldots, \mu_{1p} = \mu_{2p}$ hold simultaneously. Therefore, one might want to test H_0 by doing p univariate tests, but, in general, this is not a good idea, as can be seen from the discussion of univariate confidence intervals versus multivariate confidence regions in Section 6.3. A particular example is given in Exercise 7.

Now, suppose that samples of size N_1 and N_2 have been obtained from the two populations, and denote the usual sample mean vectors and covariance matrices as $\bar{\mathbf{x}}_j$ and $\mathbf{S}_j$, $j = 1, 2$. Then we can compute a pooled-sample covariance matrix

$$\mathbf{S} = \frac{(N_1 - 1)\mathbf{S}_1 + (N_2 - 1)\mathbf{S}_2}{(N_1 + N_2 - 2)} \tag{3}$$

and a multivariate standard distance between the mean vectors

$$D(\bar{\mathbf{x}}_1, \bar{\mathbf{x}}_2) = \left[(\bar{\mathbf{x}}_1 - \bar{\mathbf{x}}_2)'\mathbf{S}^{-1}(\bar{\mathbf{x}}_1 - \bar{\mathbf{x}}_2)\right]^{1/2}. \tag{4}$$

This is all we need to formulate a heuristic test principle.

Heuristic test. *In the situation described,*

$$\text{accept } H_0 : \boldsymbol{\mu}_1 = \boldsymbol{\mu}_2 \quad \text{if } D(\bar{\mathbf{x}}_1, \bar{\mathbf{x}}_2) \le c\,,$$

or

$$\text{reject } H_0 : \boldsymbol{\mu}_1 = \boldsymbol{\mu}_2 \quad \text{if } D(\bar{\mathbf{x}}_1, \bar{\mathbf{x}}_2) > c\,,$$

where the critical value c is to be determined such that Pr[reject $H_0 | H_0$ is true] *takes a fixed value.* Remarks apply similar to those following the formulation of the heuristic test in Section 6.2. The idea is that sample standard distance between the sample mean vectors should be used to decide whether or not to accept the hypothesis. This reflects the assumption that $D(\bar{\mathbf{x}}_1, \bar{\mathbf{x}}_2)$, in some sense, is, a good estimator of the model standard distance

$$\Delta(\boldsymbol{\mu}_1, \boldsymbol{\mu}_2) = \left[(\boldsymbol{\mu}_1 - \boldsymbol{\mu}_2)'\boldsymbol{\psi}^{-1}(\boldsymbol{\mu}_1 - \boldsymbol{\mu}_2)\right]^{1/2}, \tag{5}$$

where $\boldsymbol{\psi}$ is the common covariance matrix of both populations. It is also implicitly assumed that standard distance is a good way to summarize differences in location between two p-variate populations. Note that the assumption of a common covariance matrix is not explicitly stated in the "heuristic test" — a point to be discussed later.

The two-sample T^2-statistic is used in the same situations as the univariate t-statistic for equality of two means, except that several variables are examined simultaneously. For instance, sample 1 might consist of N_1 individuals assigned to an experimental treatment, and sample 2 of N_2 individuals in a control group. A researcher might use a T^2-test to prove that a certain treatment has an effect. However, very often a T^2-test is only the first step in an analysis of multivariate data. It ensures that a discriminant function is not just a product of random error. The student who is familiar with multiple linear regression may recall that, in any regression analysis involving more than one regressor, an overall F-test should be performed to make sure that there is something to be modelled. Similarly in linear discriminant analysis, if we estimate a discriminant function, then a preliminary overall test for mean difference may be needed to make sure that the discriminant function reflects more than just statistical error. The heuristic test based on the standard distance $D(\bar{\mathbf{x}}_1, \bar{\mathbf{x}}_2)$ serves this purpose. From Section 5.2, recall that the coefficients of the linear discriminant function are given by

$$\boldsymbol{\beta} = \boldsymbol{\psi}^{-1}(\boldsymbol{\mu}_1 - \boldsymbol{\mu}_2). \tag{6}$$

Since $\boldsymbol{\psi}^{-1}$ is assumed nonsingular, it follows that $\boldsymbol{\beta} = \mathbf{0}$ exactly if $\boldsymbol{\mu}_1 - \boldsymbol{\mu}_2 = \mathbf{0}$, which is the same as the hypothesis H_0 of equation (2). We shall elaborate on this more in Section 6.5.

Instead of the standard distance, the statistic

$$\begin{aligned} T^2 &= \frac{N_1 N_2}{N_1 + N_2} D^2(\bar{\mathbf{x}}_1, \bar{\mathbf{x}}_2) \\ &= \frac{N_1 N_2}{N_1 + N_2} (\bar{\mathbf{x}}_1 - \bar{\mathbf{x}}_2)' \mathbf{S}^{-1} (\bar{\mathbf{x}}_1 - \bar{\mathbf{x}}_2) \end{aligned} \tag{7}$$

is commonly used, as originally proposed by Hotelling (1931). Equation (7) reduces to the square of the univariate t-statistic if $p = 1$; see Exercise 1. Hotelling originally studied the distribution of T^2 under H_0, but, similarly to the one-sample case, T^2 is usually transformed to an F-statistic because of the following result.

Theorem 6.4.1 *Let $\mathbf{x}_{11}, \ldots, \mathbf{x}_{1N_1}$ and $\mathbf{x}_{21}, \ldots, \mathbf{x}_{2N_2}$ denote independent samples from two p-variate, normal distributions*

$$\mathbf{x}_{1i} \sim iid\, \mathcal{N}_p(\boldsymbol{\mu}_1, \boldsymbol{\psi}), \; i = 1, \ldots, N_1,$$

and

$$\mathbf{x}_{2i} \sim iid\, \mathcal{N}_p(\boldsymbol{\mu}_2, \boldsymbol{\psi}), \; i = 1, \ldots, N_2, \tag{8}$$

with the same covariance matrix $\boldsymbol{\psi}$ in both populations, and assume that $N_1 + N_2 > p + 1$. Let $\bar{\mathbf{x}}_1$, $\bar{\mathbf{x}}_2$, and $\mathbf{S}$ denote the sample mean vectors and the pooled sample covariance matrix. Let $\mathbf{d} = \bar{\mathbf{x}}_1 - \bar{\mathbf{x}}_2$ and $\boldsymbol{\delta} = \boldsymbol{\mu}_1 - \boldsymbol{\mu}_2$ denote the vectors of mean differences in the sample and in the model, respectively. Then the statistic

$$F = \frac{(N_1 + N_2 - p - 1)}{p(N_1 + N_2 - 2)} \cdot \frac{N_1 N_2}{N_1 + N_2} \cdot (\mathbf{d} - \boldsymbol{\delta})' \mathbf{S}^{-1} (\mathbf{d} - \boldsymbol{\delta}) \tag{9}$$

follows an F-distribution with p and $(N_1 + N_2 - p - 1)$ degrees of freedom.

Proof See Anderson (1984), Muirhead (1982), or Seber (1984). Note that the assumption $N_1 + N_2 > p + 1$ is needed to make sure that $\mathbf{S}$ is nonsingular; see Exercise 8 in Section 5.3. ■

As in Theorem 6.2.1, the distribution of F depends neither on the mean vectors $\boldsymbol{\mu}_j$ nor on the covariance matrix $\boldsymbol{\psi}$. This is, of course, highly desirable. Now, Theorem 6.4.1 can be used to construct a test for $H_0: \boldsymbol{\mu}_1 = \boldsymbol{\mu}_2$ or more generally, for the hypothesis $\boldsymbol{\delta} = \boldsymbol{\delta}_0$, where $\boldsymbol{\delta}_0$ is a prespecified vector of hypothetical mean differences; see Exercise 2. A direct application of Theorem 6.4.1 with $\boldsymbol{\delta} = \mathbf{0}$ leads to a rule based on any of the following three statistics:

(i) $D = \left[(\bar{\mathbf{x}}_1 - \bar{\mathbf{x}}_2)'\mathbf{S}^{-1}(\bar{\mathbf{x}}_1 - \bar{\mathbf{x}}_2)\right]^{1/2}$,

(ii) $T^2 = \frac{N_1 N_2}{N_1+N_2} \cdot D^2$,

(iii) $F = \frac{N_1+N_2-p-1}{(N_1+N_2-2)p} \cdot \frac{N_1 N_2}{N_1+N_2} \cdot D^2 = \frac{N_1+N_2-p-1}{(N_1+N_2-2)p} \cdot T^2$.

Rule for testing the hypothesis $\boldsymbol{\mu}_1 = \boldsymbol{\mu}_2$: Accept *the hypothesis if any of the following three equivalent conditions is satisfied*:

(i) $D \le \left[\frac{N_1+N_2}{N_1 N_2} \cdot \frac{(N_1+N_2-2)p}{N_1+N_2-p-1} \cdot f\right]^{1/2}$,

(ii) $T^2 \le \frac{(N_1+N_2-2)p}{N_1+N_2-p-1} \cdot f$,

(iii) $F \le f$,

where f is the $(1-\alpha)$-quantile of the F-distribution with p and $(N_1 + N_2 - p - 1)$ degrees of freedom. Otherwise, reject *the hypothesis.*

Theorem 6.4.1 says that this test has level α in the normal theory setup. The rule is in accordance with the heuristic principle explained earlier. The normality assumptions allow determining the critical value c as a function of the F-quantile. For large sample sizes, we may also use a chi-square approximation to the distribution of T^2; see the discussion following Theorem 6.6.1.

Example 6.4.1 For the two species of midges (see Examples 1.1, 5.2.4, and 5.3.1), the bivariate standard distance is $D = 3.94$. With $N_1 = 6$, $N_2 = 9$, and $p = 2$, we find the 95 percent quantile of the F-distribution with 2 and 12 degrees of freedom as $f = 3.89$. The corresponding critical value for D is $c = 1.53$. Hence, the null hypothesis is rejected. For a test at the significance level $\alpha = 0.01$, we find $f = 6.93$ and $c = 2.04$, and equality of the mean vectors is rejected at this higher level of significance as well. We can be fairly confident that the discriminant function discussed earlier is not just due to random error. □

In Example 6.4.1 the conclusion is clear, and, although assessing the underlying assumptions of bivariate normality and equality of the covariance matrices is almost impossible for samples as small as $N_1 = 9$ and $N_2 = 6$, most statisticians would accept the T^2-test as a valid tool in this case. However, if the significance is only marginal, then it becomes critical to understand the role of the assumptions.

As in the one-sample case, it is not really multivariate normality of the two populations that is required, but rather multivariate normality of the difference between sample mean vectors, $\mathbf{d} = \bar{\mathbf{x}}_1 - \bar{\mathbf{x}}_2$, along with the assumption that the pooled-sample covariance matrix $\mathbf{S}$ follows a Wishart distribution and is independent of $\mathbf{d}$. However, if both sample sizes are large, the assumptions on $\mathbf{S}$ become less important. At the same time, the central limit theorem ensures approximate normality of $\bar{\mathbf{x}}_1 - \bar{\mathbf{x}}_2$ if N_1 and N_2 are sufficiently large. See the discussion following Example 6.2.3.

The importance of assuming equality of the two covariance matrices is unfortunately even more difficult to assess. Although equality of covariance matrices can be assessed by a test (see Exercise 13 in Section 6.6), a decision to consider them as unequal leads to the problem that it is no longer straightforward to test for equality of the mean vectors. The problem of testing the hypothesis $\boldsymbol{\mu}_1 = \boldsymbol{\mu}_2$ based on samples from $N_p(\boldsymbol{\mu}_1, \boldsymbol{\psi}_1)$ and $N_p(\boldsymbol{\mu}_2, \boldsymbol{\psi}_2)$, respectively, is commonly referred to as the multivariate *Behrens–Fisher problem* and has about as many different solutions as authors who have published on this controversial topic. This is quite a contrast to the simplicity of the T^2-test. See Seber (1984) for a review.

Example 6.4.2 This is the continuation of Example 5.3.4 on electrodes produced by two different machines. We found the multivariate standard distance, based on $p = 5$ variables and $N_1 = N_2 = 50$ observations, as $D = 5.39$. For a test with significance level $\alpha = 0.001$, we need the 99.9 percent quantile of the F-distribution on $p = 5$ and $N_1 + N_2 - p - 1 = 94$ degrees of freedom, which is approximately $f = 4.51$. The critical value for D is then $c \approx 0.97$, and so the hypothesis of equality of the two mean vectors is clearly rejected.

Recall that we noticed some distinct differences in variability between the two machines; see Table 5.3.6 (p. 314). How could these differences affect the T^2-test? In the present example with sample sizes of 50 and a significance at the 0.1 percent level, nobody would seriously doubt the correctness of the decision to reject the null hypothesis, so these concerns would most likely be ignored. □

Although the conclusion is clear in both examples, one might still question the validity of the normality assumptions in other cases and prefer a test that does not depend on distributional assumptions. In Section 6.7 we will present randomization tests that provide a powerful alternative.

As in the one-sample case, the theoretical result that gives the distribution of the test statistic can also be used to construct confidence regions. In the two-sample case this will be a confidence region for the difference between the two mean vectors, $\boldsymbol{\delta} = \boldsymbol{\mu}_1 - \boldsymbol{\mu}_2$. Following the same logic as in Section 6.3 and assuming that the conditions of Theorem 6.4.1 hold,

$$\Pr\left[(\mathbf{d} - \boldsymbol{\delta})'\mathbf{S}^{-1}(\mathbf{d} - \boldsymbol{\delta}) \le c^2\right] = 1 - \alpha \,, \tag{10}$$

where $\mathbf{d} = \bar{\mathbf{x}}_1 - \bar{\mathbf{x}}_2$ is the vector of sample mean differences,

$$c^2 = \frac{p(N_1 + N_2 - 2)}{N_1 + N_2 - p - 1} \cdot \frac{N_1 + N_2}{N_1 N_2} \cdot f\,, \tag{11}$$

and f is the $(1-\alpha)$-quantile of the F-distribution on p and $(N_1 + N_2 - p - 1)$ degrees of freedom. Thus, all $\mathbf{x} \in \mathbb{R}^p$ inside the set

$$(\mathbf{x} - \mathbf{d})'\mathbf{S}^{-1}(\mathbf{x} - \mathbf{d}) \le c^2 \tag{12}$$

form an elliptically shaped confidence region with confidence probability $1 - \alpha$, centered at $\mathbf{d}$. The confidence region covers all values of $\boldsymbol{\delta}$ for which the hypothesis $\boldsymbol{\mu}_1 - \boldsymbol{\mu}_2 = \boldsymbol{\delta}$ would be accepted in a test with significance level α; see Exercise 3. Since the construction of confidence regions has been illustrated rather extensively in Section 6.3, no example is given here. See Exercise 7 for a practical application.

Exercises for Section 6.4

1. Show that the T^2-statistic for dimension $p = 1$ is the same as the square of a univariate t-statistic for testing equality of two means.

2. Use Theorem 6.4.1 to formulate a test for the hypothesis $\boldsymbol{\delta} = \boldsymbol{\delta}_0$, where $\boldsymbol{\delta} = \boldsymbol{\mu}_1 - \boldsymbol{\mu}_2$ is the difference between the mean vectors and $\boldsymbol{\delta}_0$ is a fixed vector in $\mathbb{R}^p$. Show that this test reduces to the one given in the text for the hypothesis $\boldsymbol{\mu}_1 = \boldsymbol{\mu}_2$ by setting $\boldsymbol{\delta}_0 = \mathbf{0}$.

3. Show that the T^2-test for equality of two mean vectors is invariant under nonsingular linear transformations. That is, show that, if all observations $\mathbf{x}_{ji}$ are transformed according to $\mathbf{y}_{ji} = \mathbf{A}\mathbf{x}_{ji} + \mathbf{b}$, where $\mathbf{A}$ is a nonsingular $p \times p$ matrix and $\mathbf{b}$ is a fixed vector, then the value of the T^2-statistic computed from the $\mathbf{y}$-data is the same as the value of the T^2-statistic computed from the $\mathbf{x}$-data. (An application is given in Exercise 7).

4. Show that the confidence region (12) consists of all vectors $\boldsymbol{\delta}_0 \in \mathbb{R}^p$ for which the hypothesis $\boldsymbol{\mu}_1 - \boldsymbol{\mu}_2 = \boldsymbol{\delta}_0$ would be accepted at a significance level of α. *Hint*: Use Exercise 1.

5. Show that, under the assumptions stated in Theorem 6.4.1, the distribution of T^2 does not depend on $\boldsymbol{\mu}_1$ and $\boldsymbol{\mu}_2$. *Hint*: Argue that the joint distribution of $\bar{\mathbf{x}}_1 - \boldsymbol{\mu}_1$, $\bar{\mathbf{x}}_2 - \boldsymbol{\mu}_2$, and $\mathbf{S}$ does not depend on the $\boldsymbol{\mu}_j$.

6. Perform a T^2-test for equality of the mean vectors of four morphometric variables in two species of flea beetles; see Example 5.3.2 and Table 5.3.2. Use a significance level of 1 percent.

7. This exercise is based on the midge data of Example 6.4.1. Define two variables Y_1 = *antenna length* + *wing length* and Y_2 = *wing length*.

(a) Compute a T^2-statistic for the hypothesis of equality of the mean vectors in the two groups, based on $\mathbf{Y}$. You should get the same value as if you use the original variables. Thus, verify the invariance result of Exercise 3 numerically for this particular example and transformation.

(b) Show that, in univariate t-tests at a significance level of 5 percent performed on each of the variables Y_1 and Y_2 separately, the hypotheses of equality of means would be accepted.

(c) Draw a scatterplot of Y_1 vs. Y_2 for the data in both groups, marking the data points of the two groups with different symbols, and explain how it can happen that, in the univariate tests, the hypotheses of equality of means would be accepted, whereas, in the multivariate test, the hypothesis of equality of both means is clearly rejected.

(d) Draw a 98 percent confidence ellipse for the vector of mean differences, using equation (12), and show that this ellipse does not cover the vector $\mathbf{0}$. In the same graph, draw a rectangle corresponding to univariate 99 percent confidence intervals for the mean differences of Y_1 and Y_2, similar to Figure 6.3.1. How does this show that it is better to do a multivariate T^2-test rather than two univariate t-tests?

8. Perform a T^2-test for equality of the mean vectors of eight morphometric variables in two species of voles; see Example 5.4.4 and Table 5.4.1. Use a significance level of 1 percent. (Only the 89 classified observations are to be used in this test).

6.5 Inference for Discriminant Function Coefficients

The two-sample T^2-test of Section 6.4 may be viewed as an overall test of significance in linear discriminant analysis. With $\boldsymbol{\mu}^{(1)}$ and $\boldsymbol{\mu}^{(2)}$ denoting the mean vectors in two populations and $\boldsymbol{\psi}$ the common covariance matrix, the vector of discriminant function coefficients is $\boldsymbol{\beta} = \boldsymbol{\psi}^{-1}(\boldsymbol{\mu}^{(1)} - \boldsymbol{\mu}^{(2)})$, and therefore, the hypothesis $\boldsymbol{\mu}^{(1)} = \boldsymbol{\mu}^{(2)}$ is equivalent to $\boldsymbol{\beta} = \mathbf{0}$, i.e., redundancy of all variables. Only if $\boldsymbol{\beta} \neq \mathbf{0}$ does the estimation of a linear classification rule make sense.

Now, suppose that a T^2-test has rejected the hypothesis $\boldsymbol{\beta} = \mathbf{0}$. Then we may proceed to a finer analysis by asking which of the variables are needed for discrimination and which ones could be omitted without loss of information. This is the problem of redundancy of variables discussed at length in Section 5.6, on which the current material is based. We will develop a test for redundancy of a variable in a discriminant function, and then present a more general test for redundancy of a subset of variables. Tests of redundancy may help to select a model that contains only a subset of the available variables. An alternative approach to model selection is based on the performance of a classification rule in classifying future observations. This has been discussed in Section 5.5, using the leave-one-out error rate.

Using the same setup as in Section 5.6, let $\mathbf{Y} = \begin{pmatrix} \mathbf{Y}_1 \\ \mathbf{Y}_2 \end{pmatrix}$ denote a p-variate random vector partitioned into subvectors $\mathbf{Y}_1$ of dimension q and $\mathbf{Y}_2$ of dimension $p - q$. Write $\boldsymbol{\mu}^{(j)} = E_j(\mathbf{Y}) = \begin{pmatrix} \boldsymbol{\mu}_1^{(j)} \\ \boldsymbol{\mu}_2^{(j)} \end{pmatrix}$ for the mean of $\mathbf{Y}$ in group j ($j = 1, 2$), and let $\boldsymbol{\psi} = \text{Cov}[\mathbf{Y}] = \begin{pmatrix} \boldsymbol{\psi}_{11} & \boldsymbol{\psi}_{12} \\ \boldsymbol{\psi}_{21} & \boldsymbol{\psi}_{22} \end{pmatrix}$ denote the common covariance matrix, all partitioned analogously to $\mathbf{Y}$. Let $\boldsymbol{\delta} = \begin{pmatrix} \boldsymbol{\delta}_1 \\ \boldsymbol{\delta}_2 \end{pmatrix} = \boldsymbol{\mu}^{(1)} - \boldsymbol{\mu}^{(2)}$ denote the vector of mean differences, and set

$$\boldsymbol{\delta}_{2.1} = \boldsymbol{\delta}_2 - \boldsymbol{\psi}_{21}\boldsymbol{\psi}_{11}^{-1}\boldsymbol{\delta}_1 \tag{1}$$

and

$$\boldsymbol{\psi}_{22.1} = \boldsymbol{\psi}_{22} - \boldsymbol{\psi}_{21}\boldsymbol{\psi}_{11}^{-1}\boldsymbol{\psi}_{12}\,. \tag{2}$$

From equations (22) and (23) of Section 5.6, we know that these can be interpreted, under appropriate conditions, as the vector of conditional mean differences of $\mathbf{Y}_2$, given $\mathbf{Y}_1$, and as the covariance matrix of the conditional distribution of $\mathbf{Y}_2$ in both groups. Also, recall from Theorem 5.6.3 that redundancy of $\mathbf{Y}_2$ is equivalent to $\boldsymbol{\delta}_{2.1} = \mathbf{0}$. Thus, if we can construct a test for the hypothesis that a conditional mean difference is zero, we automatically have a test for redundancy of subsets of variables.

From here on, assume that $\mathbf{Y}$ is p-variate normal in both populations to make sure that all conditions for Theorem 5.6.4 to apply are satisfied. Let $r_j(\mathbf{y}_1) = E_j[\mathbf{Y}_2|\mathbf{Y}_1 = \mathbf{y}_1]$ denote the regression function of $\mathbf{Y}_2$ on $\mathbf{Y}_1$ in group j, $j = 1, 2$. By Theorem 5.6.4, we can write

$$r_1(\mathbf{y}_1) = \boldsymbol{\alpha}_0 + \boldsymbol{\gamma} + \boldsymbol{\alpha}'\mathbf{y}_1\,,$$

and

$$r_2(\mathbf{y}_1) = \boldsymbol{\alpha}_0 + \boldsymbol{\alpha}'\mathbf{y}_1 \tag{3}$$

where

$$\boldsymbol{\alpha} = \boldsymbol{\psi}_{11}^{-1}\boldsymbol{\psi}_{12}\,, \quad \boldsymbol{\alpha}_0 = \boldsymbol{\mu}_2^{(2)} - \boldsymbol{\alpha}'\boldsymbol{\mu}_1^{(2)}, \quad \text{and } \boldsymbol{\gamma} = \boldsymbol{\delta}_{2.1}\,; \tag{4}$$

see Exercise 1. Using X for an indicator variable that takes value 1 in group 1 and 0 in group 2, we can rewrite the model (3) as a single equation:

$$r(x, \mathbf{y}_1) = E[\mathbf{Y}_2|X = x,\ \mathbf{Y}_1 = \mathbf{y}_1] = \boldsymbol{\alpha}_0 + \boldsymbol{\alpha}'\mathbf{y}_1 + \boldsymbol{\delta}_{2.1}x\,. \tag{5}$$

This model forms the basis for constructing a test for the hypothesis $\boldsymbol{\delta}_{2.1} = \mathbf{0}$. We will treat it in detail for the case $q = p - 1$, i.e., test for redundancy of a single variable. For $q = p - 1$, the parameters α_0 and $\delta_{2.1}$ are both scalars, and $\boldsymbol{\alpha}$ is a vector of $(p - 1)$ regression coefficients.

Specifically, suppose that we wish to test for redundancy of variable Y_j in the discriminant function. Then we have to estimate the parameters of a linear regression of Y_j on the other $(p-1)$ Y-variables and X, where X is an indicator variable for one of the two groups. We write this regression model as

$$E\left[Y_j|X, \text{ other } Y\text{-variables}\right] = \alpha_0 + \sum_{\substack{h=1 \\ h\neq j}}^{p} \alpha_h Y_h + \gamma X . \tag{6}$$

Since multivariate normality with equal covariance matrices was assumed, the conditions of the classical regression model are satisfied, including the assumption of normal errors with constant variance. Thus, we can estimate the $(p+1)$ parameters $\alpha_0, \alpha_1, \ldots, \alpha_{j-1}, \alpha_{j+1}, \ldots, \alpha_p$ and γ by least squares. Let $\hat{\gamma}$ denote the least squares estimate of γ, and $\text{se}(\hat{\gamma})$ its standard error. Then a test for the hypothesis $\gamma = 0$ can be constructed using the statistic

$$t_j = \hat{\gamma}/\text{se}(\hat{\gamma}) . \tag{7}$$

With $N = N_1 + N_2$ observations, under the hypothesis $\gamma = 0$, the statistic t_j follows a t-distribution with $(N-p-1)$ degrees of freedom. That is, at a level α, we would accept the hypothesis of redundancy if $|t_j| \leq c$ and reject it otherwise, where c is the $(1-\alpha/2)$-quantile of the t-distribution with $(N-p-1)$ degrees of freedom.

Example 6.5.1 continuation of the midge example

Let Y_1 = *antenna length* and Y_2 = *wing length*, both in mm/100 as in Table 5.3.1. In Example 5.3.1, we calculated the linear discriminant function as

$$\text{LDF} = 0.582Y_1 - 0.381Y_2 .$$

Suppose we wish to test for redundancy of Y_2, using the technique outlined. Then we have to regress Y_2 on Y_1 and X, where $X = 1$ for observations from group 1 (*Af*) and $X = 0$ for observations from group 2 (*Apf*). The regression model to be fitted is

$$E\left[Y_2|X, Y_1\right] = \alpha_0 + \alpha_1 Y_1 + \gamma X,$$

and we obtain the following least squares estimates and associated standard errors (see Exercise 2):

$$\begin{aligned} \hat{\alpha}_0 &= 84.67, & \text{se}(\hat{\alpha}_0) &= 35.6, \\ \hat{\alpha}_1 &= 0.880, & \text{se}(\hat{\alpha}_1) &= 0.288, \\ \hat{\gamma} &= -28.66, & \text{se}(\hat{\gamma}) &= 7.2 . \end{aligned}$$

This is illustrated in Figure 6.5.1, where two lines are given by

$$r_1(y_1) = \hat{\alpha}_0 + \hat{\gamma} + \hat{\alpha}_1 y_1 = 56.01 + 0.88 y_1$$

for group 1, and

$$r_2(y_1) = \hat{\alpha}_0 + \hat{\alpha}_1 y_1 = 84.67 + 0.88 y_1$$

for group 2. The parameter estimate $\hat{\gamma}$ corresponds to the vertical distance between the two lines, i.e., the estimated conditional mean difference. The t-statistic (7) for redundancy of Y_2 in the discriminant function is $t_2 = -28.66/7.2 \approx -3.98$, with $N_1 + N_2 - p - 1 = 12$ degrees of freedom, and the hypothesis of redundancy of Y_2 is rejected at any reasonable level of significance. We conclude that Y_2 cannot be omitted from the discriminant function. Note that *wing length* appears redundant in a univariate setup; see Exercise 7(b) in Section 6.4. Thus, this example is similar in nature to situation (iv) in Figure 5.6.3. □

Example 6.5.2 continuation of Example 5.3.2

Two species of flea beetles. In an earlier analysis, we saw that all four variables together do not seem to discriminate much better than subsets of three variables. Let us test if variable Y_3 = *length of the second antennal joint* is redundant. To do this test, we regress Y_3 on Y_1, Y_2, Y_4, and an indicator variable X, that is, we fit a regression

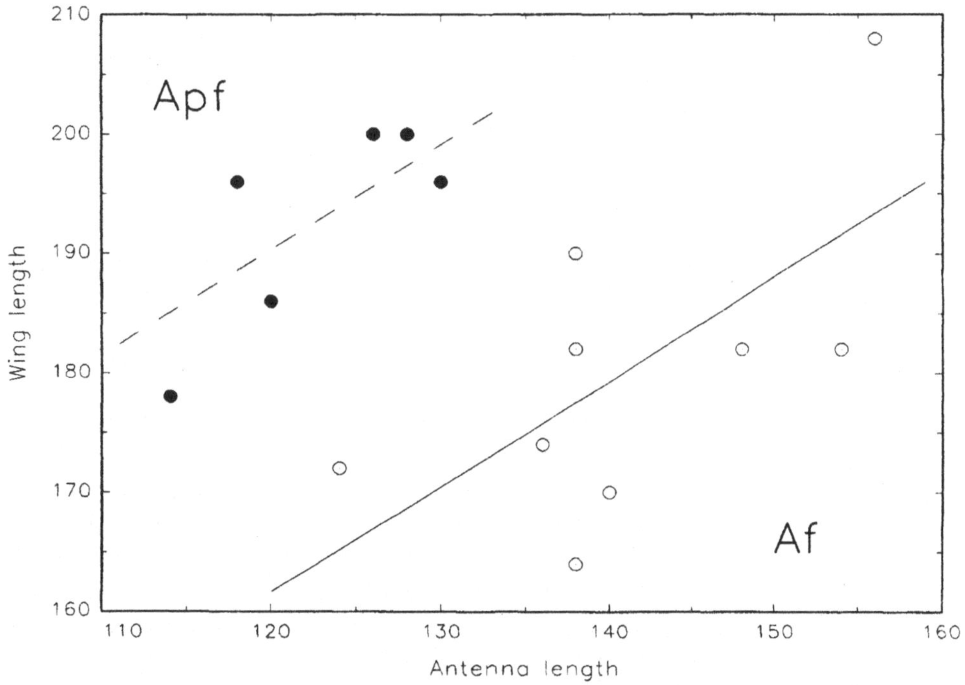

Figure 6-5-1 Scatterplot of Y_2 = *antenna length* vs. Y_1 = *wing length* in Example 6.5.1 with parallel regression lines of Y_2 on Y_1.

model

$$E[Y_3|X, Y_1, Y_2, Y_4] = \alpha_0 + \alpha_1 Y_1 + \alpha_2 Y_2 + \alpha_4 Y_4 + \gamma X.$$

Leaving numerical details to the reader (Exercise 3), we find $|t_3| = 1.332$ on 34 degrees of freedom. This corresponds to a p-value of approximately 0.19, which would usually not be considered significant. Thus we accept the hypothesis of redundancy of Y_3 and conclude that discrimination should be based only on variables Y_1, Y_2, and Y_4 or a subset of these. □

If we wish to compute a t-statistic for redundancy of each of p variables in a given set, then the procedure used in Examples 6.5.1 and 6.5.2 becomes too involved. Therefore, it will be useful to have an alternative way to compute the statistic t_j of equation (7). Such an alternative is provided by the following theorem, which, at the same time, relates the t-statistic to multivariate standard distance.

Theorem 6.5.1 *Let D denote the multivariate standard distance between the means of two p-variate samples of size N_1 and N_2, respectively. Let D_{-j} denote the multivariate standard distance between the same two samples, using all variables except Y_j. Let t_j of equation (7) denote the t-statistic for redundancy of variable Y_j in the discriminant function. Then*

$$|t_j| = \left[(N_1 + N_2 - p - 1) \cdot \frac{D^2 - D_{-j}^2}{m + D_{-j}^2}\right]^{1/2}, \tag{8}$$

where

$$m = \frac{(N_1 + N_2)(N_1 + N_2 - 2)}{N_1 N_2}. \tag{9}$$

Proof The proof is based on an analysis of the model of parallel regression. See Rao (1970). ■

As Theorem 6.5.1 shows, the t-statistic for redundancy of the jth variable depends on the data through the standard distances D (based on all variables) and D_{-j} (omitting variable Y_j). This is actually a nice result, because, in view of Theorem 5.6.3, we would expect D_{-j} to be only slightly smaller than D if Y_j is redundant.

Yet, Theorem 6.5.1 does not seem to solve the problem of complicated computation because one needs to compute D_{-j} for each variable Y_j to be tested for redundancy, which means choosing the proper subset of $p-1$ variables and computing the standard distance based on the subset. Fortunately there is a simple shortcut based on Exercise 18 in Section 5.6. Analogous to equation (26) of Section 5.6 and with $\mathbf{b} = (b_1, \ldots, b_p)' = \mathbf{S}^{-1}(\bar{\mathbf{y}}^{(1)} - \bar{\mathbf{y}}^{(2)})$ denoting the vector of discriminant function coefficients,

$$D_{-j}^2 = D^2 - b_j^2/a_{jj} \qquad j = 1, \ldots, p, \tag{10}$$

where a_{jj} is the jth diagonal entry of $\mathbf{S}^{-1}$ and $\mathbf{S}$ is the pooled-sample covariance matrix; see Exercise 4. Note that, for equation (10) to be valid, we need to use the vector of discriminant function coefficients in the form $\mathbf{b} = \mathbf{S}^{-1}(\bar{\mathbf{y}}^{(1)} - \bar{\mathbf{y}}^{(2)})$, not as an arbitrary vector proportional to $\mathbf{S}^{-1}(\bar{\mathbf{y}}^{(1)} - \bar{\mathbf{y}}^{(2)})$.

Before turning to applications, let us briefly discuss a problem closely related to the test for redundancy of a variable: standard errors of discriminant function coefficients. By equations (8) to (10), we can write

$$t_j^2 = \frac{b_j^2}{a_{jj}\left(m + D_{-j}^2\right)/(N_1 + N_2 - p - 1)}. \tag{11}$$

From (7), $t_j = \hat{\gamma}/\mathrm{se}(\hat{\gamma})$, where $\hat{\gamma}$ is the estimated conditional mean difference. Similarly, we may regard (11) as the square of the ratio of b_j to its standard error. That is, by defining

$$\mathrm{se}(b_j) = \left[\frac{a_{jj}\left(m + D_{-j}^2\right)}{N_1 + N_2 - p - 1}\right]^{1/2}, \tag{12}$$

we can write

$$t_j = \frac{b_j}{\mathrm{se}(b_j)}, \qquad j = 1, \ldots, p. \tag{13}$$

This is a bit dangerous because, by defining $\mathrm{se}(b_j)$ implicitly via equation (11), we do not really know if (12) is, in some sense, a good estimator of $\sqrt{\mathrm{var}(b_j)}$. Thus, to give a better justification for (12), we would need to find the distribution of $\mathbf{b} = \mathbf{S}^{-1}(\bar{\mathbf{y}}^{(1)} - \bar{\mathbf{y}}^{(2)})$ or at least its first and second moments $E[\mathbf{b}]$ and $\mathrm{Cov}[\mathbf{b}]$. This is beyond the level of this text, however, so the reader is asked to accept the somewhat pragmatic definition of a standard error suggested by equation (11).

There is an additional problem with the notion of standard errors of discriminant function coefficients. It is sometimes argued (see, e.g., Rao 1970) that this notion is meaningless since discriminant functions are not uniquely defined. As we have seen in Section 5.2, if $V = \boldsymbol{\beta}'\mathbf{Y}$ is a linear discriminant function, then $V^* = c \cdot V$ is an equally good linear discriminant function for any $c \neq 0$ since it yields the same standard distance between the two groups. Only *ratios* of discriminant function coefficients are well defined but not the coefficients themselves. Similarly, testing for a fixed value of a discriminant function coefficient is meaningless according to this view, unless the fixed value is zero.

Contrary to this view, if we define the vector of discriminant function coefficients as $\boldsymbol{\beta} = \boldsymbol{\psi}^{-1}\left(\boldsymbol{\mu}^{(1)} - \boldsymbol{\mu}^{(2)}\right)$, *not* as any vector proportional to $\boldsymbol{\psi}^{-1}\left(\boldsymbol{\mu}^{(1)} - \boldsymbol{\mu}^{(2)}\right)$, then the coefficients are perfectly well defined, and standard errors of the coefficients $\mathbf{b} = \mathbf{S}^{-1}\left(\bar{\mathbf{y}}^{(1)} - \bar{\mathbf{y}}^{(2)}\right)$ are a reasonable concept. This is the view taken in the current section; also see Exercise 5.

Example 6.5.3 continuation of Example 6.5.2

Applying the preceding theory to the flea beetle data with $p = 4$ variables, we obtain the results displayed in Table 6.5.1. For each variable, its "reduced standard distance" D_{-j} is computed from equation (10), and the standard error and associated t-statistic follow from equations (12) and (13). The standard distance based on all four variables is $D = 3.701$ as seen earlier in Example 5.3.2. For individual tests of redundancy of each of the four variables in the set $\{Y_1, Y_2, Y_3, Y_4\}$, we need to compare the $|t_j|$ to the $(1 - \alpha/2)$-quantile of the t-distribution with $N_1 + N_2 - p - 1 = 34$ degrees of freedom. For $\alpha = 0.05$, the critical value is 2.03, and therefore, the hypothesis of redundancy of Y_3 would be accepted. The reader should be aware, though, of the problems associated with multiple testing, which are the same here as in multiple regression analysis.

Part (b) of Table 6.5.1 displays results of an analysis after omitting Y_3. All three remaining coefficients now appear relatively stable, i.e., none of the t_j-statistics is below 3 in absolute value. □

Let us briefly discuss the relevance of the distributional assumptions for the test of redundancy. Before equation (3), we stated that **Y** will be assumed p-variate normal with equal covariance matrices in both groups. However, in the test for redundancy of Y_j we need less, namely,

(i) normality of the conditional distribution of Y_j, given all other Y-variables,

(ii) linearity of the conditional mean of Y_j, with identical regression coefficients (i.e., parallel regression lines or planes), and

(iii) equality of the conditional variance of Y_j in both groups, given the other Y-variables.

Table 6.5.1. Standard errors and t-statistics for coefficients of the discriminant function in the flea beetle example

Variable	Coefficient	Standard Error	t-Value	D_{-j}
(a) Analysis using all four variables				
Y_1	0.345	0.058	5.99	2.170
Y_2	−0.130	0.054	−2.41	3.339
Y_3	−0.106	0.080	−1.33	3.582
Y_4	−0.143	0.050	−2.87	3.206
(b) Analysis without Y_3				
Y_1	0.358	0.051	7.06	1.749
Y_2	−0.169	0.039	−4.31	2.658
Y_4	−0.149	0.047	−3.19	3.015

These conditions are weaker than multivariate normality; see, e.g., Exercise 13 in Section 5.6. In cases of concern, one might use the parallel regression technique illustrated in Examples 6.5.1 and 6.5.2 to assess the assumptions via residual analysis.

Now, we proceed to a generalization of the test for redundancy of a single variable to several variables. Suppose that $\mathbf{Y} = \begin{pmatrix} \mathbf{Y}_1 \\ \mathbf{Y}_2 \end{pmatrix}$ is partitioned into q and $(p-q)$ variables and the vector of discriminant function coefficients is partitioned analogously as $\boldsymbol{\beta} = \begin{pmatrix} \boldsymbol{\beta}_1 \\ \boldsymbol{\beta}_2 \end{pmatrix}$. Suppose that we wish to test the hypothesis of redundancy of $\mathbf{Y}_2$, i.e., $\boldsymbol{\beta}_2 = \mathbf{0}$. Let D_p and D_q denote the multivariate standard distance between the two sample mean vectors based on all p variables (D_p) and based on $\mathbf{Y}_1$ only (D_q).

Theorem 6.5.2 *Suppose that* $\mathbf{Y}$ *is p-variate normal with identical covariance matrix* ψ *in both groups. In the setup just explained, let* D_p *and* D_q *denote the standard distances based on* $\mathbf{Y}$ *and* $\mathbf{Y}_1$ *respectively. If the* $(p-q)$ *variables* $\mathbf{Y}_2$ *are redundant for discrimination, then the statistic*

$$F = \frac{N_1 + N_2 - p - 1}{p - 1} \cdot \frac{D_p^2 - D_q^2}{m + D_q^2}, \tag{14}$$

where $m = (N_1 + N_2)(N_1 + N_2 - 2)/N_1 N_2$, *follows an F-distribution with* $(p-q)$ *and* $(N_1 + N_2 - p - 1)$ *degrees of freedom.*

Proof The proof of this theorem due originally to Rao (1970), is beyond the level of this text. See Mardia et al. (1979, p. 78) or Rao (1970). ■

Theorem 6.5.2 can be applied directly to a test for redundancy of a subset of $(p-q)$ variables (or, equivalently, sufficiency of the remaining q variables) by computing the standard distances based on the full set (D_p) and the reduced set (D_q) of variates and evaluating (14). For a test of level α, let f denote the $(1-\alpha)$-quantile of the F-distribution with $(p-q)$ and $(N_1 + N_2 - p - 1)$ degrees of freedom. The hypothesis of redundancy is accepted if $F \le f$, and rejected otherwise. Again, note that the test statistic (14) depends on the data only through the two standard distances. In fact for $q = p - 1$, the F-statistic is the same as the square of the t-statistic in Theorem 6.5.1, that is, the test for redundancy of a single variable follows as a special case of Theorem 6.5.2. Interestingly, it also contains the T^2-test for equality of two mean vectors as a special case. To see this, set $q = 0$, i.e., test for simultaneous redundancy of all p variables, and define $D_0 = 0$. Then the statistic (14) becomes

$$F = \frac{N_1 + N_2 - p - 1}{mp} \cdot D_p^2, \tag{15}$$

which is the same as the F-statistic from Section 6.4. Once again, this reflects the fact that simultaneous redundancy of all p variables, i.e., $\boldsymbol{\beta} = \mathbf{0}$ is equivalent to equality of the mean vectors.

Example 6.5.4 Electrodes. In Example 5.3.4, we computed a discriminant function for two samples of size $N_1 = N_2 = 50$ of electrodes produced by two different machines. In Example 4.4.2, we applied a T^2-test based on $p = 5$ variables. The test turned out to be highly significant. It was also noted earlier that variable Y_4 by itself has a very large standard distance. Now, we will apply the test for redundancy of a subset of variables to see if the four variables $\{Y_1, Y_2, Y_3, Y_5\}$ contribute to the discrimination based on Y_4. The formal hypothesis is $\beta_1 = \beta_2 = \beta_3 = \beta_5 = 0$.

Before testing for redundancy, consider the discriminant function coefficients in some detail. Table 6.5.2 list the coefficients and their associated standard errors and t-statistics. Not all coefficients seem stable, and therefore, it is a good idea to eliminate redundant variables.

The multivariate standard distance based on all five variables is $D_5 = 5.387$, and the standard distance based on variable Y_4 alone is $D_1 = 4.186$. The value of the test statistic (14) is $F = 12.59$ on 4 and 94 degrees of freedom. This value is significant at all commonly used levels, and we conclude, therefore, that we would lose information by removing all four questionable variables from the discriminant function. This does not necessarily mean that all five variables are needed. As Table 6.5.2 shows, variable Y_3 is a good candidate for elimination. □

A final word of caution is in order regarding the tests just performed. The theory of statistical tests is based on the assumption that the hypothesis to be tested is known a priori, not generated from the very data on which the test is to be performed. We have violated this "axiom" without worrying much about how this affects the significance level. All the more, it is important to perform an overall test of significance (i.e.., a T^2-test) before any variable selection or screening is done. The overall test gives at least some assurance that the "fishing trip" is more than just a random search in the sea of white noise.

The reader may also recall from Section 5.3 that there are some doubts about the assumption of equality of the two covariance matrices in the electrode example. In turn, this casts some doubt on the validity of the assumption of a constant conditional mean difference and therefore, on the validity of the test for redundancy.

Table 6.5.2 Discriminant function coefficients in the electrode example

Variable	Coefficient	Standard Error	t-Value	D_{-j}
Y_1	−1.900	0.486	−3.91	4.940
Y_2	2.224	0.460	4.83	4.739
Y_3	−0.071	0.115	−0.62	5.374
Y_4	0.879	0.080	11.03	3.231
Y_5	0.833	0.443	1.88	5.275

Exercises for Section 6.5

1. Prove equations (3) and (4).

2. In Example 6.5.1, find the least squares estimates (and their standard errors) of the regression of $Y_2 = \textit{wing length}$ on $Y_1 = \textit{antenna length}$ and X, where X is an indicator variable for group 1. Hence, verify the numerical results reported in the text. What happens if X is an indicator variable for group 2 instead of group 1? How is the test for the hypothesis $\gamma = 0$ affected?

3. In Example 6.5.2 (flea beetles), compute the least squares estimates of the regression of Y_3 on Y_1, Y_2, Y_4, and X, where X is an indicator variable for one of the two groups. Thus, verify the results reported for the test of redundancy of Y_3 in the linear discriminant function.

4. Prove equation (10). *Hint*: Use Exercise 18 in Section 5.6.

5. Let $\mathbf{b} = \mathbf{S}^{-1}\left(\bar{\mathbf{y}}^{(1)} - \bar{\mathbf{y}}^{(2)}\right)$ denote the vector of linear discriminant function coefficients, and let $\text{se}(b_j)$ be defined as in (12). Suppose that we wish to use an equivalent version of the discriminant function with coefficients $\mathbf{b}^* = c \cdot \mathbf{b}$ for some $c \neq 0$. How should we define $\text{se}(b_j^*)$ if we want tests for redundancy based on the statistics $t_j^* = b_j^*/\text{se}(b_j^*)$ to lead to the same conclusion as tests based on $t_j = b_j/\text{se}(b_j)$?

6. In Example 6.5.4, test for simultaneous redundancy of variables Y_3 and Y_5 in the discriminant function, i.e., test the hypothesis $\beta_3 = \beta_5 = 0$. Use a significance level of $\alpha = 0.05$.

7. This exercise is based on the *Microtus* data of Example 5.4.4. Use only the 89 classified observations.
   (a) Compute the coefficients of the linear discriminant function, their standard errors, and associated t-statistics. Display the results in a table similar to Table 6.5.2.
   (b) Test for redundancy of all variables except Y_1 and Y_4. Use a significance level of one percent.
   (c) Starting with the discriminant function based on all variables, perform a stepwise elimination procedure as follows. Eliminate the variable with the smallest absolute value of t, then compute a discriminant function based on the remaining variables, with associated standard errors, etc. Continue until all remaining variables have t-values significant at the one percent level. *Note*: In part (c), it is convenient to use some canned software package that offers stepwise variable selection. Problems of multiple testing are ignored altogether in this procedure.

8. This exercise explores a rather interesting relationship between linear regression and linear discriminant functions, which is useful for statisticians stranded on a remote island, with access to software for multiple linear regression, but no other software whatsoever. Suppose we wish to compute a discriminant function based on p variables $Y_1, \ldots, Y_p$. Define X as a binary variable indicating group membership, e.g., $X = 1$ for group 1, and $X = 0$ for group 2. (Any other group code $X = c_1$ for group 1 and $X = c_2$ for group 2 will work as well, as long as $c_1 \neq c_2$). Then compute the vector

$\mathbf{a} = (a_0, a_1, \ldots, a_p)'$ of coefficients of the linear regression of X on the p regressors $Y_1, \ldots, Y_p$. Let $\mathbf{b} = (b_1, \ldots, b_p)' = \mathbf{S}^{-1}\left(\bar{\mathbf{y}}^{(1)} - \bar{\mathbf{y}}^{(2)}\right)$ denote the vector of linear discriminant function coefficients.

(a) Show that $(a_1, \ldots, a_p)'$ is proportional to $\mathbf{b}$, i.e.,

$$\begin{pmatrix} a_1 \\ \vdots \\ a_p \end{pmatrix} = c \cdot \mathbf{b} \tag{16}$$

for some $c \in \mathbb{R}$. *Hint*: The proportionality constant c in (16) may depend on $\mathbf{b}$.

(b) Let R^2 denote the coefficient of determination from the linear regression of X on $Y_1, \ldots, Y_p$, and D the p-variate standard distance between the two groups. Show that

$$R^2 = \frac{N_1 N_2 D^2}{(N_1 + N_2)(N_1 + N_2 2) + N_1 N_2 D^2} \,. \tag{17}$$

(c) Show that the partial t-statistics associated with each variable in the regression of X on $Y_1, \ldots, Y_p$ are identical with the statistics t_j for redundancy of a variable in the discriminant function, that is, the approach of this exercise gives a coefficient vector proportional to the vector of discriminant function coefficients and also gives the correct t-statistics. *Hint*: Write the partial t-statistic for variable Y_j in the regression of X on $Y_1, \ldots, Y_p$ as a function of R^2 and R^2_{-j}, where R^2_{-j} is the coefficient of determination if Y_j is omitted from the analysis.

6.6 Union-Intersection and Likelihood Ratio Testing

This is a rather technical section in which we will rederive the T^2-tests discussed earlier in this chapter, using the principles of union-intersection and likelihood ratio testing. The latter approach has been introduced in Section 4.3, and therefore, finding the likelihood ratio statistics will be a purely technical matter. Since the union-intersection approach has not been used so far, we start out with a brief general description of this technique originally proposed by Roy (1953). Various applications can be found in Roy (1957).

Unlike the likelihood machinery, the union-intersection principle is not a fully automatic procedure to generate test statistics for certain hypotheses in specified parametric families of distributions. Rather, it is a heuristic principle that can be applied to many testing problems. Different statisticians might apply the union-intersection principle in different ways and, thus generate different tests for the same hypothesis. Therefore, it is generally not correct to speak of *the* union-intersection test for a given problem, although it is common to do so.

We start the description of the union-intersection approach with an artificially simple example.

Example 6.6.1 Suppose that Y is binomial (N, π) and we wish to test the hypothesis $H : \pi = \frac{1}{2}$ against the alternative $H^c : \pi \neq \frac{1}{2}$. The alternative hypothesis can be stated as $H_1^c : \pi < \frac{1}{2}$ or $H_2^c : \pi > \frac{1}{2}$. Formally, we can write the alternative hypothesis as $\pi \in \Omega_1 \cup \Omega_2$, where $\Omega_1 = [0, \frac{1}{2})$ and $\Omega_2 = (\frac{1}{2}, 1]$. Suppose that, for a test of H against H_1^c, we would accept H if $Y \geq k_1$, and reject H if $Y < k_1$ for some integer k_1 between 0 and N. That is, in a test of H against H_1^c, the acceptance region $\mathcal{A}_1$ and the rejection region $\mathcal{R}_1$ are

$$\mathcal{A}_1 = \{k_1, k_1 + 1, \ldots, N\}$$

and

$$\mathcal{R}_1 = \mathcal{A}_1^c = \{0, \ldots, k_1 - 1\}.$$

Similarly in a test of H vs H_2^c, we might find an acceptance region $\mathcal{A}_2$ and a rejection region $\mathcal{R}_2$ as

$$\mathcal{A}_2 = \{0, 1, \ldots, k_2\}$$

and

$$\mathcal{R}_2 = \mathcal{A}_2^c = \{k_2 + 1, \ldots, N\}.$$

Define $\mathcal{A} = \mathcal{A}_1 \cap \mathcal{A}_2 = \{k_1, k_1 + 1, \ldots, k_2\}$ and $\mathcal{R} = \mathcal{A}^c = \mathcal{R}_1 \cup \mathcal{R}_2 = \{0, 1, \ldots, k_1 - 1, k_2 + 1, \ldots, N\}$. Assuming that none of the two sets $\mathcal{A}$ and $\mathcal{R}$ is empty, we can now take $\mathcal{A}$ as the acceptance region and $\mathcal{R}$ as the rejection region for the test of the hypothesis H against H^c, that is, the hypothesis is accepted if the tests of H against H_1^c and H_2^c both accept it, and the hypothesis is rejected if at least one of the tests rejects it. The term"union-intersection test" reflects the fact that the rejection region is the union of several (here, two) rejection regions, and the acceptance region is the intersection of several acceptance regions. To complete the test, we would need to determine k_1 and k_2 such that

$$\Pr\left[Y \in \mathcal{R} | \pi = \tfrac{1}{2}\right] = \alpha$$

for some fixed significance level $\alpha \in (0, 1)$. (We do not elaborate here on the problem that because of discreteness of Y it is impossible to find such integers k_1 and k_2 except for selected values of α. The purpose of the example is to illustrate the idea of unions of rejection regions and intersections of acceptance regions). □

The union-intersection principle can be formulated as follows. Suppose that $\boldsymbol{\theta}$ is the parameter of interest in a family of distributions. Let $\boldsymbol{\Omega}$ denote the parameter space and $\omega \subset \boldsymbol{\Omega}$ a subspace. We wish to test the hypothesis $H : \boldsymbol{\theta} \in \omega$ against the alternative $H^c : \boldsymbol{\theta} \notin \omega$. Suppose that the hypothesis space can be expressed as the intersection of a collection of sets, say

$$\omega = \bigcap_{s \in \mathcal{I}} \omega_s .$$

Here, $\mathcal{I}$ is an index set which may be finite or infinite. The hypothesis H holds exactly if the hypotheses $H_s : \boldsymbol{\theta} \in \omega_s$ hold simultaneously for all $s \in \mathcal{I}$, and H is wrong if at least one of the H_s is wrong, i.e., if $\boldsymbol{\theta} \in \bigcup_{s\in\mathcal{I}} \omega_s^c$. Let $\mathcal{X}$ denote a sample from the distribution under consideration, and suppose that, for each s, we can construct a test for H_s against H_s^c, with acceptance region $\mathcal{A}_s$ and rejection region $\mathcal{R}_s$. Let

$$\mathcal{A} = \bigcap_s \mathcal{A}_s$$

and

$$\mathcal{R} = \mathcal{A}^c = \bigcup_s \mathcal{R}_s ,$$

and define $\mathcal{A}$ and $\mathcal{R}$ as the acceptance and the rejection region for a test of H against H^c. Then this test is called a union-intersection test, that is, the union-intersection test accepts H exactly if H_s is accepted for all s, and rejects H otherwise.

Note that the sets Ω_s do not necessarily form a partition of Ω, i.e., their intersections may not be empty. To test multivariate hypotheses, it is often useful to choose the Ω_s as the subsets of Ω corresponding to a particular linear combination of the parameter vector $\boldsymbol{\theta}$. This is illustrated in our next example, which, at the same time, constitutes our main application of the union-intersection principle.

Example 6.6.2 A union-intersection test for the hypothesis $H : \boldsymbol{\mu} = \boldsymbol{\mu}_0$. Let $\mathcal{X} = (\mathbf{x}_1, \ldots, \mathbf{x}_N)$ be a sample from some p-variate distribution with mean vector $\boldsymbol{\mu}$, and suppose that we wish to test the hypothesis $H : \boldsymbol{\mu} = \boldsymbol{\mu}_0$, where $\boldsymbol{\mu}_0$ is a specified vector in $\mathbb{R}^p$. Ignoring the problem that there may be other parameters besides $\boldsymbol{\mu}$, let $\Omega = \mathbb{R}^p$, and $\omega = \{\boldsymbol{\mu}_0\}$. Notice that H holds true exactly if $\mathbf{a}'\boldsymbol{\mu} = \mathbf{a}'\boldsymbol{\mu}_0$ for all vectors $\mathbf{a} \in \mathbb{R}^p$; see Exercise 1. Thus, for fixed $\mathbf{a} \in \mathbb{R}^p$, let $H_\mathbf{a}$ denote the hypothesis that $\mathbf{a}'\boldsymbol{\mu} = \mathbf{a}'\boldsymbol{\mu}_0$, and $H_\mathbf{a}^c$ the alternative, $\mathbf{a}'\boldsymbol{\mu} \neq \mathbf{a}'\boldsymbol{\mu}_0$. This implicitly defines a subset $\omega_\mathbf{a} = \{\boldsymbol{\mu} \in \mathbb{R}^p : \mathbf{a}'\boldsymbol{\mu} = \mathbf{a}'\boldsymbol{\mu}_0\}$ corresponding to $H_\mathbf{a}$. The problem of testing $H_\mathbf{a} : \mathbf{a}'\boldsymbol{\mu} = \mathbf{a}'\boldsymbol{\mu}_0$ against $H_\mathbf{a}^c : \mathbf{a}'\boldsymbol{\mu} \neq \mathbf{a}'\boldsymbol{\mu}_0$ is now a univariate one. Let $y_i = \mathbf{a}'\mathbf{x}_i$ $(i = 1, \ldots, N)$ denote the linearly transformed sample. Then $H_\mathbf{a}$ is equivalent to $E[Y] = \mathbf{a}'\boldsymbol{\mu}_0$. Suppose we feel that a proper way to test $H_\mathbf{a}$ is to use the statistic

$$t_\mathbf{a} = \sqrt{N}\,\frac{\bar{y} - \mathbf{a}'\boldsymbol{\mu}_0}{s_y}, \tag{1}$$

where $\bar{y}$ and s_y are the sample mean and the sample standard deviation of $y_1, \ldots, y_N$. This generates an acceptance region

$$\mathcal{A}_\mathbf{a} = \{\mathcal{X} : |t_\mathbf{a}| \leq c\}$$

and a rejection region

$$\mathcal{R}_\mathbf{a} = \{\mathcal{X} : |t_\mathbf{a}| \geq c\} = \mathcal{A}_\mathbf{a}^c$$

for the univariate test, where c is a properly chosen constant that does not depend on **a**. Let

$$\mathcal{A} = \bigcap_{\substack{\mathbf{a} \in \mathbb{R}^p \\ \mathbf{a} \neq \mathbf{0}}} \mathcal{A}_{\mathbf{a}} = \{\mathcal{X} : |t_{\mathbf{a}}| \leq c \text{ for all } \mathbf{a} \in \mathbb{R}^p\} \tag{2}$$

and

$$\mathcal{R} = \bigcup_{\substack{\mathbf{a} \in \mathbb{R}^p \\ \mathbf{a} \neq \mathbf{0}}} \mathcal{R}_{\mathbf{a}} = \{\mathcal{X} : |t_{\mathbf{a}}| > c \text{ for at least one } \mathbf{a} \in \mathbb{R}^p\}. \tag{3}$$

The union-intersection test accepts the hypothesis $\mu = \mu_0$ if $\mathcal{X} \in \mathcal{A}$, and rejects H if $\mathcal{X} \in \mathcal{R}$. But $\mathcal{X} \in \mathcal{A}$ exactly if

$$\max_{\substack{\mathbf{a} \in \mathbb{R}^p \\ \mathbf{a} \neq \mathbf{0}}} |t_{\mathbf{a}}| \leq c\,.$$

By Lemma 5.2.3 or Theorem 5.2.4,

$$\begin{aligned} \max_{\substack{\mathbf{a} \in \mathbb{R}^p \\ \mathbf{a} \neq \mathbf{0}}} |t_{\mathbf{a}}| &= \sqrt{N}\, \left[(\mu_0 - \bar{\mathbf{x}})'\mathbf{S}^{-1}(\mu_0 - \bar{\mathbf{x}})\right]^{1/2} \\ &= \sqrt{N}\, D(\mu_0, \bar{\mathbf{x}})\,, \end{aligned} \tag{4}$$

where $\bar{\mathbf{x}}$ and $\mathbf{S}$ are the sample mean vector and covariance matrix, respectively. The constant c is to be chosen so that Pr[reject $H|H$ is true] takes a fixed value. This is the same as the heuristic test proposed in Section 6.2. □

Here are some additional comments on Example 6.6.2.

1. The critical value c for the test must be determined by the distribution of the statistic (4), not by the distribution of $t_{\mathbf{a}}$ for a fixed **a**. In fact, this union-intersection tests consists of infinitely many univariate tests. If the overall test is to have a certain level α, then the individual univariate tests cannot have the same level.
2. Multivariate normality is not assumed in the derivation of the test statistic. Explicit distributional assumptions are made only at a later stage in determining a critical value for acceptance or rejection of the hypothesis. This distinguishes the union-intersection approach from the likelihood ratio approach, in which an explicit parametric model is specified at the very beginning.
3. A different union-intersection test for the same hypothesis can be obtained if we restrict the vectors $\mathbf{a} \in \mathbb{R}^p$ to the set of p unit basis vectors, that is, compute a univariate t-statistic for each of the p variables. If t_j denotes the t-statistic for the jth variable, then the corresponding union-intersection test is based on $\max_{1 \leq j \leq p} |t_j|$. Finding a critical value for this test would be difficult though. This illustrates that there may be more than one union-intersection solution for a given testing problem. More examples of union-intersection tests are given

in Exercises 2 and 3. In particular, the construction of a union-intersection test for the problem of equality of mean vectors in two independent populations is analogous to Example 6.6.2; see Exercise 2.

Now, we turn to likelihood ratio testing, using the setup from Section 4.3. Let $L(\boldsymbol{\theta})$ denote the likelihood function for a parameter vector $\boldsymbol{\theta}$, based on a given sample of observations. Let Ω denote the unconstrained parameter space and $\omega \subset \Omega$ the constrained parameter space. Suppose that we wish to test the hypothesis $H : \boldsymbol{\theta} \in \omega$. For simplicity, assume that the likelihood function takes a unique maximum in both Ω and ω, and denote the maximum likelihood estimates of $\boldsymbol{\theta}$ by $\hat{\boldsymbol{\theta}}$ (in the unconstrained space Ω) and $\tilde{\boldsymbol{\theta}}$ (in the constrained space ω), respectively. Then the likelihood ratio statistic for the hypothesis $\boldsymbol{\theta} \in \omega$ against the alternative $\boldsymbol{\theta} \in \Omega,\ \boldsymbol{\theta} \notin \omega$, is given by

$$\text{LRS} = \frac{\max_{\boldsymbol{\theta}\in\omega} L(\boldsymbol{\theta})}{\max_{\boldsymbol{\theta}\in\Omega} L(\boldsymbol{\theta})} = \frac{L(\tilde{\boldsymbol{\theta}})}{L(\hat{\boldsymbol{\theta}})}, \tag{5}$$

and the hypothesis is rejected if LRS $< c$ for some constant c. Because of the large sample properties given in Theorem 4.3.3, we will usually prefer to use the log-likelihood ratio statistic

$$\text{LLRS} = -2\ \log(\text{LRS}) = 2\left[\ell(\hat{\boldsymbol{\theta}}) - \ell(\tilde{\boldsymbol{\theta}})\right], \tag{6}$$

where $\ell(\boldsymbol{\theta}) = \log L(\boldsymbol{\theta})$ is the log-likelihood function.

Example 6.6.3 The likelihood ratio test for equality of the mean vectors of two multivariate normal populations with identical covariance matrices. This is a classical example for which we are well prepared from Section 4.3. Suppose that $\mathbf{x}_{11}, \ldots, \mathbf{x}_{1N_1}$ and $\mathbf{x}_{21}, \ldots, \mathbf{x}_{2N_2}$ are independent samples from $\mathcal{N}_p(\boldsymbol{\mu}_1, \boldsymbol{\psi})$ and $\mathcal{N}_p(\boldsymbol{\mu}_2, \boldsymbol{\psi})$, respectively, and assume that $N_1 + N_2 \geq p + 2$. Let

$$\Omega = \{(\boldsymbol{\mu}_1, \boldsymbol{\mu}_2, \boldsymbol{\psi}) : \boldsymbol{\mu}_1 \in \mathbb{R}^p,\ \boldsymbol{\mu}_2 \in \mathbb{R}^p,\ \boldsymbol{\psi} \text{ positive definite}\}$$

denote the unconstrained parameter space, and

$$\omega = \{(\boldsymbol{\mu}_1, \boldsymbol{\mu}_2, \boldsymbol{\psi}) : \boldsymbol{\mu}_1 = \boldsymbol{\mu}_2 \in \mathbb{R}^p,\ \boldsymbol{\psi} \text{ positive definite}\}$$

the constrained parameter space. The likelihood function is given by

$$L(\boldsymbol{\mu}_1, \boldsymbol{\mu}_2, \boldsymbol{\psi}) = (2\pi)^{-(N_1+N_2)p/2}\,(\det \boldsymbol{\psi})^{-(N_1+N_2)/2}$$

$$\times \exp\left\{-\frac{1}{2}\sum_{j=1}^{2} \text{tr}\left[\boldsymbol{\psi}^{-1}\sum_{i=1}^{N_j}(\mathbf{x}_{ji} - \boldsymbol{\mu}_j)(\mathbf{x}_{ji} - \boldsymbol{\mu}_j)'\right]\right\}. \tag{7}$$

From Example 4.3.6, we know that the likelihood function takes its maximum in Ω at

$$\hat{\mu}_j = \bar{\mathbf{x}}_j = \frac{1}{N_j} \sum_{i=1}^{N_j} \mathbf{x}_{ji}, \qquad j = 1, 2, \tag{8}$$

and

$$\hat{\psi} = \frac{1}{N_1 + N_2} \sum_{j=1}^{2} \sum_{i=1}^{N_j} (\mathbf{x}_{ji} - \bar{\mathbf{x}}_j)(\mathbf{x}_{ji} - \bar{\mathbf{x}}_j)'. \tag{9}$$

Thus, the maximum of the likelihood function in Ω is given by

$$\begin{aligned} L(\hat{\mu}_1, \hat{\mu}_2, \hat{\psi}) &= (2\pi)^{-(N_1+N_2)p/2} \left(\det \hat{\psi}\right)^{-(N_1+N_2)/2} \\ &\quad \times \exp\left\{-\frac{1}{2} \sum_{j=1}^{2} \operatorname{tr}\left[(\hat{\psi})^{-1} \sum_{i=1}^{N_j} (\mathbf{x}_{ji} - \bar{\mathbf{x}}_j)(\mathbf{x}_{ji} - \bar{\mathbf{x}}_j)'\right]\right\} \qquad (10) \\ &= (2\pi e)^{-(N_1+N_2)p/2} \left(\det \hat{\psi}\right)^{-(N_1+N_2)/2}, \end{aligned}$$

as follows from (8), (9), and from $\operatorname{tr}(\mathbf{I}_p) = p$.

For $(\mu_1, \mu_2, \psi) \in \omega$, i.e., in the reduced parameter space, the two multivariate normal distributions are identical, and therefore, we can apply Theorem 4.3.1 directly. The maximum likelihood estimates in ω are given by

$$\begin{aligned} \tilde{\mu}_1 = \tilde{\mu}_2 = \bar{\mathbf{x}} &= \frac{1}{N_1 + N_2} \sum_{j=1}^{2} \sum_{i=1}^{N_j} \mathbf{x}_{ji} \\ &= \frac{1}{N_1 + N_2} (N_1 \bar{\mathbf{x}}_1 + N_2 \bar{\mathbf{x}}_2) \end{aligned} \tag{11}$$

and

$$\tilde{\psi} = \frac{1}{N_1 + N_2} \sum_{j=1}^{2} \sum_{i=1}^{N_j} (\mathbf{x}_{ji} - \bar{\mathbf{x}})(\mathbf{x}_{ji} - \bar{\mathbf{x}})'. \tag{12}$$

Therefore, the constrained maximum of the likelihood function is expressed by

$$L(\tilde{\mu}_1, \tilde{\mu}_2, \tilde{\psi}) = (2\pi\, e)^{-(N_1+N_2)p/2} \left(\det \tilde{\psi}\right)^{-(N_1+N_2)/2}, \tag{13}$$

as follows by reasoning analogous to (10). Thus, we obtain the likelihood ratio statistic

$$\text{LRS} = \frac{L(\tilde{\mu}_1, \tilde{\mu}_2, \tilde{\psi})}{L(\hat{\mu}_1, \hat{\mu}_2, \hat{\psi})} = \left(\frac{\det \tilde{\psi}}{\det \hat{\psi}}\right)^{-\frac{1}{2}(N_1+N_2)}. \tag{14}$$

Remarkably, the likelihood ratio statistic depends on the data only through the maximum likelihood estimates of ψ in Ω and ω, respectively. This may be a bit puzzling, at first, because we are testing for equality of mean vectors; see Exercise 4. The log-likelihood ratio statistic is

$$\text{LLRS} = -2\log(\text{LRS}) = (N_1 + N_2)\log\frac{\det\tilde{\psi}}{\det\hat{\psi}}. \tag{15}$$

By Theorem 4.3.3, the large sample distribution of LLRS under the hypothesis $\mu_1 = \mu_2$ is approximately chi-square with degrees of freedom equal to the difference between the dimensions of Ω and ω. Since ω is the subspace of Ω obtained by imposing the p constraints $\mu_1 = \mu_2$, this difference is p. Thus, we accept the hypothesis if LLRS $\le c$ and reject it otherwise, where c is the $(1-\alpha)$-quantile of the chi-square distribution with p degrees of freedom. This test has level approximately α. □

A similar result for the one-sample case is left to the reader; see Exercise 5. Also note that we need to assume nonsingularity of $\hat{\psi}$. This is guaranteed by the condition $N_1 + N_2 \ge p + 2$ in the normal theory setup; see Exercise 8 in Section 5.3.

Although the log-likelihood ratio statistic (15) and the T^2-statistic of Section 6.4 seem to be quite different, they are really equivalent in the sense that they are functionally related. We state this interesting result as a theorem.

Theorem 6.6.1 *The log-likelihood ratio statistic (15) for equality of two mean vectors and the two-sample T^2-statistic (7) of Section 6.4 are related by*

$$LLRS = (N_1 + N_2)\log\left(1 + \frac{1}{N_1 + N_2 - 2}T^2\right). \tag{16}$$

Proof We need to show that

$$\frac{\det\tilde{\psi}}{\det\hat{\psi}} = 1 + \frac{1}{N_1 + N_2 - 2}T^2, \tag{17}$$

where

$$T^2 = \frac{N_1 N_2}{N_1 + N_2}\mathbf{d}'\mathbf{S}^{-1}\mathbf{d}, \tag{18}$$

$\mathbf{S}$ is the pooled-sample covariance matrix, and $\mathbf{d} = \bar{\mathbf{x}}_1 - \bar{\mathbf{x}}_2$. Three partial results used in the proof are

(i) $\tilde{\psi} = \hat{\psi} + \frac{N_1 N_2}{(N_1+N_2)^2}\mathbf{dd}'$, (19)

(ii) if the symmetric $p \times p$ matrix $\mathbf{A}$ has eigenvalues $\lambda_1, \ldots, \lambda_p$, then $\mathbf{I}_p + \mathbf{A}$ has eigenvalues $1 + \lambda_1, \ldots, 1 + \lambda_p$, and

(iii) if $\mathbf{A} = \mathbf{hh}'$ for some $\mathbf{h} \in \mathbb{R}^p$, then $\mathbf{A}$ has a single nonzero eigenvalue equal to $\mathbf{h}'\mathbf{h}$.

For proofs of (i) to (iii), see Exercises 6 to 8. Recall the definition of the symmetric square root of a positive definite symmetric matrix from Appendix A.7. Then we can write

$$\frac{\det \tilde{\psi}}{\det \hat{\psi}} = \det\left[\tilde{\psi}\left(\hat{\psi}\right)^{-1}\right] = \det\left[\left(\hat{\psi}\right)^{-1/2}\tilde{\psi}\left(\hat{\psi}\right)^{-1/2}\right]. \tag{20}$$

Using (i),

$$\begin{aligned}\frac{\det \tilde{\psi}}{\det \hat{\psi}} &= \det\left[\left(\hat{\psi}\right)^{-1/2}\left(\hat{\psi} + \frac{N_1 N_2}{(N_1+N_2)^2}\mathbf{d}\mathbf{d}'\right)\left(\hat{\psi}\right)^{-1/2}\right] \\ &= \det\left[\mathbf{I}_p + \frac{N_1 N_2}{(N_1+N_2)^2}\left(\hat{\psi}\right)^{-1/2}\mathbf{d}\mathbf{d}'\left(\hat{\psi}\right)^{-1/2}\right].\end{aligned} \tag{21}$$

Let $\mathbf{h} = \frac{(N_1N_2)^{1/2}}{N_1+N_2}(\hat{\psi})^{-1/2}\mathbf{d}$, and recall that the determinant of a symmetric matrix is the product of its eigenvalues. By (ii) and (iii), the eigenvalues of $\mathbf{I}_p + \mathbf{h}\mathbf{h}'$ are 1 (with multiplicity $p-1$) and $1 + \mathbf{h}'\mathbf{h}$. Therefore,

$$\begin{aligned}\frac{\det \tilde{\psi}}{\det \hat{\psi}} &= 1 + \mathbf{h}'\mathbf{h} \\ &= 1 + \frac{N_1 N_2}{(N_1+N_2)^2}\mathbf{d}'(\hat{\psi})^{-1}\mathbf{d} \\ &= 1 + \frac{N_1 N_2}{(N_1+N_2)^2}\mathbf{d}'\left(\frac{N_1+N_2-2}{N_1+N_2}\mathbf{S}\right)^{-1}\mathbf{d} \\ &= 1 + \frac{1}{N_1+N_2-2}T^2.\end{aligned}$$

■

Theorem 6.6.1 gives strong support to our earlier heuristic justification of the T^2-statistic because now we know that T^2 is equivalent to a likelihood ratio statistic. Theorem 6.6.1 also enables us to obtain a simple asymptotic approximation to the distribution of T^2 under the hypothesis $\mu_1 = \mu_2$, as follows. From Theorem 4.3.3, we know that the asymptotic distribution of the log-likelihood ratio statistic

$$\text{LLRS} = (N_1 + N_2)\log\left(1 + \frac{1}{N_1+N_2-2}T^2\right)$$

under the hypothesis $\mu_1 = \mu_2$ is chi-square with p degrees of freedom. A first order Taylor series of the logarithm, for $|x|$ small, is $\log(1+x) \approx x$, and therefore, for $N_1 + N_2$ large,

$$\text{LLRS} \approx T^2, \tag{22}$$

that is, if $N_1 + N_2$ is large, we may use the T^2-statistic and compare it directly to the $(1-\alpha)$-quantile of the chi-square distribution with p degrees of freedom. In fact, this is the same approximation as if we assume that the common covariance matrix ψ of the two populations is known, or in other words, assume that $N_1 + N_2$ is so large that the pooled-sample covariance matrix $\mathbf{S}$ is "almost identical" to ψ. By the normality assumptions, $\bar{\mathbf{x}}_1 \sim \mathcal{N}_p(\boldsymbol{\mu}_1, \psi/N_1)$, $\bar{\mathbf{x}}_2 \sim \mathcal{N}_p(\boldsymbol{\mu}_2, \psi/N_2)$, and

$$\mathbf{d} = \bar{\mathbf{x}}_1 - \bar{\mathbf{x}}_2 \sim \mathcal{N}_p\left(\boldsymbol{\mu}_1 - \boldsymbol{\mu}_2, \frac{N_1 N_2}{N_1 + N_2}\psi\right). \tag{23}$$

Under the hypothesis $\boldsymbol{\mu}_1 = \boldsymbol{\mu}_2$, $E[\mathbf{d}] = \mathbf{0}$, and by Theorem 3.3.2,

$$\mathbf{d}'\left(\frac{N_1 N_2}{N_1 + N_2}\psi\right)^{-1}\mathbf{d} = \frac{N_1 N_2}{N_1 + N_2}(\bar{\mathbf{x}}_1 - \bar{\mathbf{x}}_2)'\psi^{-1}(\bar{\mathbf{x}}_1 - \bar{\mathbf{x}}_2)$$

follows a chi-square distribution with p degrees of freedom. If both N_1 and N_2 are large, we can even abandon the normality requirement because the multivariate central limit theorem ensures approximate normality of $\bar{\mathbf{x}}_1$ and $\bar{\mathbf{x}}_2$.

For yet a different justification of the asymptotic chi-square approximation, see Exercise 9. Similar results can also be established for the one-sample T^2-statistic, see the discussion following Example 6.2.3.

Example 6.6.4 In this example, we return to the situation of Example 6.2.1 (head length and breadth of siblings), treating it in a somewhat more general context and using the likelihood machinery. Let $\mathbf{X} = \begin{pmatrix}\mathbf{X}_{(1)}\\ \mathbf{X}_{(2)}\end{pmatrix}$ denote a $2p$-variate normal random vector, where both $\mathbf{X}_{(1)}$ and $\mathbf{X}_{(2)}$ have dimension p, and let

$$E[\mathbf{X}] = \boldsymbol{\mu} = \begin{pmatrix}\boldsymbol{\mu}_1\\ \boldsymbol{\mu}_2\end{pmatrix}$$

and

$$\text{Cov}[\mathbf{X}] = \psi = \begin{pmatrix}\psi_{11} & \psi_{12}\\ \psi_{21} & \psi_{22}\end{pmatrix}$$

be partitioned analogously. Suppose that we wish to test if $\mathbf{X}_1$ and $\mathbf{X}_2$ are exchangeable, i.e., test the hypothesis that the distribution of $\begin{pmatrix}\mathbf{X}_{(1)}\\ \mathbf{X}_{(2)}\end{pmatrix}$ is the same as the distribution of $\begin{pmatrix}\mathbf{X}_{(2)}\\ \mathbf{X}_{(1)}\end{pmatrix}$. From Exercise 12 in Section 6.2, and because first and second moments determine a multivariate normal distribution completely, we can state this hypothesis as

$$\boldsymbol{\mu}_1 = \boldsymbol{\mu}_2, \quad \psi_{11} = \psi_{22}, \text{ and } \psi_{12} \text{ is symmetric.} \tag{24}$$

Based on a sample $\mathbf{x}_1, \ldots, \mathbf{x}_N$ from $\mathcal{N}_{2p}(\boldsymbol{\mu}, \psi)$, $N > 2p$, we wish to test the hypothesis (24). Finding the maximum of the likelihood function in the unconstrained case poses no new problem, but, under the constraint (24), it seems to be difficult. Fortunately, a reparameterization makes the task easier. Let $\mathbf{Y}_{(1)} = \mathbf{X}_{(1)} + \mathbf{X}_{(2)}$, $\mathbf{Y}_{(2)} =$

$\mathbf{X}_{(1)} - \mathbf{X}_{(2)}$, and $\mathbf{Y} = \begin{pmatrix} \mathbf{Y}_{(1)} \\ \mathbf{Y}_{(2)} \end{pmatrix}$. Then $\mathbf{Y} \sim \mathcal{N}_{2p}(\boldsymbol{\nu}, \boldsymbol{\Gamma})$, where

$$\boldsymbol{\nu} = \begin{pmatrix} \boldsymbol{\nu}_1 \\ \boldsymbol{\nu}_2 \end{pmatrix} = \begin{pmatrix} \boldsymbol{\mu}_1 + \boldsymbol{\mu}_2 \\ \boldsymbol{\mu}_1 - \boldsymbol{\mu}_2 \end{pmatrix} \tag{25}$$

and

$$\boldsymbol{\Gamma} = \begin{pmatrix} \boldsymbol{\Gamma}_{11} & \boldsymbol{\Gamma}_{12} \\ \boldsymbol{\Gamma}_{21} & \boldsymbol{\Gamma}_{22} \end{pmatrix} = \begin{pmatrix} \boldsymbol{\psi}_{11} + \boldsymbol{\psi}_{21} + \boldsymbol{\psi}_{12} + \boldsymbol{\psi}_{22} & \boldsymbol{\psi}_{11} + \boldsymbol{\psi}_{21} - \boldsymbol{\psi}_{12} - \boldsymbol{\psi}_{22} \\ \boldsymbol{\psi}_{11} - \boldsymbol{\psi}_{21} + \boldsymbol{\psi}_{12} - \boldsymbol{\psi}_{22} & \boldsymbol{\psi}_{11} - \boldsymbol{\psi}_{21} - \boldsymbol{\psi}_{12} + \boldsymbol{\psi}_{22} \end{pmatrix}; \tag{26}$$

see Exercise 10. By (24), the hypothesis to be tested is now equivalent to

$$\boldsymbol{\nu}_2 = \mathbf{0}$$

and

$$\boldsymbol{\Gamma}_{12} = \boldsymbol{\Gamma}_{21} = \mathbf{0}. \tag{27}$$

To derive the likelihood ratio statistic, let $\mathbf{x}_i = \begin{pmatrix} \mathbf{x}_{i(1)} \\ \mathbf{x}_{i(2)} \end{pmatrix}$ denote the ith data vector, partitioned into subvectors of dimension p each, and set

$$\mathbf{y}_i = \begin{pmatrix} \mathbf{y}_{i(1)} \\ \mathbf{y}_{i(2)} \end{pmatrix} = \begin{pmatrix} \mathbf{x}_{i(1)} + \mathbf{x}_{i(2)} \\ \mathbf{x}_{i(1)} - \mathbf{x}_{i(2)} \end{pmatrix}, \qquad i = 1, \ldots, N. \tag{28}$$

With no constraints on the parameter space other than positive definiteness of $\boldsymbol{\Gamma}$, the maximum likelihood estimates of $\boldsymbol{\nu}$ and $\boldsymbol{\Gamma}$ are given by

$$\hat{\boldsymbol{\nu}} = \bar{\mathbf{y}} = \frac{1}{N}\sum_{i=1}^{N} \mathbf{y}_i$$

and

$$\hat{\boldsymbol{\Gamma}} = \frac{1}{N}\sum_{i=1}^{N} (\mathbf{y}_i - \bar{\mathbf{y}})(\mathbf{y}_i - \bar{\mathbf{y}})',$$

and the maximum of the likelihood function is expressed by

$$L(\hat{\boldsymbol{\nu}}, \hat{\boldsymbol{\Gamma}}) = (2\pi\, e)^{-Np} (\det \hat{\boldsymbol{\Gamma}})^{-N/2}, \tag{29}$$

see Exercise 11. In the constrained case, $\boldsymbol{\Gamma}_{12} = \mathbf{0}$ implies independence of $\mathbf{Y}_{(1)}$ and $\mathbf{Y}_{(2)}$, and the likelihood function factorizes as $L(\boldsymbol{\nu}, \boldsymbol{\Gamma}) = L_1(\boldsymbol{\nu}_1, \boldsymbol{\Gamma}_{11}) \cdot L_2(\boldsymbol{\Gamma}_{22})$ because the $\mathbf{y}_{i(1)}$ are a sample from $\mathcal{N}_p(\boldsymbol{\nu}_1, \boldsymbol{\Gamma}_{11})$ and the $\mathbf{y}_{i(2)}$ are an independent sample from $\mathcal{N}_p(\mathbf{0}, \boldsymbol{\Gamma}_{22})$. Specifically,

$$\begin{aligned} L_1(\boldsymbol{\nu}_1, \boldsymbol{\Gamma}_{11}) = {} & (2\pi)^{-Np/2} (\det \boldsymbol{\Gamma}_{11})^{-N/2} \\ & \times \exp\left\{ -\frac{1}{2} \operatorname{tr}\left[\boldsymbol{\Gamma}_{11}^{-1} \sum_{i=1}^{N} \left(\mathbf{y}_{i(1)} - \boldsymbol{\nu}_1\right) \left(\mathbf{y}_{i(1)} - \boldsymbol{\nu}_1\right)' \right] \right\}, \end{aligned}$$

and

$$L_2(\Gamma_{22}) = (2\pi)^{-Np/2} (\det \Gamma_{22})^{-N/2} \exp\left\{-\frac{1}{2}\mathrm{tr}\left[\Gamma_{22}^{-1} \sum_{i=1}^{N} \mathbf{y}_{i(2)}\mathbf{y}_{i(2)}'\right]\right\}.$$

Maximizing L_1 and L_2 individually, it follows from Theorem 4.3.1 that the constrained maximum likelihood estimates are

$$\begin{aligned} \tilde{\nu}_1 &= \bar{\mathbf{y}}_{(1)} = \frac{1}{N}\sum_{i=1}^{N} \mathbf{y}_{i(1)}, \\ \tilde{\Gamma}_{11} &= \frac{1}{N}\sum_{i=1}^{N} \left[\mathbf{y}_{i(1)} - \bar{\mathbf{y}}_{(1)}\right]\left[\mathbf{y}_{i(1)} - \bar{\mathbf{y}}_{(1)}\right]', \end{aligned} \tag{30}$$

and

$$\tilde{\Gamma}_{22} = \frac{1}{N}\sum_{i=1}^{N} \mathbf{y}_{i(2)}\mathbf{y}_{i(2)}'. \tag{31}$$

This gives the constrained maximum likelihood estimates of ν and Γ as

$$\tilde{\nu} = \begin{pmatrix} \tilde{\nu}_1 \\ \mathbf{0} \end{pmatrix}$$

and

$$\tilde{\Gamma} = \begin{pmatrix} \tilde{\Gamma}_{11} & \mathbf{0} \\ \mathbf{0} & \tilde{\Gamma}_{22} \end{pmatrix}. \tag{32}$$

Finally, the constrained maximum is given by

$$\begin{aligned} L(\tilde{\nu}, \tilde{\Gamma}) &= L_1(\tilde{\nu}_1, \tilde{\Gamma}_{11})\, L_2(\tilde{\Gamma}_{22}) \\ &= (2\pi\, e)^{-Np/2} (\det \tilde{\Gamma}_{11})^{-N/2} (2\pi\, e)^{-Np/2} (\det \tilde{\Gamma}_{22})^{-N/2} \\ &= (2\pi\, e)^{-Np}\left[(\det \tilde{\Gamma}_{11})\,(\det \tilde{\Gamma}_{22})\right]^{-N/2}. \end{aligned} \tag{33}$$

Thus, the log-likelihood ratio statistic for testing the hypothesis (27) is expressed by

$$\begin{aligned} \mathrm{LLRS} &= -2 \log \frac{L(\tilde{\nu}, \tilde{\Gamma})}{L(\hat{\nu}, \hat{\Gamma})} \\ &= N \log \frac{(\det \tilde{\Gamma}_{11})(\det \tilde{\Gamma}_{22})}{\det \hat{\Gamma}}. \end{aligned} \tag{34}$$

This is also the log-likelihood ratio statistic for testing the original hypothesis (24); see Exercise 12(b). Finally, from Theorem 4.3.3 it follows that, for large N, the distribution of LLRS under the hypothesis is approximately chi-square with $m = p(p+1)$ degrees of freedom; see Exercise 12(c) for the calculation of m. □

Example 6.6.4 is a good illustration of the power and the potential of likelihood ratio testing and of likelihood based methods in general. We will encounter other interesting applications in Sections 7.5 and 7.6 (logistic regression) and in Chapter 9 (finite mixtures). Further normal theory applications are given in Exercises 13 to 15.

We conclude this section with a practical application of the test developed in Example 6.6.4 to the data which motivated the study of this particular model.

Example 6.6.5 continuation of Examples 6.2.1 and 6.6.4

In the head length and breadth of siblings example, let $\mathbf{X}_{(1)} = \begin{pmatrix} X_1 \\ X_2 \end{pmatrix}$ denote the two variables measured on the older brother, and $\mathbf{X}_{(2)} = \begin{pmatrix} X_3 \\ X_4 \end{pmatrix}$ those measured on the younger brother. Thus, we have $p = 2$ and $N = 25$ observations. Transforming the X-variables to $\mathbf{Y}_{(1)} = \mathbf{X}_{(1)} + \mathbf{X}_{(2)}$ and $\mathbf{Y}_{(2)} = \mathbf{X}_{(1)} - \mathbf{X}_{(2)}$, we obtain a new four-dimensional random vector

$$\mathbf{Y} = \begin{pmatrix} \mathbf{Y}_{(1)} \\ \mathbf{Y}_{(2)} \end{pmatrix} = \begin{pmatrix} Y_1 \\ Y_2 \\ Y_3 \\ Y_4 \end{pmatrix} = \begin{pmatrix} X_1 + X_3 \\ X_2 + X_4 \\ X_1 - X_3 \\ X_2 - X_4 \end{pmatrix}.$$

The maximum likelihood estimates of $\nu = E[\mathbf{Y}]$ and $\Gamma = \text{Cov}[\mathbf{Y}]$ in the unconstrained case are

$$\hat{\nu} = \begin{pmatrix} 370.72 \\ 300.48 \\ 4.08 \\ 1.76 \end{pmatrix},$$

and

$$\hat{\Gamma} = \begin{pmatrix} 327.80 & 199.93 & 4.66 & -0.51 \\ 199.93 & 171.37 & 15.68 & 5.40 \\ 4.66 & 15.68 & 57.75 & 13.82 \\ -0.51 & 5.40 & 13.82 & 16.98 \end{pmatrix}.$$

Under the constraint (27), maximum likelihood estimates are $\tilde{\nu}_1 = \begin{pmatrix} 370.72 \\ 300.48 \end{pmatrix}$, $\tilde{\Gamma}_{11} = \begin{pmatrix} 327.80 & 199.93 \\ 199.93 & 171.37 \end{pmatrix}$, and $\tilde{\Gamma}_{22} = \begin{pmatrix} 74.40 & 21.00 \\ 21.00 & 20.08 \end{pmatrix}$, and thus,

$$\tilde{\nu} = \begin{pmatrix} 370.72 \\ 300.48 \\ 0 \\ 0 \end{pmatrix},$$

and

$$\tilde{\Gamma} = \begin{pmatrix} 327.80 & 199.93 & 0 & 0 \\ 199.93 & 171.37 & 0 & 0 \\ 0 & 0 & 74.40 & 21.00 \\ 0 & 0 & 21.00 & 20.08 \end{pmatrix}.$$

The log-likelihood ratio statistic (34) is LLRS = 8.99 with 6 degrees of freedom, corresponding to a p-value of 0.17 in the approximating chi-square distribution. Thus, we would accept the hypothesis (27) at any of the commonly used levels of significance and consider $\mathbf{X}_{(1)}$ and $\mathbf{X}_{(2)}$ as exchangeable.

Having reached this conclusion, the logical consequence is that we should estimate the parameters of the distribution of $\mathbf{X}$ accordingly. From (25) and (26) and if the hypothesis of exchangeability (24) holds, then

$$E[\mathbf{X}] = \mu = \frac{1}{2}\begin{pmatrix} \nu_1 \\ \nu_2 \end{pmatrix}, \tag{35}$$

and

$$\mathrm{Cov}[\mathbf{X}] = \psi = \frac{1}{4}\begin{pmatrix} \Gamma_{11} + \Gamma_{22} & \Gamma_{11} - \Gamma_{22} \\ \Gamma_{11} - \Gamma_{22} & \Gamma_{11} + \Gamma_{22} \end{pmatrix}, \tag{36}$$

see Exercise 12. By the invariance property of maximum likelihood estimates (Exercise 20 in Section 4.3), the constrained estimates of μ and ψ are obtained by using $\tilde{\nu}_1$, $\tilde{\Gamma}_{11}$, and $\tilde{\Gamma}_{22}$ in (35) and (36), giving

$$\tilde{\mu} = \begin{pmatrix} 185.36 \\ 150.24 \\ 185.36 \\ 150.24 \end{pmatrix},$$

and

$$\tilde{\psi} = \begin{pmatrix} 100.55 & 55.23 & 63.35 & 44.73 \\ 55.23 & 47.86 & 44.73 & 37.82 \\ 63.35 & 44.73 & 100.55 & 55.23 \\ 44.73 & 37.82 & 55.23 & 47.86 \end{pmatrix}.$$

These estimates reflect the exchangeability property, and are close to te unconstrained estimates of μ and ψ; for details see Exercise 12(d). □

Exercises for Section 6.6

1. This exercise refers to Example 6.6.2. Show that the hypothesis $\mu = \mu_0$ is true exactly if $\mathbf{a}'\mu = \mathbf{a}'\mu_0$ for all vectors $\mathbf{a} \in \mathbb{R}^p$.

2. Derive a union-intersection test for the hypothesis $\mu_1 = \mu_2$ in the situation where $\mathbf{x}_{11}, \ldots, \mathbf{x}_{1N_1}$ and $\mathbf{x}_{21}, \ldots, \mathbf{x}_{2N_2}$ are independent samples from some p-variate distributions with mean vectors μ_1 and μ_2, respectively. Show that $\mu_1 = \mu_2$ exactly if $\mathbf{a}'\mu_1 = \mathbf{a}'\mu_2$ for all $\mathbf{a} \in \mathbb{R}^p$, and use the univariate, two-sample t-statistic for testing the hypothesis $H_\mathbf{a} : \mathbf{a}'\mu_1 = \mathbf{a}'\mu_2$. Show that this leads to the two-sample T^2-statistic of Section 6.4.

3. A union-intersection test for correlation between a random variable Y and a p-variate random vector $\mathbf{X}$. Let $\begin{pmatrix} Y \\ \mathbf{X} \end{pmatrix}$ denote $(p+1)$ jointly distributed random variables, where $\mathbf{X}$ has dimension p. Let

$$\text{Cov}\begin{pmatrix} Y \\ \mathbf{X} \end{pmatrix} = \begin{pmatrix} \sigma_Y^2 & \sigma_{Y\mathbf{X}} \\ \sigma_{\mathbf{X}Y} & \Sigma_{\mathbf{XX}} \end{pmatrix},$$

and suppose that we wish to test the hypothesis that Y is uncorrelated with all X-variables, i.e., $\sigma_{\mathbf{X}Y} = \mathbf{0} \in \mathbb{R}^p$.

(a) Show that the hypothesis $\sigma_{\mathbf{X}Y} = \mathbf{0}$ is equivalent to $\text{corr}(Y, \mathbf{a}'\mathbf{X}) = 0$ for all $\mathbf{a} \in \mathbb{R}^p$.

(b) For a fixed vector $\mathbf{a} \in \mathbb{R}^p$, $\mathbf{a} \neq \mathbf{0}$, let $H_\mathbf{a}$ be the hypothesis that $\text{corr}(Y, \mathbf{a}'\mathbf{X}) = 0$, and let $r_\mathbf{a}$ denote the observed correlation between Y and $\mathbf{a}'\mathbf{X}$ based on a sample of size $N > 2$. Then a test for $H_\mathbf{a}$ may be based on the statistic

$$t_\mathbf{a} = \sqrt{N-2}\,\frac{r_\mathbf{a}}{\sqrt{1-r_\mathbf{a}^2}}, \tag{37}$$

and the hypothesis is rejected if $|t_\mathbf{a}| > c$ for some critical value $c > 0$. Use the statistic (37) and part (a) of this exercise to construct a union-intersection test for the hypothesis $\sigma_{\mathbf{X}Y} = \mathbf{0}$. *Hint*: First, show that maximizing $|t_\mathbf{a}|$ over all $\mathbf{a} \in \mathbb{R}^p$, $\mathbf{a} \neq \mathbf{0}$, is equivalent to maximizing $r_\mathbf{a}^2$. Then show that

$$\max_{\substack{\mathbf{a}\in\mathbb{R}^p \\ \mathbf{a}\neq\mathbf{0}}} r_\mathbf{a}^2 = \mathbf{s}_{Y\mathbf{X}}\mathbf{S}_{\mathbf{XX}}^{-1}\mathbf{s}_{\mathbf{X}Y}/s_{YY}, \tag{38}$$

where

$$\mathbf{S} = \begin{pmatrix} s_{YY} & \mathbf{s}_{Y\mathbf{X}} \\ \mathbf{s}_{\mathbf{X}Y} & \mathbf{S}_{\mathbf{XX}} \end{pmatrix}$$

is the observed covariance matrix from a sample of size N. What assumptions do you need to make on $\mathbf{S}$ and N?

(c) Show that the union-intersection statistic from part (b) is a function of the overall F-statistic for significance of the linear regression of Y on $\mathbf{X}$.

4. By equation (14), the likelihood ratio statistic for equality of two mean vectors depends on the data only through the maximum likelihood estimates $\tilde{\psi}$ and $\hat{\psi}$ of the covariance matrix. How does it depend on the estimates of the mean vectors? Explain. (Also see Exercise 6).

5. This example deals with the one-sample analog of Example 6.6.3. Let $\mathbf{x}_1, \ldots, \mathbf{x}_N$ be a sample from $\mathcal{N}_p(\mu, \psi)$, and consider testing the hypothesis $\mu = \mu_0$, where $\mu_0 \in \mathbb{R}^p$ is a fixed vector and ψ is unknown but assumed positive definite.

(a) Find the unconstrained parameter space Ω and the constrained parameter space ω.

(b) Find the maximum likelihood estimates of $\hat{\mu}$ and $\hat{\psi}$ in Ω. Show that the maximum likelihood estimates in the constrained parameter space are $\tilde{\mu} = \mu_0$ and $\tilde{\psi} = \sum_{i=1}^{N}(\mathbf{x}_i - \mu_0)(\mathbf{x}_i - \mu_0)'$. What assumptions do you need to make on the sample size N?

(c) Show that the log-likelihood ratio statistic for testing the hypothesis $\mu = \mu_0$ is given by

$$\text{LLRS} = N \log \frac{\det \tilde{\psi}}{\det \hat{\psi}}. \tag{39}$$

(d) Show that the log-likelihood ratio statistic (39) can be written as LLRS $= N \log\left[\det(\mathbf{I}_p + \mathbf{hh}')\right]$, where $\mathbf{h} = \left(\hat{\psi}\right)^{-1/2}(\mu_0 - \bar{\mathbf{x}})$.

(e) Show that the log-likelihood ratio statistic (39) and the one-sample T^2-statistic from Section 6.2, $T^2 = N(\mu_0 - \bar{\mathbf{x}})'\mathbf{S}^{-1}(\mu_0 - \bar{\mathbf{x}})$, are related by

$$\text{LLRS} = N \log\left(1 + \frac{1}{N-1} T^2\right). \tag{40}$$

(f) By Theorem 4.3.3, the large sample distribution of LLRS under the hypothesis $\mu = \mu_0$ is approximately chi-square with m degrees of freedom. What is m?

(g) Use part (f) and a first-order Taylor series expansion of the natural logarithm to argue that the distribution of T^2 under the hypothesis $\mu = \mu_0$ is approximately chi-square with p degrees of freedom.

6. Prove equation (19). *Hint*: Define a p-variate discrete mixture distribution by assigning probability $1/(N_1 + N_2)$ to each of the observed data vectors $\mathbf{x}_{ji}$, with mixing weights $\pi_j = N_j/(N_1 + N_2)$, $j = 1, 2$. Find the mean vector and the covariance matrix of the mixture as functions of the mean vectors and covariance matrices of the components.

7. Prove result (ii) in the proof of Theorem 6.6.1.

8. Prove result (iii) in the proof of Theorem 6.6.1.

9. Let T^2 denote the two-sample T^2-statistic of equation (7) in Section 6.4, and

$$F = \frac{N_1 + N_2 - p - 1}{(N_1 + N_2 - 2)p} T^2 \tag{41}$$

the corresponding F-statistic. From Theorem 6.4.1, we know that, under normality assumptions and if the hypothesis $\mu_1 = \mu_2$ is true, the statistic (41) follows an F-distribution with p and $(N_1 + N_2 - p - 1)$ degrees of freedom. Use this to show that the distribution of T^2 for large values of $N_1 + N_2$ is approximately chi-square with p degrees of freedom. *Hint*: Read the discussion following Example 6.2.3.

10. Prove equations (25) and (26).

11. Prove equation (29).

12. This exercise expands on Examples 6.6.4 and 6.6.5.

(a) Show that the maximum likelihood estimates of $\boldsymbol{\mu}$ and $\boldsymbol{\psi}$ under the hypothesis (24) are given by

$$\tilde{\boldsymbol{\mu}} = \frac{1}{2}\begin{pmatrix}\hat{\boldsymbol{\nu}}_1 \\ \hat{\boldsymbol{\nu}}_1\end{pmatrix}$$

and

$$\hat{\boldsymbol{\psi}} = \frac{1}{4}\begin{pmatrix}\hat{\boldsymbol{\Gamma}}_{11} + \hat{\boldsymbol{\Gamma}}_{22} & \hat{\boldsymbol{\Gamma}}_{11} - \hat{\boldsymbol{\Gamma}}_{22} \\ \hat{\boldsymbol{\Gamma}}_{11} - \hat{\boldsymbol{\Gamma}}_{22} & \hat{\boldsymbol{\Gamma}}_{11} + \hat{\boldsymbol{\Gamma}}_{22}\end{pmatrix}.$$

(b) Let $L^*(\boldsymbol{\mu}, \boldsymbol{\psi})$ be the likelihood function for $\boldsymbol{\mu}$ and $\boldsymbol{\psi}$, based on the observed data vectors $\mathbf{x}_1, \ldots, \mathbf{x}_N$. Denote by $(\hat{\boldsymbol{\mu}}, \hat{\boldsymbol{\psi}})$ and $(\tilde{\boldsymbol{\mu}}, \tilde{\boldsymbol{\psi}})$ the unconstrained and the constrained maximum likelihood estimates, respectively. Show that $-2 \log\left[L^*(\tilde{\boldsymbol{\mu}}, \tilde{\boldsymbol{\psi}})/L^*(\hat{\boldsymbol{\mu}}, \hat{\boldsymbol{\psi}})\right]$ is the same as the log-likelihood ratio statistic (34) computed from the reparameterized model.

(c) Find the degrees of freedom m of the asymptotic chi-square distribution of the log-likelihood ratio statistic (34). *Hint*: Count the number of independent constraints imposed by the hypothesis. Both (24) and (27) should give the same answer.

(d) In Example 6.6.5, find the unconstrained maximum likelihood estimates of $\boldsymbol{\nu}$ and $\boldsymbol{\Gamma}$ numerically. Transform them to unconstrained maximum likelihood estimates of $\boldsymbol{\mu}$ and $\boldsymbol{\psi}$, using equations (25) and (26). Verify that these are the same as the unconstrained maximum likelihood estimates computed directly from the $\mathbf{x}_i$.

13. A likelihood ratio test for equality of covariance matrices. Let $\mathbf{x}_{j1}, \ldots, \mathbf{x}_{jN_j}$ $(j = 1, \ldots, k)$ denote k independent samples from $\mathcal{N}_p(\boldsymbol{\mu}_j, \boldsymbol{\psi}_j)$, respectively, where all $N_j > p$. Consider testing the hypothesis $\boldsymbol{\psi}_1 = \cdots = \boldsymbol{\psi}_k$ against the alternative of arbitrary positive definite covariance matrices. (The mean vectors are unknown, and no constraints are imposed on them).

(a) Show that the maximum likelihood estimates of the covariance matrices are given by

$$\hat{\boldsymbol{\psi}}_j = \frac{1}{N_j}\sum_{i=1}^{N_j}(\mathbf{x}_{ji} - \bar{\mathbf{x}}_j)(\mathbf{x}_{ji} - \bar{\mathbf{x}}_j)', \quad j = 1, \ldots, k, \tag{42}$$

in the unconstrained case, and by

$$\tilde{\boldsymbol{\psi}} := \tilde{\boldsymbol{\psi}}_1 = \cdots = \tilde{\boldsymbol{\psi}}_k = \frac{1}{N_1 + \cdots + N_k}\sum_{j=1}^{k}\sum_{i=1}^{N_j}(\mathbf{x}_{ji} - \bar{\mathbf{x}}_j)(\mathbf{x}_{ji} - \bar{\mathbf{x}}_j)' \tag{43}$$

in the constrained case.

(b) Show that the log-likelihood ratio statistic for testing the hypothesis of equality of covariance matrices is

$$\text{LLRS} = \sum_{j=1}^{k} N_j \log\frac{\det \tilde{\boldsymbol{\psi}}_j}{\det \hat{\boldsymbol{\psi}}_j} = \left(\sum_{j=1}^{k} N_j\right)\log(\det \tilde{\boldsymbol{\psi}}) - \sum_{j=1}^{k} N_j \log(\det \hat{\boldsymbol{\psi}}_j). \tag{44}$$

(c) Find the large-sample distribution of the log-likelihood ratio statistic under the hypothesis, using Theorem 4.3.3. Show that the number of degrees of freedom in the asymptotic approximation equals $(k-1)p(p+1)/2$.

(d) Apply this procedure to the electrode data of Example 6.4.2, with all five variables, using a significance level of 0.05. Also see Example 6.7.4,

14. This exercise concerns testing for equality of two multivariate normal distributions. Let $\mathbf{x}_{11}, \ldots, \mathbf{x}_{1N_1}$ and $\mathbf{x}_{21}, \ldots, \mathbf{x}_{2N_2}$ denote two independent samples from $\mathcal{N}_p(\boldsymbol{\mu}_1, \boldsymbol{\psi}_1)$ and $\mathcal{N}_p(\boldsymbol{\mu}_2, \boldsymbol{\psi}_2)$, respectively, where both $N_j > p$.

(a) Find the maximum likelihood estimates of the $\boldsymbol{\mu}_j$ and the $\boldsymbol{\psi}_j$ with no constraints on the parameters (except positive definiteness of $\boldsymbol{\psi}_1$ and $\boldsymbol{\psi}_2$) and under the constraint of equality of the two distributions, i.e., $\boldsymbol{\mu}_1 = \boldsymbol{\mu}_2$ and $\boldsymbol{\psi}_1 = \boldsymbol{\psi}_2$.

(b) Show that the log-likelihood ratio statistic for testing the hypothesis H : $(\boldsymbol{\mu}_1, \boldsymbol{\psi}_1) = (\boldsymbol{\mu}_2, \boldsymbol{\psi}_2)$ is

$$\text{LLRS} = \sum_{j=1}^{2} N_j \log \frac{\det \bar{\boldsymbol{\psi}}}{\det \hat{\boldsymbol{\psi}}_j},$$

where

$$\bar{\boldsymbol{\psi}} = \frac{1}{N_1 + N_2} \sum_{j=1}^{2} \sum_{i=1}^{N_j} (\mathbf{x}_{ji} - \bar{\mathbf{x}})(\mathbf{x}_{ji} - \bar{\mathbf{x}})', \tag{45}$$

the $\hat{\boldsymbol{\psi}}_j$ are as in equation (42), and $\bar{\mathbf{x}} = (N_1\bar{\mathbf{x}}_1 + N_2\bar{\mathbf{x}}_2)/(N_1 + N_2)$. Find its asymptotic distribution, using Theorem 4.3.3. Show that the number of degrees of freedom equals $p(p+3)/2$.

(c) Let V, V_1, and V_2 denote the log-likelihood ratio statistics for testing the following hypothesis:

- $V : (\boldsymbol{\mu}_1, \boldsymbol{\psi}_1) = (\boldsymbol{\mu}_2, \boldsymbol{\psi}_2)$; see part (b).
- $V_1 : \boldsymbol{\psi}_1 = \boldsymbol{\psi}_2$, with no constraint on the means; see Exercise 13 with $k = 2$.
- $V_2 : \boldsymbol{\mu}_1 = \boldsymbol{\mu}_2$, given that $\boldsymbol{\psi}_1 = \boldsymbol{\psi}_2$; see Example 6.6.3.

Show that $V = V_1 + V_2$. *Note*: Part (c) gives a decomposition of the log-likelihood ratio statistic for equality of two normal distributions into a part due to differences in variability (V_1) and a part due to differences in location (V_2), given that there are no differences in variability. Use the notation of equations (42), (43), and (45).

(d) Apply the test from part (b) to the electrode data of Example 6.4.2, using a significance level of 0.05. Also see Exercise 13 and Example 6.7.2.

15. Show that the test statistic for redundancy of $(p-q)$ variables in a linear discriminant function (equation 14 in Section 6.5) is equivalent to the normal theory likelihood ratio statistic for this hypothesis. More precisely, let $\mathbf{x}_{11}, \ldots, \mathbf{x}_{1N_1}$ and $\mathbf{x}_{21}, \ldots, \mathbf{x}_{2N_2}$ be independent samples from $\mathcal{N}_p(\boldsymbol{\mu}_1, \boldsymbol{\psi})$ and $\mathcal{N}_p(\boldsymbol{\mu}_2, \boldsymbol{\psi})$, respectively, let $\boldsymbol{\beta} = \boldsymbol{\psi}^{-1}(\boldsymbol{\mu}_1 - \boldsymbol{\mu}_2)$ denote the vector of discriminant function coefficients, and partition β into q and $(p-q)$ components as $\beta = \begin{pmatrix} \beta_1 \\ \beta_2 \end{pmatrix}$. Find the log-likelihood ratio statistic for the hypothesis $\beta_2 = \mathbf{0}$, and show that it is a function of the statistic (14) of Section 6.5. *Note*: This is a more difficult exercise for the advanced student. We suggest solving it using the equivalence between redundancy of the last $(p-q)$ variables ($\beta_2 = \mathbf{0}$) and equality of conditional means. See Giri (1964).

16. Write the log-likelihood ratio statistic for equality of two covariance matrices (see Exercise 13) as a function of the eigenvalues of $\mathbf{S}_1^{-1}\mathbf{S}_2$, where $\mathbf{S}_1$ and $\mathbf{S}_2$ are the two sample covariance matrices.

17. Let $\mathbf{x}_{j1}, \ldots, \mathbf{x}_{jN_j}$, $(j = 1, \ldots, k)$ denote k independent samples from $\mathcal{N}_p(\boldsymbol{\mu}_j, \boldsymbol{\psi})$. Find a union-intersection test for the hypothesis $\boldsymbol{\mu}_1 = \cdots = \boldsymbol{\mu}_k$. *Note*: You might prefer to solve this exercise after reading Section 7.4 on multivariate analysis of variance.

6.7 Resampling-Based Testing

Now, we will return to questions raised earlier in this Chapter: What can we do in cases where we doubt that some assumptions made for a test are valid? If the outcome of a test is totally clear, such as in the electrode data set of Example 6.4.2, we would usually ignore such concerns. In cases where the significance is marginal, such as Example 6.2.1, it would be preferable to replace parametric theory by a procedure that does not make explicit distributional assumptions. Thus, we will take a short leave from parametric theory now and present resampling-based alternatives to the T^2-tests of Sections 6.2 and 6.4. We will focus on two specific methods, a permutation test for the two-sample problem and a bootstrap test for the one-sample problem. We will justify these methods with heuristic arguments, without elaborating much on the mathematical theory. The interested reader is referred to the books by Edgington (1980), Efron and Tibshirani (1993), Good (1994), and Manly (1991).

Broadly speaking, the approach to testing a hypothesis will be as follows. Let H denote a statistical hypothesis, such as equality of two mean vectors, and let V denote a statistic that we feel provides a good summary of all evidence against the hypothesis. For instance, in the two-group situation, the hypothesis will be $\boldsymbol{\mu}_1 = \boldsymbol{\mu}_2$, and we could base a test on the standard distance between the two sample mean vectors,

$$V = D(\bar{\mathbf{x}}_1, \bar{\mathbf{x}}_2) = \left[(\bar{\mathbf{x}}_1 - \bar{\mathbf{x}}_2)'\mathbf{S}^{-1}(\bar{\mathbf{x}}_1 - \bar{\mathbf{x}}_2)\right]^{1/2}, \tag{1}$$

as justified by the properties of multivariate standard distance. The parametric theory outlined in Section 6.4 allows finding the exact distribution of (1), or equivalently of the associated T^2 and F-statistics, if H holds true, assuming that the two populations are multivariate normal with identical covariance matrices. (For a more careful assessment of the importance of the normality assumptions, see Sections 6.2 and 6.4).

The approach to be discussed here avoids the normality part of the assumptions by computing the distribution of the test statistic V using a resampling procedure.

First, we will focus on the two-sample case, presenting a *permutation* or *randomization* procedure, which is conceptually simple and whose logic appears quite convincing. Suppose that we are sampling from two populations with distribution functions $F(\mathbf{x};\ \boldsymbol{\theta}_1)$ and $F(\mathbf{x};\ \boldsymbol{\theta}_2)$, respectively, and we wish to test the hypothesis $H : \boldsymbol{\theta}_1 = \boldsymbol{\theta}_2$. It is implicitly assumed here that if $\boldsymbol{\theta}_1 = \boldsymbol{\theta}_2$, then the two distributions are actually identical, as in the important case where the $\boldsymbol{\theta}_j$ represent mean vectors and the two populations differ in location but not in variability or any other aspect of their distribution. Let $\mathcal{X}_1 = (\mathbf{x}_{11}, \ldots, \mathbf{x}_{1N_1})$ and $\mathcal{X}_2 = (\mathbf{x}_{21}, \ldots, \mathbf{x}_{2N_2})$ denote independent samples from $F(\mathbf{x};\ \boldsymbol{\theta}_1)$ and $F(\mathbf{x};\ \boldsymbol{\theta}_2)$, respectively, and let

$$V = v(\mathcal{X}_1, \mathcal{X}_2) \tag{2}$$

denote a test statistic that we consider a good summary of all evidence against the hypothesis $H : \boldsymbol{\theta}_1 = \boldsymbol{\theta}_2$. Suppose that relatively large values of V are unlikely if H holds, i.e., we would reject H if the observed value of V is in the extreme right tail of the null distribution. The key word here is "null distribution". It is the distribution of the test statistic if the null hypothesis is true. Parametric theory allows finding this distribution exactly at the cost of making assumptions that may be difficult or impossible to verify. Resampling tests compute a null distribution appropriate for the specific data at hand, without making distributional assumptions. That means we are not really performing an alternative test. Rather, we choose a different method of computing the null distribution associated with a given test statistic. Specifically, for the randomization test in the two-sample situation, the idea is that, if $\boldsymbol{\theta}_1 = \boldsymbol{\theta}_2$, then each of the observed data vectors in sample 1 might as well have occurred in sample 2, and vice versa. In other words, if H holds, then partitioning the combined sample of $N_1 + N_2$ observations into N_1 observations, forming sample $\mathcal{X}_1$, and N_2 observations, forming sample $\mathcal{X}_2$, is artificial, and any other assignment of the observations to groups of size N_1 and N_2 might as well have been obtained.

The randomization procedure looks at all possible ways to split the combined data into subsets of N_1 and N_2 observations. There are exactly

$$\binom{N_1 + N_2}{N_1} = \frac{(N_1 + N_2)!}{N_1!\ N_2!} \tag{3}$$

different ways to partition a set of $N_1 + N_2$ objects into subsets of size N_1 and N_2. If the assignment is random, i.e., if the N_1 objects to form subset 1 are chosen randomly from the total set, then each of the possible assignments has probability $1/\binom{N_1+N_2}{N_1}$. This result is sometimes called the permutation lemma, and its proof is left to the reader; see Exercise 1. Let $\mathcal{X}_1^*$ and $\mathcal{X}_2^*$ denote two samples obtained by such a random assignment. Then

$$V^* = v(\mathcal{X}_1^*, \mathcal{X}_2^*) \tag{4}$$

is a discrete, random variable, taking each of its $\binom{N_1+N_2}{N_1}$ values with probability $1/\binom{N_1+N_2}{N_1}$. The distribution of V^* is called the *permutation distribution* or *randomization distribution* of V, and serves as the null distribution. If the actually observed value of V is in the extreme tail of the randomization distribution, then we will conclude that the evidence against the hypothesis $\theta_1 = \theta_2$ is strong enough to reject the hypothesis, that is, the randomization distribution plays exactly the role of the null distribution usually derived by parametric theory.

Before giving more details, let us illustrate this with an example.

Example 6.7.1 continuation of Example 6.4.1

Testing the hypothesis of equality of mean vectors in the midge data set. We choose standard distance between the two groups as a test statistic, but we could as well use the T^2 or the equivalent F-statistic. With $N_1 = 9$ and $N_2 = 6$, there are $\binom{15}{9} = 5,005$ different ways to assign the 15 observations to two groups. Suppose that the observations are indexed 1 to 15. The randomization procedure systematically generates all 5005 partitions, one of which will coincide with the original partition. The five first and five last partitions obtained in such a procedure are given in Table 6.7.1. The very last one in this example coincides with the actual partition into species *Af* and *Apf*. For each partition, the bivariate standard distance was computed. The resulting distribution is shown as a histogram in Figure 6.7.1. The standard distance obtained from the partition given by the actual two species is in the extreme right tail of the distribution, and therefore the null hypothesis is rejected at any reasonable level of significance. □

As mentioned earlier, the randomization distribution of D is discrete, and the histogram representation is, therefore, not completely appropriate, but it is quite useful anyway for comparison with the null distribution obtained from parametric theory. We will elaborate more on this in Example 6.7.4. Although a total of $\binom{N_1+N_2}{N_1}$ partitions is created to compute the values of the test statistic V, the actual number of discrete values that V takes is typically much smaller, particularly if there are identical observations in the data set. Although no two observations are identical in the midge data set, the randomization distribution of D has positive probability on "only" about 2,400 discrete points. Of these, the observed value $D = 3.940$ is the largest, leaving little doubt about the decision to reject the hypothesis.

Example 6.7.1 shows the main difficulty encountered with the randomization procedure: the computations become quite involved, even for small sample sizes. For instance, if $N_1 = N_2 = 10$, there are $\binom{20}{10} = 184,756$ possible ways to assign them to samples of size 10 each, which is quite a large number, even for high speed computers. We can cut this number in half by observing that, for $N_1 = N_2$, interchanging the two samples gives identical values of the test statistic, but that does not help much for large sample sizes. The good news is that it is really not necessary to compute the exact

Table 6.7.1 Generating the randomization distribution in the midge example. The table displays the first five and the last five partitions of the combined data set, obtained by systematically constructing all 5005 different partitions. The partitions were generated using $N_2 = 6$ indices $i_1, \ldots, i_6$ corresponding to the observations to be assigned to group 2 and letting i_1 go from 1 to 10, i_2 from $i_1 + 1$ to 11, ..., i_6 from $i_5 + 1$ to 15. For each partition the value of the standard distance between the two groups is shown.

Partition Number	Observations in Group 1	Observations in Group 2	D
1	7, 8, 9, 10, 11, 12, 13, 14, 15	1, 2, 3, 4, 5, 6	2.451
2	6, 8, 9, 10, 11, 12, 13, 14, 15	1, 2, 3, 4, 5, 7	2.659
3	6, 7, 9, 10, 11, 12, 13, 14, 15	1, 2, 3, 4, 5, 8	1.743
4	6, 7, 8, 10, 11, 12, 13, 14, 15	1, 2, 3, 4, 5, 9	1.315
5	6, 7, 8, 9, 11, 12, 13, 14, 15	1, 2, 3, 4, 5, 10	2.588
⋮	⋮	⋮	⋮
5001	1, 2, 3, 4, 5, 6, 7, 8, 13	9, 10, 11, 12, 14, 15	2.349
5002	1, 2, 3, 4, 5, 6, 7, 8, 12	9, 10, 11, 13, 14, 15	1.886
5003	1, 2, 3, 4, 5, 6, 7, 8, 11	9, 10, 12, 13, 14, 15	2.660
5004	1, 2, 3, 4, 5, 6, 7, 8, 10	9, 11, 12, 13, 14, 15	3.343
5005	1, 2, 3, 4, 5, 6, 7, 8, 9	10, 11, 12, 13, 14, 15	3.940

randomization distribution based on all partitions. Instead, we can generate a large number M (say $M = 10^3$ or $M = 10^4$) of random partitions. Quantiles of the discrete distribution generated by this procedure, then serve as approximate critical values for the test. Of course, this still means heavy computations, but at least the amount of work is finite. All we need is efficient software to compute standard distance and a good algorithm for generating random partitions. The latter task is surprisingly simple, as the student who has had a course on stochastic simulation may know. The interested reader is referred to the Suggested Further Readings Section at the end of this chapter and to Exercises 3 and 10. A convenient way to create a random partition is to shuffle (permute) the $N_1 + N_2$ observations randomly and assign the first N_1 observations to group 1. This explains why we use the terms "randomization test" and "permutation test" interchangeably.

Example 6.7.2 Two species of voles. In Example 5.4.4, we introduced eight-dimensional morphometric data measured on two species of voles, *Microtus multiplex* and *M. subterraneus*. There are $N_1 = 43$ classified specimens from *M. multiplex* and $N_2 = 46$ from *M. subterraneus*. The unclassified observations are omitted in the context of this example. The multivariate standard distance between the two groups, using all eight variables, is 3.117, which would be highly significant with regard to the null distribution. To

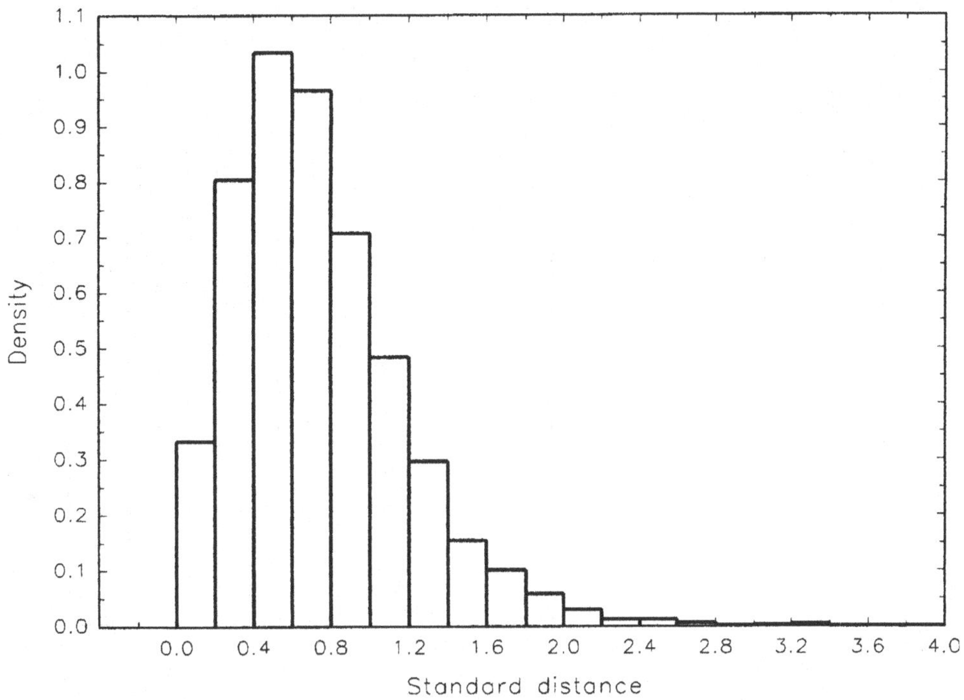

Figure 6-7-1 Randomization distribution of the standard distance D in the midge example. The distribution is summarized as a histogram.

obtain the exact randomization distribution of the standard distance D in this example, we would need to generate approximately 4.97×10^{25} randomizations. Suppose that we have good software that allows generating 1000 randomizations per second and computing the associated values of D. (This is a rather optimistic assumption because each computation of a value of D involves inversion of a matrix of dimension 8×8.) Then it would take about 1.57×10^{15} years to do all the computations. By the time the results become available, the two species of voles compared in this example will, most likely, be extinct, and multivariate statistics will either be obsolete or in the first grade curriculum.

While a comparison of the two species is still of interest, we generated $M = 10,000$ random partitions and computed the corresponding values of the standard distance D. Figure 6.7.2 shows the randomization distribution of D in form of a histogram. It turns out that the observed value 3.117 is larger than all 10,000 values. In fact, the largest of the 10,000 values generated in the randomization procedure is 1.366. This is a rather drastic confirmation of the decision reached by the parametric criterion. See Exercise 7 for more details on this analysis. □

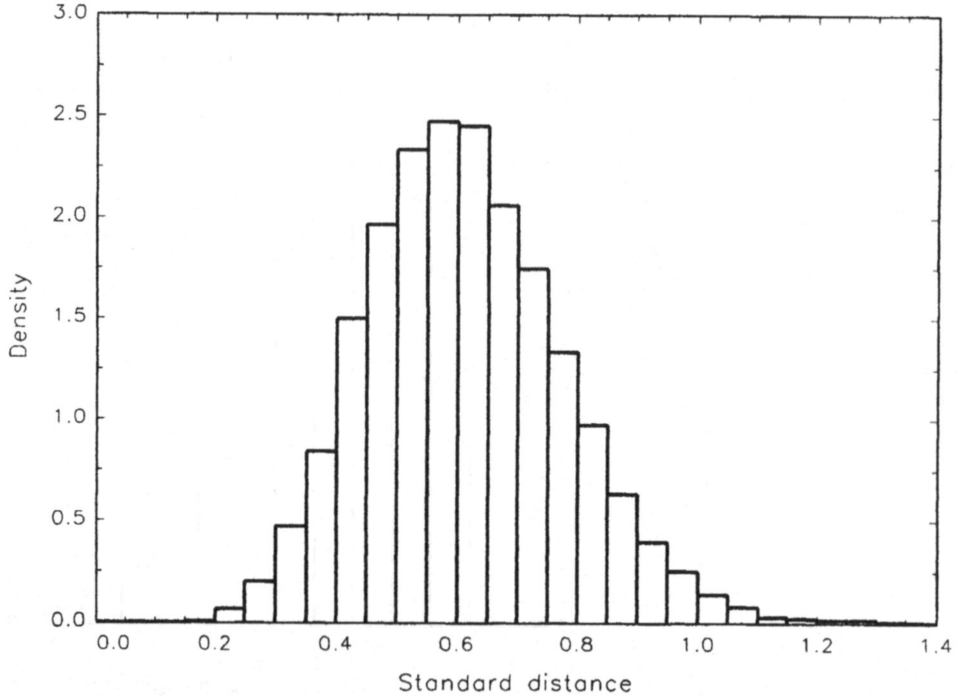

Figure 6-7-2 Randomization distribution of the standard distance D in the *Microtus* example, shown as a histogram. The distribution is based on 10,000 random partitions of the combined sample into subsamples of size 50 each. The observed value of D for the actual two groups is 3.117.

Although the randomization procedure allows relaxing the normality assumptions, it does not necessarily allow dropping all distributional assumptions. For instance, a randomization test cannot protect against the effects of dependence between observations. Also, recall that the randomization procedure for the hypothesis of equality of two mean vectors implicitly assumes that the two populations differ in location but not in any other aspect. We will return to this point in Example 6.7.4.

Results of statistical tests are often reported as p-values, or *achieved significance levels*, abbreviated as ASL. For simplicity, assume, again, that large values of the test statistic V indicate strong evidence against the hypothesis. Let v_{obs} denote the actually observed value of the test statistic, and let V_0 denote a random variable that follows the null distribution of V. Considering v_{obs} as fixed, the achieved significance level is defined as

$$\text{ASL} = \Pr[V_0 \geq v_{\text{obs}}]. \tag{5}$$

The parametric test procedure and the randomization procedure differ only in the way they compute the null distribution of V but not in the choice of the test statistic itself. In the case of Examples 6.7.1 and 6.7.2, the parametric ASL can be computed using the F-distribution. The ASL of the randomization distribution is

$$\text{ASL}_{\text{rand}} = \frac{\#\text{ of partitions with } V^* \geq v_{\text{obs}}}{\binom{N_1+N_2}{N_1}} . \tag{6}$$

If the randomization distribution is approximated using M random partitions rather than computed exactly, as will typically be the case, we get an approximate ASL

$$\widehat{\text{ASL}}_{\text{rand}} = \frac{\#\text{ of partitions with } V^* \geq v_{\text{obs}}}{M} , \tag{7}$$

but we will usually not distinguish in notation between (6) and (7), assuming that M is very large.

Example 6.7.3 continuation of Example 6.7.1

In the midge data set, the normal theory ASL is approximately 4.5×10^{-5}; see Exercise 4. The randomization ASL is $1/5005 \approx 2 \times 10^{-4}$. Both of these indicate very clearly that the hypothesis of equality of mean vectors should be rejected. □

Now, we return to a problem raised after Example 6.7.2. What if the two distributions differ considerably in variability? Although the result from the permutation test in Example 6.7.2 is clear enough to ignore such concerns, in other cases we might want instead to perform a test for equality of two distributions without assuming equality of covariance matrices. In Exercise 14 of Section 6.6 we studied the problem of testing for equality of two p-variate, normal distributions $\mathcal{N}_p(\boldsymbol{\mu}_1, \boldsymbol{\psi}_1)$ and $\mathcal{N}_p(\boldsymbol{\mu}_2, \boldsymbol{\psi}_2)$, based on independent samples $\mathbf{x}_{11}, \ldots, \mathbf{x}_{1N_1}$ and $\mathbf{x}_{21}, \ldots, \mathbf{x}_{2N_2}$. The log-likelihood ratio statistic for the hypothesis $H : (\boldsymbol{\mu}_1, \boldsymbol{\psi}_1) = (\boldsymbol{\mu}_2, \boldsymbol{\psi}_2)$ is expressed by

$$V = \sum_{j=1}^{2} N_j \log \frac{\det \bar{\boldsymbol{\psi}}}{\det \hat{\boldsymbol{\psi}}_j} = (N_1 + N_2) \log(\det \bar{\boldsymbol{\psi}}) - \sum_{j=1}^{2} N_j \log(\det \hat{\boldsymbol{\psi}}_j) , \tag{8}$$

where

$$\hat{\boldsymbol{\psi}}_j = \frac{1}{N_j} \sum_{i=1}^{N_j} (\mathbf{x}_{ji} - \bar{\mathbf{x}}_j)(\mathbf{x}_{ji} - \bar{\mathbf{x}}_j)' , \quad j = 1, 2, \tag{9}$$

$$\bar{\boldsymbol{\psi}} = \frac{1}{N_1 + N_2} \sum_{j=1}^{2} \sum_{i=1}^{N_j} (\mathbf{x}_{ji} - \bar{\mathbf{x}})(\mathbf{x}_{ji} - \bar{\mathbf{x}})' , \tag{10}$$

and $\bar{\mathbf{x}} = (N_1\bar{\mathbf{x}}_1 + N_2\bar{\mathbf{x}}_2)/(N_1 + N_2)$ is the overall mean vector. The asymptotic null distribution of V, under normality assumptions, is chi-square with $p(p+3)/2$ degrees of freedom. Although this test was derived using the normal theory likelihood methodology, we can still consider V a reasonable summary of all evidence against the hypothesis of no differences in location and variability, without making any specific distributional assumptions; see Exercise 5. But if we distrust the normality assumptions, then we should generate a null distribution of V using the randomization principle rather than relying on asymptotic theory. This is illustrated in the next example.

Example 6.7.4 continuation of Example 6.4.2

Electrodes produced by two machines. In the earlier analysis of this data set, we noticed that there are considerable differences in variability between the two machines. However, the standard distance between the two mean vectors of the five variables is so large that there is no reasonable doubt that the two distributions are different. To avoid yet another example where a randomization test is superfluous because the decision is clear anyway, we restrict ourselves to a subset of variables and a subset of observations: We use only the first 12 observations in each group and only the two variables labeled X_2 and X_5 in Table 5.3.5. Thus, $N_1 = N_2 = 12$, and $p = 2$. Because of the suspected differences in variability, we will perform a test of equality of the two distributions based on the statistic V of equation (8). Leaving details of the computations to the reader (Exercise 6), we obtain $V = 21.51$. This corresponds to an ASL of 6×10^{-4} of the asymptotic chi-square distribution with $p(p+3)/2 = 5$ degrees of freedom, which is significant by all standards. The randomization test, based on $M = 10,000$ random partitions of the data into groups of 12 each, tells a different story: 788 out of the 10,000 values of V generated by the randomization procedure exceeded the actually observed value of 21.51, which gives a randomization ASL of 0.0788. Usually, this would not be considered small enough to justify rejection of the null hypothesis of equality of the two distributions. Figure 6.7.3(a) shows the randomization distribution of V as a histogram, along with the approximating density of a chi-square distribution with five degrees of freedom. Clearly, the two distributions are very different, casting serious doubts about the use of any asymptotic approximation. There are two reasons why the normal theory approach fails so badly in this case. First, sample sizes of 12 each are not exactly "large"; secondly, contrary to tests on means, asymptotic results for tests on variances and covariances may be heavily affected by violations of the normality assumptions. Whatever the reason may be in this particular case, it is clear that the asymptotic theory cannot be trusted.

Fortunately the failure of asymptotic theory is not always as bad as in the case just illustrated, as we will show now on the same small data set. Recall from Exercise 14 in Section 6.6 that we can decompose the log-likelihood ratio statistic V of equation (8) as $V = V_1 + V_2$, where V_1 is the log-likelihood ratio statistic for equality of the two covariance matrices and V_2 is the log-likelihood ratio statistic for equality of the mean vectors, assuming equality of covariance matrices. The value of the V_2 for the two samples of 12 observations each is 5.465; see Exercise 6. From Section 6.6 we

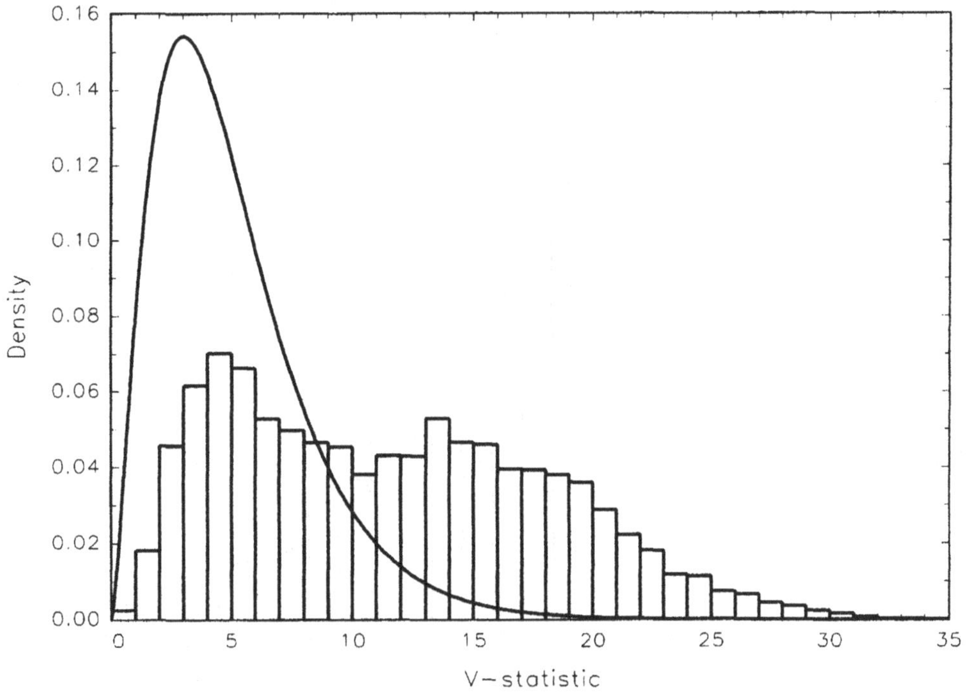

Figure 6-7-3A

know that the asymptotic distribution of V_2 under the null hypothesis of equality of mean vectors is chi-square on $p = 2$ degrees of freedom; this yields a parametric theory ASL of 0.0651. From the same 10,000 random partitions of the 24 observations used before, the randomization distribution of V_2 was computed; the randomization ASL is 0.0873. Although this is still not the same as the parametric theory ASL, the difference is not very large. Part (b) of Figure 6.7.3 shows a histogram of the randomization distribution of V_2, along with the approximating chi-square distribution with two degrees of freedom. Although not perfect, the approximation is surprisingly good, considering the fact that there are only 12 observations in each sample. An even better agreement between the parametric and the randomization null distributions is reached if we use the exact distribution theory for V_2 as follows. Recall from Theorem 6.6.1 that the statistic V_2 is a function of the two-sample T^2-statistic. That is, we can transform the 10,000 randomization values of V_2 to values of T^2 according to

$$T^2 = (N_1 + N_2 - 2)\exp\left(\frac{V_2}{N_1 + N_2}\right)$$

and compare the randomization distribution of T^2 to the exact normal theory null distribution of T^2. See Exercise 8. □

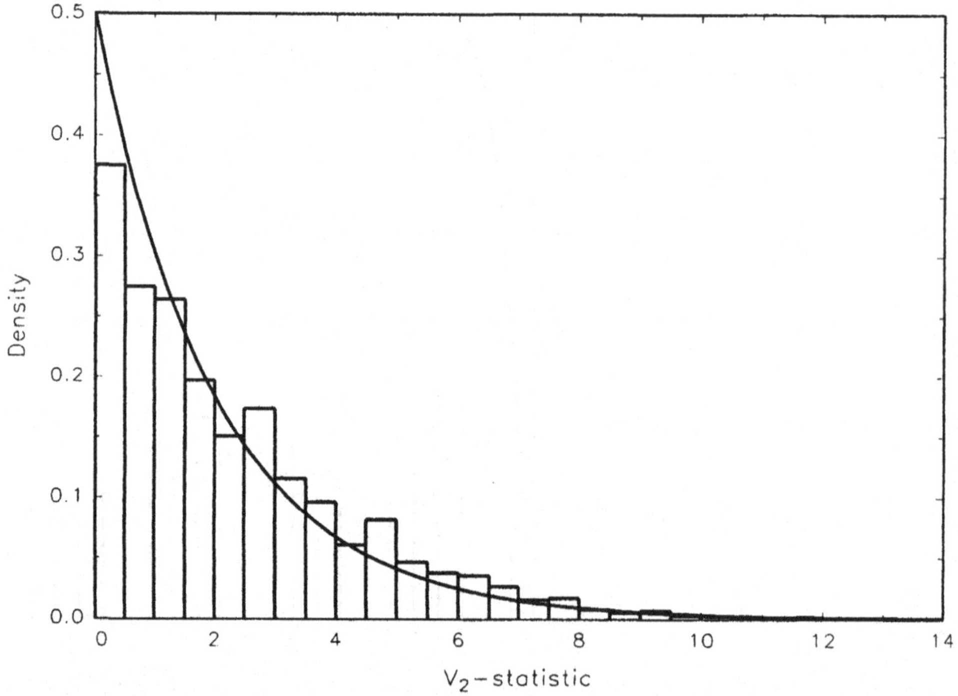

Figure 6-7-3B Randomization and approximating normal theory distributions in Example 6.7.4; (a) histogram of the randomization distribution of V, the log-likelihood ratio statistic for equality of two normal distributions, along with the asymptotic approximation by a chi-square distribution with five degrees of freedom; (b) histogram of the randomization distribution of V_2, the log-likelihood ratio statistic for equality of the mean vectors of two normal distributions, along with the asymptotic approximation by a chi-square distribution with two degrees of freedom.

Now, we return to the one-sample case of Section 6.2. With $\mathbf{x}_1, \ldots, \mathbf{x}_N$ denoting a sample from some p-variate distribution with mean vector $\boldsymbol{\mu}$, we wish to test the hypothesis $\boldsymbol{\mu} = \boldsymbol{\mu}_0$, where $\boldsymbol{\mu}_0$ is a specified vector in $\mathbb{R}^p$. The randomization idea is not going to help us here. Instead, we will use a bootstrap test as proposed in Efron and Tibshirani (1993, Chapter 16.4). We will outline its logic only on heuristic grounds, without much formal justification. As in the nonparametric bootstrap approach to estimating the distribution of a statistic (see Section 4.2), the idea is to sample from the empirical distribution of the observed data. However, there is a difficulty here: To find the null distribution of a test statistic, we need to do the calculations based on a model that satisfies the null hypothesis, in this case, the hypothesis that the mean is $\boldsymbol{\mu}_0$. The mean of the empirical distribution is $\bar{\mathbf{x}} = \sum_{i=1}^N \mathbf{x}_i/N$, which is almost certainly not equal to $\boldsymbol{\mu}_0$ (and if it were, any test should automatically come out non-significant). An easy way to create such a reference distribution with the desired property, but which is otherwise determined by the empirical distribution of the sample $\mathbf{x}_1, \ldots, \mathbf{x}_N$, is to shift the empirical distribution by $\mu_0 - \bar{\mathbf{x}}$. That is, we will take bootstrap samples from the empirical distribution of the shifted data

$$\mathbf{z}_i = \mathbf{x}_i - \bar{\mathbf{x}} + \mu_0, \quad i = 1, \ldots N. \tag{11}$$

The empirical distribution of the $\mathbf{z}_i$ has mean μ_0. We will denote this distribution by $\hat{G}$.

Suppose we use a statistic

$$V = v(\mathbf{x}_1, \ldots, \mathbf{x}_N) \tag{12}$$

to test the hypothesis $\mu = \mu_0$. Then the bootstrap null distribution of V is generated by taking a large number M of samples of size N from $\hat{G}$. Let $\mathbf{z}_1^*, \ldots, \mathbf{z}_N^*$ denote a typical bootstrap sample; then we compute $V^* = v(\mathbf{z}_1^*, \ldots, \mathbf{z}_N^*)$ and tabulate the distribution of V^* for the M bootstrap samples. (As in the permutation test, we would really like to know the *exact* bootstrap distribution of V rather than just a Monte Carlo approximation, but this involves N^N bootstrap samples, which is typically too much). Suppose large values of V indicate strong evidence against the hypothesis; then the bootstrap ASL is

$$\widehat{\text{ASL}}_{\text{boot}} = \frac{\#\text{ of bootstrap samples with } V^* \geq v_{\text{obs}}}{M}, \tag{13}$$

where v_{obs} is the observed value of V in the original sample.

Suppose the test statistic to be used is the standard distance between $\bar{\mathbf{x}}$ and μ_0,

$$D = \left[(\bar{\mathbf{x}} - \mu_0)'\mathbf{S}^{-1}(\bar{\mathbf{x}} - \mu_0)\right]^{1/2}.$$

For a given bootstrap sample $\mathbf{z}_1^*, \ldots, \mathbf{z}_N^*$, we have to compute the average

$$\bar{\mathbf{z}}^* = \frac{1}{N}\sum_{i=1}^{N}\mathbf{z}_i^*$$

and the sample covariance matrix

$$\mathbf{S}^* = \frac{1}{N-1}\sum_{i=1}^{N}(\mathbf{z}_i^* - \bar{\mathbf{z}}^*)(\mathbf{z}_i^* - \bar{\mathbf{z}}^*)'$$

and, finally, evaluate

$$D^* = \left[(\bar{\mathbf{z}}^* - \mu_0)'\,(\mathbf{S}^*)^{-1}\,(\bar{\mathbf{z}}^* - \mu_0)\right]^{1/2}. \tag{14}$$

Then the M values of D^* obtained from the M bootstrap samples define the bootstrap null distribution of the standard distance from μ_0.

Example 6.7.5 Head length and breadth of siblings. In the setup of Example 6.2.1, let $Y_1 = X_1 - X_3$ denote the difference in head length, and $Y_2 = X_2 - X_4$ the difference in head breadth between the first and second son. Setting $\mathbf{Y} = \begin{pmatrix} Y_1 \\ Y_2 \end{pmatrix}$ and $\mu = E[\mathbf{Y}]$, we wish to test

Table 6.7.2 Distribution $\hat{G}$ from which bootstrap samples are taken in Example 6.7.5. The distribution $\hat{G}$ puts probability 1/25 on each of the data points $\mathbf{z}_i' \in \mathbb{R}^2$ given by the rows of the table.

Observation #	Z_1	Z_2
1	7.92	8.24
2	−10.08	−4.76
3	−8.08	−2.76
4	−9.08	2.24
5	0.92	0.24
6	11.92	3.24
7	−5.08	−0.76
8	3.92	5.24
9	−13.08	−8.76
10	0.92	−2.76
11	2.92	1.24
12	−8.08	−1.76
13	16.92	1.24
14	7.92	−0.76
15	1.92	1.24
16	−2.08	5.24
17	7.92	−4.76
18	8.92	3.24
19	−5.08	−2.76
20	5.92	1.24
21	2.92	0.24
22	−8.08	−5.76
23	−4.08	−5.76
24	−7.08	7.24
25	−1.08	1.24

the hypothesis $H: \boldsymbol{\mu} = \mathbf{0}$. As in Example 6.2.1, we might use the standard distance from $\mathbf{0}$,

$$D = \left[\bar{\mathbf{y}}'\mathbf{S}^{-1}\bar{\mathbf{y}}\right]^{1/2},$$

as a test statistic. To facilitate the comparison of the bootstrap null distribution with the normal theory null distribution, we will use instead the statistic

$$F = \frac{N}{N-1} \cdot \frac{N-p}{p} D^2 . \tag{15}$$

Since $\boldsymbol{\mu}_0 = \mathbf{0}$, we use shifted data

$$\mathbf{z}_i = \mathbf{y}_i - \bar{\mathbf{y}}, \quad i = 1, \ldots, N . \tag{16}$$

The $N = 25$ data pairs $\mathbf{z}_i = \begin{pmatrix} z_{i1} \\ z_{i2} \end{pmatrix}$ are the rows of the data matrix displayed in Table 6.7.2; this table defines the distribution $\hat{G}$ from which bootstrap samples are taken to find the null distribution of the test statistic. Table 6.7.3 identifies the first five of $M = 10,000$ bootstrap samples taken from $\hat{G}$ by the indexes of the observations chosen. For each of the 10,000 bootstrap samples, the value F^* was computed using equations (14) and (15). The bootstrap distribution of F, thus obtained, is shown as a histogram in Figure 6.7.4. From Example 6.2.1, the observed value of F is 3.831, corresponding to an ASL of 0.037 computed from the F-distribution with 2 and 23 degrees of freedom. Since 418 values of F^* are larger than 3.831, the bootstrap ASL is approximately 0.042. This is in good agreement with the parametric result. In fact, the bootstrap null distribution and the F-distribution with 2 and 23 degrees of freedom, shown as the continuous curve in Figure 6.7.4, agree surprisingly well. This sheds a somewhat more positive light on the value of parametric theory than Figure 6.7.3(a). □

As the examples show, randomization and bootstrap tests are powerful tools for inference. They cannot replace parametric theory completely because the choice of a test statistic will typically still be based on criteria rooted in multivariate normal theory. Yet, they allow avoiding some of the assumptions made in the classical calculation of null distributions and achieved significance levels. In the author's experience, users of statistical tests seem to appreciate the logic of randomization tests as particularly convincing. In fact, in cases where a randomization test can be applied, one may well consider the calculation of a null distribution by parametric theory as nothing more than a convenient approximation to the "true null distribution" given by the randomization procedure. Edgington (1980) elaborates on this in detail.

Table 6.7.3 The first five bootstrap samples from $\hat{G}$ in Example 6.7.5. Data points selected more than once in a given bootstrap sample are identified by an appropriate power. For instance, "18^3" in the first line means that observation no. 18 from Table 6.7.2 was represented three times in the first bootstrap sample.

Bootstrap Sample	Observations Included	F^*
1	$1, 2, 3, 4^2, 7, 8^2, 9^2, 10, 11^2, 15, 16, 17, 18^3, 19^2, 21^3, 25$	0.2369
2	$2^2, 3, 4, 5, 6, 8^2, 9^2, 10^2, 11, 13^2, 16, 18, 21^2, 22, 24^4, 25$	0.7118
3	$1^3, 5, 7, 10, 12, 13, 14, 15^2, 16, 17, 18, 19, 20^3, 21^2, 22, 24^2, 25^2$	2.5708
4	$4^2, 5, 6^2, 8, 9^5, 11^3, 13, 14^2, 16, 17^3, 18, 21, 22, 25$	1.7804
5	$3^2, 5, 6, 7, 8^2, 9^2, 10, 11^2, 12, 13^2, 14, 15, 16^3, 19, 22^2, 23, 24$	0.1515
⋮	⋮	⋮

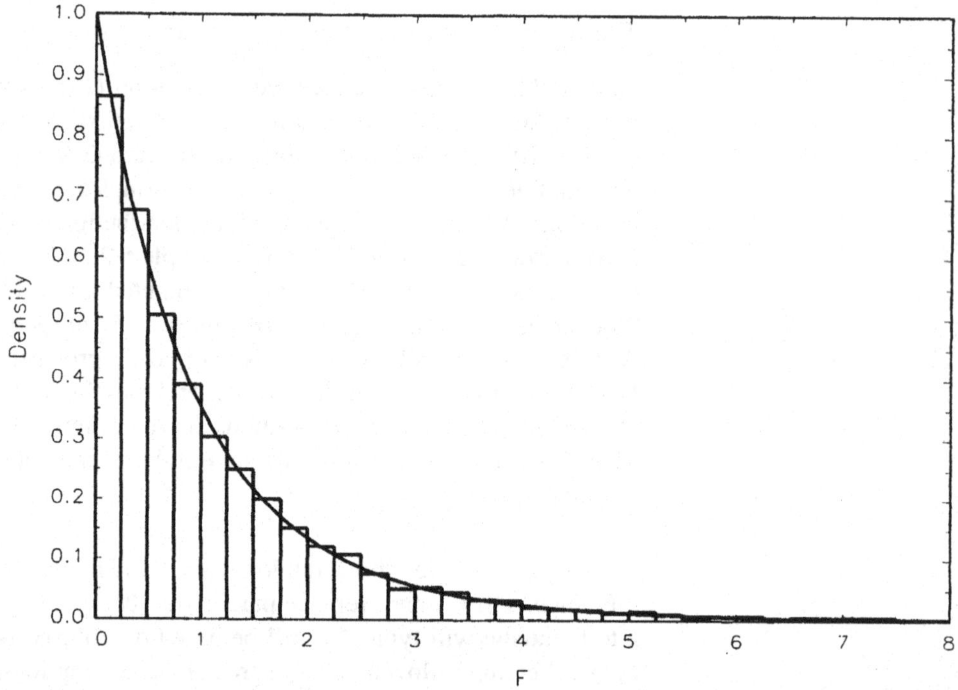

Figure 6-7-4 Histogram of the bootstrap null distribution of F in Example 6.7.5. Also shown as a solid curve is the density function of an F-distribution with 2 and 23 degrees of freedom, which serves as the parametric null distribution. The observed value from the original sample is $F = 3.831$. The 34 values of the bootstrap distribution that are larger than 8 are not shown in this graph.

Exercises for Section 6.7

1. Prove the following result (randomization lemma): From a set with N items, draw $n < N$ items sequentially, such that, at the tth draw, each of the $N - t + 1$ items not yet selected has the same probability of being drawn. Show that each subset of size n has the same probability $1/\binom{N}{n}$ of being selected.

2. Let V denote a two-sample test statistic and v_{obs} the observed value from a particular set of two samples. Let V^* be another two-sample test statistic, which is a monotonically increasing function of V. (For instance, V could be the standard distance, and V^* the corresponding T^2-statistic). Show that the ASL of v_{obs} in the randomization distribution of V is the same as the ASL of v^*_{obs} in the randomization distribution of V^*.

3. Show that the following procedure can be used to create a random partition of a set of N objects $x_1, \ldots, x_N$ into subsets of size n and $N - n$: Permute the integers $1, \ldots, N$ randomly, and choose the subset corresponding to the first n and the last $N - n$ numbers in the permuted list.

4. Find the normal theory ASL (achieved significance level) of the test for equality of mean vectors in Example 6.7.1. *Hint*: Transform the D-statistic into an F-statistic, according to the procedure explained in Section 6.4. Denote this value as f_{obs}. Compute the integral $\int_{f_{\text{obs}}}^{\infty} g(y)\,dy$ numerically, where g is the density function of an F-distribution with 2 and 12 degrees of freedom. (You need some software that allows computing tail probabilities of F-distributions to do this exercise).

5. Show that the test statistic V of equation (8) is zero if and only if the two sample mean vectors and the two sample covariance matrices are equal, and larger than zero otherwise. *Hint*: Use the decomposition $V = V_1 + V_2$ from Exercise 14 in Section 6.6.

6. This exercise is based on Example 6.7.4. Extract the first twelve observations of each of the two groups (and variables X_2 and X_5) from the electrode data set of Table 5.3.5. Verify the calculation of the test statistic V. Also calculate the values of the statistics V_1 and V_2 in the decomposition of V from Exercise 14 in Section 6.6.

7. In this exercise, we show how to compare the randomization distribution of T^2 to its exact and approximate normal theory null distributions. We will need the chi-square and F densities. The *pdf* of a random variable Y distributed as chi-square with r degrees of freedom is given by

$$g(y) = \frac{1}{\Gamma(\frac{1}{2}r)\, 2^{\frac{1}{2}r}}\, y^{\frac{1}{2}r-1}\, e^{-y/2} \quad \text{for } y > 0. \tag{17}$$

A random variable Y follows an F-distribution with r_1 and r_2 degrees of freedom if its *pdf* is given by

$$h(y) = \frac{\Gamma\left[\frac{1}{2}(r_1 + r_2)\right] \left(\frac{r_1}{r_2}\right)^{\frac{1}{2}r_1}}{\Gamma\left(\frac{1}{2}r_1\right)\, \Gamma\left(\frac{1}{2}r_2\right)} \cdot \frac{y^{\frac{1}{2}r_1-1}}{\left(1 + \frac{r_1}{r_2}y\right)^{\frac{1}{2}(r_1+r_2)}} \quad \text{for } y > 0. \tag{18}$$

(a) From Theorem 6.4.1, we know that, under the hypothesis $\mu_1 = \mu_2$ and under normality assumptions,

$$F = \frac{N_1 + N_2 - p - 1}{(N_1 + N_2 - 2)p}\, T^2$$

follows an F-distribution with $r_1 = p$ and $r_2 = (N_1 + N_2 - p - 1)$ degrees of freedom. Use this result to show that the exact distribution of T^2 under the hypothesis is given by the *pdf*

$$h^*(y) = c \cdot h(cy), \tag{19}$$

where h is the *pdf* in equation (18) and $c = \frac{(N_1+N_2-p-1)}{(N_1+N_2-2)p}$.

(b) Show that, as $N_1 + N_2$ tends to infinity while p remains constant, the *pdf* h^* of equation (19) converges to the *pdf* of a chi-square random variable with p degrees of freedom.

Note: Figure 6.7.5 illustrates an application of this exercise to the *Microtus* data of Example 6.7.2. The solid curve and the broken curve represent the exact null

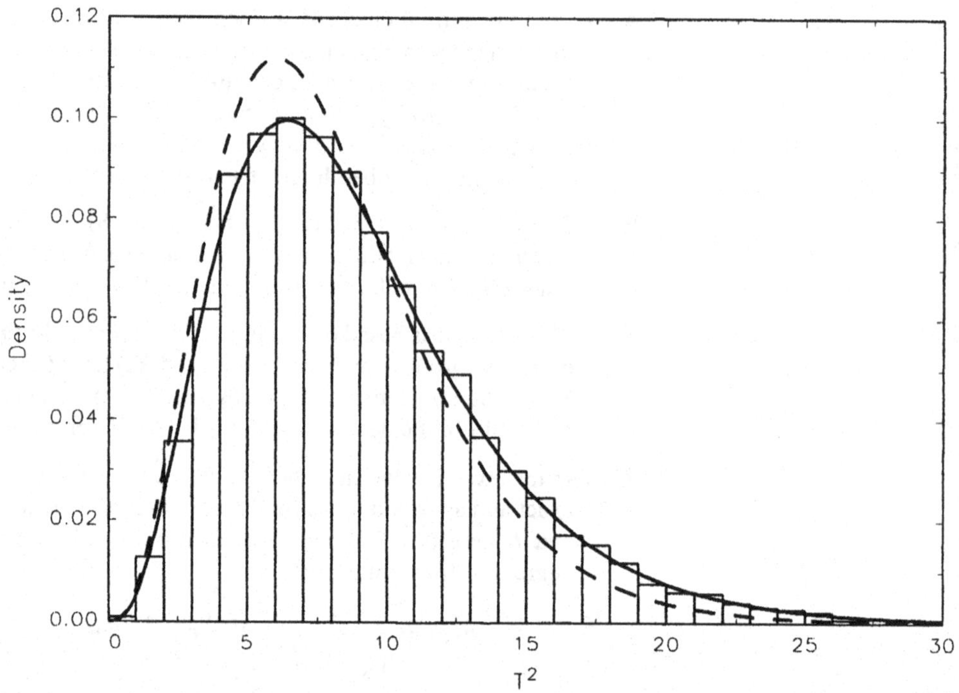

Figure 6-7-5 Histogram of the randomization distribution of T^2 in Example 6.7.2 using $M = 10,000$ randomizations. The 28 values of the randomization distribution that are larger than 30 are not included. The broken curve is the *pdf* of the chi-square approximation to the null distribution of T^2, i.e., a chi-square density with 8 degrees of freedom. The solid curve is the *pdf* of the exact normal theory null distribution of T^2, obtained from Exercise 7(a).

distribution and its large sample approximation by a chi-square distribution. The randomization distribution and the exact parametric distribution are in very close agreement.

8. This exercise refers to Example 6.7.4.
   (a) Compute a randomization distribution of T^2 in the electrode example with $p = 2$ variables and $N_1 = N_2 = 12$ observations per group. Use $M = 10,000$ randomizations. Graph the randomization distribution as a histogram.
   (b) In the same graph, plot the approximate normal theory null distribution of T^2, i.e., the *pdf* of a chi-square, random variable with 2 degrees of freedom.
   (c) Again in the same graph, plot the exact normal theory null distribution of T^2, using the method of Exercise 7.
   (d) Compare the three distributions visually and comment.

9. This exercise concerns the computation of the bootstrap null distribution of the standard distance D from a hypothetical mean vector. Show that, instead of sampling

from the distribution of the $\mathbf{z}_i$ (11) and computing D^* as in equation (14), we can equivalently take bootstrap samples $\mathbf{x}_1^*, \ldots, \mathbf{x}_N^*$ from the empirical distribution of $\mathbf{x}_1, \ldots, \mathbf{x}_N$, and calculate

$$D^* = \left[(\bar{\mathbf{x}}^* - \bar{\mathbf{x}})' \left(\mathbf{S}_\mathbf{x}^*\right)^{-1} (\bar{\mathbf{x}}^* - \bar{\mathbf{x}}) \right]^{1/2}.$$

Here, $\bar{\mathbf{x}}^*$ is the mean of the bootstrap sample, $\mathbf{S}_\mathbf{x}^*$ is its sample covariance matrix, and $\bar{\mathbf{x}}$ is the mean of the original sample $\mathbf{x}_1, \ldots, \mathbf{x}_N$.

10. The following is an algorithm for choosing a random subset with n elements from a set of N objects $\{x_1, \ldots, x_N\}$:

Set $k = 0$. Then repeat the following step for $i = 1$ to N. Select object x_i with probability $(n - k)/(N - i + 1)$. If x_i is selected, replace k by $k - 1$. Otherwise, leave k unchanged.

Show that every subset of size n has the same probability $1/\binom{N}{n}$ of being selected. *Hint*: Let $i_1 < i_2 < \cdots < i_n$ be the indexes of the elements of some subset of size n. Compute the probability that this subset will be selected, using conditional probabilities. If you find this difficult, first solve the problem for $n = 3$ and for the following two sets of indexes: $\{1, 2, 3\}$ and $\{N - 2, N - 1, N\}$.

11. This exercise refers to Example 6.2.2, cork deposits on trees. Compute a bootstrap null distribution of the statistic D for testing the hypothesis $E[\mathbf{Y}] = \mathbf{0}$. Use 10,000 bootstrap replications. Compute the bootstrap ASL and the normal theory ASL using the F-distribution. Does the bootstrap test support or contradict the decision made in Example 6.2.2?

12. Perform a randomization test with $M = 10,000$ random assignments to test for equality of the mean vectors of four morphometric variables in two species of flea beetles; see Example 5.3.2 and Table 5.3.2. Use either D, T^2, or F as a test statistic, whichever you prefer. Compute the normal theory ASL and the randomization ASL, and compare.

13. This exercise refers to Example 6.6.4, in which we studied a model of exchangeability of two p-variate, random vectors $\mathbf{X}_{(1)}$ and $\mathbf{X}_{(2)}$. In equation (34) of Section 6.6, we gave the log-likelihood ratio statistic for the hypothesis of exchangeability and stated that its large sample null distribution is approximately chi-square.

(a) Design a "randomization" test for the hypothesis of exchangeability of $\mathbf{X}_{(1)}$ and $\mathbf{X}_{(2)}$, based on the idea that, if the hypothesis holds, then for each of the N observations the values $\begin{pmatrix} \mathbf{x}_{i(1)} \\ \mathbf{x}_{i(2)} \end{pmatrix}$ and $\begin{pmatrix} \mathbf{x}_{i(2)} \\ \mathbf{x}_{i(1)} \end{pmatrix}$ are equally likely. That is, for each of the N observations, interchange $\mathbf{x}_{i(1)}$ and $\mathbf{x}_{i(2)}$ with probability $1/2$.

(b) How many different such "randomization samples" can be constructed if, for each observation, the two components can either be left as they are in the original sample or interchanged? If the interchanges are made independently and with probability $1/2$ each, what is the probability of each particular randomization sample?

(c) Apply this procedure to the siblings' data (Example 6.6.5), with $M = 10,000$ randomization samples. For each of the 10,000 samples, compute the value of the log-likelihood ratio statistic. Graph the randomization distribution, and compute the ASL of the observed value of 8.99 from Example 6.6.5. How does this compare to the asymptotic theory ASL of 0.17?

Suggested Further Reading

General: Casella and Berger (1990), Silvey (1975).
Sections 6.2 and 6.4: Hotelling (1931), Rao (1948), Takemura (1985).
Section 6.3: Fairley (1986).
Section 6.5: Mardia et al. (1979), Rao (1970).
Section 6.6: Roy (1957).
Section 6.7: Efron and Tibshirani (1993), Ripley (1987), Ross (1996), Westfall and Young (1993).

7 Discrimination and Classification, Round 2

7.1 Optimal Classification

In this chapter we continue the theory of classification developed in Chapter 5 on a somewhat more general level. We start out with some basic consideration of optimality. In the notation introduced in Section 5.4, **Y** will denote a p-variate random vector measured in k groups (or populations). Let X denote a discrete random variable that indicates group membership, i.e., takes values $1, \ldots, k$. The probabilities

$$\pi_j = \Pr[X = j] \qquad j = 1, \ldots, k, \tag{1}$$

will be referred to as *prior probabilities*, as usual. Suppose that the distribution of **Y** in the jth group is given by a *pdf* $f_j(\mathbf{y})$, which may be regarded as the conditional *pdf* of **Y**, given $X = j$. Assume for simplicity that **Y** is continuous with sample space $\mathbb{R}^p$ in each group. Then the joint *pdf* of X and **Y**, as seen from Section 2.8, is

$$f_{XY}(j, \mathbf{y}) = \begin{cases} \pi_j f_j(\mathbf{y}) & for j = 1, \ldots, k,\ \mathbf{y} \in \mathbb{R}^p, \\ 0 & \text{otherwise.} \end{cases} \tag{2}$$

The marginal *pdf* of **Y** is the *mixture density*

$$f_{\mathbf{Y}}(\mathbf{y}) = \sum_{j=1}^{k} \pi_j f_j(\mathbf{y}) \qquad \mathbf{y} \in \mathbb{R}^p. \tag{3}$$

The conditional probability of an observation being from group j, given $\mathbf{Y} = \mathbf{y}$, then is given by

$$\pi_{jy} = \Pr[X = j \mid \mathbf{Y} = \mathbf{y}] = \frac{\pi_j f_j(\mathbf{y})}{\sum_{h=1}^{k} \pi_h f_h(\mathbf{y})}, \qquad j = 1, \ldots, k. \tag{4}$$

The π_{jy} will often be referred to as *posterior probabilities*. By construction, $\sum_{j=1}^{k} \pi_{jy} = 1$ for all $\mathbf{y}$.

Let $C_1, \ldots, C_k$ be a partition of the sample space of $\mathbf{Y}$ (i.e., usually a partition of $\mathbb{R}^p$), and let $x^*(\mathbf{y})$ be a function of $\mathbf{y} \in \mathbb{R}^p$ that takes value j if $\mathbf{y} \in C_j$. A formal definition of a classification rule, extending Definition 5.5.1, can be given as follows.

Definition 7.1.1 A *classification rule* is a random variable

$$X^* = x^*(\mathbf{Y}) \tag{5}$$

that takes value j exactly if $\mathbf{Y} \in C_j$, $j = 1, \ldots, k$. □

We may regard X^* as the *predicted group membership*, as opposed to the actual group membership X.

The theory of optimal classification attempts to find classification rules that perform "best" in some sense to be defined. Ideally, $X^* = X$ with probability 1, but since X^* is a function of $\mathbf{Y}$ this will, in general, not be possible. For a given classification rule X^* with classification regions $C_1, \ldots, C_k$, the probability of assigning an observation $\mathbf{Y}$ to group i, given that it actually comes from population j, is

$$\begin{aligned} q_{ij} &= \Pr\left[X^* = i \mid X = j\right] \\ &= \Pr\left[\mathbf{Y} \in C_i \mid X = j\right] \\ &= \int_{C_i} f_j(\mathbf{y})\, d\mathbf{y} \qquad i, j = 1, \ldots, k. \end{aligned} \tag{6}$$

In some cases it may be possible to assign a cost d_{ij} to the classification of an observation from group j in group i; one would then want to find X^* such that the expected cost of classification is minimal. This is explored in some detail in Exercises 1 and 2. Here, we will focus on the overall probability of misclassification, or *error rate* of a classification procedure.

Definition 7.1.2 The *error rate* of a classification rule X^* is given by

$$\gamma(X^*) = \Pr[X^* \neq X]. \tag{7}$$

□

For a given classification rule X^* with classification regions $C_1, \ldots, C_k$, the error rate can be computed as

$$\begin{aligned}\gamma(X^*) &= \sum_{j=1}^{k} \Pr[X = j, X^* \neq j] \\ &= 1 - \sum_{j=1}^{k} \Pr[X = j, X^* = j] \\ &= 1 - \sum_{j=1}^{k} \pi_j q_{jj}.\end{aligned} \tag{8}$$

We will call a classification rule *optimal* and denote it by X^*_{opt} if $\gamma(X^*_{\text{opt}}) \leq \gamma(X^*)$ for all classification rules X^*. Equivalently to minimizing $\gamma(X^*)$, we can maximize the probability of correct classification, $1 - \gamma(X^*) = \sum_{j=1}^{k} \pi_j q_{jj}$. The solution to this maximization problem is given as a theorem.

Theorem 7.1.1 *Any optimal classification rule has classification regions*

$$C_j = \left\{\mathbf{y} \in \mathbb{R}^p : \pi_j f_j(\mathbf{y}) > \pi_h f_h(\mathbf{y}) \text{ for all } h \neq j\right\}, \qquad j = 1, \ldots, k. \tag{9}$$

The sets $\{\pi_j f_j(\mathbf{y}) = \pi_h f_h(\mathbf{y})\}$ *can be assigned arbitrarily to either* C_j *or* C_h.

Proof Let $D_1, \ldots, D_k$ denote a partition of $\mathbb{R}^p$, and define a function $h(\mathbf{y})$ as

$$h(\mathbf{y}) = \pi_j f_j(\mathbf{y}) \text{ if } \mathbf{y} \in D_j, \qquad j = 1, \ldots, k. \tag{10}$$

Then the classification rule X^* given by the partition $D_1, \ldots, D_k$ has error rate $\gamma(X^*)$ with

$$\begin{aligned}1 - \gamma(X^*) &= \sum_{j=1}^{k} \pi_j q_{jj} \\ &= \sum_{j=1}^{k} \int_{D_j} \pi_j f_j(\mathbf{y})\, d\mathbf{y} \\ &= \int_{\mathbb{R}^p} h(\mathbf{y})\, d\mathbf{y}.\end{aligned}$$

Since $\pi_j f_j(\mathbf{y}) \geq 0$ for all $\mathbf{y} \in \mathbb{R}^p$ and all $j = 1, \ldots, k$,

$$h(\mathbf{y}) \leq \max_{1 \leq j \leq k} \pi_j f_j(\mathbf{y}),$$

and therefore,

$$1 - \gamma(X^*) \leq \int_{\mathbb{R}^p} \max\{\pi_1 f_1(\mathbf{y}), \ldots, \pi_k f_k(\mathbf{y})\}\, d\mathbf{y}.$$

Hence, the probability of correct classification is maximized if we choose classification regions as in equation (9). For points $\mathbf{y} \in \mathbb{R}^p$ such that the maximum is attained by two or several of the $\pi_h f_h(\mathbf{y})$, it does not matter to which of the corresponding classification regions they are assigned. ■

An immediate consequence of Theorem 7.1.1 is that optimal classification in terms of minimal error rate is the same as classification according to highest posterior probability. By definition of the optimal classification regions C_j in (9), $\mathbf{y} \in \mathbb{R}^p$ is classified into group j if $\pi_j f_j(\mathbf{y}) > \pi_i f_i(\mathbf{y})$ for all $i \neq j$. This is equivalent to classifying $\mathbf{y}$ into group j if

$$\frac{\pi_j f_j(\mathbf{y})}{\sum_{h=1}^k \pi_h f_h(\mathbf{y})} > \frac{\pi_i f_i(\mathbf{y})}{\sum_{h=1}^k \pi_h f_h(\mathbf{y})} \qquad \text{for all } i \neq j. \tag{11}$$

This may be viewed as a late justification of the principle used in Section 5.4.

Classification rules that minimize the error rate (or more generally, the expected cost) are often referred to as *Bayes rules*, presumably because the computation of posterior probabilities reflects the Bayes formula from probability theory.

Example 7.1.1 Two univariate normal distributions. Suppose $Y \sim \mathcal{N}(\mu_j, \sigma_j^2)$ in groups $j = 1, 2$, with prior probabilities π_1 and $\pi_2 = 1 - \pi_1$. The classification regions are given by

$$C_1 = \{y \in \mathbb{R} : ay^2 + by + c > 0\}$$

and

$$C_2 = \{y \in \mathbb{R} : ay^2 + by + c < 0\}, \tag{12}$$

where

$$a = \frac{1}{2}\left(\frac{1}{\sigma_2^2} - \frac{1}{\sigma_1^2}\right),$$

$$b = \frac{\mu_1}{\sigma_1^2} - \frac{\mu_2}{\sigma_2^2} \tag{13}$$

and

$$c = \log\frac{\pi_1}{\pi_2} + \frac{1}{2}\log\frac{\sigma_2^2}{\sigma_1^2} + \frac{1}{2}\left(\frac{\mu_2^2}{\sigma_2^2} - \frac{\mu_1^2}{\sigma_1^2}\right),$$

see Exercise 3. The cutoff points between C_1 and C_2 are the solutions to the quadratic equation $ay^2 + by + c = 0$. Depending on the parameter values, this equation may have zero, one, two, or infinitely many real roots; see Exercise 4. Figure 7.1.1 illustrates a situation with two roots; the classification region C_1 consists of two half-lines, and region C_2 is a finite interval. In the same graph, the function $h(y) =$

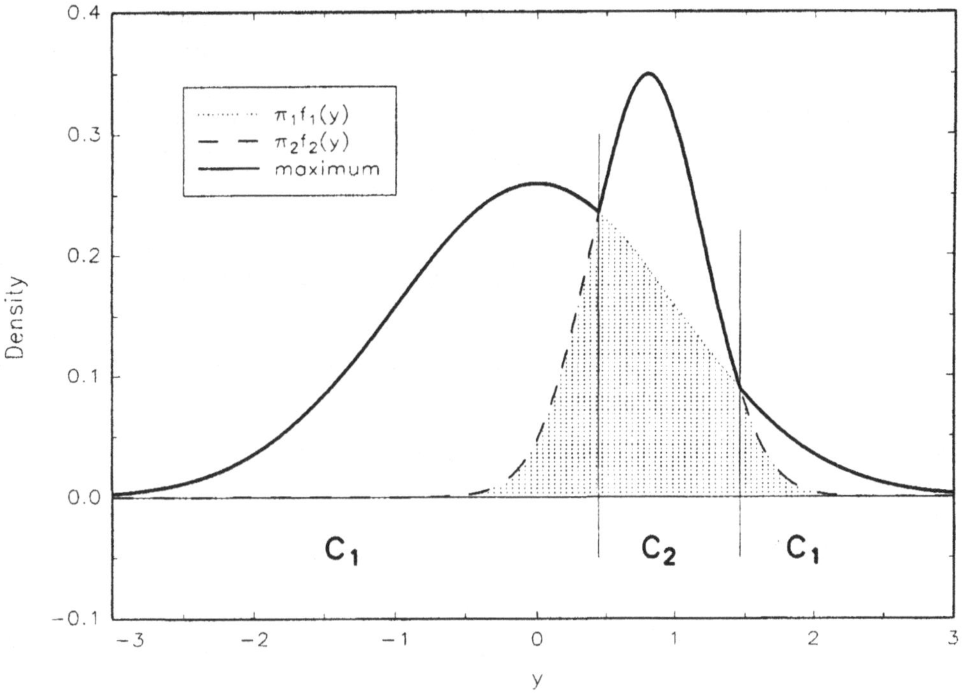

Figure 7-1-1 Optimal classification for two univariate normal distributions with different means and variances. The error rate corresponds to the shaded area. The probability of correct classification is the area under the thick curve.

$\max\{\pi_1 f_1(\mathbf{y}), \pi_2 f_2(\mathbf{y})\}$ is shown as a thick curve. The area under this curve corresponds to the probability of correct classification. Similarly, the shaded area in Figure 7.1.1 gives the error rate.

The numerical values used in Figure 7.1.1 are $\pi_1 = 0.65$, $\mu_1 = 0$, $\sigma_1 = 1$, $\pi_2 = 0.35$, $\mu_2 = 0.8$, and $\sigma_2 = 0.4$. The cutoffs are $v_1 \approx 0.444$ and $v_2 \approx 1.461$. Thus, the optimal classification rule is

$$X^*_{\text{opt}} = \begin{cases} 1 & \text{if } y < v_1 \text{ or } y > v_2 \\ 2 & \text{if } v_1 < y < v_2, \end{cases}$$

and the associated error rate $\gamma(X^*_{\text{opt}}) \approx 0.249$; see Exercise 5. □

In Example 7.1.1, the classification regions C_1 and C_2 are bounded by two cutoff points v_1 and v_2. Since the probability of a normal random variable taking a fixed value is zero (unless the normal random variable is itself constant), $\Pr[Y = v_1 \text{ or } v_2 | X = 1] = \Pr[Y = v_1 \text{ or } v_2 | X = 2] = 0$. That is, in the normal model, we do not worry about the possibility of an observation falling exactly on the boundary, and the optimal classification rule is unique except for a subset of the sample space that has probability zero. This need not always be so, as illustrated by our next example and by Exercise 6.

Example 7.1.2 Suppose $\mathbf{Y} = (Y_1, \ldots, Y_p)'$ is multinomial with $Y_1 + \cdots + Y_p = N$ and with parameters $\theta_{j1}, \ldots, \theta_{jp}$ in group j, $j = 1, 2$, that is, the probability function of $\mathbf{Y}$ in group j $(j = 1, 2)$ is given by

$$f_j(\mathbf{y}) = \binom{N}{y_1 \; y_2 \; \cdots \; y_p} \theta_{j1}^{y_1} \theta_{j2}^{y_2} \cdots \theta_{jp}^{y_p}, \text{ where all } y_i \geq 0, \; \sum_{i=1}^{p} y_i = N.$$

With prior probabilities π_1 and π_2 $(\pi_1 + \pi_2 = 1)$, we would classify $\mathbf{y}$ into group 1 if $\pi_1 f_1(\mathbf{y}) > \pi_2 f_2(\mathbf{y})$. This condition is equivalent to setting the classification regions as

$$C_1 = \left\{ \mathbf{y} : \log \frac{\pi_1 f_1(\mathbf{y})}{\pi_2 f_2(\mathbf{y})} > 0 \right\}$$

and

$$C_2 = \left\{ \mathbf{y} : \log \frac{\pi_2 f_2(\mathbf{y})}{\pi_1 f_1(\mathbf{y})} > 0 \right\}.$$

But the condition $\log [\pi_1 f_1(\mathbf{y}) / \pi_2 f_2(\mathbf{y})] > 0$ is equivalent to

$$\log \frac{\pi_1}{\pi_2} + \sum_{i=1}^{p} y_i \log \left(\frac{\theta_{1i}}{\theta_{2i}} \right) > 0.$$

Thus, classification depends on the observed counts $y_1, \ldots, y_p$ through a linear function of the y_i.

In this setup, it may well happen (with positive probability) that $\mathbf{Y}$ falls on the boundary set $\{\pi_1 f_1(\mathbf{y}) = \pi_2 f_2(\mathbf{y})\}$. For instance, if $\pi_1 = \pi_2 = \frac{1}{2}$, with $p = 2$ categories, suppose that $\theta_{11} = 2/3$, $\theta_{12} = 1/3$, $\theta_{21} = 1/3$, and $\theta_{22} = 2/3$. Then the boundary of the classification regions is given by $y_1 \log 2 - y_2 \log 2 = 0$. That is, if we observe $Y_1 = Y_2 = N/2$, then we can classify the observation into either group 1 or group 2, because

$$\Pr \left[Y = \frac{N}{2} \right] = \binom{N}{N/2} 2^{N/2} 3^{-N}$$

conditionally on both $X = 1$ and $X = 2$. A similar example is given in Exercise 10 of Section 5.4. □

In some applications, it may not be possible to assign prior probabilities to the k groups. In such a case, one might argue that, lacking any information on prior probabilities, an observation should be classified in the group with the largest value of $f_j(\mathbf{y})$. That is, the classification region for group j would then be given by

$$C_j = \left\{ \mathbf{y} \in \mathbb{R}^p : f_j(\mathbf{y}) > f_h(\mathbf{y}) \text{ for all } h \neq j \right\}. \tag{14}$$

This is sometimes called the maximum likelihood rule because it corresponds to choosing the value of a parameter (group membership) such that, under all models considered, the chosen one has the highest probability. In fact, this principle is equivalent to choosing prior probabilities $\pi_j = 1/k$ for all k groups, as seen from equations (9) and (14).

Exercises for Section 7.1

1. In a classification problem with $k = 2$ groups, suppose that classifying an observation in group i, when it actually comes from group j, costs s_{ij} dollars, where $s_{11} = s_{22} = 0$, $s_{12} > 0$, $s_{21} > 0$. Let S be a random variable that takes value s_{ij} if $X = j$ and $X^* = i$. Show that a classification rule that minimizes $E[S]$ is given by the classification regions

$$D_1 = \left\{ \mathbf{y} \in \mathbb{R}^p : \frac{f_1(\mathbf{y})}{f_2(\mathbf{y})} \geq \frac{s_{12}\pi_2}{s_{21}\pi_1} \right\}$$

and

$$D_2 = \left\{ \mathbf{y} \in \mathbb{R}^p : \frac{f_1(\mathbf{y})}{f_2(\mathbf{y})} < \frac{s_{12}\pi_2}{s_{21}\pi_1} \right\}. \tag{15}$$

Hint: For classification regions D_1 and D_2, show that $E[S] = \int_{D_2} s_{21}\pi_1 f_1(\mathbf{y}) + \int_{D_1} s_{12}\pi_2 f_2(\mathbf{y})$. Then use a technique similar to the one used in the proof of Theorem 7.1.1 to show that $E[S]$ is a minimum if D_1 and D_2 are chosen as in (15).

2. In a classification problem with k groups, suppose that the cost of classification is

$$s_{ij} = \begin{cases} s & \text{if } i \neq j, \\ 0 & \text{otherwise.} \end{cases}$$

Here, $s > 0$ is a fixed constant. Show that the classification rule that minimizes expected cost is identical to the one given in Theorem 7.1.1.

3. Prove equations (12) and (13).

4. This exercise expands on Example 7.1.1.

   (a) Show that, if $(\mu_1, \sigma_1^2) = (\mu_2, \sigma_2^2)$ and $\pi_1 = \pi_2 = \frac{1}{2}$, then the quadratic equation $ax^2 + bx + c = 0$ has infinitely many solutions and every classification rule has error rate $1/2$.

   (b) Find conditions under which the quadratic equation has no real solution. Find particular parameter values π_j, μ_j, and σ_j^2, such that there is no solution, and graph the functions $\pi_1 f_1(\mathbf{y})$ and $\pi_2 f_2(\mathbf{y})$.

   (c) Show that, if $\sigma_1 = \sigma_2$ but $\mu_1 \neq \mu_2$, the quadratic equation has a unique root.

   (d) What are the optimal classification regions if $\mu_1 = \mu_2$?

5. Find the error rate of the optimal classification procedure in Example 7.1.1 numerically, by computing

$$\gamma(X^*_{\text{opt}}) = \int_{-\infty}^{v_1} \pi_2 f_2(y)dy + \int_{v_1}^{v_2} \pi_1 f_1(y)dy + \int_{v_2}^{\infty} \pi_2 f_2(y)dy.$$

6. Suppose that X takes values 1 and 2 with probabilities π_1 and $\pi_2 = 1 - \pi_1$, and, conditionally on $X = j$, Y is uniform in the interval $[\alpha_j, \beta_j]$, $j = 1, 2$. Find values of the parameters π_j, α_j, and β_j such that the optimal classification rule is not uniquely defined, i.e., $\Pr[\pi_1 f_1(Y) = \pi_2 f_2(Y) | X = j] > 0$.

7. This exercise investigates optimal classification for spherically symmetric distributions with $k = 2$ groups and prior probabilities π_1, π_2.
   (a) Suppose that $\mathbf{Y} \sim \mathcal{N}_p(\boldsymbol{\mu}_j, \sigma^2 \mathbf{I}_p)$ in groups $j = 1, 2$. Show that the optimal classification rule depends on $\mathbf{y}$ only through the Euclidean distances $||\mathbf{y} - \boldsymbol{\mu}_1||$ and $||\mathbf{y} - \boldsymbol{\mu}_2||$ between $\mathbf{y}$ and the mean vectors.
   (b) Consider the following generalization of part (a): Suppose that $\mathbf{Z}$ is a spherically symmetric random vector as in Definition 3.4.1 and $\mathbf{Y}$ has the same distribution as $\boldsymbol{\mu}_j + \mathbf{Z}$ in groups $j = 1, 2$, that is, $\mathbf{Y}$ has a spherical distribution shifted to $\boldsymbol{\mu}_1$ and $\boldsymbol{\mu}_2$, respectively. Show that the optimal classification rule in this situation depends on $\mathbf{y}$ only through $||\mathbf{y} - \boldsymbol{\mu}_1||$ and $||\mathbf{y} - \boldsymbol{\mu}_2||$, or give a counterexample.

8. Partition the p-variate random vector $\mathbf{Y} = \begin{pmatrix} \mathbf{Y}_1 \\ \mathbf{Y}_2 \end{pmatrix}$ into q variables $\mathbf{Y}_1$ and $(p - q)$ variables $\mathbf{Y}_2$. Assume that $\mathbf{Y}_1$ and $\mathbf{Y}_2$ are independent in all k groups to be considered in a classification problem and that $\mathbf{Y}_2$ has the same *pdf* $h(\mathbf{y}_2)$ in all k groups. Show that optimal classification depends only on $\mathbf{Y}_1$.

9. (continuation of problem 8). Find an example of discrimination between two groups, where $\mathbf{Y}_2$ has the same distribution in both groups, but the optimal classification rule depends on both $\mathbf{Y}_1$ and $\mathbf{Y}_2$ ($\mathbf{Y}_1$ and $\mathbf{Y}_2$ are not independent in this case). *Hint*: Look into Section 5.6.

10. Do Exercise 19 in Section 8.4.

11. Prove the following result. Suppose that $\mathbf{Y}$ has *pdf* $f_j(\mathbf{y})$ in groups $j = 1, \ldots, k$, with prior probabilities π_j. Let $\mathbf{Z} = g(\mathbf{Y})$, where $g(\mathbf{y})$ is a one-to-one transformation defined on the sample space of $\mathbf{Y}$. Then the optimal classification rule based on $\mathbf{Z}$ has the same error rate as the optimal classification rule based on $\mathbf{Y}$.

7.2 Normal Theory Classification Revisited: Linear vs Quadratic

In Section 5.4 we saw that optimal classification into two multivariate normal distributions with identical covariance matrices is based on the linear discriminant function. In Section 7.3 we will generalize this to the case of several normal distributions, all with the same covariance matrix. In the current section, we focus on the case of general multivariate normal distributions and discuss the relative merits of both linear and quadratic classification rules.

Suppose that $\mathbf{Y}$ is p-variate normal with mean vector $\boldsymbol{\mu}_j$ and covariance matrix $\boldsymbol{\psi}_j$ in group j, $j = 1, \ldots, k$. As before, let X denote a group indicator variable, taking value j with probability π_j. Assuming that $\boldsymbol{\psi}_j$ is positive-definite, the conditional *pdf* of $\mathbf{Y}$, given $X = j$, is

$$f_j(\mathbf{y}) = (2\pi)^{-p/2}(\det \boldsymbol{\psi}_j)^{-1/2} \exp\left[-\frac{1}{2}(\mathbf{y} - \boldsymbol{\mu}_j)'\boldsymbol{\psi}_j^{-1}(\mathbf{y} - \boldsymbol{\mu}_j)\right], \quad \mathbf{y} \in \mathbb{R}^p. \quad (1)$$

By Theorem 7.1.1, the optimal classification rule X^*_{opt} uses classification regions

$$C_j = \{\mathbf{y} \in \mathbb{R}^p : \pi_j f_j(\mathbf{y}) > \pi_h f_h(\mathbf{y}) \text{ for all } h \neq j\}, \qquad j = 1, \ldots, k. \quad (2)$$

Equivalently, C_j consists of all $\mathbf{y} \in \mathbb{R}^p$ such that $\log\left[\pi_j f_j(\mathbf{y})\right] > \log\left[\pi_h f_h(\mathbf{y})\right]$ for all $h \neq j$. Let

$$q_j(\mathbf{y}) = \log[\pi_j f_j(\mathbf{y})] + \frac{p}{2}\log(2\pi), \quad j = 1, \ldots, k. \quad (3)$$

Exercise 1 shows that

$$q_j(\mathbf{y}) = \mathbf{y}'\mathbf{A}_j\mathbf{y} + \mathbf{b}_j'\mathbf{y} + c_j\,, \quad (4)$$

where

$$\mathbf{A}_j = -\frac{1}{2}\boldsymbol{\psi}_j^{-1},$$
$$\mathbf{b}_j = \boldsymbol{\psi}_j^{-1}\boldsymbol{\mu}_j, \quad (5)$$

and

$$c_j = \log \pi_j - \frac{1}{2}\log(\det \boldsymbol{\psi}_j) - \frac{1}{2}\boldsymbol{\mu}_j'\boldsymbol{\psi}_j^{-1}\boldsymbol{\mu}_j.$$

For a given $\mathbf{y} \in \mathbb{R}^p$, we need to evaluate the k quadratic functions $q_1(\mathbf{y}), \ldots, q_k(\mathbf{y})$, and classify $\mathbf{y}$ in group j if $q_j(\mathbf{y})$ is the largest of the k values. The q_j are called *classification functions*. It also follows from equation (3) that the posterior probability of membership in group j is

$$\pi_{jy} = \Pr[X = j | \mathbf{Y} = \mathbf{y}] = \frac{\exp\left[q_j(\mathbf{y})\right]}{\sum_{h=1}^k \exp\left[q_h(\mathbf{y})\right]}, \qquad j = 1, \ldots, k. \quad (6)$$

Thus, the optimal classification rule can be formulated in terms of the posterior probabilities or in terms of the classification functions.

In general, it is not easy to visualize the classification regions, but some insight can be gained by studying the case of two groups in more detail. For $k = 2$, the classification regions are given by

$$C_1 = \{\mathbf{y} \in \mathbb{R}^p : q_1(\mathbf{y}) - q_2(\mathbf{y}) > 0\}$$

and

$$C_2 = \{\mathbf{y} \in \mathbb{R}^p : q_1(\mathbf{y}) - q_2(\mathbf{y}) < 0\}. \tag{7}$$

Define

$$Q(\mathbf{y}) = q_1(\mathbf{y}) - q_2(\mathbf{y}) \tag{8}$$

as the *discriminant function.* Then $Q(\mathbf{y})$ is itself a quadratic function

$$Q(\mathbf{y}) = \mathbf{y}'\mathbf{A}\mathbf{y} + \mathbf{b}'\mathbf{y} + c, \tag{9}$$

where

$$\begin{aligned} \mathbf{A} &= \frac{1}{2}\left(\psi_2^{-1} - \psi_1^{-1}\right), \\ \mathbf{b} &= \psi_1^{-1}\mu_1 - \psi_2^{-1}\mu_2, \end{aligned} \tag{10}$$

and

$$c = \log\frac{\pi_1}{\pi_2} + \frac{1}{2}\log\frac{\det\psi_2}{\det\psi_1} + \frac{1}{2}\left[\mu_2'\psi_2^{-1}\mu_2 - \mu_1'\psi_1^{-1}\mu_1\right].$$

The boundary of the classification regions consists of all $\mathbf{y} \in \mathbb{R}^p$ such that $Q(\mathbf{y}) = 0$. In the special case $\psi_1 = \psi_2$, the matrix $\mathbf{A}$ in (10) is zero, and the quadratic discriminant function reduces to the familiar linear discriminant function; see Exercise 2. If $\psi_1 \neq \psi_2$ there will always be quadratic terms in $Q(\mathbf{y})$.

Example 7.2.1 Two bivariate normal distributions with unequal covariance matrices. This is the continuation of Example 5.4.1, with parameters

$$\mu_1 = \begin{pmatrix} -1.0 \\ 0.5 \end{pmatrix}, \qquad \psi_1 = \begin{pmatrix} 1 & 0.6 \\ 0.6 & 1 \end{pmatrix},$$

$$\mu_2 = \begin{pmatrix} 1.5 \\ -0.5 \end{pmatrix}, \qquad \psi_2 = \begin{pmatrix} 1 & -0.8 \\ -0.8 & 1 \end{pmatrix},$$

and prior probabilities $\pi_1 = 0.75$ and $\pi_2 = 0.25$. Figure 7.2.1 shows two ellipses corresponding to sets of constant standard distance 1 from the respective group means. Classification region C_1 is the contiguous region bounded by the two branches of the hyperbola $\mathbf{y}'\mathbf{A}\mathbf{y} + \mathbf{b}'\mathbf{y} + c = 0$, where $\mathbf{A}$, $\mathbf{b}$, and c are computed from equation (10). Classification region C_2 consists of two disjoint parts. □

As we will see shortly, the boundary of the two classification regions is not necessarily a hyperbola. To understand the various possibilities better, it will be convenient to transform classification problems into a "standardized form" which is easier to analyze. By Exercise 11 of Section 7.1, any nonsingular linear or affine transformation of the random vector $\mathbf{Y}$ will leave the error rate of the optimal classification procedure unchanged. Thus, we will seek a transformation that reduces the quadratic discriminant function to a particularly simple form.

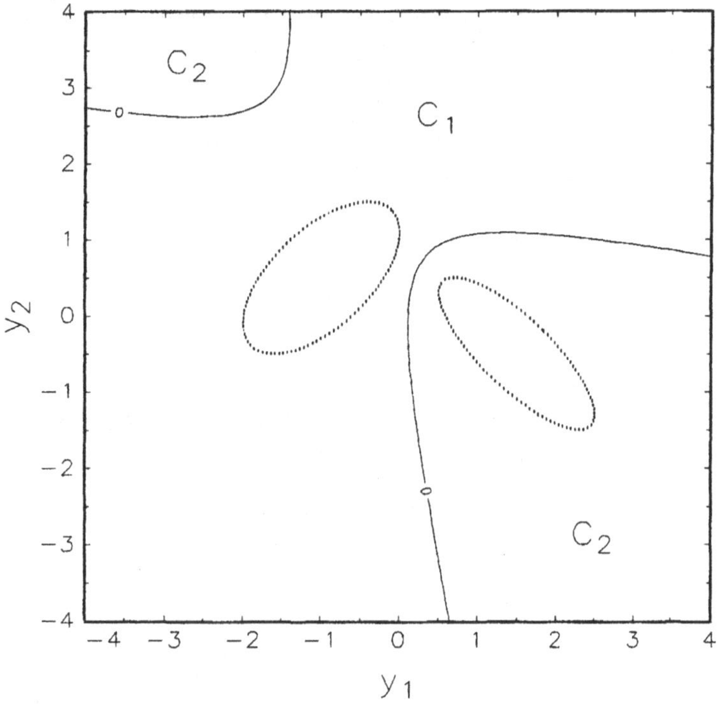

Figure 7-2-1 Classification regions in Example 7.2.1.

Definition 7.2.1 Suppose that $\mathbf{Y} \sim \mathcal{N}_p(\boldsymbol{\mu}_j, \boldsymbol{\psi}_j)$ in groups $j = 1, 2$. Then $\mathbf{Y}$ is in *canonical form* if $\boldsymbol{\mu}_1 = \mathbf{0}$, $\boldsymbol{\psi}_1 = \mathbf{I}_p$, and $\boldsymbol{\psi}_2$ is diagonal. □

The following theorem shows that a multivariate normal discriminant problem can always be transformed into an equivalent one which is in canonical form.

Theorem 7.2.1 *Suppose that $\mathbf{Y} \sim \mathcal{N}_p(\boldsymbol{\mu}_j, \boldsymbol{\psi}_j)$ in groups $j = 1,2$, where both covariance matrices are positive-definite. Then there exists a nonsingular $p \times p$ matrix $\boldsymbol{\Gamma}$ such that*

$$\mathbf{Z} = \boldsymbol{\Gamma}'(\mathbf{Y} - \boldsymbol{\mu}_1) \tag{11}$$

is in canonical form.

Proof By the simultaneous spectral decomposition theorem of Appendix A.9, there exists a nonsingular matrix $\boldsymbol{\Gamma}$ such that $\boldsymbol{\Gamma}'\boldsymbol{\psi}_1\boldsymbol{\Gamma} = \mathbf{I}_p$ and $\boldsymbol{\Gamma}'\boldsymbol{\psi}_2\boldsymbol{\Gamma} = \boldsymbol{\Lambda} = \text{diag}(\lambda_1, \lambda_2, \ldots, \lambda_p)$. The random vector $\mathbf{Z}$ is itself p-variate normal with mean $\mathbf{0}$ and covariance matrix $\mathbf{I}_p$ in group 1 and with mean $\boldsymbol{\Gamma}'(\boldsymbol{\mu}_2 - \boldsymbol{\mu}_1)$ and covariance matrix $\boldsymbol{\Lambda}$ in group 2. ■

Suppose now that the p-variate normal random vector $\mathbf{Y}$ is in canonical form. Writing $\boldsymbol{\mu}_1 = \mathbf{0}$, $\boldsymbol{\psi}_1 = \mathbf{I}_p$,

$$\boldsymbol{\mu}_2 = \boldsymbol{\delta} = \begin{pmatrix} \delta_1 \\ \vdots \\ \delta_p \end{pmatrix},$$

and

$$\boldsymbol{\psi}_2 = \boldsymbol{\Lambda} = \begin{pmatrix} \lambda_1 & 0 & \cdots & 0 \\ 0 & \lambda_2 & \cdots & 0 \\ \vdots & \vdots & \ddots & \vdots \\ 0 & 0 & \cdots & \lambda_p \end{pmatrix},$$

the coefficient matrix $\mathbf{A}$ from (10) is itself diagonal, and the quadratic discriminant function is given by

$$Q(\mathbf{y}) = \sum_{i=1}^{p} a_{ii} y_i^2 + \sum_{i=1}^{p} \mathbf{b}_i y_i + c, \tag{12}$$

where

$$a_{ii} = \frac{1}{2}\left(\frac{1}{\lambda_i} - 1\right), \qquad i = 1, \ldots, p,$$

$$b_i = -\frac{\delta_i}{\lambda_i}, \qquad i = 1, \ldots, p,$$

and

$$c = \log\frac{\pi_1}{\pi_2} + \frac{1}{2}\sum_{h=1}^{p} \log\lambda_h + \frac{1}{2}\sum_{h=1}^{p} \frac{\delta_h^2}{\lambda_h}. \tag{13}$$

The main benefit of the canonical form is that all terms of the form $a_{ij} y_i y_j$ (for $i \neq j$) vanish, which makes the classification boundary easier to understand. Our next example illustrates this for the bivariate case.

Example 7.2.2 Bivariate normal classification. Suppose that $\mathbf{Y}$ is in canonical form with mean vector $\mathbf{0}$ and covariance matrix $\mathbf{I}_p$ in group 1 and with mean vector and covariance matrix

$$\boldsymbol{\delta} = \begin{pmatrix} \delta_1 \\ \delta_2 \end{pmatrix},$$

and

$$\boldsymbol{\Lambda} = \begin{pmatrix} \lambda_1 & 0 \\ 0 & \lambda_2 \end{pmatrix}$$

in group 2. Assume equal prior probabilities $\pi_1 = \pi_2 = \frac{1}{2}$. Figure 7.2.2 illustrates four cases, labeled A to D. In each case, the dotted ellipses in Figure 7.2.2 indicate sets of constant standard distance 1 from the respective group mean (for group 1, this is just the unit circle centered at the origin). A solid curve gives the boundary between the classification regions. The numerical setup for each case is given in Table 7.2.1. A hyperbola with two branches has already been seen in Example 7.2.1 (albeit not in canonical form) and is not repeated here.

Case A: This is a situation where the boundary curve is an ellipse. It is characterized by the fact that a_{11} and a_{22} have the same sign, which, in turn, occurs whenever either $\psi_1 - \psi_2$ or $\psi_2 - \psi_1$ is positive-definite; see Exercise 3. Depending on the other parameter values, it may also happen that the equation $Q(\mathbf{y}) = 0$ has no solution, in which case one of the classification regions is empty; see Exercise 4.

Case B: This is a rather unexpected case in which the classification regions are bounded by two intersecting straight lines. The straight lines may be viewed as limiting cases of a hyperbola. This situation occurs when the quadratic discriminant function can be factorized as $Q(\mathbf{y}) = (c_{11}y_1 + c_{12}y_2 + c_{13})(c_{21}y_1 + c_{22}y_2 + c_{23})$ for some constants c_{ij}. See Exercises 5 and 6. In the numerical setup of this case (see Table 7.2.1), the quadratic discriminant function is given by

$$\begin{aligned} Q(\mathbf{y}) &= -\frac{3}{8}y_1^2 + \frac{3}{2}y_2^2 - \frac{1}{2}y_1 + 4y_2 + \frac{5}{2} \\ &= \left(-\frac{1}{2}y_1 + y_2 + 1\right)\left(\frac{3}{4}y_1 + \frac{3}{2}y_2 + \frac{5}{2}\right), \end{aligned}$$

and setting either of the two terms in parentheses equal to zero, gives one of the straight lines.

Case C: Because $\lambda_2 = 1$, one of the quadratic coefficients vanishes ($a_{22} = 0$), and therefore, the boundary curve is a parabola. Both classification regions are contiguous.

Case D: This is a rather interesting special case of **C**, in which not only a quadratic coefficient but also the associated linear coefficient vanishes. In the specific numerical setup, $a_{22} = b_2 = 0$, and optimal classification is based entirely on the value of the first variable. The classification regions consist of an infinite strip and two half-planes. In fact, this is a particular case where Exercise 8 of Section 7.1 applies: Because of

Table 7.2.1 Numerical setup for four situations of quadratic classification in canonical form, illustrated in Figure 7.2.2. Group 1 follows a bivariate standard normal distribution. In group 2 the means are δ_1 and δ_2, and the variances are λ_1 and λ_2. The coefficients a_{ii}, b_i, and c of the quadratic discriminant function are defined in equation (13).

Case	δ_1	δ_2	λ_1	λ_2	a_{11}	a_{22}	b_1	b_2	c
A	2	1	0.4	0.1	0.75	4.5	−5	−10	8.391
B	2	−1	4	0.25	−0.375	1.5	−0.5	4	2.5
C	3	1	4	1	−0.375	0	−0.75	−1	2.318
D	2	0	10	1	−0.45	0	−0.2	0	1.351

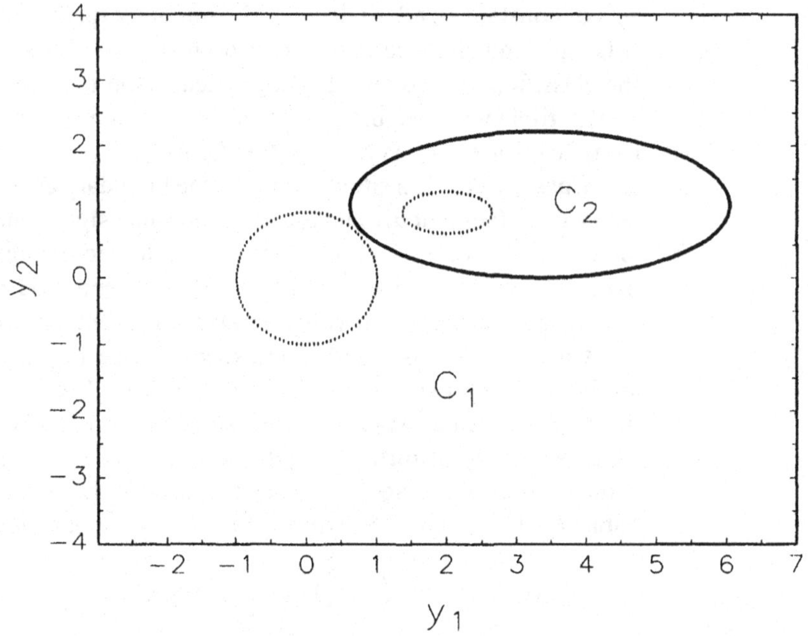

Figure 7-2-2A

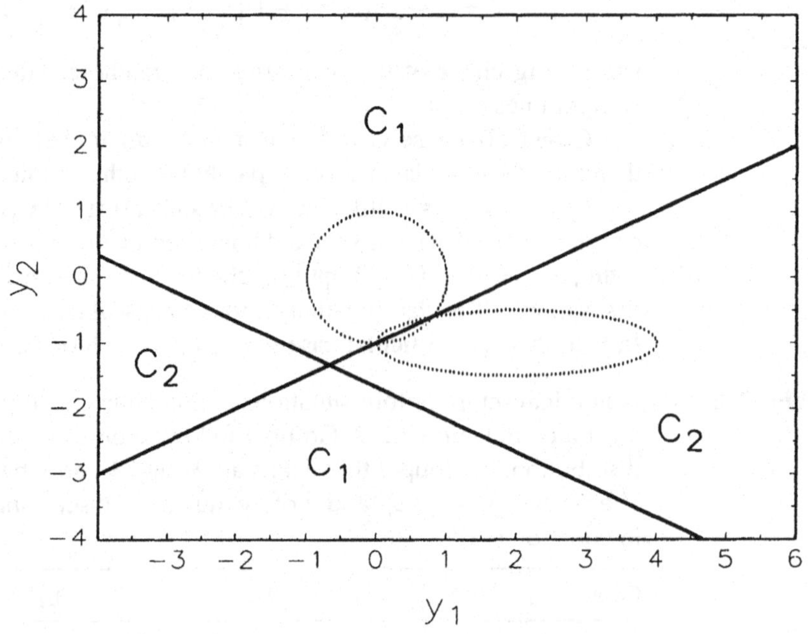

Figure 7-2-2B

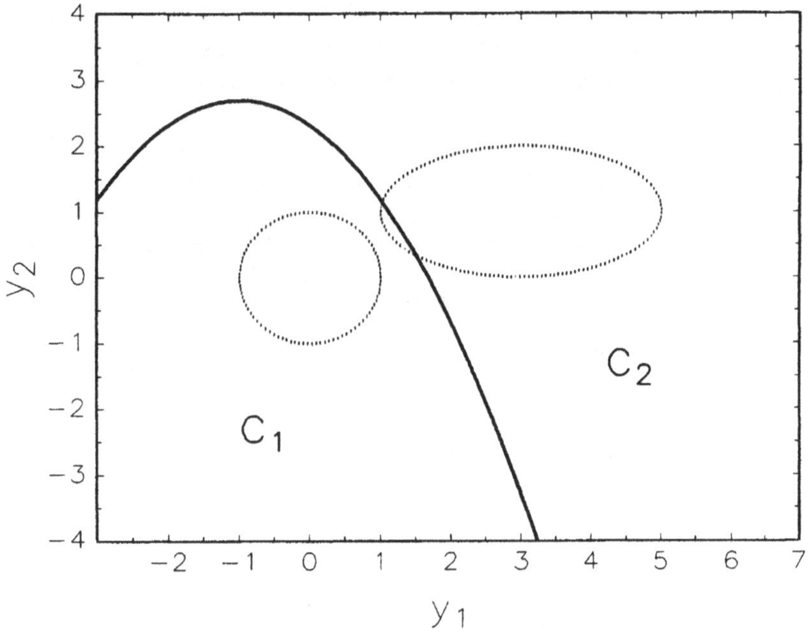

Figure 7-2-2C

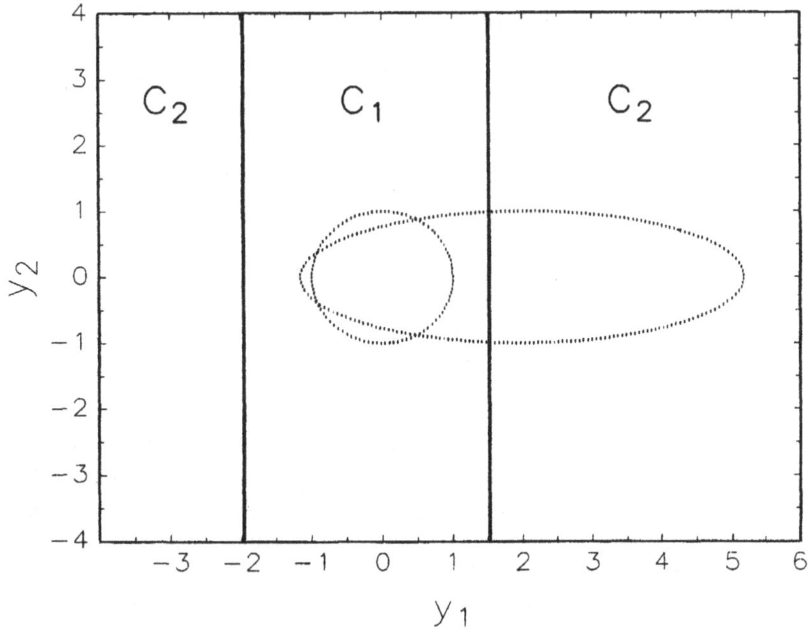

Figure 7-2-2D Bivariate normal classification in canonical form. The numerical setup for situations A to D is given in Table 7.2.1. Dotted ellipses indicate sets of constant standard distance from the means of the respective groups. Solid curves are boundaries of the classification regions.

joint normality of Y_1 and Y_2 and because the covariance matrices are diagonal, Y_1 and Y_2 are independent in both groups. Moreover, the distribution of Y_2 is the same in both groups, and therefore, optimal classification is based on Y_1 alone. □

Standard applications of normal theory quadratic discrimination are straightforward: Simply replace the parameters by plug-in estimates or maximum likelihood estimates. We will follow the convention of using the unbiased-sample covariance matrices $\mathbf{S}_j$. Thus, the jth classification function is estimated as

$$\hat{q}_j(\mathbf{y}) = \mathbf{y}'\hat{\mathbf{A}}_j\mathbf{y} + \hat{\mathbf{b}}_j'\mathbf{y} + \hat{c}_j, \tag{14}$$

where

$$\begin{aligned}\hat{\mathbf{A}}_j &= -\frac{1}{2}\mathbf{S}_j^{-1},\\ \hat{\mathbf{b}}_j &= \mathbf{S}_j^{-1}\bar{\mathbf{y}}_j,\end{aligned} \tag{15}$$

and

$$\hat{c}_j = \log\hat{\pi}_j - \frac{1}{2}\log(\det\mathbf{S}_j) - \frac{1}{2}\bar{\mathbf{y}}_j'\mathbf{S}_j^{-1}\bar{\mathbf{y}}_j.$$

The prior probabilities $\hat{\pi}_j$ may be estimated from the data or may be assigned fixed values. Estimated posterior probabilities are computed as

$$\hat{\pi}_{jy} = \frac{\exp\left[\hat{q}_j(\mathbf{y})\right]}{\sum_{h=1}^{k}\exp\left[\hat{q}_h(\mathbf{y})\right]}, \qquad j = 1,\ldots,k. \tag{16}$$

In the case of $k = 2$ groups, we estimate the quadratic discriminant function as

$$\hat{Q}(\mathbf{y}) = \hat{q}_1(\mathbf{y}) - \hat{q}_2(\mathbf{y}). \tag{17}$$

An illustrative example follows.

Example 7.2.3 Electrodes produced by two machines. In Example 5.3.4, we noticed that there are substantial differences in variances and covariances between the two groups. For the sake of illustration, we will now use two variables Y_1 and Y_5 that show particularly large differences in variability. Without giving numerical details (see Exercise 7), Figure 7.2.3 shows a scatterplot of Y_1 vs. Y_5, along with the classification boundaries for linear and for quadratic classification, using prior probabilities $\pi_1 = \pi_2 = \frac{1}{2}$. The quadratic classification boundary is an ellipse whose interior is the classification region for machine 2.

As seen from Figure 7.2.3, both classification rules would work identically for the 100 observations in the training samples, i.e., exactly the same observations would be misclassified. Thus, the quadratic rule has little or no advantage over the much simpler linear rule. In fact, as we will see in Example 7.2.4, using the quadratic

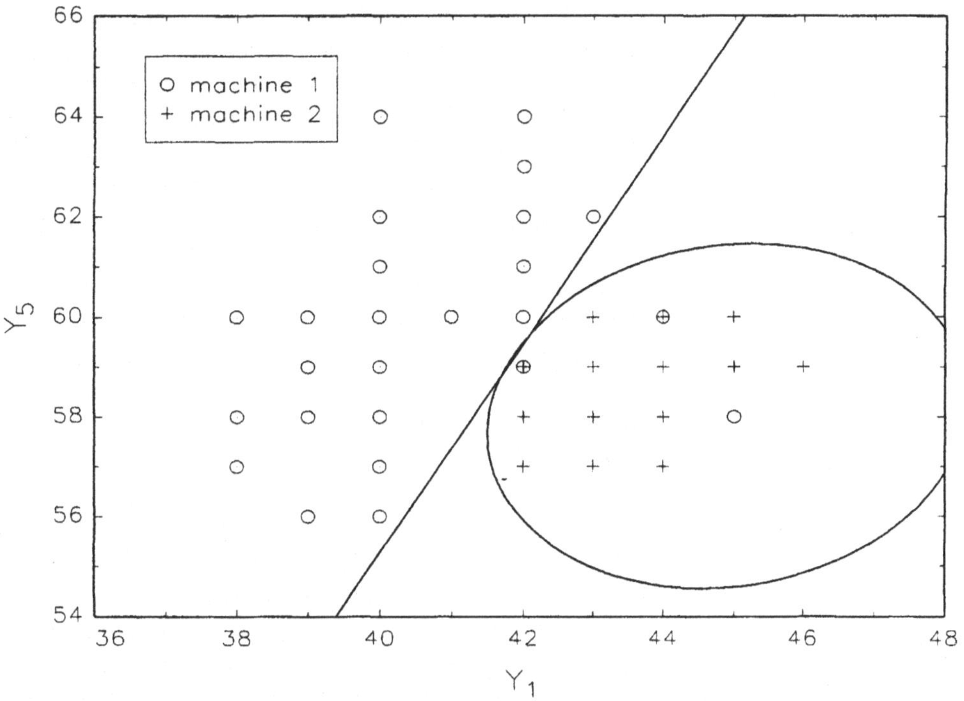

Figure 7-2-3 Scatter plot of Y_1 vs. Y_5 in the electrode example, along with the normal theory classification regions for both linear and quadratic discrimination.

rule may actually be worse than using the linear rule. The graph also points toward a frequently observed phenomenon: the quadratic classification boundary may be reasonable in the area where the two observed distributions actually overlap but appears almost entirely artificial elsewhere, i.e., in the lower right-hand corner of Figure 7.2.3. □

In the general multivariate case, it is not as easy as in Example 7.2.3 to compare the performance of the linear and quadratic classification rules. Too often the choice between the two rules is made following a test for equality of covariance matrices. That is, if a test of the hypothesis $\psi_1 = \psi_2$ is significant, then quadratic discrimination is used. See Exercise 13 in Section 6.6 for the likelihood ratio test for equality of covariance matrices. Although such a decision appears reasonable in view of the fact that linear discrimination is optimal exactly if the covariance matrices are identical, a considerable amount of literature shows that the linear rule is (in a sense to be discussed later) better in many cases even when the assumption $\psi_1 = \psi_2$ is clearly violated; see Seber (1984, pp. 299–300) for a review of the older literature on this topic.

Naive users of discriminant analysis tend to fall into the trap of simply choosing the method that produces the smaller plug-in error rate. This is dangerous, as we have already seen in Section 5.5. A better assessment of the relative merits of classification procedures would be based on actual error rates rather than on plug-in error rates. Many studies comparing expected actual error rates of linear and quadratic classification rules have reached the conclusion that linear discrimination performs better than quadratic discrimination as long as the differences in variability are only moderate, and the sample sizes are relatively small. This is due to the fact that the quadratic rule estimates more parameters (two covariance matrices instead of one), which gives a better fit to the observed data but less stable estimates. Unfortunately, the calculation of expected actual error rates is rather complicated and, typically, can be done only by simulation. We will therefore use a different approach to demonstrate the danger of quadratic discrimination: a comparison based on leave-one-out error rates; see Section 5.5.

Example 7.2.4 Two species of voles. In Example 5.4.4 we studied linear discrimination between *Microtus multiplex* and *M. subterraneus* using two of the eight variables from the data set reproduced in Table 5.4.1. Now, we will compare the linear and the quadratic classification rules using all eight variables and training samples of size $N_1 = 43$ (*M. multiplex*) and $N_2 = 46$ (*M. subterraneus*). For the linear classification rule with prior probabilities estimated by the relative sample sizes $\hat{\pi}_1 = N_1/(N_1 + N_2)$ and $\hat{\pi}_2 = N_2/(N_1 + N_2)$, the error rates are $\hat{\gamma}_{\text{plug-in}} = 5/89 = 5.62$ percent and $\hat{\gamma}_{\text{leave-one-out}} = 6/89 = 6.74$ percent. Using the quadratic classification rule, the plug-in error rate is $4/89 = 4.49$ percent, suggesting a slight advantage of quadratic over linear discrimination. However, the leave-one-out analysis tells a different story: 11 out of the 89 observations were misclassified, giving an error rate of $\hat{\gamma}_{\text{leave-one-out}} =$ 12.36 percent.

Table 7.2.2 gives some numerical details on the 11 misclassified observations, leaving a complete analysis to the reader (Exercise 8). In many cases, the estimated posterior probability of an observation changes drastically when the observation is omitted in the estimation procedure. This indicates that estimation of the quadratic classification rule is heavily affected by single observations. In view of the leave-one-out error rates, the linear rule should be preferred. □

Example 7.2.4 shows the dangers of quadratic discrimination rather drastically. In some cases, however, quadratic discrimination may be justified on substantial grounds, as illustrated in the following example.

Example 7.2.5 Zero mean discrimination. Occasionally, we may have previous knowledge about either the means or the variances and covariances, which might be considered in the construction of a classification rule. A particular case is discrimination between

Table 7.2.2 Details of the computation of error rates in the *Microtus* example. The symbols used are x_i, actual group of observation i; $\hat{z}_i$ = value of the quadratic discriminant function using all observations; $\hat{\pi}_{1i}$ = estimated posterior probability (percent) in group 1 using the quadratic discriminant function based on all observations; $\hat{x}_i^*$ = predicted group membership using the quadratic discriminant function based on all observations; $\hat{\pi}_{1i,-i}$ and $\hat{x}_{i,-i}^*$ = analogous quantities using the quadratic discriminant function based on all but the ith observation. Only results for the 11 observations that are misclassified in the leave-one-out analysis are reported.

i	x_i	$\hat{z}_i$	$\hat{\pi}_{1i}$	$\hat{x}_i^*$	$\hat{\pi}_{1i,-i}$	$\hat{x}_{i,-i}^*$
3	1	-2.03	11.63	2	0.97	2
21	1	1.61	83.30	1	46.06	2
24	1	-2.52	7.48	2	1.26	2
29	1	-0.22	44.61	2	31.86	2
52	2	-0.43	39.42	2	63.98	1
58	2	0.66	66.04	1	95.71	1
66	2	-2.51	7.53	2	100.00	1
72	2	-0.03	49.17	2	78.14	1
76	2	-1.18	23.55	2	76.58	1
82	2	-0.60	35.54	2	61.47	1
87	2	-1.43	19.32	2	51.63	1

monozygotic (identical) and dizygotic (fraternal) twins. Suppose that a morphometric variable is measured on both twins of a pair and Y denotes the difference between the first and second twin, both of the same gender. Assuming that the labels "first twin" and "second twin" are not based on morphometric characteristics, then it is reasonable to assume $E[Y] = 0$ for both monozygotic and dizygotic twins. Moreover, an intuitively reasonable classification rule would assign a pair of twins with observed difference y to be

$$\text{monozygotic, if } |y| < a,$$

$$\text{dizygotic, if } |y| \geq a,$$

for some constant $a \geq 0$. This is exactly the normal theory rule for this situation; see Exercise 9.

Now, suppose now that $\mathbf{Y}$ is the vector of differences between first and second twin measured in p morphometric variables. Assuming $E[\mathbf{Y}] = \mathbf{0}$ for both monozygotic and dizygotic twins leads to the quadratic discriminant function $Q(\mathbf{y}) = \mathbf{y}'\mathbf{A}\mathbf{y} + c$, where $\mathbf{A} = \frac{1}{2}\left(\psi_2^{-1} - \psi_1^{-1}\right)$ and $c = \log(\pi_1/\pi_2) + \frac{1}{2}\log(\det \Psi_2/\det \psi_1)$. However, one might argue that even stronger assumptions should be made. For instance, because dizygotic twins are expected to differ more than monozygotic twins, one might argue that the covariance matrices should be proportional, i.e., $\psi_2 = \gamma\psi_1$ for some constant $\gamma > 0$. (Presumably γ would be larger than 1 if group 2 is the dizygotic twins). Some motivation is given in Exercise 10. Writing the covariance matrices as $\psi_1 = \psi$ and $\psi_2 = \gamma\psi$, the quadratic discriminant function is given by $Q(\mathbf{y}) = \mathbf{y}'\mathbf{A}\mathbf{y} + c$, where

$$\mathbf{A} = \frac{1}{2}\left[(\gamma\psi)^{-1} - \psi^{-1}\right] = \frac{1-\gamma}{2\gamma}\psi^{-1}$$

and

$$c = \log\frac{\pi_1}{\pi_2} + \frac{1}{2}\log\frac{\det(\gamma\psi)}{\det\psi} = \log\frac{\pi_1}{\pi_2} + \frac{p}{2}\log\gamma. \tag{18}$$

Since ψ^{-1} is positive definite, the classification regions are bordered by an ellipse centered at the origin, with the same shape as ellipses of constant density of either of the two normal distributions. □

The model introduced in Example 7.2.5 is appealing because it is parsimonious: instead of two mean vectors and two covariance matrices, there is only one covariance matrix (ψ) and a proportionality constant (γ) to be estimated. This should reduce the variability of the parameter estimates, and thus, lead to better performance of the classification rule compared to ordinary quadratic classification. The price to be paid is that parameter estimation is no longer as simple as in the ordinary linear or quadratic case. The plug-in principle fails to give a solution to the estimation problem in this case, and therefore we will resort to maximum likelihood. Some technical details in the following example are left to the reader; see Exercise 11.

Example 7.2.6 A quadratic classification model with zero mean vectors and proportional covariance matrices. Suppose that $\mathbf{Y}_{11}, \ldots, \mathbf{Y}_{1N_1}$ and $\mathbf{Y}_{21}, \ldots, \mathbf{Y}_{2N_2}$ are independent samples, such that

$$\mathbf{Y}_{1i} \sim \mathcal{N}_p(\mathbf{0}, \psi) \qquad i = 1, \ldots, N_1,$$

and

$$\mathbf{Y}_{2i} \sim \mathcal{N}_p(\mathbf{0}, \gamma\psi) \qquad i = 1, \ldots, N_2, \tag{19}$$

where ψ and γ are unknown parameters, ψ is assumed positive definite, and $\gamma > 0$. By equation (16) in Section 4.3, the log-likelihood function based on observed data

$\mathbf{y}_{11}, \ldots, \mathbf{y}_{1N_1}$ and $\mathbf{y}_{21}, \ldots, \mathbf{y}_{2N_2}$ is given by

$$\begin{aligned}
\ell(\psi, \gamma) = & -\frac{N_1 p}{2}\log(2\pi) - \frac{N_1}{2}\log(\det\psi) - \frac{1}{2}\sum_{i=1}^{N_1}\mathbf{y}_{1i}'\psi^{-1}\mathbf{y}_{1i} \\
& -\frac{N_2 p}{2}\log(2\pi) - \frac{N_2}{2}\log[(\det\gamma\psi)] - \frac{1}{2}\sum_{i=1}^{N_2}\mathbf{y}_{2i}'(\gamma\psi)^{-1}\mathbf{y}_{2i} \\
= & -\frac{(N_1+N_2)p}{2}\log(2\pi) + \frac{(N_1+N_2)}{2}\log\left(\det\psi^{-1}\right) \\
& -\frac{N_1}{2}\operatorname{tr}\left(\psi^{-1}\mathbf{M}_1\right) - \frac{N_2}{2}\left[p\log\gamma + \frac{1}{\gamma}\operatorname{tr}\left(\psi^{-1}\mathbf{M}_2\right)\right],
\end{aligned} \tag{20}$$

where

$$\mathbf{M}_1 = \frac{1}{N_1}\sum_{i=1}^{N_1}\mathbf{y}_{1i}\mathbf{y}_{1i}',$$

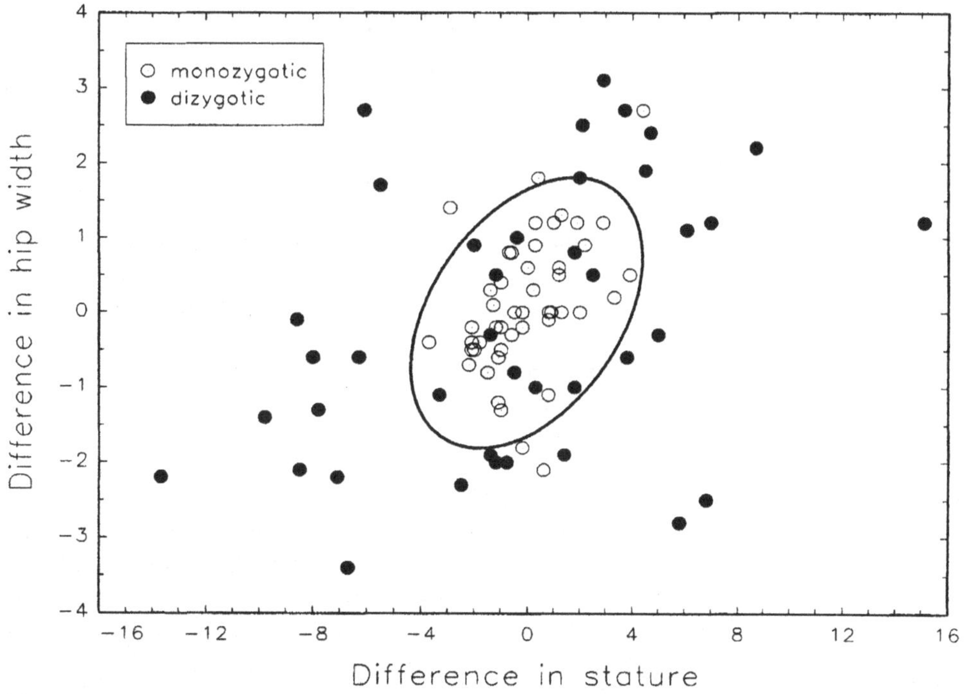

Figure 7-2-4 Scatterplot of Y_1 (difference in stature) vs. Y_2 (difference in hip width) for 89 pairs of male twins. The interior of the ellipse is the classification region for monozygotic twins obtained from the model with zero means and proportional covariance matrices treated in Examples 7.2.5 and 7.2.6.

and

$$\mathbf{M}_2 = \frac{1}{N_2} \sum_{i=1}^{N_2} \mathbf{y}_{2i}\mathbf{y}'_{2i} \,. \tag{21}$$

To find the values of ψ and γ that maximize the likelihood, we will need two results from matrix differential calculus given in appendix A.4. If $\mathbf{X}$ is a $p \times p$ matrix of p^2 variables and $\mathbf{M}$ is a fixed $p \times p$ matrix, then we will use the notation

$$\frac{\partial \log(\det \mathbf{X})}{\partial \mathbf{X}} = \begin{pmatrix} \frac{\partial \log(\det \mathbf{X})}{\partial x_{11}} & \cdots & \frac{\partial \log(\det \mathbf{X})}{\partial x_{1p}} \\ \vdots & \ddots & \vdots \\ \frac{\partial \log(\det \mathbf{X})}{\partial x_{p1}} & \cdots & \frac{\partial \log(\det \mathbf{X})}{\partial x_{pp}} \end{pmatrix}$$

and

$$\frac{\partial \operatorname{tr}(\mathbf{XM})}{\partial \mathbf{X}} = \begin{pmatrix} \frac{\partial \operatorname{tr}(\mathbf{XM})}{\partial x_{11}} & \cdots & \frac{\partial \operatorname{tr}(\mathbf{XM})}{\partial x_{1p}} \\ \vdots & \ddots & \vdots \\ \frac{\partial \operatorname{tr}(\mathbf{XM})}{\partial x_{p1}} & \cdots & \frac{\partial \operatorname{tr}(\mathbf{XM})}{\partial x_{pp}} \end{pmatrix}$$

for the matrices of derivatives of the functions $\log(\det \mathbf{X})$ and $\operatorname{tr}(\mathbf{XM})$. Using rules (iii) and (iv) from Appendix A.4, we obtain

$$\begin{aligned} \frac{\partial \ell(\psi, \gamma)}{\partial \psi^{-1}} &= \frac{N_1}{2} \left\{ 2\psi - \operatorname{diag}(\psi) - [2\mathbf{M}_1 - \operatorname{diag}(\mathbf{M}_1)] \right\} \\ &\quad + \frac{N_2}{2} \left\{ 2\psi - \operatorname{diag}(\psi) - \frac{1}{\gamma}[2\mathbf{M}_2 - \operatorname{diag}(\mathbf{M}_2)] \right\}. \end{aligned} \tag{22}$$

Note that (22) is a $p \times p$ matrix. Setting (22) equal to the zero matrix gives

$$\psi = \frac{1}{N_1 + N_2} \left(N_1\mathbf{M}_1 + \frac{1}{\gamma} N_2\mathbf{M}_2 \right). \tag{23}$$

Thus for given γ, equation (23) gives the value of ψ that maximizes the likelihood. Taking the derivative of $\ell(\psi, \gamma)$ with respect to γ and setting it equal to zero gives

$$\gamma = \frac{1}{p} \operatorname{tr}(\psi^{-1}\mathbf{M}_2), \tag{24}$$

providing the optimal value of γ for given ψ.

Equations (23) and (24) need to be solved simultaneously to find stationary points of the log-likelihood function. (It turns out that there is a unique solution, provided that $\mathbf{M}_1$ and $\mathbf{M}_2$ are positive definite, but a proof is beyond the scope of this book). A convenient way to solve the equation system is to iterate between equations (23) and (24), starting with an arbitrary initial value of γ, for instance, $\gamma = 1$. That is, for given γ equation (23) is evaluated, and, with the matrix ψ thus obtained, a new value of γ is computed using equation (24), and so on. Once two consecutive values of γ are identical (up to numerical inaccuracy), the maximum likelihood estimates have been found. We will denote them by $\hat{\psi}$ and $\hat{\gamma}$, as usual. See Exercise 12 for more details on the computational procedure.

Table 7.2.3 gives anthropometric data for 89 pairs of male twins measured in the 1950s at the University of Hamburg, Germany. Of the 89 pairs, $N_1 = 49$ are monozygotic, and $N_2 = 40$ are dizygotic. There are six variables for each pair of twins: stature, hip width, and chest circumference for each of the two brothers. Taking differences between first and second twins, we will use the variables

$$Y_1 = \text{difference in stature}$$

and

$$Y_2 = \text{difference in hip width}$$

for illustration. Leaving numerical details to the reader (Exercise 12), the maximum likelihood estimates are

$$\hat{\psi} = \begin{pmatrix} 3.8166 & 0.6462 \\ 0.6462 & 0.6473 \end{pmatrix}$$

and

$$\hat{\gamma} = 7.070.$$

Figure 7.2.4 shows a scatterplot of Y_1 vs. Y_2 for both groups, along with the classification boundary $\mathbf{y}'\hat{\mathbf{A}}\mathbf{y} + \hat{c} = 0$, where $\hat{\mathbf{A}}$ and $\hat{c}$ are obtained by substituting maximum likelihood estimates for parameters in (18) and using relative sample sizes $N_j/(N_1 + N_2)$ as prior probabilities. Without giving a formal assessment of the performance of the classification rule, the result seems quite reasonable. The inside of the ellipse is the classification region for monozygotic twins. Thus, pairs of twins with relatively small differences would be classified as identical. □

Whether or not the model explored in Examples 7.2.5 and 7.2.6 is ultimately appropriate, it illustrates a case where a quadratic classification rule is justified by the setup of the problem. In many cases, however, there is no prior information available that would imply quadratic discrimination, yet, the data itself suggests that linear discrimination is not sufficient because there are substantial differences in variability. At the same time, ordinary quadratic classification based on equations (9) and (10) may lead to unsatisfactory results, as seen in Example 7.2.4. To overcome the disadvantages of the quadratic rule, many authors have suggested "compromises" between linear and quadratic discrimination, mainly by imposing constraints on the parameter space. A review of this topic, which goes beyond the scope of this introduction, can be found in the book by McLachlan (1992, Chapter 5). We will briefly discuss just one particular attempt to improve quadratic discrimination, called the Discrimination Subspace Model.

Table 7.2.3 Male twin data. Variables are *Type*, type of pair (1 = monozygotic, 2 = dizygotic); *STA1* = stature of first twin; *HIP1* = hip width of first twin; *CHE1* = chest circumference of first twin; *STA2, HIP2, CHE2,* analogous measurements for second twin. All variables are in centimeters.

Type	STA1	HIP1	CHE1	STA2	HIP2	CHE2
1	143.4	22.5	65.5	143.6	24.3	65.0
1	141.1	23.2	65.0	142.6	24.0	65.0
1	147.2	25.8	75.0	146.9	24.6	71.0
1	172.4	32.2	87.0	173.4	32.7	87.0
1	170.5	32.8	89.0	170.1	31.0	87.0
1	186.4	32.7	89.0	186.6	32.9	88.0
1	178.0	28.5	78.0	177.2	28.5	77.0
1	181.4	33.6	93.0	183.2	34.0	92.0
1	152.0	27.5	75.0	154.1	27.9	75.0
1	170.6	30.0	84.0	169.3	28.7	82.0
1	151.9	24.6	72.0	155.6	25.0	72.0
1	167.6	30.4	89.0	169.6	30.9	88.0
1	127.0	19.7	59.0	125.7	19.7	58.0
1	134.1	21.0	62.0	136.1	21.5	63.0
1	132.0	23.0	64.0	132.5	23.0	64.0
1	139.0	22.7	65.0	135.7	22.5	64.0
1	134.0	21.5	67.0	132.0	21.5	68.0
1	180.0	30.8	84.0	180.6	30.0	88.0
1	135.1	21.7	67.0	133.9	21.2	65.0
1	142.5	26.7	73.0	139.6	25.5	75.0
1	127.0	21.6	60.0	127.0	21.0	60.5
1	134.4	21.7	65.0	133.6	22.8	65.5
1	167.1	31.5	82.0	169.2	32.0	86.0
1	141.7	25.0	69.0	144.6	23.6	70.0
1	151.5	26.5	78.0	151.7	26.5	78.0
1	129.5	21.2	62.0	130.5	22.5	61.0
1	151.6	24.6	65.0	153.8	25.3	66.0
1	136.9	21.2	66.0	136.7	20.9	65.0
1	159.7	25.9	75.0	159.4	25.0	75.0
1	139.4	22.9	72.0	140.5	23.5	72.0
1	141.0	25.7	72.0	140.0	24.5	71.0
1	180.7	33.1	94.0	181.8	34.3	92.0
1	137.3	23.5	70.0	136.5	23.6	68.0
1	173.9	28.0	83.0	173.0	28.0	83.0
1	170.0	29.0	84.0	170.7	28.2	77.5
1	141.3	23.4	65.0	142.3	23.6	65.0
1	153.2	24.9	69.0	153.7	24.9	70.5
1	172.9	32.8	84.0	169.0	32.3	84.0
1	169.3	29.9	80.5	164.9	27.2	81.0
1	175.7	30.5	86.0	173.5	29.6	85.5
1	166.2	27.3	82.0	167.5	27.2	82.0

Table 7.2.3 *Continued*

Type	STA1	HIP1	CHE1	STA2	HIP2	CHE2
1	142.7	24.7	73.0	140.8	23.5	61.0
1	160.5	27.6	80.0	159.3	27.0	75.5
1	133.9	22.3	62.0	136.0	22.5	63.0
1	157.3	27.0	70.0	158.5	27.2	72.0
1	162.4	26.5	77.0	163.0	26.8	76.0
1	158.5	26.1	77.0	159.5	25.7	76.0
1	171.9	30.5	86.0	171.3	32.6	84.0
1	163.6	24.4	73.0	165.0	24.1	73.0
2	149.5	26.0	78.0	155.0	24.3	71.5
2	148.0	25.7	72.0	150.0	24.8	71.0
2	146.5	23.1	63.0	144.7	24.1	65.0
2	148.0	24.6	68.0	149.2	26.6	70.0
2	146.7	22.6	67.0	147.5	24.6	70.5
2	140.8	22.5	68.0	144.1	23.6	71.0
2	160.0	23.5	71.0	167.8	24.8	74.0
2	143.7	24.0	67.0	144.9	23.5	68.0
2	172.0	27.6	77.0	167.3	25.2	77.0
2	177.6	30.7	91.0	168.9	28.5	91.0
2	173.5	30.5	82.0	166.7	33.0	82.0
2	163.8	27.4	77.0	170.9	29.6	82.0
2	174.5	29.3	76.5	181.2	32.7	86.0
2	183.0	33.7	94.0	176.9	32.6	93.0
2	138.6	23.7	64.0	134.1	21.8	62.0
2	172.1	33.5	97.0	178.2	30.8	90.0
2	168.4	30.1	79.0	161.4	28.9	79.0
2	162.0	30.8	85.0	170.0	31.4	86.0
2	124.2	21.4	60.0	132.7	23.5	64.0
2	137.6	24.5	66.0	138.0	23.5	65.5
2	149.7	24.8	68.0	143.9	27.6	70.0
2	139.8	24.4	68.0	138.0	23.6	64.0
2	156.6	27.0	74.5	152.9	24.3	71.0
2	133.2	21.8	63.0	133.7	22.6	61.0
2	171.6	27.7	80.5	174.1	30.0	82.5
2	138.2	22.3	63.5	136.8	24.2	68.0
2	143.3	24.5	66.5	141.3	22.7	65.0
2	143.0	22.7	66.0	138.0	23.0	68.0
2	170.0	32.2	86.0	176.3	32.8	87.0
2	139.8	23.6	65.0	137.3	23.1	65.0
2	161.4	28.6	80.0	159.3	26.1	73.0
2	159.9	30.1	74.0	157.0	27.0	73.0
2	166.9	27.2	82.0	151.8	26.0	75.0
2	173.6	30.7	86.0	175.0	32.6	90.0
2	146.4	24.7	68.5	155.0	24.8	68.0
2	173.4	32.0	86.0	169.6	32.6	87.0
2	124.6	21.3	58.0	124.3	22.3	59.0
2	181.6	31.2	82.0	191.4	32.6	86.0

Table 7.2.3 *Continued*

Type	STA1	HIP1	CHE1	STA2	HIP2	CHE2
2	134.2	22.8	66.0	135.6	23.1	71.0
2	163.3	28.7	76.0	177.0	30.9	87.0

Data courtesy of the Institute of Anthropology, University of Hamburg

Example 7.2.7 Quadratic classification based on a single linear combination. Suppose that $\mathbf{Y} \sim \mathcal{N}_2(\boldsymbol{\mu}_j, \boldsymbol{\psi}_j)$ in groups $j = 1, 2$, with

$$\boldsymbol{\mu}_1 = \begin{pmatrix} 1 \\ 2 \end{pmatrix},$$

$$\boldsymbol{\psi}_1 = \begin{pmatrix} 10 & -5 \\ -5 & 5 \end{pmatrix},$$

and

$$\boldsymbol{\mu}_2 = \begin{pmatrix} 4 \\ 5 \end{pmatrix},$$

$$\boldsymbol{\psi}_2 = \begin{pmatrix} 17 & 2 \\ 2 & 12 \end{pmatrix}.$$

Figure 7.2.5 shows ellipses of constant standard distance 1 from the respective means in both groups, and optimal classification regions assuming equal prior probabilities. The classification regions are an infinite strip for group 1 and two half-planes for group 2. Numerical calculations (see Exercise 16) show that the quadratic discriminant function can be written as

$$Q(\mathbf{y}) = az^2 + bz + c, \qquad \text{with } z = 0.4(y_1 - 1) + 0.6(y_2 - 2), \tag{25}$$

where $a = -7/16, b = -3/8$, and $c = \log(\pi 1/\pi 2) + \frac{1}{2}\log 8 + 9/16$. This means that, for all values of the prior probabilities π_j, optimal classification depends on $\mathbf{Y}$ only through the value of the linear combination $Z = 0.4Y_1 + 0.6Y_2$. Irrespective of the prior probabilities, the optimal classification regions are bounded by two parallel lines. This example corresponds to situation D in Figure 7.2.2. That is, in canonical form, the optimal classification rule depends only on the first variable; see Exercise 16. □

Although Example 7.2.7 may appear to be just a curious artifact, unlikely to be applicable to practical discrimination problems, the same idea has considerable appeal in high-dimensional classification problems. Suppose that $\mathbf{Y} \sim \mathcal{N}_p(\boldsymbol{\mu}_j, \boldsymbol{\psi}_j)$ in groups $j = 1, 2$. By Theorem 7.2.1, we can obtain a canonical form of the same

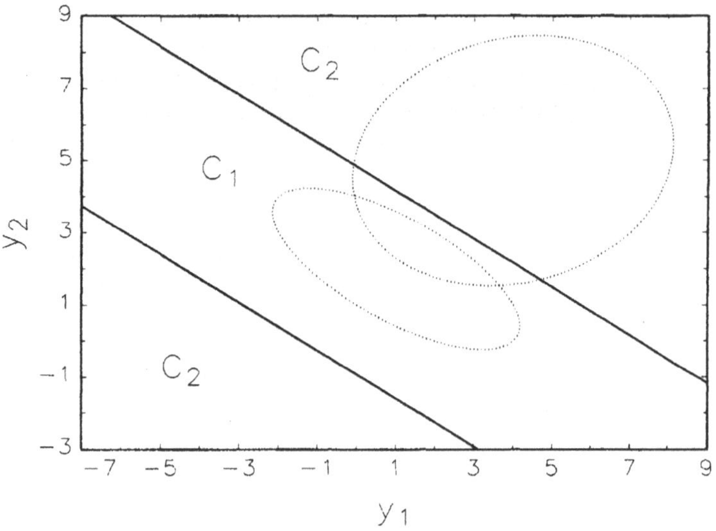

Figure 7-2-5 Normal theory classification regions in Example 7.2.7.

classification problem by the transformation

$$\mathbf{Z} = \begin{pmatrix} Z_1 \\ \vdots \\ Z_p \end{pmatrix} = \boldsymbol{\Gamma}'(\mathbf{Y} - \boldsymbol{\mu}_1), \tag{26}$$

with a properly chosen matrix $\boldsymbol{\Gamma}$. Recall that canonical form means $E[\mathbf{Z}] = \mathbf{0}$, $\text{Cov}[\mathbf{Z}] = \mathbf{I}_p$ in group 1 and $\text{Cov}[\mathbf{Z}] = \text{diag}(\lambda_1, \lambda_2, \ldots, \lambda_p)$ in group 2. Let $\boldsymbol{\delta} = (\delta_1, \delta_2, \ldots, \delta_p)'$ be the mean vector of $\mathbf{Z}$ in group 2. If $\lambda_2 = \cdots = \lambda_p = 1$ and $\delta_2 = \cdots = \delta_p = 0$, then optimal classification depends on $\mathbf{Z}$ only through Z_1, since Z_2 to Z_p are independent standard normal random variables in both groups. This is called a one-dimensional *Discrimination Subspace Model*. The main appeal of this model is that, similar to linear discrimination, all differences between the two groups are summarized by a single linear combination of $\mathbf{Y}$ with coefficients given by the first column of the matrix $\boldsymbol{\Gamma}$ in (26). Example 7.2.7 is a particular case with $p = 2$ and $\boldsymbol{\Gamma} = \begin{pmatrix} 0.4 & 0.2 \\ 0.6 & -0.2 \end{pmatrix}$. The discrimination subspace model is mostly useful if the number of variables is large, but the problem of parameter estimation is difficult, and therefore, is not treated in this text.

Exercises for Section 7.2

1. Prove equations (4) and (5).

2. Show that the quadratic discriminant function (9) reduces to the linear discriminant function of Chapter 5 exactly if $\boldsymbol{\psi}_1 = \boldsymbol{\psi}_2$.

3. Let ψ_1 and ψ_2 be the covariance matrices, assumed positive definite, in a bivariate normal classification problem. If $\psi_1 - \psi_2$ or $\psi_2 - \psi_1$ is positive definite, show that either the boundary curve $Q(\mathbf{y}) = 0$ for quadratic discrimination is an ellipse, or else the classification region for one of the two groups is empty. *Hint*: Transform the problem to canonical form.

4. Suppose that $\mathbf{Y} \sim \mathcal{N}_2(\boldsymbol{\mu}_j, \psi_j)$ in groups $j = 1, 2$, where

$$\boldsymbol{\mu}_1 = \begin{pmatrix} 1 \\ 2 \end{pmatrix},$$

$$\psi_1 = \begin{pmatrix} 10 & -5 \\ -5 & 5 \end{pmatrix},$$

$$\boldsymbol{\mu}_2 = \begin{pmatrix} 10 \\ 1 \end{pmatrix},$$

and

$$\psi_2 = \begin{pmatrix} 40 & -20 \\ -20 & 20 \end{pmatrix}.$$

For which values of the prior probability π_1 does the equation $Q(\mathbf{y}) = 0$ not have a solution? If there is a solution, what is the shape of the classification boundary? *Hint*: Transform the problem to canonical form.

5. This exercise refers to case B in Example 7.2.2.

(a) Prove the factorization of the quadratic discriminant function given in the example.

(b) In the same numerical setup, what are the optimal classification regions if $\pi_1 = 0.49$ or $\pi_1 = 0.51$? Graph these two cases, and describe what happens when the value of the prior probability π_1 varies between 0 and 1.

6. Suppose that $\mathbf{Y} \sim \mathcal{N}_2(\boldsymbol{\mu}_j, \psi_j)$ in groups $j = 1, 2$, where

$$\boldsymbol{\mu}_1 = \begin{pmatrix} 4 \\ 1 \end{pmatrix},$$

$$\psi_1 = \begin{pmatrix} 4 & 0 \\ 0 & 1 \end{pmatrix},$$

$$\boldsymbol{\mu}_2 = \begin{pmatrix} 1 \\ 4 \end{pmatrix},$$

and

$$\psi_2 = \begin{pmatrix} 1 & 0 \\ 0 & 4 \end{pmatrix}.$$

Assume equal prior probabilities.

(a) Sketch the ellipses $(\mathbf{y} - \boldsymbol{\mu}_1)'\psi_j^{-1}(\mathbf{y} - \boldsymbol{\mu}_j) = 1$ for $j = 1, 2$, and guess at the shape of the optimal classification regions. *Hint*: Since $\psi_1 \neq \psi_2$, optimal classification is quadratic.

(b) Use (9) and (10) to find the optimal classification regions, and graph them.

7. This exercise expands on Example 7.2.3.

(a) Compute mean vectors and covariance matrices for both groups of electrodes, using variables X_1 and X_5 from Table 5.3.5.

(b) Find the quadratic classification functions (14), using equal prior probabilities. Estimate posterior probabilities for all 100 observations using equation (16), and identify observations that are misclassified.

8. This exercise refers to Example 7.2.4.

(a) Compute mean vectors and covariance matrices for both species, using all eight variables, and training samples of size $N_1 = 43$ and $N_2 = 46$. The raw data are in Table 5.4.1.

(b) Use (15) to estimate the parameters of the quadratic classification rule with relative sample sizes as prior probabilities.

(c) Estimate posterior probabilities for all 89 observations in the training samples.

(d) Similar to parts (a) to (c), perform a leave-one-out analysis, and summarize the results in a table. For 11 of the 89 observations in the training samples, you should obtain the same results as those reproduced in Table 7.2.2.

9. Suppose that $Y \sim \mathcal{N}(0, \sigma_j^2)$ in groups $j = 1, 2$, with prior probabilities π_j. Assume that $\sigma_2^2 > \sigma_1^2$. Show that the optimal classification rule is

$$X^* = \begin{cases} 1 & \text{if } |Y| < a, \\ 2 & \text{otherwise,} \end{cases}$$

for some constant $a > 0$. Find the constant as a function of the variances and the prior probabilities.

10. This exercise gives some rather crude motivation for the model studied in Example 7.2.5. It is not meant to be genetically correct.

(a) Let $\mathbf{Z}_1$ and $\mathbf{Z}_2$ denote the vectors of p morphometric variables measured on the first and the second siblings in a pair of identical twins. Suppose $\mathbf{Z}_1 = \mathbf{Z} + \mathbf{F}_1$ and $\mathbf{Z}_2 = \mathbf{Z} + \mathbf{F}_2$, where $\mathbf{Z}$, $\mathbf{F}_1$, and $\mathbf{F}_2$ are mutually independent random vectors with $\text{Cov}[\mathbf{Z}] = \boldsymbol{\Omega}$, $E[\mathbf{F}_1] = E[\mathbf{F}_2] = \mathbf{0}$, and $\text{Cov}[\mathbf{F}_1] = \text{Cov}[\mathbf{F}_2] = \boldsymbol{\Sigma}$. Find $E[\mathbf{Y}]$ and $\text{Cov}[\mathbf{Y}]$, where $\mathbf{Y} = \mathbf{Z}_1 - \mathbf{Z}_2$. (The vectors $\mathbf{F}_j$ represent "errors" added to the first and the second twin's "true" dimensions).

(b) Suppose that a similar model for fraternal twins can be written as $\mathbf{Z}_1 = \mathbf{Z} + \gamma\mathbf{F}_1$ and $\mathbf{Z}_2 = \mathbf{Z} + \gamma\mathbf{F}_2$, where $\gamma > 1$ is a constant, and $\mathbf{Z}$, $\mathbf{F}_1$, and $\mathbf{F}_2$ are as in part (a). Find the mean vector and the covariance matrix of $\mathbf{Y} = \mathbf{Z}_1 - \mathbf{Z}_2$, and compare these to the result of part (a).

11. This exercise refers to Example 7.2.6.

(a) Prove equations (20) and (21).

(b) Derive equation (23) explicitly for the case of $p = 2$ variables by writing the bivariate normal densities in scalar notation and taking partial derivatives with respect to the three parameters (two variances, one covariance) that determine the covariance matrix.

(c) Show that setting equation (22) equal to zero implies (23). *Hint*: Argue that, for two $p \times p$ matrices $\mathbf{A}$ and $\mathbf{B}$, $2\mathbf{A} - \text{diag}\mathbf{A} = 2\mathbf{B} - \text{diag}\mathbf{B}$ implies $\mathbf{A} = \mathbf{B}$.

12. Write a computer program for solving equations (23) and (24) iteratively. The input to your program should consist of two positive definite symmetric $p \times p$ matrices $\mathbf{M}_1$ and $\mathbf{M}_2$ and two positive integers N_1 and N_2. Starting with $\gamma^{(0)} = 1$, compute

$$\psi^{(i)} = \frac{1}{N_1 + N_2}\left[N_1\mathbf{M}_1 + \frac{1}{\gamma^{(i)}}N_2\mathbf{M}_2\right]$$

and

$$\gamma^{(i+1)} = \frac{1}{p}\,\text{tr}\left\{\left[\psi^{(i)}\right]^{-1}\mathbf{M}_2\right\}$$

iteratively for $i = 0, 1, 2, \ldots$. Stop the calculations when $|\gamma^{(i+1)} - \gamma^{(i)}| < \epsilon$ for some small positive constant ϵ, e.g., $\epsilon = 10^{-10}$. Test your program using the data of Example 7.2.6.

13. Apply your program from Exercise 12 to all three variables (i.e., differences between first and second twins) in the male twin data set to find maximum likelihood estimates. Estimate posterior probabilities for all 89 pairs of twins, and calculate a plug-in error rate. Is classification better than if you use only variables Y_1 and Y_2?

14. In Example 7.2.6 with two variables, graph the classification regions estimated from ordinary quadratic discrimination using (15) and (17), and compare your graph to Figure 7.2.4. Which of the two classification rules would you prefer?

15. Table 7.2.4 reports data measured on 79 pairs of female twins, obtained from the same study as the data of the male twins. Do the same analysis as in Example 7.2.6 for the female twins. Then repeat Exercises 13 and 14 for the female twins. How do you feel about the assumption of proportionality of the covariance matrices of monozygotic and dizygotic twins?

16. This exercise expands on Example 7.2.7.
    (a) Prove equation (25).
    (b) Transform the classification problem to canonical form, and show that optimal classification depends only on the first variable.

17. Suppose that $\mathbf{Y} \sim \mathcal{N}_3(\mu_j, \psi_j)$ in groups $j = 1, 2$, with

$$\mu_1 = \begin{pmatrix} 1 \\ 2 \\ 3 \end{pmatrix},$$

$$\psi_1 = \begin{pmatrix} 1 & -2 & 1 \\ -2 & 5 & -4 \\ 1 & -4 & 6 \end{pmatrix},$$

$$\mu_2 = \begin{pmatrix} 4 \\ -4 \\ 6 \end{pmatrix},$$

Table 7.2.4 Female twin data. Variables are defined as in Table 7.2.3.

TYPE	STA1	HIP1	CHE1	STA2	HIP2	CHE2
1	159.3	28.0	79.0	160.3	27.9	76.5
1	157.9	28.5	77.0	158.0	28.5	76.5
1	128.3	20.5	61.5	126.6	20.5	63.0
1	136.0	28.0	79.0	137.1	29.4	80.0
1	163.6	32.2	74.0	164.3	31.6	78.0
1	155.1	32.1	74.0	157.5	32.4	76.0
1	142.2	23.7	70.0	142.3	23.2	68.0
1	127.2	20.5	61.0	124.9	20.3	60.0
1	151.5	31.6	81.0	154.7	32.4	81.0
1	158.9	29.8	76.0	158.4	30.0	77.0
1	137.1	24.3	68.0	137.5	24.3	67.5
1	172.3	31.5	79.0	172.8	30.0	76.0
1	156.0	31.8	84.0	152.8	32.1	86.0
1	148.9	28.6	68.0	149.6	26.6	68.0
1	158.0	32.8	81.5	158.3	31.0	83.0
1	154.2	28.7	70.0	155.8	28.9	72.0
1	138.5	23.2	65.0	138.1	24.0	65.0
1	149.2	26.5	72.0	148.9	28.0	76.0
1	134.0	21.6	62.0	135.2	22.1	62.0
1	147.7	24.0	70.5	145.7	23.2	69.0
1	165.6	32.7	82.0	164.7	33.3	85.0
1	153.4	29.5	75.0	152.2	28.5	75.0
1	133.8	21.0	60.0	132.0	20.4	60.0
1	158.9	28.3	79.0	158.8	28.5	78.5
1	181.8	33.1	82.0	175.2	32.6	82.0
1	157.2	29.5	76.0	156.9	29.0	76.0
1	155.0	26.3	72.0	160.6	26.5	77.5
1	153.0	26.0	72.5	153.6	27.0	73.0
1	147.4	26.2	75.0	148.0	26.5	76.0
1	153.7	27.6	70.0	156.3	28.3	73.0
1	165.9	28.1	78.0	167.2	28.8	81.0
1	162.3	27.3	81.0	162.9	27.3	81.0
1	159.3	32.1	83.0	159.1	32.1	83.5
1	151.3	29.5	80.0	151.3	30.5	78.0
1	129.7	22.3	61.5	128.8	21.5	63.0
1	157.8	27.3	69.0	163.1	28.6	73.0
2	139.8	22.0	64.0	137.2	21.2	61.0
2	130.1	22.0	60.0	126.8	24.0	63.0
2	134.1	22.8	67.0	132.3	23.0	67.5
2	151.5	24.6	70.0	151.7	24.5	72.0
2	153.5	29.8	72.0	152.3	28.7	70.0
2	153.1	26.0	76.5	159.3	26.5	72.0
2	150.7	28.2	72.0	157.7	30.4	74.0
2	148.8	29.0	74.5	140.7	23.9	64.5
2	151.6	29.9	83.0	153.7	29.3	79.5
2	159.0	29.9	70.0	158.5	28.9	74.5
2	146.0	29.0	69.0	159.4	28.8	71.0

Table 7.2.4 *Continued*

TYPE	STA1	HIP1	CHE1	STA2	HIP2	CHE2
2	154.2	26.1	75.0	154.4	27.6	75.5
2	147.1	24.4	64.0	154.0	26.0	68.0
2	148.0	27.5	82.0	147.1	24.9	72.0
2	127.8	20.0	63.0	135.7	21.0	64.0
2	166.8	31.5	80.5	163.8	28.9	74.0
2	162.5	32.0	80.0	168.6	29.2	79.0
2	161.1	30.7	81.0	156.6	32.5	87.0
2	163.4	33.0	83.0	164.4	32.0	80.0
2	155.9	30.6	75.0	160.0	31.0	82.0
2	159.8	27.5	85.5	157.6	28.0	85.0
2	116.1	20.0	55.0	121.3	21.0	57.0
2	134.7	21.9	67.0	145.0	26.4	74.0
2	155.6	27.2	70.0	159.3	30.0	72.5
2	134.0	21.0	59.0	131.0	22.2	65.0
2	146.5	25.2	75.0	143.5	23.7	69.0
2	142.0	25.0	67.0	140.0	23.5	64.0
2	148.3	26.6	79.0	152.4	25.0	74.5
2	134.4	23.2	67.0	139.6	27.6	77.0
2	168.0	30.5	82.0	168.5	32.2	81.0
2	145.2	22.7	65.5	138.7	20.9	64.0
2	128.4	23.6	64.0	128.5	23.4	63.0
2	144.5	23.5	66.0	144.1	26.0	66.0
2	148.5	23.7	70.0	146.0	24.9	64.0
2	170.9	33.9	84.0	167.8	32.1	82.0
2	159.9	29.9	77.5	150.5	25.0	68.0
2	167.0	32.7	79.0	172.0	34.2	81.0
2	157.3	27.0	73.0	159.0	28.2	77.0
2	120.0	21.5	60.0	115.4	20.6	56.0
2	165.2	29.4	87.0	157.6	26.0	76.0
2	168.2	32.8	89.0	163.1	31.1	79.0
2	159.9	28.5	79.0	165.5	27.4	75.5
2	150.1	32.6	82.0	159.5	31.9	82.0

Data courtesy of the Institute of Anthropology, University of Hamburg.

and

$$\psi_2 = \begin{pmatrix} 9 & -18 & 9 \\ -18 & 37 & -20 \\ 9 & -20 & 14 \end{pmatrix}.$$

Show that optimal classification depends on $\mathbf{Y}$ only through Y_1. *Hint*: The matrix Γ in the transformation to canonical form is

$$\Gamma = \begin{pmatrix} 1 & 2 & 3 \\ 0 & 1 & 2 \\ 0 & 0 & 1 \end{pmatrix}.$$

Use the canonical form, or else show the result directly by evaluating (10).

7.3 Canonical Discriminant Functions

In this section we will study the problem of linear discrimination between $k \geq 2$ groups, assuming equality of all covariance matrices. This generalizes the theory developed in Chapter 5. Analogously to the derivation of the linear discriminant function, we will define canonical discriminant functions using at first a heuristic principle, and then show that the same classification rule is obtained from the normal theory approach.

We start with a preliminary result in which equality of covariance matrices across groups is not yet assumed. In the setup and notation introduced in Section 7.1, suppose that, conditionally on $X = j$, $\mathbf{Y}$ follows some p-variate distribution with mean vector $\boldsymbol{\mu}_j$ and covariance matrix $\boldsymbol{\psi}_j$, $j = 1, \ldots, k$. As before, let $\pi_j = \Pr[X = j]$, $j = 1, \ldots, k$, denote the prior probabilities. Let $\bar{\boldsymbol{\mu}}$ and $\bar{\boldsymbol{\psi}}$ be the mean vector and the covariance matrix of the mixture distribution. Then we have the following result which generalizes Theorem 2.8.1:

Theorem 7.3.1 *The mean vector $\bar{\boldsymbol{\mu}}$ and the covariance matrix $\bar{\boldsymbol{\psi}}$ of the mixture distribution are related to the mean vectors $\boldsymbol{\mu}_j$ and the covariance matrices $\boldsymbol{\psi}_j$ of the conditional distributions by*

$$\bar{\boldsymbol{\mu}} = \sum_{j=1}^{k} \pi_j \boldsymbol{\mu}_j \tag{1}$$

and

$$\bar{\boldsymbol{\psi}} = \sum_{j=1}^{k} \pi_j \boldsymbol{\psi}_j + \sum_{j=1}^{k} \pi_j (\boldsymbol{\mu}_j - \bar{\boldsymbol{\mu}})(\boldsymbol{\mu}_j - \bar{\boldsymbol{\mu}})'. \tag{2}$$

Proof Equation (1) follows directly from the analogous univariate result, i.e., equation (18) in Section 2.8. To show (2), notice that

$$E[\mathbf{Y}\mathbf{Y}'] = \sum_{j=1}^{k} \pi_j E[\mathbf{Y}\mathbf{Y}' | X = j] = \sum_{j=1}^{k} \pi_j \left[\boldsymbol{\psi}_j + \boldsymbol{\mu}_j \boldsymbol{\mu}_j'\right].$$

Moreover, as seen from Exercise 1,

$$\sum_{j=1}^{k} \pi_j \boldsymbol{\mu}_j \boldsymbol{\mu}_j' = \sum_{j=1}^{k} \pi_j (\boldsymbol{\mu}_j - \bar{\boldsymbol{\mu}})(\boldsymbol{\mu}_j - \bar{\boldsymbol{\mu}})' + \bar{\boldsymbol{\mu}}\bar{\boldsymbol{\mu}}', \tag{3}$$

and hence,

$$\bar{\boldsymbol{\psi}} = E\left[(\mathbf{Y} - \bar{\boldsymbol{\mu}})(\mathbf{Y} - \bar{\boldsymbol{\mu}})'\right]$$

$$= E[\mathbf{Y}\mathbf{Y}'] - \bar{\mu}\bar{\mu}'$$

$$= \sum_{j=1}^{k} \pi_j \psi_j + \sum_{j=1}^{k} \pi_j (\mu_j - \bar{\mu})(\mu_j - \bar{\mu})'.$$

■

An alternative proof of Theorem 7.3.1 is given in Exercise 2.

Equation (2) is a theoretical version of the so-called *multivariate analysis of variance equality*, which is used in the construction of tests for equality of several mean vectors; see Section 7.4. It partitions the covariance matrix of the mixture into a part that depends only on the conditional (i.e., within-group) covariance matrices and a part that depends only on the mean vectors. Setting

$$\psi = \sum_{j=1}^{k} \pi_j \psi_j ,$$

and

$$\psi_B = \sum_{j=1}^{k} \pi_j (\mu_j - \bar{\mu})(\mu_j - \bar{\mu})', \tag{4}$$

we can write (2) symbolically as

$$\bar{\psi} = \psi + \psi_B, \tag{5}$$

which is often interpreted as

$$\text{total variability} = \text{variability within groups} + \text{variability between groups}. \tag{6}$$

In the general setup, the "within" term $\psi = \sum_{j=1}^{k} \pi_j \psi_j$ is a weighted average of the conditional covariance matrices and may lack an easy interpretation. However, if we assume that all ψ_j are equal, then they must be identical to their weighted average, and the interpretation (6) becomes sensible. The matrix ψ_B is itself a covariance matrix (see Exercise 3), and therefore, is positive semidefinite. Thus, equation (5) also shows that the covariance matrix $\bar{\psi}$ of the mixture is equal to the weighted average of the conditional covariance matrices exactly if all mean vectors μ_j are identical. This fact underlies the construction of tests for equality of several mean vectors presented in Section 7.4.

In the following derivation of canonical discriminant functions we will assume for simplicity that all ψ_j are equal and, therefore, identical with the weighted average ψ. This assumption facilitates the understanding, although mathematically the only assumption made will be positive definiteness of $\psi = \sum_{j=1}^{k} \pi_j \psi_j$. Similar to the heuristic principle used in Section 5.2, we will study linear combinations that are "interesting" in the sense that they separate groups as much as possible. For any linear combination $Z = \mathbf{a}'\mathbf{Y}$,

$$\text{var}[Z] = \mathbf{a}'\bar{\psi}\mathbf{a} = \mathbf{a}'\psi\mathbf{a} + \mathbf{a}'\psi_B\mathbf{a}. \tag{7}$$

Therefore, we can decompose the variance of Z into a part "within" and a part "between." A linear combination for which the unconditional variance $\mathbf{a}'\bar{\psi}\mathbf{a}$ is much larger than the conditional (within-group) variance $\mathbf{a}'\psi\mathbf{a}$ will be particularly interesting because it tells us that the within-groups variability is inflated by differences in location. More precisely, we will maximize the ratio

$$\frac{\mathrm{var}[\mathbf{a}'\mathbf{Y}]}{\mathrm{var}[\mathbf{a}'\mathbf{Y}|X=j]} = \frac{\mathbf{a}'\bar{\psi}\mathbf{a}}{\mathbf{a}'\psi\mathbf{a}} \tag{8}$$

over $\mathbf{a} \in \mathbb{R}^p$. By equation (7),

$$\frac{\mathbf{a}'\bar{\psi}\mathbf{a}}{\mathbf{a}'\psi\mathbf{a}} = 1 + \frac{\mathbf{a}'\psi_B\mathbf{a}}{\mathbf{a}'\psi\mathbf{a}},$$

so we can equivalently maximize $\mathbf{a}'\psi_B\mathbf{a}/\mathbf{a}'\psi\mathbf{a}$, the ratio of variability between groups to variability within groups.

First, we show first how this is related to the two-group case studied in Chapter 5. By Exercise 4, in the case $k = 2$,

$$\psi_B = \pi_1\pi_2(\mu_1 - \mu_2)(\mu_1 - \mu_2)', \tag{9}$$

and maximization of (8) is equivalent to maximization of $[\mathbf{a}'(\mu_1 - \mu_2)]^2/\mathbf{a}'\psi\mathbf{a}$, which led to the definition of the linear discriminant function. Thus, our current problem generalizes the theory of Section 5.2.

Now, we proceed now to maximization of

$$r(\mathbf{a}) = \frac{\mathbf{a}'\psi_B\mathbf{a}}{\mathbf{a}'\psi\mathbf{a}} \tag{10}$$

over $\mathbf{a} \in \mathbb{R}^p$, in the general case $k \geq 2$, assuming that ψ is positive definite. By the simultaneous spectral decomposition theorem for two matrices (Appendix A.9), there exists a nonsingular matrix $\mathbf{H}$ of dimension $p \times p$ and a diagonal matrix $\Lambda = \mathrm{diag}(\lambda_1, \ldots, \lambda_p)$, all $\lambda_i \geq 0$, such that

$$\psi = \mathbf{H}\mathbf{H}',$$

and

$$\psi_B = \mathbf{H}\Lambda\mathbf{H}'. \tag{11}$$

Let $m = \mathrm{rank}(\psi_B)$, then exactly m of the diagonal entries of Λ are strictly positive, and we can arrange the columns of $\mathbf{H}$ according to decreasing order of magnitude of the λ_i, i.e., assume $\lambda_1 \geq \cdots \geq \lambda_m > 0 = \lambda_{m+1} = \cdots = \lambda_p$. Using (11),

$$r(\mathbf{a}) = \frac{\mathbf{a}'\mathbf{H}\Lambda\mathbf{H}'\mathbf{a}}{\mathbf{a}'\mathbf{H}\mathbf{H}'\mathbf{a}} = \frac{\mathbf{b}'\Lambda\mathbf{b}}{\mathbf{b}'\mathbf{b}}, \tag{12}$$

where $\mathbf{b} = \mathbf{H}'\mathbf{a}$. This ratio is unchanged if we multiply $\mathbf{b}$ by any nonzero constant c, and therefore, we can restrict maximization to normalized vectors $\mathbf{b}$, i.e., assume

$\mathbf{b}'\mathbf{b} = 1$. Writing $\mathbf{b} = (b_1, \ldots, b_p)'$, it remains to maximize $\mathbf{b}'\Lambda\mathbf{b} = \sum_{i=1}^m \lambda_i b_i^2$ over all vectors of unit length in $\mathbb{R}^p$. Since $\lambda_1 \geq \lambda_i$ for all $i \neq 1$,

$$\sum_{i=1}^{m} \lambda_i b_i^2 = \sum_{i=1}^{p} \lambda_i b_i^2 \leq \lambda_1 \sum_{i=1}^{p} b_i^2 = \lambda_1 . \tag{13}$$

The maximum λ_1 is attained for $\mathbf{b} = \mathbf{e}_1 = (1, 0, \ldots, 0)'$ because $\mathbf{e}_1'\Lambda\mathbf{e}_1 = \lambda_1$. This means that the function $r(\mathbf{a})$ is maximized if we choose

$$\mathbf{a} = (\mathbf{H}')^{-1}\mathbf{e}_1 \tag{14}$$

or any vector proportional to (14). Setting

$$\Gamma = (\gamma_1, \ldots, \gamma_p) = (\mathbf{H}')^{-1}, \tag{15}$$

a linear combination which maximizes $r(\mathbf{a})$ is given by

$$Z_1 = \gamma_1'\mathbf{Y}. \tag{16}$$

The linear combination (16) is called the *first canonical variate*. It is unique only up to a constant of proportionality, and, for simplicity, we will always take the proportionality constant equal to 1. Further nonuniqueness arises if $\lambda_1 = \lambda_2$, a problem that we will discuss later. By construction, Z_1 gives the best separation between the k groups in the sense of maximum between-group variability relative to within-group variability.

In the case of $k = 2$ groups, the vector γ_1 is proportional to $\psi^{-1}(\mu_1 - \mu_2)$; see Exercise 5. In Exercise 12 of Section 5.2, we have seen that any linear combination uncorrelated with $Z_1 = \gamma_1'\mathbf{Y}$ has identical means in both groups. However, in the case of $k > 2$ groups, we may be able to extract more information about group differences by studying further linear combinations. Thus, in a second step we will attempt, again to maximize the ratio $r(\mathbf{a}) = \mathbf{a}'\psi_B\mathbf{a}/\mathbf{a}'\psi\mathbf{a}$ but now subject to the constraint that the new linear combination $Z = \mathbf{a}'\mathbf{Y}$ be uncorrelated with the first canonical variate $Z_1 = \gamma_1'\mathbf{Y}$. Now, the problem is to maximize

$$r(\mathbf{a}) = \frac{\mathbf{a}'\psi_B\mathbf{a}}{\mathbf{a}'\psi\mathbf{a}}, \quad \text{subject to} \quad \mathbf{a}'\psi\gamma_1 = 0. \tag{17}$$

Writing $\psi_B = \mathbf{H}\Lambda\mathbf{H}'$ and $\psi = \mathbf{H}\mathbf{H}'$ as before, and setting $\mathbf{b} = \mathbf{H}'\mathbf{a}$, we have to maximize $\mathbf{b}'\Lambda\mathbf{b}/\mathbf{b}'\mathbf{b}$, and we can assume again, without loss of generality, that $\mathbf{b}'\mathbf{b} = 1$. Because $\mathbf{H}'\gamma_1 = \mathbf{e}_1$, the constraint of uncorrelatedness reads

$$0 = \mathbf{a}'\psi\gamma_1 = \mathbf{a}'\mathbf{H}\mathbf{H}'\gamma_1 = \mathbf{b}'\mathbf{e}_1, \tag{18}$$

i.e., orthogonality of $\mathbf{b}$ to $\mathbf{e}_1$. Thus, we have to maximize

$$\mathbf{b}'\Lambda\mathbf{b} \quad \text{subject to} \quad \mathbf{b}'\mathbf{b} = 1 \quad \text{and} \quad \mathbf{b}'\mathbf{e}_1 = 0. \tag{19}$$

But $\mathbf{b}'\mathbf{e}_1 = b_1$, and therefore,

$$\mathbf{b}'\boldsymbol{\Lambda}\mathbf{b} = \sum_{i=2}^{m} \lambda_i b_i^2 = \sum_{i=2}^{p} \lambda_i b_i^2 \le \lambda_2 \sum_{i=2}^{p} b_i^2 = \lambda_2. \tag{20}$$

The value λ_2 is attained if we choose $\mathbf{b} = \mathbf{e}_2 = (0, 1, 0, \dots, 0)'$, which means that a maximum is attained for

$$\mathbf{a} = \left(\mathbf{H}'\right)^{-1} \mathbf{e}_2 = \boldsymbol{\Gamma}\mathbf{e}_2 = \boldsymbol{\gamma}_2. \tag{21}$$

The linear combination $Z_2 = \boldsymbol{\gamma}_2'\mathbf{Y}$, thus obtained, is called the *second canonical variate*.

It is now clear how this process continues. Suppose that $r - 1 < m$ canonical discriminant functions have been found; these are the linear combinations $Z_j = \boldsymbol{\gamma}_j'\mathbf{Y}$, $j = 1, \dots, r-1$, where $\boldsymbol{\gamma}_j$ is the jth column of $\boldsymbol{\Gamma} = (\mathbf{H}')^{-1}$. In the next step, we wish to find a new linear combination $Z = \mathbf{a}'\mathbf{Y}$ so as to maximize $r(\mathbf{a}) = \mathbf{a}'\boldsymbol{\psi}_B\mathbf{a}/\mathbf{a}'\boldsymbol{\psi}\mathbf{a}$ subject to uncorrelatedness of Z with $Z_1, \dots, Z_{r-1}$. The constraints of uncorrelatedness (see Exercise 6) read

$$\text{cov}[\mathbf{a}'\mathbf{Y},\ Z_j] = \mathbf{a}'\boldsymbol{\psi}\boldsymbol{\gamma}_j \quad \text{for } j = 1, \dots, r-1. \tag{22}$$

Writing $\boldsymbol{\psi}_B = \mathbf{H}\boldsymbol{\Lambda}\mathbf{H}'$, $\boldsymbol{\psi} = \mathbf{H}\mathbf{H}'$, and $\mathbf{b} = \mathbf{H}'\mathbf{a}$, as before, we have to maximize $\mathbf{b}'\boldsymbol{\Lambda}\mathbf{b}/\mathbf{b}'\mathbf{b}$, and we can constrain the maximization problem to vectors of unit length, i.e., assume $\mathbf{b}'\mathbf{b} = 1$. The constraints (22) now read

$$0 = \mathbf{a}'\boldsymbol{\psi}\boldsymbol{\gamma}_j = \mathbf{a}'\mathbf{H}\mathbf{H}'\boldsymbol{\gamma}_j = \mathbf{b}'\mathbf{e}_j\,, \quad j = 1, \dots, r-1. \tag{23}$$

Thus, we have to maximize

$$\mathbf{b}'\boldsymbol{\Lambda}\mathbf{b} \quad \text{subject to } \mathbf{b}'\mathbf{b} = 1 \quad \text{and} \quad \mathbf{b}'\mathbf{e}_j = 0 \quad \text{for } j = 1, \dots, r-1. \tag{24}$$

Since $\mathbf{b}'\mathbf{e}_i = b_i$,

$$\mathbf{b}'\boldsymbol{\Lambda}\mathbf{b} = \sum_{i=r}^{m} \lambda_i b_i^2 = \sum_{i=r}^{p} \lambda_i b_i^2 \le \lambda_r \sum_{i=r}^{p} b_i^2 = \lambda_r\,, \tag{25}$$

and the value λ_r is attained if we choose $\mathbf{b} = \mathbf{e}_r$. Equivalently, the maximum is attained by choosing

$$\mathbf{a} = (\mathbf{H}')^{-1}\mathbf{e}_r = \boldsymbol{\Gamma}\mathbf{e}_r = \boldsymbol{\gamma}_r\,. \tag{26}$$

The linear combination $Z_r = \boldsymbol{\gamma}_r\mathbf{Y}$, thus obtained, is called the rth canonical discriminant function.

This process continues until we have found m canonical discriminant functions $Z_j = \boldsymbol{\gamma}_j'\mathbf{Y}$, $j = 1, \dots, m$. For any linear combination $Z = \mathbf{a}'\mathbf{Y}$, which is uncorrelated with $Z_1, Z_2, \dots, Z_m$, the ratio $\mathbf{a}'\boldsymbol{\psi}_B\mathbf{a}/\mathbf{a}'\boldsymbol{\psi}\mathbf{a}$ is zero, meaning that the means $\mathbf{a}'\boldsymbol{\mu}_j$ of variable Z in the k groups are all identical; see Exercise 7. Thus, all differences in location among the k groups are contained in the $m \le p$ variables $Z_1, \dots, Z_m$.

What are the possible values of m? Because the random vector $\mathbf{Y}$ has dimension p, m is at most equal to p. But also because ψ_B is the covariance matrix of a discrete random vector which can take k different values (see exercise 3), the rank of ψ_B is at most $k-1$. Therefore,

$$m \le \min(p, k-1). \tag{27}$$

Example 7.3.1 Suppose that we have $k = 3$ groups and $p = 2$ variables; then m can be at most 2. If the three mean vectors fall on a single line, e.g., if $\boldsymbol{\mu}_1 = \begin{pmatrix} 0 \\ 1 \end{pmatrix}$, $\boldsymbol{\mu}_2 = \begin{pmatrix} 1 \\ 2 \end{pmatrix}$, and $\boldsymbol{\mu}_2 = \begin{pmatrix} 2 \\ 3 \end{pmatrix}$, then the mean vector $\bar{\boldsymbol{\mu}}$ of the mixture lies on the same line, and the between-groups covariance matrix ψ_B has rank $m = 1$, irrespective of the probabilities π_1, π_2, and π_3. That is, a single linear combination summarizes all information about differences in location. If the three mean vectors do not fall on a line, then $m = 2$, and the transformation from $p = 2$ variables to $m = 2$ canonical variates does not reduce the dimensionality of the problem. See Exercise 9 for further numerical examples. □

Summarizing the theory developed so far, we have seen that, if the between-groups covariance matrix ψ_B has rank m, then we can find m canonical discriminant functions $Z_1, \ldots, Z_m$ which contain all the differences in location among the k groups. The transformation from $\mathbf{Y}$ to $Z_1, \ldots, Z_m$ is particularly appealing if m is small and p is large.

The transformation to canonical discriminant functions may also be understood in terms of a nonsingular linear transformation of the original variables $\mathbf{Y}$ into a new p-variate vector $\mathbf{Z}$ for which it is obvious that only the first m components are important for discrimination. In the parametrization of equation (11) and with $\boldsymbol{\Gamma} = (\mathbf{H}')^{-1}$, partition the matrix $\boldsymbol{\Gamma}$ as

$$\boldsymbol{\Gamma} = (\boldsymbol{\Gamma}_1 \quad \boldsymbol{\Gamma}_2), \tag{28}$$

where $\boldsymbol{\Gamma}_1 = (\boldsymbol{\gamma}_1 \quad \cdots \quad \boldsymbol{\gamma}_m)$ has dimension $p \times m$. Let $\mathbf{Z} = \boldsymbol{\Gamma}'\mathbf{Y}$. Assuming, as before, that the within-groups covariance matrix ψ is identical across all k groups, the common within-groups covariance matrix of $\mathbf{Z}$ is given by

$$\text{Cov}[\mathbf{Z}|X = j] = \boldsymbol{\Gamma}'\psi\boldsymbol{\Gamma} = \boldsymbol{\Gamma}'\mathbf{H}\mathbf{H}'\boldsymbol{\Gamma} = \mathbf{I}_p, \tag{29}$$

that is, in the transformed coordinate system all variables are uncorrelated and have variance 1. Let $\boldsymbol{\nu}_j = \boldsymbol{\Gamma}'\boldsymbol{\mu}_j$ denote the mean vector of $\mathbf{Z}$ in the jth group, and $\bar{\boldsymbol{\nu}} = \boldsymbol{\Gamma}'\bar{\boldsymbol{\mu}} = \sum_{j=1}^{k} \pi_j \boldsymbol{\nu}_j$ the mean vector of the mixture distribution. Then the between-group covariance matrix for the random vector $\mathbf{Z}$ is

$$\begin{aligned}\boldsymbol{\Gamma}'\boldsymbol{\psi}_B\boldsymbol{\Gamma} &= \sum_{j=1}^{k} \pi_j \boldsymbol{\Gamma}'(\boldsymbol{\mu}_j - \bar{\boldsymbol{\mu}})(\boldsymbol{\mu}_j - \bar{\boldsymbol{\mu}})'\boldsymbol{\Gamma} \\ &= \sum_{j=1}^{k} \pi_j (\boldsymbol{\nu}_j - \bar{\boldsymbol{\nu}})(\boldsymbol{\nu}_j - \bar{\boldsymbol{\nu}})'.\end{aligned} \tag{30}$$

On the other hand,

$$\boldsymbol{\Gamma}'\boldsymbol{\psi}_B\boldsymbol{\Gamma} = \boldsymbol{\Gamma}'\mathbf{H}\boldsymbol{\Lambda}\mathbf{H}'\boldsymbol{\Gamma} = \boldsymbol{\Lambda} = \begin{pmatrix} \boldsymbol{\Lambda}_1 & \mathbf{O} \\ \mathbf{O} & \mathbf{O} \end{pmatrix}, \tag{31}$$

where $\boldsymbol{\Lambda}_1 = \text{diag}(\lambda_1, \ldots, \lambda_m)$. From (30) and (31), it follows that all vectors $\boldsymbol{\nu}_j$ are identical in their last $p - m$ entries. After the transformation $\mathbf{Z} = \boldsymbol{\Gamma}'\mathbf{Y}$, all differences in location among the k groups occur only in the first m coordinates, which correspond to the m canonical discriminant functions.

The vectors $\boldsymbol{\gamma}_j$ are unique only up to multiplication by -1. Also, if two or more λ_j are identical, the associated $\boldsymbol{\gamma}_j$ are not uniquely defined. See Exercise 11 for more details and a numerical example.

Having seen that only the first $m \leq p$ canonical variates are of interest for classification, let

$$\mathbf{Z}_1 = \begin{pmatrix} Z_1 \\ \vdots \\ Z_m \end{pmatrix} = \begin{pmatrix} \boldsymbol{\gamma}_1' \\ \vdots \\ \boldsymbol{\gamma}_m' \end{pmatrix} \mathbf{Y} = \boldsymbol{\Gamma}_1'\mathbf{Y} \tag{32}$$

denote the vector of canonical discriminant functions. Then by (29),

$$\text{Cov}[\mathbf{Z}_1 | X = j] = \mathbf{I}_m \quad \text{for } j = 1, \ldots, k. \tag{33}$$

Let

$$\boldsymbol{\nu}_j^{(1)} = E[\mathbf{Z}_1 | X = j] = \boldsymbol{\Gamma}_1'\boldsymbol{\mu}_j \tag{34}$$

denote the mean vector of the canonical variates in the jth group. Since the within-group covariance matrix of $\mathbf{Z}_1$ is the identity, it is appropriate to use Euclidean distance for classification. In a heuristic approach ignoring prior probabilities we define a classification region for group j as

$$C_j = \left\{ \mathbf{z}_1 \in \mathbb{R}^m : ||\mathbf{z}_1 - \boldsymbol{\nu}_j^{(1)}|| < ||\mathbf{z}_1 - \boldsymbol{\nu}_h^{(1)}|| \text{ for all } h \neq j \right\}. \tag{35}$$

Note that this rule applies to the transformed variables $\mathbf{Z}_1 = \boldsymbol{\Gamma}_1'\mathbf{Y}$. That is, for an observation $\mathbf{y} \in \mathbb{R}^p$ in the original coordinate system we need to compute $\mathbf{z}_1 = \boldsymbol{\Gamma}_1'\mathbf{y}$ and find the particular mean vector among the $\boldsymbol{\nu}_j^{(1)}$ from which it has the smallest Euclidean distance. This is equivalent to classifying $\mathbf{y}$ in group j if the standard distance of $\mathbf{y}$ to $\boldsymbol{\mu}_j$ is smaller than the standard distance to all other mean vectors; see Exercise 12.

For $m = 1$, classification based on the single canonical discriminant function Z_1 corresponds to partitioning the real line into intervals. For $m = 2$, the classification rule generates a partition of $\mathbb{R}^2$ in which each area of allocation is bounded by line segments, half-lines, or lines. We will illustrate this in Example 7.3.2.

The application of the foregoing theory to observed data is straightforward, using the plug-in principle of estimation from Section 4.2. Let $\mathbf{y}_{ji}$ denote the ith observation in group j, $j = 1, \ldots, k$, $i = 1, \ldots, N_j$, and set $N = \sum_{j=1}^{k} N_j$. The plug-in estimates of the groupwise mean vectors and covariance matrices are given by

$$\hat{\mu}_j = \bar{\mathbf{y}}_j = \frac{1}{N_j} \sum_{i=1}^{N_j} \mathbf{y}_{ji} , \qquad j = 1, \ldots, k, \tag{36}$$

and

$$\hat{\psi}_j = \frac{1}{N_j} \sum_{i=1}^{N_j} (\mathbf{y}_{ji} - \bar{\mathbf{y}}_j)(\mathbf{y}_{ji} - \bar{\mathbf{y}}_j)' , \qquad j = 1, \ldots, k. \tag{37}$$

Using the observed proportions

$$\hat{\pi}_j = N_j / N, \qquad j = 1, \ldots, k \tag{38}$$

as estimates of the prior probabilities, the plug-in estimates of mean vector and covariance matrix of the mixture distribution are given by

$$\hat{\bar{\mu}} = \frac{1}{N} \sum_{j=1}^{k} \sum_{i=1}^{N_j} \mathbf{y}_{ji} = \sum_{j=1}^{k} \hat{\pi}_j \bar{\mathbf{y}}_j \tag{39}$$

and

$$\hat{\bar{\psi}} = \frac{1}{N} \sum_{j=1}^{k} \sum_{i=1}^{N_j} (\mathbf{y}_{ji} - \bar{\mathbf{y}})(\mathbf{y}_{ji} - \bar{\mathbf{y}})'. \tag{40}$$

The plug-in estimates correspond to parameters of the empirical distribution that puts probability $1/N$ on each of the N observed data vectors $\mathbf{y}_{ji}$. Therefore, Theorem 7.3.1 gives

$$\begin{aligned} \hat{\bar{\psi}} &= \sum_{j=1}^{k} \hat{\pi}_j \hat{\psi}_j + \sum_{j=1}^{k} \hat{\pi}_j (\bar{\mathbf{y}}_j - \bar{\mathbf{y}})(\bar{\mathbf{y}}_j - \bar{\mathbf{y}})' \\ &= \hat{\psi} + \hat{\psi}_B , \end{aligned} \tag{41}$$

where $\hat{\psi} = \sum_{j=1}^{k} \hat{\pi}_j \hat{\psi}_j$ and $\hat{\psi}_B = \sum_{j=1}^{k} \hat{\pi}_j (\bar{\mathbf{y}}_j - \bar{\mathbf{y}})(\bar{\mathbf{y}}_j - \bar{\mathbf{y}})'$. Notice that, in this approach, we do not necessarily assume equality of covariance matrices within groups, and $\hat{\psi}$ is merely a weighted average of the groupwise sample covariance matrices. If we multiply equation (41) by N and use $N_j = N\hat{\pi}_j$, we obtain

$$\sum_{j=1}^{k}\sum_{i=1}^{N_j}(\mathbf{y}_{ji}-\bar{\mathbf{y}})(\mathbf{y}_{ji}-\bar{\mathbf{y}})' = \sum_{j=1}^{k}\sum_{i=1}^{N_j}(\mathbf{y}_{ji}-\bar{\mathbf{y}}_j)(\mathbf{y}_{ji}-\bar{\mathbf{y}}_j)' + \sum_{j=1}^{k} N_j(\bar{\mathbf{y}}_j-\bar{\mathbf{y}})(\bar{\mathbf{y}}_j-\bar{\mathbf{y}})'. \quad (42)$$

Equation (42) is commonly called the MANOVA (Multivariate ANalysis Of VAriance) identity, often written as

$$\text{SSP(total)} = \text{SSP(within)} + \text{SSP(between)}, \quad (43)$$

where "SSP" stands for *Sums of Squares and Products*. The author of this book prefers to refer to equation (41) as the MANOVA identity because it reflects the relationship between parameter estimates.

The sample canonical discriminant functions can now be defined analogously to the theoretical ones. We need to determine the vectors $\hat{\gamma}_h$, $h = 1, \ldots, m$, that successively maximize the ratio

$$r(\mathbf{a}) = \frac{\mathbf{a}'\hat{\psi}_B\mathbf{a}}{\mathbf{a}'\hat{\psi}\mathbf{a}}, \quad (44)$$

subject to the same types of constraints as before. Here, $m \le \max(p, k-1)$ is the rank of the between-group covariance matrix $\hat{\psi}_B$.

As usual, calculations become more complicated by the unfortunate convention of using unbiased-sample covariance matrices rather than the more natural plug-in estimates. If $\mathbf{S}_j$ denotes the sample covariance matrix in group j, then $\hat{\psi}_j = (N_j - 1)\mathbf{S}_j/N_j$, and the within-group estimate $\hat{\psi}$ is computed as

$$\hat{\psi} = \frac{1}{N}\sum_{j=1}^{k}(N_j - 1)\mathbf{S}_j. \quad (45)$$

Example 7.3.2 Leaf dimensions of three species of flowers. This is, perhaps, the most famous of all data sets used in multivariate statistics, originally due to E. Anderson (1935) and analyzed by Fisher (1936) in his seminal paper on discriminant analysis. Table 7.3.1 displays data measured on three species of flowers called *Iris Setosa*, *Iris Versicolor*, and *Iris Virginica*. There are 50 specimens of each species and four variables: Y_1 = sepal length, Y_2 = sepal width, Y_3 = petal length, and Y_4 = petal width, all measured in mm.

Numerical results are as follows.

- *Setosa*: $\bar{\mathbf{y}}_1 = \begin{pmatrix} 50.06 \\ 34.28 \\ 14.62 \\ 2.46 \end{pmatrix}$, $\hat{\psi}_1 = \begin{pmatrix} 12.18 & 9.72 & 1.60 & 1.01 \\ 9.72 & 14.08 & 1.15 & 0.91 \\ 1.60 & 1.15 & 2.96 & 0.59 \\ 1.01 & 0.91 & 0.59 & 1.09 \end{pmatrix}$.

Table 7.3.1 Iris data from Fisher (1936). Species code is 1 = *Setosa*, 2 = *Versicolor*, 3 = *Virginica*. Variables are measured in mm.

Species	Sepal Length	Sepal Width	Petal Length	Petal Width
1	51	35	14	2
1	49	30	14	2
1	47	32	13	2
1	46	31	15	2
1	50	36	14	2
1	54	39	17	4
1	46	34	14	3
1	50	34	15	2
1	44	29	14	2
1	49	31	15	1
1	54	37	15	2
1	48.	34	16	2
1	48	30	14	1
1	43	30	11	1
1	58	40	12	2
1	57	44	15	4
1	54	39	13	4
1	51	35	14	3
1	57	38	17	3
1	51	38	15	3
1	54	34	17	2
1	51	37	15	4
1	46	36	10	2
1	51	33	17	5
1	48	34	19	2
1	50	30	16	2
1	50	34	16	4
1	52	35	15	2
1	52	34	14	2
1	47	32	16	2
1	48	31	16	2
1	54	34	15	4
1	52	41	15	1
1	55	42	14	2
1	49	31	15	2
1	50	32	12	2
1	55	35	13	2
1	49	36	14	1
1	44	30	13	2
1	51	34	15	2
1	50	35	13	3
1	45	23	13	3
1	44	32	13	2
1	50	35	16	6
1	51	38	19	4

Table 7.3.1 *Continued*

Species	Sepal Length	Sepal Width	Petal Length	Petal Width
1	48	30	14	3
1	51	38	16	2
1	46	32	14	2
1	53	37	15	2
1	50	33	14	2
2	70	32	47	14
2	64	32	45	15
2	69	31	49	15
2	55	23	40	13
2	65	28	46	15
2	57	28	45	13
2	63	33	47	16
2	49	24	33	10
2	66	29	46	13
2	52	27	39	14
2	50	20	35	10
2	59	30	42	15
2	60	22	40	10
2	61	29	47	14
2	56	29	36	13
2	67	31	44	14
2	56	30	45	15
2	58	27	41	10
2	62	22	45	15
2	56	25	39	11
2	59	32	48	18
2	61	28	40	13
2	63	25	49	15
2	61	28	47	12
2	64	29	43	13
2	66	30	44	14
2	68	28	48	14
2	67	30	50	17
2	60	29	45	15
2	57	26	35	10
2	55	24	38	11
2	55	24	37	10
2	58	27	39	12
2	60	27	51	16
2	54	30	45	15
2	60	34	45	16
2	67	31	47	15
2	63	23	44	13
2	56	30	41	13
2	55	25	40	13
2	55	26	44	12

Table 7.3.1 *Continued*

Species	Sepal Length	Sepal Width	Petal Length	Petal Width
2	61	30	46	14
2	58	26	40	12
2	50	23	33	10
2	56	27	42	13
2	57	30	42	12
2	57	29	42	13
2	62	29	43	13
2	51	25	30	11
2	57	28	41	13
3	63	33	60	25
3	58	27	51	19
3	71	30	59	21
3	63	29	56	18
3	65	30	58	22
3	76	30	66	21
3	49	25	45	17
3	73	29	63	18
3	67	25	58	18
3	72	36	61	25
3	65	32	51	20
3	64	27	53	19
3	68	30	55	21
3	57	25	50	20
3	58	28	51	24
3	64	32	53	23
3	65	30	55	18
3	77	38	67	22
3	77	26	69	23
3	60	22	50	15
3	69	32	57	23
3	56	28	49	20
3	77	28	67	20
3	63	27	49	18
3	67	33	57	21
3	72	32	60	18
3	62	28	48	18
3	61	30	49	18
3	64	28	56	21
3	72	30	58	16
3	74	28	61	19
3	79	38	64	20
3	64	28	56	22
3	63	28	51	15
3	61	26	56	14
3	77	30	61	23
3	63	34	56	24

Table 7.3.1 *Continued*

Species	Sepal Length	Sepal Width	Petal Length	Petal Width
3	64	31	55	18
3	60	30	48	18
3	69	31	54	21
3	67	31	56	24
3	69	31	51	23
3	58	27	51	19
3	68	32	59	23
3	67	33	57	25
3	67	30	52	23
3	63	25	50	19
3	65	30	52	20
3	62	34	54	23
3	59	30	51	18

- *Versicolor*: $\bar{\mathbf{y}}_2 = \begin{pmatrix} 59.36 \\ 27.70 \\ 42.60 \\ 13.26 \end{pmatrix}, \qquad \hat{\psi}_2 = \begin{pmatrix} 26.11 & 8.35 & 17.92 & 5.47 \\ 8.35 & 9.65 & 8.10 & 4.04 \\ 17.92 & 8.10 & 21.64 & 7.16 \\ 5.47 & 4.04 & 7.16 & 3.83 \end{pmatrix}.$

- *Virginica*: $\bar{\mathbf{y}}_3 = \begin{pmatrix} 65.88 \\ 29.74 \\ 55.52 \\ 20.26 \end{pmatrix}, \qquad \hat{\psi}_3 = \begin{pmatrix} 39.63 & 9.19 & 29.72 & 4.81 \\ 9.19 & 10.19 & 7.00 & 4.67 \\ 29.72 & 7.00 & 29.85 & 4.78 \\ 4.81 & 4.67 & 4.78 & 7.39 \end{pmatrix}.$

Note that the $\hat{\psi}_j$ are plug-in estimates, not the usual unbiased ones. With $N_1 = N_2 = N_3 = 50$, $\hat{\pi}_j = 1/3$ for all three groups, and we obtain the estimates of within-group and between-group covariance matrices as

$$\hat{\psi} = \begin{pmatrix} 25.97 & 9.09 & 16.42 & 3.76 \\ 9.09 & 11.31 & 5.41 & 3.21 \\ 16.42 & 5.41 & 18.15 & 4.18 \\ 3.76 & 3.21 & 4.18 & 4.10 \end{pmatrix}$$

and

$$\hat{\psi}_B = \begin{pmatrix} 42.14 & -13.30 & 110.17 & 47.52 \\ -13.30 & 7.56 & -38.16 & -15.29 \\ 110.17 & -38.16 & 291.40 & 124.52 \\ 47.52 & -15.29 & 124.52 & 53.61 \end{pmatrix}.$$

The rank of $\hat{\psi}_B$ is $m = \min(p, k-1) = 2$, and the vectors of canonical discriminant function coefficients are

$$\hat{\gamma}_1 = \begin{pmatrix} -0.0838 \\ -0.1550 \\ 0.2224 \\ 0.2839 \end{pmatrix},$$

and

$$\hat{\gamma}_2 = \begin{pmatrix} 0.0024 \\ 0.2186 \\ -0.0941 \\ 0.2868 \end{pmatrix},$$

with $\hat{\lambda}_1 = 32.192$ and $\hat{\lambda}_2 = 0.285$. Thus, the two canonical variates are

$$Z_1 = -0.0838\, Y_1 - 0.1550\, Y_2 + 0.2224\, Y_3 + 0.2839\, Y_4 \,,$$

and

$$Z_2 = 0.0024\, Y_1 + 0.2186\, Y_2 - 0.0941\, Y_3 + 0.2868\, Y_4 \,.$$

The value $\hat{\lambda}_1 = 32.192$ means that, for variable Z_1, the between-group variance is about 32 times the within-group variance, indicating a good separation between at least two of the three groups. Contrary to this, $\hat{\lambda}_2$ is quite small, and we might reasonably ask if classification should be based on Z_1 alone. Figure 7.3.1 shows a scatterplot of Z_1 vs Z_2; the good separation of the three groups is clearly visible. The graph also shows classification regions according to the criterion of minimum (Euclidean) distance to the respective group mean in the (Z_1, Z_2)-plane. The computation of error rates is left to the reader; see Exercise 14.

In the derivation of the canonical discriminant functions, we imposed the constraint that they are uncorrelated with each other, with respect to the within-group variability. If the assumption of equal covariance matrices in all groups is correct, we would expect the observed covariance matrices of the two canonical variates to be approximately equal to the identity matrix of dimension 2×2. Calculations (see Exercise 13) show a rather strong negative correlation between Z_1 and Z_2 for group *Setosa* and moderate positive correlations for the other two species. Without doing a formal test for equality of the three covariance matrices, this seems to indicate that linear discrimination is not optimal; see Section 7.2. However, the differences in location are so big that a nonlinear rule would probably not have any advantage for classification. □

As the Iris example shows, much of the appeal of canonical discriminant functions is that they allow us to get a low-dimensional representation of possibly high-dimensional data. If $m > 2$ then one may construct scatterplots of pairs of canonical discriminant functions, but typically the plot of Z_1 vs. Z_2 will be the most informative

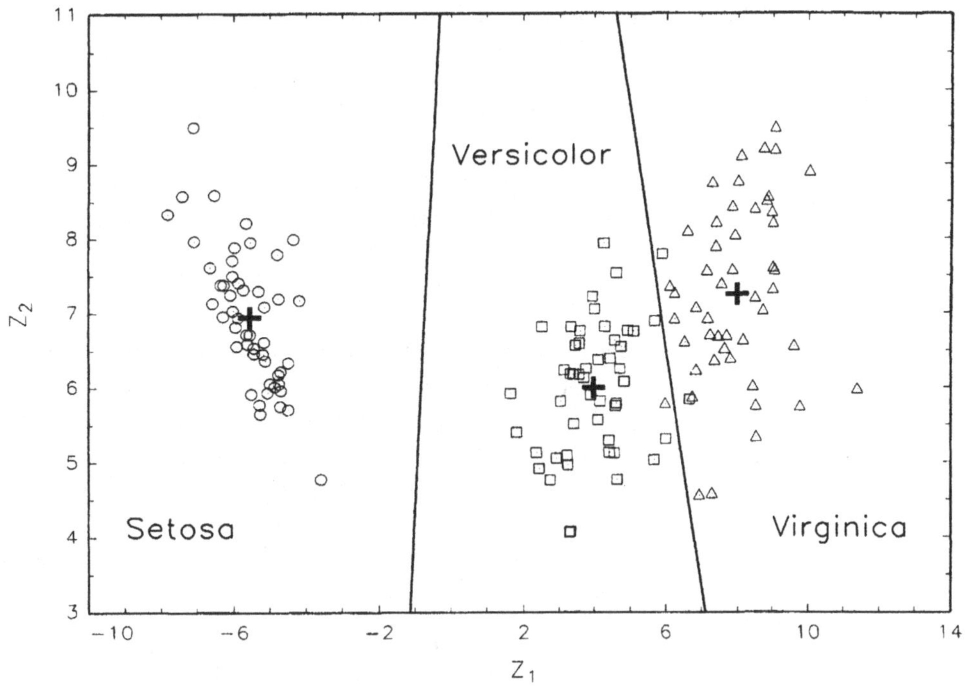

Figure 7-3-1 Scatterplot of the two canonical discriminant functions in the Iris example. Groupwise mean vectors are marked as solid plus signs. Straight lines are boundaries of the areas of classification.

one. On the other hand, if $p = 2$, then no reduction in dimensionality is achieved by the transformation to canonical discriminant functions, and one may as well graph the data in the original coordinate system. The only simplification achieved by the transformation in the case $m = p$ is a somewhat simplified calculation of classification functions because the common within-group covariance matrix is the identity matrix.

Now, we will show that the theory of optimal classification, assuming multivariate normality and equal covariance matrices in all groups, leads to the use of canonical discriminant functions. Suppose that $\mathbf{Y} \sim \mathcal{N}_p(\boldsymbol{\mu}_j, \boldsymbol{\psi})$ in groups $j = 1, \ldots, k$, with prior probabilities π_j. Define $\boldsymbol{\psi}_B$ as in equation (4) and $\mathbf{H}$ as in (11). With $\boldsymbol{\Gamma} = (\mathbf{H}')^{-1}$,

$$\boldsymbol{\Gamma}'\boldsymbol{\psi}\boldsymbol{\Gamma} = \mathbf{I}_p ,$$

and

$$\boldsymbol{\Gamma}'\boldsymbol{\psi}_B\boldsymbol{\Gamma} = \boldsymbol{\Lambda} = \text{diag}(\lambda_1, \ldots, \lambda_m, 0, \ldots, 0), \tag{46}$$

where $m = \text{rank}(\psi_B)$ and $\lambda_1 \geq \cdots \geq \lambda_m > 0$. Partition $\boldsymbol{\Gamma}$ as $(\boldsymbol{\Gamma}_1 \quad \boldsymbol{\Gamma}_2)$, where $\boldsymbol{\Gamma}_1$ has m columns, and let

$$\mathbf{Z} = \boldsymbol{\Gamma}'\mathbf{Y} = \begin{pmatrix} \boldsymbol{\Gamma}_1'\mathbf{Y} \\ \boldsymbol{\Gamma}_2'\mathbf{Y} \end{pmatrix} = \begin{pmatrix} \mathbf{Z}_1 \\ \mathbf{Z}_2 \end{pmatrix}. \tag{47}$$

Then $\mathbf{Z} \sim \mathcal{N}_p(\boldsymbol{\nu}_j, \mathbf{I}_p)$ in group j, and by Exercise 11 of Section 7.1, optimal classification may be based on $\mathbf{Z}$ instead of $\mathbf{Y}$. In partitioned form,

$$\boldsymbol{\nu}_j = \begin{pmatrix} \boldsymbol{\nu}_j^{(1)} \\ \boldsymbol{\nu}^* \end{pmatrix}, \qquad j = 1, \ldots, k, \tag{48}$$

that is, the mean vector of $\mathbf{Z}_2$ is identical for all k groups; see Exercise 7. Since $\mathbf{Z}$ is multivariate normal, diagonality of the covariance matrix implies independence of $\mathbf{Z}_1$ and $\mathbf{Z}_2$ in all k groups. Moreover, the distribution of $\mathbf{Z}_2$ is identical across groups. Therefore, by Exercise 8 of Section 7.1, optimal classification is based on $\mathbf{Z}_1$ alone. But $\mathbf{Z}_1 \sim \mathcal{N}_p(\boldsymbol{\nu}_j^{(1)}, \mathbf{I}_m)$ in group j, and therefore we can use the classification functions of equations (4) and (5) in Section 7.2. As the quadratic terms are identical and the common covariance matrix is $\mathbf{I}_m$, we can use the simplified classification functions

$$q_j(\mathbf{z}_1) = \left[\boldsymbol{\nu}_j^{(1)}\right]' \left[\mathbf{z}_1 - \frac{1}{2}\boldsymbol{\nu}_j^{(1)}\right] + \log \pi_j\,, \qquad j = 1, \ldots, k. \tag{49}$$

Optimal classification regions are defined by

$$C_j = \{\mathbf{z}_1 \in \mathbb{R}^m : q_j(\mathbf{z}_1) > q_h(\mathbf{z}_1) \text{ for all } h \neq j\}, \qquad j = 1, \ldots, k. \tag{50}$$

The posterior probability of membership in group j can be computed from the classification function using

$$\Pr[X = j | \mathbf{Z}_1 = \mathbf{z}_1] = \frac{\exp\left[q_j(\mathbf{z}_1)\right]}{\sum_{h=1}^k \exp\left[q_h(\mathbf{z}_1)\right]}, \qquad j = 1, \ldots, k, \tag{51}$$

see Exercise 15. Classification according to regions (50) is equivalent to assigning $\mathbf{z}_1$ to the group for which it has the highest posterior probability.

In applications to observed data, all parameters are replaced by their respective estimates from the training samples. Also, for classification of new observations, the user may choose the prior probabilities different from the $\hat{\pi}_j = N_j/N$ used in the estimation of the within- and between- group covariance matrices. Let $\pi_1^0, \ldots, \pi_k^0$ be prior probabilities chosen by the investigator; then the jth classification function would be computed as

$$\hat{q}_j(\mathbf{z}_1) = \left[\hat{\boldsymbol{\nu}}_j^{(1)}\right]' \left[\mathbf{z}_1 - \frac{1}{2}\hat{\boldsymbol{\nu}}_j^{(1)}\right] + \log \pi_j^0\,. \tag{52}$$

In particular, if we choose all $\pi_j^0 = 1/k$, then the classification regions are the same as the heuristic ones given in equation (35).

Example 7.3.3 Origin of wines. Marais, Versini, van Wyk, and Rapp (1992) collected data on the chemical composition of Weisser Riesling wines from South Africa, Germany, and Italy. Table 7.3.2 displays concentrations of 15 free monoterpenes and C_{13}-norisoprenoids, which are important constituents of wine aroma. For simplicity, we label the concentrations Y_1 to Y_{15}; for a detailed description see Marais et al. (1992). A total of 26 wines were compared, nine from South Africa, seven from Pfalz (Germany), and ten from Trentino Alto Adige and Friuli (Northern Italy).

Because the number of variables is quite large ($p = 15$), we leave numerical details to the student; see Exercise 16. With $k = 3$ groups, the rank of the between-group covariance matrix is $m = \min(p, k - 1) = 2$. Numerical values of the two vectors of canonical discriminant function coefficients are given in Exercise 16; the associated ratios of between-groups variance to within-group variance are $\hat{\lambda}_1 = 10.498$ and $\hat{\lambda}_2 = 3.299$. This indicates that at least two of the groups are well separated, and in contrast to the Iris example, the second canonical discriminant function seems to be fairly important. Figure 7.3.2 shows a scatterplot of the canonical discriminant functions, confirming the excellent separation of the three groups. Table 7.3.3 gives numerical results for the classification, using equal prior probabilities $\pi_j^0 = 1/3$ in the calculation of the classification functions (52). Remarkably, all posterior probabilities are very close to 0 or 1, which might give us a lot of confidence about the ability of the canonical discriminant functions to tell the origin of the wines based on the chemical indicators.

The posterior probabilities displayed in Table 7.3.3 are based explicitly on normality assumptions, which may not be reasonable for concentrations. However, there is a worse problem: estimation of the discriminant functions is based on 15 variables and 26 observations, that is, we have a very small sample size compared to the dimensionality of the problem. As we have seen in earlier examples, estimating many parameters leads to an overly optimistic assessment of the actual error rates. Thus, it is important to perform a leave-one-out analysis in this example. Results are also displayed in Table 7.3.3: eight of the 26 observations are misclassified, giving a leave-one-out error rate of $8/26 \approx 31$ percent. Remarkably, the normal theory posterior probabilities (calculated again using equal prior probabilities) are mostly very close to 0 and 1, illustrating rather dramatically how dangerous it is to use too many variables. We leave it to the reader (see Exercise 16) to analyze this example for a subset of variables, which gives a group separation not quite as nice as in Figure 7.3.2, but a much better leave-one-out error rate, and thus a better method to actually identify the origin of a wine based on the chemical indicators. □

In this section, we have treated the problem of linear discrimination between $k \geq 2$ groups mostly on a descriptive level, similar to the treatment of the two-group case in Chapter 5. Although, in both data examples of this section, there is very little doubt that the distributions differ between groups (which, in turn justifies any attempt

Table 7.3.2 Wine data. The table displays C = country of origin, j = sequential number, and 15 chemical indicators labeled Y_1 to Y_{15} for 26 wines from South Africa, Germany, and Italy. Data courtesy of Dr. G. Versini, Istituto Agrario Provinciale, 38010 San Michele all'Adige (Trento), Italy. Country codes are 1 for South Africa, 2 for Germany, 3 for Italy.

C	j	Y_1	Y_2	Y_3	Y_4	Y_5	Y_6	Y_7	Y_8
		Y_9	Y_{10}	Y_{11}	Y_{12}	Y_{13}	Y_{14}	Y_{15}	
1	1	36.306	13.409	17.796	114.573	135.282	159.252	15.614	6.165
		5.402	4.420	54.464	22.652	3.747	15.483	11.875	
1	2	65.271	24.934	41.455	1.326	50.701	74.265	14.336	6.200
		1.067	1.077	43.856	28.722	24.687	30.592	40.412	
1	3	26.730	17.453	7.921	102.013	45.168	162.280	17.603	6.337
		5.344	6.198	28.728	29.514	0.000	17.132	13.201	
1	4	58.361	29.623	27.650	3.558	40.265	100.269	16.690	6.266
		3.073	3.022	26.612	37.529	20.903	39.599	40.365	
1	5	81.553	40.751	40.267	0.000	8.031	26.847	13.231	6.480
		1.765	2.819	20.994	23.216	29.998	41.060	45.959	
1	6	16.362	10.760	7.118	50.855	35.033	85.345	9.514	4.413
		2.497	2.866	26.038	15.361	0.000	9.879	8.180	
1	7	27.675	15.119	17.100	3.909	46.370	49.159	7.800	5.569
		0.000	1.886	23.830	13.463	8.868	18.950	18.323	
1	8	18.860	11.554	7.740	65.129	28.458	85.923	3.631	2.801
		2.807	3.098	25.569	17.543	0.000	6.673	7.028	
1	9	53.517	21.557	32.041	0.000	10.066	12.947	4.443	4.269
		1.000	1.000	14.707	14.548	3.696	45.976	39.473	
2	10	14.179	3.499	4.135	37.994	68.698	31.480	15.099	2.234
		6.149	0.746	51.522	3.436	0.804	7.505	1.199	
2	11	25.031	6.427	8.885	4.068	34.642	22.374	11.323	2.322
		1.000	1.239	36.928	6.811	3.713	8.237	1.618	
2	12	32.698	6.983	12.346	89.364	75.210	91.384	21.646	4.664
		5.948	3.624	64.769	11.640	2.809	13.424	3.696	
2	13	41.050	13.375	24.546	3.202	51.694	41.759	18.795	4.432
		1.000	2.252	39.232	8.695	6.473	16.730	12.238	
2	14	25.662	6.129	10.008	166.594	77.726	90.224	20.675	3.697
		2.694	4.688	141.407	7.660	1.480	9.420	1.179	
2	15	49.556	15.650	29.004	51.721	133.495	102.959	18.606	7.011
		1.042	3.494	134.595	20.645	2.743	25.348	8.161	
2	16	77.110	25.527	71.980	0.000	84.141	42.724	24.272	3.252
		1.000	2.643	97.858	20.553	16.048	25.369	22.256	
3	17	103.021	8.520	6.532	98.743	40.621	40.914	38.364	6.659
		3.350	2.626	146.784	3.319	0.000	4.192	1.010	
3	18	89.719	10.236	6.906	104.507	60.151	58.143	41.501	6.969
		3.803	3.571	157.712	4.404	0.000	5.187	2.397	
3	19	93.047	8.112	6.019	89.888	68.510	47.985	45.329	7.977
		2.929	3.454	139.754	3.439	0.000	4.620	2.120	
3	20	14.691	4.934	0.811	41.746	15.121	25.868	5.595	2.480
		3.516	2.842	29.086	1.977	0.000	0.000	1.638	

Table 7.3.2 *Continued*

C	j	Y_1 / Y_9	Y_2 / Y_{10}	Y_3 / Y_{11}	Y_4 / Y_{12}	Y_5 / Y_{13}	Y_6 / Y_{14}	Y_7 / Y_{15}	Y_8
3	21	34.227	8.910	16.681	5.807	12.256	25.310	5.776	3.323
		0.734	1.568	20.249	7.579	0.000	0.578	1.303	
3	22	40.591	10.546	132.637	0.000	9.998	12.913	5.460	1.681
		0.000	4.760	17.042	6.579	7.300	3.808	2.960	
3	23	19.076	8.084	2.510	60.979	58.099	28.012	28.508	5.412
		3.888	1.992	44.238	1.714	0.000	1.232	0.917	
3	24	39.375	10.009	16.472	21.583	55.422	75.098	26.287	5.466
		1.193	1.629	76.699	13.784	4.000	2.693	4.396	
3	25	16.381	10.475	2.100	110.954	23.115	52.587	8.940	2.850
		6.414	5.390	46.343	3.173	0.000	0.854	1.115	
3	26	42.682	16.922	23.330	12.761	40.772	151.610	14.427	2.875
		0.000	2.125	40.919	29.387	12.000	6.695	3.859	

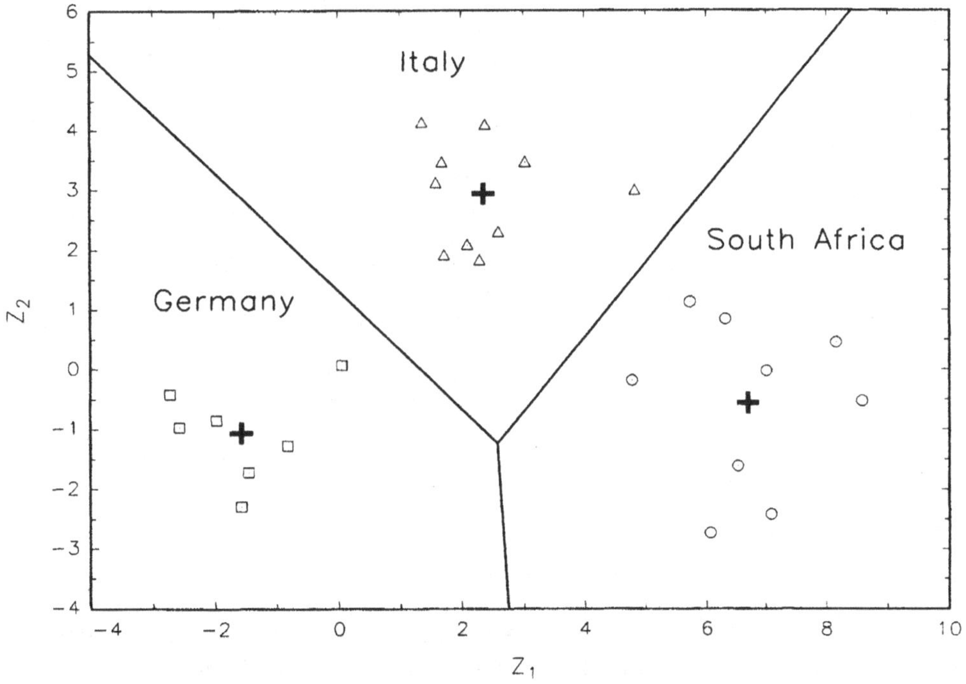

Figure 7-3-2 Scatterplot of the two canonical discriminant functions in the Wine example. Groupwise mean vectors are marked as solid plus-signs. Straight lines are boundaries of the areas of classification.

Table 7.3.3 Details of the computation of error rates in the wine example. The symbols used are x_i, actual group of observation i; $\hat{\pi}_{ji}$ = estimated posterior probability (percent) of observation i in group j using the classification functions based on all observations; $\hat{x}_i^*$ = predicted group membership using the classification functions based on all observations; $\hat{\pi}_{ji,-i}$ and $\hat{x}_{i,-i}^*$ = analogous quantities using the classification functions based on all but the ith observation. Incorrectly predicted group membership is marked with an asterisk (*).

i	x_i	$\hat{\pi}_{1i}$	$\hat{\pi}_{2i}$	$\hat{\pi}_{3i}$	$\hat{x}_i^*$	$\hat{\pi}_{1i,-i}$	$\hat{\pi}_{2i,-i}$	$\hat{\pi}_{3i,-i}$	$\hat{x}_{i,-i}^*$
1	1	100.00	0.00	0.00	1	54.80	45.20	0.00	1
2	1	100.00	0.00	0.00	1	100.00	0.00	0.00	1
3	1	100.00	0.00	0.00	1	100.00	0.00	0.00	1
4	1	100.00	0.00	0.00	1	99.95	0.05	0.00	1
5	1	100.00	0.00	0.00	1	0.21	0.00	99.79	3*
6	1	99.56	0.00	0.44	1	96.39	0.00	3.61	1
7	1	99.72	0.00	0.28	1	6.02	0.00	93.98	3*
8	1	99.99	0.00	0.01	1	94.52	0.00	5.48	1
9	1	100.00	0.00	0.00	1	100.00	0.00	0.00	1
10	2	0.00	100.00	0.00	2	0.00	100.00	0.00	2
11	2	0.00	100.00	0.00	2	0.00	100.00	0.00	2
12	2	0.00	100.00	0.00	2	0.00	99.99	0.01	2
13	2	0.00	99.17	0.83	2	0.00	0.00	100.00	3*
14	2	0.00	100.00	0.00	2	99.03	0.93	0.04	1*
15	2	0.00	100.00	0.00	2	6.68	93.32	0.00	2
16	2	0.00	100.00	0.00	2	0.00	96.71	3.29	2
17	3	0.00	0.00	100.00	3	0.00	0.00	100.00	3
18	3	0.00	0.00	100.00	3	0.00	0.00	100.00	3
19	3	0.00	0.00	100.00	3	0.00	0.01	99.99	3
20	3	0.00	0.00	100.00	3	0.01	0.24	99.75	3
21	3	0.67	0.00	99.33	3	100.00	0.00	0.00	1*
22	3	0.00	0.00	100.00	3	0.00	100.00	0.00	2*
23	3	0.00	0.00	100.00	3	0.68	99.32	0.00	2*
24	3	0.00	0.00	100.00	3	0.00	0.00	100.00	3
25	3	0.00	0.00	100.00	3	0.76	0.02	99.22	3
26	3	0.00	0.01	99.99	3	0.00	100.00	0.00	2*

to estimate a classification rule), we will in general be concerned about the possibility that the observed mean vectors might differ only because of sampling variation. That is, we will want to test the hypothesis of equality of several mean vectors, a problem we have discussed in detail for the case $k = 2$ in Section 6.4. For the general case $k \geq 2$, appropriate tests are introduced in Section 7.4. As we have seen, the number m of canonical discriminant functions is at most $r = \min(p, k-1)$. If this minimum is at least 2, we might further ask if all canonical discriminant functions are really needed. More formally, we may want to find the smallest value of m such that the hypothesis $\text{rank}(\psi_B) = m$ is acceptable. In particular, the hypothesis $m = 0$ is the same as equality of all k mean vectors. For instance, in a classification problem with $k = 4$ groups, we might find that the four observed mean vectors actually lie almost exactly in a plane, which would make it possible to use only two instead of three canonical discriminant functions. In the spirit of a first introduction, the interested reader is referred to more advanced texts for answers to such questions. However, a pragmatic answer can always be given using a leave-one-out procedure. If the leave-one-out error rate of a classification rule based on a reduced set of canonical variates is as good or better than the leave-one-out error rate based on all canonical variates, then one should use the reduced set.

Similar questions may be asked about the importance of variables, a problem which we have treated in detail in Section 6.5 for the case of two groups. In view of cases like the wine example, the question of redundancy of variables should not be ignored. As seen in Example 7.3.3, the leave-one-out method may again be used for variable selection. The reader is referred to more advanced texts for formal testing procedures.

In the two-group case, much emphasis was put on the notion of standard distance (or Mahalanobis distance). In the setup of this section, it is straightforward to compute a $k \times k$ matrix of standard distances between pairs of mean vectors; see Exercise 17. The interpretation of the variance ratios λ_i is not as straightforward, and therefore, a graph of the canonical discriminant functions is important for the user to see how good or how bad group separation is.

Finally, here is a cautionary remark. As mentioned earlier, equation (42) is commonly referred to as the MANOVA identity, and, accordingly, most commercially available software will display the sums of squares and products matrices instead of the plug-in estimates $\hat{\boldsymbol{\psi}}$ and $\hat{\boldsymbol{\psi}}_B$. Also, for reasons of unbiasedness, most authors use the pooled within-group covariance matrix

$$\mathbf{S} = \frac{1}{N_1 + \cdots + N_k - k} \sum_{j=1}^{k} (N_j - 1)\mathbf{S}_j = (N - k)\hat{\boldsymbol{\psi}}/N, \tag{53}$$

instead of the more natural $\hat{\boldsymbol{\psi}}$, as an estimator of the common $\boldsymbol{\psi}$. Consequently, the columns $\mathbf{b}$ of the matrix $\boldsymbol{\Gamma}$ are normalized such that $\mathbf{b}'\mathbf{S}\mathbf{b} = 1$, not $\mathbf{b}'\hat{\boldsymbol{\psi}}\mathbf{b} = 1$ as in this chapter. Although this does not affect the actual classification rule, it may change the values of the classification functions and the estimated posterior probabilities considerably in cases where the number of groups k is large compared to the total number of observations N. It will also affect the values of the pairwise standard distances between groups; see Exercise 17.

Exercises for Section 7.3

1. Prove equation (3).

2. This exercise generalizes equations (17) and (18) from Section 2.7. Here, **X** and **Y** denote two jointly distributed random vectors.

   (a) Show that $E[\mathbf{Y}] = E\big[E[\mathbf{Y}|\mathbf{X}]\big]$.

   (b) Show that $\text{Cov}[\mathbf{Y}] = E\big[\text{Cov}[\mathbf{Y}|\mathbf{X}]\big] + \text{Cov}\big[E[\mathbf{Y}|\mathbf{X}]\big]$.

   (c) Use parts (a) and (b) to prove Theorem 7.3.1.

3. Let $\mathbf{Y}^*$ be a p-variate discrete random vector that takes values $\boldsymbol{\mu}_1, \ldots, \boldsymbol{\mu}_k$ with probabilities $\pi_1, \ldots, \pi_k$. Show that $\text{Cov}[\mathbf{Y}^*] = \boldsymbol{\psi}_B$ (equation 4).

4. Prove equation (9).

5. Show that, if $k = 2$, then the vector $\boldsymbol{\gamma}_1$ of the first canonical variate is proportional to $\boldsymbol{\psi}^{-1}(\boldsymbol{\mu}_1 - \boldsymbol{\mu}_2)$.

6. Verify equation (22).

7. Let $Z_1, \ldots, Z_m$ denote the m canonical discriminant functions in a discrimination problem with p variables and $\text{rank}(\boldsymbol{\Gamma}) = m < p$. Assume, for simplicity, that no two λ_i are identical. Show that any linear combination $Z = \mathbf{a}'\mathbf{Y}$ which is uncorrelated with $Z_1, \ldots, Z_m$ provides no information about differences in location, i.e., $\mathbf{a}'\boldsymbol{\mu}_1 = \cdots = \mathbf{a}'\boldsymbol{\mu}_k$.

8. Let $\boldsymbol{\gamma}_1, \ldots, \boldsymbol{\gamma}_p$ be defined as in equation (15).

   (a) Show that the $\boldsymbol{\gamma}_j$ are eigenvectors of $\boldsymbol{\psi}^{-1}\boldsymbol{\psi}_B$. What are the associated eigenvalues?

   (b) Show that the $\boldsymbol{\gamma}_j$ are eigenvectors of $\boldsymbol{\psi}^{-1}\bar{\boldsymbol{\psi}}$. What are the associated eigenvalues?

9. In a discrimination problem with $k = 4$ groups and $p = 3$ variables, suppose that the prior probabilities π_j are all 1/4, the mean vectors are $\boldsymbol{\mu}_j$, and the common within-group covariance matrix is $\boldsymbol{\psi}$. Compute the between-group covariance matrix $\boldsymbol{\psi}_B$, the rank m of $\boldsymbol{\psi}_B$, the simultaneous decomposition $\boldsymbol{\psi} = \mathbf{H}\mathbf{H}'$ and $\boldsymbol{\psi}_B = \mathbf{H}\boldsymbol{\Lambda}\mathbf{H}'$, and the columns $\boldsymbol{\gamma}_j$ of $\boldsymbol{\Gamma} = (\mathbf{H}')^{-1}$. Thus, describe the transformation from the original variables $\mathbf{Y}$ to m canonical variates for each of the following three cases:

   (a) $\boldsymbol{\psi} = \begin{pmatrix} 1 & 2 & 3 \\ 2 & 5 & 10 \\ 3 & 10 & 26 \end{pmatrix}$, $\boldsymbol{\mu}_1 = \begin{pmatrix} -2 \\ -4 \\ -6 \end{pmatrix}$, $\boldsymbol{\mu}_2 = \begin{pmatrix} 2 \\ 4 \\ 6 \end{pmatrix}$, $\boldsymbol{\mu}_3 = \begin{pmatrix} 4 \\ 8 \\ 12 \end{pmatrix}$, and $\boldsymbol{\mu}_4 = \begin{pmatrix} 0 \\ 0 \\ 0 \end{pmatrix}$.

   (b) Same as part (a), except $\boldsymbol{\mu}_3 = (0 \quad 2 \quad 8)'$ and $\boldsymbol{\mu}_4 = (0 \quad 6 \quad 24)'$.

   (c) Same as part (b), except $\boldsymbol{\mu}_3 = (0 \quad 2 \quad 12)'$.

10. An appealing property of the linear discriminant function, in the case of two groups, is that it does not depend on the probabilities of group membership π_1 and π_2. Show that, in general, this is not true for canonical discriminant functions when $m = \text{rank}(\boldsymbol{\psi}_B) \geq 2$. *Hint*: Construct a numerical example with $p = 2$ and $k = 3$, and compute the matrix $\boldsymbol{\Gamma}$ for two different choices of the probabilities π_1, π_2, and π_3.

11. This exercise refers to nonuniqueness of the canonical discriminant functions if some of the λ_i are identical. Let $\mathbf{W}$ be a positive definite symmetric matrix of dimension $p \times p$, and $\mathbf{A}$ a symmetric matrix of the same dimension. By Appendix A.9 there exists a nonsingular $p \times p$ matrix $\mathbf{H}$ such that $\mathbf{W} = \mathbf{HH}'$ and $\mathbf{A} = \mathbf{H}\Lambda\mathbf{H}'$, where $\Lambda = \text{diag}(\lambda_1, \ldots, \lambda_p)$. Show that, if some of the λ_i are identical, then this decomposition is not unique. *Hint*: Let $\Gamma = (\gamma_1 \ \cdots \ \gamma_p) = (\mathbf{H}')^{-1}$. Then $\Gamma'\mathbf{W}\Gamma = \mathbf{I}_p$ and $\Gamma'\mathbf{A}\Gamma = \Lambda$. Assume $\lambda_1 = \cdots = \lambda_r$ for some $r \geq 2$, and set

$$\Gamma^* = \Gamma \begin{pmatrix} \mathbf{M} & \mathbf{O} \\ \mathbf{O} & \mathbf{I}_{p-r} \end{pmatrix}$$

for an arbitrary orthogonal matrix $\mathbf{M}$ of dimension $r \times r$. Show that Γ^* diagonalizes $\mathbf{W}$ and $\mathbf{A}$ simultaneously.

12. Suppose that $\mathbf{Y}$ is a p-variate random vector with mean $\boldsymbol{\mu}_j$ in group j and common covariance matrix $\boldsymbol{\psi}$ in all groups. Show that the classification rule based on classification regions

$$C_j^* = \left\{\mathbf{y} \in \mathbb{R}^p : (\mathbf{y} - \boldsymbol{\mu}_j)'\boldsymbol{\psi}^{-1}(\mathbf{y} - \boldsymbol{\mu}_j) < (\mathbf{y} - \boldsymbol{\mu}_h)'\boldsymbol{\psi}^{-1}(\mathbf{y} - \boldsymbol{\mu}_h) \ \text{ for all } h \neq j\right\}$$

is equivalent to the rule based on (35).

13. This exercise expands on Example 7.3.2 (Iris data).
    (a) Verify the calculations of Example 7.3.2. In particular, compute a matrix $\mathbf{H}$ such that $\hat{\boldsymbol{\psi}} = \mathbf{HH}'$ and $\hat{\boldsymbol{\psi}}_B = \mathbf{H}\Lambda\mathbf{H}'$, find its inverse $\Gamma' = \mathbf{H}^{-1}$, and verify that two of the columns of Γ are the vectors $\hat{\gamma}_j$ given in Example 7.3.2. *Note*: Depending on the normalization of eigenvectors used by your software, you might not get exactly these two vectors, but rather two vectors proportional to $\hat{\gamma}_1$ and $\hat{\gamma}_2$. However, you should get the same eigenvalues $\hat{\lambda}_1 = 32.192$ and $\hat{\lambda}_2 = 0.285$.
    (b) Compute the mean vector and the covariance matrix of the two canonical discriminant functions (Z_1, Z_2) in all three groups. Compute also $\bar{\mathbf{z}}_j = \Gamma'\bar{\mathbf{y}}_j$ and $\Gamma'\hat{\boldsymbol{\psi}}_j\Gamma$ for $j = 1, 2, 3$, and verify numerically that the three mean vectors $\bar{\mathbf{z}}_j$ are identical in their third and fourth components.
    (c) Either from Figure 7.3.1 or by calculation, the plug-in error rate for equal prior probabilities is $3/150 = 2$ percent. Calculate a leave-one-out error rate (similarly to Example 7.5.3), and comment on the ability of the two canonical discriminant functions to identify the three species of Iris flowers.

14. This exercise concerns the computation of error rates in Example 7.3.2.
    (a) Construct a "confusion matrix" as follows: The three rows of the matrix correspond to actual group membership, and the three columns to predicted group membership according to the classification rule of Example 7.3.2 (assignment to the closest of the means in Euclidean distance, based on the two canonical discriminant functions). From this matrix, compute a plug-in estimate of the actual error rate.
    (b) Repeat part (a), but use a leave-one-out procedure, as introduced in Section 5.5.

15. Prove equation (51).

16. This exercise expands on Example 7.3.3 (wine data).

(a) Compute the within-groups covariance matrix $\hat{\psi}$, the between-groups covariance matrix $\hat{\psi}_B$, and a matrix Γ that diagonalizes both matrices simultaneously. For the purpose of verifying your numerical results, the two columns of Γ associated with the nonzero roots $\hat{\lambda}_1$ and $\hat{\lambda}_2$ should be

$$\hat{\gamma}_1' = (0.0293,\ 0.1276,\ 0.0359,\ 0.0422,\ -0.0172,\ 0.0152,\ -0.1447,\ 1.1328,$$
$$-0.0742,\ -0.8768,\ -0.0312,\ 0.0655,\ -0.3110,\ -0.3609,\ 0.3953),$$

and

$$\hat{\gamma}_2' = (-0.0006,\ 0.2185,\ 0.0360,\ -0.0272,\ -0.0421,\ 0.0486,\ -0.0320,\ 0.4725,$$
$$0.3073,\ -0.4115,\ 0.0288,\ -0.2171,\ -0.1255,\ -0.2459,\ 0.1083).$$

Note: Depending on the normalization of eigenvectors used by your software, you might not get these two vectors, but rather two vectors proportional to $\hat{\gamma}_1$ and $\hat{\gamma}_2$. However, you should get the same eigenvalues $\hat{\lambda}_1 = 10.498$ and $\hat{\lambda}_2 = 3.299$.

(b) Repeat the analysis of the wine data using only the four variables labeled Y_{12} to Y_{15}. Graph the two canonical discriminant functions similarly to Figure 7.3.2, and compute posterior probabilities for all 26 observations. Finally, perform a leave-one-out analysis similar to the one presented in Table 7.3.3. (The author got three misclassifications in this analysis, using equal prior probabilities in the computation of the classification functions).

(c) Why is it not possible to find a normal theory quadratic classification rule (as in equations (14) to (16) of Section 7.2) for this example? If you cannot think of the answer right away, try it out.

17. This exercise concerns the computation of standard distances between groups. Assume that $\mathbf{Y}$ has a p-variate distribution with mean vector $\boldsymbol{\mu}_j$ and (common) covariance matrix ψ in groups $j = 1, \ldots, k$. Let

$$\delta_{jh} = \left[(\boldsymbol{\mu}_j - \boldsymbol{\mu}_h)'\psi^{-1}(\boldsymbol{\mu}_j - \boldsymbol{\mu}_h)\right]^{1/2}, \quad j, h = 1, \ldots, k, \tag{54}$$

be the standard distance between groups j and h, and $\boldsymbol{\Delta} = (\delta_{jh})$ the $k \times k$ matrix of pairwise standard distances.

(a) Show that the standard distance δ_{jh} is the Euclidean distance between $\boldsymbol{\nu}_j^{(1)}$ and $\boldsymbol{\nu}_h^{(1)}$ of equation (34).

(b) Define sample standard distances between groups by replacing parameters in (54) with plug-in estimates. How do the standard distances change if the unbiased (pooled)-sample covariance matrix $\mathbf{S}$ of equation (53) is used instead of the plug-in estimate $\hat{\psi}$?

(c) Compute standard distances between all three groups of Iris flowers, using part (a).

(d) Using only the first two groups of Iris flowers (*Setosa, Versicolor*) and the methodology of Section 5.3, compute a standard distance between the two groups. Explain why the numerical result is not the same as the one obtained in part (c).

7.4 Multivariate Analysis of Variance

Multivariate analysis of variance, or MANOVA, for short, is treated extensively in most multivariate books. The current section gives a rather limited introduction to this topic, presenting only the case of one-way MANOVA and omitting most of the distributional results with which the multivariate literature abounds. An easy review, including tables of critical values, is given in Rencher (1995). We will focus on one-way MANOVA as an overall test of significance associated with canonical discriminant functions. If the k mean vectors $\boldsymbol{\mu}_1, \ldots, \boldsymbol{\mu}_k$ are identical, then all differences in location found by the canonical discriminant functions are purely random, and linear discrimination would then not be worthwhile.

We will work with the same setup and terminology as in Section 7.3, where we have seen that the between-group covariance matrix ψ_B is always positive semidefinite, and is zero exactly if all $\boldsymbol{\mu}_j$ are identical. Moreover, by equation (11) of Section 7.3, $\psi_B = \mathbf{O}$ exactly if all λ_h are zero. In the coordinate system given by the canonical discriminant functions, the λ_h are ratios of between-group variance to within-group variance for m uncorrelated (and in a normal theory setup, independent) variables. Because univariate ANOVA tests are based on ratios of between-group variability to within-group variability, we might attempt to construct a test for equality of all $\boldsymbol{\mu}_j$ by somehow combining the $\hat{\lambda}_h$ into a single test statistic. Intuition tells us that a reasonable test should reject the hypothesis if some $\hat{\lambda}_h$ are large, but what exactly the shape of the rejection region should be is not obvious.

Before we proceed to the construction of test statistics, some algebra will be helpful.

Lemma 7.4.1 *Let ψ be a positive definite symmetric matrix of dimension $p \times p$, ψ_B a symmetric matrix of the same dimension, and set $\bar{\psi} = \psi + \psi_B$ as in equation (5) of Section 7.3.*

(i) *The matrices $\psi^{-1}\psi_B$, $\psi^{-1}\bar{\psi}$, and $\bar{\psi}^{-1}\psi_B$ have the same p real eigenvectors.*

(ii) *Let the eigenvalues of $\psi^{-1}\psi_B$ be $\lambda_1, \ldots, \lambda_p$. Then the eigenvalues of $\psi^{-1}\bar{\psi}$ are $1 + \lambda_h$, $h = 1, \ldots, p$, and the eigenvalues of $\bar{\psi}^{-1}\psi_B$ are $\theta_h = \lambda_h/(1 + \lambda_h)$, $h = 1, \ldots, p$.*

Proof Let $\psi = \mathbf{H}\mathbf{H}'$, $\psi_B = \mathbf{H}\boldsymbol{\Lambda}\mathbf{H}'$ be a simultaneous decomposition of ψ and ψ_B, as given in Appendix A.9. Here, $\boldsymbol{\Lambda}$ is diagonal, with diagonal entries λ_h. Set $\boldsymbol{\Gamma} = (\mathbf{H}')^{-1}$, and let $\gamma_1, \ldots, \gamma_p$ be the columns of $\boldsymbol{\Gamma}$. Then

$$\begin{aligned} \psi^{-1}\psi_B\boldsymbol{\Gamma} &= (\mathbf{H}\mathbf{H}')^{-1}\mathbf{H}\boldsymbol{\Lambda}\mathbf{H}'\boldsymbol{\Gamma} \\ &= \boldsymbol{\Gamma}\boldsymbol{\Gamma}'\mathbf{H}\boldsymbol{\Lambda}\mathbf{H}'\boldsymbol{\Gamma} \qquad (1) \\ &= \boldsymbol{\Gamma}\boldsymbol{\Lambda}. \end{aligned}$$

The jth column of equation (1) reads

$$\psi^{-1}\psi_B\gamma_h = \lambda_h\gamma_h\,, \tag{2}$$

that is, γ_h is an eigenvector of $\psi^{-1}\psi_B$, with associated eigenvalue λ_h.

Since $\bar{\psi} = \psi + \psi_B = \mathbf{H}(\mathbf{I}_p + \Lambda)\mathbf{H}'$, it follows similarly to (1) that

$$\psi^{-1}\bar{\psi}\Gamma = \Gamma(\mathbf{I}_p + \Lambda), \tag{3}$$

i.e., $\psi^{-1}\psi_B\gamma_h = (1 + \lambda_h)\gamma_h$ for $h = 1, \ldots, p$. Finally,

$$\bar{\psi}^{-1}\psi_B\Gamma = \Gamma(\mathbf{I}_p + \Lambda)^{-1}\Lambda. \tag{4}$$

Since $(\mathbf{I}_p + \Lambda)^{-1}\Lambda$ is diagonal with diagonal entries $\theta_h = \lambda_h/(1 + \lambda_h)$, it follows that $\bar{\psi}^{-1}\psi_B$ has eigenvectors γ_h and associated eigenvalues θ_h, $h = 1, \ldots, p$. ∎

We now proceed to the likelihood ratio test for equality of the means of k p-variate normal distributions, assuming equality of all covariance matrices. Let $\mathbf{y}_{j1}, \ldots, \mathbf{y}_{jN_j}$, $j = 1, \ldots, k$, be independent samples from multivariate normal distributions $\mathcal{N}_p(\boldsymbol{\mu}_j, \psi)$, respectively. The parameter space is

$$\Omega = \left\{(\boldsymbol{\mu}_1, \ldots, \boldsymbol{\mu}_k, \psi) : \text{ all } \boldsymbol{\mu}_j \in \mathbb{R}^p,\ \psi \text{ positive definite symmetric}\right\}, \tag{5}$$

and the hypothesis $\boldsymbol{\mu}_1 = \cdots = \boldsymbol{\mu}_k$ imposed on the means reduces Ω to

$$\omega = \{(\boldsymbol{\mu}_1, \ldots, \boldsymbol{\mu}_k, \psi) : \boldsymbol{\mu}_1 = \cdots = \boldsymbol{\mu}_k \in \mathbb{R}^p,\ \psi \text{ positive definite symmetric}\}. \tag{6}$$

The joint log-likelihood function of the k samples, by equation (18) of Section 4.3, is given by

$$\ell(\boldsymbol{\mu}_1, \ldots, \boldsymbol{\mu}_k, \psi) = -\frac{1}{2}\sum_{j=1}^{k} N_j\left[p\log(2\pi) + \log\det\psi + (\bar{\mathbf{y}}_j - \boldsymbol{\mu}_j)'\psi^{-1}(\bar{\mathbf{y}}_j - \boldsymbol{\mu}_j) + \operatorname{tr}(\psi^{-1}\mathbf{A}_j)\right], \tag{7}$$

where $\bar{\mathbf{y}}_j = \sum_{i=1}^{N_j}\mathbf{y}_{ji}$ and $\mathbf{A}_j = \sum_{i=1}^{N_j}(\mathbf{y}_{ji} - \bar{\mathbf{y}}_j)(\mathbf{y}_{ji} - \bar{\mathbf{y}}_j)'/N_j$. Let $N = \sum_{j=1}^{k} N_j$. Assuming that $N \geq p + k$ (why?), Exercise 12 of Section 4.3 shows that the log-likelihood function takes a unique maximum in Ω at

$$\hat{\boldsymbol{\mu}}_j = \bar{\mathbf{y}}_j, \quad j = 1, \ldots, k,$$
$$\hat{\psi} = \sum_{j=1}^{k} N_j\mathbf{A}_j/N. \tag{8}$$

The maximum of the log-likelihood function in Ω is therefore given by

$$\ell(\bar{\mathbf{y}}_1, \ldots, \bar{\mathbf{y}}_k, \hat{\psi}) = -\frac{1}{2}\sum_{j=1}^{k} N_j\left[p\log(2\pi) + \log\det\hat{\psi} + p\right]. \tag{9}$$

In the constrained parameter space ω, the N observations form a sample from a single, p-variate normal distribution. Writing $\boldsymbol{\mu}$ for the common mean, the maximum is therefore taken at

$$\tilde{\mu} = \bar{\mathbf{y}} = \frac{1}{N}\sum_{j=1}^{k} N_j \bar{\mathbf{y}}_j ,$$
$$\tilde{\psi} = \frac{1}{N}\sum_{j=1}^{k}\sum_{i=1}^{N_j} (\mathbf{y}_{ji} - \bar{\mathbf{y}})(\mathbf{y}_{ji} - \bar{\mathbf{y}})'. \tag{10}$$

Notice that $\bar{\mathbf{y}}$ and $\tilde{\psi}$ are the plug-in estimators of the mean vector and the covariance matrix of the mixture distribution, as explained in Section 7.3. Thus we will now return to the notation of Section 7.3, writing $\hat{\bar{\psi}}$, instead of $\tilde{\psi}$, and using $\hat{\psi}_B = \sum_{j=1}^{k} N_j(\bar{\mathbf{y}}_j - \bar{\mathbf{y}})(\bar{\mathbf{y}}_j - \bar{\mathbf{y}})'/N$ as the estimate of the between-group covariance matrix, as in equation (41) of Section 7.3. The maximum of the log-likelihood function in ω is given by

$$\ell(\bar{\mathbf{y}}, \ldots, \bar{\mathbf{y}}, \hat{\bar{\psi}}) = -\frac{1}{2}\sum_{j=1}^{k} N_j \left[p \log(2\pi) + \log\det\hat{\bar{\psi}} + p\right]. \tag{11}$$

Thus, the log-likelihood ratio statistic for the hypothesis of a common mean vector is

$$\begin{aligned} \text{LLRS} &= 2\left[\ell(\bar{\mathbf{y}}_1, \ldots, \bar{\mathbf{y}}_k, \hat{\psi}) - \ell(\bar{\mathbf{y}}, \ldots, \bar{\mathbf{y}}, \hat{\bar{\psi}})\right] \\ &= N \log \frac{\det\hat{\bar{\psi}}}{\det\hat{\psi}} \\ &= N \log\det\left(\hat{\psi}^{-1}\hat{\bar{\psi}}\right); \end{aligned} \tag{12}$$

see Exercise 1. By Lemma 7.4.1 and Exercise 2 and writing $\hat{\lambda}_h$ for the eigenvalues of $\hat{\psi}^{-1}\hat{\psi}_B$, we obtain

$$\text{LLRS} = N \log\left[\prod_{h=1}^{p}(1+\hat{\lambda}_h)\right] = N\sum_{h=1}^{p}\log(1+\hat{\lambda}_h) = N\sum_{h=1}^{m}\log(1+\hat{\lambda}_h)\,, \tag{13}$$

where m is the rank of $\hat{\psi}_B$. The last equality in (13) follows from the usual convention of ordering the eigenvalues as $\hat{\lambda}_1 \geq \hat{\lambda}_2 \geq \ldots$, i.e., the last $(p-m)$ eigenvalues are $\hat{\lambda}_{m+1} = \cdots = \hat{\lambda}_p = 0$. From the large-sample theory for likelihood ratio tests (see Section 4.3), it follows that the distribution of LLRS under the hypothesis of equality of all mean vectors is asymptotically chi-square with $(k-1)p$ degrees of freedom. However, this result is hardly ever used because tables of the exact distribution of an equivalent statistic, the so-called Wilks' Lambda, are available. We will return to this point later. For current purposes, we notice that the log-likelihood ratio statistic depends on the data through the eigenvalues $\hat{\lambda}_h$, as conjectured earlier.

Example 7.4.1 continuation of Example 7.3.2

Three species of Iris flowers. There are two nonzero eigenvalues, $\hat{\lambda}_1 = 32.192$ and $\hat{\lambda}_2 = 0.285$, yielding LLRS = 563.01. With $k = 3$ and $p = 4$, the number of degrees of freedom of the asymptotic chi-square distribution is 8, and therefore the hypothesis of equality of all mean vectors is rejected at any reasonable level of significance. In fact, in view of Figure 7.3.1, a formal test of significance is not necessary in this example. □

Four test statistics commonly used in MANOVA are all based on the m nonzero eigenvalues $\hat{\lambda}_1, \ldots, \hat{\lambda}_m$ of $\hat{\psi}^{-1}\hat{\psi}_B$. The four statistics are Wilks' Lambda (which we will call W), the Lawley-Hotelling trace (LH), Pillai's trace (P), and Roy's largest root (R) defined as follows:

$$W = \prod_{h=1}^{m} \frac{1}{1 + \hat{\lambda}_h},$$

$$LH = \sum_{h=1}^{m} \hat{\lambda}_h,$$

$$P = \sum_{h=1}^{m} \frac{\hat{\lambda}_h}{1 + \hat{\lambda}_h}, \tag{14}$$

and

$$R = \frac{\hat{\lambda}_1}{1 + \hat{\lambda}_1}.$$

In the definition of these four statistics, it is assumed that the eigenvalues are ordered as $\hat{\lambda}_1 \geq \cdots \geq \hat{\lambda}_m > 0$, and the remaining eigenvalues are all zero. Thus, the sums or products in (14) can be taken to go from 1 to p. For tests based on LH, P, R, and on the log-likelihood ratio statistic, the hypothesis is rejected if the value of the statistic exceeds some critical value, whereas for Wilks' statistic, the rejection region is an interval from 0 to some critical value.

The first of these four statistics, Wilks' Lambda, is, perhaps, the most popular. By equation (13), $W = \exp\left(-\frac{1}{N}\text{LLRS}\right)$, or, equivalently, $\text{LLRS} = -N \log W$. Since W is a monotonically decreasing function of the log-likelihood ratio statistic, the rejection region for a test of level α consists of all values smaller than a critical value w_α. Tables of quantiles of the exact normal theory distribution of W are available in many multivariate books, e.g., in Seber (1984) or Rencher (1995), and are not reproduced here. In special cases, there exists a transformation of Wilks' statistic to an F-statistic as follows. (Here, $X \sim F(\text{df}_1, \text{df}_2)$ means that X follows an F-distribution with df_1 degrees of freedom in the numerator and df_2 degrees of freedom in the denominator):

- If $p = 1$, then

$$\frac{N-k}{k-1} \cdot \frac{1-W}{W} \sim F(k-1,\ N-k). \tag{15}$$

- If $p = 2$, then

$$\frac{N-k-1}{k-1} \cdot \frac{1-\sqrt{W}}{\sqrt{W}} \sim F\left[2(k-1),\ 2(N-k-1)\right]. \tag{16}$$

- If $k = 2$, then

$$\frac{N-p-1}{p} \cdot \frac{1-W}{W} \sim F(p,\ N-p-1). \tag{17}$$

- If $k = 3$, then

$$\frac{N-p-2}{p} \cdot \frac{1-\sqrt{W}}{\sqrt{W}} \sim F\left[2p,\ 2(N-p-2)\right]. \tag{18}$$

Equation (15) refers to the univariate case $p = 1$, in which MANOVA reduces to ordinary ANOVA; see Exercise 4. Result (17) follows from the equivalence of Hotelling's T^2 and Wilks' statistic in the case of $k = 2$ groups; see Exercise 5. Results (16) and (18) are more difficult to prove; see Anderson (1984, Section 8.4). The transformation to an F-statistic is often useful because tables of the F-distribution are widely available, and many software packages have built-in functions to evaluate the distribution function of F-variates. Instead of transforming W into an F-variable as in equations (15) to (18), the same formulas may be used to transform upper quantiles of F to lower quantiles of W. For values of $p > 2$ or $k > 3$, which are not covered by the above four cases, the reader is referred to tables in books such as Seber (1984, Appendix D13) or Rencher (1995, Table A.9). Some software may also calculate an approximate F-statistic and corresponding p-values.

The Lawley-Hotelling trace (LH) is the sum of the eigenvalues of $\hat{\psi}^{-1}\hat{\psi}_B$ and seems rather natural in view of our derivation of canonical discriminant functions of Section 7.3. Pillai's trace (P) is the sum of the eigenvalues $\hat{\theta}_h$ of $\hat{\bar{\psi}}^{-1}\hat{\psi}_B$; see Lemma 7.4.1. Each $\hat{\theta}_h$ is the ratio of variance between groups to variance of the mixture for a canonical discriminant function. Finally, Roy's largest root statistic (R) is the largest eigenvalue of $\hat{\bar{\psi}}^{-1}\hat{\psi}_B$. Roy's criterion follows naturally from the union-intersection approach to MANOVA; see Exercise 17 in Section 6.6. The justification for Roy's largest root is that the hypothesis $\mu_1 = \cdots = \mu_k$ holds true exactly if the largest eigenvalue θ_1 of $\bar{\psi}^{-1}\psi_B$ is zero. The derivation of the union-intersection statistic R parallels the derivation of the first canonical discriminant function given in Section 7.3. Instead of $\hat{\theta}_1$, we might equivalently use $\hat{\lambda}_1$, but tables of the largest root distribution are commonly given for $\hat{\theta}_1$.

There is an extensive literature on the advantages and disadvantages of the four statistics in terms of their power to detect various types of deviations from the hypothesis of equality of all mean vectors, and robustness against violations of the assumptions of multivariate normality and equality of the within-group covariance matrices. For instance, Roy's largest root statistic has the highest power among the four statistics if the k mean vectors are almost on a straight line. For summaries of the relevant results, see again the books by Seber (1984) and Rencher (1995). We will discuss a few aspects in Example 7.4.3. In the case of $k = 2$ groups or dimension $p = 1$, all four statistics are equivalent, and, in turn, equivalent to Hotelling's T^2 statistic; see Exercise 5.

Example 7.4.2 continuation of Example 7.3.3

Wines from three countries. With eigenvalues $\hat{\lambda}_1 = 10.498$, $\hat{\lambda}_2 = 3.299$, and $N = 26$ we obtain LLRS $= 101.4$ and $W = 0.0202$. For a test using the asymptotic chi–square approximation to the distribution of the log-likelihood ratio statistic, we need quantiles $c_{1-\alpha}$ of the chi-square distribution with $(k-1)p = 30$ degrees of freedom; for $\alpha = 0.05$ and $\alpha = 0.01$, these are $c_{0.95} = 43.77$ and $c_{0.99} = 50.89$. Using the transformation $W = \exp(-\text{LLRS}/N)$, we obtain approximate lower quantiles of the distribution of W: $w_{0.05} \approx 0.1857$, and $w_{0.01} \approx 0.1412$. Thus, the hypothesis of equality of the three mean vectors would clearly be rejected. However, a sample size of $N = 26$ is very small compared to the number of variables ($p = 15$) used in this example, and asymptotic theory should therefore be avoided. By equation (17), we can use critical values $f_{1-\alpha}$ of the $F(30, 26)$-distribution, $f_{0.95} = 1.901$ and $f_{0.99} = 2.503$. The corresponding lower quantiles of the distribution of W are $w_{0.05} = 0.0576$ and $w_{0.01} = 0.0374$; see Exercise 6. Using the exact distribution theory, we still reject the hypothesis at the usual levels of significance but not as clearly. □

Example 7.4.2 illustrates the danger of the uncritical use of large-sample approximation rather dramatically. But can we really trust the "exact" distribution theory? For 15-dimensional data and $N = 26$ observations, it is practically impossible to assess the assumptions of multivariate normality and equality of the covariance matrices. Therefore, we might prefer to use a randomization technique as outlined in Section 6.7. For samples of size $N_1, \ldots, N_k$, $\sum_{j=1}^{k} N_j = N$, there are

$$\binom{N}{N_1 \quad N_2 \quad \cdots \quad N_k} = \frac{N!}{N_1!\, N_2!\, \cdots\, N_k!} \tag{19}$$

different ways to partition the N observations into groups of size $N_1, \ldots, N_k$. For each partition, we calculate the value of our statistic of interest, say "V." Since (19) is typically a very large number and computing the value of V for all partitions might exceed your lifetime, we will generate a large number M of random partitions and thus calculate an approximate randomization distribution. If V_{obs} denotes the value of the statistic V calculated for the original k samples, the achieved significance level (ASL) of the randomization procedure is estimated by

$$\widehat{\text{ASL}}_{\text{rand}} = \frac{\#\text{ of random partitions with value of } V \geq V_{\text{obs}}}{M} . \tag{20}$$

Also, recall also that the randomization test is not really a new test; it is nothing but a calculation of the null distribution of a test statistic using the randomization approach rather than distributional calculus. See Section 6.7 for further discussion.

Example 7.4.3 continuation of Example 7.4.2

Wines from three countries. We generated $M = 1000$ random partitions of the wine data into samples of size $N_1 = 9$, $N_2 = 7$, and $N_3 = 10$, using all 15 variables, and thus generated an approximate joint randomization distribution of the two eigenvalues $\hat{\lambda}_1$ and $\hat{\lambda}_2$. For each of the four test statistics, the achieved significance level was estimated using equation (20). Results are as follows:

- Wilks: $W_{\text{obs}} = 0.0202$, $\widehat{\text{ASL}}_{\text{rand}} = 0.002$.
- Lawley-Hotelling: $LH_{\text{obs}} = 13.796$, $\widehat{\text{ASL}}_{\text{rand}} = 0.006$.
- Pillai: $P_{\text{obs}} = 1.6804$, $\widehat{\text{ASL}}_{\text{rand}} = 0.001$.
- Roy: $R_{\text{obs}} = 0.9130$, $\widehat{\text{ASL}}_{\text{rand}} = 0.014$.

Thus Pillai's statistic would reject the hypothesis most strongly, whereas Roy's statistic gives an achieved significance level of 1.4 percent.

We can use this example to illustrate how the four test statistics give different weight to the eigenvalues $\hat{\lambda}_h$ in constructing the rejection region. Figure 7.4.1 shows a scatterplot of the 1000 joint values $(\hat{\lambda}_{1i}, \hat{\lambda}_{2i})$, $i = 1, \ldots, 1000$, of the randomization distribution, along with the observed value (10.498, 3.299). The 95th percentile of the randomization distribution of W is $w_{0.95} = 0.0537$. The curve marked "W" in Figure 7.4.1 is the set of all $(\hat{\lambda}_1, \hat{\lambda}_2)$ such that

$$\frac{1}{(1+\hat{\lambda}_1)} \cdot \frac{1}{(1+\hat{\lambda}_2)} = w_{0.95} .$$

All points above this curve form the rejection region for the randomization test. By construction, exactly 50 points may be counted in this region, whereas 949 are in the acceptance region and one is on the curve. Similar curves can be constructed for the other three statistics. The rejection region of the Lawley-Hotelling statistic is bounded by a straight line because LH is the sum of the two eigenvalues. Similarly, the rejection region for Roy's largest root test is a vertical line because R uses only the largest eigenvalue. Notice that the shape of these regions would be the same if we use parametric theory for the computation of the cutoff value, that is, the shape

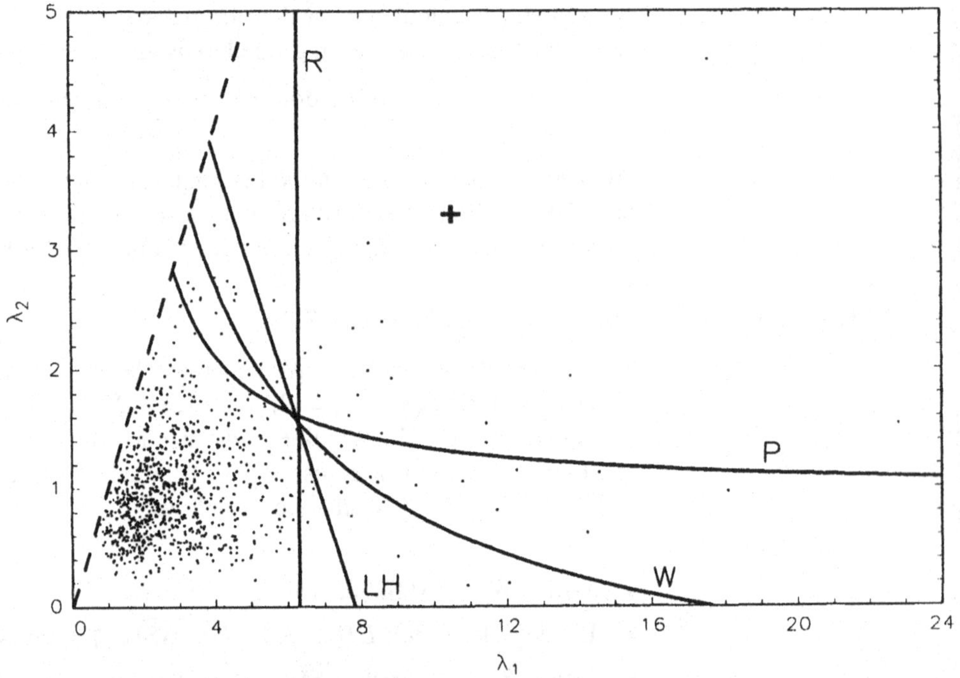

Figure 7-4-1 Joint randomization distribution of the two eigenvalues $\hat{\lambda}_1$ and $\hat{\lambda}_2$ in the wine example with $M = 1000$ randomizations. The solid plus sign represents the observed value $(\hat{\lambda}_1, \hat{\lambda}_2) = (10.498, 3.299)$. The dashed line corresponds to $\hat{\lambda}_1 = \hat{\lambda}_2$; there are no dots above this line because the eigenvalues are ordered. The four curves mark the border of the rejection region for a test of level $\alpha = 0.05$ using the test statistics W, LH, P, and R, respectively.

of the rejection regions reflects the construction of the test statistic and is in no way tied to the randomization test.

Figure 7.4.1 shows that the four statistics use the information contained in the eigenvalues $\hat{\lambda}_h$ quite differently. Whereas Roy's largest root statistic ignores all but the largest eigenvalue, Pillai's statistic gives considerable weight to the second (and in general, the second to mth) eigenvalue. Figure 7.4.1 is not a graph that one would usually want to construct in a practical application of MANOVA, but it illustrates the differences among the four statistics rather nicely. □

In this section, we have treated MANOVA only for the special case of a one-way analysis, i.e., as a test for equality of the means of k distributions. To the reader who is familiar with experimental design and with ANOVA, it will be clear that, for any design used for a single response variable, such as a two-way factorial design with or without interactions, a similar analysis can be done on multivariate response data. Most multivariate books give reasonably comprehensive reviews; see e.g., Seber (1984), Rencher (1995), or Hand and Taylor (1987).

Exercises for Section 7.4

1. Verify equations (9), (11), and (12).

2. In the setup of Lemma 7.4.1, show that $\det(\psi^{-1}\bar{\psi}) = \prod_{h=1}^{p}(1+\lambda_h)$. *Note*: The matrix $\psi^{-1}\bar{\psi}$ is in general not symmetric.

3. In Example 7.4.3 (Iris data), compute exact normal theory quantiles of Wilks' statistic W for tests of levels $\alpha = 0.05$ and $\alpha = 0.01$, using the transformation (18). Are the conclusions the same as if the asymptotic chi-square approximation is used?

4. In univariate ANOVA, the F-statistic for equality of k means is defined as

$$F = \frac{\text{SS(between)}/(k-1)}{\text{SS(within)}/(N-k)},$$

where $\text{SS(between)} = \sum_{j=1}^{k} N_j(\bar{y}_j - \bar{y})^2$ and $\text{SS(within)} = \sum_{j=1}^{k}\sum_{i=1}^{N_j}(y_{ji} - \bar{y}_j)^2$. How is this related to Wilks' statistic if $p = 1$?

5. If the number of groups in a one-way MANOVA problem is $k = 2$, then the number of nonzero eigenvalues is $m = 1$. In this case all four statistics (14) are equivalent and, in turn, equivalent to the two-sample T^2-statistic of Section 6.4. Find explicit formulas for T^2 and for the multivariate standard distance D as functions of W, LH, P, and R.

6. This exercise expands on the wine example.
   (a) Using the exact normal distribution theory, find the critical value of W for a test of significance level $\alpha = 0.05$. You may use the F-transformation of equation (18). How different is this critical value from the cutoff found by the randomization procedure?
   (b) On a photocopy of Figure 7.4.1, graph the curve corresponding to the normal theory cutoff found in part (a).

7. In the setup of the likelihood ratio test for equality of the mean vectors of k multivariate normal distributions with identical covariance matrices, assuming that ψ is positive definite symmetric and $N \geq p + k$, show that

$$\Pr[\text{rank}(\hat{\psi}_B) = \min(p,\ k-1)] = 1.$$

Hint: See Exercise 10 in Section 4.2.

8. Table 7.4.1 gives data on apple trees of different rootstocks for $k = 6$ rootstocks and $p = 4$ variables. We will be interested in comparing the six groups.
   (a) Test the hypothesis of equality of the mean vectors of all $k = 6$ groups using Wilks' statistic and a significance level of $\alpha = 0.05$. (The exact normal theory critical value of W is 0.455).
   (b) Repeat (a) using the randomization test. Is the conclusion the same?
   (c) Use all four canonical discriminant functions to find the plug-in error rate of the normal theory classification procedure with equal prior probabilities.

Table 7.4.1 Rootstock data from Andrews and Herzberg 1985, pp. 357-360. The table display variables measured on apple trees of six different rootstocks. For each rootstock, there are eight trees. Variables are Y_1 = trunk girth at 4 years, in units of 10 cm; Y_2 = extension growth at 4 years, in m; Y_3 = trunk girth at 15 years, in units of 10 cm; Y_4 = weight of tree above ground at 15 years, in units of 1000 pounds.

Rootstock	Y_1	Y_2	Y_3	Y_4
1	1.11	2.569	3.58	0.760
1	1.19	2.928	3.75	0.821
1	1.09	2.865	3.93	0.928
1	1.25	3.844	3.94	1.009
1	1.11	3.027	3.60	0.766
1	1.08	2.336	3.51	0.726
1	1.11	3.211	3.98	1.209
1	1.16	3.037	3.62	0.750
2	1.05	2.074	4.09	1.036
2	1.17	2.885	4.06	1.094
2	1.11	3.378	4.87	1.635
2	1.25	3.906	4.98	1.517
2	1.17	2.782	4.38	1.197
2	1.15	3.018	4.65	1.244
2	1.17	3.383	4.69	1.495
2	1.19	3.447	4.40	1.026
3	1.07	2.505	3.76	0.912
3	0.99	2.315	4.44	1.398
3	1.06	2.667	4.38	1.197
3	1.02	2.390	4.67	1.613
3	1.15	3.021	4.48	1.476
3	1.20	3.085	4.78	1.571
3	1.20	3.308	4.57	1.506
3	1.17	3.231	4.56	1.458
4	1.22	2.838	3.89	0.944
4	1.03	2.351	4.05	1.241
4	1.14	3.001	4.05	1.023
4	1.01	2.439	3.92	1.067
4	0.99	2.199	3.27	0.693
4	1.11	3.318	3.95	1.085
4	1.20	3.601	4.27	1.242
4	1.08	3.291	3.85	1.017
5	0.91	1.532	4.04	1.084
5	1.15	2.552	4.16	1.151
5	1.14	3.083	4.79	1.381
5	1.05	2.330	4.42	1.242
5	0.99	2.079	3.47	0.673
5	1.22	3.366	4.41	1.137
5	1.05	2.416	4.64	1.455

Table 7.4.1 *Continued*

Rootstock	Y_1	Y_2	Y_3	Y_4
5	1.13	3.100	4.57	1.325
6	1.11	2.813	3.76	0.800
6	0.75	0.840	3.14	0.606
6	1.05	2.199	3.75	0.790
6	1.02	2.132	3.99	0.853
6	1.05	1.949	3.34	0.610
6	1.07	2.251	3.21	0.562
6	1.13	3.064	3.63	0.707
6	1.11	2.469	3.95	0.952

(d) Use a leave-one-out procedure to estimate the expected actual error rate of the normal theory classification procedure.

(e) Graph the data of the first two canonical discriminant functions using different symbols for trees from different rootstocks. Comment on the ability of the classification rule to correctly identify the six different rootstocks. Is it worth constructing graphs that involve the third and/or fourth canonical discriminant function? Why or why not?

9. Show that all four MANOVA statistics (14) are invariant under nonsingular linear transformations of the data. That is, show the following: If the data $\mathbf{y}_{ji}$ are transformed to $\mathbf{y}^*_{ji} = \mathbf{A}\mathbf{y}_{ji}$, where $\mathbf{A}$ is an arbitrary nonsingular matrix of dimension $p \times p$, then the values of the test statistics are the same whether the original data $\mathbf{y}_{ji}$ or the transformed data $\mathbf{y}^*_{ji}$ are used. *Hint*: Show invariance of the eigenvalues $\hat{\lambda}_h$.

7.5 Simple Logistic Regression

All classification methods studied so far were based on the joint distribution of a group indicator variable X and a vector $\mathbf{Y}$ of p variables whose conditional distribution, given $X = j$, is assumed to follow some *pdf* $f_j(\mathbf{y})$. The values taken by X may themselves be random (if the investigator samples from the mixture distribution), or they may be fixed (if sampling is from the conditional distributions of $\mathbf{Y}$, given $X = j$). Whether X is random or fixed, a crucial assumption has always been that $\mathbf{Y}$ is random. However, in many practical situations the experimenter may be able to choose a value of $\mathbf{Y}$, and then observe the "group indicator" X. In such cases, we will be interested in modeling the conditional distribution of X, given $\mathbf{Y} = \mathbf{y}$, without making explicit distributional assumptions on $\mathbf{Y}$. Some preliminary discussion of this has already been given in the text following Example 5.4.3. An introductory example may help further to clarify things.

Example 7.5.1 Basketball. On April 18, 1996, the author and his sons visited Bryan Park in Bloomington, Indiana, to perform an experiment. From distances of one to twenty feet from the hoop, each participant shot the ball repeatedly, aiming carefully, and recorded hit or miss. Table 7.5.1 reports a part of Steve's data: for distances of 1, 2, ..., 20 feet, success or failure in the first shot out of six shots is given. Setting Y = distance from the hoop, $X = 1$ for hit, and $X = 0$ for miss, the data consists of 20 pairs (y_i, x_i). A discriminant analysis approach would split the data into two groups according to the value of X and estimate the conditional distributions of Y, given $X = 0$ and $X = 1$. This would not be very meaningful because the distances were chosen systematically by the experimenter. A better approach will attempt to model the conditional probabilities

$$\Pr[X = 1 | Y = y] = \Pr[\text{hit}|\text{distance}]$$

directly as a function of y, without any distributional assumptions on Y. Figure 7.5.1 shows Steve's 20 data points with Y = distance on the horizontal axis, and X = hit or miss on the vertical axis. In anticipation of our discussion of the logistic regression model, Figure 7.5.1 also shows a smooth curve that estimates the success probability as a function of distance. Without knowing the details of the model or the estimation process at this time, notice that the curve is in no way meant to be an approximation to the data points. Rather, it gives the (estimated) value of a parameter as a function of distance. □

As illustrated by the preceding example, we will now take a reverse view to discriminant analysis: what was formerly the group indicator (X) will now be the dependent variable in a regression setup, and the former response variable Y will now be the regressor, whose values can ideally be chosen by the investigator. Following the convention used throughout most of the statistical literature, we will therefore change notation: The dependent variable will be denoted by Y, and the regressor variable by X, or by $\mathbf{X}$ if there are several regressor variables. We will study almost exclusively the case where Y can take only two values, typically labeled "success" and "failure" and represented by the numerical values 1 and 0, respectively. That is, Y will be a Bernoulli random variable whose success probability π depends on the value of a regressor X.

Now, suppose that we perform N experiments. In the ith experiment, we choose $X = x_i$ and observe $Y_i = 0$ or 1. Then the *pdf* of Y_i is given by

$$f_i(y_i) = \pi_i^{y_i}(1 - \pi_i)^{1-y_i} \quad \text{for } y_i = 0, 1,$$

Table 7.5.1 Steve's basketball data. Steve shot the ball six times from each of the distances 1 to 20 feet. Here, only the result of the first attempt from each distance is given.

Distance:	1	2	3	4	5	6	7	8	9	10
Hit or Miss:	H	H	H	H	H	M	H	M	M	M
Distance:	11	12	13	14	15	16	17	18	19	20
Hit or Miss:	M	M	M	H	H	M	M	M	M	M

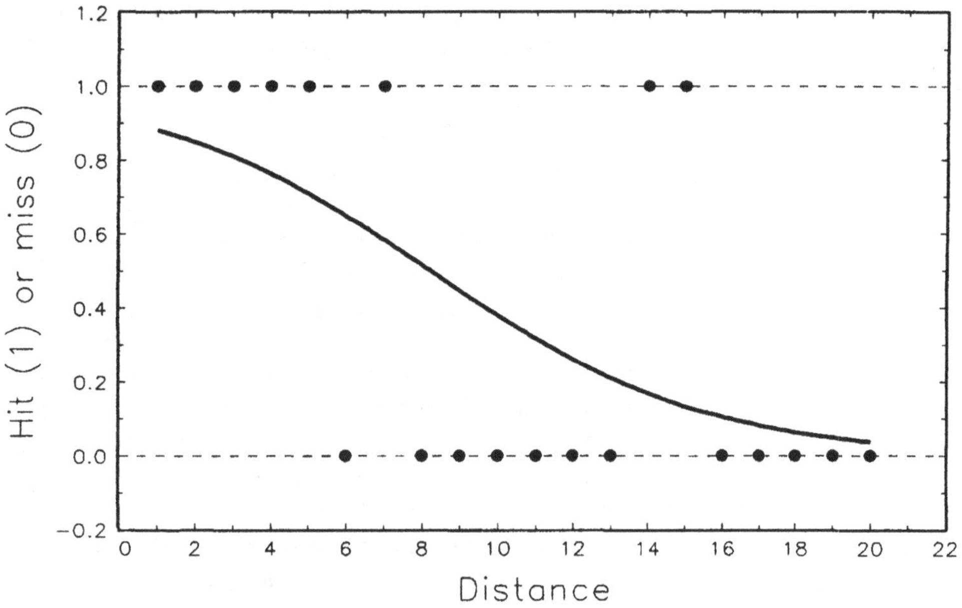

Figure 7-5-1 Steve's basketball data for the first shot from distances of 1 to 20 feet, along with an estimated logistic regression curve.

where π_i depends on x_i. How can we model this dependency? In Example 7.5.1, it is reasonable to assume that the success probability is a continuous, monotonically decreasing function of distance, taking values close to 1 for small distances and approaching 0 as the distance increases. Obviously, a linear function ($\pi_i = \alpha + \beta x_i$) is not a good candidate as it could take values outside the interval [0, 1]. For various reasons (one of which is explored in Exercise 9), the *logistic regression function*

$$\pi = g(x) = \frac{\exp(\alpha + \beta x)}{1 + \exp(\alpha + \beta x)} \tag{1}$$

is a popular choice. Here, α and β are parameters that determine the exact shape of the curve. The function $g(x)$ of equation (1) has the following properties (see Exercise 1):

(i) Unless $\beta = 0$, it maps the real line onto the interval (0, 1).

(ii) The function $g(x)$ is monotonically increasing if $\beta > 0$ and monotonically decreasing if $\beta < 0$.

(iii) The function takes the value $\frac{1}{2}$ at $x = -\alpha/\beta$ and is symmetric to the point $(-\alpha/\beta,\ 1/2)$, i.e.,

$$g(-\frac{\alpha}{\beta} - u) = 1 - g(-\frac{\alpha}{\beta} + u)$$

for all $u \in \mathbb{R}$.

It is customary to write equation (1) in a different form. Solving for the expression in the exponent (exercise 2) gives

$$\log \frac{\pi}{1-\pi} = \alpha + \beta x. \tag{2}$$

The function $\log \frac{\pi}{1-\pi}$ is called the *logit* of π; it maps the unit interval onto the real line. Yet another term used for the same function is *log-odds*; this comes from the use of the word "odds" for the ratio $\pi/(1-\pi)$ when π is the probability of an event. Equation (2) expresses the basic assumption of the logistic regression model: The logit of the success probability π is a linear function of the regressor variable X. If the logit of π is zero, then $\pi = 1/2$. Probabilities larger and smaller than 1/2 correspond to positive and negative values, respectively, on the logit scale. Finally, probabilities of 1 and 0 correspond to $\pm\infty$ on the logit scale.

More generally, we may have several Bernoulli trials for each selected value x_i of the regressor variable. For instance, the basketball player may shoot the ball M_i times from distance x_i. Let Y_i be the number of successes out of the M_i trials conducted at $X = x_i$. Assuming independence of the individual Bernoulli trials, Y_i has a binomial distribution with M_i trials and success probability π_i.

Now we have now collected all the building blocks for a formal definition of the logistic regression model.

Definition 7.5.1 The simple logistic regression model consists of $N \geq 2$ independent random variables $Y_1, \ldots, Y_N$, such that

(a)

$$Y_i \sim \text{binomial}(M_i,\ \pi_i) \quad \text{for } i = 1, \ldots, N, \tag{3}$$

where the $M_i \geq 1$ are fixed integers, and

(b)

$$\text{logit}(\pi_i) = \log \frac{\pi_i}{1-\pi_i} = \alpha + \beta x_i \quad \text{for } i = 1, \ldots, N. \tag{4}$$

Here, $\alpha \in \mathbb{R}$ and $\beta \in \mathbb{R}$ are parameters, and $x_1, \ldots, x_N$ are known values of a regressor variable. □

In Example 7.5.1, all M_i are equal to 1, but, in general, the M_i may be larger than 1 and differ from each other. The main problem will be to estimate the parameters α and β. Note that equations (3) and (4) define a regression problem in the sense of Section 2.7 because we are estimating the conditional mean of Y, given $X = x$:

$$E[Y_i|x_i] = M_i\pi_i = M_i \cdot \frac{\exp(\alpha + \beta x_i)}{1 + \exp(\alpha + \beta x_i)}. \tag{5}$$

We are now going to find the maximum likelihood estimators of the parameters α and β, using the techniques of Section 4.3. The likelihood function for observed counts

$y_1, \ldots, y_N$ is given by

$$L(\alpha, \beta; y_1, \ldots, y_N) = \prod_{i=1}^{N} \binom{M_i}{y_i} \pi_i^{y_i} (1 - \pi_i)^{M_i - y_i}, \tag{6}$$

for $(\alpha, \beta) \in \mathbb{R}^2$, where the success probabilities π_i are functions of α and β, according to

$$\pi_i = \frac{\exp(\alpha + \beta x_i)}{1 + \exp(\alpha + \beta x_i)}, \qquad i = 1, \ldots, N. \tag{7}$$

The log-likelihood function is

$$\ell(\alpha, \beta) = C + \sum_{i=1}^{N} \left[y_i \log \pi_i + (M_i - y_i) \log(1 - \pi_i) \right], \tag{8}$$

where C is a constant that does not depend on the unknown parameters. Then rearranging terms in (8) and using (4) gives

$$\ell(\alpha, \beta) = C + \sum_{i=1}^{N} \left[y_i(\alpha + \beta x_i) + M_i \log(1 - \pi_i) \right]. \tag{9}$$

Exercise 3 shows that the derivatives of π_i with respect to α and β are

$$\frac{\partial \pi_i}{\partial \alpha} = \pi_i(1 - \pi_i)$$

and

$$\frac{\partial \pi_i}{\partial \beta} = x_i \pi_i (1 - \pi_i). \tag{10}$$

Therefore, the score functions are

$$\begin{aligned} S_1(\alpha, \beta) = \frac{\partial \ell(\alpha, \beta)}{\partial \alpha} &= \sum_{i=1}^{N} \left[y_i + M_i \frac{\partial \log(1 - \pi_i)}{\partial \pi_i} \cdot \frac{\partial \pi_i}{\partial \alpha} \right] \\ &= \sum_{i=1}^{N} (y_i - M_i \pi_i) \end{aligned} \tag{11}$$

and

$$\begin{aligned} S_2(\alpha, \beta) = \frac{\partial \ell(\alpha, \beta)}{\partial \beta} &= \sum_{i=1}^{N} \left[x_i y_i + M_i \frac{\partial \log(1 - \pi_i)}{\partial \pi_i} \cdot \frac{\partial \pi_i}{\partial \beta} \right] \\ &= \sum_{i=1}^{N} x_i (y_i - M_i \pi_i). \end{aligned} \tag{12}$$

To find maximum likelihood estimates, we need to solve the two equations $S_1(\alpha, \beta) = 0$ and $S_2(\alpha, \beta) = 0$ simultaneously. Since the π_i depend on α and β in a complicated way, it is in general not possible to find an explicit solution. For numerical maximization and to find the asymptotic distribution of the maximum likelihood estimators, we will need the information function:

$$\mathbf{I}(\alpha, \beta) = \begin{pmatrix} I_{11}(\alpha, \beta) & I_{12}(\alpha, \beta) \\ I_{21}(\alpha, \beta) & I_{22}(\alpha, \beta) \end{pmatrix}$$
$$= -\begin{pmatrix} \frac{\partial S_1}{\partial \alpha} & \frac{\partial S_1}{\partial \beta} \\ \frac{\partial S_2}{\partial \alpha} & \frac{\partial S_2}{\partial \beta} \end{pmatrix} \tag{13}$$
$$= \begin{bmatrix} \sum_{i=1}^{N} M_i \pi_i (1 - \pi_i) & \sum_{i=1}^{N} x_i M_i \pi_i (1 - \pi_i) \\ \sum_{i=1}^{N} x_i M_i \pi_i (1 - \pi_i) & \sum_{i=1}^{N} x_i^2 M_i \pi_i (1 - \pi_i) \end{bmatrix}.$$

Exercise 4 shows that $\mathbf{I}(\alpha, \beta)$ is positive definite provided there are at least two different values x_i and provided that all π_i are different from 0 and 1. The latter condition holds by (7), as long as both α and β are finite. We will explore exceptions in a later example. For now, positive definiteness of $\mathbf{I}(\alpha, \beta)$ means that the log-likelihood function is concave and can therefore take at most one maximum, corresponding to the maximum likelihood estimate.

Before developing the theory further, let us return to our introductory example.

Example 7.5.2 continuation of Example 7.5.1

In the notation introduced, $N = 20$, $M_i = 1$ for $i = 1, \ldots, 20$, $x_i = i$ for $i = 1, \ldots, 20$, and $y_1 = 1$, $y_2 = 1$, $\ldots$, $y_{20} = 0$. Figure 7.5.2 shows a contour graph of the log-likelihood function. The contours are nearly concentric ellipses, an aspect to which we will return in the context of numerical maximization. The log-likelihood function takes a unique maximum at $\alpha = \hat{\alpha} = 2.271$ and $\beta = \hat{\beta} = -0.276$, that is, $(\hat{\alpha}, \hat{\beta})$ is the unique solution to the equation system $S_1(\alpha, \beta) = 0$ and $S_2(\alpha, \beta) = 0$. The logistic curve in Figure 7.5.1 was computed using these estimates, i.e., it is defined by

$$g(x; \hat{\alpha}, \hat{\beta}) = \frac{\exp(\hat{\alpha} + \hat{\beta}x)}{1 + \exp(\hat{\alpha} + \hat{\beta}x)}.$$ □

As always in estimation problems, we will be interested now in the distribution of the estimated parameters. The powerful large-sample theory for maximum likelihood estimators (Section 4.3) allows us to find the approximate distribution of the maximum likelihood estimators $\hat{\alpha}$ and $\hat{\beta}$, despite the fact that we do not have explicit formulas. Actually, Theorem 4.3.2 is not directly applicable in this case because $Y_1, \ldots, Y_N$ are not identically distributed. However, a generalization of Theorem 4.3.2 shows that, for large N, the joint distribution of $\hat{\alpha}$ and $\hat{\beta}$ is approximately bivariate normal

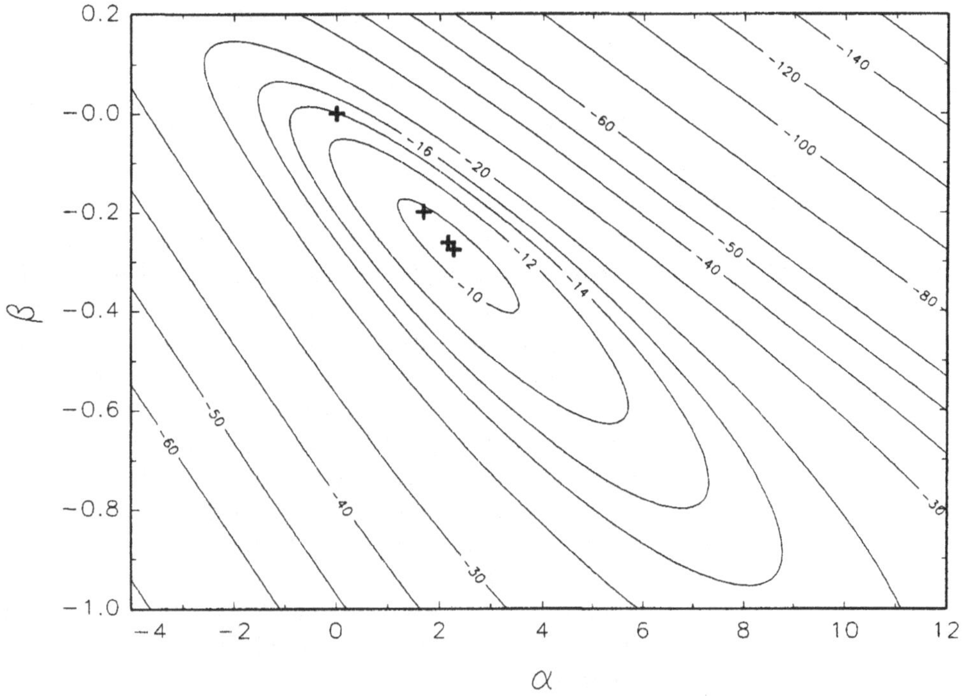

Figure 7-5-2 Contour plot of the log-likelihood function in Example 7.5.2. Plus signs indicate parameter values in the iterations of the numerical maximization procedure, starting at $\alpha^{(0)} = 0$ and $\beta^{(0)} = 0$.

with means α and β and with covariance matrix $\boldsymbol{\Sigma}$, where $\boldsymbol{\Sigma} = \{E\,[\mathbf{I}(\alpha, \beta)]\}^{-1}$. As seen from (13), the random variables Y_i do not appear in $\mathbf{I}(\alpha, \beta)$, and therefore, $E\,[\mathbf{I}(\alpha, \beta)] = \mathbf{I}(\alpha, \beta)$. However, $\boldsymbol{\Sigma}$ depends on the unknown parameters α and β and needs to be estimated. Therefore, we define

$$\hat{\boldsymbol{\Sigma}} = \left[\mathbf{I}(\hat{\alpha}, \hat{\beta})\right]^{-1} = \begin{bmatrix} \hat{\sigma}_{11} & \hat{\sigma}_{12} \\ \hat{\sigma}_{21} & \hat{\sigma}_{22} \end{bmatrix}$$
$$= \begin{pmatrix} \sum_{i=1}^{N} M_i \hat{\pi}_i (1 - \hat{\pi}_i) & \sum_{i=1}^{N} x_i M_i \hat{\pi}_i (1 - \hat{\pi}_i) \\ \sum_{i=1}^{N} x_i M_i \hat{\pi}_i (1 - \hat{\pi}_i) & \sum_{i=1}^{N} x_i^2 M_i \hat{\pi}_i (1 - \hat{\pi}_i) \end{pmatrix}^{-1} \tag{14}$$

as the estimated covariance matrix of $\hat{\alpha}$ and $\hat{\beta}$.

Furthermore, the square roots of the diagonal elements of $\hat{\boldsymbol{\Sigma}}$ will be referred to as standard errors of the parameter estimates, i.e.,

$$\text{se}\,(\hat{\alpha}) = \sqrt{\hat{\sigma}_{11}}, \quad \text{and se}\,(\hat{\beta}) = \sqrt{\hat{\sigma}_{22}}. \tag{15}$$

In many applications, it will also be useful to report the estimated success probabilities

$$\hat{\pi}_i = g(x_i; \hat{\alpha}, \hat{\beta}) = \frac{\exp(\hat{\alpha} + \hat{\beta} x_i)}{1 + \exp(\hat{\alpha} + \hat{\beta} x_i)}, \qquad i = 1, \ldots, N, \tag{16}$$

as well as the estimated expected frequencies

$$\hat{y}_i = M_i \hat{\pi}_i, \qquad i = 1, \ldots, N. \tag{17}$$

This is illustrated in the following classical example.

Example 7.5.3 Beetle mortality. An important question in chemical and pharmaceutical experiments often is to determine the dosage of some substance necessary to get a desired result. A typical example is from Bliss (1935). Table 7.5.2 shows numbers of insects dead after five hours' exposure to gaseous carbon disulphide at various concentrations. Experiments were done at eight different levels x_i of the dose. At dose x_i, M_i insects were tested, and the number y_i of insects killed was recorded. (As is often the case in applied statistics, death is considered as success, and survival is called failure). How does the probability of killing an insect depend on the dose of the poison? And for what value of the dose can we expect to kill half of the insects exposed?

Maximum likelihood estimates are $\hat{\alpha} = -60.717$, $\hat{\beta} = 34.270$. The logistic regression curve is shown in Figure 7.5.3, along with the observed proportions of success y_i/M_i. The logistic curve fits the observed proportions quite well, although a formal test, to be discussed in Section 7.6, indicates some lack of fit. In addition to the original data, Table 7.5.2 also lists the estimated probability of success $\hat{\pi}_i$ for each value of the dose and the estimated expected frequencies computed from equation (17). The information function evaluated at $(\hat{\alpha}, \hat{\beta})$ is

$$\mathbf{I}(\hat{\alpha}, \hat{\beta}) = \begin{pmatrix} 58.48 & 104.10 \\ 104.01 & 185.09 \end{pmatrix},$$

Table 7.5.2 Beetle mortality data from Bliss (1935). The data used as input for the logistic regression analysis are x_i = dose in units of $\log_{10} CS_2$mf/l; M_i = number of insects tested; y_i = number of insects killed. The remaining columns are $\hat{y}_i$ = estimated expected number of insects killed; y_i/M_i = observed proportion of insects killed; $\hat{\pi}_i$ = estimated success probability.

i	x_i	M_i	y_i	$\hat{y}_i$	y_i/M_i	$\hat{\pi}_i$
1	1.6907	59	6	3.5	0.102	0.059
2	1.7242	60	13	9.8	0.217	0.164
3	1.7552	62	18	22.5	0.290	0.362
4	1.7842	56	28	33.9	0.500	0.605
5	1.8113	63	52	50.1	0.825	0.795
6	1.8369	59	53	53.3	0.898	0.903
7	1.8610	62	61	59.2	0.984	0.955
8	1.8839	60	60	58.7	1.000	0.979

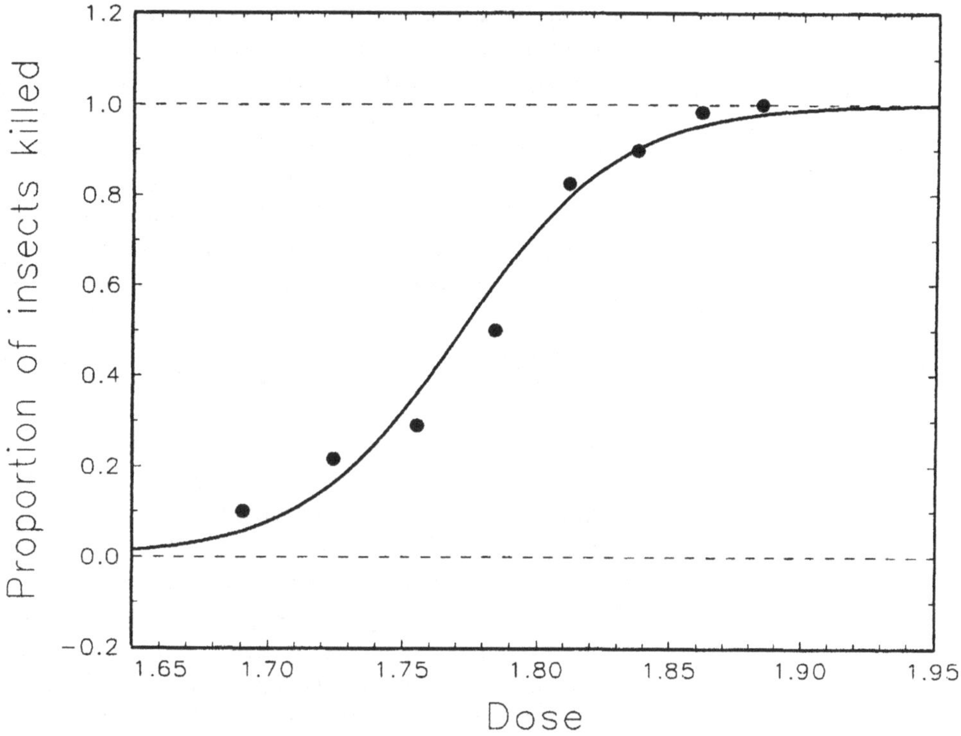

Figure 7-5-3 Proportion of insects killed as a function of dose in Example 7.5.3 and the estimated logistic regression curve.

giving an estimate of the covariance matrix of $\hat{\alpha}$ and $\hat{\beta}$,

$$\hat{\boldsymbol{\Sigma}} = \left[\mathbf{I}(\hat{\alpha}, \hat{\beta})\right]^{-1} = \begin{pmatrix} 26.84 & -15.08 \\ -15.08 & 8.48 \end{pmatrix}.$$

The standard errors of the two parameter estimates are the square roots of the diagonal elements of $\hat{\boldsymbol{\Sigma}}$, i.e., $\text{se}(\hat{\alpha}) = 5.181$ and $\text{se}(\hat{\beta}) = 2.912$. Without doing a formal test at this point, it appears that the parameter estimates are quite stable since the standard errors are relatively small. Finally, by property (iii) of the logistic regression function, we estimate the dose at which the success probability is 50 percent as $-\hat{\alpha}/\hat{\beta} = 1.772$. □

Questions of testing hypothesis arise naturally from the logistic regression model, the single most important one being the test for the hypothesis $\beta = 0$, because this hypothesis implies that the probability of success is a constant, independent of the regressor. See Exercise 6. We will treat the likelihood ratio tests for various hypotheses in Section 7.6 in a more general context, and therefore skip the problem at this point. Instead, let us return to the problem of numerical maximization of the log-likelihood function. A highly efficient procedure for the logistic regression model is the Newton-Raphson algorithm, presented without proof. The algorithm requires first and second derivatives of the function to be maximized, which we already know from equations (11) to (13). The algorithm does essentially the following. At the current value (α, β) of the arguments of the log-likelihood function $\ell(\alpha, \beta)$, it computes a second-order Taylor series approximation, whose maximum can be computed explicitly. This leads to new values of α and β, for which in turn a new Taylor series approximation is calculated, and so on. Specifically, if $\alpha^{(j)}$ and $\beta^{(j)}$ are the current parameter values after the jth step in the algorithm, the next pair of values $\left(\alpha^{(j+1)}, \beta^{(j+1)}\right)$ is computed as

$$\begin{pmatrix} \alpha^{(j+1)} \\ \beta^{(j+1)} \end{pmatrix} = \begin{bmatrix} \alpha^{(j)} \\ \beta^{(j)} \end{bmatrix} + \left[\mathbf{I}\left(\alpha^{(j)}, \beta^{(j)}\right)\right]^{-1} \begin{pmatrix} S_1\left(\alpha^{(j)}, \beta^{(j)}\right) \\ S_2\left(\alpha^{(j)}, \beta^{(j)}\right) \end{pmatrix}. \tag{18}$$

Starting with arbitrary values $\alpha^{(0)}$ and $\beta^{(0)}$, the repeated evaluations of equation (18) produce a sequence $\left(\alpha^{(1)}, \beta^{(1)}\right), \left(\alpha^{(2)}, \beta^{(2)}\right), \ldots,$ which converges to the maximum likelihood estimate $(\hat{\alpha}, \hat{\beta})$, provided the maximum exists. Since the maximum is unique (provided it exists), the choice of the initial values $\alpha^{(0)}$ and $\beta^{(0)}$ is not important. A reasonable and simple choice is to set $\alpha^{(0)} = \beta^{(0)} = 0$, which means that all probabilities π_i are $\frac{1}{2}$ at the beginning of the algorithm. As seen from (18), each iteration requires evaluating the score and information functions at the current parameter values. Although this may seem to be a lot of numerical work, convergence is typically very fast, as illustrated by the following example.

Example 7.5.4 continuation of Examples 7.5.1 and 7.5.2

Starting with initial values $\alpha^{(0)} = \beta^{(0)} = 0$, five iterations of equation (18) produced the final values $\hat{\alpha} = 2.271$, $\hat{\beta} = -0.276$, as seen from Table 7.5.3; the last three of these points are so close together that they are indistinguishable in Figure 7.5.2. Convergence of the numerical algorithm is quite fast. To the expert in numerical analysis, this is not a surprise because the log-likelihood function is approximated well by a quadratic function in the neighborhood of its maximum, as seen from the almost elliptical contours in Figure 7.5.2. □

At the beginning of this section, we argued that the logistic function (1) is a suitable model for probabilities because it maps the real line into the unit interval. More generally, we could take any cumulative distribution function F and fit a model based on

$$\pi = g(x) = F(\alpha + \beta x), \tag{19}$$

with parameters α and β to be estimated. A popular choice is to take F as the *cdf* of the standard normal, which leads to so-called *probit models*. Whichever function is taken, the advantage of these models compared to discriminant analysis is that distributional assumptions on X are avoided. Any classification problem based

Table 7.5.3 Performance of the Newton-Raphson algorithm in the basketball example. Displayed are the values $\alpha^{(j)}$ and $\beta^{(j)}$ in the jth iteration and the associated value of the log-likelihood function $\ell(\alpha^{(j)}, \beta^{(j)})$.

Iteration (j)	$\alpha^{(j)}$	$\beta^{(j)}$	$\ell(\alpha^{(j)}, \beta^{(j)})$
0	0.0000000	0.00000000	-13.862944
1	1.6842105	-0.19849624	-9.8206298
2	2.1652963	-0.26182802	-9.5698573
3	2.2666729	-0.27583952	-9.5614973
4	2.2710045	-0.27645260	-9.5614826
5	2.2710125	-0.27645374	-9.5614826

on posterior probabilities can, in principle, be tackled with a logistic regression approach, or a similar approach based on some alternative distribution function F. (So far, of course, we could handle only univariate problems). For instance, to classify midges based on wing length or antenna length (Example 1.1), we could use a logistic regression approach, setting $Y = 1$ for one species and $Y = 0$ for the other species. See Exercise 10. This way it seems we could estimate posterior probabilities without making distributional assumptions. However, surprises are sometimes waiting, as we will see in the next example.

Example 7.5.5 Classification of midges. In our introductory example from Chapter 1, set $Y = 0$ for *Af* midges and $Y = 1$ for *Apf* midges. As the regressor variable, use X = wing length − antenna length in mm/100. That is, each of the $N = 15$ midges is considered as a Bernoulli trial, success being defined as *Apf.* When trying to estimate the parameters of a logistic regression model, the Newton-Raphson algorithm aborted after 35 iterations, producing an error message that the information function $\mathbf{I}(\alpha, \beta)$, evaluated at the current values of the parameters, was not positive definite. What happened?

Part (a) of Figure 7.5.4 shows a contour plot of the log-likelihood function, with plus signs marking the values of the parameters at the initialization $\left(\alpha^{(0)} = \beta^{(0)} = 0\right)$ and after each of the first three iterations. Afterwards, the algorithm continued walking along an ever flatter ridge extending up and to the left in Figure 7.5.4(a).

Why does this happen? The reason is that, as shown in part (b) of Figure 7.5.4, the data of the two groups of midges do not overlap. The numerical maximization algorithm produced a sequence of parameter values corresponding to steeper and steeper logistic curves. When the algorithm finally failed after 35 iterations, the parameter values were $\alpha^{(35)} = -327.923$ and $\beta^{(35)} = 5.649$. The corresponding logistic curve is almost a step function. Although it approximates the data well, it is useless for practical purposes because it estimates the probabilities of midges being *Af* or *Apf*, respectively, as practically 0 and 1, except in the very narrow interval where the function jumps. If this curve were reasonable, we would have to conclude that the two species can be classified without error, based just on the difference between

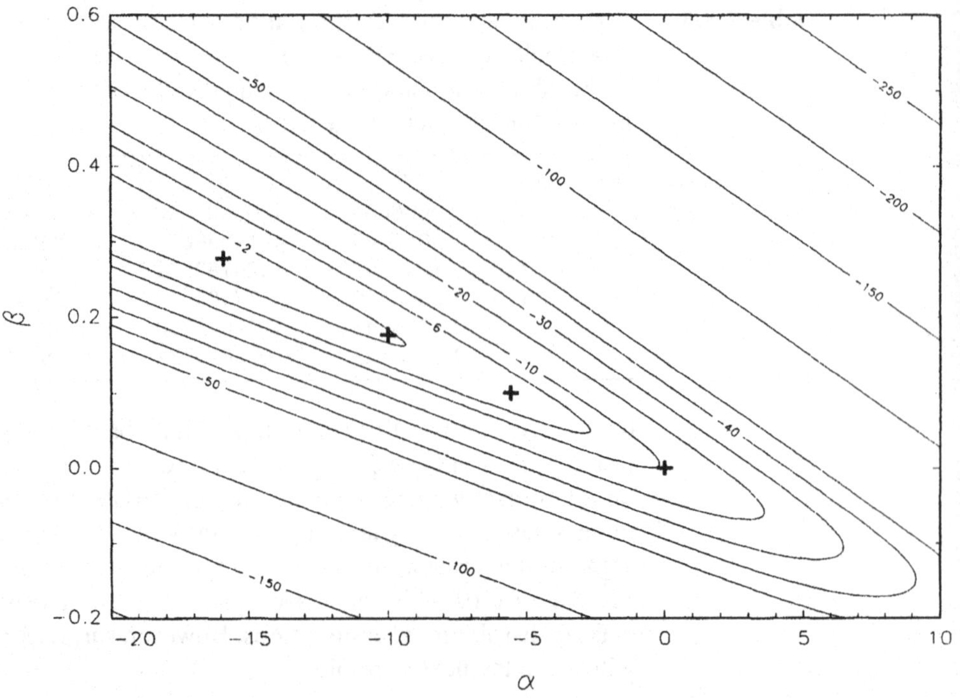

Figure 7-5-4A

wing length and antenna length. The assessment of error rates from normal theory discriminant analysis is more reasonable in this case. □

The phenomenon observed in Example 7.5.5 has nothing to do with the fact that X is random. It will occur whenever successes and failures are completely separated along the x-axis, i.e., whenever there is a number $c \in \mathbb{R}$ such that $y_i = 0$ for all $x_i < c$ and $y_i = M_i$ for all $x_i > c$, or vice-versa. (We ignore the possibility $x_i = c$ here; see Exercise 11). For convenience, assume that $y_i = 0$ for all $x_i < c$ and $y_i = M_i$ for all $x_i > c$. Then the likelihood function (6) is

$$L(\alpha, \beta) = \left[\prod_{\{i: x_i < c\}} (1 - \pi_i)^{M_i} \right] \left[\prod_{\{i: x_i > c\}} \pi_i^{M_i} \right], \tag{20}$$

where the π_i depend on α and β through (7). If there were no constraints on the π_i, $L(\alpha, \beta)$ would be maximized by setting $\pi_i = 0$ if $x_i < c$, and $\pi_i = 1$ if $x_i > c$, and the maximum would be 1. We can get arbitrarily close to this value by letting β tend

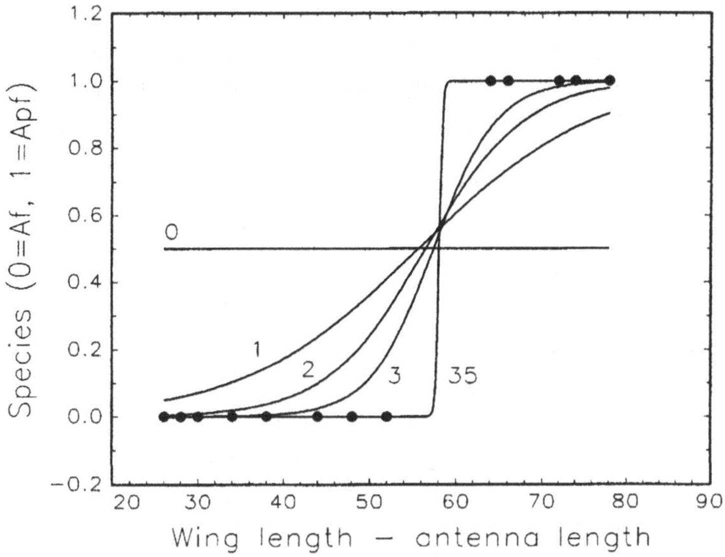

Figure 7-5-4B Maximum likelihood estimation in Example 7.5.5: (a) contour plot of the log-likelihood function, with plus signs indicating parameter values at the initialization of the numerical maximization procedure ($\alpha^{(0)} = 0$, $\beta^{(0)} = 0$) and after each of the first three iterations; (b) logistic regression curves corresponding to parameter values after a number of iterations of the numerical maximization procedure as indicated. The horizontal line labeled "0" refers to the initialization of the algorithm.

to infinity and setting $\alpha = -c\beta$. Then

$$\lim_{\substack{\beta\to\infty\\ \alpha=-c\beta}} \pi_i(\alpha, \beta) = \lim_{\beta\to\infty} \frac{\exp[\beta(x_i - c)]}{1 + \exp[\beta(x_i - c)]}$$

$$= \begin{cases} 0 & \text{if } x_i < c, \\ 1 & \text{if } x_i > c. \end{cases}$$

Therefore, as β tends to infinity and $\alpha = -c\beta$, the likelihood function takes larger and larger values without actually reaching the limiting value of 1. This explains the flat ridge seen in Figure 7.5.4(a).

We argued earlier that one might use some other distribution function, rather than the logistic, to model the dependency of the success probability on the regressor variable. A formal assessment of the fit of a logistic curve can be made using the notion of deviance, treated in the next section. Alternatively, we may also try transformations of the regressor variable illustrated in the next example.

Example 7.5.6 Basketball again. In the same experiment reported in Example 7.5.1, the author and his sons Steve, Andy, and Chris shot the ball six times from each of the distances 1, 2, . . . , 20 feet. Numerical results are given in Table 7.5.4. For the current purpose, we look only at Andy's data. Let Y_i = number of hits from distance x_i, $i = 1, \ldots, 20$, where $x_i = i$ and $M_i = 6$ for $i = 1, \ldots, 20$. Figure 7.5.5 shows a graph of distance vs. the observed proportions of success. Assuming $Y_i \sim$ binomial (M_i, π_i), with logit $(\pi_i) = \alpha + \beta x_i$, maximum likelihood estimates and their standard errors are

$$\hat{\alpha} = 0.800, \qquad \text{se}(\hat{\alpha}) = 0.418,$$

$$\hat{\beta} = -0.170, \quad \text{and} \quad \text{se}(\hat{\beta}) = 0.042.$$

Table 7.5.4 Basketball data. Variables are X = distance from hoop (in feet); S = Steve's score; A = Andy's score; C = Chris' score; B = author's score (please don't laugh – for someone who can hardly distinguish between a touchdown and a homerun, these results are really not that bad). All four players had six shots from each of the distances; the author did not try from any distances more than 10 feet. All data were obtained in the afternoon of April 18, 1996, at Bryan Park, Bloomington, Indiana.

X	S	A	C	B
1	6	5	5	4
2	5	4	5	3
3	2	3	3	3
4	3	2	1	3
5	4	4	5	0
6	1	2	1	2
7	3	2	3	1
8	1	3	1	1
9	1	3	2	1
10	1	0	5	1
11	4	1	2	-
12	0	3	2	-
13	1	0	2	-
14	2	0	3	-
15	4	0	3	-
16	1	1	1	-
17	0	1	1	-
18	1	2	0	-
19	0	1	3	-
20	2	0	0	-

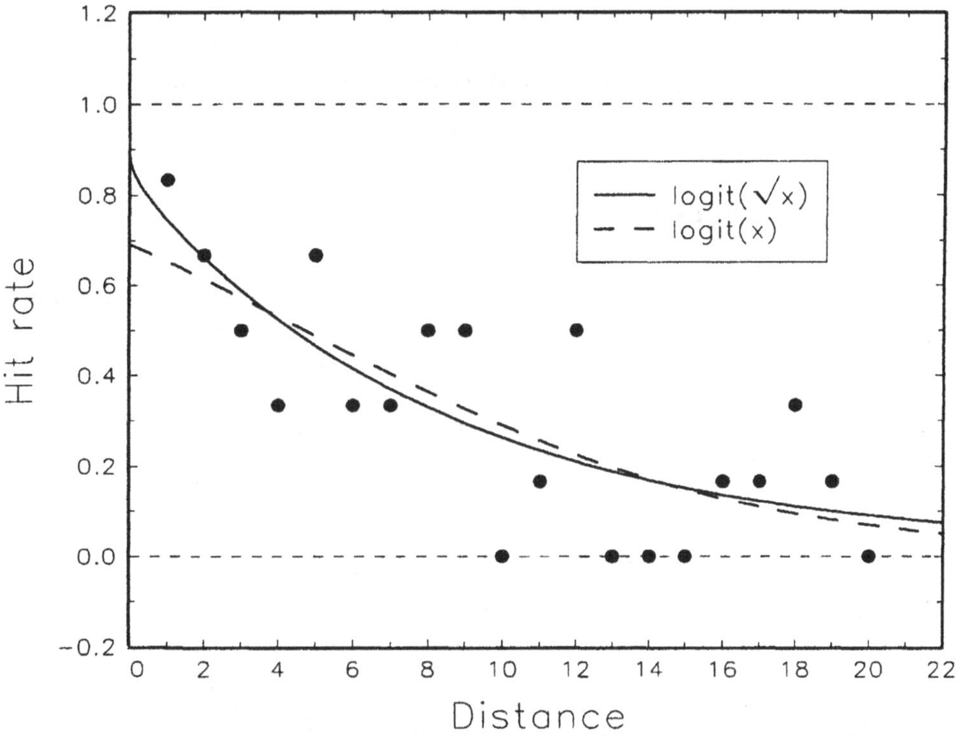

Figure 7-5-5 Andy's basketball data. Data points are proportions of hits from distances ranging from 1 to 20 feet. The two curves are estimated logistic regression curves with distance and the square root of distance, respectively, as regressors.

One might argue that the probability of a hit for small distances should be larger than estimated by this logistic regression curve. A possible remedy is to transform x, for instance, by taking its square root. The maximum likelihood estimates of α^* and β^* in the model

$$\text{logit}(\pi_i) = \alpha^* + \beta^* \sqrt{x_i}, \qquad i = 1, \ldots, N$$

are

$$\hat{\alpha}^* = 2.036, \qquad \text{se}(\hat{\alpha}^*) = 0.684, \quad \text{and}$$

$$\hat{\beta}^* = -0.971, \qquad \text{se}(\hat{\beta}^*) = 0.230.$$

The corresponding logistic regression curve

$$\hat{\pi}_i = \frac{\exp\left(2.036 - 0.971\sqrt{x_i}\right)}{1 + \exp\left(2.036 - 0.971\sqrt{x_i}\right)}$$

seems marginally better than the original one, as seen in Figure 7.5.5. Differences between the two curves are most pronounced for distances close to zero. See Exercise 15 for further analyses of the basketball data. □

Exercises for Section 7.5

1. Prove properties (i) to (iii) of the logistic regression function; see the text following equation (1).

2. Prove equation (2).

3. Prove equations (10). *Hint*: First, show that
$$1-\pi_i = \frac{1}{1+\exp[\alpha+\beta x_i]}.$$

4. This exercise concerns the information function $\mathbf{I}(\alpha, \beta)$.
   (a) Show that $\mathbf{I}(\alpha, \beta)$ has the form given in equation (13).
   (b) Show that $\mathbf{I}(\alpha, \beta)$ is positive definite for all finite values of α and β, provided there are at least two different x_i. *Hint*: Apply the Cauchy-Schwartz inequality (Lemma 5.2.2) to the vectors $\mathbf{u} = (u_1, \ldots, u_N)'$ and $\mathbf{v} = (v_1, \ldots, v_N)'$, where $u_i = \sqrt{M_i \pi_i (1-\pi_i)}$ and $v_i = u_i x_i$.

5. In Example 7.5.2, compute $\mathbf{I}(\hat{\alpha}, \hat{\beta})$, its inverse, and the standard errors of the parameter estimates.

6. In the simple logistic regression model, show that $\beta = 0$ is equivalent to equality of all π_i. Under the hypothesis $\beta = 0$, show that the maximum likelihood estimate of α is $\log\left[\hat{\pi}/(1-\hat{\pi})\right]$, where $\hat{\pi} = \left(\sum_{i=1}^N y_i\right) / \left(\sum_{i=1}^N M_i\right)$.

7. This exercise gives some insight into the Newton-Raphson algorithm for maximizing or minimizing a function of a single variable.
   (a) Let $h(x) = ax^2 + bx + c$, where $a \neq 0$. Suppose that we know $h(x_0)$, $h'(x_0)$, and $h''(x_0)$ for some $x_o \in \mathbb{R}$. Show that the function $h(x)$ takes its maximum or minimum at
$$x = x_0 - \frac{h'(x_0)}{h''(x_0)}.$$
   (b) Explain why this suggests an iterative procedure
$$x_{j+1} = x_j - \frac{h'(x_j)}{h''(x_j)}, \qquad j = 0, 1, 2, \ldots \tag{21}$$
   for finding the minimum/maximum of a function $h(x)$ that is twice continuously differentiable. Here, x_0 is an arbitrary initial value, preferably a value close to a

relative maximum or minimum of $h(x)$. Show that $x_{j+1} = x_j$ exactly if x_j is a stationary point.

(c) The function $h(x) = \frac{1}{3}x^3 - bx$, where $b > 0$, takes a unique minimum for $x > 0$ at $x = \sqrt{b}$. Show that the Newton-Raphson algorithm for finding the square root of a positive number b consists of iterations of the equation

$$x_{j+1} = \frac{1}{2}\left(x_j + \frac{b}{x_j}\right), \quad j = 0, 1, \ldots,$$

where $x_0 > 0$ is an arbitrary starting point.

(d) Argue that equation (18) is analogous to (21) for a real-valued function of two variables.

8. Program the Newton-Raphson algorithm (18) in a programming language of your choice. Use your program to verify the results of Example 7.5.3 numerically.

9. Suppose that $Y \sim$ Bernoulli(θ) and, conditionally on $Y = j$, X is normal with mean μ_j and variance σ^2. Let $\pi_{jx} = \Pr[Y = j | X = x]$. Show that

$$\text{logit}(\pi_{1x}) = \alpha + \beta x$$

with

$$\alpha = \log \frac{\theta}{1-\theta} - \frac{1}{2\sigma^2}\left(\mu_1^2 - \mu_0^2\right)$$

and

$$\beta = \frac{\mu_1 - \mu_0}{\sigma^2}.$$

That is, show that the conditional distributions of Y, given $X = x$, satisfy the logistic regression model.

10. In the midge data set (Example 1.1), set $Y = 0$ for *Af* midges and $Y = 1$ for *Apf* midges, and let X = antenna length. Estimate the parameters of a logistic regression model and graph the results similarly to Figure 7.5.1.

11. In a simple logistic regression problem, suppose that there is a number $c \in \mathbb{R}$ such that

$$y_i = 0 \text{ for all } x_i < c$$

and

$$y_i = M_i \text{ for all } x_i > c.$$

For $x_i = c$ (provided there is such a value of the regressor), we may have any value of y_i between 0 and M_i. Show that the same phenomenon occurs as in Example 7.5.5, i.e., the likelihood function does not take a maximum for finite parameter values.

12. In a study relating the outcome of a psychiatric diagnosis (patient needs or does not need psychiatric treatment) to the score on a 12 item General Health Questionnaire (GHQ), the data of Table 7.5.5 was obtained. Logistic regression may be used to

Table 7.5.5 General Health Questionnaire Data from Silvapulle (1981). Variables are Gender (0 = male, 1 = female); GHQ = score on questionnaire; M_i = number of patients for the particular combination of gender and GHQ score; y_i = number of patients needing psychiatric treatment.

GENDER	GHQ	M_i	y_i
0	0	18	0
0	1	8	0
0	2	2	1
0	4	1	1
0	5	3	3
0	7	2	2
0	10	1	1
1	0	44	2
1	1	16	2
1	2	9	4
1	3	4	3
1	4	3	2
1	5	3	3
1	6	1	1
1	7	1	1
1	8	3	3
1	9	1	1

estimate the probability that a person needs psychiatric treatment as a function of the GHQ score.

(a) Estimate the parameters of a logistic regression model for the 85 females only, and graph the results similarly to Figure 7.5.3. For which value of $X = GHQ$-score would you estimate the probability of "success" (= probability of needing treatment) to be $\frac{1}{2}$?

(b) Repeat the analysis of part (a), using only the 35 males. What happens when you try to estimate the parameters? Explain. *Hint*: See Exercise 11.

13. In a logistic regression problem with $Y_i \sim \text{binomial}(M_i, \pi_i)$ and $\text{logit}(\pi_i) = \alpha + \beta x_i$, suppose that the results of the individual Bernoulli trials are known for some or all y_i. Therefore, we might estimate a logistic regression model with y_i replaced by M_i binary observations $y_{i1}, \ldots, y_{iN_i}$, exactly y_i of which are equal to 1 and $M_i - y_i$ are equal to zero. Show that the maximum likelihood estimates based on the Bernoulli data are identical to those based on the y_i. *Hint*: It suffices to show that the likelihood function based on the y_{ij} is proportional to (6).

14. In the Challenger disaster of January 20, 1986, the cause of the disaster was determined to be the sensitivity of O-rings to low temperature. Table 7.5.6 gives data on numbers of damaged O-rings on each of 23 previous flights, along with the ambient temperature

at the time of launching. A detailed description can be found in Chatterjee et al. (1995, pp. 33–35). Let Y_i = number of damaged O-rings in the ith flight and x_i = temperature. There are six O-rings total, so $M_i = 6$ for $i = 1, \ldots, 23$.

(a) Estimate the parameters of a logistic regression model, and graph the results similarly to Figure 7.5.3. Graph the curve for temperatures between 30 and 90° F.

(b) Using the parameter values from part (a), estimate the probability that an O-ring will be damaged if the ambient temperature at launching time is 31° F., as on the day of the fatal flight. Estimate the expected number of O-rings (out of 6) that will be damaged.

(c) How do you feel about the prediction made in part (b)?

15. Repeat the analysis of Example 7.5.6 for each of the other three players, including graphs like Figure 7.5.5. Do you think that the model with x (distance) as regressor is more appropriate, or the model with the square root of x as regressor? Do you think that combining the data of all four players in a single model (i.e., estimating a common logistic regression curve for all four players) would be reasonable, or is

Table 7.5.6 Challenger data from Chatterjee et al. (1995, p. 33). Variables are X = ambient temperature (° F.) at launching time; Y = number of damaged O-rings (out of 6), for each of 23 previous flights.

X	Y
53	2
57	1
58	1
63	1
66	0
67	0
67	0
67	0
68	0
69	0
70	0
70	0
70	1
70	1
72	0
73	0
75	0
75	2
76	0
76	0
78	0
79	0
81	0

there evidence that each player should be modeled individually? (Only an informal answer is expected at this time; a formal test will be treated in Section 7.6).

16. Suppose that $N = 2$ in a logistic regression model. Find the maximum likelihood estimates of α and β explicitly as functions of the x_i, y_i, and M_i, and show that $\hat{\pi}_i = y_i/M_i$ for $i = 1, 2$.

17. In the midge example, set X = wing length − antenna length. Estimate normal theory posterior probabilities for both groups, using equal prior probabilities. Graph the posterior probabilities, and compare the curve to Figure 7.5.4(b). You may use either the general theory of Section 5.4 or Exercise 9 in the current section.

7.6 Multiple Logistic Regression

The multiple logistic regression model generalizes simple logistic regression to the case of an arbitrary number $p \geq 1$ of regressors $X_1, \ldots, X_p$. Let $x_{i1}, x_{i2}, \ldots, x_{ip}$ denote the values of the p regressor variables for the ith observation, and let Y_i be the ith response. The multiple logistic regression model assumes that the Y_i are independent,

$$Y_i \sim \text{binomial}\,(M_i, \pi_i) \qquad i = 1, \ldots, N, \tag{1}$$

and

$$\text{logit}(\pi_i) = \log \frac{\pi_i}{1 - \pi_i} = \beta_0 + \beta_1 x_{i1} + \cdots + \beta_p x_{ip}, \qquad i = 1, \ldots, N, \tag{2}$$

for $p + 1$ unknown parameters $\beta_0, \beta_1, \ldots, \beta_p$. We will first discuss parameter estimation, and then turn to statistical inference.

As in ordinary multiple regression, matrix notation will be useful. (In fact, knowing multiple regression will help considerably to understand the following technical part). Let

$$\mathbf{x}_i' = (1, x_{i1}, \ldots, x_{ip}), \qquad i = 1, \ldots, N, \tag{3}$$

be the vector of regressor values of the ith observation, and

$$\boldsymbol{\beta}' = (\beta_0, \beta_1, \ldots, \beta_p) \tag{4}$$

the vector of $p + 1$ unknown parameters. Then equation (2) can be written as

$$\text{logit}(\pi_i) = \mathbf{x}_i'\boldsymbol{\beta}, \qquad i = 1, \ldots, N, \tag{5}$$

or equivalently,

$$\pi_i = \frac{\exp\left(\mathbf{x}_i'\boldsymbol{\beta}\right)}{1 + \exp\left(\mathbf{x}_i'\boldsymbol{\beta}\right)}. \tag{6}$$

As these equations show, the only generalization compared to the simple logistic regression model is in the number of regressor variables; the distributional assumptions on the Y_i and the transformation from the linear function of the regressor variables to probabilities remain the same.

The commonly used method of estimation is again maximum likelihood. As in equation (8) of Section 7.5, the log-likelihood-function for observed counts $y_1, \ldots, y_N$ is given by

$$\begin{aligned} \ell(\boldsymbol{\beta}) &= C + \sum_{i=1}^{N} \left[y_i \log \pi_i + (M_i - y_i) \log(1 - \pi_i) \right] \\ &= C + \sum_{i=1}^{N} \left[y_i \mathbf{x}_i' \boldsymbol{\beta} + M_i \log(1 - \pi_i) \right], \end{aligned} \tag{7}$$

where C is a constant that does not depend on the parameters, and the π_i depend on $\boldsymbol{\beta}$ through (6). The score function $\mathbf{S}(\boldsymbol{\beta})$ is a vector of $p + 1$ functions

$$\mathbf{S}(\boldsymbol{\beta}) = \begin{bmatrix} S_0(\boldsymbol{\beta}) \\ \vdots \\ S_p(\boldsymbol{\beta}) \end{bmatrix} = \begin{bmatrix} \dfrac{\partial}{\partial \beta_0} \ell(\boldsymbol{\beta}) \\ \vdots \\ \dfrac{\partial}{\partial \beta_p} \ell(\boldsymbol{\beta}) \end{bmatrix}. \tag{8}$$

Let $\partial \pi_i / \partial \boldsymbol{\beta}$ denote the column vector of partial derivatives of π_i with respect to the elements of $\boldsymbol{\beta}$. In Exercise 1, it is shown that

$$\frac{\partial \pi_i}{\partial \boldsymbol{\beta}} = \pi_i (1 - \pi_i) \mathbf{x}_i. \tag{9}$$

Using the rules for vector differentiation from Appendix A.4, the score function is given by

$$\begin{aligned} \mathbf{S}(\boldsymbol{\beta}) = \frac{\partial}{\partial \boldsymbol{\beta}} \ell(\boldsymbol{\beta}) &= \sum_{i=1}^{N} \frac{\partial}{\partial \boldsymbol{\beta}} \left[y_i \mathbf{x}_i' \boldsymbol{\beta} + M_i \log(1 - \pi_i) \right] \\ &= \sum_{i=1}^{N} \left(\mathbf{x}_i y_i - M_i \frac{1}{1 - \pi_i} \frac{\partial \pi_i}{\partial \boldsymbol{\beta}} \right) \\ &= \sum_{i=1}^{N} \mathbf{x}_i (y_i - M_i \pi_i). \end{aligned} \tag{10}$$

The equation system $\mathbf{S}(\boldsymbol{\beta}) = \mathbf{0} \in \mathbb{R}^{p+1}$ usually has no explicit solution because the π_i depend on $\boldsymbol{\beta}$ in a complicated way. Therefore, second derivatives will be useful for numerical maximization of the likelihood function. The information function is a $(p+1) \times (p+1)$ matrix

$$\mathbf{I}(\boldsymbol{\beta}) = -\frac{\partial}{\partial \boldsymbol{\beta}'}\mathbf{S}(\boldsymbol{\beta}) = -\begin{pmatrix} \frac{\partial S_0(\boldsymbol{\beta})}{\partial \beta_0} & \cdots & \frac{\partial S_0(\boldsymbol{\beta})}{\partial \beta_p} \\ \vdots & \ddots & \vdots \\ \frac{\partial S_p(\boldsymbol{\beta})}{\partial \beta_0} & \cdots & \frac{\partial S_p(\boldsymbol{\beta})}{\partial \beta_p} \end{pmatrix}. \tag{11}$$

Using results from Appendix A.4, we obtain

$$\begin{aligned} \mathbf{I}(\boldsymbol{\beta}) &= -\frac{\partial}{\partial \boldsymbol{\beta}'}\left[\sum_{i=1}^{N} \mathbf{x}_i (y_i - M_i \pi_i)\right] \\ &= \sum_{i=1}^{N} M_i \mathbf{x}_i \left(\frac{\partial \pi_i}{\partial \boldsymbol{\beta}}\right)' \\ &= \sum_{i=1}^{N} M_i \pi_i (1-\pi_i) \mathbf{x}_i \mathbf{x}_i', \end{aligned} \tag{12}$$

see Exercise 2. We leave it to the reader to verify that equations (10) and (12) reduce to equations (11) to (13) of Section 7.5 in the case $p = 1$.

It is worth writing the score and information functions in concise matrix notation. Let

$$\mathbf{X} = \begin{pmatrix} \mathbf{x}_1' \\ \mathbf{x}_2' \\ \vdots \\ \mathbf{x}_N' \end{pmatrix} \tag{13}$$

denote the $N \times (p+1)$ *design matrix*, whose first column is a vector of ones as in least squares multiple regression. Moreover, let

$$\mathbf{y} = \begin{pmatrix} y_1 \\ \vdots \\ y_N \end{pmatrix}$$

and

$$\mathbf{e} = \begin{pmatrix} M_1 \pi_1 \\ \vdots \\ M_N \pi_N \end{pmatrix} \tag{14}$$

be the vectors of observed frequencies and the vector of expected frequencies for given probabilities π_i. Then the score function can be written as

$$\mathbf{S}(\boldsymbol{\beta}) = \mathbf{X}'(\mathbf{y} - \mathbf{e}). \tag{15}$$

At the maximum of the likelihood function, $\mathbf{X}'(\mathbf{y}-\mathbf{e}) = \mathbf{0}$. But $\mathbf{y}-\mathbf{e}$ is the vector of deviations of the observed from the expected frequencies. As in least square regression, $\mathbf{S}(\boldsymbol{\beta}) = \mathbf{0}$ means that the parameters have to be chosen so that the vector of deviations, usually called residuals, is uncorrelated with all regressor variables. However, an explicit solution is not possible because of the complicated dependency of $\mathbf{e}$ on the parameters. Defining $\mathbf{W}$ as the $N \times N$ diagonal matrix with diagonal entries $M_i\pi_i(1-\pi_i)$, i.e.,

$$\mathbf{W} = \text{diag}\left(M_1\pi_1(1-\pi_1), \ldots, M_N\pi_N(1-\pi_N)\right), \tag{16}$$

the information function can be written as

$$\mathbf{I}(\boldsymbol{\beta}) = \mathbf{X}'\mathbf{W}\mathbf{X}. \tag{17}$$

For all finite values of the parameter vector $\boldsymbol{\beta}$, the diagonal elements of $\mathbf{W}$ are strictly positive. Thus, $\mathbf{I}(\boldsymbol{\beta})$ is positive definite for all finite $\boldsymbol{\beta} \in \mathbb{R}^{p+1}$, provided the design matrix $\mathbf{X}$ has full rank $p+1$, i.e., has $p+1$ linearly independent rows. This is exactly the same condition as encountered in least squares regression. Positive definiteness of $\mathbf{I}(\boldsymbol{\beta})$ means that the log-likelihood function is concave, and therefore, has at most one maximum.

Numerical maximization of the likelihood function can be done efficiently with the Newton-Raphson algorithm. Starting with an arbitrary initial vector $\boldsymbol{\beta}^{(0)} \in \mathbb{R}^{p+1}$, e.g., $\boldsymbol{\beta}^{(0)} = \mathbf{0}$, the equation

$$\boldsymbol{\beta}^{(j+1)} = \boldsymbol{\beta}^{(j)} + \left[\mathbf{I}\left(\boldsymbol{\beta}^{(j)}\right)\right]^{-1} \mathbf{S}\left(\boldsymbol{\beta}^{(j)}\right) \qquad j = 0, 1, \ldots \tag{18}$$

is iterated until a suitable convergence criterion is met. No further details on the numerical algorithm are given here. As usual, we will denote the parameter estimates by $\hat{\boldsymbol{\beta}} = \left(\hat{\beta}_0, \hat{\beta}_1, \ldots, \hat{\beta}_p\right)'$, from which estimated probabilities and estimated expected frequencies are obtained by

$$\hat{\pi}_i = \frac{\exp\left(\mathbf{x}_i'\hat{\boldsymbol{\beta}}\right)}{1+\exp\left(\mathbf{x}_i'\hat{\boldsymbol{\beta}}\right)}$$

and

$$\hat{y}_i = M_i\hat{\pi}_i, \qquad i = 1, \ldots, N, \tag{19}$$

respectively.

The second important reason for computing the information function is, again, its use in approximating the distribution of the maximum likelihood estimators. By the same reasoning as in Section 7.5, the distribution of $\hat{\boldsymbol{\beta}}$ for large N is approximately $(p+1)$-variate normal with mean vector $\boldsymbol{\beta}$ and covariance matrix $\boldsymbol{\Sigma} = \{E\,[\mathbf{I}(\boldsymbol{\beta})]\}^{-1}$. But the random vectors $\mathbf{Y}_i$ do not appear in $\mathbf{I}(\boldsymbol{\beta})$, and therefore $E\,[\mathbf{I}(\boldsymbol{\beta})] = \mathbf{I}(\boldsymbol{\beta})$.

Replacing parameters by their maximum likelihood estimates, we define

$$\hat{\Sigma} = \left[\mathbf{I}\left(\hat{\beta}\right)\right]^{-1} \tag{20}$$

as the estimated covariance matrix of $\hat{\beta}$. In particular, the square roots of the diagonal elements of $\hat{\Sigma}$ are the standard errors $\text{se}(\hat{\beta}_0), \ldots, \text{se}(\hat{\beta}_p)$.

Example 7.6.1 A physiological experiment. In a study of the effect of the rate and volume of air inspired by human subjects on the occurrence or nonoccurrence of vasoconstriction in the skin of the fingers, Finney (1947) reports the data reproduced in Table 7.6.1. There are 39 observations of three variables. The regressor variables are VOLUME = volume of air inspired and RATE = rate of air inspired. The response variable is binary with $Y = 1$ indicating occurrence and $Y = 0$ indicating nonoccurrence of vasoconstriction. All M_i are equal to 1, and therefore we assume

$$Y_i \sim \text{Bernoulli}(\pi_i), \qquad i = 1, \ldots, N,$$

where

$$\text{logit}(\pi_i) = \beta_0 + \beta_1 \text{VOLUME}_i + \beta_2 \text{RATE}_i.$$

However, several of the measurements were obtained on the same human subjects, which casts some doubt on the validity of the assumption of independence of all Y_i. We will illustrate the use of multiple logistic regression ignoring this concern; see Finney (1947) or Aitkin et al. (1989, pp. 167 ff) for some discussion.

Table 7.6.2 displays numerical results of the Newton-Raphson algorithm, starting at $\beta^{(0)} = \mathbf{0} \in \mathbb{R}^3$. Again, convergence is quite fast. The information function, evaluated at $\hat{\beta}$, is

$$\mathbf{I}(\hat{\beta}) = \begin{pmatrix} 4.8217 & 6.0519 & 8.5115 \\ 6.0519 & 8.7765 & 9.2721 \\ 8.5115 & 9.2721 & 17.9079 \end{pmatrix},$$

with inverse

$$\hat{\Sigma} = \left[\mathbf{I}\left(\hat{\beta}\right)\right]^{-1} = \begin{pmatrix} 10.4536 & -4.3250 & -2.7292 \\ -4.3250 & 2.0409 & 0.9989 \\ -2.7292 & 0.9989 & 0.8358 \end{pmatrix}.$$

Standard errors of estimated logistic regression coefficients are the square roots of the diagonal elements in $\hat{\Sigma}$. We have the following numerical results:

$$\hat{\beta}_0 = -9.530 \qquad \text{se}(\hat{\beta}_0) = 3.323$$

$$\hat{\beta}_1 = 3.882 \qquad \text{se}(\hat{\beta}_1) = 1.429$$

$$\hat{\beta}_2 = 2.649 \qquad \text{se}(\hat{\beta}_2) = 0.914.$$

Table 7.6.1 Vasoconstriction data from Finney (1947). Variables are as explained in Example 7.6.1.

Volume	Rate	Y
3.70	0.825	1
3.50	1.090	1
1.25	2.500	1
0.75	1.500	1
0.80	3.200	1
0.70	3.500	1
0.60	0.750	0
1.10	1.700	0
0.90	0.750	0
0.90	0.450	0
0.80	0.570	0
0.55	2.750	0
0.60	3.000	0
1.40	2.330	1
0.75	3.750	1
2.30	1.640	1
3.20	1.600	1
0.85	1.415	1
1.70	1.060	0
1.80	1.800	1
0.40	2.000	0
0.95	1.360	0
1.35	1.350	0
1.50	1.360	0
1.60	1.780	1
0.60	1.500	0
1.80	1.500	1
0.95	1.900	0
1.90	0.950	1
1.60	0.400	0
2.70	0.750	1
2.35	0.030	0
1.10	1.830	0
1.10	2.200	1
1.20	2.000	1
0.80	3.330	1
0.95	1.900	0
0.75	1.900	0
1.30	1.625	1

Table 7.6.2 Performance of the Newton-Raphson algorithm in the vasoconstriction example. Displayed are the values $\beta^{(j)}$ in the jth iteration and the associated value of the log-likelihood function $\ell(\beta^{(j)})$.

Iteration (j)	$\beta_0^{(j)}$	$\beta_1^{(j)}$	$\beta_2^{(j)}$	$\ell(\beta^{(j)})$
0	0.0000000	0.0000000	0.0000000	-27.0327400
1	-4.4504521	1.6044453	1.3737081	-17.0121217
2	-7.0285972	2.7221447	2.0464592	-15.2969115
3	-8.8445766	3.5680216	2.4830076	-14.9119003
4	-9.4761167	3.8582065	2.6358622	-14.8862967
5	-9.5292591	3.8820066	2.6490359	-14.8861523
6	-9.5295856	3.8821504	2.6491183	-14.8861522
7	-9.5295856	3.8821504	2.6491183	-14.8861522

Thus, an increase in either of the variables VOLUME and RATE appears to increase the probability of vasoconstriction.

Instead of a list of estimated probabilities (19), Figure 7.6.1 shows a scatterplot of VOLUME vs RATE with circles marking "successes" and black squares marking "failures." Superimposed on the scatterplot are lines of equal estimated probability of success, in percent. This graph allows a rough, informal assessment of the fit of the model, which appears satisfactory. □

As in least squares regression and in linear discriminant analysis, it is often useful to assess the contribution of individual variables or subsets of variables to the logistic regression function. The asymptotic likelihood theory of Section 4.3 provides relatively straightforward testing procedures. Note, however, that we need to invoke a generalization of Theorem 4.3.3 because the Y_i are not identically distributed.

For given maximum likelihood estimates $\hat{\beta}$, we can evaluate the log-likelihood function (7) at its maximum. Writing the maximum as a function of the estimated probabilities $\hat{\pi}' = (\hat{\pi}_1, \ldots, \hat{\pi}_M)$,

$$\ell(\hat{\pi}) = C + \sum_{i=1}^{N} \left[y_i \log \hat{\pi}_i + (M_i - y_i) \log(1 - \hat{\pi}_i) \right]. \tag{21}$$

It is customary to compare $\ell(\hat{\pi})$ to the value of the log-likelihood function for a "perfect" model in which each of the estimated success probabilities is

$$\tilde{\pi}_i = \frac{y_i}{M_i}, \qquad i = 1, \ldots, N. \tag{22}$$

The $\tilde{\pi}_i$ may be regarded as maximum likelihood estimates in a logistic regression model that has as many parameters as observations, often called the *saturated model* or *maximal model*. However, there is no need to formally write such a model (see Exercise 3), because all we need are the $\tilde{\pi}_i$. The value of the log-likelihood function

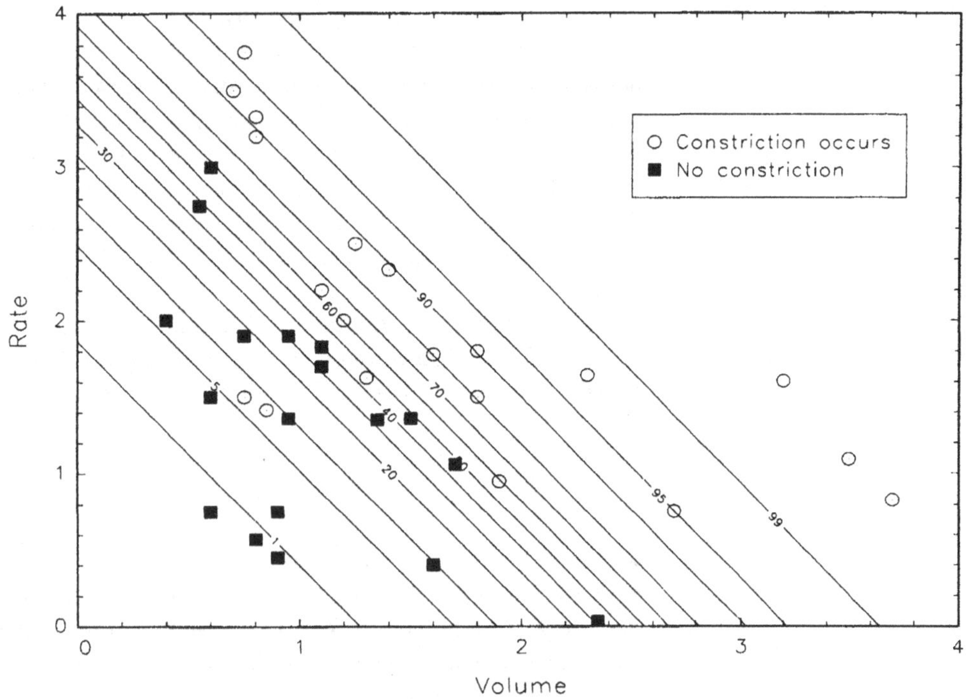

Figure 7-6-1 Plot of Volume vs. Rate in Example 7.6.1. Open circles and black squares mark observations where vasoconstriction did and did not occur, respectively. Parallel lines indicate sets of constant estimated probability of vasoconstriction.

for the "perfect" model with $\tilde{\pi}' = (\tilde{\pi}_1, \ldots, \tilde{\pi}_N) = (y_1/M_1, \ldots, y_N/M_N)$ is given by

$$\ell(\tilde{\pi}) = C + \sum_{i=1}^{N} \left[y_i \log \frac{y_i}{M_i} + (M_i - y_i) \log \frac{M_i - y_i}{M_i} \right]. \tag{23}$$

Some care must be taken in evaluating (23); if $y_i = 0$, then $\log(y_i/M_i)$ is not defined and the term $y_i \log(y_i/M_i)$ is to be interpreted as zero. Similarly, if $y_i = M_i$, then the second term in the sum of (23) is to be interpreted as zero. See Exercise 4 for some mathematical justification. In fact, if $M_i = 1$, then y_i can only be equal to 0 or M_i.

The *deviance* D of a model with parameter estimates $\hat{\beta}$ and estimated success probabilities $\hat{\pi}_i$ is defined as twice the difference between (23) and (21), i.e.,

$$\begin{aligned} D &= 2\left[\ell(\tilde{\pi}) - \ell(\hat{\pi})\right] \\ &= 2\sum_{i=1}^{N} \left[y_i \left(\log \frac{y_i}{M_i} - \log \hat{\pi}_i \right) + (M_i - y_i) \left(\log \frac{M_i - y_i}{M_i} - \log(1 - \hat{\pi}_i) \right) \right] \\ &= 2\sum_{i=1}^{N} \left[y_i \log \frac{y_i}{\hat{y}_i} + (M_i - y_i) \log \frac{M_i - y_i}{M_i - \hat{y}_i} \right], \end{aligned} \tag{24}$$

where $\hat{y}_i = M_i \hat{\pi}_i$ as in equation (19). The same note applies as explained in the text following (23) if $y_i = 0$ or $y_i = M_i$. Since $\ell(\tilde{\pi}) \geq \ell(\hat{\pi})$, the deviance is always nonnegative (see Exercise 5) and may be viewed as a measure of fit of a model, similar to the residual sum of squares in least squares regression. Asymptotic theory shows that if all M_i are large and the specified logistic regression model is correct, the distribution of D is approximately chi-square with $(N - p - 1)$ degrees of freedom, that is, a large value of D may indicate that the model is misspecified. However, we will not elaborate on this since we will use deviance mostly as a tool to compare different models.

The question of redundancy of variables or subsets of variables arises naturally, just as in linear discriminant analysis; see Sections 5.6 and 6.5. Formally, let us partition the vector of logistic regression coefficients into $(q+1)$ and $(p-q)$ components as

$$\beta = \begin{pmatrix} \beta^{(1)} \\ \beta^{(2)} \end{pmatrix}, \tag{25}$$

where $\beta^{(1)} = (\beta_0, \ldots, \beta_q)'$ and $\beta^{(2)} = (\beta_{q+1}, \ldots, \beta_p)'$. Here, q can be any number between 0 and $p - 1$. Then the hypothesis of simultaneous redundancy of variables X_{q+1} to X_p can be $\beta^{(2)} = \mathbf{0} \in \mathbb{R}^{p-q}$. We will refer to the model with all p variables included as the *full model* and to the model containing only variables X_1 to X_q as the *reduced model*, respectively.

Let $\hat{\pi}^{(p)}$ and $\hat{\pi}^{(q)}$ be the vectors of estimated success probabilities for the full model and the reduced model, respectively, and let D_p and D_q be the associated values of the deviance. Because the reduced model is obtained from the full model by imposing a constraint, $D_q \geq D_p$ holds (see Exercise 6). The log-likelihood ratio statistic for testing the hypothesis $\beta^{(2)} = \mathbf{0}$ is

$$\begin{aligned} \text{LLRS} &= 2\left[\ell(\hat{\pi}^{(p)}) - \ell(\hat{\pi}^{(q)})\right] \\ &= 2\left\{\left[\ell(\hat{\pi}^{(p)}) - \ell(\tilde{\pi})\right] - \left[\ell(\hat{\pi}^{(q)}) - \ell(\tilde{\pi})\right]\right\} \\ &= D_q - D_p, \end{aligned} \tag{26}$$

that is, the log-likelihood ratio statistic for redundancy of a subset of variables is the difference between the two deviances. It follows from the large sample theory of likelihood ratio tests that, assuming that the logistic regression model is correct and the hypothesis holds true, LLRS follows approximately a chi square distribution with $(p-q)$ degrees of freedom. We will accept the hypothesis if LLRS $\leq c$ and reject the hypothesis if LLRS $> c$, where c is the $(1-\alpha)$-quantile of the chi-square distribution with $(p-q)$ degrees of freedom, and α is the significance level. The condition that all M_i be large does not apply to the test for redundancy of variables because p and q are fixed in the current context. In contrast, in a saturated model, the number of parameters estimated is N, that is, it grows with the number of observations. A formal proof of this statement is beyond the level of this text.

An important special case of the test for redundancy is the overall test of significance, i.e., the test of the hypothesis $\beta_1 = \cdots = \beta_p = 0$. In this particular reduced model, it is assumed that

$$\text{logit}(\pi_i) = \beta_0 \qquad \text{for } i = 1, \ldots, N, \tag{27}$$

and the common maximum likelihood estimate of all π_i is given by $\bar{\pi} = \left(\sum_{i=1}^N y_i\right)/\left(\sum_{i=1}^N M_i\right)$; see Exercise 6 in Section 7.5. Thus, the estimated expected frequencies under the reduced model are

$$\hat{y}_i = M_i\bar{\pi} \qquad i = 1, \ldots, N, \tag{28}$$

and the deviance of the reduced model with no regressor variables included is

$$D_0 = 2\sum_{i=1}^{N}\left[y_i \log \frac{y_i}{M_i\bar{\pi}} + (M_i - y_i)\log \frac{M_i - y_i}{M_i(1-\bar{\pi})}\right]. \tag{29}$$

The log-likelihood ratio statistic for the overall hypothesis of redundancy of all variables is LLRS $= D_0 - D_p$, and the value of LLRS is to be compared to critical values of the chi-square distribution with p degrees of freedom. This test parallels the overall test of significance in least squares regression and the test for redundancy of all variables in linear discriminant analysis.

Example 7.6.2 continuation of Example 7.6.1

Since there are two regressor variables, there are four models that can be considered, including the one with no variables included. Numerical results for the full model with two variables have been given in Example 7.6.1; the deviance is $D_2 = 29.772$. Numerical results for the remaining models follow.

- model $\text{logit}(\pi_i) = \beta_0 + \beta_1\text{VOLUME}_i$: Maximum likelihood estimates are $\hat{\beta}_0 = -1.664$ and $\hat{\beta}_1 = 1.336$ with standard errors $\text{se}(\hat{\beta}_0) = 0.812$ and $\text{se}(\hat{\beta}_1) = 0.616$. The deviance is $D_1 = 46.989$. The log-likelihood ratio test statistic for redundancy of RATE is LLRS $= D_1 - D_2 = 17.22$, to be compared with quantiles of the chi-square distribution with one degree of freedom. The hypothesis of redundancy of RATE would be rejected at all reasonable levels of significance.
- model $\text{logit}(\pi_i) = \beta_0 + \beta_2\text{RATE}_i$: Maximum likelihood estimates are $\hat{\beta}_0 = -1.353$ and $\hat{\beta}_2 = 0.844$ with standard errors $\text{se}(\hat{\beta}_0) = 0.793$ and $\text{se}(\hat{\beta}_2) = 0.439$. The deviance is $D_1 = 49.655$, and the log-likelihood ratio statistic for redundancy of VOLUME is LLRS $= D_1 - D_2 = 19.88$. Thus, the hypothesis of redundancy of VOLUME is rejected.
- model $\text{logit}(\pi_i) = \beta_0$: The maximum likelihood estimate of the single parameter is $\hat{\beta}_0 = 0.051$ with $\text{se}(\hat{\beta}_0) = 0.320$. This corresponds to a common estimate of the success probability $\hat{\pi}_i = \frac{1}{39}\sum_{i=1}^{39} y_i = 0.503$. The deviance is $D_0 =$

54.040. The log-likelihood ratio statistic for simultaneous redundancy of both variables is LLRS $= D_0 - D_2 = 24.27$, to be compared with quantiles of the chi-square distribution with two degrees of freedom. The hypothesis is rejected at any reasonable level of significance.

Thus, we conclude that both variables should be kept in the model (which, looking at Figure 7.6.1, is really not a surprise). Notice however, that we have totally ignored the earlier concerns about the possible lack of independence of the Y_i, whose effect on the distribution of LLRS would be difficult to assess. By computing several test statistics in the same analysis, we have also ignored concerns about multiple testing. In the current case, these concerns are unimportant, but they should be taken seriously in applications with many regressor variables or marginal significance. □

Tests for redundancy of a single variable are often performed using a somewhat simpler approach. For parameter estimates $\hat{\beta}_j$ and standard errors se$(\hat{\beta}_j)$, the ratios

$$w_j = \hat{\beta}_j / \text{se}(\hat{\beta}_j) \tag{30}$$

are called *Wald statistics*. For large N and if the hypothesis of redundancy of the jth variable holds true, w_j is approximately standard normal, a result whose proof is a little too technical to be given in this text. Thus we may use the following procedure for testing the hypothesis $\beta_j = 0$: Accept the hypothesis if $|w_j| \le c$, and reject otherwise, where c is the $(1 - \alpha/2)$-quantile of the standard normal distribution. Alternatively, one may construct a confidence interval for β_j and accept the hypothesis if the interval covers 0. Notice that the Wald-statistic for redundancy of X_j is not simply a transformation of the corresponding likelihood ratio statistic as in least squares regression. However, both the log-likelihood ratio statistic and w_j^2 have approximately a chi-square distribution with one degree of freedom under the hypothesis $\beta_j = 0$, and, in fact, if the hypothesis is true, then the two statistics are approximately the same. The interested reader is referred to Silvey (1975) for a good discussion of the relationship between log-likelihood ratio and Wald statistics.

As in least squares regression, categorical regressor variables can be entered in a logistic regression model by introducing indicator variables. We illustrate this with an example.

Example 7.6.3 Prediction of the need for psychiatric treatment. In Exercise 12 of Section 7.5, we introduced a data set relating the outcome of a psychiatric diagnosis to the score of patients on a 12 item General Health Questionnaire (GHQ); the data is reproduced in Table 7.5.5. Let X = GHQ score and G = gender, with $G = 0$ for males and $G = 1$ for females. For the data of males and females combined, let us study the model

$$\text{logit}(\pi_i) = \beta_0 + \beta_1 X_i + \beta_2 G_i, \qquad i = 1, \ldots, 17. \tag{31}$$

This may be interpreted as a model of two parallel logistic regression curves: since $G_i = 0$ for males, and $G_i = 1$ for females, equation (31) reads

$$\text{logit}(\pi_i) = \beta_0 + \beta_1 X_i \quad \text{for males}$$

and

$$\text{logit}(\pi_i) = \beta_0 + \beta_2 + \beta_1 X_i \quad \text{for females.}$$

Let $f_0(x) = \beta_0 + \beta_1 x$ and $f_1(x) = \beta_0 + \beta_2 + \beta_1 x$. Then $f_1(x) = f_0(x + \beta_2/\beta_1)$, that is, the logistic regression curve for females is shifted by an amount β_2/β_1 compared to the curve for males. Or, in other words, for all values of X, the probability that a female will need treatment is the same as the probability that a male will need treatment at $X - \beta_2/\beta_1$.

The following maximum likelihood estimates were obtained for model (31):

Parameter β_j	Estimate $\hat{\beta}_j$	Std. Error $\text{se}(\hat{\beta}_j)$	Wald-Statistic $\hat{\beta}_j/\text{se}(\hat{\beta}_j)$
β_0	−4.072	0.976	−4.17
β_1	1.433	0.291	4.96
β_2	0.794	0.929	0.85

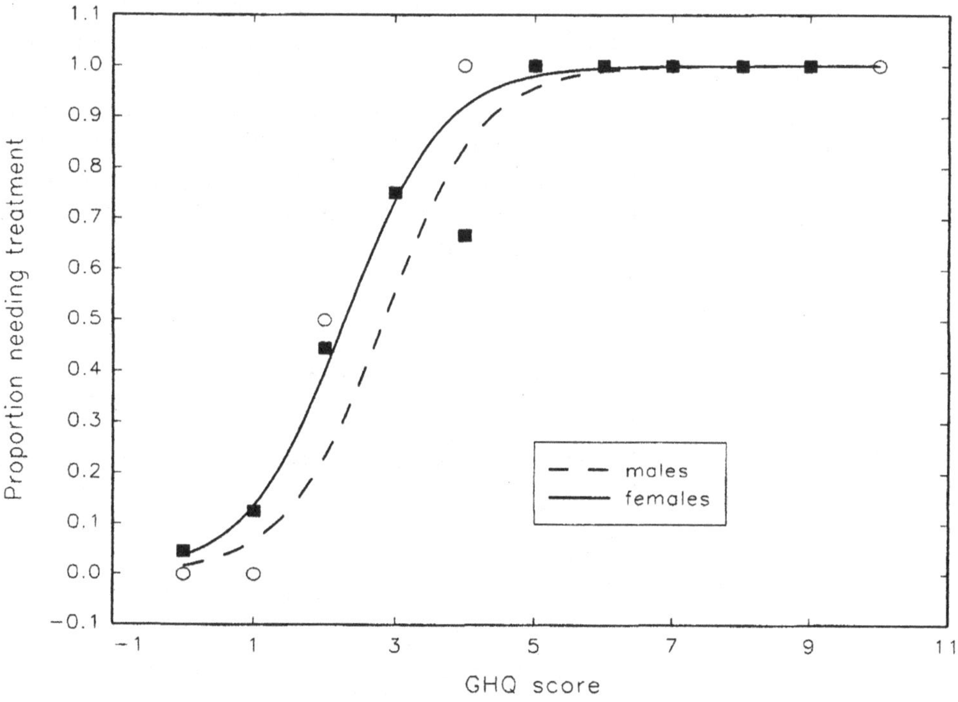

Figure 7-6-2
Simultaneous analysis of male and female data in Example 7.6.3, with an indicator variable for gender. Open circles indicate males; black squares, females.

Figure 7.6.2 shows observed proportions of subjects needing psychiatric treatment, along with the logistic regression curves

$$\hat{\pi} = \frac{\exp\left(\hat{\beta}_0 + \hat{\beta}_1 x\right)}{1 + \exp\left\{\hat{\beta}_0 + \hat{\beta}_1 x\right\}} \quad \text{(for males)},$$

and

$$\hat{\pi} = \frac{\exp\left(\hat{\beta}_0 + \hat{\beta}_2 + \hat{\beta}_1 x\right)}{1 + \exp\left(\hat{\beta}_0 + \hat{\beta}_2 + \hat{\beta}_1 x\right)} \quad \text{(for females)}.$$

The curve for females is shifted by an amount of $\hat{\beta}_2/\hat{\beta}_1 \approx 0.55$ compared to the curve for males. According to the maximum likelihood estimates, females would be somewhat more likely to need psychiatric treatment than males for all values of the GHQ score. However, the Wald statistic for variable G is $\hat{\beta}_2/\text{se}(\hat{\beta}_2) = 0.85$, which is far from significant at any customary level. Omitting variable G from the analysis, and fitting a logistic model

$$\text{logit}(\pi_i) = \beta_0 + \beta_1 X_i$$

for the data of both genders combined, the parameter estimates $\hat{\beta}_0 = -3.454$ and $\hat{\beta}_1 = 1.441$ were obtained. This corresponds to a single logistic regression curve for both genders. The same decision about redundancy of GENDER is reached with the log-likelihood ratio statistic; see Exercise 7.

Instead of collapsing data across gender, we might also consider estimating gender-specific logistic regression curves, one for males and one for females. This can be done in two ways: (1) by splitting the data file in two files and estimating the parameters separately for the male and the female data or (2) by introducing a third regressor variable XG, the product of X and G, That is, XG will be zero for all males and will be equal to X for all females. We can then estimate the parameters of the model

$$\text{logit}(\pi_i) = \beta_0 + \beta_1 X_i + \beta_2 G_i + \beta_3 (XG)_i.$$

Both methods lead to the same estimated probabilities. However, in this particular example, a problem will arise with the numerical maximization procedure; see Exercise 7. □

In a more general setup, suppose that a regressor variable C is categorical, taking values $1, \ldots, k$. Define k indicator variables $I_1, \ldots, I_k$ by

$$I_j = \begin{cases} 1 & \text{if } C = j \\ 0 & \text{otherwise.} \end{cases} \tag{32}$$

Let $X_1, \ldots, X_r$ be the other regressors to be used. We may then consider a logistic model

$$\text{logit}(\pi) = \sum_{h=1}^{r} \beta_h X_h + \sum_{j=1}^{k} \alpha_j I_j \,, \tag{33}$$

where the index (i) for observations has been omitted for simplicity. This model can be interpreted as k shifted logistic regression surfaces:

$$\text{logit}(\pi) = \sum_{h=1}^{r} \beta_h X_h + \alpha_j \quad \text{if } C = j. \tag{34}$$

Model (33) has no intercept term, and therefore most commercial software will not be able to estimate its parameters since the presence of a first column of ones in the design matrix is automatically assumed. To get around this difficulty, we may choose $k - 1$ out of the k indicator variables and include an intercept term. Omitting the kth indicator variable, the model reads

$$\text{logit}(\pi) = \beta_0 + \sum_{h=1}^{r} \beta_h X_h + \sum_{j=1}^{k-1} \alpha_j I_j. \tag{35}$$

This model can equivalently be interpreted as

$$\text{logit}(\pi) = \begin{cases} \beta_0 + \gamma_j + \sum_{h=1}^{r} \beta_h X_h & \text{if } C = 1, \ldots, k-1, \\ \beta_0 + \sum_{h=1}^{r} \beta_h X_h & \text{if } C = k, \end{cases} \tag{36}$$

that is, the kth category serves as a reference in this parameterization, and the γ_j are "corrections" to the intercept. The parameters α_j in models (33) and γ_j in model (35) are related by

$$\alpha_j = \begin{cases} \beta_0 + \gamma_j & \text{if } C = 1, \ldots, k-1, \\ \beta_0 & \text{if } C = k. \end{cases} \tag{37}$$

Of course, in practical applications, we may choose any of the k categories of C as the reference category — the choice of the kth is purely for mathematical convenience.

Example 7.6.4 continuation of Example 7.5.6

Basketball — the final analysis. In the basketball data of Table 7.5.4, let X = distance from the hoop, and define four indicator variables I_S, I_A, I_C, and I_B, one for each of the players. The data matrix has to be set up differently from Table 7.5.4, with 70 rows, as illustrated in Table 7.6.3. As suggested in Example 7.5.6, we will use the square root of distance as a regressor, instead of distance.

Table 7.6.3 Data matrix for the combined analysis of all four players in the basketball example. This matrix has 70 rows: 20 for each of Steve, Andy, and Chris, and 10 for the author. Only the first two and the last two rows are shown for each player.

Distance	I_S	I_A	I_C	I_B	Y	M
1	1	0	0	0	6	6
2	1	0	0	0	5	6
⋮	⋮	⋮	⋮	⋮	⋮	⋮
19	1	0	0	0	0	6
20	1	0	0	0	2	6
1	0	1	0	0	5	6
2	0	1	0	0	4	6
⋮	⋮	⋮	⋮	⋮	⋮	⋮
19	0	1	0	0	1	6
20	0	1	0	0	0	6
1	0	0	1	0	5	6
2	0	0	1	0	5	6
⋮	⋮	⋮	⋮	⋮	⋮	⋮
19	0	0	1	0	3	6
20	0	0	1	0	0	6
1	0	0	0	1	4	6
2	0	0	0	1	3	6
⋮	⋮	⋮	⋮	⋮	⋮	⋮
9	0	0	0	1	1	6
10	0	0	0	1	1	6

First, consider the model

$$\text{logit}(\pi) = \beta_0 + \beta_1\sqrt{X} + \gamma_1 I_S + \gamma_2 I_A + \gamma_3 I_C. \tag{38}$$

In this model the author serves as a reference category. The following parameter estimates were obtained:

Parameter	Estimate	Std. Error	Wald-Statistic
β_0	1.068	0.385	2.775
β_1	−0.842	0.120	−6.990
γ_1	0.823	0.366	1.641
γ_2	0.604	0.368	1.641
γ_3	1.073	0.366	2.930

The deviance of this model is $D_4 = 89.791$. The parameter estimates $\hat{\gamma}_j$ are all positive, reflecting the fact that the author's sons are better at hitting the basket. Of the three $\hat{\gamma}_j$'s, $\hat{\gamma}_3$ is the largest, which means that Chris would be considered as the best of the four players. However, the three coefficients are similar in magnitude. Therefore, we consider a model in which Steve, Andy, and Chris have the same curve, whereas the author's curve is shifted:

$$\text{logit}(\pi) = \beta_0^* + \beta_1^* \sqrt{X} + \gamma^* I_B. \tag{39}$$

Parameter estimates are as follows:

Parameter	Estimate	Std. Error	Wald-Statistic
β_0^*	1.888	0.369	5.118
β_1^*	−0.835	0.120	−6.975
γ^*	−0.833	0.325	−2.563

The deviance is $D_2 = 92.372$. Although model (39) is not obtained directly from model (38) by setting some parameters equal to zero, the log-likelihood ratio statistic (26) can be used to compare the two models (see Exercise 9), giving LLRS $= D_2 - D_4 = 2.58$. For comparison, the 95 percent quantile of the chi-square distribution with two degrees of freedom is 5.99, and the simpler model (39) would therefore be accepted. Figure 7.6.3 shows the estimated logistic regression curves according to model (39); it suggests a further modification in which the authors curve would be steeper. See Exercise 9. □

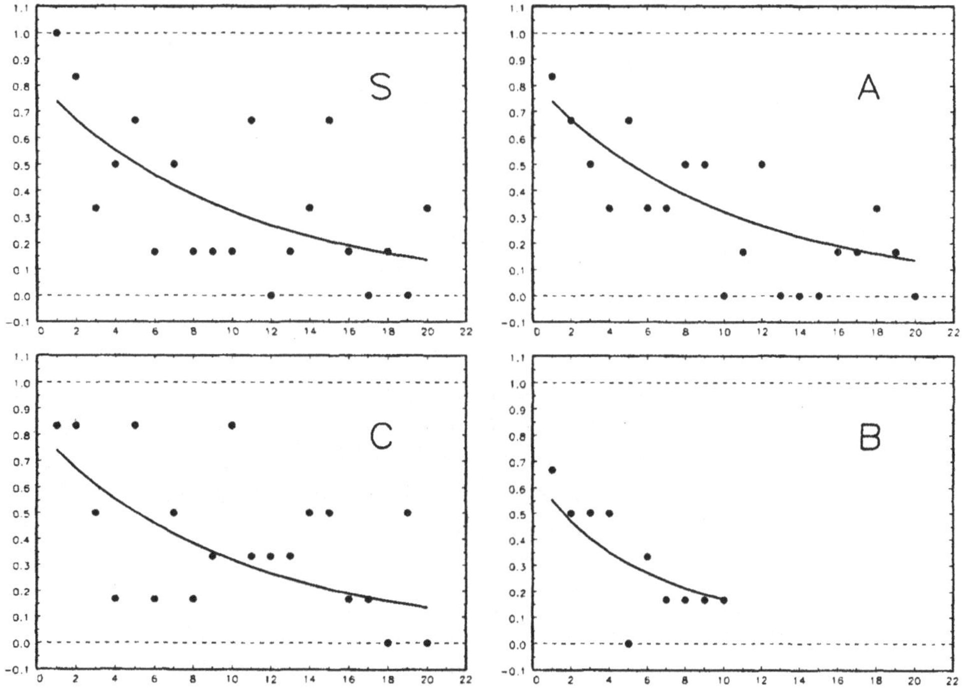

Figure 7-6-3
Simultaneous analysis of the results of all four basketball players. In the model shown here, Steve (S), Andy (A), and Chris (C) have the same logistic regression curve, whereas the author's curve (B) is shifted.

In Example 7.6.4 we used a transformation of the regressor variable (the square root of distance instead of distance) because this transformation seemed to give a better fitting model. In some situations, transformations are suggested by the very setup of the problem, as illustrated in the next example.

Example 7.6.5 Snails. Table 7.6.4 displays data on aquatic snails of the species *Potamopyrgus antipodarum* recorded by E. Levri and C. Lively from the Indiana University Department of Biology. During the day snails are less likely to be found on tops of rocks than at night to avoid predation. Levri and Lively collected snails at various times of the day and registered whether they were found on top of a rock or hiding underneath. In addition, the snails were classified in one of three mutually exclusive classes: infected by a disease, juvenile, or brooding female. The primary purpose of the study was to assess the influence of infection on the behavior of the snails. In Table 7.6.4, three indicator variables identify the groups: I_1 is the indicator variable for the infected group, I_2 for juvenile snails, and I_3 for brooding females. Time is given in decimal hours, i.e., 17.75 means 5:45pm.

First, consider first the data of group 3 only, i.e., the last nine rows in Table 7.6.4. Observed proportions of snails at the various time of day are shown graphically in Figure 7.6.4, along with a curve whose construction we are now going to describe. With Y_i denoting the number of snails found on top at time T_i, $i = 1, \ldots, 9$, assume $Y_i \sim$ binomial (π_i, M_i), where $\text{logit}(\pi_i)$ is a function of T_i. It would not be a good idea to fit the model $\text{logit}(\pi_i) = \alpha + \beta T_i$ since the probabilities should follow a 24-hour cyclic pattern. A simple model, some justification for which is given in Exercise 10, is to assume

$$\text{logit}(\pi_i) = h(T_i) = \beta_0 + \beta_1 C_i + \beta_2 S_i, \qquad i = 1, \ldots, 9, \tag{40}$$

where

$$\begin{aligned} C_i &= \cos\left(\frac{\Pi}{12} T_i\right), \\ S_i &= \sin\left(\frac{\Pi}{12} T_i\right), \qquad i = 1, \ldots, 9. \end{aligned} \tag{41}$$

(Here, $\Pi \approx 3.1415926535897$, not to be confused with the probability parameters). The function (40) has a 24-hour period, and may actually be considered as a first-order Fourier series approximation to an unknown periodic function.

Maximum likelihood estimates for the data of group 3 shown in Figure 7.6.4 were obtained as follows:

Parameter	Estimate	Std. Error	Wald-Statistic
β_0	0.386	0.115	3.351
β_1	1.858	0.178	10.445
β_2	0.399	0.155	2.575

Table 7.6.4 Snail data. I_1, I_2, and I_3 are indicator variables for the three types of snails (infected, juvenile, brooding female). Time = time of day in decimal hours; M = number of snails collected; and Y = number found on top of a rock. Data courtesy of Drs. Ed Levri and Curt Lively, Department of Biology, Indiana University.

I_1	I_2	I_3	TIME	Y	M
1	0	0	6.25	10	36
1	0	0	8.17	6	80
1	0	0	10.25	7	81
1	0	0	12.50	2	108
1	0	0	14.25	2	60
1	0	0	15.50	6	35
1	0	0	17.75	14	69
1	0	0	20.25	14	68
1	0	0	21.75	36	68
0	1	0	6.25	32	62
0	1	0	8.17	19	111
0	1	0	10.25	20	110
0	1	0	12.50	21	104
0	1	0	14.25	11	102
0	1	0	15.50	24	81
0	1	0	17.75	33	93
0	1	0	20.25	53	118
0	1	0	21.75	227	299
0	0	1	6.25	22	32
0	0	1	8.17	21	54
0	0	1	10.25	18	76
0	0	1	12.50	17	54
0	0	1	14.25	3	39
0	0	1	15.50	8	35
0	0	1	17.75	33	72
0	0	1	20.25	62	79
0	0	1	21.75	85	102

The curve in Figure 7.6.4 corresponds to these parameter estimates. The deviance is $D_2 = 11.696$. Since all M_i are fairly large, varying from 32 to 102, it is appropriate to assess the fit of the model using the large sample distribution of deviance mentioned in the text following (24). The observed value 11.696 is approximately the 93 percent quantile of the chi-square distribution with six degrees of freedom, and therefore we conclude that the fit is satisfactory but not perfect.

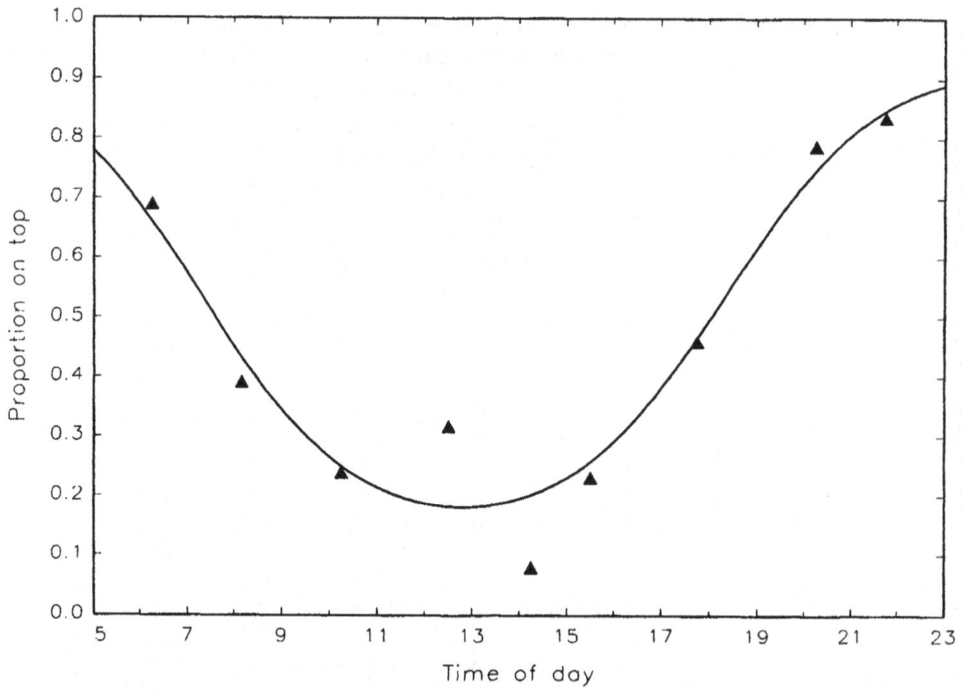

Figure 7-6-4 A periodic logistic regression model for the snail data. Only snails of type 3 are represented here.

Next we try to model the data of all three groups simultaneously. Similarly to Example 7.6.4, we use indicator variables for the different groups. For the model

$$\text{logit}(\pi) = \beta_0 + \beta_1 C + \beta_2 S + \gamma_2 I_2 + \gamma_3 I_3 \tag{42}$$

applied to all 27 observations, the following maximum likelihood estimates were obtained:

Parameter	Estimate	Std. Error	Wald-Statistic
β_0	−1.387	0.122	−11.389
β_1	1.633	0.087	18.744
β_2	0.240	0.084	2.852
γ_2	1.108	0.139	7.962
γ_3	1.694	0.155	10.911

Figure 7.6.5 shows the data of all three groups, along with the curves obtained from the above parameter estimates. For instance, the curve for group 2 corresponds to $\text{logit}(\hat{\pi}) = \hat{\beta}_0 + \hat{\beta}_1 C + \hat{\beta}_2 S + \hat{\gamma}_2$, where C and S are the cosine and sine functions as in equation (41). In earlier examples, an indicator variable caused a shift of the logistic response curve; here, the three periodic curves take their maxima and minima at the same time of day. See Exercise 11 for further refinements.

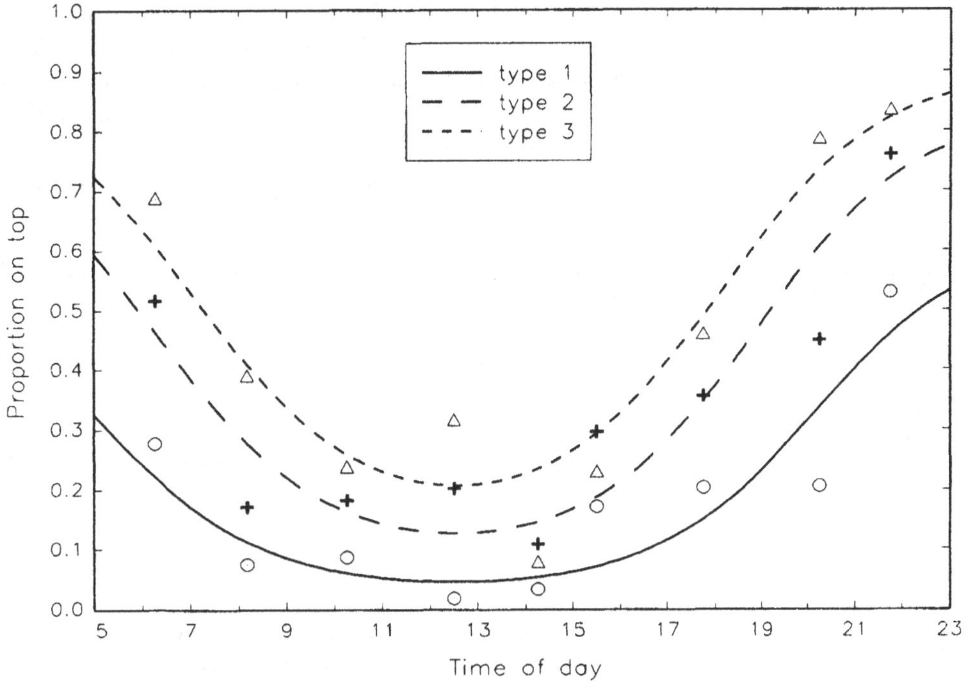

Figure 7-6-5 Simultaneous analysis of all three types of snails using periodic logistic regression and indicator variables. Circles, plus signs, and triangles mark data for groups 1, 2, and 3, respectively.

The deviance of the joint model for all three groups is 64.941, which would be highly significant (there are 27 observations and five parameters, so critical values are taken from the chi-square distribution with 22 degrees of freedom). This constitutes a distinct lack of fit, mostly due to groups 1 and 2 (see Exercise 11). Nevertheless, the curves shown in Figure 7.6.5 are a nice example of the power and flexibility of the logistic regression model. See Levri and Lively (1996) for further analysis. □

There are many more aspects of logistic regression we would need to discuss in a comprehensive treatment of the method. However, in the spirit of a first introduction, we will quit at this point, referring the interested student to the books listed in the Suggested Further Readings Section following the Bibliography. Among the omitted topics are residual analysis and generalizations of logistic regression to ordered response categories. Logistic regression is a special case of a class of models called *generalized linear models*, which also includes least squares regression, Poisson regression, and log-linear models that are used frequently in categorical data analysis.

Users of classification techniques are sometimes not sure whether they should use discriminant analysis or logistic regression for a given problem. A rough guideline is as follows. In the notation of this section, if **X** is fixed by the design of the experiment and Y is random, then discriminant analysis is not appropriate, as seen for instance in the basketball example. Conversely, if Y is fixed and $\mathbf{X}$ is random, then a discriminant analysis approach is appropriate. If both $\mathbf{X}$ and Y are random, then either method may be used. In the special case where $\mathbf{X}$ follows a multivariate normal mixture model with equal covariance matrices in both components, the logistic regression function coincides with the posterior probabilities of Y, given $\mathbf{X} = \mathbf{x}$, as seen from Exercise 12.

Exercises for Section 7.6

1. Prove equation (9). *Hint*: First, show that $1 - \pi_i = 1/[1 + \exp(\mathbf{x}_i'\beta)]$.
2. Verify equation (12).
3. A saturated logistic regression model can be written as

$$\text{logit}(\pi_i) = \alpha_i \qquad i = 1, \ldots, N,$$

where there are no constraints on the α_i. Show that the maximum likelihood estimates of the π_i are y_i/M_i.

4. Expressions of the form "0 log(0)" occur in the computation of deviance; see equations (23) and (24). Such terms are set equal to zero. Justify this by showing that $\lim_{x \to 0} x \log(x) = 0$.
5. Show that the deviance D of a logistic regression model is always nonnegative, and $D = 0$ exactly if all $\hat{\pi}_i = y_i/M_i$.
6. Suppose that D_p and D_q are the deviances for a full model and a reduced model, where the reduced model is obtained from the full model by setting some $(p - q)$ parameters equal to zero. Show that $D_q \geq D_p$, with equality exactly if all probabilities are estimated identically in both models.
7. This exercise expands on Example 7.6.3.
   (a) In the model $\text{logit}(\pi) = \beta_0 + \beta_1 X + \beta_2 G$ from equation (31), test the hypothesis $\beta_2 = 0$ using the log-likelihood ratio statistic.
   (b) Consider the model $\text{logit}(\pi) = \beta_0 + \beta_1 X + \beta_2 G + \beta_3 XG$. Prove formally that the estimated probabilities $\hat{\pi}_i$ are the same as if separate logistic regressions were performed for males and females.
   (c) Try to estimate the parameters in the model from part (b). Explain what happens. *Hint*: See Exercises 11 and 12 in Section 7.5.
8. Consider a logistic regression model with a categorical variable C that takes values 1, ..., k. Define k indicator variables $I_1, \ldots, I_k$ for the categories. For simplicity, suppose that there is only one additional regressor variable X, and study the model

$$\text{logit}(\pi) = \beta_0 + \beta_1 X + \sum_{h=1}^{k-1} \gamma_h I_h + \sum_{h=1}^{k-1} \delta_h X I_h \,, \tag{43}$$

where $XI_h = X$ if $I_h = 1$, and $XI_h = 0$ if $I_h = 0$. (These regressor variables are sometimes called interactions). Show that the estimated probabilities $\hat{\pi}_i$ are the same as if separate logistic regression models were fitted to the data in each category, i.e., the parameters of the model

$$\text{logit}(\pi) = \alpha_{h0} + \alpha_{h1}X$$

are estimated using the data of category h only.

9. This exercise expands on Example 7.6.4.

(a) Why can the log-likelihood ratio statistic be applied to the comparison of models (38) and (39)? Explain. *Hint*: In model (38), use a different set of indicator variables.

(b) Estimate the parameters of model (39) with an additional regressor variable $\sqrt{X}I_B$. Graph the results similarly to Figure 7.6.3. Verify numerically that the estimated probabilities from this model are the same as when separate logistic regression models are fitted for the author and for the pooled data of Steve, Andy, and Chris. Test for significance of the newly introduced "interaction" variable, using either the Wald-statistic or the log-likelihood ratio statistic.

10. Consider the periodic logistic regression model

$$\text{logit}(\pi) = \alpha_0 + \alpha_1 \cos(X - \alpha_2).$$

Show that this model can be written as

$$\text{logit}(\pi) = \beta_0 + \beta_1 \cos X + \beta_2 \sin X$$

for some parameters β_0, β_1, and β_2. How are the α_j and the β_j related? *Hint*: $\cos(u - v) = \cos(u)\cos(v) + \sin(u)\sin(v)$.

11. This exercise expands on Example 7.6.5.

(a) Estimate the parameters of a logistic regression model with variables C, S, I_2, I_3, CI_2, CI_3, SI_2, and SI_3. Compute estimated probabilities $\hat{\pi}_i$ for all 27 observations.

(b) Estimate the parameters of a periodic logistic regression model separately for each group of snails, using C and S as regressors. How are the parameter estimates related to those of part (a)?

(c) Test for simultaneous redundancy of the last four variables in the list of part (a), using the log-likelihood ratio statistic, and argue that this can be considered a test of the hypothesis that the curves of all three groups reach a minimum at the same time of day.

(d) Write the deviance of model (42) as $D = D^{(1)} + D^{(2)} + D^{(3)}$, where $D^{(j)}$ is the part of the deviance due to the data in group j. That is, write equation (42) as a sum of three terms, corresponding to nine observations each. Thus, verify numerically that the large deviance of model (42) is mostly due to groups 1 and 2.

12. Suppose that $Y \sim \text{Bernoulli}(\theta)$, and, conditionally on $Y = j$, $\mathbf{X}$ is p-variate normal with mean vector $\boldsymbol{\mu}_j$ and covariance matrix $\boldsymbol{\psi}$. Let $\pi_{j\mathbf{x}} = \Pr[Y = j|\mathbf{X} = \mathbf{x}]$. Show that

$$\text{logit}(\pi_{j\mathbf{x}}) = \beta_0 + \beta_1'\mathbf{x}$$

for some $\beta_0 \in \mathbb{R}$ and $\beta_1 \in \mathbb{R}^p$. Find β_0 and β_1 explicitly, and thus, show that β_1 is proportional to the vector of linear discriminant function coefficients.

13. This exercise shows how logistic regression can be used in categorical data analysis. In January 1975, the Committee on Drugs of the American Academy of Pediatrics recommended that tetracycline drugs not be used for children under age eight. A two-year study was conducted in Tennessee to investigate the extent to which physicians prescribed this drug between 1973 and 1975. In one part of the study, family practice physicians were categorized according to their county of practice (urban, intermediate, rural), and according to whether or not they had prescribed tetracycline to at least one patient under age eight. The following table summarizes data from Ray (1977). We might be interested mostly in estimating the proportion of doctors who prescribe tetracycline and in testing if the probability that a doctor prescribes tetracycline varies by type of county.

	Urban	Intermediate	Rural
Tetracycline (y_i)	65	90	172
No Tetracycline	149	136	158
Total (M_i)	214	226	330

Assume that the number of doctors in county type i who prescribe tetracycline is binomial(M_i, π_i), the M_i's being the column totals in the frequency table. Define two indicator variables X_1 and X_2 as

$$X_1 = \begin{cases} 1 & \text{for Urban,} \\ 0 & \text{otherwise} \end{cases}$$

and

$$X_2 = \begin{cases} 1 & \text{for Intermediate} \\ 0 & \text{otherwise.} \end{cases}$$

(a) Estimate the parameters of the model $\text{logit}(\pi_i) = \beta_0 + \beta_1 x_{i1} + \beta_2 x_{i2}, i = 1, 2, 3$. (This is a saturated model).

(b) Test the hypothesis $\beta_1 = \beta_2 = 0$, using the log-likelihood ratio statistic. Show that this hypothesis is equivalent to the hypothesis that the probability of prescribing tetracycline does not depend on the type of county.

14. Table 7.6.5 displays data recorded by K. Dorman and D. Whitehead of the Indiana University Department of Biology on the frequency of flowering dogwood trees in Hoosier National Forest in Southern Indiana. For 39 equally sized plots in the forest, Dorman and Whitehead counted the total number of trees and the number of flowering dogwood trees, and recorded the aspect (direction of steepest descent, in degrees) of the plot.

(a) Let Y = number of flowering dogwood trees and M = number of trees total. Estimate the parameters of a periodic logistic regression model, similar to Ex-

Table 7.6.5 Flowering dogwood tree data. Aspect is the orientation of a plot, i.e., the direction of steepest descent, in degrees measured clockwise from north. M is the total number of trees on the plot, and Y is the number of flowering dogwood trees.

Aspect	M	Y
15	29	0
36	68	0
45	68	0
45	30	0
47	45	0
50	33	0
60	41	0
63	30	0
64	17	0
73	24	0
80	16	0
80	12	0
82	13	0
96	31	0
98	31	0
99	32	0
106	16	0
110	58	0
130	42	0
142	60	3
146	57	0
150	42	1
185	64	0
191	31	0
198	32	3
199	53	17
210	57	1
210	36	0
217	32	0
218	31	1
226	40	9
230	42	0
231	36	0
235	32	8
279	98	0
328	38	0
339	49	0
340	21	0
355	28	0

Unpublished data on vegetation structure; courtesy of Ms. Karin Dorman and Dr. Don Whitehead, Department of Biology, Indiana University.

ample 7.6.5, to estimate the probability that a tree is a flowering dogwood, as a function of aspect. Use the cosine and sine of aspect as regressors. Graph the results similarly to Figure 7.6.4.

(b) When you look at the data of Table 7.6.5 or at the graph from part (a), you should have serious doubts about the validity of the logistic regression model. Which assumption is violated?

(c) Estimate the parameters of a periodic model as in part (a), with the following modification: Consider each of the 39 plots as a single Bernoulli experiment (that is, set $M_i = 1$ for all observations), and consider a plot a "success" if it contains at least one flowering dogwood tree, and a "failure" otherwise. In this setup π_i is the probability that a plot has at least one black chestnut tree, as a function of aspect. Graph the response curve again.

Suggested Further Reading

General: Hand (1981), McLachlan (1992).
Section 7.1: Anderson (1984, Chapter 6), Krzanowski (1988, Chapter 12).
Section 7.2: Bartlett and Please (1963), Bensmail and Celeux (1996), Flury et al. (1997), Friedman (1989).
Section 7.3: Fisher (1936), Johnson and Wichern (1988).
Section 7.4: Hand and Taylor (1987), Rencher (1995, Chapter 6).
Sections 7.5 and 7.6: Dobson (1990), Efron (1975), Hosmer and Lemeshow (1989), Lavine (1991), McCullagh and Nelder (1989).

8 Linear Principal Component Analysis

8.1 Introduction

Principal component analysis is a classical multivariate technique dating back to publications by Pearson (1901) and Hotelling (1933). Pearson focused on the aspect of approximation: Given a p-variate random vector (or a "system of points in space," in Pearson's terminology), find an optimal approximation in a linear subspace of lower dimension. More specifically, Pearson studied the problem of fitting a line to multivariate data so as to minimize the sum of squared deviations of the points from the line, deviation being measured orthogonally to the line. We will discuss Pearson's approach in Section 8.3; however, it will be treated in a somewhat more abstract way by studying approximations of multivariate random vectors using the criterion of mean-squared error.

Hotelling's reinvention of principal components focused on a different aspect: Suppose that $\mathbf{X}$ is a p-variate random vector. Then how can we transform $\mathbf{X}$ linearly into a new random vector $\mathbf{U}$ whose components are uncorrelated or even independent? Expanding on this theme, if the linear transformation from $\mathbf{X}$ to $\mathbf{U}$ is nonsingular, then we should be able to write $\mathbf{X}$ as a linear transformation of $\mathbf{U}$, i.e., express the p measured variables as functions of p or fewer uncorrelated and unobserved variables. The latter aspect moves Hotelling's approach into the neighborhood of factor analysis about which the current author has mixed feelings. Therefore, it is rather downplayed in this introduction, although Section 8.4 and, in particular, Example 8.4.5 contain some relevant material.

Our treatment of principal component analysis will be in the spirit of approximation, both in the sense of Pearson's orthogonal least squares and in the sense of self-consistency, as proposed by Tarpey and Flury (1996), based on earlier work by Hastie and Stuetzle (1989). The notion of self-consistency appears, to the author's knowledge, for the first time in a textbook on multivariate statistics. Since self-consistency is a relatively abstract concept, we will give it a detailed treatment in Section 8.2. The reader who feels uncomfortable with it may prefer the more classical derivation in terms of orthogonal least squares given in the second part of Section 8.3. Whichever approach is chosen, it will be important to have a solid understanding of notions like orthogonal matrices, eigenvectors, and eigenvalues; see Appendix A.5. A particularly important technical tool used in this section is the projection matrix associated with the orthogonal projection on a line or plane; this topic is outlined in Appendix A.8.

The reader who is interested mostly in practical applications of principal component analysis might find the theoretical introduction given by Sections 8.2 to 8.4 difficult. Therefore, we start with a practical, albeit artificially simple, example that should help the student to find sufficient motivation for going through the possible hardship of the technical derivations.

Example 8.1.1 Shell dimensions of turtles. This is a rather well-known and classical example due to Jolicoeur and Mosimann (1960), which we have encountered first in Exercise 4 of Chapter 1. For the current purpose, we use only two variables, namely, $X_1 = 10\log(\text{carapace length})$, and $X_2 = 10\log(\text{carapace width})$ of 24 male turtles of the species *Chrysemys picta marginata* (painted turtle). The raw data is given in Table 1.4. The log transformation is common in morphometric studies; we will return to it in Section 8.6. The multiplication of both log-transformed variables by 10 is purely for numerical convenience and is not essential, but notice that the same factor is used for both variables.

Figure 8.1.1(a) shows a scatterplot of X_1 vs. X_2, along with a straight line that appears to approximate the scatterplot rather well. Actually, the straight line was obtained by minimizing the sum of squared deviations of the points from the line, distance being measured orthogonally to the line. This illustrates the orthogonal least squares principle introduced by Pearson (1901). Notice that this line is not a regression line. In fact, the regression line of X_2 on X_1 would be flatter, and the regression line of X_1 on X_2 would be steeper than the line shown in Figure 8.1.1(a); see Exercise 1. In the current context, none of the variables is declared as "dependent" or "independent." This constitutes an essential difference from regression analysis.

In what sense does the straight line approximate the scatterplot? Perhaps a better way to formulate the idea of approximation is to say that we approximate the bivariate data by their projection on a straight line, that is, for each data point $\mathbf{x}_i = (x_{i1}, x_{i2})'$,

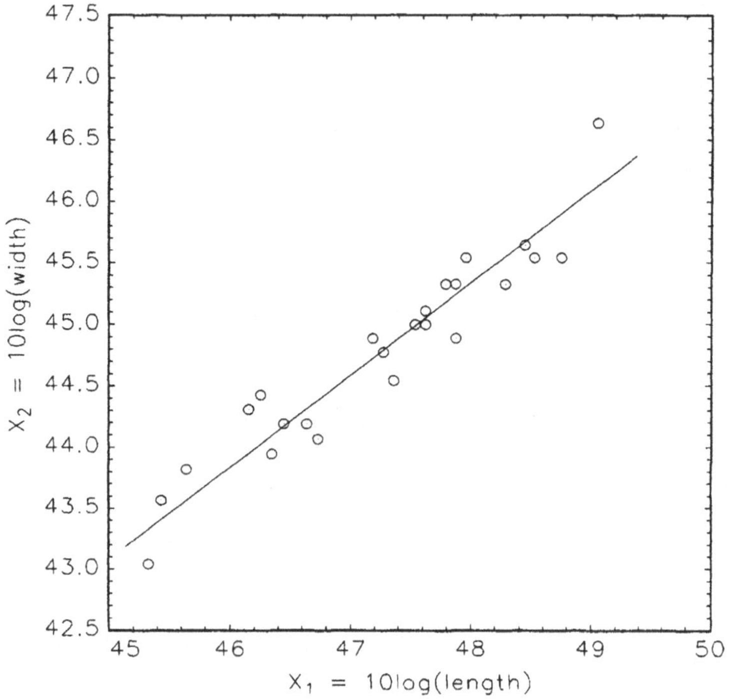

Figure 8-1-1A

we will find the coordinates of its orthogonal projection on the line. Denote the coordinates of the projected points by $\mathbf{y}_i = (y_{i1}, y_{i2})'$, $i = 1, \ldots, 24$. Then $\mathbf{y}_i$ is the point on the line closest to $\mathbf{x}_i$. Now, we will give some numerical details of this projection without theoretical justification. Let $\bar{\mathbf{x}} = (\bar{x}_1, \bar{x}_2)' = (47.25, 44.78)'$ denote the vector of sample means, and let $\mathbf{b} = (b_1, b_2)' = (0.80, 0.60)'$ (which for now has fallen as manna from heaven). Define for $i = 1, \ldots, 24$,

$$\begin{aligned} u_i &= \mathbf{b}'(\mathbf{x}_i - \bar{\mathbf{x}}) \\ &= b_1(x_{i1} - \bar{x}_1) + b_2(x_{i2} - \bar{x}_2) \\ &= 0.8(x_{i1} - 47.25) + 0.6(x_{i2} - 44.78). \end{aligned} \tag{1}$$

Then for $i = 1, \ldots, 24$,

$$\begin{aligned} \mathbf{y}_i &= \begin{pmatrix} y_{i1} \\ y_{i2} \end{pmatrix} \\ &= \bar{\mathbf{x}} + \mathbf{b}u_i \\ &= \begin{pmatrix} \bar{x}_1 \\ \bar{x}_2 \end{pmatrix} + \begin{pmatrix} b_1 u_i \\ b_2 u_i \end{pmatrix} \\ &= \begin{pmatrix} 47.25 + 0.8u_i \\ 44.78 + 0.6u_i \end{pmatrix}. \end{aligned} \tag{2}$$

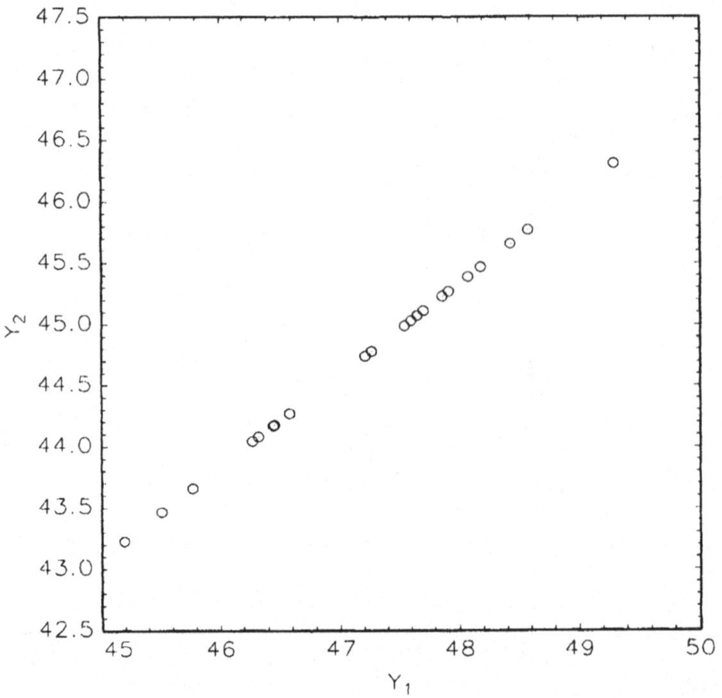

Figure 8-1-1B A linear principal component approximation in the turtle shell dimensions example; (a) scatterplot of X_1 vs. X_2, with a straight line found by orthogonal least squares; (b) scatterplot of Y_1 vs. Y_2, illustrating the one-dimensional principal component approximation to the bivariate data.

Using equation (1) in the second line of (2), we can also write

$$\begin{aligned}\mathbf{y}_i &= \bar{\mathbf{x}} + \mathbf{b}\mathbf{b}'(\mathbf{x}_i - \bar{\mathbf{x}}) \\ &= \begin{pmatrix} \bar{x}_1 \\ \bar{x}_2 \end{pmatrix} + \begin{pmatrix} b_1^2 & b_1 b_2 \\ b_2 b_1 & b_2^2 \end{pmatrix} \begin{pmatrix} x_{i1} - \bar{x}_1 \\ x_{i2} - \bar{x}_2 \end{pmatrix} \\ &= \begin{pmatrix} 47.25 \\ 44.78 \end{pmatrix} + \begin{pmatrix} 0.64 & 0.48 \\ 0.48 & 0.36 \end{pmatrix} \begin{pmatrix} x_{i1} - 47.25 \\ x_{i2} - 44.78 \end{pmatrix}.\end{aligned} \tag{3}$$

The first line of equation (3) is actually a formula that we will encounter many times in the current chapter. It allows computing the coordinates of the approximating points $\mathbf{y}_i$ directly from the $\mathbf{x}_i$. The matrix $\mathbf{b}\mathbf{b}' = \begin{pmatrix} 0.64 & 0.48 \\ 0.48 & 0.36 \end{pmatrix}$ is a *projection matrix*, and the transformation given by equation (3) is the desired orthogonal projection of the data points on the best fitting straight line. Figure 8.1.1(b) shows a scatterplot of the data pairs (y_{i1}, y_{i2}); by construction, all data points are exactly on a straight line.

In anticipation of formal definitions to be given in Section 8.3, the variable U defined by (1) is called the *first linear principal component* of X_1 and X_2. The variables Y_1 and Y_2, defined in (2) or (3), are a one-dimensional principal component approximation to the joint distribution of X_1 and X_2. Both Y_1 and Y_2 are determined entirely by U, together with the vector $\mathbf{b}$ and the mean vector $\bar{\mathbf{x}}$. That is, if we know $\bar{\mathbf{x}}$, $\mathbf{b}$, and the values $u_1, \ldots, u_{24}$, then we can generate the approximate data of Figure 8.1.1(b). This points towards the main theme of this chapter: approximation of high-dimensional data by lower dimensional data. In this particular case, we approximate bivariate data by univariate data, which is not a very big gain, but, nevertheless, the example illustrates the purpose of linear principal component analysis well. To clarify the idea of approximation by lower dimensional data, Table 8.1.1 reports the original data (x_{i1}, x_{i2}), the values u_i of the first principal component, and the approximate data (y_{i1}, y_{i2}). The data in the last two columns are generated from the data in the column labeled u_i, using the transformation (2).

As seen in Figure 8.1.1(a), the 24 data points are distributed about evenly on both sides of the straight line. This points towards a concept called *self-consistency* that we will introduce in Section 8.2. If the shape of the scatterplot showed some curvature, then we would find that in some segments of the line there are too many points on one side and too few on the other side, which, in turn, would render the linear approximation questionable. More details on this example are given in Exercise 2. □

Table 8.1.1 Data used in Example 8.1.1, along with values u_i of the first principal component and values (y_{i1}, y_{i2}) of the one-dimensional principal component approximation

i	x_{i1}	x_{i2}	u_i	y_{i1}	y_{i2}
1	45.326	43.041	−2.584	45.188	43.224
2	45.433	43.567	−2.182	45.509	43.465
3	45.643	43.820	−1.862	45.766	43.658
4	46.151	44.308	−1.163	46.325	44.077
5	46.250	44.427	−1.013	46.444	44.167
6	46.347	43.944	−1.225	46.275	44.040
7	46.444	44.188	−1.001	46.454	44.175
8	46.634	44.188	−0.848	46.576	44.266
9	46.728	44.067	−0.846	46.578	44.268
10	47.185	44.886	0.011	47.263	44.782
11	47.274	44.773	0.014	47.266	44.784
12	47.362	44.543	−0.053	47.212	44.744
13	47.536	44.998	0.359	47.541	44.991
14	47.622	44.998	0.427	47.596	45.032
15	47.622	45.109	0.494	47.649	45.072
16	47.791	45.326	0.760	47.862	45.232
17	47.875	44.886	0.563	47.704	45.114
18	47.875	45.326	0.827	47.915	45.272
19	47.958	45.539	1.021	48.071	45.389
20	48.283	45.326	1.153	48.176	45.468
21	48.442	45.643	1.471	48.430	45.659
22	48.520	45.539	1.470	48.430	45.659
23	48.752	45.539	1.656	48.578	45.770
24	49.053	46.634	2.554	49.297	46.310

Exercises for Section 8.1

1. In Example 8.1.1, calculate the linear regression of X_2 on X_1 and the linear regression of X_1 on X_2. Graph them in a scatterplot of X_1 vs. X_2, together with the orthogonal least squares line of Figure 8.1.1. Verify that all three lines pass through the average $\bar{\mathbf{x}}$ and that the orthogonal least squares line is between the two regression lines.

2. This exercise gives some further results on Example 8.1.1.
   (a) Verify that $\bar{\mathbf{y}} = \bar{\mathbf{x}}$.
   (b) Calculate the sample covariance matrices $\mathbf{S_X}$ and $\mathbf{S_Y}$ of the X-variables and the Y-variables, respectively. Verify that the correlation between Y_1 and Y_2 is 1 and that the variances of the Y-variables are smaller than the variances of the corresponding X-variables.
   (c) Show that, for any vector $\mathbf{a} \in \mathbb{R}^2$, the (sample) variance of $\mathbf{a}'\mathbf{X}$ is larger or equal to the variance of $\mathbf{a}'\mathbf{Y}$. *Hint*: Show that $\mathbf{S_X} - \mathbf{S_Y}$ is positive semi-definite.

8.2 Self-Consistent Approximations

The mean $E[\mathbf{X}]$ of a random vector $\mathbf{X}$ is the simplest summary of the distribution of $\mathbf{X}$ and may be regarded as a crude approximation. A less crude approximation might be obtained by replacing $\mathbf{X}$ with a random vector $\mathbf{Y}$ which in some sense is "simpler" than $\mathbf{X}$ but reflects important properties of $\mathbf{X}$. For instance, in a high-dimensional case, "simpler" might mean that $\mathbf{Y}$ takes values only in a linear subspace of lower dimension. Or, as we will discuss briefly in this section, "simpler" may mean discrete with only a small number of values that $\mathbf{Y}$ takes with positive probability. Let us illustrate this with two examples.

Example 8.2.1 Suppose that we wish to approximate the standard normal distribution by a discrete distribution which puts positive probability on exactly two points. How should this be done? One possibility is to "chop" the distribution in half and replace each of the "half-normals" by its expected value. More precisely, suppose that $X \sim \mathcal{N}(0, 1)$. Then conditionally on $X \geq 0$ and $X < 0$, X has the *pdf*

$$f_+(x) = \begin{cases} \sqrt{2/\pi}\, e^{-x^2/2} & \text{if } x \geq 0 \\ 0 & \text{otherwise,} \end{cases} \tag{1}$$

and

$$f_-(x) = \begin{cases} \sqrt{2/\pi}\, e^{-x^2/2} & \text{if } x < 0 \\ 0, & \text{otherwise,} \end{cases} \tag{2}$$

respectively (see Exercise 1). Then

$$E[X|X \geq 0] = \int_0^\infty x\, f_+(x)\, dx = \sqrt{2/\pi}, \tag{3}$$

and

$$E[X|X<0] = \int_{-\infty}^{0} x\, f_{-}(x)\, dx = -\sqrt{2/\pi}. \tag{4}$$

Now, define a discrete random variable Y by

$$Y = \begin{cases} -\sqrt{2/\pi} & \text{if } X < 0 \\ \sqrt{2/\pi} & \text{if } X \geq 0, \end{cases} \tag{5}$$

as illustrated in Figure 8.2.1. Thus, Y takes the values $\pm\sqrt{2/\pi}$ with probability $1/2$ each. In fact, Y may be regarded as a *discretization* of X. Note that $E[Y] = 0 = E[X]$, so both X and its discrete approximation Y have the same mean. For a general univariate normal random variable X with mean μ and variance σ^2, we may similarly define

$$Y = \begin{cases} \mu - \sigma\sqrt{2/\pi} & \text{if } X < \mu \\ \mu + \sigma\sqrt{2/\pi} & \text{if } X \geq \mu \end{cases} \tag{6}$$

(see Exercise 2) as a simple approximation. Again, notice that $E[Y] = E[X]$. □

Example 8.2.2 Suppose that X_1 and X_2 are jointly distributed random variables with $E[X_1] = \mu_1$ and $E[X_2] = \mu_2$. Let $\mathbf{X} = \begin{pmatrix} X_1 \\ X_2 \end{pmatrix}$, and define an approximation $\mathbf{Y}$ for $\mathbf{X}$ as

$$\mathbf{Y} = \begin{pmatrix} Y_1 \\ Y_2 \end{pmatrix} = \begin{pmatrix} \mu_1 \\ X_2 \end{pmatrix}. \tag{7}$$

Thus, the first variable is replaced by its mean, and the second variable remains unchanged. Whether or not $\mathbf{Y}$ is a good approximation of $\mathbf{X}$, in some sense, depends

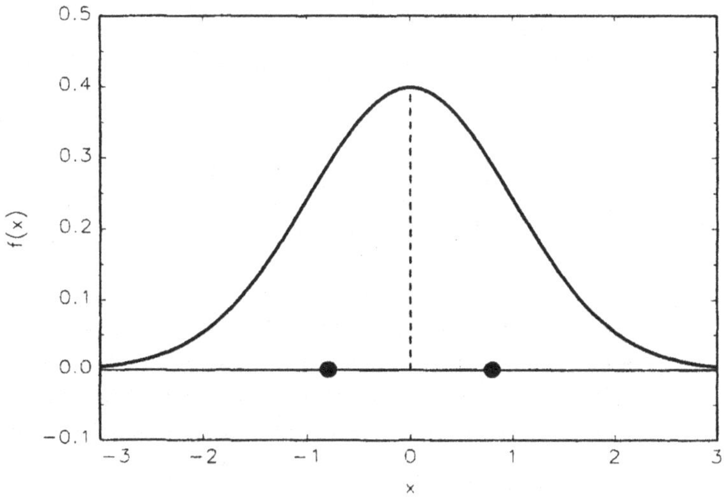

Figure 8-2-1 Density of a $\mathcal{N}(0,1)$ random variable along with an approximation by a discrete variable which takes values $\pm\sqrt{2/\pi}$ with probability 1/2 each. The two values are shown as solid dots.

on the exact distribution of **X** and remains to be seen. However, we notice that

$$E[\mathbf{Y}] = \begin{pmatrix} E[Y_1] \\ E[Y_2] \end{pmatrix} = \begin{pmatrix} E[\mu_1] \\ E[X_2] \end{pmatrix} = \begin{pmatrix} \mu_1 \\ \mu_2 \end{pmatrix} = E[\mathbf{X}], \tag{8}$$

similar to what we found in Example 8.2.1. Notice that the dimension of **Y** is the same as the dimension of **X**, i.e., both are bivariate. However, **Y** is "simpler" in the sense that one of the random variables (X_1) has been replaced by its expected value. We might justify such an approximation if the variability of X_1 is, in a sense to be defined later, negligible. □

Example 8.2.2 may look artifically simple at this point, but it is actually quite illustrative for the aspect of dimension reduction, as we will see in principal component approximations.

We will return to the two examples just discussed in the light of *self-consistency*, the main concept to be introduced in this section. Self-consistency is based on conditional expectation, and the student is advised to review Section 2.7, if necessary. In particular, it will be useful to view conditional expectations as random variables. Suppose that **U** and **V** are jointly distributed, random vectors, and suppose that the regression function of **U** on **V** is

$$r(\mathbf{v}) := E[\mathbf{U}|\mathbf{V} = \mathbf{v}]. \tag{9}$$

Then we define the conditional expectation of **U**, given **V**, as the random vector

$$E[\mathbf{U}|\mathbf{V}] = r(\mathbf{V}). \tag{10}$$

For instance, if $\begin{pmatrix} \mathbf{X}_1 \\ \mathbf{X}_2 \end{pmatrix}$ is multivariate normal with mean vector $\begin{pmatrix} \mu_1 \\ \mu_2 \end{pmatrix}$ and positive-definite covariance matrix $\begin{pmatrix} \psi_{11} & \psi_{12} \\ \psi_{21} & \psi_{22} \end{pmatrix}$, then $E[\mathbf{X}_1|\mathbf{X}_2] = \mu_1 + \psi_{12}\psi_{22}^{-1}(\mathbf{X}_2 - \mu_2)$; see Chapter 3.

Now, we are ready to define the notion of self-consistency.

Definition 8.2.1 Suppose that **X** and **Y** are two jointly distributed random vectors, both of the same dimension. Then **Y** is *self-consistent* for **X** if

$$E[\mathbf{X}|\mathbf{Y}] = \mathbf{Y}. \tag{11}$$

□

As self-consistency seems to be a rather abstract notion, we give some explanation of its meaning in the context of approximation. Suppose that we define Y as a function $h(X)$ for a random variable X. For a given fixed value y that Y can take, let $\mathcal{S}(y)$ denote the set of all x-values that are mapped into y, i.e.,

$$\mathcal{S}(y) = \{x \in \mathbb{R} : h(x) = y\}. \tag{12}$$

Then self-consistency of Y for X means that, for each such set, the mean of all x-values that are mapped into y is y itself—indeed, a natural concept. Or formally, self-consistency means that

$$E[X|h(X) = y] = y \tag{13}$$

for all values of y that Y can take.

Basic examples for self-consistent approximations are as follows. Any random vector $\mathbf{X}$ is self-consistent for itself, i.e., $\mathbf{Y} = \mathbf{X}$ is self-consistent for $\mathbf{X}$. This is indeed a "perfect," but rather uninteresting, approximation. Also, for any random vector $\mathbf{X}$ with finite mean, $\mathbf{Y} = E[\mathbf{X}]$ is self-consistent for $\mathbf{X}$ (see Exercise 5). This is an approximation of $\mathbf{X}$ by a degenerated random variable which takes the value $\mu = E[\mathbf{X}]$ with probability one. More interesting self-consistent approximations are somewhere "between" these two extreme possibilities.

Example 8.2.3 continuation of Example 8.2.1

By construction, $Y = \sqrt{2/\pi}$ exactly if $X \geq 0$, and $Y = -\sqrt{2/\pi}$ if $X < 0$. Therefore,

$$E\left[X|Y = \sqrt{2/\pi}\right] = E[X|X \geq 0] = \sqrt{2/\pi},$$

and

$$E[X|Y = -\sqrt{2/\pi}] = E[X|X < 0] = -\sqrt{2/\pi}.$$

This means that Y is self-consistent for X. See Exercises 6 and 7 for related material. □

Example 8.2.4 continuation of Example 8.2.2

Recall that X_1 and X_2 are two jointly distributed random variables with means μ_1 and μ_2, respectively, and $\mathbf{Y} = \begin{pmatrix} Y_1 \\ Y_2 \end{pmatrix} = \begin{pmatrix} \mu_1 \\ X_2 \end{pmatrix}$. Is $\mathbf{Y}$ self-consistent for $\mathbf{X}$? For a fixed value x_2^* of X_2, the set of all $\begin{pmatrix} x_1 \\ x_2 \end{pmatrix}$ that are mapped into $\begin{pmatrix} \mu_1 \\ x_2^* \end{pmatrix}$ is

$$S = \left\{\begin{pmatrix} x_1 \\ x_2 \end{pmatrix} : x_1 \in \mathbb{R},\ x_2 = x_2^*\right\}.$$

Hence,

$$\begin{aligned} E\left\{\mathbf{X}|\mathbf{Y} = \begin{pmatrix} \mu_1 \\ x_2^* \end{pmatrix}\right\} &= E[\mathbf{X}|X_2 = x_2^*] \\ &= \begin{Bmatrix} E[X_1|X_2 = x_2^*] \\ E[X_2|X_2 = x_2^*] \end{Bmatrix} \\ &= \begin{Bmatrix} E[X_1|X_2 = x_2^*] \\ x_2^* \end{Bmatrix}. \end{aligned}$$

If $\mathbf{Y}$ is to be self-consistent for $\mathbf{X}$, then we must require that $E[X_1|X_2 = x_2] = \mu_1$ irrespective of the value of x_2, which is generally not true. However, if X_1 and X_2 are independent, then $E[X_1|X_2] = E[X_1] = \mu_1$, and $\mathbf{Y} = \begin{pmatrix} \mu_1 \\ X_2 \end{pmatrix}$ is self-consistent for $\mathbf{X}$. Independence is a sufficient, but not a necessary condition; see Exercise 8. □

In anticipation of the derivation of principal components in Section 8.3, notice that the transformation from $\mathbf{X}$ to $\mathbf{Y}$ in Example 8.2.4 is actually a *projection* on the line $x_1 = \mu_1$ in the real plane. Setting $\boldsymbol{\mu} = \begin{pmatrix} \mu_1 \\ \mu_2 \end{pmatrix}$ and $\mathbf{P} = \begin{pmatrix} 0 & 0 \\ 0 & 1 \end{pmatrix}$, we can write $\mathbf{Y}$ as

$$\mathbf{Y} = \boldsymbol{\mu} + \mathbf{P}(\mathbf{X} - \boldsymbol{\mu}). \tag{14}$$

Self-consistent approximations have a simple but important property: they "preserve location," as shown in a first theoretical result.

Lemma 8.2.1 *Suppose that* $\mathbf{Y}$ *is self-consistent for* $\mathbf{X}$. *Then* $E[\mathbf{Y}] = E[\mathbf{X}]$.

Proof From Section 2.7, we know that, for any jointly distributed random vectors $\mathbf{U}$ and $\mathbf{V}$,

$$E[\mathbf{U}] = E\big[E[\mathbf{U}|\mathbf{V}]\big]. \tag{15}$$

Setting $\mathbf{X} = \mathbf{U}$ and $\mathbf{Y} = \mathbf{V}$, $E\big[E[\mathbf{X}|\mathbf{Y}]\big] = E[\mathbf{X}]$ by (15). On the other hand, by self-consistency $E[\mathbf{X}|\mathbf{Y}] = \mathbf{Y}$, and hence, $E\big[E[\mathbf{X}|\mathbf{Y}]\big] = E[\mathbf{Y}]$, which completes the proof. ■

Thus, a self-consistent approximation $\mathbf{Y}$ has the same mean as $\mathbf{X}$, but the variability of $\mathbf{Y}$ is in general smaller than the variability of $\mathbf{X}$, as we shall see later.

It will often be important to assess how well a random vector $\mathbf{Y}$ approximates $\mathbf{X}$, whether $\mathbf{Y}$ is self-consistent for $\mathbf{X}$ or not. A convenient measure is the average squared distance between $\mathbf{X}$ and $\mathbf{Y}$, as in the following definition.

Definition 8.2.2 Let $\mathbf{X}$ and $\mathbf{Y}$ denote two jointly distributed, p-variate, random vectors. Then the *mean squared difference* between $\mathbf{X}$ and $\mathbf{Y}$ is given by

$$\begin{aligned} \text{MSD}(\mathbf{X}, \mathbf{Y}) &= E\Big[\,||\mathbf{X} - \mathbf{Y}||^2\Big] = E\Big[(\mathbf{X} - \mathbf{Y})'(\mathbf{X} - \mathbf{Y})\Big] \\ &= \sum_{i=1}^{p} E[(X_i - Y_i)^2], \end{aligned} \tag{16}$$

i.e., the expected squared distance of $\mathbf{X} - \mathbf{Y}$ from $\mathbf{0}$. □

In the context of approximating a random vector $\mathbf{X}$ by another random vector $\mathbf{Y}$, we will often use the terminology "mean squared error" of $\mathbf{Y}$, and write $\text{MSE}(\mathbf{Y}; \mathbf{X})$ instead of $\text{MSD}(\mathbf{X}, \mathbf{Y})$.

Example 8.2.5 continuation of Examples 8.2.1 and 8.2.3

Since X is univariate, we obtain $\text{MSE}(Y; X) = E\big[(X - Y)^2]\big)$. But

$$X - Y = \begin{cases} X + \sqrt{2/\pi} & \text{if } X < 0 \\ X - \sqrt{2/\pi} & \text{if } X \geq 0, \end{cases}$$

and each possibility occurs with probability 1/2. Hence,

$$\begin{aligned} \text{MSE}(Y; X) &= \frac{1}{2}E\big[(X + \sqrt{2/\pi})^2 | X < 0\big] + \frac{1}{2}E\big[(X - \sqrt{2/\pi})^2 | X \geq 0\big] \\ &= 1 - \frac{2}{\pi} \\ &\approx 0.3634; \end{aligned}$$

see Exercise 9. □

Example 8.2.6 continuation of Example 8.2.4

Suppose that X_1 and X_2 are two jointly distributed random variables with $E[X_i] = \mu_i$ and $\text{var}[X_i] = \sigma_i^2$; $i = 1, 2$. Let $\mathbf{Y} = \begin{pmatrix} Y_1 \\ Y_2 \end{pmatrix} = \begin{pmatrix} \mu_1 \\ X_2 \end{pmatrix}$, then

$$\begin{aligned} \text{MSE}(\mathbf{Y}; \mathbf{X}) &= E\big[||\mathbf{X} - \mathbf{Y}||^2\big] \\ &= E\big[(X_1 - Y_1)^2 + (X_2 - Y_2)^2\big] \\ &= E\big[(X_1 - \mu_1)^2 + (X_2 - X_2)^2\big] \\ &= \sigma_1^2. \end{aligned}$$

Thus, the "loss" incurred by replacing X_1 by its mean and leaving X_2 unchanged is σ_1^2, the variance of X_1. This holds whether $\mathbf{Y}$ is self-consistent for X or not. □

Our next result connects mean squared error with self-consistency. We will use the symbol "tr" for the trace of a square matrix.

Theorem 8.2.2 *Let* $\mathbf{X}$ *and* $\mathbf{Y}$ *denote p-variate random vectors with covariance matrices* $\text{Cov}[\mathbf{X}] = \psi_{\mathbf{X}}$ *and* $\text{Cov}[\mathbf{Y}] = \psi_{\mathbf{Y}}$, *and assume that* $\mathbf{Y}$ *is self-consistent for* $\mathbf{X}$. *Then*

$$\begin{aligned} \text{MSE}(\mathbf{Y}; \mathbf{X}) &= \text{tr}(\psi_{\mathbf{X}}) - \text{tr}(\psi_{\mathbf{Y}}) \\ &= \sum_{i=1}^{p} \big\{\text{var}[X_i] - \text{var}[Y_i]\big\}. \end{aligned} \tag{17}$$

Proof First, notice that, because of self-consistency, $E[\mathbf{Y}] = E[\mathbf{X}]$, and hence $\text{MSE}(\mathbf{Y}; \mathbf{X}) = E\big(||\mathbf{X} - \mathbf{Y}||^2\big)$ does not depend on the means. Therefore assume, without loss of generality, that $E[\mathbf{X}] = \mathbf{0}$, which implies $\text{Cov}[\mathbf{X}] = E[\mathbf{XX}']$ and $\text{Cov}[\mathbf{Y}] = E[\mathbf{YY}']$. Now,

$$\begin{aligned}\text{MSE}(\mathbf{Y};\mathbf{X}) &= E\Big[(\mathbf{X}-\mathbf{Y})'(\mathbf{X}-\mathbf{Y})\Big] \\ &= E\big[\mathbf{X}'\mathbf{X}\big] + E\big[\mathbf{Y}'\mathbf{Y}\big] - 2E\big[\mathbf{Y}'\mathbf{X}\big] \\ &= \text{tr}(\psi_{\mathbf{X}}) + \text{tr}(\psi_{\mathbf{Y}}) - 2E\big[\mathbf{Y}'\mathbf{X}\big],\end{aligned} \tag{18}$$

by Exercise 13 and since $\mathbf{X}'\mathbf{Y} = \mathbf{Y}'\mathbf{X}$ is scalar. Recall the rule

$$E[\mathbf{U}] = E\big[E[\mathbf{U}|\mathbf{V}]\big] \tag{19}$$

for any jointly distributed random vectors $\mathbf{U}$ and $\mathbf{V}$. Setting $\mathbf{U} = \mathbf{Y}'\mathbf{X}$ and $\mathbf{V} = \mathbf{Y}$,

$$E[\mathbf{Y}'\mathbf{X}] = E\big[E[\mathbf{Y}'\mathbf{X}|\mathbf{Y}]\big]. \tag{20}$$

But $E[\mathbf{Y}'\mathbf{X}|\mathbf{Y}] = \mathbf{Y}'\ E[\mathbf{X}|\mathbf{Y}]$ (see Exercise 14), and hence by self-consistency of $\mathbf{Y}$ for $\mathbf{X}$,

$$\begin{aligned}E[\mathbf{Y}'\mathbf{X}] &= E\big[\mathbf{Y}'E[\mathbf{X}|\mathbf{Y}]\big] \\ &= E[\mathbf{Y}'\mathbf{Y}] \\ &= \text{tr}(\psi_{\mathbf{Y}}).\end{aligned} \tag{21}$$

Therefore, from (18),

$$\text{MSE}(\mathbf{Y};\mathbf{X}) = \text{tr}(\psi_{\mathbf{X}}) + \text{tr}(\psi_{\mathbf{Y}}) - 2\text{tr}(\psi_{\mathbf{Y}}), \tag{22}$$

and the result follows. ■

We leave it to the student to verify the results of Examples 8.2.5 and 8.2.6 using the preceding theorem; see Exercises 15 and 16. It is typically easier to compute the mean squared error using Theorem 8.2.2 than by direct calculation. The theorem also shows that self-consistent approximations usually involve a reduction in variability. More precisely, in the univariate case, suppose that Y is self-consistent for X, then

$$\text{MSE}(Y;X) = \text{var}[X] - \text{var}[Y] \geq 0 \tag{23}$$

because $\text{MSE}(Y;X)$ is the expected value of a nonnegative random variable. Equality in (23) is achieved only if $\text{var}[X-Y] = 0$, i.e., if X and Y are identical (except perhaps on a set of probability zero). The other extreme, $\text{MSE}(Y;X) = \text{var}[X]$, is achieved if we choose the self-consistent approximation $Y = E[X]$. A good self-consistent approximation will have a mean-squared error close to zero, yet be "relatively simple."

We conclude this section with a final example which is pertinent to principal component approximations.

Example 8.2.7 Suppose that $X_1, \ldots, X_p$ are p independent variables with means μ_i and variances σ_i^2, $i = 1, \ldots, p$, and let q $(1 \leq q < p)$ be a fixed number. Suppose that we wish to find a self-consistent approximation $\mathbf{Y} = (Y_1, \ldots, Y_p)$ to $\mathbf{X} = (X_1, \ldots, X_p)$, as follows. For q of the X_i, set $Y_i = X_i$; for the remaining $(p-q)$ variables set $Y_i = \mu_i$. Then self-consistency of $\mathbf{Y}$ for $\mathbf{X}$ follows, as in Example 8.2.4. For simplicity, take the case that $Y_i = X_i$ for $i = 1, \ldots, q$ and $Y_i = \mu_i$ for $i = q+1, \ldots, p$. Then $\text{var}[Y_i] = \sigma_i^2$ for $i = 1, \ldots, q$ and $\text{var}[Y_i] = 0$ for $i = q+1, \ldots, p$. Hence, by Theorem 8.2.2,

$$\text{MSE}(\mathbf{Y}; \mathbf{X}) = \sum_{i=q+1}^{p} \sigma_i^2.$$

According to the criterion of choosing a self-consistent approximation with small mean squared error, we should set $Y_i = X_i$ for the q variables X_i with the largest variances and set $Y_i = \mu_i$ for the remaining $(p-q)$ variables. □

Exercises for Section 8.2

1. If $X \sim \mathcal{N}(0, 1)$,

   (a) find the conditional distributions of X, given $X \geq 0$ and given $X < 0$, as well as the associated density functions. Thus, verify equations (1) and (2).

   (b) find $E[X|X \geq 0]$ and $E[X|X < 0]$.

2. If $X \sim \mathcal{N}(\mu, \sigma^2)$, find $E[X|X \geq \mu]$ and $E[X|X < \mu]$.

3. Let X_1 and X_2 denote two independent random variables, both with mean zero. Show that X_1 is self-consistent for $X_1 + X_2$.

4. Let X_1 and X_2 denote the numbers shown by two independent, regular dice. Show that $Y = X_1 + 3.5$ is self-consistent for $X = X_1 + X_2$. More generally, if $\mathbf{X}_1$ and $\mathbf{X}_2$ are independent random vectors of dimension p each, with $E[\mathbf{X}_2] = \mu_2$, show that $\mathbf{Y} = \mathbf{X}_1 + \mu_2$ is self-consistent for $\mathbf{X} = \mathbf{X}_1 + \mathbf{X}_2$. *Hint*: Write $\mathbf{X} = \mathbf{Y} + (\mathbf{X}_2 - \mu_2)$, and notice that $\mathbf{X}_2 - \mu_2$ is independent of $\mathbf{Y}$.

5. For a random vector $\mathbf{X}$, show that the following random vectors $\mathbf{Y}_1$ and $\mathbf{Y}_2$ are self-consistent:

   (a) $\mathbf{Y} = \mathbf{X}$.

   (b) $\mathbf{Y} = E[\mathbf{X}]$, provided $E[\mathbf{X}]$ exists.

6. Let $\mathbf{X}$ denote a p-variate random vector, and let $\{\mathcal{A}_1, \ldots, \mathcal{A}_k\}$ denote a partition of $\mathbb{R}^p$, i.e., $\bigcup_{i=1}^{k} \mathcal{A}_i = \mathbb{R}^p$, and $\mathcal{A}_i \cap \mathcal{A}_j = \emptyset$ if $i \neq j$. Assume that $\Pr[\mathbf{X} \in \mathcal{A}_i] > 0$ for $i = 1, \ldots, k$, and define a p-variate, random vector $\mathbf{Y}$ by $\mathbf{Y} = E[\mathbf{X}|\mathbf{X} \in \mathcal{A}_i]$ if $\mathbf{X} \in \mathcal{A}_i$. Show that $\mathbf{Y}$ is self-consistent for $\mathbf{X}$.

7. Let U denote a random variable with uniform distribution in the interval $[0, 1]$. Let $m_1 = E[U|U \leq c]$ and $m_2 = E[U|U > c]$. Define a random variable Y by

$$Y = \begin{cases} m_1 & \text{if } U \leq c \\ m_2 & \text{if } U > c. \end{cases}$$

(a) Show that Y is self-consistent for U.

(b) Suppose that we wish to choose c such that the above definition of Y is equivalent to

$$Y = \begin{cases} m_1 & \text{if } |U - m_1| \leq |U - m_2| \\ m_2 & \text{if } |U - m_1| > |U - m_2|. \end{cases}$$

Show that $c = \frac{1}{2}$ is the only solution.

8. Let X_1 and X_2 denote two jointly distributed random variables such that $E[X_1] = \mu_1$ and $E[X_2] = \mu_2$.

(a) If $E[X_1|X_2 = x_2]$ does not depend on x_2, show that $\mathbf{Y} = \begin{pmatrix} \mu_1 \\ X_2 \end{pmatrix}$ is self-consistent for $\mathbf{X}$.

(b) Find an explicit example of two jointly distributed, random variables X_1 and X_2 that are not independent, yet $\mathbf{Y} = \begin{pmatrix} \mu_1 \\ X_2 \end{pmatrix}$ is self-consistent for $\mathbf{X}$. *Hint*: Read Section 3.4.

9. In Example 8.2.5, show that $\text{MSE}(Y; X) = 1 - 2/\pi$ without using Theorem 8.2.2. *Hint*: This can be done by straightforward integration or using the following trick. Consider X as a mixture of two half-normals X_- and X_+ with densities

$$f_-(x) = \begin{cases} \sqrt{2/\pi}\, e^{-x^2/2} & \text{if } x < 0, \\ 0 & \text{otherwise} \end{cases}$$

and

$$f_+(x) = \begin{cases} \sqrt{2/\pi}\, e^{-x^2/2} & \text{if } x \geq 0, \\ 0 & \text{otherwise,} \end{cases}$$

respectively, and with mixing weights $p_1 = p_2 = \frac{1}{2}$. Then use the fact that $\text{var}[X_-] = \text{var}[X_+] = E\big[(X - \sqrt{2/\pi})^2 | X > 0\big]$ and Theorem 2.8.1.

10. Let U denote a random variable with a uniform distribution in the interval $[0, 1]$. For a fixed integer $k \geq 1$, define

$$Y_k = \frac{2j-1}{2k} \quad \text{if} \quad \frac{j-1}{k} < U \leq \frac{j}{k}, \qquad j = 1, \ldots, k.$$

(a) Show that Y_k is self-consistent for U.

(b) Find $\text{MSE}(Y_k; X)$.

11. Let X_1 and X_2 denote two independent random variables, both normal with mean zero and with $\text{var}[X_1] = 1$, $\text{var}[X_2] = \sigma^2 > 0$. Let

$$\mathbf{Y} = \begin{cases} \begin{pmatrix} \sqrt{2/\pi} \\ 0 \end{pmatrix} & \text{if } X_1 \ge 0 \\ \begin{pmatrix} -\sqrt{2/\pi} \\ 0 \end{pmatrix} & \text{if } X_1 < 0 \end{cases}$$

and

$$\mathbf{Z} = \begin{cases} \begin{pmatrix} 0 \\ \sigma\sqrt{2/\pi} \end{pmatrix} & \text{if } X_2 \ge 0 \\ \begin{pmatrix} 0 \\ -\sigma\sqrt{2/\pi} \end{pmatrix} & \text{if } X_2 < 0. \end{cases}$$

(a) Show that both $\mathbf{Y}$ and $\mathbf{Z}$ are self-consistent for $\mathbf{X} = \begin{pmatrix} X_1 \\ X_2 \end{pmatrix}$.

(b) Find $\text{MSE}(\mathbf{Y}; \mathbf{X})$ and $\text{MSE}(\mathbf{Z}; \mathbf{X})$. For which values of σ is $\text{MSE}(\mathbf{Y}; \mathbf{X})$ smaller than $\text{MSE}(\mathbf{Z}; \mathbf{X})$?

12. In Exercise 7, compute $\text{MSE}(Y; U)$ as a function of c, $0 < c < 1$. For which value of c is the mean squared error minimal?

13. If $\mathbf{X}$ is a p-variate random vector with $E[\mathbf{X}] = \boldsymbol{\mu}$ and $\text{Cov}[\mathbf{X}] = \boldsymbol{\psi}$, show that $E\big[(\mathbf{X} - \boldsymbol{\mu})'(\mathbf{X} - \boldsymbol{\mu})\big] = \text{tr}(\boldsymbol{\psi})$.

14. If U and V are jointly distributed random variables, show that $E[UV|U] = U\ E[V|U]$. More generally, if $\mathbf{U}$ and $\mathbf{V}$ are jointly distributed random vectors of dimension p, show that $E[\mathbf{U}'\mathbf{V}|\mathbf{U}] = \mathbf{U}'\ E[\mathbf{V}|\mathbf{U}]$.

15. Use Theorem 8.2.2 to show that $\text{MSE}(Y; X) = 1 - 2/\pi$ in Example 8.2.5.

16. Use Theorem 8.2.2 to show that $\text{MSE}(Y; X) = \sigma^2$ in Example 8.2.6, with X_1 and X_2 assumed independent.

17. Suppose that $\mathbf{X}$ is a p-variate random vector with $\text{Cov}[\mathbf{X}] = \boldsymbol{\psi}_{\mathbf{X}}$ and $\mathbf{Y}$ is self-consistent for $\mathbf{X}$ with $\text{Cov}[\mathbf{Y}] = \boldsymbol{\psi}_{\mathbf{Y}}$. Use the technique in the proof of Theorem 8.2.2 to show that the matrix $\boldsymbol{\psi}_{\mathbf{X}} - \boldsymbol{\psi}_{\mathbf{Y}}$ is positive semidefinite. *Hint*: First, show that $\text{Cov}[\mathbf{X} - \mathbf{Y}] = E\big[(\mathbf{X} - \mathbf{Y})(\mathbf{X} - \mathbf{Y})'\big]$.

18. Let $\mathbf{X} = \begin{pmatrix} X_1 \\ X_2 \end{pmatrix}$ follow a bivariate standard normal distribution, i.e., X_1 and X_2 are independent standard normal random variables. Let

$$\mathbf{Y} = \begin{bmatrix} Y_1 \\ Y_2 \end{bmatrix} = \sqrt{2/\pi} \begin{bmatrix} \text{sgn}(X_1) \\ \text{sgn}(X_2) \end{bmatrix},$$

where "sgn" is the sign, and let

$$\mathbf{Z} = \begin{pmatrix} X_1 \\ 0 \end{pmatrix}.$$

(a) Show that both $\mathbf{Y}$ and $\mathbf{Z}$ are self-consistent for $\mathbf{X}$.

(b) Which one of the self-consistent approximations $\mathbf{Y}$ and $\mathbf{Z}$ has the smaller mean squared error for $\mathbf{X}$?

19. Repeat Exercise 7 for the standard normal distribution. Show that $c = 0$ is the only solution. *Note*: This will be more difficult than Exercise 7.

20. Let $X_1, \ldots, X_N$ be independent, identically distributed Poisson random variables with parameter $\lambda > 0$, and denote by $\bar{X}$ and S^2 the sample mean and (unbiased) sample variance. Show that $\bar{X}$ is self-consistent for S^2. *Hint*: See Exercise 25 in Section 4.2.

8.3 Self-Consistent Projections and Orthogonal Least Squares

In this section we will derive principal components of a random vector $\mathbf{X}$ in two ways: first, as self-consistent projections of $\mathbf{X}$ in a subspace of lower dimension; and second, according to the principle of orthogonal least squares. Both derivations are based on orthogonal projections on lines, planes, or hyperplanes, and the student is advised to review the material in appendix A.8 first. Also crucial for the understanding of the material in this section is a solid knowledge of eigenvectors and eigenvalues of symmetric matrices; see appendix A.5.

Let us start out non-technically, to develop some intuition. Figure 8.3.1(a) shows contours of a bivariate normal distribution with mean vector $\boldsymbol{\mu} = \begin{pmatrix} 0 \\ 0 \end{pmatrix}$ and covariance matrix $\boldsymbol{\psi} = \begin{pmatrix} 2 & 1 \\ 1 & 2 \end{pmatrix}$. The solid line indicates the linear subspace $\mathcal{S} = \{\mathbf{x} \in \mathbb{R}^2 : x_1 = x_2\}$. Several broken lines, all orthogonal to the line $x_1 = x_2$, represent sets of points in $\mathbb{R}^2$ that are mapped into the same point in an orthogonal projection on $\mathcal{S}$. For instance, all points on the rightmost broken line in Figure 8.3.1(a) are mapped into the point $\begin{pmatrix} 2 \\ 2 \end{pmatrix} \in \mathcal{S}$.

Consider now the bivariate random vector $\mathbf{Y}$ that corresponds to the orthogonal projection of $\mathbf{X} = \begin{pmatrix} X_1 \\ X_2 \end{pmatrix}$ into the subspace $\mathcal{S}$. This projection can be written as

$$\mathbf{Y} = \begin{pmatrix} Y_1 \\ Y_2 \end{pmatrix} = \frac{1}{2} \begin{pmatrix} X_1 + X_2 \\ X_1 + X_2 \end{pmatrix} \tag{1}$$

(see Exercise 1). By definition, $\mathbf{Y}$ is self-consistent for $\mathbf{X}$ if $E[\mathbf{X}|\mathbf{Y}] = \mathbf{Y}$. Or in other words, $\mathbf{Y}$ is self-consistent for $\mathbf{X}$ if, for each $\mathbf{y} \in \mathcal{S}$, $\mathbf{y}$ is the mean of all $\mathbf{x} \in \mathbb{R}^2$ that project orthogonally on $\mathbf{y}$. For a given $\mathbf{y} = \begin{pmatrix} c \\ c \end{pmatrix} \in \mathcal{S}$, the set of all $\mathbf{x} \in \mathbb{R}^2$ that project on $\mathbf{y}$ is

$$\left\{\mathbf{x} \in \mathbb{R}^2 : \frac{1}{2}(x_1 + x_2) = c\right\}. \tag{2}$$

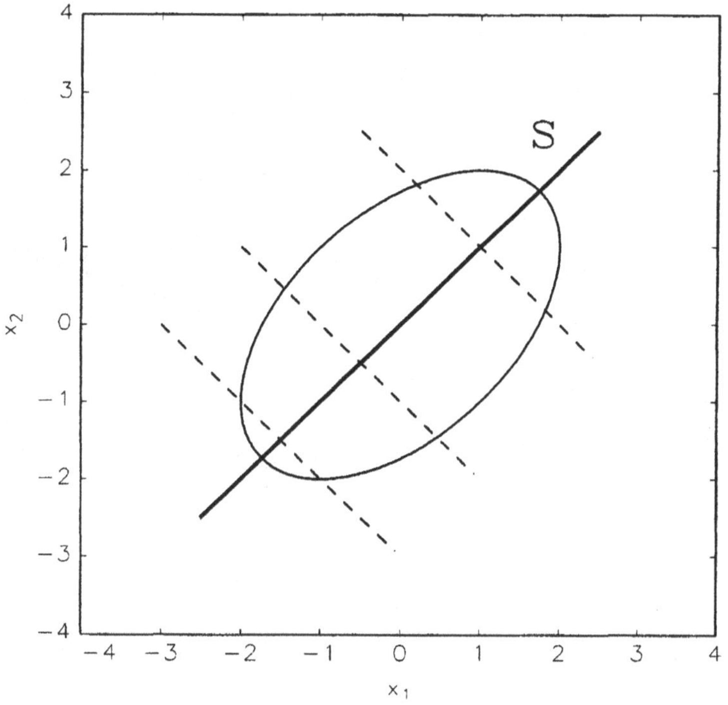

Figure 8-3-1A

From symmetry considerations, intuition would tell us that **Y** is indeed self-consistent because each of the broken lines in Figure 8.3.1(a) "cuts a slice out of the normal mountain," such that the mean on each broken line lies on the solid line. Formally,

$$E[\mathbf{X}|\mathbf{Y}=\mathbf{y}] = \begin{pmatrix} E[X_1|X_1+X_2=2c] \\ E[X_2|X_1+X_2=2c] \end{pmatrix}. \tag{3}$$

A little trick helps to evaluate the two conditional expectations. Notice that X_1+X_2 and X_1-X_2 are independent (Exercise 1), and therefore, $E[X_1-X_2|X_1+X_2=2c]=0$, irrespective of c. Also, $E[X_1+X_2|X_1+X_2=2c]=2c$, and therefore,

$$E[X_1|X_1+X_2=2c] = \frac{1}{2}E\big[(X_1+X_2)+(X_1-X_2)|X_1+X_2=2c\big] = c,$$

and

$$E[X_2|X_1+X_2=2c] = \frac{1}{2}E\big[(X_1+X_2)-(X_1-X_2)|X_1+X_2=2c\big] = c.$$

Thus

$$E\left[\mathbf{X}|\mathbf{Y}=\begin{pmatrix} c \\ c \end{pmatrix}\right] = \begin{pmatrix} c \\ c \end{pmatrix}. \tag{4}$$

Since this holds for all values of c, **Y** is self-consistent for **X**.

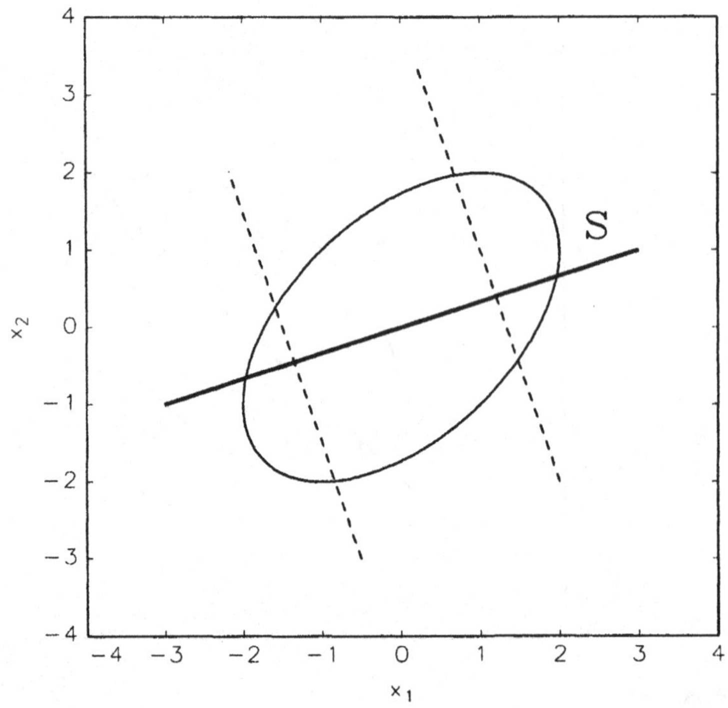

Figure 8-3-1B Contour plot of a bivariate normal distribution and orthogonal projections on a line. The solid line marked S represents a one-dimensional linear subspace; broken lines indicate sets of points that are projected into the same point in S; (a) the projection is self-consistent; (b) the projection is not self-consistent.

Part (b) of Figure 8.3.1 shows contours of the same bivariate normal distribution, this time with a line S corresponding to $x_2 = \frac{1}{3}x_1$. Broken lines represent sets of points that project orthogonally onto the same point in S. Intuition would tell us that the random vector **Y** obtained by projecting **X** orthogonally on S is not self-consistent because the average of all points on a given broken line is in general not the point on the solid line. A formal proof is left to the student; see Exercise 2.

In the preceding discussion, we have studied only projections on lines through the origin, which, in this case, was also the mean of the distribution. It seems intuitive that a line must contain the mean of the distribution if the orthogonal projection on the line is to be self-consistent. Actually, this forms part of the first important result of this section, as given in the following theorem.

Theorem 8.3.1 *Let* $\mathbf{X}$ *denote a p-variate random vector with mean* μ *and covariance matrix* ψ. *Let* $\mathbf{Y}$ *be the orthogonal projection of* $\mathbf{X}$ *on a line in* $\mathbb{R}^p$. *If* $\mathbf{Y}$ *is self-consistent for* $\mathbf{X}$, *then* $\mathbf{Y} = \mu + \mathbf{P}(\mathbf{X} - \mu)$, *where* $\mathbf{P} = \mathbf{aa}'$, *and* $\mathbf{a}$ *is a normalized eigenvector of* ψ.

Proof Since **Y** is a projection on a line, it can be written as

$$\mathbf{Y} = \mathbf{x}_0 + \mathbf{P}(\mathbf{X} - \mathbf{x}_0) \tag{5}$$

for some $\mathbf{x}_0 \in \mathbb{R}^p$, where $\mathbf{P} = \mathbf{aa}'$ for some $\mathbf{a} \in \mathbb{R}^p$, $\mathbf{a}'\mathbf{a} = 1$; see Appendix A.8. Taking the expectations on both sides of (5),

$$E[\mathbf{Y}] = \mathbf{x}_0 + \mathbf{P}\big(E[\mathbf{X}] - \mathbf{x}_0\big). \tag{6}$$

By Lemma 8.2.1, $E[\mathbf{Y}] = E[\mathbf{X}]$, and therefore,

$$\boldsymbol{\mu} = \mathbf{x}_0 + \mathbf{P}(\boldsymbol{\mu} - \mathbf{x}_0), \tag{7}$$

that is, $\boldsymbol{\mu}$ is on the line. Without loss of generality we can therefore take $\mathbf{x}_0 = \boldsymbol{\mu}$, i.e., the projection has the form

$$\mathbf{Y} = \boldsymbol{\mu} + \mathbf{P}(\mathbf{X} - \boldsymbol{\mu}). \tag{8}$$

For the remainder of the proof and for simplicity, assume $\boldsymbol{\mu} = \mathbf{0}$, otherwise replace $\mathbf{X}$ by $\mathbf{X}^* = \mathbf{X} - \boldsymbol{\mu}$, which does not change the covariance matrix. Thus, $\mathbf{Y} = \mathbf{PX}$, where $\mathbf{P} = \mathbf{aa}'$ for some $\mathbf{a} \in \mathbb{R}^p$, $\mathbf{a}'\mathbf{a} = 1$. Then $\mathbf{Y}$ will be self-consistent for $\mathbf{X}$ if $E[\mathbf{X}|\mathbf{Y}] = \mathbf{Y}$. But

$$\begin{aligned} E[\mathbf{X}|\mathbf{Y}] &= E[\mathbf{X}|\mathbf{PX}] \\ &= E\big[\mathbf{PX} + (\mathbf{I}_p - \mathbf{P})\mathbf{X}|\mathbf{PX}\big] \\ &= E[\mathbf{PX}|\mathbf{PX}] + E\big[(\mathbf{I}_p - \mathbf{P})\mathbf{X}|\mathbf{PX}\big] \\ &= \mathbf{PX} + E\big[(\mathbf{I}_p - \mathbf{P})\mathbf{X}|\mathbf{PX}\big]. \end{aligned} \tag{9}$$

By self-consistency, this expectation is $\mathbf{PX}$, and therefore,

$$E\big[(\mathbf{I}_p - \mathbf{P})\mathbf{X}|\mathbf{PX}\big] = \mathbf{0}. \tag{10}$$

Both $\mathbf{PX}$ and $(\mathbf{I} - \mathbf{P})\mathbf{X}$ are p-variate random vectors. Since

$$\begin{matrix} \mathbf{PX} \\ (\mathbf{I}_p - \mathbf{P})\mathbf{X} \end{matrix} = \begin{pmatrix} \mathbf{P} \\ \mathbf{I}_p - \mathbf{P} \end{pmatrix} \mathbf{X}, \tag{11}$$

their joint $2p$-variate distribution has covariance matrix

$$\begin{bmatrix} \mathbf{P} \\ \mathbf{I}_p - \mathbf{P} \end{bmatrix} \boldsymbol{\psi} \, (\mathbf{P} : \mathbf{I}_p - \mathbf{P}) = \begin{bmatrix} \mathbf{P}\boldsymbol{\psi}\mathbf{P} & \mathbf{P}\boldsymbol{\psi}(\mathbf{I}_p - \mathbf{P}) \\ (\mathbf{I}_p - \mathbf{P})\boldsymbol{\psi}\mathbf{P} & (\mathbf{I}_p - \mathbf{P})\boldsymbol{\psi}(\mathbf{I}_p - \mathbf{P}) \end{bmatrix}, \tag{12}$$

where we have used the fact that both $\mathbf{P}$ and $\mathbf{I}_p - \mathbf{P}$ are symmetric. For any jointly distributed random vectors $\mathbf{U}$ and $\mathbf{V}$, $E[\mathbf{U}|\mathbf{V}] = \mathbf{0}$ implies that all covariances between $\mathbf{U}$ and $\mathbf{V}$ are zero; see Exercise 17. Thus,

$$\mathbf{P}\boldsymbol{\psi}(\mathbf{I}_p - \mathbf{P}) = \mathbf{0}, \tag{13}$$

or

$$\mathbf{P}\boldsymbol{\psi} = \mathbf{P}\boldsymbol{\psi}\mathbf{P}. \tag{14}$$

Multiplying both sides of (14) from the left by $\mathbf{a}'$ and recalling that $\mathbf{P} = \mathbf{aa}'$, we obtain

$$\mathbf{a}'(\mathbf{aa}')\psi = \mathbf{a}'(\mathbf{aa}')\psi(\mathbf{aa}'), \tag{15}$$

or

$$(\mathbf{a}'\mathbf{a})\mathbf{a}'\psi = (\mathbf{a}'\mathbf{a})(\mathbf{a}'\psi\mathbf{a})\mathbf{a}'. \tag{16}$$

But $\mathbf{a}'\mathbf{a} = 1$, and $\mathbf{a}'\psi\mathbf{a}$ is a scalar, say, $\mathbf{a}'\psi\mathbf{a} = \lambda$. Thus, (16) reads $\mathbf{a}'\psi = \lambda\mathbf{a}'$, or, in the more familiar form,

$$\psi\mathbf{a} = \lambda\mathbf{a}, \tag{17}$$

that is, $\mathbf{a}$ is an eigenvector of ψ, and $\lambda = \mathbf{a}'\psi\mathbf{a}$ is the associated eigenvalue. This concludes the proof. ∎

A symmetric $p \times p$ matrix has p real eigenvalues, often labeled as λ_i and ordered such that $\lambda_1 \geq \cdots \geq \lambda_p$, and associated eigenvectors $\beta_1, \ldots, \beta_p$. If the eigenvalues are distinct (i.e., no two eigenvalues are equal), then the eigenvectors are unique up to multiplication by a nonzero constant, and it is convenient to normalize them, i.e., scale them such that $\beta_i'\beta_i = 1$ for $i = 1, \ldots, p$. Since the eigenvectors of a symmetric matrix ψ are mutually orthogonal, or can be chosen so in case of multiple eigenvalues, they form an orthonormal basis of $\mathbb{R}^p$.

In the context of Theorem 8.3.1, this means the following. Provided that the eigenvalues λ_i of the covariance matrix ψ are all unique, there are exactly p "candidates" for self-consistent projections of the p-variate random vector $\mathbf{X}$, namely, those that can be written as

$$\mathbf{Y} = \boldsymbol{\mu} + \mathbf{P}_i(\mathbf{X} - \boldsymbol{\mu}), \tag{18}$$

where $\mathbf{P}_i = \beta_i\beta_i'$ and β_i is a normalized eigenvector of ψ. Note, however, that the projection (18) is not automatically self-consistent: β_i being an eigenvector of ψ is a necessary but not a sufficient condition for self-consistency.

Example 8.3.1 Suppose that $\mathbf{X} = \begin{pmatrix} X_1 \\ X_2 \end{pmatrix}$ is bivariate normal with mean $\boldsymbol{\mu} = \mathbf{0}$ and covariance matrix $\psi = \begin{pmatrix} 2 & 1 \\ 1 & 2 \end{pmatrix}$, as in the introductory example to this section. The covariance matrix has two eigenvectors $\beta_1 = \frac{1}{\sqrt{2}}\begin{pmatrix} 1 \\ 1 \end{pmatrix}$ and $\beta_2 = \frac{1}{\sqrt{2}}\begin{pmatrix} 1 \\ -1 \end{pmatrix}$ and associated eigenvalues $\lambda_1 = 3,\ \lambda_2 = 1$. The two associated projections

$$\mathbf{Y}_{(1)} = \beta_1\beta_1'\mathbf{X} = \frac{1}{2}\begin{pmatrix} X_1 + X_2 \\ X_1 + X_2 \end{pmatrix}$$

and

$$\mathbf{Y}_{(2)} = \boldsymbol{\beta}_2\boldsymbol{\beta}_2'\mathbf{X} = \frac{1}{2}\begin{pmatrix} X_1 - X_2 \\ X_1 - X_2 \end{pmatrix}$$

are both self-consistent. For the first one, self-consistency has been shown before; for the latter, a proof can be constructed similarly; see Exercise 3. The two lines associated with the projections $\mathbf{Y}_{(1)}$ and $\mathbf{Y}_{(2)}$ correspond to the larger and smaller diameter of the ellipses of constant density; see Exercise 4. How good are the two projections $\mathbf{Y}_{(1)}$ and $\mathbf{Y}_{(2)}$ as approximations for $\mathbf{X}$? Using Theorem 8.2.2 and $\psi_{\mathbf{X}} = \begin{pmatrix} 2 & 1 \\ 1 & 2 \end{pmatrix}$,

$$\begin{aligned} \operatorname{tr}\bigl(\operatorname{Cov}[\mathbf{X}]\bigr) &= \operatorname{tr}(\psi_{\mathbf{X}}) = 4, \\ \operatorname{tr}\bigl(\operatorname{Cov}[\mathbf{Y}_{(1)}]\bigr) &= \operatorname{tr}\bigl(\boldsymbol{\beta}_1\boldsymbol{\beta}_1'\psi_{\mathbf{X}}\boldsymbol{\beta}_1\boldsymbol{\beta}_1'\bigr) = 3, \end{aligned}$$

and

$$\operatorname{tr}\bigl(\operatorname{Cov}[\mathbf{Y}_{(2)}]\bigr) = \operatorname{tr}\bigl(\boldsymbol{\beta}_2\boldsymbol{\beta}_2'\psi_{\mathbf{X}}\boldsymbol{\beta}_2\boldsymbol{\beta}_2'\bigr) = 1,$$

yielding

$$\operatorname{MSE}(\mathbf{Y}_{(1)}; \mathbf{X}) = 1 \quad \text{and} \quad \operatorname{MSE}(\mathbf{Y}_{(2)}; \mathbf{X}) = 3.$$

Thus, the self-consistent approximation $\mathbf{Y}_{(1)}$, corresponding to a projection on the longer axis in Figure 8.3.1(a), is better than $\mathbf{Y}_{(2)}$. □

Example 8.3.2 Suppose that $\mathbf{X} \sim \mathcal{N}_p(\boldsymbol{\mu}, \psi)$, where $\psi = \boldsymbol{\beta}\boldsymbol{\Lambda}\boldsymbol{\beta}'$, $\boldsymbol{\beta} = (\boldsymbol{\beta}_1, \ldots, \boldsymbol{\beta}_p)$ is an orthogonal $p \times p$ matrix and $\boldsymbol{\Lambda} = \operatorname{diag}(\lambda_1, \ldots, \lambda_p)$, $\lambda_1 > \lambda_2 > \cdots > \lambda_p > 0$. That is, ψ has p positive eigenvalues, all distinct, and associated eigenvectors $\boldsymbol{\beta}_1, \ldots, \boldsymbol{\beta}_p$ of unit length. Let $\mathbf{P}_i = \boldsymbol{\beta}_i\boldsymbol{\beta}_i'$. Then the projections

$$\mathbf{Y}_{(i)} = \boldsymbol{\mu} + \mathbf{P}_i(\mathbf{X} - \boldsymbol{\mu})$$

are all self-consistent; a proof of self-consistency will be given in Section 8.4. For the ith projection, the covariance matrix is given by $\operatorname{Cov}[\mathbf{Y}_{(i)}] = \mathbf{P}_i\psi\mathbf{P}_i = \boldsymbol{\beta}_i\boldsymbol{\beta}_i'\psi\boldsymbol{\beta}_i\boldsymbol{\beta}_i'$, and the mean squared error is

$$\begin{aligned} \operatorname{MSE}(\mathbf{Y}_{(i)}; \mathbf{X}) &= \operatorname{tr}(\psi) - \operatorname{tr}(\boldsymbol{\beta}_i\boldsymbol{\beta}_i'\psi\boldsymbol{\beta}_i\boldsymbol{\beta}_i') \\ &= \operatorname{tr}(\psi) - \boldsymbol{\beta}_i'\psi\boldsymbol{\beta}_i \\ &= \sum_{\substack{j=1 \\ j \neq i}}^{p} \lambda_j \end{aligned}$$

(see Exercise 5). Thus the best self-consistent projection on a line, in the sense of mean squared error, is $\mathbf{Y}_{(1)} = \boldsymbol{\mu} + \mathbf{P}_1(\mathbf{X} - \boldsymbol{\mu})$. □

Example 8.3.3 Suppose that $\mathbf{X} = \begin{pmatrix} X_1 \\ X_2 \end{pmatrix}$ has a uniform distribution inside the triangle with corner points $\begin{pmatrix} -1 \\ 0 \end{pmatrix}$, $\begin{pmatrix} 1 \\ 0 \end{pmatrix}$, and $\begin{pmatrix} 0 \\ 1 \end{pmatrix}$. Then $\boldsymbol{\mu} = E[\mathbf{X}] = \begin{pmatrix} 0 \\ 1/3 \end{pmatrix}$, and $\mathbf{Y} = \text{Cov}[\mathbf{X}] = \text{diag}(\frac{1}{6}, \frac{1}{18})$. Since ψ is diagonal, its eigenvectors are $\mathbf{e}_1 = \begin{pmatrix} 1 \\ 0 \end{pmatrix}$ and $\mathbf{e}_2 = \begin{pmatrix} 0 \\ 1 \end{pmatrix}$, and the associated eigenvalues are $\lambda_1 = \frac{1}{6}$, $\lambda_2 = \frac{1}{18}$. Thus the two candidates for self-consistent projections on a line are $\mathbf{Y}_{(1)} = \begin{pmatrix} X_1 \\ \frac{1}{3} \end{pmatrix}$ and $\mathbf{Y}_{(2)} = \begin{pmatrix} 0 \\ X_2 \end{pmatrix}$, but only $\mathbf{Y}_{(2)}$ is self-consistent; see Exercise 6. Further examples are given in Exercises 7 to 9. □

Theorem 8.3.1 tells us that, in view of self-consistency, orthogonal projections $\mathbf{Y} = \boldsymbol{\mu} + \mathbf{P}(\mathbf{X} - \boldsymbol{\mu})$ should be chosen with $\mathbf{P} = \boldsymbol{\beta}_i\boldsymbol{\beta}_i'$, where $\boldsymbol{\beta}_i$ is an eigenvector of the covariance matrix. The same result is obtained if we focus on the mean squared error of the projection on a line, irrespective of self-consistency, as shown in the next theorem. Actually, this is a version of Pearson's (1901) original approach in which he minimized the sum of squared orthogonal distances of N points in $\mathbb{R}^p$ from a line. See Exercise 10.

Theorem 8.3.2 *Let* $\mathbf{X}$ *denote a p-variate random vector with* $E[\mathbf{X}] = \boldsymbol{\mu}$ *and* $\text{Cov}[\mathbf{X}] = \psi$, *and let* $\mathbf{Y}$ *denote an orthogonal projection of* $\mathbf{X}$ *on a line in* $\mathbb{R}^p$. *Then* $\text{MSE}(\mathbf{Y};\mathbf{X})$ *is minimal for*

$$\mathbf{Y} = \mathbf{Y}_{(1)} = \boldsymbol{\mu} + \boldsymbol{\beta}_1\boldsymbol{\beta}_1'(\mathbf{X} - \boldsymbol{\mu}), \tag{19}$$

where $\boldsymbol{\beta}_1$ *is a normalized eigenvector associated with the largest eigenvalue* λ_1 *of* ψ, *and*

$$\text{MSE}(\mathbf{Y}_{(1)}; \mathbf{X}) = \text{tr}(\psi) - \lambda_1. \tag{20}$$

Proof Since $\mathbf{Y}$ is an orthogonal projection on a line, it can be written as $\mathbf{Y} = \mathbf{x}_0 + \mathbf{b}\mathbf{b}'(\mathbf{X} - \mathbf{x}_0)$ for some $\mathbf{x}_0 \in \mathbb{R}^p$ and some $\mathbf{b} \in \mathbb{R}^p$, $\mathbf{b}'\mathbf{b} = 1$. We need to determine $\mathbf{x}_0$ and $\mathbf{b}$ so as to minimize $\text{MSE}(\mathbf{Y}; \mathbf{X})$. With $\mathbf{P} = \mathbf{b}\mathbf{b}'$ denoting the projection matrix,

$$\begin{aligned} \text{MSE}(\mathbf{Y}; \mathbf{X}) &= E\left[||\mathbf{X} - \mathbf{Y}||^2\right] \\ &= E\left[||(\mathbf{X} - \mathbf{x}_0) - \mathbf{P}(\mathbf{X} - \mathbf{x}_0)||^2\right] \\ &= E\left[||(\mathbf{I}_p - \mathbf{P})(\mathbf{X} - \mathbf{x}_0)||^2\right]. \end{aligned} \tag{21}$$

For any p-variate random vector $\mathbf{U}$ with finite second moments and for any fixed matrix $\mathbf{A}$ of dimension $s \times p$, it is true that

$$E\left[||\mathbf{A}(\mathbf{U} - E[\mathbf{U}])||^2\right] \le E\left[||\mathbf{A}(\mathbf{U} - \mathbf{m})||^2\right] \tag{22}$$

for all $\mathbf{m} \in \mathbb{R}^p$; see Exercise 11. Hence,

$$\begin{aligned} \text{MSE}(\mathbf{Y}; \mathbf{X}) &= E\left[||(\mathbf{I}_p - \mathbf{P})(\mathbf{X} - \mathbf{x}_0)||^2\right] \\ &\geq E\left[||(\mathbf{I}_p - \mathbf{P})(\mathbf{X} - \boldsymbol{\mu})||^2\right] \\ &= \text{tr}\left\{\text{Cov}\left[(\mathbf{I}_p - \mathbf{P})\mathbf{X}\right]\right\} \\ &= \text{tr}\left[(\mathbf{I}_p - \mathbf{P})\boldsymbol{\psi}(\mathbf{I}_p - \mathbf{P})\right]. \end{aligned} \tag{23}$$

Using idempotence of $\mathbf{I}_p - \mathbf{P}$, we obtain

$$\begin{aligned} \text{MSE}(\mathbf{Y}; \mathbf{X}) &\geq \text{tr}\left[(\mathbf{I}_p - \mathbf{P})^2\boldsymbol{\psi}\right] \\ &= \text{tr}\left[(\mathbf{I}_p - \mathbf{P})\boldsymbol{\psi}\right] \\ &= \text{tr}(\boldsymbol{\psi}) - \text{tr}(\mathbf{P}\boldsymbol{\psi}). \end{aligned} \tag{24}$$

Therefore, to minimize MSE$(\mathbf{Y}; \mathbf{X})$, we have to maximize $\text{tr}(\mathbf{P}\boldsymbol{\psi})$. But

$$\begin{aligned} \text{tr}(\mathbf{P}\boldsymbol{\psi}) &= \text{tr}(\mathbf{b}\mathbf{b}'\boldsymbol{\psi}) \\ &= \text{tr}(\mathbf{b}'\boldsymbol{\psi}\mathbf{b}) \\ &= \mathbf{b}'\boldsymbol{\psi}\mathbf{b}. \end{aligned} \tag{25}$$

Thus, we have to maximize $\text{var}[\mathbf{b}'\mathbf{X}] = \mathbf{b}'\boldsymbol{\psi}\mathbf{b}$ over all vectors $\mathbf{b} \in \mathbb{R}^p$ of unit length. Let $h(\mathbf{b}) = \mathbf{b}'\boldsymbol{\psi}\mathbf{b}$ and $h^*(\mathbf{b}, \lambda) = h(\mathbf{b}) - \lambda(\mathbf{b}'\mathbf{b} - 1)$, where λ is a Lagrange multiplier. Then the maximum of the function h under the constraint $\mathbf{b}'\mathbf{b} = 1$ is obtained at a critical point of the function h^*. Using the rules for differentiation from appendix A.4, we obtain

$$\frac{\partial h^*}{\partial \mathbf{b}} = 2\boldsymbol{\psi}\mathbf{b} - 2\lambda\mathbf{b}. \tag{26}$$

Setting (26) equal to zero, it follows that

$$\boldsymbol{\psi}\mathbf{b} = \lambda\mathbf{b}, \tag{27}$$

that is, critical points are given by the eigenvectors $\boldsymbol{\beta}_i$ and associated eigenvalues λ_i of $\boldsymbol{\psi}$. Since

$$\text{var}[\boldsymbol{\beta}_i'\mathbf{X}] = \boldsymbol{\beta}_i'\boldsymbol{\psi}\boldsymbol{\beta}_i = \boldsymbol{\beta}_i'\lambda_i\boldsymbol{\beta}_i = \lambda_i \tag{28}$$

for all eigenvectors $\boldsymbol{\beta}_i$, the variance of $\mathbf{b}'\mathbf{X}$ is maximized by choosing $\mathbf{b} = \boldsymbol{\beta}_1$, a normalized eigenvector associated with the largest root.

Finally, setting $\mathbf{x}_0 = \boldsymbol{\mu}$, $\mathbf{P} = \boldsymbol{\beta}_1\boldsymbol{\beta}_1'$, and $\mathbf{Y}_{(1)} = \boldsymbol{\mu} + \mathbf{P}(\mathbf{X} - \boldsymbol{\mu})$,

$$\begin{aligned} \text{MSE}(\mathbf{Y}_{(1)}; \mathbf{X}) &= \text{tr}(\boldsymbol{\psi}) - \text{tr}(\mathbf{P}\boldsymbol{\psi}) \\ &= \text{tr}(\boldsymbol{\psi}) - \lambda_1. \end{aligned} \tag{29}$$

This concludes the proof. ■

Example 8.3.4 continuation of Example 8.3.2

For $\mathbf{X} \sim \mathcal{N}_p(\boldsymbol{\mu}, \boldsymbol{\psi})$, where $\boldsymbol{\psi}$ has a distinct largest eigenvalue λ_1, the best orthogonal projection is $\mathbf{Y}_{(1)} = \boldsymbol{\mu} + \boldsymbol{\beta}_1\boldsymbol{\beta}_1'(\mathbf{X} - \boldsymbol{\mu})$. Here, $\boldsymbol{\beta}_1$ is the normalized eigenvector associated with λ_1, and $\boldsymbol{\beta}_1$ is unique up to multiplication by -1. □

Example 8.3.5 continuation of Example 8.3.3

In this case, $\mathbf{Y}_{(1)} = \begin{pmatrix} X_1 \\ 1/3 \end{pmatrix}$ is the projection that minimizes the mean squared error, with $\text{MSE}(\mathbf{Y}_{(1)}; \mathbf{X}) = 1/18$, but $\mathbf{Y}_{(1)}$ is not self-consistent for $\mathbf{X}$. □

Example 8.3.6 Suppose that $\mathbf{X}$ is bivariate uniform in the unit square. Then $E[\mathbf{X}] = \frac{1}{2}\begin{pmatrix} 1 \\ 1 \end{pmatrix}$, and $\text{Cov}[\mathbf{X}] = \frac{1}{12}\mathbf{I}_2$, with two identical eigenvalues of $1/12$. Hence, any projection on a line through the mean has the same mean squared error for $\mathbf{X}$, but only four of the projections are self-consistent; see Exercise 7. □

The third theorem of this section generalizes both Theorems 8.3.1 and 8.3.2 to projections on planes or hyperplanes. The proof is left to the technically inclined student; see Exercises 13 and 14.

Theorem 8.3.3 *Suppose that $\mathbf{X}$ is a p-variate random vector with mean vector $\boldsymbol{\mu}$ and covariance matrix $\boldsymbol{\psi}$. Let $\mathbf{Y}$ denote the orthogonal projection of $\mathbf{X}$ on a line, plane, or hyperplane of dimension q, $1 \le q \le p$.*

(a) *If $\mathbf{Y}$ is self-consistent for $\mathbf{X}$, then*

$$\mathbf{Y} = \boldsymbol{\mu} + \mathbf{P}(\mathbf{X} - \boldsymbol{\mu}), \tag{30}$$

where $\mathbf{P} = \mathbf{A}\mathbf{A}'$ and $\mathbf{A}$ is a matrix of dimension p by q whose columns are q distinct orthonormal eigenvectors of $\boldsymbol{\psi}$.

(b) *The mean squared error of $\mathbf{Y}$ for $\mathbf{X}$ is minimal if we define $\mathbf{Y}$ as in equation (30), where the columns of $\mathbf{A}$ are q orthonormal eigenvectors of $\boldsymbol{\psi}$ associated with the q largest roots of $\boldsymbol{\psi}$. With $\lambda_1 \ge \cdots \ge \lambda_p \ge 0$ denoting the ordered eigenvalues, the corresponding mean squared error is $\sum_{i=q+1}^{p} \lambda_i$.*

Part (a) of Theorem 8.3.3 generalizes the earlier result on self-consistent projections on lines to self-consistent projections on planes or hyperplanes of any dimension. Essentially it says that, if we look for a self-consistent projection in a subspace of dimension q, then it suffices to look at subspaces spanned by q different eigenvectors of the covariance matrix. However, such a projection is not automatically self-consistent, as we have seen in Example 8.3.3. Rather, self-consistency depends on the exact form of the multivariate distribution.

Part (b) of Theorem 8.3.3 generalizes Theorem 8.3.2. It says that, if we want to approximate **X** by an orthogonal projection on a plane or hyperplane, then according to the principle of minimizing mean squared error, we have to choose a subspace spanned by q eigenvectors of the covariance matrix associated with the q largest roots. This holds irrespective of the exact distributional shape and irrespective of whether or not the projection is self-consistent. A remarkable consequence is that, as we increase q, the optimal subspace is always obtained by "adding another dimension" to the previous subspace. More precisely, for $q = 1$, the optimal subspace is spanned by β_1; for $q = 2$, by β_1 and β_2, and so on. This is ultimately the justification for the popular definition of principal components in terms of a stepwise maximization procedure, in each step maximizing the variance of a normalized linear combination of **X**, subject to constraints of uncorrelatedness. This is outlined in Exercise 15.

We are now finally ready for a definition of principal components.

Definition 8.3.1 Let **X** denote a p-variate random vector with $E[\mathbf{X}] = \mu$ and $\text{Cov}[\mathbf{X}] = \psi$. Let $\lambda_1 \geq \cdots \geq \lambda_p \geq 0$ denote the eigenvalues of ψ and $\beta_1, \ldots, \beta_p$ associated orthonormal eigenvectors, such that $\mathbf{B} = [\beta_1, \ldots, \beta_p]$ is orthogonal. For $1 \leq q \leq p$, let $\mathbf{A}_q = [\beta_1, \ldots, \beta_q]$. Then

(a) the *linear principal components* of **X** are the p jointly distributed random variables

$$\mathbf{U} = \begin{bmatrix} U_1 \\ \vdots \\ U_p \end{bmatrix} = \mathbf{B}'(\mathbf{X} - \mu) = \begin{bmatrix} \beta_1'(\mathbf{X} - \mu) \\ \vdots \\ \beta_p'(\mathbf{X} - \mu) \end{bmatrix}, \tag{31}$$

and

(b) the p-variate random vector

$$\mathbf{Y}_{(q)} = \mu + \mathbf{A}_q \mathbf{A}_q'(\mathbf{X} - \mu) \tag{32}$$

is called a *q-dimensional principal component approximation* of **X**. □

It is important to understand this definition correctly. The transformation $\mathbf{U} = \mathbf{B}'(\mathbf{X} - \mu)$ maps the p variables X_i into p new variables U_i called the principal components. It is a nonsingular transformation, whose inverse is

$$\mathbf{X} = \mu + \mathbf{B}\mathbf{U}, \tag{33}$$

as follows from orthogonality of **B**. This can be illustrated graphically. Figure 8.3.2, once again, shows a bivariate normal distribution in the form of a contour plot. The transformation (31) from **X** to **U** corresponds to a shift of the origin of the coordinate system to the center of the distribution, followed by a rotation by an angle corresponding to the orthogonal matrix. Thus, the p principal components are merely a representation of the distribution in a different coordinate system. In contrast, the one-dimensional principal component approximation $\mathbf{Y}_{(1)}$ is a two-dimensional random vector, corresponding to the projection of $\mathbf{X}$ on the line labeled U_1. It takes values only on the straight line, but nevertheless it is a bivariate random vector. The coordinate system of $\mathbf{Y}_{(1)}$ is the same as the coordinate system of $\mathbf{X}$. Actually one can think of $\mathbf{Y}_{(1)}$ as a two-dimensional random vector obtained by the following procedure: First, transform $\mathbf{X}$ to principal components. Then replace the second principal component (U_2) by its mean 0, and then apply the inverse transformation (33) to the vector $\begin{pmatrix} U_1 \\ 0 \end{pmatrix}$. This follows from the fact that $\mathbf{Y}_{(q)}$ can be written as

$$\mathbf{Y}_{(q)} = \boldsymbol{\mu} + \mathbf{A}_q \begin{pmatrix} U_1 \\ \vdots \\ U_q \end{pmatrix}. \tag{34}$$

See also Exercise 16.

In the definition of q-dimensional principal component approximations, the case $q = p$ is included. Since in this case $\mathbf{A}_q = \mathbf{B}$, and $\mathbf{B}$ is orthogonal,

$$\mathbf{Y}_{(p)} = \boldsymbol{\mu} + \mathbf{BB}'(\mathbf{X} - \boldsymbol{\mu}) = \mathbf{X}, \tag{35}$$

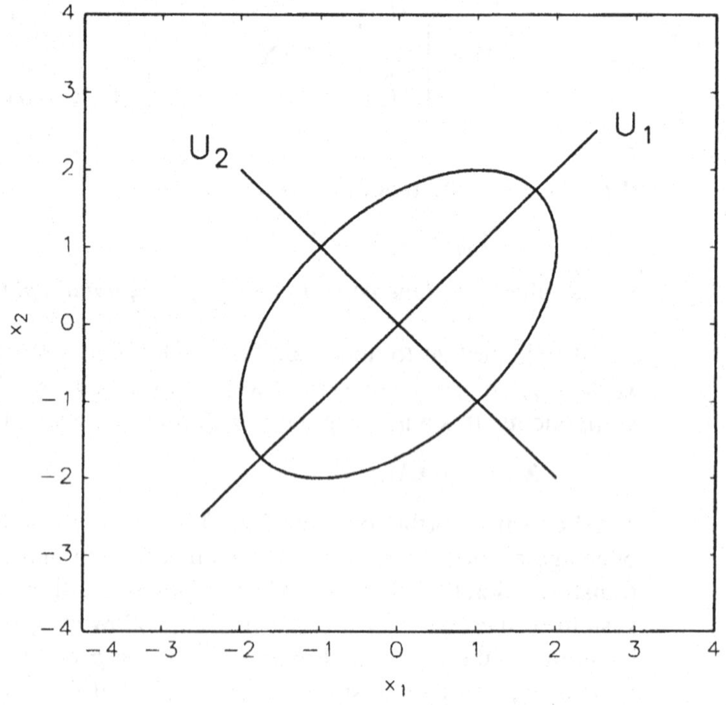

Figure 8-3-2
Contour plot of a bivariate normal distribution, with coordinate system marked (U_1, U_2) of the principal components. The one-dimensional principal component approximation is the orthogonal projection on the line marked U_1.

i.e., the p-dimensional, principal component approximation of $\mathbf{X}$ is $\mathbf{X}$ itself. Formally, even the case $q = 0$ may be included, in which the matrix $\mathbf{A}_q$ disappears, by declaring $\mathbf{Y}_{(0)} = \boldsymbol{\mu}$ to be the zero-dimensional principal component approximation of $\mathbf{X}$.

Definition 8.3.1 attaches the word *linear* to the principal components. Most textbooks would call the U_i just principal components. The word "linear" has been added to stress the fact that only linear transformations are considered, which may not always be appropriate. Actually, one might argue that a transformed variable U_i, as in equation (31), should be called a principal component only if the associated projection $\boldsymbol{\mu} + \boldsymbol{\beta}_i \boldsymbol{\beta}_i'(\mathbf{X} - \boldsymbol{\mu})$ is self-consistent for $\mathbf{X}$.

Exercises for Section 8.3

1. Verify equation (1).

2. This exercise concerns the projection illustrated in Figure 8.3.1(b).

(a) Find the projection matrix $\mathbf{P}$ associated with the projection on the line given by $x_2 = \frac{1}{3}x_1$.

(b) Show that $\mathbf{Y} = \mathbf{PX}$ is not self-consistent for $\mathbf{X}$. *Hint*: Find a vector $\begin{pmatrix} c \\ c/3 \end{pmatrix}$ such that

$$E\left[\mathbf{X} | \mathbf{PX} = \begin{pmatrix} c \\ c/3 \end{pmatrix}\right] \neq \begin{pmatrix} c \\ c/3 \end{pmatrix}.$$

3. In Example 8.3.1, show that $\mathbf{Y}_{(2)}$ is self-consistent for $\mathbf{X}$ without using the results from Section 8.4.

4. In Figure 8.3.1, an ellipse of constant density has the form $\mathbf{x}'\psi^{-1}\mathbf{x} = c^2$ for some $c > 0$, where $\psi = \begin{pmatrix} 2 & 1 \\ 1 & 2 \end{pmatrix}$. Show that the principal axes of the ellipse are along the lines $x_1 = x_2$ and $x_1 = -x_2$, and find the lengths of these axes (i.e., the longest and the shortest diameter) as functions of c. More generally, let an ellipse in $\mathbb{R}^2$ be given by $(\mathbf{x}-)'\mathbf{A}^{-1}(\mathbf{x}-) = \mathbf{c}^2$, where $\in \mathbb{R}^2$ is fixed and $\mathbf{A}$ is a positive-definite symmetric matrix of dimension 2×2. Find the directions and lengths of the principal axes of the ellipse.

5. In Example 8.3.2, verify that

$$\text{MSE}(\mathbf{Y}_{(i)}; \mathbf{X}) = \sum_{\substack{j=1 \\ j \neq i}}^{p} \lambda_j.$$

6. In Example 8.3.3, show that $\mathbf{Y}_{(2)}$ is self-consistent for $\mathbf{X}$, but $\mathbf{Y}_{(1)}$ is not.

7. Let $\mathbf{X}$ denote a bivariate random vector with uniform distribution inside the rectangle $[0, a] \times [0, b]$, where $a > 0$ and $b > 0$.

(a) If $a \neq b$, show that there are exactly two self-consistent projections on lines. Which one has the smaller mean squared error?

(b) If $a = b$, show that there are exactly four self-consistent projections on lines. Also show that all projections on a line through the point $\begin{pmatrix} a/2 \\ b/2 \end{pmatrix}$ have the same mean squared error for $\mathbf{X}$.

8. Define a discrete bivariate random vector $\mathbf{X}$ as follows. For a fixed number $k > 1$, choose k equally spaced points on a circle in $\mathbb{R}^2$, and put probability $1/k$ on each of the k points. Show that there are exactly k self-consistent projections on lines. *Note*: Distinguish between the cases k odd and k even.

9. Suppose that $\mathbf{X}$ follows a p-variate spherical distribution, with $E[\mathbf{X}] = \mathbf{0}$. Show that all projections on lines through the origin are self-consistent.

10. This exercise illustrates Pearson's original approach to principal components in terms of orthogonal least squares. Let $\mathbf{x}_1, \ldots, \mathbf{x}_N$ denote N data points in $\mathbb{R}^p$, let $\bar{\mathbf{x}} = \frac{1}{N}\sum_{i=1}^N \mathbf{x}_i$ and $\mathbf{S} = \frac{1}{N}\sum_{i=1}^N(\mathbf{x}_i - \bar{\mathbf{x}})(\mathbf{x}_i - \bar{\mathbf{x}})'$. Let $\mathbf{y}_i$ denote the orthogonal projections of the $\mathbf{x}_i$ on a straight line. Show that the projection which minimizes

$$\sum_{i=1}^N ||\mathbf{x}_i - \mathbf{y}_i||^2$$

is given by $\mathbf{y}_i = \bar{\mathbf{x}} + \mathbf{b}_1\mathbf{b}_1'(\mathbf{x}_i - \bar{\mathbf{x}})$, where $\mathbf{b}_1$ is an eigenvector associated with the largest root of $\mathbf{S}$. *Hint*: this can be done using Theorem 8.3.2, by formulating an appropriate p-variate distribution.

11. Prove equation (22).

12. This exercise gives an alternative technique to the last part of the proof of Theorem 8.3.2, which does not involve calculus. In the setup and notation of the theorem, suppose we wish to maximize $\mathbf{b}'\boldsymbol{\psi}\mathbf{b}$ with respect to $\mathbf{b} \in \mathbb{R}^p$ and subject to the constraint $\mathbf{b}'\mathbf{b} = 1$. Let $\boldsymbol{\psi} = \mathbf{B}\boldsymbol{\Lambda}\mathbf{B}$ denote a spectral decomposition of $\boldsymbol{\psi}$, where $\mathbf{B} = (\boldsymbol{\beta}_1, \ldots, \boldsymbol{\beta}_p)$ is orthogonal and $\boldsymbol{\Lambda} = \text{diag}(\lambda_1 \ldots, \lambda_p)$, the eigenvalues being in descending order. Without using derivatives, show that the maximum is taken for $\mathbf{b} = \boldsymbol{\beta}_1$. *Hint*: Define $\mathbf{y} = \mathbf{B}'\mathbf{b}$ and notice that $\mathbf{y}'\mathbf{y} = 1$.

13. Prove part (a) of Theorem 8.3.3 along the following lines.

(a) Using the same technique as in the proof of Theorem 8.3.1, derive an equation generalizing the one in the text following (16),

$$\mathbf{A}'\boldsymbol{\psi} = \mathbf{A}'\boldsymbol{\psi}\mathbf{A}\mathbf{A}'.$$

(b) Let $\mathbf{H}$ denote an orthogonal matrix of dimension $q \times q$ such that $\mathbf{H}'(\mathbf{A}'\boldsymbol{\psi}\mathbf{A})\mathbf{H}$ is diagonal. Show that the columns of $\mathbf{A}^* = \mathbf{A}\mathbf{H}$ are q eigenvectors of $\boldsymbol{\psi}$.

(c) Show that the columns of $\mathbf{A}$ span the same subspace as the columns of $\mathbf{A}^*$.

14. Prove part (b) of Theorem 8.3.3 along the following lines.

(a) Using the technique in the proof of Theorem 8.3.2, show that the problem reduces to maximizing $\text{tr}(\mathbf{A}'\psi\mathbf{A})$ over all matrices $\mathbf{A}$ of dimension $p \times q$ such that $\mathbf{A}'\mathbf{A} = \mathbf{I}_q$.

(b) Writing $\mathbf{a}_1, \ldots, \mathbf{a}_q$ for the q columns of $\mathbf{A}$, introduce Lagrange multipliers λ_i for the constraints $\mathbf{a}_i'\mathbf{a}_i = 1$ $(i = 1, \ldots, q)$ and Lagrange multipliers λ_{ij} $(1 \le i < j \le q)$ for the constraints $\mathbf{a}_i'\mathbf{a}_j = 0$. Take derivatives with respect to the vectors $\mathbf{a}_i$, set them equal to zero, and show, by algebraic manipulation, that critical points are given by substituting eigenvectors of ψ for the $\mathbf{a}_i$.

(c) Argue that $\mathbf{a}_1, \ldots, \mathbf{a}_q$ should be chosen as the eigenvectors associated with the q largest roots of ψ to minimize the mean squared error.

15. Let $\mathbf{X}$ denote a p-variate random vector with mean $\boldsymbol{\mu} = \mathbf{0}$ and covariance matrix ψ. In the proof of Theorem 8.3.2, it has implicitly been shown that the first principal component U_1 can be thought of as a linear combination $\mathbf{b}'\mathbf{X}$, which has maximum variance among all normalized linear combinations of $\mathbf{X}$.

(a) Show that the second principal component U_2 can be obtained as a linear combination $\mathbf{b}'\mathbf{X}$ which has maximum variance among all normalized linear combinations of $\mathbf{X}$, subject to the constraint of uncorrelatedness with U_1, i.e., subject to $\text{cov}[\mathbf{b}'\mathbf{X}, U_1] = 0$. *Hint*: Show that this constraint is equivalent to $\mathbf{b}'\boldsymbol{\beta}_1 = 0$, where $\boldsymbol{\beta}_1$ is an eigenvector associated with the largest root of ψ.

(c) Show, more generally, that the hth principal component U_h $(2 \le h \le p)$ can be obtained as the linear combination of $\mathbf{X}$ which has maximum variance among all normalized linear combinations of $\mathbf{X}$ that are uncorrelated with all previously obtained principal components, i.e., uncorrelated with $U_1, \ldots, U_{h-1}$.

16. Show that the q-dimensional principal component approximation $\mathbf{Y}_{(q)}$ defined in equation (32) can be obtained from the following procedure.

- Transform the random vector $\mathbf{X}$ to principal components $\mathbf{U}$ according to equation (31).
- Replace the last $(p - q)$ principal components by their mean 0, i.e., set

$$\mathbf{U}^* = \begin{pmatrix} U_1 \\ \vdots \\ U_q \\ 0 \\ \vdots \\ 0 \end{pmatrix}.$$

- Transform back to original coordinates, i.e., set

$$\mathbf{Y}_{(q)} = \boldsymbol{\mu} + \mathbf{B}\mathbf{U}^*.$$

17. Show the following result. For two jointly distributed random variables U and V with finite second moments, $E[U|V] = E[U]$ implies $\text{cov}[U, V] = 0$. More generally, for jointly distributed, random vectors $\mathbf{U}$ and $\mathbf{V}$, show that $E[\mathbf{U}|\mathbf{V}] = \mathbf{0}$ implies that all covariances between $\mathbf{U}$ and $\mathbf{V}$ are zero. *Hint*: Assume, without loss of generality, that $E[U] = 0$ and $E[V] = 0$, and use $E[UV] = E\big[E[UV|V]\big]$.

8.4 Properties of Linear Principal Components

In this section, we present some properties of linear principal components, and comment on them. Throughout the section, it will be assumed that $\mathbf{X}$ is a p-variate random vector with $E[\mathbf{X}] = \boldsymbol{\mu}$ and $\mathrm{Cov}[\mathbf{X}] = \boldsymbol{\psi}$. Using the spectral decomposition of $\boldsymbol{\psi}$, we will write

$$\boldsymbol{\psi} = \mathbf{B}\boldsymbol{\Lambda}\mathbf{B}', \tag{1}$$

where $\mathbf{B} = [\boldsymbol{\beta}_1, \ldots, \boldsymbol{\beta}_p]$ is orthogonal with eigenvectors of $\boldsymbol{\psi}$ as columns and $\boldsymbol{\Lambda} = \mathrm{diag}(\lambda_1, \ldots, \lambda_p)$. It will always be assumed that the eigenvalues λ_i are ordered, i.e., $\lambda_1 \geq \cdots \geq \lambda_p \geq 0$.

Property 1 *The principal components* $\mathbf{U} = \mathbf{B}'(\mathbf{X} - \boldsymbol{\mu})$ *have mean vector* $\mathbf{0}$ *and covariance matrix* $\boldsymbol{\Lambda}$, *that is, principal components are centered at the origin and uncorrelated.*

Property 1 follows directly from the definition of principal components. If $\mathbf{X}$ is multivariate normal, then the principal components are even independent. Inverting the transformation gives

$$\mathbf{X} = \boldsymbol{\mu} + \mathbf{B}\mathbf{U}, \tag{2}$$

that is, we can write the p measured variables as linear combinations of the p uncorrelated random variables $\mathbf{U}$. This may have an interpretation if some meaning can be attached to principal components, or if the variances of the last principal components are so small that, in fact, the measured variables $\mathbf{X}$ are mostly determined by the first few principal components. An illustrative example follows.

Example 8.4.1 Suppose that two observers independently measure a random variable V, but each of them makes some measurement error. Thus, the actual measurements obtained are $X_1 = V + \epsilon_1$ and $X_2 = V + \epsilon_2$, where ϵ_1 and ϵ_2 are the errors. Let $E[V] = \mu$, $\mathrm{var}[V] = \sigma^2$, and assume that $E[\epsilon_i] = 0$, $\mathrm{var}[\epsilon_i] = \sigma_\epsilon^2$, $i = 1, 2$. Moreover, suppose that V, ϵ_1, and ϵ_2 are mutually independent. Setting $\mathbf{X} = \begin{pmatrix} X_1 \\ X_2 \end{pmatrix}$, it follows that

$$E[\mathbf{X}] = \begin{pmatrix} \mu \\ \mu \end{pmatrix}, \quad \text{and} \quad \mathrm{Cov}[\mathbf{X}] = \boldsymbol{\psi} = \begin{pmatrix} \sigma^2 + \sigma_\epsilon^2 & \sigma^2 \\ \sigma^2 & \sigma^2 + \sigma_\epsilon^2 \end{pmatrix},$$

see Exercise 1. The eigenvectors of $\boldsymbol{\psi}$ are $\boldsymbol{\beta}_1 = \frac{1}{\sqrt{2}}\begin{pmatrix} 1 \\ 1 \end{pmatrix}$ and $\boldsymbol{\beta}_2 = \frac{1}{\sqrt{2}}\begin{pmatrix} -1 \\ 1 \end{pmatrix}$; the associated eigenvalues are $\lambda_1 = 2\sigma^2 + \sigma_\epsilon^2$ and $\lambda_2 = \sigma_\epsilon^2$. Thus the principal components and their variances are given by

$$U_1 = \frac{1}{\sqrt{2}}(X_1 + X_2 - 2\mu), \quad \text{var}[U_1] = 2\sigma^2 + \sigma_\epsilon^2$$

and

$$U_2 = \frac{1}{\sqrt{2}}(X_2 - X_1), \quad \text{var}[U_2] = \sigma_\epsilon^2.$$

If σ_ϵ^2 is relatively small compared to σ^2, we may be willing to use a one-dimensional principal component approximation to the joint distribution of X_1 and X_2, which has the form

$$\begin{aligned}\mathbf{Y}_{(1)} &= \begin{pmatrix} \mu \\ \mu \end{pmatrix} + \frac{1}{2}\begin{pmatrix} 1 & 1 \\ 1 & 1 \end{pmatrix}\begin{pmatrix} X_1 - \mu \\ X_2 - \mu \end{pmatrix} \\ &= \frac{1}{2}\begin{pmatrix} X_1 + X_2 \\ X_1 + X_2 \end{pmatrix}.\end{aligned}$$

That is, substituting the average of X_1 and X_2 for both measurements gives us the orthogonal projection with the smallest mean squared error. Also note that by substituting the original random variables V and ϵ_i in X_1 and X_2, we obtain

$$U_1 = \frac{1}{\sqrt{2}}(2V + \epsilon_1 + \epsilon_2 - 2\mu)$$

and

$$U_2 = \frac{1}{\sqrt{2}}(\epsilon_2 - \epsilon_1).$$

Thus, the second principal component is determined entirely by the errors of measurement. The first principal component may be regarded as being "mostly" a function of V, always provided the errors of measurement have small variability. The inverse transformation is given by

$$X_1 = \mu + \frac{1}{\sqrt{2}}(U_1 - U_2)$$

and

$$X_2 = \mu + \frac{1}{\sqrt{2}}(U_1 + U_2).$$

If $\lambda_2 = \sigma_\epsilon^2$ is small, we might argue that both X_1 and X_2 are "mostly determined" by the first principal component U_1. However, one has to be careful with such interpretations because they might lead to erroneous conclusions. In fact, both U_1 and U_2 are constructed variables, and there is no guarantee that there exists a physical quantity V that determines U_1. Or in other words, if we observe a mean vector and covariance matrix with the structure given in this example, we cannot conclude that the model underlying the data is $X_1 = V + \epsilon_1$, $X_2 = V + \epsilon_2$, with independent errors ϵ_i. Only the reverse conclusion is correct: If the model as stated holds, then the mean vector and the covariance matrix will have the given form, and the principal components will exhibit the pattern shown in this example. See also Exercises 2 and 3. □

Property 2 *If* $\mathbf{X}$ *is multivariate normal or elliptical, then the* q*-dimensional principal component approximation* $\mathbf{Y}_{(q)}$ *is self-consistent for* $\mathbf{X}$.

In the case $q = p$, self-consistency follows directly and does not even require normality or ellipticity, since $\mathbf{Y}_{(p)} = \mathbf{X}$. If $1 \le q < p$, suppose that $\mathbf{X}$ is p-variate normal, and let $\mathbf{A}_q = [\boldsymbol{\beta}_1, \ldots, \boldsymbol{\beta}_q]$ and $\mathbf{B}_q = [\boldsymbol{\beta}_{q+1}, \ldots, \boldsymbol{\beta}_p]$. Then $\mathbf{Y}_{(q)} = \boldsymbol{\mu} + \mathbf{A}_q\mathbf{A}_q'(\mathbf{X} - \boldsymbol{\mu})$. We need to show that $E[\mathbf{X}|\mathbf{Y}_{(q)}] = \mathbf{Y}_{(q)}$. Notice that, for a p-variate random vector $\mathbf{Z}$, the following holds: If $\mathbf{Y}$ is self-consistent for $\mathbf{Z}$, and $\mathbf{m} \in \mathbb{R}^p$ is a fixed vector, then $\mathbf{Y} + \mathbf{m}$ is self-consistent for $\mathbf{Z} + \mathbf{m}$; see Exercise 4. Hence we need to show that $\mathbf{Y}_{(q)} - \boldsymbol{\mu} = \mathbf{A}_q\mathbf{A}_q'(\mathbf{X} - \boldsymbol{\mu})$ is self-consistent for $\mathbf{X} - \boldsymbol{\mu}$. Notice that

$$\begin{aligned} E\left[\mathbf{X} - \boldsymbol{\mu}|\mathbf{A}_q\mathbf{A}_q'(\mathbf{X} - \boldsymbol{\mu})\right] &= E\left[\mathbf{A}_q\mathbf{A}_q'(\mathbf{X} - \boldsymbol{\mu}) + \mathbf{B}_q\mathbf{B}_q'(\mathbf{X} - \boldsymbol{\mu})|\mathbf{A}_q\mathbf{A}_q'(\mathbf{X} - \boldsymbol{\mu})\right] \\ &= \mathbf{A}_q\mathbf{A}_q'(\mathbf{X} - \boldsymbol{\mu}) + E\left[\mathbf{B}_q\mathbf{B}_q'(\mathbf{X} - \boldsymbol{\mu})|\mathbf{A}_q\mathbf{A}_q'(\mathbf{X} - \boldsymbol{\mu})\right]. \end{aligned} \tag{3}$$

But $\mathbf{A}_q'(\mathbf{X} - \boldsymbol{\mu}) = \begin{pmatrix} U_1 \\ \vdots \\ U_q \end{pmatrix}$, and $\mathbf{B}_q'(\mathbf{X} - \boldsymbol{\mu}) = \begin{pmatrix} U_{q+1} \\ \vdots \\ U_p \end{pmatrix}$, which by property 1 are independent. Hence the last conditional expectation in (3) is zero, and the result follows. In the general elliptical case independence does not hold but the result is still true; see Exercise 5.

Example 8.4.2 continuation of Example 8.4.1

If we assume that V, ϵ_1, and ϵ_2 are all normal, then the joint distribution of X_1 and X_2 is bivariate normal, and the one-dimensional principal component approximation

$$\mathbf{Y}_{(1)} = \frac{1}{2}\begin{pmatrix} X_1 + X_2 \\ X_1 + X_2 \end{pmatrix}$$

is self-consistent for $\mathbf{X}$. However, normality assumptions are not needed in this case; see Exercise 6. □

Property 2 does not have any direct practical applications, but it may serve as a theoretical justification for linear principal component analysis, based on the principle of self-consistency. It also says that, if we find a linear principal component which is not self-consistent, then we might want to search the data for nonlinear structure. See Exercises 17 and 18 and Example 8.7.5.

Property 3 *The covariance matrix of the q-dimensional principal component approximation of* **X** *is*

$$\mathrm{Cov}[\mathbf{Y}_{(q)}] = \sum_{i=1}^{q} \lambda_i \boldsymbol{\beta}_i \boldsymbol{\beta}_i' = \boldsymbol{\psi} - \sum_{i=q+1}^{p} \lambda_i \boldsymbol{\beta}_i \boldsymbol{\beta}_i'. \tag{4}$$

This property is obtained by a direct calculation:

$$\begin{aligned}
\mathrm{Cov}\left[\mathbf{Y}_{(q)}\right] &= \mathrm{Cov}\left[\mathbf{A}_q \mathbf{A}_q' (\mathbf{X} - \boldsymbol{\mu})\right] \\
&= \mathbf{A}_q \mathbf{A}_q' \boldsymbol{\psi} \mathbf{A}_q \mathbf{A}_q' \\
&= \mathbf{A}_q \mathrm{diag}(\lambda_1, \ldots, \lambda_q) \mathbf{A}_q' \\
&= (\boldsymbol{\beta}_1, \ldots, \boldsymbol{\beta}_q) \begin{pmatrix} \lambda_1 & \ldots & 0 \\ \vdots & \ddots & \vdots \\ 0 & \ldots & \lambda_q \end{pmatrix} \begin{pmatrix} \boldsymbol{\beta}_1' \\ \vdots \\ \boldsymbol{\beta}_q' \end{pmatrix} \\
&= \sum_{i=1}^{q} \lambda_i \boldsymbol{\beta}_i \boldsymbol{\beta}_i'.
\end{aligned} \tag{5}$$

The last equality in equation (4) then follows from

$$\boldsymbol{\psi} = \mathbf{B} \boldsymbol{\Lambda} \mathbf{B}' = \sum_{i=1}^{p} \lambda_i \boldsymbol{\beta}_i \boldsymbol{\beta}_i'. \tag{6}$$

Property 3 is of considerable intuitive appeal. It says that the principal component approximations for increasing q "build" the covariance matrix $\boldsymbol{\psi}$ by adding successively, and in order of decreasing importance, projections into one-dimensional subspaces spanned by eigenvectors of the covariance matrix. If the last $(p-q)$ eigenvalues are relatively small, then the covariance matrix of $\mathbf{Y}_{(q)}$ will be similar to $\boldsymbol{\psi}$. Equation (4) also shows that if the $(p-q)$ smallest eigenvalues are zero, then the approximation is perfect.

Example 8.4.3 continuation of Example 8.4.1

Since $\lambda_1 = 2\sigma^2 + \sigma_\epsilon^2$ and $\boldsymbol{\beta}_1 = \frac{1}{\sqrt{2}} \binom{1}{1}$,

$$\mathrm{Cov}\left[\mathbf{Y}_{(1)}\right] = \begin{pmatrix} \sigma^2 + \frac{1}{2}\sigma_\epsilon^2 & \sigma^2 + \frac{1}{2}\sigma_\epsilon^2 \\ \sigma^2 + \frac{1}{2}\sigma_\epsilon^2 & \sigma^2 + \frac{1}{2}\sigma_\epsilon^2 \end{pmatrix}.$$

Of course, we could have obtained this result directly by computing the variance of $X_1 + X_2$. Again, we see that if σ_ϵ^2 is small, then the approximation is good. □

Property 4 *The mean squared error of the q-dimensional-principal component approximation is given by*

$$\mathrm{MSE}\left(\mathbf{Y}_{(q)}; \mathbf{X}\right) = \sum_{i=q+1}^{p} \lambda_i. \tag{7}$$

This property can be shown as follows. By Definition 8.2.2,

$$\begin{aligned}
\mathrm{MSE}\left(\mathbf{Y}_{(q)}; \mathbf{X}\right) &= E\left[||\mathbf{X} - \mathbf{Y}_{(q)}||^2\right] \\
&= E\left[||(\mathbf{X} - \boldsymbol{\mu}) - \mathbf{A}_q\mathbf{A}_q'(\mathbf{X} - \boldsymbol{\mu})||^2\right] \\
&= E\left[||(\mathbf{I}_p - \mathbf{A}_q\mathbf{A}_q')(\mathbf{X} - \boldsymbol{\mu})||^2\right] \\
&= E\left[||\mathbf{B}_q\mathbf{B}_q'(\mathbf{X} - \boldsymbol{\mu})||^2\right] \\
&= \mathrm{tr}\left(\mathrm{Cov}[\mathbf{B}_q\mathbf{B}_q'\mathbf{X}]\right).
\end{aligned} \tag{8}$$

Using the same technique as in the proof of property 3, we see that

$$\mathrm{Cov}[\mathbf{B}_q\mathbf{B}_q'\mathbf{X}] = \sum_{i=q+1}^{p} \lambda_i \boldsymbol{\beta}_i \boldsymbol{\beta}_i', \tag{9}$$

and property 4 follows by taking the trace of (9). Note that, if self-consistency is assumed, then the result follows directly from Theorem 8.2.2.

It is usually more informative to assess the quality of a principal component approximation by the *relative mean squared error*

$$\mathrm{RMSE}(\mathbf{Y}_{(q)}; \mathbf{X}) := \frac{\mathrm{MSE}(\mathbf{Y}_{(q)}; \mathbf{X})}{\mathrm{MSE}(\boldsymbol{\mu}; \mathbf{X})} \tag{10}$$

rather than by the mean squared error itself. In (10), the mean squared error of $\mathbf{Y}_{(q)}$ is compared to the mean squared error obtained from the approximation of $\mathbf{X}$ by $\boldsymbol{\mu}$. It follows from equation (7) that

$$\mathrm{RMSE}(\mathbf{Y}_{(q)}; \mathbf{X}) = \frac{\sum_{i=q+1}^{p} \lambda_i}{\sum_{i=1}^{p} \lambda_i} = 1 - \frac{\sum_{i=1}^{q} \lambda_i}{\sum_{i=1}^{p} \lambda_i}. \tag{11}$$

Since $\sum_{i=1}^{p} \lambda_i = \mathrm{tr}(\boldsymbol{\psi}) = \sum_{i=1}^{p} \mathrm{var}[X_i]$, the ratio

$$\frac{\sum_{i=1}^{q} \lambda_i}{\mathrm{tr}(\boldsymbol{\psi})} = 1 - \mathrm{RMSE}(\mathbf{Y}_{(q)}; \mathbf{X}) \tag{12}$$

is often used to measure the quality of the approximation, with values close to one indicating a good fit. In fact, many authors call (12) the "proportion of the total variance explained by the first q principal components." This is rather bad terminology because the word "explain" might suggest a causal interpretation. Therefore it is better to call it the proportion of the total variance accounted for by the first q principal components, but a truly meaningful interpretation is given only by its relationship to mean squared error.

It follows from property 4 that, if $\lambda_{q+1} = \cdots = \lambda_p = 0$, i.e., if the smallest $(p-q)$ eigenvalues are all zero, then the mean squared error of $\mathbf{Y}_{(q)}$ is zero, and thus the approximation is perfect. This can also be seen from property 3 and is illustrated in the following example.

Example 8.4.4 The singular multivariate normal distribution. Suppose that $\mathbf{X} \sim \mathcal{N}_p(\boldsymbol{\mu}, \boldsymbol{\psi})$, where $\boldsymbol{\psi}$ has rank $q < p$, i.e., $\boldsymbol{\psi}$ is singular. Then $\lambda_1 \geq \lambda_2 \geq \cdots \geq \lambda_q > 0$, and $\lambda_{q+1} = \cdots = \lambda_p = 0$. Let $\boldsymbol{\beta}_1, \ldots, \boldsymbol{\beta}_p$ denote the normalized eigenvectors. Then

$$\boldsymbol{\psi} = \sum_{i=1}^{p} \lambda_i \boldsymbol{\beta}_i \boldsymbol{\beta}_i' = \sum_{i=1}^{q} \lambda_i \boldsymbol{\beta}_i \boldsymbol{\beta}_i', \tag{13}$$

see equation (6). Set $\mathbf{A}_q = [\boldsymbol{\beta}_1, \ldots, \boldsymbol{\beta}_q]$ and $\mathbf{B}_q = [\boldsymbol{\beta}_{q+1}, \ldots, \boldsymbol{\beta}_p]$, and write the vector of principal components in partitioned form as $\mathbf{U} = \begin{pmatrix} \mathbf{U}_{(1)} \\ \mathbf{U}_{(2)} \end{pmatrix}$, where $\mathbf{U}_{(1)}$ contains the first q principal components. Then

$$\mathbf{U}_{(1)} = \mathbf{A}_q'(\mathbf{X} - \boldsymbol{\mu}) \tag{14}$$

and

$$\mathbf{U}_{(2)} = \mathbf{B}_q'(\mathbf{X} - \boldsymbol{\mu}). \tag{15}$$

But $\text{Cov}[\mathbf{U}_{(2)}] = \mathbf{0}$, and therefore $\mathbf{U}_{(2)} = \mathbf{0} \in \mathbb{R}^{p-q}$ with probability one (see Exercise 7). Thus,

$$\begin{aligned} \mathbf{X} &= \boldsymbol{\mu} + \mathbf{A}_q\mathbf{A}_q'(\mathbf{X} - \boldsymbol{\mu}) + \mathbf{B}_q\mathbf{B}_q'(\mathbf{X} - \boldsymbol{\mu}) \\ &= \boldsymbol{\mu} + \mathbf{A}_q\mathbf{A}_q'(\mathbf{X} - \boldsymbol{\mu}) \\ &= \boldsymbol{\mu} + \mathbf{A}_q\mathbf{U}_{(1)} \\ &= \mathbf{Y}_{(q)}, \end{aligned} \tag{16}$$

that is, the p-variate random vector $\mathbf{X}$ is determined entirely by the first q principal components. It also follows from property 3 that $\text{Cov}[\mathbf{Y}_{(q)}] = \boldsymbol{\psi} = \text{Cov}[\mathbf{X}]$. Since the largest q eigenvalues are positive, the matrix

$$\text{Cov}[\mathbf{U}_{(1)}] = \text{diag}(\lambda_1, \ldots, \lambda_q) \tag{17}$$

is positive-definite, and therefore $\mathbf{U}_{(1)}$ has a density function, although $\mathbf{X}$ itself does not have a density. □

Example 8.4.5 This is a more general version of Example 8.4.1 and Exercise 3. Suppose that $Z_1, \ldots, Z_q$ are q independent random variables, all with mean 0 and variance 1. Let $\mathbf{C}$ denote a matrix of dimension $p \times q$, where $p > q$, and assume $\text{rank}(\mathbf{C}) = q$.

Let $\epsilon_1, \ldots, \epsilon_p$ denote p independent random variables, independent also of $\mathbf{Z}$, with $E[\epsilon_i] = 0$ and $\text{var}[\epsilon_i] = \sigma_\epsilon^2$. Finally, let $\boldsymbol{\mu} \in \mathbb{R}^p$ denote a fixed vector, and set

$$\mathbf{X} = \boldsymbol{\mu} + \mathbf{CZ} + \boldsymbol{\epsilon}. \tag{18}$$

Thus $\mathbf{X}$ is a p-variate random vector, and each X_i can be written as

$$X_i = \mu_i + \sum_{j=1}^{q} c_{ij} Z_j + \epsilon_i \qquad i = 1, \ldots, p. \tag{19}$$

It follows that $E[\mathbf{X}] = \boldsymbol{\mu}$ and

$$\text{Cov}[\mathbf{X}] = \boldsymbol{\psi} = \mathbf{CC}' + \sigma_\epsilon^2 \mathbf{I}_p, \tag{20}$$

see Exercise 8. What are the principal components of $\mathbf{X}$? Let $\mathbf{M} = \mathbf{CC}'$. Then $\mathbf{M}$ is positive semidefinite and has rank q, that is, $\mathbf{M}$ has q positive eigenvalues $\gamma_1 \geq \cdots \geq \gamma_q > 0$ and associated eigenvectors $\boldsymbol{\beta}_1, \ldots, \boldsymbol{\beta}_q$. But $\boldsymbol{\psi} = \mathbf{M} + \sigma_\epsilon^2 \mathbf{I}_p$, so

$$\begin{aligned} \boldsymbol{\psi}\boldsymbol{\beta}_i &= (\mathbf{M} + \sigma_\epsilon^2 \mathbf{I}_p)\boldsymbol{\beta}_i \\ &= \mathbf{M}\boldsymbol{\beta}_i + \sigma_\epsilon^2 \boldsymbol{\beta}_i \\ &= (\gamma_i + \sigma_\epsilon^2)\boldsymbol{\beta}_i, \qquad i = 1, \ldots q. \end{aligned} \tag{21}$$

Hence, $\boldsymbol{\beta}_i$ is also an eigenvector of $\boldsymbol{\psi}$, with associated eigenvalue $\gamma_i + \sigma_\epsilon^2$. Now let $\mathbf{b} \in \mathbb{R}^p$ denote any vector orthogonal to the columns of $\mathbf{C}$, i.e., $\mathbf{C}'\mathbf{b} = 0$. Then

$$\begin{aligned} \boldsymbol{\psi}\mathbf{b} &= (\mathbf{CC}' + \sigma_\epsilon^2 \mathbf{I}_p)\mathbf{b} \\ &= \mathbf{CC}'\mathbf{b} + \sigma_\epsilon^2 \mathbf{b} \\ &= \sigma_\epsilon^2 \mathbf{b}. \end{aligned} \tag{22}$$

Thus, any vector $\mathbf{b} \in \mathbb{R}^p$ which is orthogonal to the columns of $\mathbf{C}$ is also an eigenvector of $\boldsymbol{\psi}$, with associated eigenvalue σ_ϵ^2. Since $\mathbf{C}'\mathbf{b} = \mathbf{0}$ implies $\boldsymbol{\beta}_i'\mathbf{b} = 0$ for $i = 1, \ldots, q$ (see Exercise 8), we have the following result: The eigenvalues of $\boldsymbol{\psi}$ are $\lambda_i = \gamma_i + \sigma_\epsilon^2$ for $i = 1, \ldots, q$ and $\lambda_{q+1}, \ldots, \lambda_p = \sigma_\epsilon^2$. The associated eigenvectors are $\boldsymbol{\beta}_1, \ldots, \boldsymbol{\beta}_q$ (i.e., the eigenvectors of $\mathbf{CC}'$ associated with the positive roots), and $(p-q)$ eigenvectors $\boldsymbol{\beta}_{q+1}, \ldots, \boldsymbol{\beta}_p$, which are not uniquely defined unless $q = p - 1$.

This suggests using a principal component approximation of dimension q. Setting $\mathbf{A}_q = [\boldsymbol{\beta}_1, \ldots, \boldsymbol{\beta}_q]$, we obtain

$$\mathbf{Y}_{(q)} = \boldsymbol{\mu} + \mathbf{A}_q \mathbf{A}_q'(\mathbf{X} - \boldsymbol{\mu}), \tag{23}$$

which, by property 3, has covariance matrix

$$\begin{aligned} \text{Cov}[\mathbf{Y}_{(q)}] &= \boldsymbol{\psi} - \sigma_\epsilon^2 \sum_{i=q+1}^{p} \boldsymbol{\beta}_i \boldsymbol{\beta}_i' \\ &= \mathbf{CC}' + \sigma_\epsilon^2 \sum_{i=1}^{q} \boldsymbol{\beta}_i \boldsymbol{\beta}_i', \end{aligned} \tag{24}$$

see Exercise 8. Note that the term $\sum_{i=1}^{q} \boldsymbol{\beta}_i \boldsymbol{\beta}_i'$ is the projection matrix associated with the projection in the subspace spanned by the columns of $\mathbf{C}$. Moreover, $\text{MSE}(\mathbf{Y}_{(q)}; \mathbf{X}) = \sum_{i=q+1}^{p} \lambda_i = (p-q)\sigma_\epsilon^2$, and therefore the relative mean squared error of $\mathbf{Y}_{(q)}$ is given by

$$\begin{aligned} \text{RMSE}(\mathbf{Y}_{(q)}; \mathbf{X}) &= \frac{(p-q)\sigma_\epsilon^2}{\sum_{i=1}^{q} \lambda_i + p\sigma_\epsilon^2} \\ &= 1 - \frac{\text{tr}(\mathbf{CC}') + q\sigma_\epsilon^2}{\text{tr}(\mathbf{CC}') + p\sigma_\epsilon^2}. \end{aligned} \tag{25}$$

Thus, the q-dimensional principal component approximation will be relatively good if σ_ϵ^2 is small and p is not too large.

It is characteristic for this example that the last $(p-q)$ eigenvalues of ψ are identical and equal to σ_ϵ^2. A possible interpretation of the model (18) is that the p measured variables X_i are all functions of a smaller number q of independent variables Z_h, plus some errors with variance σ_ϵ^2. It is tempting to call the Z_h "factors" or "latent variables." However, as in Example 8.4.1, the investigator should be careful not to draw incorrect conclusions. If a principal component analysis gives the result that the last q principal components all have the same variance (here, σ_ϵ^2), then we still cannot conclude that the data was generated from a model like (18). Only the reverse conclusion is correct: The model implies that the last q eigenvalues are identical. Thus, it is not possible to conclude that there are q "factors" in the p-variate random vector $\mathbf{X}$. □

Example 8.4.6 The equicorrelation model. This example gives a more straightforward generalization of Example 8.4.1 in terms of a repeated measurements model. Suppose p measurements are taken of a variable V with mean μ and variance σ_V^2, but the ith measurement is affected by an additive error ϵ_i, $i = 1, \ldots, p$. Assume that all ϵ_i are independent, and independent also of V, with $E[\epsilon_i] = 0$ and $\text{var}[\epsilon_i] = \sigma_\epsilon^2$. Thus, the measurements obtained are $X_i = V + \epsilon_i$, $i = 1, \ldots, p$. Setting $\mathbf{X} = (X_1, \ldots, X_p)'$ and denoting a vector with 1 in each position by $\mathbf{1}_p \in \mathbb{R}^p$, we obtain

$$E[\mathbf{X}] = \mu \mathbf{1}_p \tag{26}$$

and

$$\begin{aligned} \psi = \text{Cov}[\mathbf{X}] &= \sigma_V^2 \mathbf{1}_p \mathbf{1}_p' + \sigma_\epsilon^2 \mathbf{I}_p \\ &= \begin{pmatrix} \sigma_V^2 + \sigma_\epsilon^2 & \sigma_V^2 & \cdots & \sigma_V^2 \\ \sigma_V^2 & \sigma_V^2 + \sigma_\epsilon^2 & \cdots & \sigma_V^2 \\ \vdots & \vdots & \ddots & \vdots \\ \sigma_V^2 & \sigma_V^2 & \cdots & \sigma_V^2 + \sigma_\epsilon^2 \end{pmatrix}, \end{aligned} \tag{27}$$

see Exercise 9. The correlation between each pair of variables X_i and X_j, $i \neq j$, is $\rho = \sigma_V^2/(\sigma_V^2 + \sigma_\epsilon^2)$. Setting $\tau^2 = \sigma_V^2 + \sigma_\epsilon^2$, we can write

$$\begin{aligned}
\psi &= \begin{pmatrix} \tau^2 & \rho\tau^2 & \cdots & \rho\tau^2 \\ \rho\tau^2 & \tau^2 & \cdots & \rho\tau^2 \\ \vdots & \vdots & \ddots & \vdots \\ \rho\tau^2 & \rho\tau^2 & \cdots & \tau^2 \end{pmatrix} \\
&= \tau^2 \begin{pmatrix} 1 & \rho & \cdots & \rho \\ \rho & 1 & \cdots & \rho \\ \vdots & \vdots & \ddots & \vdots \\ \rho & \rho & \cdots & 1 \end{pmatrix} \\
&= \tau^2\left[(1-\rho)\mathbf{I}_p + \rho\mathbf{1}_p\mathbf{1}_p'\right].
\end{aligned} \tag{28}$$

The matrix $\mathbf{R} = (1-\rho)\mathbf{I}_p + \rho\mathbf{1}_p\mathbf{1}_p'$ is called the *equicorrelation matrix* of dimension $p \times p$, with correlation coefficient ρ. Since $\psi = \tau^2\mathbf{R}$, the eigenvectors of ψ are identical to those of $\mathbf{R}$, and the eigenvalues of ψ are eigenvalues of $\mathbf{R}$ multiplied by τ^2. Thus, we need to find the spectral decomposition of the equicorrelation matrix $\mathbf{R}$.

In Exercise 10 it is shown that the equicorrelation matrix has one eigenvalue $\lambda_1^* = 1+(p-1)\rho$, associated with the eigenvector $\beta_1 = \frac{1}{\sqrt{p}}\mathbf{1}_p$, and $p-1$ eigenvalues $\lambda_2^* = \cdots = \lambda_p^* = 1-\rho$, with associated eigenvectors that are not uniquely determined but orthogonal to $\mathbf{1}_p$. Actually, writing $\lambda_1^* = 1+(p-1)\rho$ assumes that ρ is positive; otherwise, the multiple eigenvalue $1-\rho$ will be larger than λ_1^*. It is also shown in Exercise 10 that ρ must be in the interval $(-1/(p-1), 1)$ for $\mathbf{R}$ to be positive definite. Assume, for now, that $\rho > 0$. Then $\psi = \tau^2\mathbf{R}$ has eigenvalues $\lambda_1 = \tau^2[1+(p-1)\rho]$ and $\lambda_2 = \cdots = \lambda_p = \tau^2(1-\rho)$, and the associated eigenvector is $\beta_1 = \frac{1}{\sqrt{p}}\mathbf{1}_p$. The first principal component is given by

$$\mathrm{U}_1 = \frac{1}{\sqrt{p}}\mathbf{1}_p'(\mathbf{X} - \mu\,\mathbf{1}_p) = \frac{1}{\sqrt{p}}\left(\sum_{i=1}^p X_i - p\mu\right), \tag{29}$$

that is, it is essentially the sum of the X_i. The one-dimensional principal component approximation is given by

$$\begin{aligned}
\mathbf{Y}_{(1)} &= \mu\,\mathbf{1}_p + \frac{1}{\sqrt{p}}\mathbf{1}_p\mathrm{U}_1 \\
&= \mu\,\mathbf{1}_p + \frac{1}{p}\mathbf{1}_p\mathbf{1}_p'(\mathbf{X} - \mu\,\mathbf{1}_p) \\
&= \mu\,\mathbf{1}_p + \frac{1}{p}\mathbf{1}_p\mathbf{1}_p'\mathbf{X} - \mu\frac{1}{p}\mathbf{1}_p\mathbf{1}_p'\mathbf{1}_p.
\end{aligned} \tag{30}$$

But $\mathbf{1}_p'\mathbf{X} = \sum_{i=1}^p X_i$ and $\mathbf{1}_p'\mathbf{1}_p = p$, and therefore,

$$\mathbf{Y}_{(1)} = \frac{1}{p}\mathbf{1}_p \sum_{i=1}^{p} X_i = \begin{pmatrix} \bar{X} \\ \vdots \\ \bar{X} \end{pmatrix}, \tag{31}$$

where $\bar{X} = \frac{1}{p}\sum_{i=1}^p X_i$. Thus the one-dimensional principal component approximation replaces each X_i by the average of all p measurements. Also, see Exercise 30 in the appendix.

The relative mean squared error of $\mathbf{Y}_{(1)}$ is

$$\begin{aligned} \text{RMSE}(\mathbf{Y}_{(1)}; \mathbf{X}) &= \frac{\sum_{i=2}^p \lambda_i}{\sum_{i=1}^p \lambda_i} \\ &= (1-\rho)(1-\frac{1}{p}), \end{aligned} \tag{32}$$

so the approximation is good if the correlation ρ is close to one.

One remarkable aspect of the equicorrelation model is that the eigenvectors of ψ do not depend on the parameters τ^2 and ρ. That is, we can write the principal component transformation without knowing the parameters τ^2 and ρ, by just using the assumptions of the repeated measurements model. This can be used in parameter estimation; see Exercise 11. Suppose we know τ^2 and ρ. Then the original parameters σ_V^2 and σ_ϵ^2 are given by

$$\sigma_V^2 = \rho\tau^2$$

and

$$\sigma_\epsilon^2 = (1-\rho)\tau^2. \tag{33}$$

There is a somewhat disturbing aspect in this transformation: If the correlation coefficient ρ is negative, then $\sigma_V^2 < 0$, which is not possible. Thus, not every equicorrelation matrix can be associated with a repeated measurements model; we have to require that the correlation ρ is nonnegative, see Exercise 11. □

Property 5 *Principal components are not scale-invariant.*

This is really a "nonproperty" rather than a property. What it means is that, if we change the unit of measurement of one or several of the variables, then the analysis may change drastically. We illustrate this with two examples.

Example 8.4.7 Let $\mathbf{X} = \begin{pmatrix} X_1 \\ X_2 \end{pmatrix}$ denote a bivariate random vector with mean $\mathbf{0}$ and covariance matrix $\psi = \begin{pmatrix} \sigma^2 & 0 \\ 0 & 1 \end{pmatrix}$. Assume that $\sigma^2 > 1$. Then the principal components are $U_1 = X_1$

and $U_2 = X_2$. The one-dimensional principal component approximation is $Y_{(1)} = X_1$, and the associated relative mean squared error is $1/(1+\sigma^2)$. Now, suppose that we change the scale of measurement of X_1 by a factor of $1/a$ for some $a > 0$, i.e., set $X_1^* = aX_1$ and $X_2^* = X_2$. Then $\text{Cov}[\mathbf{X}^*] = \begin{pmatrix} a^2\sigma^2 & 0 \\ 0 & 1 \end{pmatrix}$. As long as $a^2\sigma^2 > 1$, the principal components will be $U_1^* = X_1^*$ and $U_2^* = X_2$, but the relative mean squared error of the one-dimensional principal component approximation is $1/(1 + a^2\sigma^2)$, which approaches zero as a becomes large. Thus, we may make a principal component approximation arbitrarily good by changing the scale of a variable. If we choose a such that $a^2\sigma^2 < 1$, then the first principal component will be X_2, and the relative mean squared error of the one-dimensional principal component approximation approaches again zero as a tends to 0. □

Example 8.4.8 In the setup of Example 8.4.7, suppose that $\psi = \begin{pmatrix} 1 & \rho \\ \rho & 1 \end{pmatrix}$, where $0 < \rho < 1$. Then the larger eigenvalue of ψ is $\lambda_1 = 1 + \rho$, the associated eigenvector is $\boldsymbol{\beta}_1 = \frac{1}{\sqrt{2}}\begin{pmatrix} 1 \\ 1 \end{pmatrix}$, and $\text{RMSE}(Y_{(1)}; \mathbf{X}) = (1-\rho)/2$. Again, suppose that we change the scale of X_1 by a factor $1/a$, $a > 0$, that is, for the rescaled variables $X_1^* = aX_1$ and $X_2^* = X_2$, the covariance matrix is $\psi^* = \begin{pmatrix} a^2 & a\rho \\ a\rho & 1 \end{pmatrix}$. As a becomes large, the coefficients of the first eigenvector approach $\begin{pmatrix} 1 \\ 0 \end{pmatrix}$, and the relative mean squared error of the one-dimensional, principal component approximation goes to zero; see Exercise 13. □

Thus, any principal component analysis depends crucially on the units of measurement. As a general rule, one should use principal components only if all variables have the same unit of measurement (the reader who is not yet convinced is asked to do Exercise 14). Ultimately the reason for scale-dependency is that we use the Euclidean metric, meaning that a distance of 1 means the same in all directions of p-dimensional space. Under changes of scale, or under general nonsingular linear transformations, Euclidean distances are not preserved, and therefore an orthogonal projection, in general, does not remain orthogonal; see Exercise 15.

Sometimes users of principal component analysis try to escape from the problem of scale-dependency by standardizing variables, i.e., by first replacing the random vector $\mathbf{X}$ by

$$\mathbf{Z} = (\text{diag}\,\psi)^{-1/2}(\mathbf{X} - \boldsymbol{\mu}). \tag{34}$$

Here, $\boldsymbol{\mu} = E[\mathbf{X}]$, $\psi = \text{Cov}[\mathbf{X}]$, and $\text{diag}(\psi)^{1/2}$ is the diagonal matrix with the standard deviations of the X_i on the diagonal. Then the covariance matrix of $\mathbf{Z}$ is the same as the correlation matrix of $\mathbf{X}$. It is open to debate if such a standardization is reasonable. Another way of avoiding scale-dependency is to compute principal components on the logarithm of the variables. However, this is feasible only if all X_i

are strictly positive random variables, and it usually changes the nature and purpose of the analysis drastically; see Exercise 16, Example 8.5.4, and the material on allometric growth in Section 8.6.

Exercises for Section 8.4

1. Find $E[\mathbf{X}]$ and $\text{Cov}[\mathbf{X}]$ in Example 8.4.1.

2. In Example 8.4.1, find the correlation between V and U_1 and the correlation between V and U_2, as functions of σ^2 and σ_ϵ^2.

3. This exercise gives a generalization of Example 8.4.1. Suppose that $X_i = a_i V + \epsilon_i$, $i = 1, \ldots, p$, where V and $\epsilon_1, \ldots, \epsilon_p$ are independent random variables, with $E[V] = \mu$, $\text{var}[V] = \sigma^2 > 0$, $E[\epsilon_i] = 0$ and $\text{var}[\epsilon_i] = \sigma_\epsilon^2 > 0$, $i = 1, \ldots, p$. Here, the a_i are constants, that is, the ith observed random variable X_i is a multiple of V plus some "error of measurement." Set $\mathbf{X} = (X_1, \ldots, X_p)'$ and $\mathbf{a}' = (a_1, \ldots, a_p)'$.

(a) Find $E[\mathbf{X}]$, and show that $\psi = \text{Cov}[\mathbf{X}] = \sigma^2 \mathbf{a}\mathbf{a}' + \sigma_\epsilon^2 \mathbf{I}_p$.

(b) Show that $\beta_1 = \mathbf{a}/\sqrt{\mathbf{a}'\mathbf{a}}$ is a normalized eigenvector of ψ, and find the associated eigenvalue λ_1.

(c) Show that any vector $\mathbf{b}$ orthogonal to β_1 is an eigenvector of ψ, with associated eigenvalue σ_ϵ^2. Thus, show that ψ has a distinct largest eigenvalue λ_1 and $(p-1)$ identical eigenvalues $\lambda_2 = \cdots = \lambda_p = \sigma_\epsilon^2$.

(d) Find the correlations between V and the principal components of $\mathbf{X}$.

(e) Find the one-dimensional principal component approximation $\mathbf{Y}_{(1)}$ of $\mathbf{X}$, as well as its mean vector, covariance matrix, and mean squared error.

4. If $\mathbf{Z}$ is a p-variate random vector, $\in \mathbb{R}^p$, and $\mathbf{Y}$ is self-consistent for $\mathbf{Z}$, show that $\mathbf{Y}+$ is self-consistent for $\mathbf{Z}+$.

5. Prove property 2 if $\mathbf{X}$ follows a p-variate elliptical distribution. *Hint*: Use the fact that $\mathbf{A}_q'(\mathbf{X} - \mu)$ and $\mathbf{B}_q'(\mathbf{X} - \mu)$ have a joint elliptical distribution.

6. In Example 8.4.1, show that the one-dimensional principal component approximation $\mathbf{Y}_{(1)}$ is self-consistent for $\mathbf{X}$ under the stated assumptions, irrespective of the exact distribution of V, ϵ_1, and ϵ_2.

7. In Example 8.4.4, show that $\Pr[\mathbf{U}_{(2)} = \mathbf{0}] = 1$.

8. This exercise is based on Example 8.4.5.

(a) Prove equation (20).

(b) Let $\mathbf{C}$ denote a $p \times q$ matrix of rank q, and let $\beta_1, \ldots, \beta_q$ denote the q eigenvectors of $\mathbf{CC}'$ associated with the positive roots. Show that $\mathbf{C}'\mathbf{b} = \mathbf{0}$ implies $\beta_i'\mathbf{b} = 0$ $(i = 1, \ldots, q)$ for any $\mathbf{b} \in \mathbb{R}^p$.

(c) Prove equation (24).

9. Prove equation (27),

(a) directly, by computing $\text{var}[X_i]$ and $\text{cov}[X_i, X_j]$.

(b) using the formula $\text{Cov}[\mathbf{X}] = E\big[\text{Cov}[\mathbf{X}|\mathbf{V}]\big] + \text{Cov}\big[E[\mathbf{X}|\mathbf{V}]\big]$.

10. Let $\mathbf{R}$ denote the equicorrelation matrix of dimension $p \times p$, with correlation coefficient ρ.

(a) Find the eigenvalues and eigenvectors of $\mathbf{R}$.

(b) Show that $\mathbf{R}$ is positive definite exactly if $\frac{-1}{p-1} < \rho < 1$.

(c) Show that

$$\mathbf{R}^{-1} = \frac{1}{1-\rho}\left[\mathbf{I}_p - \frac{\rho}{1+(p-1)\rho}\mathbf{1}_p\mathbf{1}_p'\right].$$

Hint: Use the spectral decomposition of $\mathbf{R}$.

11. This exercise concerns parameter estimation in the equicorrelation model. Suppose that $\mathbf{X} \sim \mathcal{N}_p(\mathbf{0}, \boldsymbol{\psi})$, where $\boldsymbol{\psi} = \tau^2\mathbf{R}$ and $\mathbf{R}$ is the equicorrelation matrix with correlation coefficient ρ. Let $\mathbf{X}_1, \ldots, \mathbf{X}_N$ denote a sample of size $N \geq p$ from $\mathbf{X}$.

(a) Find the maximum likelihood estimator of $\boldsymbol{\psi}$. *Hint*: Since $\boldsymbol{\psi}$ is determined by τ^2 and ρ, find the maximum likelihood estimators of these two parameters. A convenient way to do this is to transform the $\mathbf{X}_i$ to $\mathbf{Y}_i = \mathbf{B}'\mathbf{X}_i$, where $\boldsymbol{\psi} = \mathbf{B}\boldsymbol{\Lambda}\mathbf{B}'$ is a spectral decomposition of $\boldsymbol{\psi}$. Then use known results on maximum likelihood estimation in the univariate normal distribution.

(b) Using the results from part (a), find $\Pr[\hat{\rho} < 0]$ if $\rho = 0$.

(c) Find the maximum likelihood estimators of σ_V^2 and σ_ϵ^2 if $\boldsymbol{\psi}$ is written in the form of equation (27). *Hint*: What happens if $\hat{\rho} < 0$?

12. Suppose that $\boldsymbol{\Sigma}$ is a positive definite symmetric matrix of dimension $p \times p$ and ρ ($|\rho| < 1$) is a constant.

(a) Find the eigenvectors and eigenvalues of the matrix

$$\boldsymbol{\psi} = \begin{pmatrix} \boldsymbol{\Sigma} & \rho\boldsymbol{\Sigma} \\ \rho\boldsymbol{\Sigma} & \boldsymbol{\Sigma} \end{pmatrix}$$

in terms of ρ and the eigenvectors and eigenvalues of $\boldsymbol{\Sigma}$.

(b) Generalize the result to a matrix of dimension $pk \times pk$,

$$\boldsymbol{\psi} = \begin{pmatrix} \boldsymbol{\Sigma} & \rho\boldsymbol{\Sigma} & \cdots & \rho\boldsymbol{\Sigma} \\ \rho\boldsymbol{\Sigma} & \boldsymbol{\Sigma} & \cdots & \rho\boldsymbol{\Sigma} \\ \vdots & \vdots & \ddots & \vdots \\ \rho\boldsymbol{\Sigma} & \rho\boldsymbol{\Sigma} & \cdots & \boldsymbol{\Sigma} \end{pmatrix}.$$

What are the values of ρ for which $\boldsymbol{\psi}$ is positive-definite?

13. In Example 8.4.8, show that the first eigenvector tends to $\begin{pmatrix} 1 \\ 0 \end{pmatrix}$ and the relative mean squared error of $\mathbf{Y}_{(1)}$ tends to zero as a goes to infinity.

14. Let $\mathbf{X} = \begin{pmatrix} X_1 \\ X_2 \end{pmatrix}$, with $E[\mathbf{X}] = \mathbf{0}$ and $\text{Cov}[\mathbf{X}] = \begin{pmatrix} 1 & 1/2 \\ 1/2 & 1 \end{pmatrix}$. Denote the principal components of $\mathbf{X}$ by U_1 and U_2. Let $\mathbf{X}^* = \begin{pmatrix} X_1^* \\ X_2^* \end{pmatrix} = \begin{pmatrix} 2X_1 \\ X_2 \end{pmatrix}$, and denote the principal components of $\mathbf{X}^*$ by U_1^*, U_2^* Write U_1^* and U_2^* as functions of X_1 and X_2, and verify that they are not identical to U_1 and U_2.

15. Let $\mathbf{P}$ denote a $p \times p$ projection matrix of rank $q \le p$, and let $\mathbf{A}$ denote a nonsingular matrix of dimension $p \times p$. For $\mathbf{x} \in \mathbb{R}^p$, let $\mathbf{x}^* = \mathbf{A}\mathbf{x}$, $\mathbf{y} = \mathbf{P}\mathbf{x}$, and $\mathbf{y}^* = \mathbf{A}\mathbf{y}$. Show that $\mathbf{y}^*$ is a linear function of $\mathbf{x}^*$ but in general not an orthogonal projection of $\mathbf{x}^*$.

16. Let $X_1, \ldots, X_p$ denote p jointly distributed positive random variables, and let $a_1, \ldots, a_p$ be positive constants. Let $L_i = \log(a_i X_i)$, $i = 1, \ldots, p$, and set $\mathbf{L} = (L_1, \ldots, L_p)'$. Show that the covariance matrix of $\mathbf{L}$ does not depend on the a_i.

17. Let $\mathbf{X}$ denote a bivariate random vector with a uniform distribution in the half-circle $\mathbf{x}'\mathbf{x} \le 1, x_2 > 0$.
    (a) Find the linear principal components of $\mathbf{X}$.
    (b) Show that the orthogonal projection on the first principal component axis (i.e., the one-dimensional principal component approximation) is not self-consistent. *Hint*: Graph the conditional mean of X_2, given X_1.
    (c) Show that the projection on the second principal component axis is self-consistent.

18. Repeat Exercise 17 for the random vector $\mathbf{X} = \begin{pmatrix} X_1 \\ X_2 \end{pmatrix}$, where X_1 has a uniform distribution in the interval $[-2, 2]$ and $X_2 = X_1^2$.

19. Suppose that $\mathbf{X}$ is a p-variate random vector with mean vectors μ_1 and μ_2 in groups 1 and 2, respectively, and with common covariance matrix $\psi = \sigma^2\mathbf{R}$, where $\mathbf{R}$ is the equicorrelation matrix of dimension $p \times p$ with correlation coefficient ρ. Let $\delta = \mu_1 - \mu_2 = (\delta_1, \ldots, \delta_p)'$ and $\bar{\delta} = \frac{1}{p}\sum_{j=1}^{p} \delta_j$.
    (a) Show that the vector of linear discriminant function coefficients between the two groups is
    $$\beta = \frac{1}{\sigma^2(1-\rho)}\left[\delta - \bar{\delta}\frac{p\rho}{1+(p-1)\rho}\mathbf{1}_p\right].$$
    Hint: Use Exercise 10.
    (b) Show that the the linear discriminant function is proportional to $\sum_{j=1}^{p} X_i$ exactly if $\delta_1 = \cdots = \delta_p$.

8.5 Applications

Having developed a fair amount of theory, we turn to practical applications of principal component analysis. Most examples found in the literature constitute a straightforward transfer of the theoretical results to empirical distributions, that is, sample mean vectors and sample covariance matrices are substituted in all formulas for their theoretical counterparts. This can be justified by the plug-in principle of estimation (Section 4.2) or by normal theory maximum likelihood. Either way, applied principal component analysis consists most often of a mere computation of eigenvectors and eigenvalues of a sample covariance matrix or correlation matrix. However, thinking a bit about the particular data being analyzed often leads to more challenging problems, as we shall see in Example 8.5.3.

All examples presented in this section are from the life sciences, which reflects the author's practical experience and should not be misinterpreted as a statement that principal component analysis is applicable only to this type of data.

Throughout this section, we will denote the eigenvalues of the sample covariance matrix by $\ell_1 \geq \ell_2 \geq \cdots \geq \ell_p$ and the associated eigenvectors by $\mathbf{b}_1, \ldots, \mathbf{b}_p$. Writing $\mathbf{B} = (\mathbf{b}_1, \ldots, \mathbf{b}_p)$, the ith data point $\mathbf{x}_i = (x_{i1}, \ldots, x_{ip})'$ can then be transformed to principal component scores $\mathbf{u}_i = (u_{i1}, \ldots, u_{ip})'$ by

$$\mathbf{u}_i = \mathbf{B}'(\mathbf{x}_i - \bar{\mathbf{x}}) \quad i = 1, \ldots, N. \tag{1}$$

Example 8.5.1 Tooth size of voles. This example is based on the data from Table 5.4.1. In the current context, we use only the $N = 43$ observations of the species *Microtus multiplex* and only the first three variables (length of the upper left molars no. 1, 2, and 3), which we will call X_1 to X_3 for the current purpose. For numerical convenience, the data used here are in mm/100.

The sample mean vector and covariance matrix are

$$\bar{\mathbf{x}} = \begin{pmatrix} 205.45 \\ 163.65 \\ 181.99 \end{pmatrix},$$

and

$$\mathbf{S} = \begin{pmatrix} 171.51 & 97.41 & 121.22 \\ 97.41 & 102.31 & 110.37 \\ 121.22 & 110.37 & 232.57 \end{pmatrix}.$$

The eigenvalues of $\mathbf{S}$ are $\ell_1 = 399.70$, $\ell_2 = 78.33$, and $\ell_3 = 28.37$. The associated eigenvectors are given by

$$\mathbf{b}_1 = \begin{pmatrix} 0.561 \\ 0.443 \\ 0.699 \end{pmatrix}, \qquad \mathbf{b}_2 = \begin{pmatrix} -0.727 \\ -0.141 \\ 0.672 \end{pmatrix},$$

and

$$\mathbf{b}_3 = \begin{pmatrix} 0.397 \\ -0.885 \\ 0.243 \end{pmatrix}.$$

The first sample principal component, $U_1 = 0.561X_1 + 0.443X_2 + 0.699X_3$, is a rather typical example of a so-called *size variable*, with all coefficients positive. In morphometric studies, it is usually the case that the first principal component has all coefficients positive (or all negative, depending on the particular choice of the first eigenvector). Actually, this is a direct consequence of the fact that all covariances are positive; see Exercise 1. A possible explanation is that size differences among the objects being measured are the dominant source of variation, and one hopes that the first principal component is an adequate description of size. Indeed, a skull with large values in all three variables would score high in U_1, and a skull with small values would score low. In contrast, the second and third principal components might be interpreted as *shape* variables. In particular, the second component is practically a contrast between the first tooth and the third tooth.

According to this interpretation of the first principal component, a one-dimensional principal component approximation should reproduce the data maintaining *size* but discarding *shape* and (hopefully) error of measurement. Denoting by $\mathbf{x}_i$ the data vector of the ith observation, the corresponding value of the one-dimensional principal component approximation is given by

$$\mathbf{y}_{(1)i} = \bar{\mathbf{x}} + \mathbf{b}_1\mathbf{b}_1'(\mathbf{x}_i - \bar{\mathbf{x}}) = \bar{\mathbf{x}} + \mathbf{b}_1 u_{i1} \qquad i = 1, \ldots, 43.$$

We leave it to the reader to do the actual computations; see Exercise 2. The relative mean squared error associated with this approximation is $\text{RMSE}(\mathbf{Y}_{(1)}; \mathbf{X}) = (\ell_2 + \ell_3)/(\ell_1 + \ell_2 + \ell_3) = 0.211$. About one fifth of the variability is lost in the one-dimensional approximation. For the two-dimensional principal component approximation, the relative mean squared error is only 0.056, but of course, a reduction from dimension 3 to dimension 2 is not exactly spectacular.

Is the one-dimensional principal component approximation self-consistent for the three-dimensional distribution? A naive answer would be "no," because exact self-consistency will almost never hold for an empirical distribution. A more sophisticated answer is that self-consistency may well hold for the model that generates the data, so the question really refers to the distribution from which we are sampling. Different graphs may be used to assess self-consistency informally, similar to residual plots used in regression to assess assumptions of linearity. This is explored in some detail in Exercise 3.

We might also consider a model that exhibits some special structure, like Example 8.4.1 or the model discussed in Exercise 3 of Section 8.4, in which there is one single variable ("size") that determines the measured variables, except for error of measurement. In both cases, the smallest two eigenvalues of the covariance matrix should be identical. How far apart do the sample eigenvalues ℓ_2 and ℓ_3 need to be to allow the conclusion that the eigenvalues in the model are not identical? This is a problem of *sphericity* to be discussed briefly in Section 8.6. □

Example 8.5.2 continuation of Example 1.2

Head dimensions of young men. Table 8.5.1 gives the mean vector and the covariance matrix of the six variables presented in Example 1.2 for $N = 200$ observations. Table 8.5.2 displays eigenvectors and eigenvalues of the covariance matrix. As in the preceding example, all covariances are positive, and hence the first principal component has all coefficients positive. However, the relative mean squared error of the one-dimensional principal component approximation is quite large: $\text{RMSE}(\mathbf{Y}_{(1)}; \mathbf{X}) = 0.571$. Even the three-dimensional principal component approximation is still bad, with a relative mean squared error of 22.1 percent. This reflects the fact that there is considerable independent variability in human heads and faces. Whether or not one is willing to consider the principal components as "sources of variation," this result shows that much more information is required to adequately describe a human head than for a more primitive morphometric unit, such as a turtle shell or an animal skull. Also recall from the description of the original purpose of this investigation in Example 1.2 that a small number k of "typical representatives" were to be identified, using (hopefully) a good one-dimensional approximation to the six-dimensional data. In view of the results just found, the reduction to a single dimension would probably not be a good idea. □

Sometimes it is useful to compute standard distances of all observations from the mean vector to find potential outliers. The computation of standard distances is straightforward using principal component scores. This is based on the following lemma, the proof of which is left to the reader (Exercise 5).

Lemma 8.5.1 *Let* $\mathbf{X}$ *denote a* p*-variate random vector with mean* $\boldsymbol{\mu}$ *and covariance matrix* $\boldsymbol{\psi}$*. Let* $\boldsymbol{\psi} = \mathbf{B}\boldsymbol{\Lambda}\mathbf{B}'$ *denote a spectral decomposition of* $\boldsymbol{\psi}$*, where* $\mathbf{B} = (\boldsymbol{\beta}_1, \ldots, \boldsymbol{\beta}_p)$ *is*

Table 8.5.1 Sample statistics for the head dimensions example

Mean Vector

MFB	BAM	TFH	LGAN	LTN	LTG
114.72	115.91	123.05	57.99	122.23	138.83

Covariance Matrix

	MFB	BAM	TFH	LGAN	LTN	LTG
MFB	26.90	12.62	5.38	2.93	8.18	12.11
BAM	12.62	27.25	2.88	2.06	7.13	11.44
TFH	5.38	2.88	35.23	10.37	6.03	7.97
LGAN	2.93	2.06	10.37	17.85	2.92	4.99
LTN	8.18	7.13	6.03	2.92	15.37	14.52
LTG	12.11	11.44	7.97	4.99	14.52	31.84

Table 8.5.2 Eigenvectors and eigenvalues of the covariance matrix in the head dimension example

	Eigenvectors					
	U_1	U_2	U_3	U_4	U_5	U_6
MFB	0.445	0.263	−0.421	0.731	0.126	−0.067
BAM	0.411	0.375	−0.500	−0.656	−0.100	0.000
TFH	0.395	−0.799	−0.194	−0.000	−0.404	−0.063
LGAN	0.207	−0.345	−0.039	−0.174	0.896	0.056
LTN	0.347	0.072	0.309	0.036	−0.087	0.878
LTG	0.560	0.163	0.662	−0.057	−0.020	−0.467
	Eigenvalues					
	66.33	34.42	19.63	14.33	12.96	6.77

orthogonal and $\Lambda = \text{diag}(\lambda_1, \ldots, \lambda_p)$. *Then the squared standard distance of* $\mathbf{x} \in \mathbb{R}^p$ *from* μ *is given by*

$$\Delta^2_{\mathbf{X}}(\mathbf{x}, \mu) = (\mathbf{x} - \mu)'\psi^{-1}(\mathbf{x} - \mu) = \sum_{j=1}^{p} u_j^2/\lambda_j, \tag{2}$$

where $\mathbf{u} = (u_1, \ldots, u_p)' = \mathbf{B}'(\mathbf{x} - \mu)$.

For a given q-dimensional principal component approximation, Lemma 8.5.1 can be used to decompose the squared standard distance into a part $\Delta_1^2 = \sum_{j=1}^{q} u_j^2/\lambda_j$ due to the first q components, and a part $\Delta_2^2 = \sum_{j=q+1}^{p} u_j^2/\lambda_j$ due to the last $(p-q)$ components. If $\mathbf{X}$ is p-variate normal, then the random variable $\Delta^2_{\mathbf{X}}(\mathbf{X}, \mu)$ follows a chi-square distribution on p degrees of freedom; see Theorem 3.3.2. Let $U_j = \beta_j'(\mathbf{X} - \mu)$, and set $\Delta_1^2(\mathbf{X}, \mu) = \sum_{j=1}^{q} U_j^2/\lambda_j$ and $\Delta_2^2(\mathbf{X}, \mu) = \sum_{j=q+1}^{p} U_j^2/\lambda_j$. Then $\Delta^2(\mathbf{X}, \mu) = \Delta_1^2(\mathbf{X}, \mu) + \Delta_2^2(\mathbf{X}, \mu)$, and the two terms in the sum are independent chi-square random variables on q and $(p-q)$ degrees of freedom, respectively.

In practical applications, we can use the principal component scores $\mathbf{u}_i$ of equation (1) to decompose the observed squared standard distance of the ith observation as

$$\begin{aligned} D_i^2 &= D_i^2(\mathbf{x}_i, \bar{\mathbf{x}}) = (\mathbf{x}_i - \bar{\mathbf{x}})'\mathbf{S}^{-1}(\mathbf{x}_i - \bar{\mathbf{x}}) \\ &= \sum_{j=1}^{q} u_{ij}^2/\ell_j + \sum_{j=q+1}^{p} u_{ij}^2/\ell_j \\ &= D_{i1}^2 + D_{i2}^2, \qquad i = 1, \ldots, N. \end{aligned} \tag{3}$$

If the distribution from which the sample has been obtained is normal and if the sample size is large enough, we would expect a frequency plot of the observed D_i^2 to look similar to the graph of a chi-square distribution on p degrees of freedom. A graph of the D_{i1}^2 vs. D_{i2}^2 will reveal which observations are well represented by the given

q-dimensional principal component approximation and which ones are not, because large values of D_{i2} indicate that the associated observations are not represented well by the q-dimensional principal component approximation.

Example 8.5.3 continuation of Example 8.5.2

For the head dimension data, Figure 8.5.1 shows a histogram of the squared standard distances D_i^2 for all 200 observations, along with the density function of a chi-square random variable V with six degrees of freedom,

$$f_V(v) = \frac{1}{16} v^2 e^{-v/2}, \quad v \geq 0.$$

Without doing a formal test, the agreement between the observed and the theoretical curve seems rather good, and thus no objections are raised against the assumptions of multivariate normality. In particular, there appear to be no obvious outliers. Of course, this informal assessment does not actually *prove* normality. It just makes it plausible as a model for the data. Also see Exercises 7 and 8. □

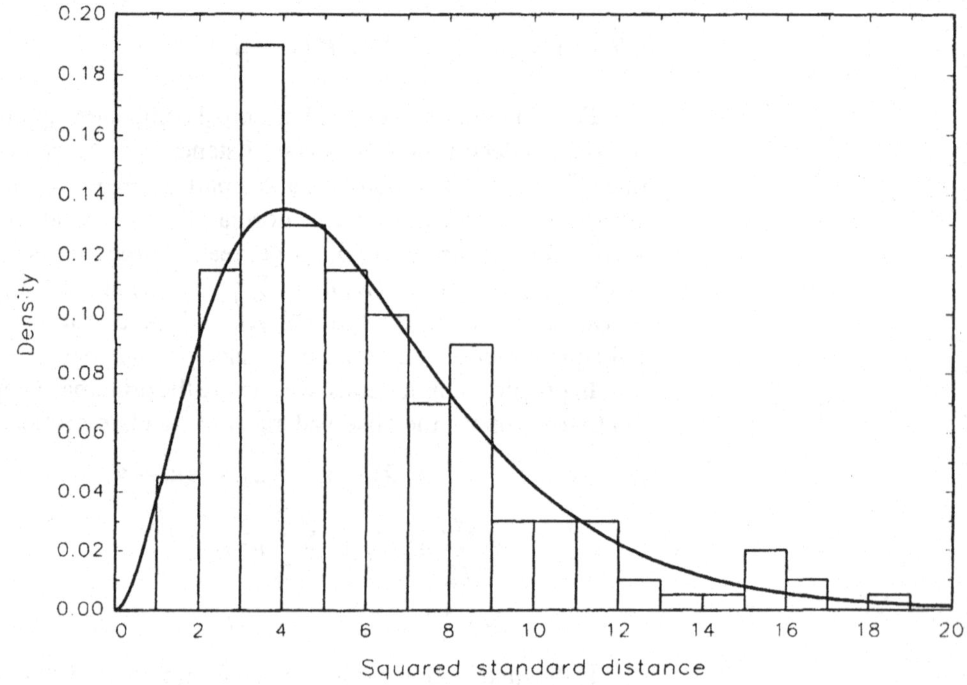

Figure 8-5-1 Histogram of squared standard distances of all 200 observations from the mean in the head dimensions example and density function of a chi-square, random variable on six degrees of freedom.

Example 8.5.4 Skeletal dimensions of water striders. Water striders are small insects living on the surface of lakes and ponds. They grow in six discrete stages called instars. At the transition from one instar to the next, they shed their skin, which is, at the same time, their skeleton. The deposited skin can be measured accurately under a microscope, and the same animal can be followed throughout its six stages of growth until it has reached maturity. Table 8.5.3 gives data for $N = 88$ female water striders of the species *Limnoporus canaliculatus* that were raised by Dr. C.P. Klingenberg at the University of Alberta, Edmonton, Canada. The six variables measured are average length of the femur (F) and average length of the tibia (T) of the hind legs for the first three instars. With indices 1 to 3 denoting instars, we have a six-dimensional random vector $(F_1, T_1, F_2, T_2, F_3, T_3)'$. This random vector can be naturally partitioned into three parts according to stage of growth or into two parts according to body part. As usual in studies of growth, we will use log-transformed variables. The actual analysis was done on the variables $F_i^* = 100 \log F_i$ and $T_i^* = 100 \log T_i$. Now, we will present the results of an ordinary principal component analysis on all six variables and then discuss some possible modifications and extensions.

Table 8.5.4 gives the usual sample statistics. As can be expected in this type of data, there are high correlations between all variables, ranging from 0.577 to 0.842. Eigenvectors and eigenvalues of the sample covariance matrix are displayed in Table 8.5.5. The mean squared error of the one-dimensional principal component approximation is 21.4 percent; hence, such an approximation reproduces the six-dimensional data quite well. As seen from the coefficients in the first eigenvector, the approximation will again reproduce "size" but ignore "shape." Thus, we may reasonably reduce the data of each observation to its score in the first principal component, $u_{i1} = \mathbf{b}_1'(\mathbf{x}_i - \bar{\mathbf{x}})$. The values of the one-dimensional principal component approximation $\mathbf{y}_{(1)i} = \bar{\mathbf{x}} + \mathbf{b}_1 u_{i1}$ are generally close to the observed data vectors $\mathbf{x}_i$; see Exercise 9. □

The coefficients of the first eigenvector have a remarkable property: all three coefficients associated with the *Femur* are practically identical, as are the coefficients associated with the *Tibia*. Thus, we might ask if the differences are due to random error only. More importantly, we might consider models like the one studied in Exercise 12(b) of Section 8.4. (The practical application of such a model is beyond the purpose of this book because the estimation problem is too difficult). We will return to this data set briefly in Example 8.7.4. For now, Exercise 10 explores some more aspects of this highly interesting data set.

Exercises for Section 8.5

1. Let $\mathbf{X}$ be a p-variate random vector with covariance matrix ψ, where all covariances are positive. Let β denote a normalized eigenvector of ψ associated with the largest

Table 8.5.3 Skeletal dimensions of 88 female water striders *Limnoporus canaliculatus*. All measurements are in mm/1000. F_i and T_i are average femur length and average tibia length of the hind legs at instar i, $i = 1, 2, 3$. Data courtesy of Dr. C.P. Klingenberg, University of Alberta.

F_1	T_1	F_2	T_2	F_3	T_3
577	440	943	609	1488	919
579	440	907	595	1547	949
578	435	912	585	1499	883
577	443	909	609	1505	917
578	434	936	629	1517	943
568	438	929	616	1508	943
584	430	930	622	1512	946
582	440	917	600	1522	930
579	448	958	637	1558	958
565	427	902	597	1548	958
570	425	894	572	1494	908
595	431	927	613	1542	935
501	391	806	542	1349	832
594	462	996	666	1631	1014
555	417	893	576	1478	888
564	436	907	607	1492	929
597	441	952	603	1490	914
587	452	947	631	1512	917
570	413	897	591	1485	887
581	420	953	609	1552	951
574	419	920	585	1493	897
597	444	949	626	1554	936
593	447	950	619	1519	937
546	406	902	586	1470	867
565	422	929	598	1484	897
571	428	940	618	1505	923
584	443	923	611	1516	936
571	423	922	613	1485	928
573	424	913	602	1506	944
527	401	880	574	1424	886
571	431	933	612	1577	976
522	396	894	570	1468	867
576	436	916	602	1497	916
564	402	902	583	1451	852
580	431	939	599	1551	925
562	412	911	589	1484	883
580	425	906	597	1467	900
537	391	873	562	1404	838
543	422	853	573	1408	891
569	421	884	576	1434	865

Table 8.5.3 *Continued*

F_1	T_1	F_2	T_2	F_3	T_3
591	431	967	608	1567	933
566	422	912	582	1491	903
551	400	878	567	1476	882
588	422	942	595	1526	907
588	449	965	630	1593	967
572	423	915	580	1508	902
580	424	922	590	1548	918
561	414	950	613	1545	936
570	416	923	602	1511	919
565	436	906	612	1488	911
576	426	910	582	1465	859
558	420	920	593	1531	904
549	427	883	600	1440	899
562	423	921	615	1551	971
549	410	882	577	1474	899
569	411	917	573	1502	889
567	413	903	572	1520	898
569	419	904	578	1491	893
597	439	960	619	1571	958
548	414	899	574	1501	882
579	430	927	610	1520	958
537	396	884	583	1468	878
589	450	921	617	1517	933
559	428	905	606	1503	958
574	423	926	576	1523	894
576	422	910	601	1468	910
566	426	931	611	1521	942
582	438	949	618	1560	930
566	420	900	592	1498	923
552	412	894	575	1501	913
567	424	907	593	1498	917
585	435	958	612	1539	935
568	423	902	579	1462	874
592	430	931	605	1507	901
582	427	930	617	1528	948
569	421	902	585	1467	892
572	419	917	594	1475	896
551	418	896	591	1451	908
586	443	931	613	1487	898
571	410	907	591	1464	902
550	402	894	563	1461	867
567	429	923	619	1493	963
570	414	905	583	1503	902
566	425	927	613	1499	935
583	437	939	607	1547	948
575	426	925	602	1500	903

Table 8.5.3 *Continued*

F_1	T_1	F_2	T_2	F_3	T_3
577	413	901	583	1499	888
585	448	949	616	1568	957

Table 8.5.4 Sample statistics for the water strider example

Mean Vector

F_1^*	T_1^*	F_2^*	T_2^*	F_3^*	T_3^*
634.515	605.194	682.102	639.299	731.473	681.790

Covariance Matrix

	F_1^*	T_1^*	F_2^*	T_2^*	F_3^*	T_3^*
F_1^*	9.247	7.972	7.274	6.894	6.110	6.252
T_1^*	7.972	11.248	7.032	9.016	6.190	8.521
F_2^*	7.274	7.032	8.782	8.071	7.128	7.400
T_2^*	6.894	9.016	8.071	11.082	6.663	9.862
F_3^*	6.110	6.190	7.128	6.663	8.171	8.232
T_3^*	6.252	8.521	7.400	9.862	8.232	12.680

root. Show that all coefficients of $\boldsymbol{\beta}$ are positive (or all are negative). *Hint*: Use the fact that $\boldsymbol{\beta}$ maximizes $\mathbf{b}'\psi\mathbf{b}$ over all $\mathbf{b} \in \mathbb{R}^p$, $\mathbf{b}'\mathbf{b} = 1$.

2. In Example 8.5.1, compute the numerical values of all three principal components and of the one-dimensional principal component approximation, similar to Table 8.1.1. For each observation, compute the distance between the observed value and the approximation, i.e., the length of the vector $\mathbf{x}_i - \mathbf{y}_{(1)i}$. How are these distances related to the principal component scores of equation (1)?

Table 8.5.5 Eigenvectors and eigenvalues of the covariance matrix in the water strider example

Eigenvectors

	U_1	U_2	U_3	U_4	U_5	U_6
F_1^*	0.368	−0.628	0.165	0.200	0.625	−0.113
T_1^*	0.427	−0.302	−0.603	0.339	−0.485	0.113
F_2^*	0.386	−0.177	0.417	−0.472	−0.258	0.597
T_2^*	0.443	0.168	−0.320	−0.655	0.114	−0.481
F_3^*	0.359	0.156	0.573	0.310	−0.397	−0.513
T_3^*	0.455	0.656	−0.062	0.313	0.371	0.351

Eigenvalues

48.102	5.248	3.987	2.261	1.145	0.465

3. In Example 8.5.1, construct scatterplots of the first vs. the second and the third principal components. What do you expect these scatterplots to look like if the one-dimensional principal component approximation is self-consistent?

4. In Example 8.5.2, construct scatterplots of the first vs. the other principal components. Is there any evidence against the assumption that the one-dimensional principal component approximation is self-consistent?

5. Prove Lemma 8.5.1.

6. For a data set with N p-variate observations, let the D_i^2, D_{i1}^2, and D_{i2}^2 be defined as in (3), using the sample covariance matrix with denominator $N-1$. Show that $\sum_{i=1}^N D_i^2 = (N-1)p$, $\sum_{i=1}^N D_{i1}^2 = (N-1)q$, and $\sum_{i=1}^N D_{i2}^2 = (N-1)(p-q)$.

7. Suppose that a data set contains outliers that are not extreme in any of the one-dimensional marginal distributions. (No formal definition of "outlier" is given here). Explain in words and graphically why you would expect such outliers to show up by large values of D_{i2}^2 in a q-dimensional principal component approximation.

8. This exercise is based on Example 8.5.2 with q=3.
   (a) Graph the pairs (D_{i1}^2, D_{i2}^2), $i = 1, \ldots, 200$.
   (b) If V_1 and V_2 are two independent chi-square random variables with three degrees of freedom each, find the number c such that $\Pr[V_1 + V_2 \le c] = 0.9$. What is the shape of this region in the support of the joint distribution of V_1 and V_2?
   (c) Under normality assumptions, the D_{i1}^2 and D_{i2}^2 from part (a) should roughly follow the joint distribution of two independent chi-square random variables on three degrees of freedom each. Use part (b) to graph a region in the scatterplot of part (a) that should contain about 90 percent of all observations.

9. This exercise is based on the water strider data of Example 8.5.4, using log-transformed data.
   (a) Compute standard distances from the mean for all observations, and graph the squared standard distances in a histogram similar to Figure 8.5.1, along with the density curve of a chi-square distribution on six degrees of freedom. Comment on the fit of the curve.
   (b) Identify the observation with the largest standard distance. Is this particular observation well represented in the one-dimensional principal component approximation?
   (c) Compute the $\mathbf{y}_{(1)i}$, i.e., the data of the one-dimensional principal component approximation for all 88 observations.
   (d) Figure 8.5.2 shows a graphical representation of the six-dimensional data vectors of observations 11 to 15. Each curve represents one water strider, and the symbols mark the data values (F_i^*, T_i^*), $i = 1, 2, 3$. Among these observations is also the "suspicious" observation from part (b). Draw the same graph for the approximate data from part (c), and comment. What changes, if any, does the approximation seem to make?

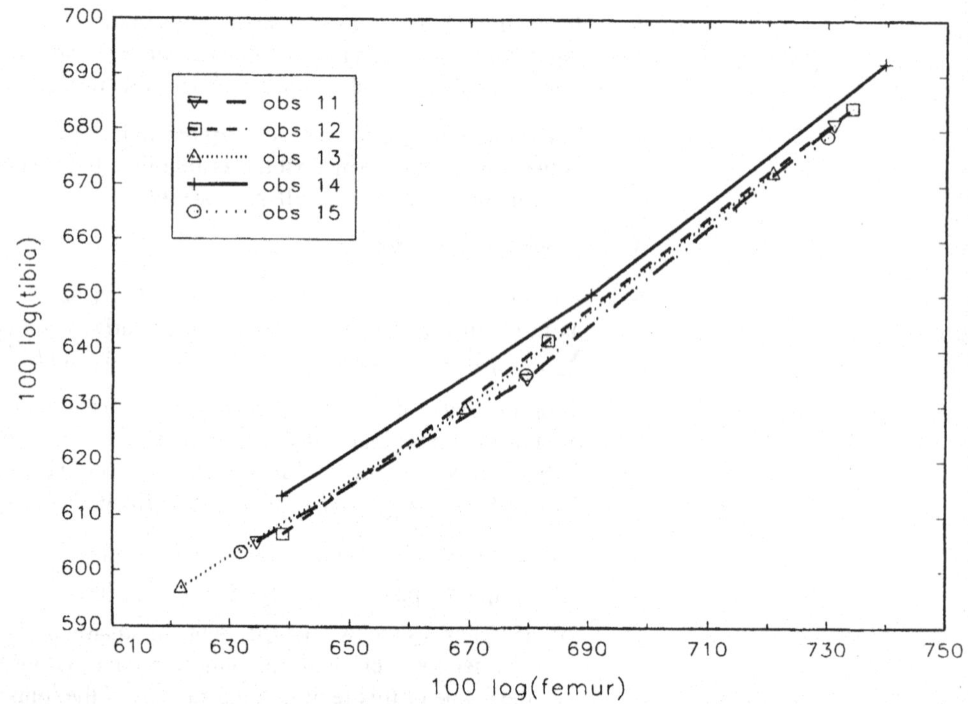

Figure 8-5-2 Growth of water striders no. 11 to 15, illustrated as curves for variables *femur* and *tibia*.

10. This exercise explores the water strider data of Example 8.5.4 in more detail. Let $\mathbf{S}$ denote the sample covariance matrix from table 8.5.4, and partition $\mathbf{S}$ into submatrices of dimension 2×2 as

$$\mathbf{S} = \begin{pmatrix} \mathbf{S}_{11} & \mathbf{S}_{12} & \mathbf{S}_{13} \\ \mathbf{S}_{21} & \mathbf{S}_{22} & \mathbf{S}_{23} \\ \mathbf{S}_{31} & \mathbf{S}_{32} & \mathbf{S}_{33} \end{pmatrix}.$$

(a) Compute the eigenvectors of each $\mathbf{S}_{ii}$, and call the orthogonal 2×2 matrices $\mathbf{B}_i$, $i = 1, 2, 3$. Comment on similarities (if any) among the three $\mathbf{B}_i$.

(b) Let $\mathbf{b}_i^*$ denote the first eigenvector of $\mathbf{S}_{ii}$. Construct a vector $\mathbf{b} \in \mathbb{R}^6$ as

$$\mathbf{b} = \frac{1}{\sqrt{3}} \begin{pmatrix} \mathbf{b}_1^* \\ \mathbf{b}_2^* \\ \mathbf{b}_3^* \end{pmatrix}.$$

Compare this vector to the first eigenvector of $\mathbf{S}$ from Table 8.5.5, and comment.

(c) Using the matrices $\mathbf{B}_i$ from part (a), construct a 6×6 matrix

$$\mathbf{B} = \begin{pmatrix} \mathbf{B}_1 & \mathbf{O} & \mathbf{O} \\ \mathbf{O} & \mathbf{B}_2 & \mathbf{O} \\ \mathbf{O} & \mathbf{O} & \mathbf{B}_3, \end{pmatrix},$$

and compute the matrix $\mathbf{M} = \mathbf{B}'\mathbf{S}\mathbf{B}$. What is the meaning of the matrix $\mathbf{M}$? Does $\mathbf{M}$ show any particular structure?

(d) If a matrix $\mathbf{B}$ is constructed as in part (c), where all $\mathbf{B}_i$ are orthogonal matrices, show that $\mathbf{B}$ is itself orthogonal.

11. Let $\boldsymbol{\Sigma}$ denote a positive definite symmetric $p \times p$ matrix, and let $\mathbf{R} = (r_{ij})$ denote a positive-definite symmetric $k \times k$ matrix. Define a $kp \times kp$ matrix $\boldsymbol{\psi}$ as

$$\boldsymbol{\psi} = \begin{pmatrix} r_{11}\boldsymbol{\Sigma} & r_{12}\boldsymbol{\Sigma} & \cdots & r_{1k}\boldsymbol{\Sigma} \\ r_{21}\boldsymbol{\Sigma} & r_{22}\boldsymbol{\Sigma} & \cdots & r_{2k}\boldsymbol{\Sigma} \\ \vdots & \vdots & \ddots & \vdots \\ r_{k1}\boldsymbol{\Sigma} & r_{k2}\boldsymbol{\Sigma} & \cdots & r_{kk}\boldsymbol{\Sigma} \end{pmatrix}.$$

(a) Show that $\boldsymbol{\psi}$ is positive-definite symmetric.

(b) Let $\mathbf{A}$ denote an orthogonal $p \times p$ matrix such that $\mathbf{A}'\boldsymbol{\Sigma}\mathbf{A} = \boldsymbol{\Lambda}$ is diagonal. Let $\mathbf{B}$ denote the block-diagonal matrix of dimension $pk \times pk$ with each diagonal block equal to $\mathbf{A}$. Show that $\mathbf{B}$ is orthogonal, and comment on the structure of the matrix $\mathbf{B}'\boldsymbol{\psi}\mathbf{B}$.

(c) Find the eigenvalues and eigenvectors of $\boldsymbol{\psi}$ in terms of the eigenvalues and eigenvectors of $\boldsymbol{\Sigma}$ and $\mathbf{R}$. *Hint*: If you find this difficult, choose some specific numerical examples for $\boldsymbol{\Sigma}$ and $\mathbf{R}$, and solve the problem numerically. Then let the numerical results inspire you.

(d) How is this exercise related to Example 8.5.4? How is it related to Exercise 12 in Section 8.4?

8.6 Sampling Properties

In this section, we will discuss some sampling properties of principal components, but we will mostly skip proofs because they are beyond the level of this text. The material is included because the problems discussed here form an important part of principal component analysis, yet they are ignored by almost all textbooks on multivariate statistics.

Sometimes a principal component analysis may have no other purpose than approximation of a high-dimensional data set in lower dimension, as illustrated repeatedly in the examples of Section 8.5. However, sometimes the investigator may have reasons to attach some meaning to the coefficients of the principal components or even to perform a formal test of a hypothesis. The most important case, in the author's experience, occurs in the estimation of constants of allometric growth. The model of allometric growth assumes that various parts of an organism grow at constant relative rates. For instance, the length of an organism may grow at twice the rate as the width. Then the relative growth rate of length to width, or allometric constant, would be 2:1. In general, there may be p parts of an organism that are being measured. The constants of allometric growth can then be summarized in the form of a vector $\mathbf{r} = (r_1, \ldots, r_p)'$, such that the relative growth rate of part i, compared to part j, is r_i/r_j. Thus, only ratios of the r_h are well defined, and therefore one may use some arbitrary rule for normalization of the vector $\mathbf{r}$. A special case of allometry is *isometric* growth, defined by the assumption that all r_j are equal, and therefore, all ratios r_i/r_j are equal to 1. This means that, in all variables measured, the organisms grow at the same rate, thus maintaining their shape while changing size. Jolicoeur (1963), who developed much of the theory of allometric growth, suggested estimating $\mathbf{r}$ as the eigenvector of the covariance matrix of log-transformed variables associated with the largest root. This is ultimately the reason for the log-transformations in Examples 8.1.1 and 8.5.4. For a review of allometry, see Hills (1982) or Klingenberg (1996).

If constants of allometric growth are to be estimated using principal component analysis, the question naturally arises how stable the estimates are. Just as in estimation of a linear regression function or discriminant function, standard errors are a useful tool for assessing the stability of estimated coefficients. The result on which the calculation of standard errors in principal component analysis is based is now given as a theorem.

Theorem 8.6.1. *Suppose that $\mathbf{X}_1, \ldots, \mathbf{X}_N$ is a sample of size $N > p$ from $\mathcal{N}_p(\boldsymbol{\mu}, \boldsymbol{\psi})$, where $\boldsymbol{\psi}$ is positive definite, and let $\mathbf{S}$ denote the usual (unbiased) sample covariance matrix. Let $\lambda_1 \ge \lambda_2 \ge \cdots \ge \lambda_p > 0$ denote the eigenvalues of $\boldsymbol{\psi}$, and $\boldsymbol{\beta}_1, \ldots, \boldsymbol{\beta}_p$ the associated normalized eigenvectors. Let $\ell_1 > \ell_2 > \cdots > \ell_p > 0$ denote the eigenvalues of $\mathbf{S}$, and $\mathbf{b}_1, \ldots, \mathbf{b}_p$ the associated eigenvectors. If λ_h is a simple root (i.e., $\lambda_{h-1} > \lambda_h > \lambda_{h+1}$), then*

(a) *the asymptotic distribution of $\sqrt{N}(\ell_h - \lambda_h)$ as N tends to infinity is normal with mean zero and variance $2\lambda_h^2$,*

(b) *the ℓ_h are asymptotically independent of each other and independent of the $\mathbf{b}_j$, and*

(c) *the asymptotic distribution of $\sqrt{N}(\mathbf{b}_h - \boldsymbol{\beta}_h)$ is p-variate normal with mean vector $\mathbf{0}$ and covariance matrix*

$$\mathbf{V}_h = \sum_{j=1}^{p} \theta_{hj} \boldsymbol{\beta}_j \boldsymbol{\beta}_j' \qquad h = 1, \ldots, p, \tag{1}$$

where

$$\theta_{hj} = \begin{cases} 0 & \text{if } h = j \\ \dfrac{\lambda_h \lambda_j}{(\lambda_h - \lambda_j)^2} & \text{if } h \neq j \end{cases} \qquad h, j = 1, \ldots, p. \tag{2}$$

Proof see Anderson (1963) or Flury (1988, Chapter 2.4). ■

The proof of part (a) is actually not too difficult and is stated as Exercise 1 for the student who knows asymptotic theory. Parts (b) and (c) are considerably more involved.

The assumption of simplicity of λ_h in Theorem 8.6.1 is crucial, as is seen from equation (2): If λ_h is not simple, then some of the θ_{hj} will be infinite, which expresses the fact that the eigenvectors associated with multiple roots are not uniquely defined. The same formula shows that the coefficients of $\mathbf{b}_h$ will be relatively unstable if λ_h is not well separated from λ_{h-1} or λ_{h+1}. We will discuss this in some more detail in Example 8.6.2.

The main practical applications of Theorem 8.6.1 concern testing of hypotheses about eigenvectors and eigenvalues and computing standard errors. We will ignore testing problems here; the interested student is again referred to Anderson (1963) and Flury (1988). For the computation of standard errors, we will replace all parameters in (1) and (2) by the respective sample counterparts, i.e., estimate $\mathbf{V}_h$ by

$$\hat{\mathbf{V}}_h = \sum_{h=1}^{p} \hat{\theta}_{hj} \mathbf{b}_j \mathbf{b}'_j, \tag{3}$$

where

$$\hat{\theta}_{hj} = \begin{cases} 0 & \text{if } h = j \\ \dfrac{\ell_h \ell_j}{(\ell_h - \ell_j)^2} & \text{if } h \neq j. \end{cases} \tag{4}$$

Writing

$$\mathbf{B} = [\mathbf{b}_1, \mathbf{b}_2, \ldots, \mathbf{b}_p] = \begin{pmatrix} b_{11} & \cdots & b_{1p} \\ \vdots & \ddots & \vdots \\ b_{p1} & \cdots & b_{pp} \end{pmatrix} \tag{5}$$

for the matrix with eigenvectors of $\mathbf{S}$ as columns, we will estimate the variance of b_{mh} as the mth diagonal entry of $\hat{\mathbf{V}}_h/N$. In scalar notation we define the standard error of b_{mh} as

$$\text{se}(b_{mh}) = \left[\frac{1}{N} \ell_h \sum_{\substack{j=1 \\ j \neq h}}^{p} \frac{\ell_j}{(\ell_j - \ell_h)^2} b_{mj}^2 \right]^{1/2}, \qquad m, h = 1, \ldots, p, \tag{6}$$

and the standard error of ℓ_h as

$$\text{se}(\ell_h) = \sqrt{2/N}\, \ell_h, \quad h = 1, \ldots, p. \tag{7}$$

These standard errors depend on the data only through the eigenvectors and eigenvalues of the sample covariance matrix. Also, the standard error of each coefficient depends on *all* eigenvalues and eigenvectors, although if ℓ_j is much larger or much smaller than ℓ_h, then the effect of the jth eigenvector on the standard errors of the hth eigenvector will be small.

This method of computing standard errors is quick and relatively simple, but it suffers from the disadvantages that it is based explicitly on normality assumptions and it is approximate only because it is derived from an asymptotic result. The obvious alternative is to use computer power, i.e., the bootstrap or jackknife estimates of standard error. We will return to this point in the examples.

Example 8.6.1 continuation of Example 8.1.1

Shell dimensions of turtles. This time we will use variables $X_1 = 10\log(\text{length})$, $X_2 = 10\log(\text{width})$, and $X_3 = 10\log(\text{height})$, measured on $N = 24$ male turtles of the species *Chrysemys picta marginata*. Sample statistics are

$$\bar{\mathbf{x}} = \begin{pmatrix} 47.25 \\ 44.78 \\ 37.03 \end{pmatrix},$$

and

$$\mathbf{S} = \begin{pmatrix} 1.107 & 0.802 & 0.816 \\ 0.802 & 0.642 & 0.601 \\ 0.816 & 0.601 & 0.677 \end{pmatrix}.$$

The eigenvectors are

$$\mathbf{b}_1 = \begin{pmatrix} 0.683 \\ 0.510 \\ 0.523 \end{pmatrix}, \mathbf{b}_2 = \begin{pmatrix} -0.159 \\ -0.594 \\ 0.788 \end{pmatrix},$$

and

$$\mathbf{b}_3 = \begin{pmatrix} 0.713 \\ -0.622 \\ -0.324 \end{pmatrix}.$$

The eigenvalues are $\ell_1 = 2.330$, $\ell_2 = 0.060$, and $\ell_3 = 0.036$ with standard errors $\text{se}(\ell_1) = 0.673$, $\text{se}(\ell_2) = 0.017$, and $\text{se}(\ell_3) = 0.010$.

The entries of $\mathbf{b}_1$ provide estimates of constants of allometric growth; accordingly, growth in length is somewhat faster than in width or height. The estimated standard errors of the coefficients of $\mathbf{b}_1$ computed from equation (6) are as follows:

$$\text{se}(\mathbf{b}_1) = \begin{pmatrix} \text{se}(b_{11}) \\ \text{se}(b_{21}) \\ \text{se}(b_{31}) \end{pmatrix} = \begin{pmatrix} 0.019 \\ 0.026 \\ 0.028 \end{pmatrix}.$$

This means that the parameter estimates are fairly stable. Without doing a formal test, this result seems to exclude the hypothesis of isometric growth, which implies equality of all coefficients in β_1, i.e., $\beta_1' = \frac{1}{\sqrt{3}}(1, 1, 1) \approx (0.577, 0.577, 0.577)$. However, a sample size of $N = 24$ is hardly large enough to justify the use of asymptotic theory, and therefore, we might look for an alternative assessment of stability.

A bootstrap analysis of this example, based on 10,000 bootstrap samples from the empirical distribution (see Efron and Tibshirani, 1993), gave the following standard errors:

$$\text{se}(\mathbf{b}_1) = \begin{pmatrix} 0.019 \\ 0.035 \\ 0.033 \end{pmatrix}.$$

The standard error of the first coefficient is (to the third decimal digit) exactly the one obtained from asymptotic theory, whereas the standard errors of the second and third coefficients have increased. However, the conclusion about good stability of the estimated coefficients remains unchanged. See Exercise 4 for a more detailed analysis. □

Here is a cautionary remark about the use of the bootstrap in principal component analysis. The remark refers to the nonuniqueness of eigenvectors due to multiplication by -1, which has not bothered us so far, but which may pose a serious problem in the bootstrap. Generally speaking, if we calculate the eigenvectors of two almost identical matrices $\mathbf{M}_1$ and $\mathbf{M}_2$ (such as the covariance matrices of two different bootstrap samples), then the eigenvectors will be "close" in some sense. However, an eigenvector may at times appear with signs reversed, which reflects nothing but the particular convention used by the software. Of course, such reversals of signs will seriously affect the bootstrap distribution of the coefficients, but they can usually be detected by graphing their distributions. It is better to catch such reversals by checking the result of each bootstrap replication. In Example 8.6.1 the following rule was used: The first coefficient in $\mathbf{b}_1$, the third coefficient in $\mathbf{b}_2$, and the first coefficient in $\mathbf{b}_3$ must always be positive. If one of these coefficients turns out negative in any bootstrap replication, multiply the respective eigenvector by -1. In the 10,000 bootstrap replications of Example 8.6.1, no sign reversals were necessary for $\mathbf{b}_1$, but roughly 700 for each of $\mathbf{b}_2$ and $\mathbf{b}_3$.

Example 8.6.2 continuation of Example 8.5.2

Eigenvectors and eigenvalues of the sample covariance matrix of the head dimension data were given in Table 8.5.2. Table 8.6.1 displays standard errors of the coefficients of all eigenvectors computed from equation (6). For the first eigenvector, all standard errors are quite small. In the fourth and fifth eigenvectors, large standard errors show that some coefficients are highly unstable. Recall that, because of the normalization of eigenvectors, all coefficients must be between -1 and 1; therefore, a standard error of 0.5 or more automatically means that the corresponding coefficient could be practically anything. This means in turn that it is impossible to attach any interpretation to a principal component whose coefficients have large standard errors. The large standard errors in the coefficients of components 4 and 5 are due to the closeness of eigenvalues ℓ_4 and ℓ_5, as seen from Table 8.5.2.

Table 8.6.1 Standard errors of principal component coefficients in the head dimensions example, arranged as in Table 8.5.2

	U_1	U_2	U_3	U_4	U_5	U_6
MBF	0.048	0.091	0.170	0.132	0.518	0.075
BAM	0.056	0.092	0.157	0.136	0.468	0.071
TFH	0.087	0.056	0.123	0.294	0.067	0.056
LGAM	0.051	0.068	0.163	0.628	0.125	0.098
LTM	0.029	0.064	0.062	0.124	0.111	0.024
LTG	0.042	0.103	0.049	0.157	0.131	0.045

For comparison, bootstrap standard errors of the coefficients of the first eigenvector based on 10,000 bootstrap samples were (0.047, 0.049, 0.078, 0.048, 0.030, and 0.044), which is in quite good agreement with the standard errors given by asymptotic theory; also see Exercise 10. □

The phenomenon of instability of the fourth and fifth eigenvectors in Example 8.6.2 is commonly referred to as the problem of sphericity. We will formulate it more precisely on the theoretical level. For a p-variate random vector $\mathbf{X}$ with $\text{Cov}[\mathbf{X}] = \psi$, the principal components U_h and U_{h+1} are called *spherical* if $\lambda_h = \lambda_{h+1}$, i.e., if the hth and $(h+1)$st eigenvalues are identical. The terminology "sphericity" derives from the fact that, under normality assumptions, the joint distribution of U_h and U_{h+1} is spherical. The bad thing about sphericity is that if $\lambda_h = \lambda_{h+1}$, then the two associated eigenvectors are not uniquely defined. In fact, they can be chosen arbitrarily subject to orthogonality to the other eigenvectors (Exercise 5). Only the subspace spanned by the eigenvectors associated with the multiple root is well defined.

Before attaching any interpretation to a principal component, it is important to make sure that it is well defined. This can be done using a test for sphericity given in the following theorem.

Theorem 8.6.2 *Under the same assumptions as in Theorem 8.6.1, the log-likelihood ratio statistic for sphericity of components $r+1, \ldots, r+h$, i.e., for the hypothesis $\lambda_{r+1} = \cdots = \lambda_{r+h}$ against the alternative $\lambda_{r+1} > \cdots > \lambda_{r+h}$ is given by*

$$S(r,h) = Nh \log \frac{\frac{1}{h}\sum_{j=r+1}^{r+h} \ell_j}{\left(\prod_{j=r+1}^{r+h} \ell_j\right)^{1/h}}. \tag{8}$$

The asymptotic distribution of $S(r,h)$ as N tends to infinity is chi-square on $h(h+1)/2 - 1$ degrees of freedom under the null hypothesis.

Proof See Anderson (1963, 1984). ■

The problem of estimating a covariance matrix under the assumption of sphericity of specified components is closely related to testing for sphericity. We will not expand on this theme here, but the interested student is referred to Flury (1988) for a detailed discussion.

Example 8.6.3 continuation of Example 8.6.1

Shell dimensions of turtles. With $\ell_1 = 2.320$, $\ell_2 = 0.060$, and $\ell_3 = 0.036$, the three possible sphericity statistics and the associated degrees of freedom (df) of the asymptotic chi-square distribution are

- $S(0, 3) = 111.71$ for sphericity of all three components (df = 5),
- $S(0, 2) = 55.77$ for sphericity of components 1 and 2 (df = 2), and
- $S(1, 3) = 1.55$ for sphericity of components 2 and 3 (df = 2).

Comparing these to quantiles of the respective chi-square distributions, at any reasonable level, we would reject sphericity of the first and second component as well as sphericity of all three, but accept sphericity of the second and third. Consequently, no interpretation should be attached to $\mathbf{U}_2$ and $\mathbf{U}_3$. In fact, this result might suggest that the variation in turtle shells can be adequately described in a single dimension and a one-dimensional principal component approximation is appropriate. Also see Example 8.4.5. □

Generally, users of principal component analysis tend to overinterpret the results by trying to attach a meaning to the eigenvectors. Without testing for sphericity or computing standard errors of coefficients, this is dangerous because the results are often far less stable than one would hope. Therefore, the computation of standard errors should be an integral part of any principal component analysis unless the only purpose of the analysis is approximation of p-variate data in a lower dimension.

Other problems of statistical inference in principal component analysis concern testing of hypotheses about eigenvectors (e.g., testing for isometric growth), confidence intervals for eigenvalues, tests on eigenvalues, and confidence intervals for the relative mean squared error of a q-dimensional principal component approximation. In a multivariate normal theory setup, these problems can be treated using Theorem 8.6.1. The interested reader is referred to Flury (1988).

Exercises for Section 8.6

1. Prove part (a) of Theorem 8.6.1.
2. Show that the matrix $\mathbf{V}_h$ of equation (1) has rank $p - 1$.

3. Use Theorem 8.6.1 to show that an approximate confidence interval for coefficient β_{hj}, with coverage probability $1-\alpha$, is given by $\mathbf{b}_{hj} \pm z_{1-\alpha/2}\,\text{se}(\mathbf{b}_{hj})$, where z_ϵ is the ϵ-quantile of the standard normal distribution.

4. This exercise is based on the turtle data of Example 8.6.1.

   (a) Do a bootstrap analysis of the principal components, using a large number B ($B \geq 1000$) of bootstrap samples drawn from the empirical distribution. *Note*: Make sure to take care of the problem discussed in the text following Example 8.6.1.

   (b) Graph the bootstrap distributions of the three estimated constants of allometric growth, and find 95 percent confidence intervals using the 2.5% quantile and the 97.5% quantile of the bootstrap distributions.

   (c) Compute confidence intervals for all three coefficients of allometric growth, using Exercise 3 and the normal theory standard errors given in the example. Compare them to the bootstrap confidence intervals from part (b), and comment.

   (d) Graph the joint bootstrap distribution of the first and the second coefficients of allometric growth in a scatterplot. Would this scatterplot rather support or contradict the hypothesis of isometric growth?

5. Let ψ denote a symmetric $p \times p$ matrix with an eigenvalue λ^* of multiplicity q, and $(p-q)$ other eigenvalues different from λ^*. Let $\mathbf{B} = [\mathbf{B}_1, \mathbf{B}_2]$ denote an orthogonal matrix such that $\mathbf{B}'\psi\mathbf{B}$ is diagonal, where $\mathbf{B}_1$ has dimension $p \times q$, with rows corresponding to a set of eigenvectors associated with the multiple root λ^*. Let $\mathbf{H}$ denote an arbitrary orthogonal matrix of dimension $q \times q$, and set $\mathbf{C} = (\mathbf{B}_1\mathbf{H} \quad \mathbf{B}_2)$. Show that $\mathbf{C}$ is orthogonal and $\mathbf{C}'\psi\mathbf{C}$ is diagonal.

6. Show that the log-likelihood ratio statistic for sphericity in equation (8), defined for $\ell_j > 0$, can take only nonnegative values. Under what condition is $S(r, h) = 0$?

7. Show that sphericity of all p principal components (i.e., equality of all p eigenvalues) is equivalent to $\text{Cov}[\mathbf{X}] = \sigma^2\mathbf{I}_p$ for some $\sigma^2 > 0$.

8. Let $\mathbf{X}_1, \ldots \mathbf{X}_N$ denote a sample from $\mathcal{N}_p(\mu, \psi)$, ψ positive definite symmetric, $N > p$. Find the log-likelihood ratio statistic for sphericity of all p principal components. Then compare it to equation (8) and show that it is a special case for $r = 0$, $h = p$.

9. Show that for $h = 2$, the sphericity statistic depends on the eigenvalues ℓ_{r+1} and ℓ_{r+2} only through their ratio.

10. In Example 8.5.2/8.6.2 (head dimensions), compute the sphericity statistic for all five pairs of adjacent principal components, and compute the associated p-values from the asymptotic distribution of the test statistic. How do these results relate to the standard errors in Table 8.6.1?

11. Table 8.6.2 gives head dimensions of 59 young Swiss women, age 20, who were measured in the same investigation as the 200 men of Example 8.6.2.

    (a) Compute eigenvectors and eigenvalues of the sample covariance matrix. Comment on the coefficients of the first eigenvector.

(b) Compute standard errors of all principal component coefficients, and comment on them.

(c) Calculate the log likelihood ratio test statistics $S(0, 2), \ldots, S(4, 2)$ for all five adjacent pairs of principal components. Show that, for no such pair, sphericity could be rejected at the 5 percent level of significance using the asymptotic chi-square distribution.

(d) Calculate the statistic for overall sphericity of all components, and show that overall sphericity would be clearly rejected.

12. This exercise is based on the water strider data of Example 8.5.4.

(a) Compute standard errors of all principal component coefficients, using equation (6). Comment on the stability of the coefficients in the first eigenvector.

(b) Repeat (a), using the bootstrap with at least 1000 bootstrap samples.

(c) Test for simultaneous sphericity of the second to sixth components, i.e., the hypothesis $\lambda_2 = \cdots = \lambda_6$.

8.7 Outlook

This final section, intended for the reader who would like to go deeper into principal component analysis, describes some recent developments without giving too much detail and provides appropriate references. We will motivate these developments with practical examples.

Example 8.7.1 continuation of Example 8.6.1

Shell dimensions of turtles. In the same publication by Jolicoeur and Mosimann (1960), data are provided for 24 female turtles of the species *Chrysemys picta marginata*. The data are reproduced in Table 1.4. Leaving details of the computation to the reader (Exercise 1), the constants of allometric growth for females and associated standard errors are estimated as

$$\mathbf{b}_1 = \begin{pmatrix} 0.622 \\ 0.484 \\ 0.615 \end{pmatrix}$$

and

$$\text{se}(\mathbf{b}_1) = \begin{pmatrix} 0.015 \\ 0.017 \\ 0.017 \end{pmatrix}.$$

The question immediately arises, do the coefficients of $\mathbf{b}_1$ differ from the corresponding coefficients in the analysis of the male turtles just by chance, or is there a systematic difference? Or, in other words, are the constants of allometric growth the same for both genders?

Table 8.6.2 Head dimensions of 59 young swiss women. Variables are the same as in Table 1.2 and Figure 1.5.

MBF	BAM	TFH	LGAN	LTN	LTG
109.2	120.1	103.0	47.6	120.1	132.4
101.5	117.3	109.6	55.7	123.1	137.1
109.5	111.7	93.4	48.0	105.4	127.2
96.1	117.8	115.5	56.8	115.7	126.7
107.6	117.9	96.6	50.5	111.8	125.3
101.4	102.4	108.9	56.5	108.3	122.6
106.7	123.8	111.4	53.4	116.8	117.0
85.6	120.3	112.4	61.2	117.7	136.2
99.7	122.0	110.8	61.0	122.1	124.5
95.0	116.9	102.7	53.3	127.6	146.1
83.9	105.7	115.4	65.4	113.9	131.1
80.7	116.4	115.3	66.5	131.9	133.6
89.8	116.8	111.0	64.2	119.5	129.8
95.5	110.6	118.1	58.9	111.5	117.6
107.9	125.6	114.2	61.0	126.6	134.3
87.2	112.6	114.2	60.6	126.7	127.8
96.0	112.3	112.9	57.0	119.3	130.4
107.4	113.5	114.4	56.4	125.3	131.3
100.5	111.7	114.8	63.9	115.9	133.3
94.8	111.2	124.7	53.2	116.2	127.6
107.7	116.1	117.2	57.1	124.1	139.5
111.3	111.2	119.4	52.4	112.4	130.7
109.6	121.6	124.5	52.2	111.8	128.2
106.6	134.6	116.5	58.2	122.9	140.1
108.2	119.1	119.4	59.9	123.2	131.2
105.5	113.8	132.3	54.2	123.7	127.5
106.5	116.1	112.7	57.7	115.6	130.2
101.7	116.7	120.0	56.5	119.8	137.4
110.4	111.7	122.2	56.6	117.5	128.5
106.8	107.4	113.4	56.8	126.2	115.7
104.4	108.7	109.3	51.4	115.9	121.6
102.2	106.8	120.5	57.4	117.3	130.6
93.5	116.6	118.8	56.4	113.8	133.3
103.3	114.1	119.3	59.5	113.8	127.1
86.1	111.7	116.0	52.9	119.4	129.9
97.0	108.4	108.1	49.9	114.4	123.2
113.3	111.8	125.2	51.4	124.0	134.1
109.5	119.4	117.8	51.0	121.4	130.5
103.5	110.5	130.8	57.3	125.9	136.0
104.3	108.6	116.1	53.3	119.4	137.6
108.3	122.5	123.2	52.9	124.3	124.5
99.6	112.2	103.3	52.8	115.7	128.4
97.5	122.6	118.8	50.7	114.1	117.8
96.6	111.2	115.3	54.8	117.1	130.5
109.3	124.0	113.8	51.1	115.6	119.7

Table 8.6.2 *Continued*

MBF	BAM	TFH	LGAN	LTN	LTG
112.4	121.1	111.7	53.7	115.9	129.7
105.5	123.9	111.3	50.1	119.8	122.3
105.7	111.3	124.3	56.0	117.8	132.7
101.4	105.4	112.1	53.1	113.0	128.4
107.6	124.3	112.0	48.5	118.9	132.4
103.1	117.8	119.4	47.4	117.7	137.1
103.3	111.7	112.8	54.8	122.2	131.9
105.1	117.5	116.3	56.7	119.5	131.6
90.8	113.0	115.6	50.9	114.3	130.2
114.2	121.1	120.9	54.3	122.2	140.7
114.5	121.4	115.9	57.2	123.5	134.6
110.6	119.4	115.2	53.5	119.6	128.2
110.1	114.4	116.8	51.8	123.1	135.3
104.6	110.9	113.5	49.4	117.1	130.0

Of course, if the covariance matrices in both populations are identical, then all eigenvectors must be identical across groups, and any differences between $\mathbf{b}_1$ for males and $\mathbf{b}_1$ for females are purely random. However, a test for equality of covariance matrices leads to the conclusion that they are different (Exercise 1). Hence, the question of equality of the constants of allometric growth is not trivial. □

Questions of equality of eigenvectors across groups have been treated by several authors; for reviews see Airoldi and Flury (1988), Flury (1988, 1995), or Krzanowski (1988). A theoretical setup is as follows. Suppose that a p-variate random vector $\mathbf{X}$ is measured in two groups, with $\text{Cov}[\mathbf{X}] = \psi_i$ in group i, $i = 1, 2$. In general, each ψ_i has its own spectral decomposition, i.e., there exist two orthogonal matrices $\mathbf{B}_1$ and $\mathbf{B}_2$ such that $\psi_i = \mathbf{B}_i \boldsymbol{\Lambda}_i \mathbf{B}_i'$, where $\boldsymbol{\Lambda}_i = \text{diag}(\lambda_{i1}, \ldots, \lambda_{ip})$, $i = 1, 2$. This corresponds to groupwise principal components without any relationship between the two groups. The CPC (for *C*ommon *P*rincipal *C*omponent) model assumes that the two matrices $\mathbf{B}_1$ and $\mathbf{B}_2$ are identical, that is, it is assumed that there exists an orthogonal matrix $\mathbf{B} = [\boldsymbol{\beta}_1, \ldots, \boldsymbol{\beta}_p]$ of dimension $p \times p$ such that

$$\psi_1 = \mathbf{B}\boldsymbol{\Lambda}_1\mathbf{B}' = \sum_{j=1}^{p} \lambda_{1j}\boldsymbol{\beta}_j\boldsymbol{\beta}_j'$$

and

$$\psi_2 = \mathbf{B}\boldsymbol{\Lambda}_2\mathbf{B}' = \sum_{j=1}^{p} \lambda_{2j}\boldsymbol{\beta}_j\boldsymbol{\beta}_j'. \tag{1}$$

Under the CPC model the two covariance matrices may differ in their eigenvalues but not in their eigenvectors. For sample covariance matrices $\mathbf{S}_1$ and $\mathbf{S}_2$ from populations that satisfy the CPC model, we would expect the two orthogonal matrices to be "almost" identical, i.e., to differ only by sampling error. This leads to the problem of estimating principal component coefficients under the constraint (1), a problem which is beyond the scope of this introduction. Incidentally, (1) is equivalent to commutativity of ψ_1 and ψ_2 by multiplication; see Exercise 2.

The CPC model just discussed does not correspond exactly to the situation alluded to in Example 7.1, in which we were interested only in the first eigenvector. A generalization of (1), called the *partial* CPC model, assumes that only a subset of eigenvectors is identical across groups, whereas the remaining ones are allowed to vary. For the case of exactly one common principal component, this can be written

$$\psi_1 = \lambda_{11}\beta_1\beta_1' + \sum_{j=2}^{p} \lambda_{1j}\beta_j^{(1)}\beta_j^{(1)'}$$

and

$$\psi_2 = \lambda_{21}\beta_1\beta_1' + \sum_{j=2}^{p} \lambda_{2j}\beta_j^{(2)}\beta_j^{(2)'}. \tag{2}$$

In this model, the two covariance matrices share one common eigenvector β_1, whereas all other eigenvectors and all eigenvalues are allowed to vary between groups. Notice that the matrices $\mathbf{B}_1 = [\beta_1, \beta_2^{(1)}, \ldots, \beta_p^{(1)}]$ and $\mathbf{B}_2 = [\beta_1, \beta_2^{(2)}, \ldots, \beta_p^{(2)}]$ must both be orthogonal.

Now, suppose that $\mathbf{X}$ is a p-variate random vector with mean μ_i and covariance matrix ψ_i in group i, $i = 1, 2$. Suppose that a partial CPC model holds as in equation (2), and assume that λ_{11} and λ_{21} are the largest eigenvalues in their respective groups. Then the one-dimensional principal component approximations are given by

$$\mathbf{Y}_{(1)} = \mu_1 + \beta_1\beta_1'(\mathbf{X} - \mu_1) \qquad \text{in group 1}$$

and

$$\mathbf{Y}_{(1)} = \mu_2 + \beta_1\beta_1'(\mathbf{X} - \mu_2) \qquad \text{in group 2.} \tag{3}$$

Since the projection matrix $\beta_1\beta_1'$ is identical in both cases, this corresponds to orthogonal projections on parallel lines.

Example 8.7.2 continuation of Example 8.7.1

Shell dimensions of turtles. In this case, we use only two variables, $X_1 = 10\log(\text{length})$ and $X_2 = 10\log(\text{width})$, of $N_1 = 24$ male turtles and $N_2 = 24$ female turtles. Figure 8.7.1 shows a scatterplot of the data of both genders. The two groups overlap considerably, showing more variability for females than for males, but the combined data roughly follow one straight line. If the constants of allometric growth are identical for both genders, then we would expect the two lines corresponding to the groupwise one-dimensional principal component approximations to be almost parallel. In Figure 8.7.1, the two lines (one of which is the same as the line in Figure 8.1.1) appear not only close in slope, but they nearly coincide. Thus, we might attempt to fit one single line, using the one-dimensional principal component approximation to the combined data of both genders, which would hopefully provide an estimate of the common constants of allometric growth. To what extent this is true is explored in the following text. □

Example 8.7.2 illustrates a notion in morphometrics called *allometric extension*, which, loosely speaking, means that one group is the extension of the other group along the main axis of variation (Hills, 1982). In the terminology of this chapter, two groups (or rather, two random vectors) are allometric extensions of each other if their one-dimensional principal component approximations are projections on the same line. Figure 8.7.2 gives a symbolic definition of a case of allometric extension, with ellipses indicating contours of constant density of two elliptical distributions. The major axes of the two ellipses coincide; hence, this is a case of CPC where, in addition to identical eigenvectors in both covariance matrices, the difference between the mean vectors gives the same direction as the first principal component axis.

The following is a theoretical setup to handle allometric extension. Suppose that $\mathbf{X}_1$ and $\mathbf{X}_2$ are p-variate random vectors with mean vectors μ_i and covariance matrices ψ_i, $i = 1, 2$. Let $\lambda_{11} > \lambda_{12} \geq \cdots \geq \lambda_{1p}$ and $\lambda_{21} > \lambda_{22} \geq \cdots \geq \lambda_{2p}$ denote the eigenvalues of ψ_1 and ψ_2 respectively, and $\mathbf{B}_i = \left[\beta_1^{(i)}, \ldots, \beta_p^{(i)}\right]$ the associated orthogonal matrices with eigenvectors as columns. Then the one-dimensional principal

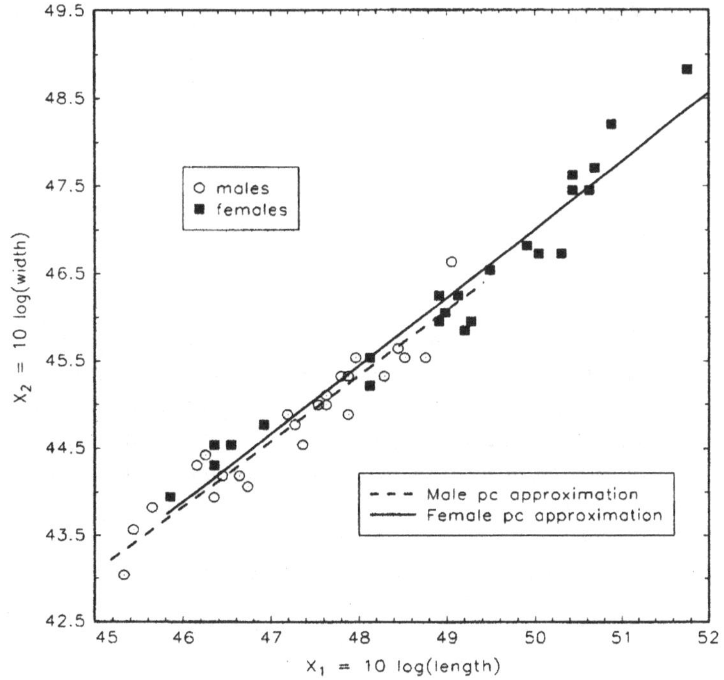

Figure 8-7-1
Scatterplot of log(carapace length) vs. log(carapace width) for 48 turtles. Open circles denote males; black squares, females. The two lines correspond to one-dimensional principal component approximations computed individually for each group.

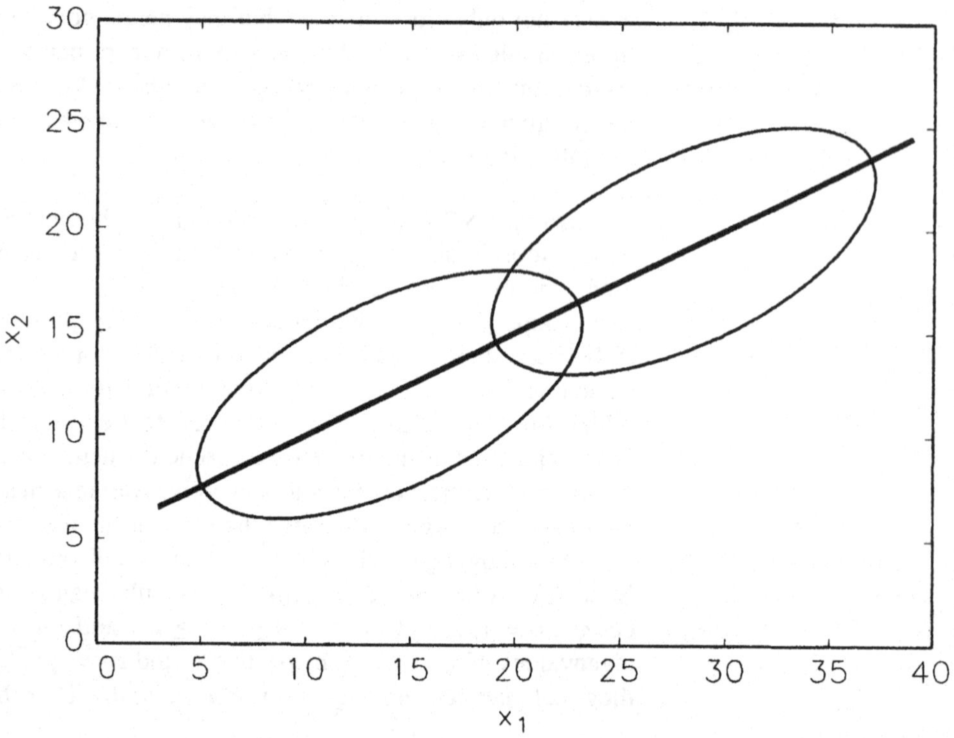

Figure 8-7-2 Illustration of two groups that are allometric extensions of each other.

component approximations of $\mathbf{X}_1$ and $\mathbf{X}_2$ are projections on the same line exactly if

$$\boldsymbol{\beta}_1^{(1)} = \boldsymbol{\beta}_1^{(2)} =: \boldsymbol{\beta}_1, \qquad \text{say,} \tag{4}$$

and

$$\boldsymbol{\mu}_2 - \boldsymbol{\mu}_1 = \delta \boldsymbol{\beta}_1 \qquad \text{for some } \delta \in \mathbb{R}. \tag{5}$$

Then we can state the following result.

Lemma 8.7.1 *Suppose that the one-dimensional principal component approximations of* $\mathbf{X}_1$ *and* $\mathbf{X}_2$ *are projections on the same line, as explained above. Let* $\mathbf{X}$ *denote a random vector whose distribution is a mixture of the distributions of* $\mathbf{X}_1$ *and* $\mathbf{X}_2$ *with arbitrary mixing weights* π_1 *and* π_2 ($\pi_1 + \pi_2 = 1$). *Then the one-dimensional, principal component approximation of* $\mathbf{X}$ *is a projection on the same line as the approximations of* $\mathbf{X}_1$ *and* $\mathbf{X}_2$.

Proof By (4), we can write the spectral decompositions of $\boldsymbol{\psi}_1$ and $\boldsymbol{\psi}_2$ as

$$\boldsymbol{\psi}_1 = \lambda_{11} \boldsymbol{\beta}_1 \boldsymbol{\beta}_1' + \sum_{j=2}^{p} \lambda_{1j} \boldsymbol{\beta}_j^{(1)} \boldsymbol{\beta}_j^{(1)\prime}$$

and

$$\psi_2 = \lambda_{21}\beta_1\beta_1' + \sum_{j=2}^{p} \lambda_{2j}\beta_j^{(2)}\beta_j^{(2)'}. \tag{6}$$

From Theorem 7.3.1, the mean vector $\bar{\mu}$ and the covariance matrix $\bar{\psi}$ of the mixture distribution with mixing weights π_1 and π_2 are given by

$$\bar{\mu} = \sum_{i=1}^{2} \pi_i \mu_i = \mu_1 + \pi_2 \delta \beta_1 = \mu_2 - \pi_1 \delta \beta_1 \tag{7}$$

and

$$\bar{\psi} = \sum_{i=1}^{2} \pi_i \psi_i + \sum_{i=1}^{2} \pi_i (\mu_i - \bar{\mu})(\mu_i - \bar{\mu})'. \tag{8}$$

In Exercise 4, it is shown that

$$\sum_{i=1}^{2} \pi_i (\mu_i - \bar{\mu})(\mu_i - \bar{\mu})' = (\pi_1\pi_2\delta)^2 \beta_1\beta_1'. \tag{9}$$

Using orthogonality of the eigenvectors of ψ_1 and ψ_2,

$$\bar{\psi}\beta_1 = \left[\pi_1\lambda_{11} + \pi_2\lambda_{21} + (\pi_1\pi_2\delta)^2\right]\beta_1. \tag{10}$$

Thus, β_1 is an eigenvector of $\bar{\psi}$ with associated eigenvalue $\bar{\lambda} = \pi_1\lambda_{11} + \pi_2\lambda_{21} + (\pi_1\pi_2\delta)^2$.

Let ξ be any other normalized eigenvector of $\bar{\psi}$ orthogonal to β_1. Then

$$\bar{\psi}\xi = \sum_{i=1}^{2} \pi_i \psi_i \xi \tag{11}$$

because $\beta_1'\xi = 0$. Hence, the eigenvalue of $\bar{\psi}$ associated with ξ is given by

$$\xi'\bar{\psi}\xi = \sum_{i=1}^{2} \pi_i \xi' \psi_i \xi. \tag{12}$$

Because of the maximization property of the eigenvector associated with the largest root, $\xi'\psi_i\xi < \lambda_{i1}$, $(i = 1, 2)$, and therefore,

$$\xi'\bar{\psi}\xi < \pi_1\lambda_{11} + \pi_2\lambda_{21} \le \bar{\lambda}. \tag{13}$$

This shows that β_1 is the eigenvector of $\bar{\psi}$ associated with the largest root. The one-dimensional principal component approximation of $\mathbf{X}$ is

$$\mathbf{Y}_{(1)} = \bar{\mu} + \beta_1\beta_1'(\mathbf{X} - \bar{\mu}), \tag{14}$$

and we leave it to the reader (Exercise 5) to show that both μ_1 and μ_2 are on the line corresponding to this projection. ■

Lemma 8.7.1 has an immediate practical consequence. Suppose that we have samples of size N_1 and N_2 from two groups (as in Example 8.7.2), and we assume that they are allometric extensions of each other. Using the combined data of both groups without group distinction corresponds to a mixture of the two distributions with mixing weights $N_1/(N_1 + N_2)$ and $N_2/(N_1 + N_2)$. Thus, the lemma tells us that the first eigenvector of the covariance matrix of the combined data may be used to estimate the common constants of allometric growth under the assumption of allometric extension.

Example 8.7.3 continuation of Example 8.7.2

For the combined data of male and female turtles, the first eigenvector of the covariance matrix is $\mathbf{b}_1 = \begin{pmatrix} 0.785 \\ 0.619 \end{pmatrix}$. For practical purposes, this is the same as the results found in groupwise analyses. We conclude that shell length and width of both male and female turtles of the species *Chrysemys picta marginata* grow at a ratio of roughly 4 to 3. □

Estimating common constants of allometric growth, as illustrated in the preceding example, is straightforward, but it is applicable only in situations where the two groups are allometric extensions of each other. Unfortunately, it is not easy to construct a formal test for allometric extension. For some relevant material, see Hills (1982), or Flury, Nel, and Pienaar (1995).

We leave the discussion of allometry here and return to the water strider data.

Example 8.7.4 continuation of Example 8.5.4

Skeletal dimensions of water striders. Recall that there are six variables partitioned in sets of two variables corresponding to the respective instars, or stages of growth. Sample statistics have been given in Table 8.5.4.

Let

$$\mathbf{S} = \begin{pmatrix} \mathbf{S}_{11} & \mathbf{S}_{12} & \mathbf{S}_{13} \\ \mathbf{S}_{21} & \mathbf{S}_{22} & \mathbf{S}_{23} \\ \mathbf{S}_{31} & \mathbf{S}_{32} & \mathbf{S}_{33} \end{pmatrix}$$

denote the sample covariance matrix partitioned into submatrices of dimension 2×2. Thus, $\mathbf{S}_{ii}$ is the covariance matrix of the two variables measured at time i, $i = 1, 2, 3$. What happens if we compute principal components individually for each instar? Writing $\mathbf{S}_{ii} = \mathbf{B}_i \mathbf{L}_i \mathbf{B}_i'$ for the spectral decomposition of $\mathbf{S}_{ii}$, we obtain

- instar 1: $\mathbf{S}_{11} = \begin{pmatrix} 9.247 & 7.972 \\ 7.972 & 11.248 \end{pmatrix}$, $\mathbf{B}_1 = \begin{pmatrix} 0.662 & 0.750 \\ 0.750 & -0.662 \end{pmatrix}$,

  $\mathbf{L}_1 = \text{diag}(18.282, 2.213)$

- instar 2: $\mathbf{S}_{22} = \begin{pmatrix} 8.728 & 8.071 \\ 8.071 & 11.082 \end{pmatrix}$, $\mathbf{B}_2 = \begin{pmatrix} 0.655 & 0.755 \\ 0.755 & -0.655 \end{pmatrix}$,
  $\mathbf{L}_2 = \text{diag}(18.084, 1.780)$
- instar 3: $\mathbf{S}_{33} = \begin{pmatrix} 8.171 & 8.232 \\ 8.232 & 12.680 \end{pmatrix}$, $\mathbf{B}_3 = \begin{pmatrix} 0.607 & 0.795 \\ 0.795 & -0.607 \end{pmatrix}$,
  $\mathbf{L}_3 = \text{diag}(18.961, 1.890)$.

Since the three orthogonal matrices are so similar, we might think of a CPC model like the one in the text following Example 8.7.1. However there is an added difficulty: the $\mathbf{S}_{ii}$ are not sample covariance matrices obtained from independent groups, but rather submatrices of one large covariance matrix, and therefore we are not quite in the same situation as in Example 8.7.1. □

Examples like the preceding one have initiated research into principal component analysis in situations where the random vector $\mathbf{X}$ is partitioned into subvectors that exhibit certain similarities. A particular model studied by Neuenschwander (1991) is as follows. Suppose that $\mathbf{X}$ is a random vector of dimension pk, partitioned into k subvectors of dimension p each. For instance, take $k = 3$ as in Example 8.7.4. Let

$$\psi = \begin{pmatrix} \psi_{11} & \psi_{12} & \psi_{13} \\ \psi_{21} & \psi_{22} & \psi_{23} \\ \psi_{31} & \psi_{32} & \psi_{33} \end{pmatrix} \tag{15}$$

denote the covariance matrix of $\mathbf{X}$ partitioned into submatrices of dimension $p \times p$. Then the *CPC model for dependent random vectors* assumes that there is an orthogonal $p \times p$ matrix $\mathbf{B}$ such that

$$\Lambda_{ij} = \mathbf{B}'\psi_i j\mathbf{B} \tag{16}$$

is diagonal for all i and j, that is, not only the submatrices along the main diagonal of ψ, but also the off-diagonal blocks are assumed to have the same eigenvectors. A variation of this model is discussed in Exercise 11 of Section 8.5. Another variation, suggested by the similarity of the $\mathbf{S}_{ii}$ in Example 8.7.4, might assume that the submatrices ψ_{ii} along the main diagonal are all identical.

Estimating the parameters in such models poses quite challenging problems that go beyond the mere computation of eigenvectors and eigenvalues. The interested reader is referred to Neuenschwander (1991), Flury and Neuenschwander (1995), and to Klingenberg, Neuenschwander, and Flury (1996). Other models for partitioned random vectors have been studied by Andersson (1975) and Szatrowski (1985).

Finally, let us turn briefly to some recent research on nonlinear generalizations of principal component analysis. A good review can be found in Krzanowski and Marriott (1994, Chapter 8). Here we are mostly going to motivate non-linear generalizations with a practical example.

Example 8.7.5 continuation of Example 8.7.3

The careful reader may have discovered in Figure 8.7.1 that the data of female turtles seem to exhibit a slight curvature. Let U_1 and U_2 denote the sample principal components of female turtles for variables X_1 and X_2, as in Example 8.7.2. Figure 8.7.3 shows a scatterplot of U_1 vs. U_2. This graph is to be viewed like a residual plot in regression. If the one-dimensional principal component approximation is self-consistent, then $E[U_2|U_1] = 0$. However, the data in Figure 8.7.3 contradict the assumption of self-consistency because, at both ends of the range of U_1, we find only positive values of U_2, whereas mostly negative values of U_2 are found in the middle. Actually, the slight non-linearity can be seen as well in the original graph of X_1 vs. X_2, but the principal component plot enhances the visual effect.

This result casts some doubt on the earlier analysis. Since the relationship between log(length) and log(width) is not quite linear, the model of allometric growth is not valid. It is a matter of judgment whether or not one is willing to consider it as approximately valid anyway, because the deviation from linearity is not very pronounced. □

Among the many attempts to generalize principal component analysis to a non-linear method, the most promising one (in the author's opinion) is the *principal curves* of Hastie and Stuetzle (1989). Principal curves are smooth curves in $\mathbb{R}^p$ that minimize the expected squared distance of the random vector $\mathbf{X}$ from the nearest point on the curve. This constitutes a generalization in the spirit of Pearson's least squares derivation of linear principal components. Determining principal curves of a random vector will be more difficult than finding principal components because of the flexibility added by allowing for non-linear curves. Typically, iterative numerical

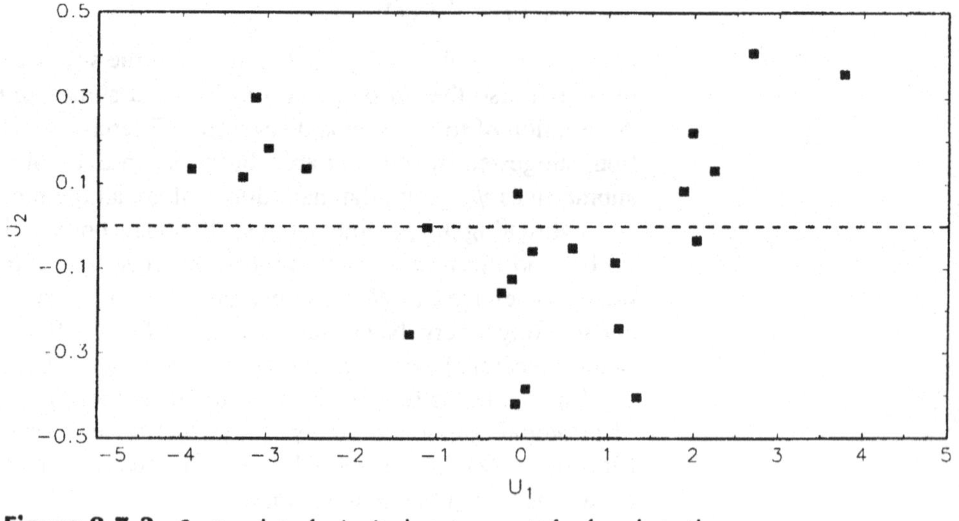

Figure 8-7-3 Scatterplot of principal components for female turtles.

calculations are required to find principal curves even in relatively simple situations. Estimating principal curves from observed data is an even more challenging problem because, without specification of an explicit parametric model, the plug-in principle of estimation cannot be applied.

We illustrate the idea of principal curves with a final example, which, at the same time, brings us back to the fundamental notion of self-consistency introduced in Section 8.2.

Example 8.7.6 Principal curves of spherical distributions (Hastie and Stuetzle 1989, Tarpey and Flury 1996). Suppose that $\mathbf{X} = \begin{pmatrix} X_1 \\ X_2 \end{pmatrix}$ follows a spherical distribution (see Section 3.4), with finite second moments. Then $E[\mathbf{X}] = \mathbf{0}$ and $\text{Cov}[\mathbf{X}] = \sigma^2\mathbf{I}_2$ for some $\sigma^2 \geq 0$, and the projection on any line through the origin gives a one-dimensional principal component approximation to $\mathbf{X}$ with mean squared error σ^2.

Consider a partition of $\mathbb{R}^2$ into $k > 2$ pie-slice-shaped areas $\mathcal{D}_1, \ldots, \mathcal{D}_k$, bordered by half-lines meeting at the origin, with angles $2\pi/k$ between successive half-lines. In each region, choose a representative point $\mathbf{y}_j$ so as to minimize the expected squared error $E[(\mathbf{X}-\mathbf{y}_j)^2|\mathbf{X} \in \mathcal{D}_j]$. Since the mean minimizes the expected squared deviation, $\mathbf{y}_j$ should be chosen as

$$\mathbf{y}_j = E[\mathbf{X}|\mathbf{X} \in \mathcal{D}_j], \quad j = 1, \ldots, k. \tag{17}$$

By sphericity, the points $\mathbf{y}_1, \ldots, \mathbf{y}_k$ are equally spaced on a circle with some radius r_k, centered at the origin. To determine r_k, arrange the k points so that $\mathbf{y}_1$ lies on the positive x_1-axis with coordinates $(r_k, 0)'$. Then the area $\mathcal{D}_1$ is given by the inequalities $x_1 > 0$ and $|x_2| < x_1 \tan(\alpha_k)$, where $\alpha_k = \pi/k$ and $r_k = E[X_1|\mathbf{X} \in \mathcal{D}_1]$. In polar coordinates, $X_1 = R\cos(U)$, and $X_2 = R\sin(U)$. The joint conditional distribution of R and U, given $\mathbf{X} \in \mathcal{D}_1$, has a *pdf* given by

$$f_{RU}(r, u) = \begin{cases} \dfrac{1}{2\alpha_k} f_R(r) & \text{if } |u| < \alpha_k \\ 0 & \text{otherwise} \end{cases} \tag{18}$$

(see Exercise 7), where f_R is the *pdf* of $R = \left(X_1^2 + X_2^2\right)^{1/2}$. Thus, R and U in (18) are independent, and

$$\begin{aligned} r_k &= E[X_1|\mathbf{X} \in \mathcal{D}_1] \\ &= E[R\,\cos(U)|\mathbf{X} \in \mathcal{D}_1] \\ &= E[R|\mathbf{X} \in \mathcal{D}_1]\, E[\cos(U)|\mathbf{X} \in \mathcal{D}_1] \\ &= \left[\int_0^\infty r\, f_R(r)\, dr\right] \times \frac{1}{2\alpha_k}\int_{-\alpha_k}^{\alpha_k} \cos(u)\, du \\ &= \frac{1}{\alpha_k}\sin(\alpha_k)\, E[R]. \end{aligned} \tag{19}$$

Since $\Pr[\mathbf{X} \in \mathcal{D}_j] = 1/k$ for all j, the random vector $\mathbf{Y}_k$, which takes values $\mathbf{y}_j$ with probability $1/k$ each, is a self-consistent approximation to $\mathbf{X}$. As k tends to infinity, the $\mathbf{y}_j$ fill up a self-consistent circle with radius

$$r_\infty = E[R] \lim_{k\to\infty} \frac{\sin(\alpha_k)}{\alpha_k} = E[R]. \tag{20}$$

The circle with radius $E[R] = E\left[\left(X_1^2 + X_2^2\right)^{1/2}\right]$ is a self-consistent curve for the spherical distribution that can be thought of as obtained by minimizing the expected squared distance orthogonal to the nearest point on the circle. For the bivariate standard normal, the radius of the circle is $\sqrt{\pi/2}$ (Exercise 8). □

Exercises for Section 8.7

1. This Exercise is based on the female turtle data given in Table 1.4 and analyzed in Example 8.7.1.
   (a) Compute the mean vector and covariance matrix of the data of female turtles (do not forget to do the log-transformation first).
   (b) Compute the eigenvectors and eigenvalues of the covariance matrix, as well as the standard errors of the eigenvector associated with the largest eigenvalue.
   (c) Perform a test, at significance level $\alpha = 0.05$, of the hypothesis of equality of the covariance matrices; see Exercise 13 in Section 6.6.
   (d) Perform a test at significance level $\alpha = 0.05$ of the hypothesis of equality of both mean vectors and covariance matrices.

2. Let ψ_1 and ψ_2 denote two positive definite symmetric matrices of dimension $p \times p$. Show that the following four conditions are equivalent.
   (i) The eigenvectors of ψ_1 and ψ_2 are identical (or can be chosen so in case of multiple roots).
   (ii) $\psi_1\psi_2 = \psi_2\psi_1$.
   (iii) $\psi_1^{-1}\psi_2$ is symmetric.
   (iv) The eigenvectors of $\psi_1^{-1}\psi_2$ are mutually orthogonal (or can be chosen so if $\psi_1^{-1}\psi_2$ has multiple roots).

3. For $p = 2$, show that the partial CPC model of equation (2) is equivalent to the CPC model of equation (1).

4. Prove equation (9).

5. In equation (14), show that μ_1 and μ_2 are projected onto themselves, i.e., the line corresponding to the projection (14) passes through μ_1 and μ_2.

6. Consider a CPC model for dependent random vectors as in equation (16), where $\mathbf{X}$ is partitioned into $k \geq 2$ subvectors of dimension p each. Show that the covariance

matrix of $\mathbf{X}$ is determined by $\frac{1}{2}p\,[p-1+k(k+1)]$ parameters. What is the number of parameters if no constraints are imposed on ψ? *Hint*: An orthogonal matrix of dimension $m \times m$ is determined by $m(m-1)/2$ numbers, except for multiplication of columns by -1.

7. Prove equation (18).

8. This exercise gives some more details for Example 8.7.6.
   (a) Show that the radius of the self-consistent circle for the bivariate normal distribution with mean $\mathbf{0}$ and covariance matrix $\sigma^2\mathbf{I}_2$ is $\sigma\sqrt{\pi/2}$.
   (b) Show that the sequence $\{r_k\}_{k=2,3,\ldots}$ is increasing.
   (c) Find MSE[$\mathbf{Y}$; $\mathbf{X}$], where $\mathbf{X}$ has a bivariate standard normal distribution and $\mathbf{Y}$ is the self-consistent approximation of $\mathbf{X}$ on a circle of radius $\sqrt{\pi/2}$. Is this approximation better (in terms of mean squared error) than a one-dimensional principal component approximation?

Suggested Further Reading

General: Flury (1988), Jolliffe (1986).
Section 8.1: Jolicoeur and Mosimann (1960), Rao (1964).
Section 8.2: Hastie and Stuetzle (1989), Tarpey and Flury (1996).
Section 8.3: Hotelling (1931), Pearson (1901).
Section 8.5 and 8.7: Hills (1982), Klingenberg (1996).
Section 8.6: Anderson (1963).

9 Normal Mixtures

9.1 Introduction

This chapter gives a brief introduction to theory and applications of finite mixtures, focusing on the most studied and reasonably well-understood case of normal components. An introductory example has already been given in Chapter 1 (Example 1.3) and has been discussed to some extent in Section 2.8, where the relevant terminology has been established. Before turning to the mathematical setup, let us look at yet another example to motivate the theory.

Example 9.1.1 In developed countries childhood diseases like rubella, mumps, and measles, are largely under control through vaccination programs. To assess the relevance and success of such programs, it is important to know the prevalence of immunity against the disease, i.e., to estimate the proportion of persons in a given age group who are immune. Immunity can be obtained by vaccination, or, naturally, by contact with the virus that causes the disease. The data reported in Table 9.1.1 are from such a study administered by the Swiss Federal Office of Public Health, Division of Epidemiology. The measured variable Y is log concentration of antibodies against mumps in blood samples of $N = 385$ unvaccinated children, all 14 years old. A histogram is shown in Figure 9.1.1, showing a somewhat asymmetric distribution with a peak near $y = 2.5$ and a "bump" on the left side around $y = 0$.

Table 9.1.1 Mumps data. Values displayed are log-concentration of antibodies against mumps in 385 children age 14. Data courtesy of Dr. Beat Neuenschwander, Swiss Federal Office of Public Health, Division of Epidemiology, 3097 Liebefeld, Switzerland.

1.2528	2.2824	0.0000	3.2809	−0.5108	2.8094	2.5096	2.7279
1.5892	2.9755	2.0541	−1.2040	2.7344	2.8094	3.2229	3.1864
−0.6931	3.8816	−1.2040	2.5802	1.2809	−0.1054	0.5306	−0.2231
0.0953	2.9232	0.0953	0.2624	3.7062	−2.3026	0.0953	1.3083
−0.5108	3.2581	3.1570	−1.2040	2.3224	0.2624	−1.2040	1.2238
3.2734	−0.1054	−0.5108	0.2624	0.3365	−0.1054	0.2624	1.6094
−0.6931	−0.3567	2.6603	3.1946	−0.2231	−1.2040	−0.3567	−1.6094
−1.2040	2.7081	−1.6094	0.5306	2.4596	2.4765	0.7419	−1.6094
0.4700	2.8959	1.6677	3.1905	3.0155	2.3702	2.6532	−0.9163
−1.6094	0.3365	0.8755	2.7344	0.4055	2.3702	3.5056	1.3350
0.7885	2.7788	1.9601	−1.6094	2.8565	3.4874	3.4340	1.7918
2.6741	2.2192	3.0155	2.2925	1.6487	0.0000	1.4816	−1.2040
2.2192	0.0000	0.1823	0.2624	2.7147	0.9933	1.7228	1.6677
−1.2040	0.5306	1.4816	−0.5108	0.7419	−2.3026	0.0000	0.3365
3.4812	0.7419	0.0953	−0.2231	0.3365	3.8416	0.2624	0.2624
−0.9163	−0.9163	−0.1054	−0.9163	−0.1054	2.0149	3.1905	−2.3026
0.9163	3.6988	2.9601	2.0794	2.8792	4.2485	2.7408	4.4415
2.4159	3.2734	−0.6931	4.2428	2.5177	1.0296	0.6931	4.2513
1.7579	4.3845	1.9601	2.0412	−0.3567	−0.3567	2.9907	3.2426
5.6682	−2.3026	0.4055	2.0669	4.6230	−1.2040	3.0634	3.0445
−2.3026	3.2308	3.7819	1.5686	1.3083	1.9601	0.0953	2.3609
2.5177	2.2300	1.8718	3.1442	0.0000	2.0015	2.2513	2.2824
−1.2040	1.1939	1.1314	0.0000	2.6027	3.1311	4.4104	−1.6094
4.0656	0.6931	2.0149	1.4110	3.3069	0.5306	−2.3026	−2.3026
2.3795	0.0000	4.3758	4.3412	3.0865	2.8565	2.2824	2.6603
1.4816	3.8133	3.4012	−0.9163	2.0281	−1.6094	3.3604	4.5768
−1.6094	4.1043	1.1314	1.0296	3.2426	2.5494	2.2083	0.6419
2.4681	2.6319	2.6174	3.7087	2.2300	0.3365	3.8712	3.9200
3.6839	3.0865	2.6027	3.6481	3.1485	3.2149	3.6610	3.3638
4.6191	3.0106	2.5337	3.0204	3.0681	3.8373	2.7014	1.7579
4.0483	4.2499	1.7918	3.1905	3.3105	3.0397	3.3499	2.5726
2.3321	2.1282	1.1632	4.7041	0.7885	2.6174	3.1612	4.4284
1.5686	3.2387	3.2465	4.1336	2.6174	3.1046	1.6292	2.6878
3.8628	2.4765	2.7344	3.7542	2.5177	4.1174	2.5096	3.0587
3.4210	3.8795	3.6082	4.0792	3.7448	0.8329	1.7228	3.5553
1.6487	1.5476	2.0015	3.0445	2.6247	2.9601	2.0919	−0.6931
1.1314	2.9857	2.3321	2.0412	2.2824	−2.3026	2.2513	2.9857
4.3605	2.8565	3.8330	0.1823	2.8391	3.5145	4.5454	2.0919
4.5315	2.5953	3.7329	2.9014	3.6082	2.9232	2.6603	2.3795
3.4468	4.2711	2.6174	2.3026	3.2619	3.7519	2.2721	0.8329
3.9338	3.8979	3.3214	2.9178	2.5572	2.8094	3.2027	3.2958
3.5525	3.5639	2.7014	2.3514	4.0893	3.1946	3.1311	4.1775
−1.2040	1.1632	1.7750	4.6161	3.6136	1.8718	2.5649	3.1091
4.3944	4.5921	2.7081	4.6308	1.7750	4.1479	2.7081	1.8083
3.1268	2.9806	4.4716	5.4528	4.3108	3.0773	4.6141	4.5860

Table 9.1.1 *Continued*

1.7579	3.0910	2.7081	2.7408	4.0843	4.2836	2.7344	2.7147
2.4248	2.9339	4.8267	3.0350	3.0397	3.6481	2.8904	1.9879
2.7279	1.9315	2.9755	3.1268	1.5686	2.8848	3.2465	2.7600
2.6946							

Experience suggests that, in a homogeneous population, such a histogram should be closer to the shape of a normal distribution. What could cause the bump? A reasonable explanation would be that the distribution shown is actually the mixture of two groups: a group of immune children and a group of susceptible children mixed in an unknown proportion. If we are willing to adopt this explanation, then fitting a mixture distribution with two normal components would make sense, and the mixing proportion of the component with the larger mean (supposedly the immune group) would serve as an estimate of the prevalence of immunity in unvaccinated children of this age. Unlike the wing length data of Example 1.3, separation of the two groups is not as clear, and there is more overlap between the two components of the distribution. Another important difference is that, in the wing length example,

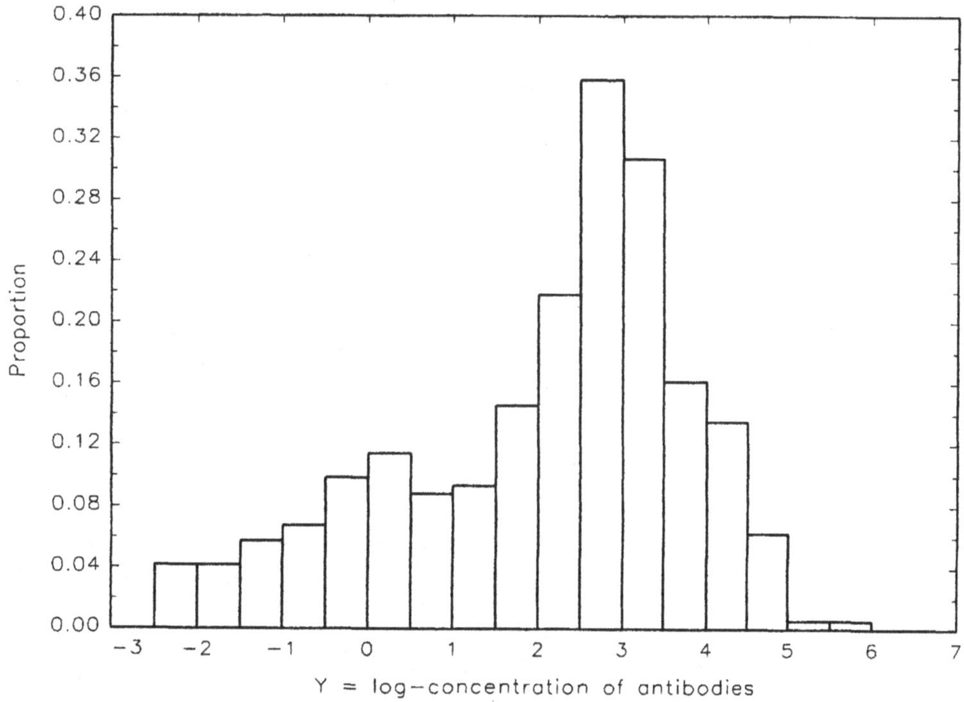

Figure 9-1-1 Histogram of log-concentration of antibodies against mumps in 385 unvaccinated children. *Note:* the vertical axis is scaled so as to make the area under the histogram unity.

we had a physically existing quantity (gender) which provides a causal interpretation of the shape of the distribution. In the current example, the existence of two distinct groups labelled "immune" and "susceptible" is suggested by experience but is not directly observable. □

It is important to distinguish finite mixture analysis from a rather popular group of classification techniques called cluster analysis. This theme has already been mentioned briefly in Example 1.3. A typical cluster-analytical technique would be to "slice" a data set into different, supposedly more homogeneous, subsets. In Example 9.1.1, suppose we partition the data into two subsets according to a cutoff value of $y = 1.0$, assuming that a classification into "susceptible" (if $y < 1.0$) and "immune" (if $y > 1.0$) would be roughly correct. Such a procedure would ignore the uncertainty inherent in any assignment of a person to a group and it would also lead to distorted estimates of means and variances. This is explained in more detail in Exercise 1.

In Chapter 2.8 we introduced mixture distributions as the marginal distribution of Y in a pair of random variables (X, Y), where X is a group code, taking values $1, \ldots, k$, and where the conditional density of Y, given $X = i$, is given for $i = 1, \ldots, k$. We are now going to change the setup by introducing k indicator variables. At first, this may seem a bit artificial and overly complicated, but it will actually be much better for the purpose of estimation. Let $\mathbf{X} = \begin{pmatrix} X_1 \\ \vdots \\ X_k \end{pmatrix}$ denote a multinomial random vector, such that exactly one of the X_j is equal to 1 and all others are zero. Let

$$\pi_j = \Pr[X_j = 1], \qquad j = 1, \ldots, k, \tag{1}$$

where $\sum_{j=1}^{k} \pi_j = 1$ and all $\pi_j > 0$. Then the *pdf* of $\mathbf{X}$ is given by

$$\begin{aligned} f_{\mathbf{X}}(x_1, \ldots, x_k) &= \begin{cases} \pi_j & \text{if } x_j = 1, \text{ all other } x_h = 0, \\ 0 & \text{otherwise} \end{cases} \\ &= \prod_{j=1}^{k} \pi_j^{x_j} \qquad \text{all } x_j = 0 \text{ or } 1, \quad \sum_{j=1}^{k} x_j = 1. \end{aligned} \tag{2}$$

Note that, in contrast to Section 2.8, now we are using the symbol "π_j" instead of "p_j" for the probability of group membership.

Let $\mathbf{Y} = \begin{pmatrix} Y_1 \\ \vdots \\ Y_p \end{pmatrix}$ denote a p-variate random vector whose distribution is given conditionally on $\mathbf{X}$, that is, conditionally on $X_j = 1$, $\mathbf{Y}$ has *pdf* $f_j(\mathbf{y})$. For instance, in the childhood disease example we have $k = 2$ components, and we define $X_1 = 1$ if the child is susceptible and $X_2 = 1$ if the child is immune. The *pdf* $f_1(y)$ is the density function of log concentration of antibodies for immune children, and $f_2(y)$ is the density for susceptible children.

Now we will assume that $\mathbf{Y}$ is continuous and all $f_j(\mathbf{y})$ are actual density functions. The joint *pdf* of $\mathbf{X}$ and $\mathbf{Y}$ can then be written as

$$f_{\mathbf{XY}}(\mathbf{x}, \mathbf{y}) = \begin{cases} \pi_j \, f_j(\mathbf{y}) & \text{if } x_j = 1, \text{ all other } x_h = 0, \mathbf{y} \in \mathbb{R}^p, \\ 0 & \text{otherwise.} \end{cases} \tag{3}$$

This is to be understood like equation (2) in Section 2.8: For any set $\mathcal{A} \subset \mathbb{R}^p$,

$$\Pr[X_j = 1, \; \mathbf{Y} \in \mathcal{A}] = \pi_j \int_{\mathcal{A}} f_j(\mathbf{y}) \, d\mathbf{y}. \tag{4}$$

It will be useful to write equation (3) in a different form, namely,

$$f_{\mathbf{XY}}(\mathbf{x}, \mathbf{y}) = \prod_{j=1}^{k} [\pi_j \, f_j(\mathbf{y})]^{x_j}, \qquad \text{all } x_j = 0 \text{ or } 1, \quad \sum_{j=1}^{k} x_j = 1, \quad \mathbf{y} \in \mathbb{R}^p. \tag{5}$$

Since exactly one of the x_j is equal to 1, this is the same as equation (3).

From the joint distribution of $\mathbf{X}$ and $\mathbf{Y}$, we can derive the marginal and conditional distributions of $\mathbf{X}$ and $\mathbf{Y}$. The marginal *pdf* of $\mathbf{X}$ is given by equation (1), and the conditional *pdf* of $\mathbf{Y}$, given $\mathbf{X} = j$, is $f_j(\mathbf{y})$; see Exercise 2. For the marginal density of $\mathbf{Y}$, we obtain

$$\begin{aligned} f_{\mathbf{Y}}(\mathbf{y}) &= \sum_{\text{all } \mathbf{x}} f_{\mathbf{XY}}(\mathbf{x}, \mathbf{y}) \\ &= \sum_{\text{all } \mathbf{x}} \prod_{j=1}^{k} [\pi_j \, f_j(\mathbf{y})]^{x_j} \\ &= \sum_{j=1}^{k} \pi_j \, f_j(\mathbf{y}). \end{aligned} \tag{6}$$

The last equality in (6) follows because the sum contains exactly k nonzero terms, the jth of which contributes $\pi_j f_j(\mathbf{y})$. The *pdf* (6) is called a *finite mixture density* and is the central object of interest in this chapter.

The conditional *pdf* of $\mathbf{X}$, given $\mathbf{Y}$, is

$$\begin{aligned} f_{\mathbf{X}|\mathbf{Y}}(\mathbf{x}|\mathbf{y}) &= \Pr[\mathbf{X} = \mathbf{x} | \mathbf{Y} = \mathbf{y}] \\ &= \frac{f_{\mathbf{XY}}(\mathbf{x}, \mathbf{y})}{f_{\mathbf{Y}}(\mathbf{y})} \\ &= \frac{\prod_{j=1}^{k} [\pi_j \, f_j(\mathbf{y})]^{x_j}}{\sum_{h=1}^{k} \pi_h \, f_h(\mathbf{y})}, \qquad x_j = 0 \text{ or } 1, \quad \sum_{j=1}^{k} x_j = 1, \end{aligned} \tag{7}$$

and is defined for all $\mathbf{y} \in \mathbb{R}^p$ such that $f_{\mathbf{Y}}(\mathbf{y}) > 0$. Equation (7) means that

$$\pi_{jy} := \Pr[X_j = 1 | \mathbf{Y} = \mathbf{y}] = \frac{\pi_j f_j(\mathbf{y})}{\sum_{h=1}^{k} \pi_h f_h(\mathbf{y})} \qquad j = 1, \ldots, k. \tag{8}$$

We will often refer to the π_{jy} in equation (8) as *posterior probabilities*. See the discussion in Section 2.8. In contrast, the mixing weights or proportions π_j are also called *prior probabilities*.

In discriminant analysis, we start with training samples, i.e., a sample from $(\mathbf{X}, \mathbf{Y})$, and the parameters of each conditional density $f_j(\mathbf{y})$ are estimated using the observations that actually belong to the jth component. Thus, discriminant analysis fits into the framework of the theory just developed. In finite mixture analysis, the $\mathbf{X}$-vector is missing, and therefore we are sampling from the mixture density $f_{\mathbf{Y}}(\mathbf{y})$. Yet, we will attempt to estimate the mixing proportions π_j and the parameters of each component, at least in the case of a mixture of normal distributions. This problem is bound to be more difficult than if group membership were known for each observation. Section 9.2 discusses maximum likelihood estimation, and Section 9.3 puts it into the framework of the *EM*-algorithm introduced in Section 4.4.

In Theorem 2.8.1, we saw how the means and variances of the components of a univariate mixture are related to the mean and variance of the mixture distribution. Similar results hold for the multivariate case; they have already been given as Theorem 7.3.1 in the context of canonical discriminant analysis, and therefore are not repeated here.

Exercises for Section 9.1

1. This exercise explains why means and variances of clusters obtained by splitting data into groups are, in general, not good estimates of means and variances of the components in a mixture.

   (a) Let $x_{(1)} \le x_{(2)} \le \cdots \le x_{(N)}$ denote an ordered sample of size N, and $y_{(1)} \le y_{(2)} \le \cdots \le y_{(M)}$ an ordered sample of size M. Denote their sample means and variances by $\bar{x}$, $\bar{y}$, s_x^2, and s_y^2, respectively. Assume that $\bar{x} < \bar{y}$ and $y_{(1)} < x_{(N)}$, i.e., the two samples overlap. Let c be any number between $y_{(1)}$ and $x_{(N)}$, and denote the pooled data by $z_1, \ldots, z_{N+M}$. Define an assignment to two clusters C_1 and C_2 as follows: All z_i below the cutoff c belong to cluster C_1, and all z_i at or above the cutoff c belong to cluster C_2. Show that the average of the data in cluster C_1 is less that $\bar{x}$, the average of the data in cluster C_2 is larger that $\bar{y}$, and the variances of the clusters are smaller than s_x^2 and s_y^2, respectively.

   (b) Let $f_Y(y)$ denote the mixture density of two normals with means μ_1 and μ_2 ($\mu_1 < \mu_2$), with common variance σ^2, and with mixing weights $\pi_1 = \pi_2 = \frac{1}{2}$. Define a "clustering" by a cutoff value of $c = (\mu_1 + \mu_2)/2$. Show that $E[Y|Y < c] < \mu_1$, $E[Y|Y > c] > \mu_2$, and $\text{var}[Y|Y < c] = \text{var}[Y|Y > c] < \sigma^2$.

2. Using the joint *pdf* (3), find the marginal *pdf* of $\mathbf{X}$ and the conditional *pdf* of $\mathbf{Y}$, given $X_j = 1$, for $j = 1, \ldots, k$.

9.2 Maximum Likelihood Estimation

In this section we will find likelihood equations for normal mixtures. Before attempting to estimate parameters, it is important to make sure that the parameters are actually identifiable. Identifiability means that any two different choices of parameters in the parameter space lead to two different distributions. With finite mixtures, we need to state precisely what we mean by "different choices of parameters" and therefore give a formal definition.

Definition 9.2.1 Let $\mathcal{H}$ denote a class of finite mixture densities. Let

$$f(\mathbf{y}) = \sum_{j=1}^{k} \pi_j f_j(\mathbf{y}) \text{ and } \qquad f^*(\mathbf{y}) = \sum_{j=1}^{k^*} \pi_j^* f_j^*(\mathbf{y}) \tag{1}$$

be any two members of $\mathcal{H}$, where both k and k^* are positive integers. Suppose that $f(\mathbf{y}) = f^*(\mathbf{y})$ exactly if $k = k^*$ and the terms in the sum can be ordered such that $\pi_j = \pi_j^*$ and $f_j = f_j^*$ for $j = 1, \ldots, k$. Then $\mathcal{H}$ is called identifiable. □

The foregoing definition assumes that no two component densities in a finite mixture are allowed to be identical, otherwise identifiability would never hold.

Example 9.2.1 Mixtures of beta densities. Let

$$f(y; \alpha, \beta) = \begin{cases} c(\alpha, \beta)\, y^{\alpha-1} (1-y)^{\beta-1} & \text{if } 0 < y < 1, \\ 0 & \text{otherwise} \end{cases} \tag{2}$$

denote the *pdf* of a univariate random variable, where $\alpha > 0$ and $\beta > 0$ are two parameters and $c(\alpha, \beta)$ is a constant. The class of mixtures of densities of the form (2) is not identifiable because

$$\frac{1}{2} f(y; 2, 1) + \frac{1}{2} f(y; 1, 2) = f(y; 1, 1),$$

see Exercise 1. □

Another example of a nonidentifiable class has already been given in Example 2.8.5 and in Exercise 4 of Section 2.8. Also see Exercises 2 and 3.

On the other hand, the class of mixtures with normal components is identifiable. We give this result without proof; the interested student is referred to any of the books listed in the Suggested Further Readings Section at the end of this chapter. If we know that a given density is a normal mixture, then the number of components, the mixing weights, and all parameters of the component densities are uniquely determined, up to permutation of the order of the components. Without identifiability, parameter estimation would not be justified.

We start our discussion of maximum likelihood estimation with a simplified problem: Assume that the component densities $f_1(\mathbf{y}), \ldots, f_k(\mathbf{y})$ are known (i.e., they do not depend on any unknown parameters) and the only parameters to be estimated are the mixing weights $\pi_1, \ldots, \pi_k$. Note that $\mathbf{Y}$ may be univariate or multivariate. The mixture density is given by

$$f_{\mathbf{Y}}(\mathbf{y}; \pi_1, \ldots, \pi_k) = \sum_{j=1}^{k} \pi_j f_j(\mathbf{y}). \tag{3}$$

The likelihood function based on observations $\mathbf{y}_1, \ldots, \mathbf{y}_N$ from (3) is

$$\begin{aligned} L(\pi_1, \ldots, \pi_k) &= \prod_{i=1}^{N} f_{\mathbf{Y}}(\mathbf{y}; \pi_1, \ldots, \pi_k) \\ &= \prod_{i=1}^{N} \left[\sum_{j=1}^{k} \pi_j f_j(\mathbf{y}_i) \right], \end{aligned} \tag{4}$$

and the log likelihood function is

$$\begin{aligned} \ell(\pi_1, \ldots, \pi_k) &= \log L(\pi_1, \ldots, \pi_k) \\ &= \sum_{i=1}^{N} \log \left[\sum_{j=1}^{k} \pi_j f_j(\mathbf{y}_i) \right]. \end{aligned} \tag{5}$$

We have to maximize (5) subject to the constraint $\sum_{j=1}^{k} \pi_j = 1$. Introducing a Lagrange multiplier λ for the constraint, we have to find stationary points of the function

$$\ell^*(\pi_1, \ldots, \pi_k, \lambda) = \ell(\pi_1, \ldots, \pi_k) - \lambda \left[\sum_{j=1}^{k} \pi_j - 1 \right]. \tag{6}$$

The partial derivative of ℓ^* with respect to π_h is

$$\frac{\partial \ell^*}{\partial \pi_h} = \sum_{i=1}^{N} \frac{f_h(\mathbf{y}_i)}{\sum_{j=1}^{k} \pi_j f_j(\mathbf{y}_i)} - \lambda, \qquad h = 1, \ldots, k. \tag{7}$$

Setting (7) equal to zero and multiplying it by π_h (assuming $\pi_h \neq 0$) yields

$$\lambda \pi_h = \sum_{i=1}^{N} \frac{\pi_h f_h(\mathbf{y}_i)}{\sum_{j=1}^{k} \pi_j f_j(\mathbf{y}_i)} \qquad h = 1, \ldots, k. \tag{8}$$

At this point, it is convenient to introduce the symbols

$$\pi_{hi} = \frac{\pi_h f_h(\mathbf{y}_i)}{\sum_{j=1}^{k} \pi_j f_j(\mathbf{y}_i)} \qquad h = 1, \ldots, k; \quad i = 1, \ldots, N, \tag{9}$$

corresponding to the posterior probability of the ith observation coming from component h. Then equation (8) reads

$$\lambda \pi_h = \sum_{i=1}^{N} \pi_{hi} \qquad h = 1, \ldots, k. \tag{10}$$

Adding equations (10) for $h = 1, \ldots, k$ and using the constraint $\sum_{j=1}^{k} \pi_j = 1$, we obtain $\lambda = N$. Inserting this into equation (10) yields

$$\pi_h = \frac{1}{N} \sum_{i=1}^{N} \pi_{hi} \qquad h = 1, \ldots, k. \tag{11}$$

This is a formula we will encounter many times, which states that the prior probability for the hth component is to be estimated as the average of the posterior probabilities of all N observations for this component.

Dividing equation (11) by π_h (assuming $\pi_h \neq 0$) and writing π_{hi} in the form (9), we obtain k equations

$$\sum_{i=1}^{N} \frac{f_h(\mathbf{y}_i)}{\sum_{j=1}^{k} \pi_j f_j(\mathbf{y}_i)} = N \qquad h = 1, \ldots, k. \tag{12}$$

Together with the constraint $\sum_{j=1}^{k} \pi_j = 1$, these equations form a complicated system with no guarantee that there exists a solution such that all π_j are in the interval [0, 1]. However, some insight can be gained by studying the special case of two components. Suppose that $\mathbf{Y}$ follows a mixture density with $k = 2$ known components and mixing weights $\pi_1 = \pi$ and $\pi_2 = 1 - \pi$. For simplicity, write $f_{1i} = f_1(\mathbf{y}_i)$, $f_{2i} = f_2(\mathbf{y}_i)$, and $m_i = \pi f_{1i} + (1 - \pi) f_{2i}$. Then the log likelihood function is

$$\ell(\pi) = \sum_{i=1}^{N} \log m_i, \tag{13}$$

and the first and second derivatives are

$$\frac{\partial \ell}{\partial \pi} = \sum_{i=1}^{N} \frac{f_{1i} - f_{2i}}{m_i} \tag{14}$$

and

$$\frac{\partial^2 \ell}{\partial \pi^2} = -\sum_{i=1}^{N} \left[\frac{f_{1i} - f_{2i}}{m_i} \right]^2. \tag{15}$$

By equation (15), the second derivative is negative, unless $f_{1i} = f_{2i}$ for $i = 1, \ldots, N$. Hence the log likelihood is concave and takes at most one maximum. However, this maximum may occur inside or outside the interval [0, 1], as the following example shows.

Example 9.2.2 Suppose that we study mixtures of the two densities

$$f_1(y) = \begin{cases} 2y & \text{if } 0 \le y \le 1, \\ 0 & \text{otherwise} \end{cases}$$

and

$$f_2(y) = \begin{cases} 2(1-y) & \text{if } 0 \le y \le 1, \\ 0 & \text{otherwise.} \end{cases}$$

For $N = 5$ data points $y_1 = 0.1$, $y_2 = 0.2$, $y_3 = 0.3$, $y_4 = 0.4$, and $y_5 = 0.9$, a graph of the log likelihood function (13) is given as a solid curve in Figure 9.2.1. A maximum occurs near $\pi = 0.21$. If the fifth data point is changed to $y_5 = 0.7$, then the log likelihood changes to the broken curve, which takes its maximum in the interval [0, 1] at $\pi = 0$. Actually, the continuation of the curve to the left of 0 shows a maximum for π near −0.28, which is outside the parameter space. Hence, in this case, the maximum likelihood estimate of π is zero. □

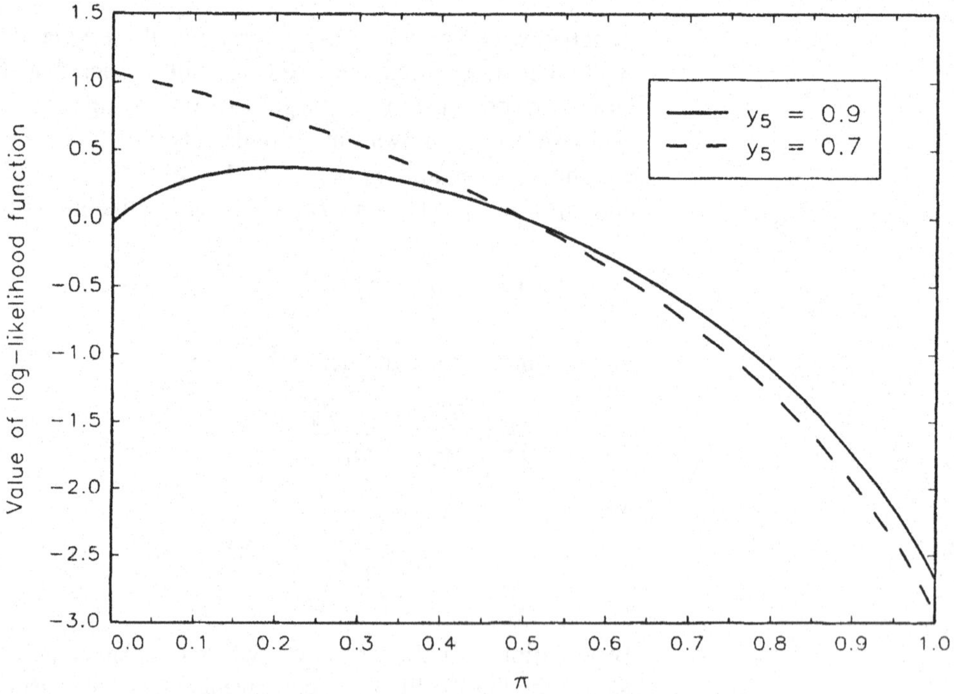

Figure 9-2-1 Graphs of log-likelihood functions in Example 9.2.2. The solid curve refers to the original data set with $y_5 = 0.9$; the broken line to the modified data set with $y_5 = 0.7$.

To find the solution of the equation $\partial \ell / \partial \pi = 0$, i.e., of

$$\sum_{i=1}^{N} (f_{1i} - f_{2i})/m_i = 0, \tag{16}$$

one would typically use some numerical procedure, such as the Newton-Raphson algorithm. An alternative follows from a modification of (16): Using $f_{1i} - f_{2i} = (f_{1i} - m_i)/(1 - \pi)$, equation (16) can be rewritten as

$$\begin{aligned} \pi &= \frac{1}{N} \sum_{i=1}^{N} \frac{\pi f_{1i}}{m_i} \\ &= \frac{1}{N} \sum_{i=1}^{N} \pi_{1i}, \end{aligned} \tag{17}$$

see Exercise 4. Here, $\pi_{1i} = \pi f_{1i}/m_i$ is the posterior probability that observation y_i is from component 1, as in equation (9). (Actually equation (17) is a special case of equation (11) for the case of $k = 2$ components.) This suggests iterating between the two equations

$$\begin{aligned} &(a) \qquad \pi_{1i} = \pi f_{1i}/m_i \qquad i = 1, \ldots, N, and \\ &(b) \qquad \pi = \frac{1}{N} \sum_{i=1}^{N} \pi_{1i}, \end{aligned} \tag{18}$$

starting with some initial value of π, say $\pi^{(0)}$. To be more precise, iteration between (a) and (b) means computing a sequence $\pi^{(1)}, \pi^{(2)}, \cdots$, according to the rule

$$\pi^{(s+1)} = \frac{1}{N} \sum_{i=1}^{N} \pi_{1i}^{(s)}, \tag{19}$$

where

$$\pi_{1i}^{(s)} = \pi^{(s)} f_{1i} / \left[\pi^{(s)} f_{1i} + (1 - \pi^{(s)}) f_{2i}\right], \qquad i = 1, \ldots, N. \tag{20}$$

Remarkably, as is seen from (18), such a procedure can never lead to a value of π outside [0, 1], provided it is started with an initial value inside the unit interval. We will study this procedure in more generality later.

Example 9.2.3 continuation of Example 9.2.2

Table 9.2.1 shows numerical details for ten iterations of the procedure outlined in (19) and (20), starting at an initial value of $\pi = 0.9$. The data used is the first set in Example 9.2.2, with $y_5 = 0.9$. Within ten iterations the successive values of π approach the point of the maximum rather closely. See also Exercise 5. □

Table 9.2.1 Numerical details for ten iterations of equations (18) in Example 9.2.3

Iteration	π	π_{11}	π_{12}	π_{13}	π_{14}	π_{15}
0	0.9000	0.5000	0.6923	0.7941	0.8571	0.9878
1	0.7663	0.2670	0.4504	0.5842	0.6861	0.9672
2	0.5910	0.1383	0.2654	0.3824	0.4907	0.9286
3	0.4411	0.0806	0.1648	0.2527	0.3447	0.8766
4	0.3439	0.0550	0.1159	0.1834	0.2589	0.8251
5	0.2877	0.0429	0.0917	0.1475	0.2121	0.7842
6	0.2557	0.0368	0.0791	0.1283	0.1864	0.7556
7	0.2372	0.0334	0.0721	0.1176	0.1717	0.7368
8	0.2263	0.0315	0.0682	0.1114	0.1632	0.7247
9	0.2198	0.0304	0.0658	0.1077	0.1581	0.7172
10	0.2158	0.0297	0.0644	0.1055	0.1550	0.7124

The computations illustrated in Example 9.2.3 are a special case of a method that can be used more generally for $k \geq 2$ components. Starting with initial values $\pi_1, \ldots, \pi_k$ of the mixing weights, one iterates between the equations

$$\pi_{ji} = \pi_j\, f_j(\mathbf{y}_i)/f(\mathbf{y}_i) \qquad j = 1, \ldots, k; \quad i = 1, \ldots, N, \tag{21}$$

and

$$\pi_j = \frac{1}{N}\sum_{i=1}^{N} \pi_{ji} \qquad j = 1, \ldots, k, \tag{22}$$

until convergence is reached, That is, one iteratively recomputes posterior probabilities as functions of the current prior probabilities, and prior probabilities as functions of the current posterior probabilities, until no further changes are obtained in some iteration. This procedure is usually fairly easy to program, unless evaluation of the component densities f_j is difficult.

Mixtures with known components, where only the mixing proportions need to be estimated, is a rather unusual problem in practice, although it may appear occasionally; see Exercise 7.

Now we turn to normal mixtures, treating, for the moment, only the univariate case. Denote the mixture density by

$$f(y) = \sum_{j=1}^{k} \pi_j\, f_j(y), \tag{23}$$

where

$$f_j(y) = \frac{1}{\sqrt{2\pi}\,\sigma_j} \exp\left[-\frac{1}{2}\left(\frac{y-\mu_j}{\sigma_j}\right)^2\right], \qquad y \in \mathbb{R}. \tag{24}$$

The unknown parameters are $\pi_1, \ldots, \pi_k,\ \mu_1, \ldots, \mu_k$, and $\sigma_1, \ldots, \sigma_k$, subject to constraints $\sum_{j=1}^{k} \pi_j = 1$, all $\pi_j \in [0, 1]$, and all $\sigma_j > 0$. The log-likelihood function based on N observations $y_1, \ldots, y_N$ from the mixture distribution (23) is given by

$$\begin{aligned} \ell &= \ell(\pi_1, \ldots, \pi_k, \mu_1, \ldots, \mu_k, \sigma_1, \ldots, \sigma_k) \\ &= \sum_{i=1}^{N} \log f(y_i). \end{aligned} \tag{25}$$

To keep the formalism under control, we will neglect writing the various functions explicitly as functions of the parameters. For instance, if we write $f_j(y_i)$, then it will always be implicitly assumed that f_j is a function of μ_j and σ_j, evaluated at y_i.

The normal log-likelihood (25) has a very disturbing feature to which we will return later. For now, we proceed as usual, taking derivatives with respect to the unknown parameters and setting them equal to zero. The derivatives with respect to the mixing proportions π_h are formally the same as in equation (7), and hence we get equations (9) and (11) again. Next, for $1 \le h \le k$,

$$\begin{aligned} \frac{\partial \ell}{\partial \mu_h} &= \sum_{i=1}^{N} \frac{\partial \log f(y_i)}{\partial \mu_h} \\ &= \sum_{i=1}^{N} \frac{1}{f(y_i)} \frac{\partial f(y_i)}{\partial \mu_h} \\ &= \sum_{i=1}^{N} \frac{1}{f(y_i)} \pi_h \frac{\partial f_h(y_i)}{\partial \mu_h}, \end{aligned} \tag{26}$$

because μ_h appears only in the hth component density. Using the fact that for a nonnegative function $g(u)$,

$$\frac{\partial g(u)}{\partial u} = g(u) \frac{\partial \log g(u)}{\partial u}, \tag{27}$$

we obtain

$$\begin{aligned} \frac{\partial \ell}{\partial \mu_h} &= \sum_{i=1}^{N} \frac{\pi_h f_h(y_i)}{f(y_i)} \frac{\partial \log f_h(y_i)}{\partial \mu_h} \\ &= \sum_{i=1}^{N} \pi_{hi} \frac{(y_i - \mu_h)}{\sigma_h^2}, \qquad h = 1, \ldots, k. \end{aligned} \tag{28}$$

Setting (28) equal to zero and multiplying by σ_h^2 yields

$$\sum_{i=1}^{N} \pi_{hi} (y_i - \mu_h) = 0, \qquad h = 1, \ldots, k. \tag{29}$$

But $\sum_{i=1}^{N} \pi_{hi} = N\pi_h$ from equation (11), and hence, (29) is the same as

$$\mu_h = \frac{1}{N\pi_h} \sum_{i=1}^{N} \pi_{hi} y_i \qquad h = 1, \ldots, k. \tag{30}$$

Similarly, one obtains (see Exercise 9)

$$\sigma_h^2 = \frac{1}{N\pi_h} \sum_{i=1}^{N} \pi_{hi} (y_i - \mu_h)^2 \qquad h = 1, \ldots, k. \tag{31}$$

Thus, the relevant equations are (9), (11), (30), and (31). If it is assumed that the variances of all k components are equal, say σ^2, then equations (31) are replaced by

$$\sigma^2 = \frac{1}{N} \sum_{h=1}^{k} \sum_{i=1}^{N} \pi_{hi} (y_i - \mu_h)^2, \tag{32}$$

which is a weighted average of the σ_h^2 in (31); see Exercise 10.

We will give a detailed discussion of this equation system in Section 9.3 and finish the current section with a disturbing aspect of the normal mixture likelihood: There are many solutions to the equation system that make no sense. Take any single sample value (say y_1), and assume that no other y_m is equal to y_1. Set $\pi_{11} = 1$ and $\pi_{1i} = 0$ for $i = 2, \ldots, N$. Then equation (11) gives $\pi_1 = \frac{1}{N}$, equation (30) gives $\mu_1 = y_1$, and equation (31) yields $\sigma_1^2 = 0$, that is, the first data point y_1 "determines" its own normal distribution, with mean $\mu_1 = y_1$ and variance 0, whereas the remaining $N-1$ points form a mixture with $k-1$ components. The same thing happens for each data point y_i. If we set one of the means equal to y_i, and let the corresponding variance go to zero, then the equation system "thinks" that there is a component sitting right at y_i with very small variance. To make things worse, all such solutions correspond to infinite peaks of the log-likelihood. To see this, write the log-likelihood function (25) in explicit form as

$$\ell = \sum_{i=1}^{N} \log \left\{ \frac{1}{\sqrt{2\pi}} \sum_{j=1}^{k} \frac{\pi_j}{\sigma_j} \exp\left[-\frac{1}{2} \left(\frac{y_i - \mu_j}{\sigma_j} \right)^2 \right] \right\}. \tag{33}$$

Now set $\mu_1 = y_1$. Then $\exp[-\frac{1}{2}(y_1 - \mu_1)^2/\sigma_1^2] = 1$. Let σ_1 tend to zero. Then for $\pi_1 > 0$, the term (π_1/σ_1) tends to infinity, and the value of the log-likelihood tends to infinity as well. It is as if the likelihood function "thought" it found an extremely nice-fitting component. Ultimately, this is the same problem as trying to estimate the mean and variance of a normal distribution using a single observation.

Example 9.2.4 A univariate normal mixture with $k = 2$ components and unequal variances has five parameters to estimate; it is therefore difficult to visualize the likelihood function. However, we can construct a simpler problem that illustrates the phenomenon of infinite peaks. Suppose that we are sampling from a normal mixture with two components, where the first component is known to be $\mathcal{N}(0, 1)$, the mixing weights are

$\pi_1 = \pi_2 = \frac{1}{2}$, and the second component is $\mathcal{N}(\mu, \sigma^2)$, both μ and σ^2 unknown. The log-likelihood function based on N observations $y_1, \ldots, y_N$ from the mixture is then given by

$$\ell(\mu, \sigma) = C + \sum_{i=1}^{N} \log \left\{ \exp\left[-\tfrac{1}{2}y_i^2\right] + \frac{1}{\sigma} \exp\left[-\tfrac{1}{2}\left(\frac{y_i - \mu}{\sigma}\right)^2\right] \right\},$$

where C is a constant that does not depend on the unknown parameters. Figure 9.2.2 shows a contour plot of this log-likelihood function for $N = 6$ data values equal to $-1, 0, 1, 3, 4,$ and 5. The log-likelihood function takes a "reasonable" local maximum near $(\mu, \sigma) = (4, 1)$, but, for μ equal to any of the six data values and σ approaching zero, the singularities appear. □

The ugly consequence of this is that the likelihood function of a normal mixture is literally "cluttered" with singularities. To find reasonable estimates, we have to somehow keep the variance estimates away from zero, hoping to find a local maximum that yields a consistent estimate of the mixture parameters. Strictly speaking, one should not even call the solution (however reasonable it may be) a maximum likelihood estimate. Despite the singularities, there is little doubt that it makes good sense to use the equation system found in this section to estimate the parameters of a

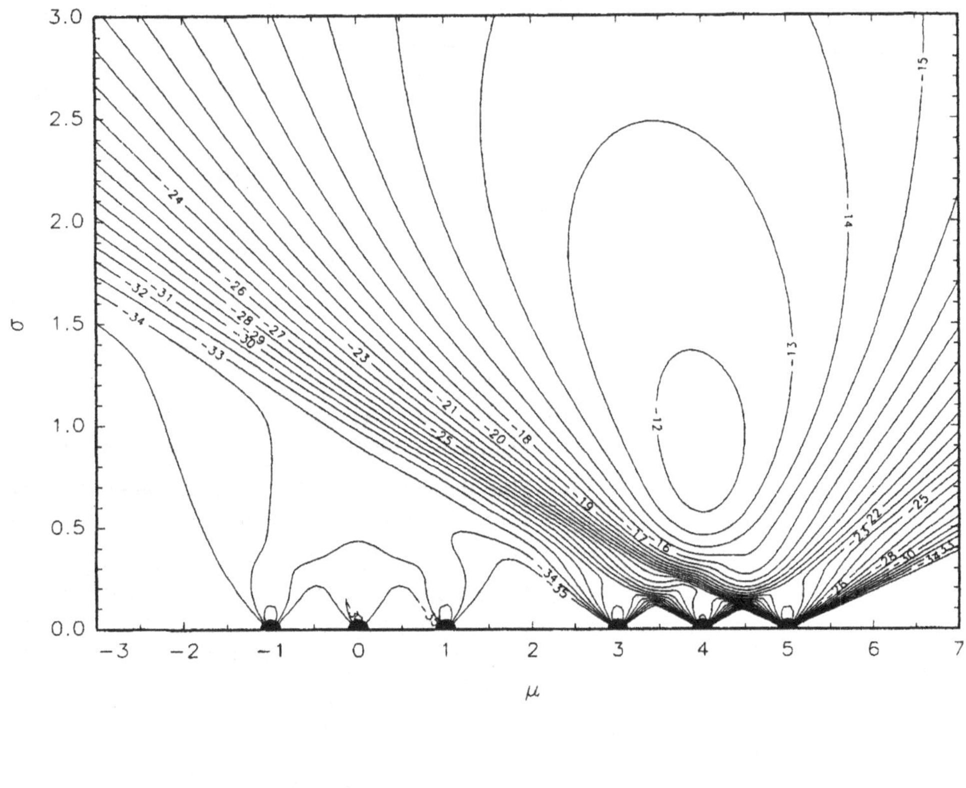

Figure 9-2-2 Contour plot of the log-likelihood function in Example 9.2.4. The log-likelihood function has six infinite peaks for $\mu = -1, 0, 1, 3, 4, 5$, and $\sigma = 0$; these peaks are marked by solid dots. A reasonable local maximum is taken near $(\mu, \sigma) = (4, 1)$.

normal mixture. The reader who is interested in more detail is once again referred to the books listed in the Suggested Further Readings Section at the end of this chapter.

One way to keep the variance estimates away from zero is to assume equality of all variances. If there are at least $k+1$ distinct data points, then the estimate of the common σ^2 cannot become arbitrarily small. Yet, the likelihood function may still have several local maxima. We will return to this point in Section 9.3.

In this section we have considered only the univariate normal distribution. The multivariate case differs little from the univariate case, except that the derivation of the multivariate analog of equation (31) is much messier. We skip it because the *EM*-algorithm of Section 9.3 will give the solution almost free. The problem with infinite peaks is just as bad as in the univariate case: Setting a mean vector equal to an observed data point $\mathbf{y}_i$ and letting the corresponding covariance matrix approach singularity will also create infinite peaks. The easiest solution, again is, to assume equality of all covariance matrices, whenever this is justifiable.

Exercises for Section 9.2

1. Show that the class of mixtures of beta distributions is not identifiable by verifying Example 9.2.1.

2. Show that the following classes of finite mixtures are not identifiable:
   (a) mixtures of Bernoulli distributions, and
   (b) mixtures of uniform distributions in intervals $[a_j, b_j]$, where both endpoints are allowed to vary.

3. Show that the class of finite mixtures of uniform distributions in intervals $[0, b_j]$ is identifiable.

4. Verify equation (17). What assumptions do you need to make on π?

5. This exercise refers to the numerical calculations of Example 9.2.2.
   (a) Write a computer program to do the iterations of equation (18) for arbitrary densities $f_1(y)$ and $f_2(y)$.
   (b) For $f_1(y)$ and $f_2(y)$ as in Example 9.2.2, use your program to verify the numerical results of Table 9.2.1 for the first ten iterations. Then do another 90 iterations.
   (c) Find the maximum of the log-likelihood function using a numerical method of your choice, e.g., the Newton–Raphson algorithm. Compare the result with the final value of π in part (b).
   (d) Repeat parts (b) and (c) for the modified data with $y_5 = 0.7$. Comment on the behavior of the iterative procedure. Does it seem to converge?
   Note: Your numerical maximization procedure may have difficulty handling the constraint $0 \le \pi \le 1$.

6. Let $\ell'(\pi)$ denote the first derivative of the log-likelihood function (13). Prove the following result. The maximum likelihood estimate for $\pi \in [0, 1]$ is

$$\hat{\pi} = \begin{cases} 0 & \text{if } \ell'(0) < 0 \\ 1 & \text{if } \ell'(1) > 0 \\ \text{the solution to } \ell'(\pi) = 0 & \text{otherwise.} \end{cases}$$

7. This exercise is based on the wing length data of Exercise 21 in Section 2.8. The data is listed in Table 2.8.1. Since only frequencies for intervals are given, use approximate data by repeating the midpoint of each interval, that is, generate 20 data values equal to 1.1, 14 data values equal to 1.3, etc.

   (a) Graph the log-likelihood function for the unknown mixing proportion similarly to Figure 9.2.1.

   (b) Find the maximum likelihood estimate of the mixing proportion using a method of your choice. *Note*: If you use the iterative procedure from Exercise 5(a), do not forget to change the definition of your density functions $f_1(y)$ and $f_2(y)$.

8. Verify equation (27) in less than 10 seconds.

9. Derive equation (31).

10. Derive equation (32).

11. Find likelihood equations for a mixture of k exponential densities, i.e., components of the form

$$f_j(y) = \begin{cases} \frac{1}{\lambda_j} e^{-y/\lambda_j} & \text{if } y \geq 0 \\ 0 & \text{otherwise,} \end{cases}$$

where the $\lambda_j > 0$ are unknown.

12. Suppose that we have a sample of size 1 (i.e., a single observation) from a mixture of two known components with unknown mixing proportion π. Show that the maximum likelihood estimate of π is either 0 or 1, or an arbitrary value in the unit interval.

13. Suppose that Y follows a mixture of two components with the density functions $f_1(y)$ and $f_2(y)$ from Example 9.2.2. Let $\hat{\pi}$ denote the maximum likelihood estimator of the unknown mixing proportion π based on a sample of size $N = 2$. If the true parameter value is $\pi = 1/2$, show that $0 < \Pr[\hat{\pi} = 0] < 4/9$ and $0 < \Pr[\hat{\pi} = 1] < 4/9$. *Hint*: Use Exercise 6.

14. Suppose we wish to estimate the parameters of a normal mixture with k components, where the means and variances of the components are unknown, but the mixing proportions π_j are known. Derive a modified equation system that takes this into account.

9.3 The EM-Algorithm for Normal Mixtures

In this section, we will show how the *EM*-algorithm introduced in Section 4.4 can be used to estimate the parameters of a p-variate normal mixture. At the same time, this will connect the theory of Section 9.1 to the equations we found in Section 9.2.

Recall from Section 9.1 that the joint *pdf* of $\mathbf{X}$ and $\mathbf{Y}$, where $\mathbf{X}$ is a k-vector of indicator variables, is given by

$$f_{\mathbf{XY}}(\mathbf{x}, \mathbf{y}) = \prod_{j=1}^{k} \left[\pi_j f_j(\mathbf{y})\right]^{x_j}, \qquad \text{all } x_j = 0 \text{ or } 1,\ \sum_{j=1}^{k} x_j = 1,\ \mathbf{y} \in \mathbb{R}^p. \quad (1)$$

Suppose that $f_j(\mathbf{y})$ is a multivariate normal density with mean vector $\boldsymbol{\mu}_j$ and positive-definite covariance matrix ψ_j. Let $\begin{pmatrix}\mathbf{X}_1\\ \mathbf{Y}_1\end{pmatrix}, \ldots, \begin{pmatrix}\mathbf{X}_N\\ \mathbf{Y}_N\end{pmatrix}$ denote a sample of size N from the joint distribution of $\mathbf{X}$ and $\mathbf{Y}$. That is, we assume for the moment that all N observations are fully categorized. Write

$$\mathbf{X}_i = \begin{pmatrix} X_{1i} \\ \vdots \\ X_{ki} \end{pmatrix} \quad (2)$$

for the vector of indicator variables for the ith observation. Formally, write $\boldsymbol{\theta}$ for the set of all unknown parameters, $\mathcal{X}$ for the matrix $(\mathbf{X}_1, \cdots, \mathbf{X}_N)$ and $\mathcal{Y}$ for the matrix $(\mathbf{Y}_1, \cdots, \mathbf{Y}_N)$. Then the log-likelihood function, denoted by ℓ_c to indicate that it is based on the complete data where all observations are categorized, is

$$\begin{aligned}
\ell_c(\boldsymbol{\theta}; \mathcal{X}, \mathcal{Y}) &= \sum_{i=1}^{N} \sum_{j=1}^{k} x_{ji} \log[\pi_j f_j(\mathbf{y}_i)] \\
&= \sum_{j=1}^{k} \sum_{i=1}^{N} \left[x_{ji} \log \pi_j + x_{ji} \log f_j(\mathbf{y}_i)\right] \qquad (3) \\
&= \sum_{j=1}^{k} \sum_{x_{ji}=1} \log \pi_j + \sum_{j=1}^{k} \sum_{x_{ji}=1} \log f_j(\mathbf{y}_i).
\end{aligned}$$

Each of the two parts in the last line of equation (3) can be maximized individually because they depend on disjoint subsets of the parameter set. Using the usual Lagrange multiplier technique, it follows (see Exercise 1) that the likelihood equations for the mixing proportions are

$$\pi_h = \frac{1}{N} \sum_{i=1}^{N} x_{hi} = \frac{N_h}{N}, \qquad h = 1, \ldots, k, \quad (4)$$

where N_h is the number of observations from the hth component. Note that the estimate of the mixing proportion does not depend, in any way, on the specific form of the densities $f_j(\mathbf{y})$ or on the data $\mathcal{Y}$.

Finding the maximum of the likelihood function with respect to the parameters μ_h and ψ_h is the same as finding the maximum likelihood estimators of the parameters of k independent normal populations based on samples of size $N_1, \ldots, N_k$. This holds because the term $\sum_{x_{ji}=1} \log f_j(\mathbf{y}_i)$ is a normal log-likelihood based on the N_j observations that come from the jth component. That means we obtain the same solution as in Section 4.2, namely,

$$\mu_h = \frac{1}{N_h} \sum_{x_{hi}=1} \mathbf{y}_i = \frac{1}{N_h} \sum_{i=1}^{N} x_{hi} \mathbf{y}_i \qquad h = 1, \ldots, k, \tag{5}$$

and

$$\begin{aligned} \psi_h &= \frac{1}{N_h} \sum_{x_{hi}=1} (\mathbf{y}_i - \mu_h)(\mathbf{y}_i - \mu_h)' \\ &= \frac{1}{N_h} \sum_{i=1}^{N} x_{hi} (\mathbf{y}_i - \mu_h)(\mathbf{y}_i - \mu_h)' \qquad h = 1, \ldots, k. \end{aligned} \tag{6}$$

We have to be careful not to forget that the N_h are random in the current setup. Also, we have implicitly assumed that $N_h > p$ for $h = 1, \ldots, k$. Thus, (4), (5), and (6) give the complete data likelihood equations. If equality of all covariance matrices is assumed, then the likelihood equation for the common ψ is

$$\psi = \frac{1}{N} \sum_{j=1}^{k} \sum_{i=1}^{N} x_{ji} (\mathbf{y}_i - \mu_j)(\mathbf{y}_i - \mu_j)', \tag{7}$$

which replaces equation (6).

Recall from Section 4.4 that the *EM*-algorithm for maximizing a likelihood function with missing data consists of iterated applications of the following two steps:

- *E*-step (Expectation): Replace the complete data log-likelihood by its expectation taken with respect to the missing data $\mathcal{X}$, given the observed data $\mathcal{Y}$, and using the current parameter values in the calculation of the expectation.
- *M*-step (Maximization): Solve the likelihood equations for the complete data likelihood, using the data substituted in the *E-step*.

From equation (6), the complete data log-likelihood function is linear in the X_{ji}, and therefore the E-step amounts to replacing the X_{ji} by their expected values, given $\mathcal{Y}$ and current parameter values. The expected value of X_{ji}, given the nonmissing data and parameters $\boldsymbol{\theta}$, is given by

$$\begin{aligned}\pi_{ji} = E\left[X_{ji}|\mathcal{Y};\boldsymbol{\theta}\right] &= E[X_{ji}|\mathbf{Y}_i;\boldsymbol{\theta}] \\ &= \Pr[X_{ji}=1|\mathbf{Y}_i;\boldsymbol{\theta}] \\ &= \frac{\pi_j\, f_j(\mathbf{y}_i;\boldsymbol{\theta})}{f(\mathbf{y}_i;\boldsymbol{\theta})} \qquad j=1,\ldots,k;\ i=1,\ldots,N,\end{aligned} \tag{8}$$

where $f(\mathbf{y}_i;\boldsymbol{\theta})$ is the mixture density evaluated at $\mathbf{y}_i$. These are exactly the posterior probabilities we have encountered before. The expected value of N_j, given the data $\mathcal{Y}$ and the parameters, is given by

$$E[N_j|\mathcal{Y};\boldsymbol{\theta}] = \sum_{i=1}^{N} \pi_{ji} \qquad j=1,\ldots,k. \tag{9}$$

Thus, the *EM*-algorithm for the mixture problem can be formulated as follows. Starting with some initial parameter vector $\boldsymbol{\theta}^{(0)}$, compute a sequence of parameter vectors $\boldsymbol{\theta}^{(1)}, \boldsymbol{\theta}^{(2)}, \ldots,$ until convergence, according to the following two steps:

E-step: set $\mathcal{X}^{(s)} = E[\mathcal{X}|\mathcal{Y};\boldsymbol{\theta}^{(s)}]$

M-step: Find $\boldsymbol{\theta}^{(s+1)}$ that maximizes $\ell_c(\boldsymbol{\theta}|\mathcal{X}^{(s)},\mathcal{Y})$.

For $s = 0, 1, \ldots,$ the two steps are applied iteratively until some convergence criterion is met.

We write this out explicitly in the form of an algorithm that is easy to program. It is understood that both the component densities

$$f_h(\mathbf{y}) = (2\pi)^{-p/2}\,(\det \boldsymbol{\psi}_h)^{-1/2}\,\exp\left[-\frac{1}{2}(\mathbf{y}-\boldsymbol{\mu}_h)'\boldsymbol{\psi}_h^{-1}(\mathbf{y}-\boldsymbol{\mu}_h)\right] \tag{10}$$

and the mixture density

$$f(\mathbf{y}) = \sum_{j=1}^{k} \pi_j\, f_j(\mathbf{y}) \tag{11}$$

are always evaluated for the most current values of the parameters.

EM-Algorithm for Normal Mixtures

Step 0 (initialization): Assign initial values to the parameters $\pi_1,\ldots,\pi_k;\boldsymbol{\mu}_1,\ldots,\boldsymbol{\mu}_k$, and $\boldsymbol{\psi}_1,\ldots,\boldsymbol{\psi}_k$.

Step 1 (*E*-step): set

$$\pi_{hi} = \frac{\pi_h\, f_h(\mathbf{y}_i)}{f(\mathbf{y}_i)} \tag{12}$$

for $h = 1, \ldots, k;\ i = 1, \ldots, N.$

Step 2 (M-step): set

$$\pi_h = \frac{1}{N}\sum_{i=1}^{N}\pi_{hi} \qquad h = 1, \ldots, k, \tag{13}$$

$$\mu_h = \frac{1}{N\pi_h}\sum_{i=1}^{N}\pi_{hi}\,\mathbf{y}_i \qquad h = 1, \ldots, k, \tag{14}$$

and

$$\psi_h = \frac{1}{N\pi_h}\sum_{i=1}^{N}\pi_{hi}(\mathbf{y}_i - \mu_h)(\mathbf{y}_i - \mu_h)' \qquad h = 1, \ldots, k. \tag{15}$$

Then return to step 1.

Note that the equal signs in equations (12) to (15) mean "assignment:" Assign to the variable on the left-hand side the value of the expression on the right-hand side. Equation (12) is the same as (8). Equation (13) is the same as (4) applied to the expected values of the X_{ji}. Finally, (14) and (15) are applications of (5) and (6) to the expected values of the X_{hi}, keeping in mind that $N_h = \sum_{i=1}^{N} X_{hi}$. If it is assumed that all covariance matrices are equal, say ψ, then equation (15) is to be replaced by

$$\psi = \frac{1}{N}\sum_{j=1}^{k}\sum_{i=1}^{N}\pi_{ji}\,(\mathbf{y}_i - \mu_j)(\mathbf{y}_i - \mu_j)'. \tag{16}$$

There are two questions: (1) How should the algorithm be initialized, and (2) how long does it need to run, i.e., what is the stopping criterion? We will return to these questions after giving an example.

Example 9.3.1 continuation of Examples 1.3 and 2.8.4

In the wing length of birds example, we started the EM-algorithm using initial values of $\pi_1 = 0.2$, $\pi_2 = 0.8$, $\mu_1 = 84$, $\mu_2 = 96$, and $\sigma_1^2 = \sigma_2^2 = 10$. The symbols σ_1^2 and σ_2^2 are used here instead of ψ_1 and ψ_2 because this is a univariate situation. Table 9.3.1 gives details of the computations for the first 24 iterations, using the version for unequal variances. The parameter estimates reach their final values practically during the first ten iterations. The procedure was continued to over 50 iterations without producing any changes in the decimal digits shown in the last line in Table 9.3.1. The final parameter estimates are $\hat{\pi}_1 = 0.486$, $\hat{\pi}_2 = 0.514$, $\hat{\mu}_1 = 86.140$, $\hat{\mu}_2 = 92.328$, $\hat{\sigma}_1 = 1.490$, and $\hat{\sigma}_2 = 1.579$. These estimates are slightly different from those reported in Example 2.8.4 because here we did not assume equality of variances. We leave to the student (Exercise 2) to verify the numerical results of Example 2.8.4. □

Table 9.3.1 The first 24 iterations of the *EM*-algorithm in Example 9.3.1.

iter	π_1	π_2	μ_1	μ_2	σ_1^2	σ_2^2	ℓ
0	0.2000	0.8000	84.0000	96.0000	10.0000	10.0000	−1315.5912
1	0.4666	0.5334	86.1962	92.0532	2.8815	3.8274	−956.4721
2	0.4725	0.5275	86.1087	92.1967	2.2828	3.0469	−948.7652
3	0.4766	0.5234	86.0951	92.2575	2.1438	2.7306	−947.5868
4	0.4797	0.5203	86.1047	92.2849	2.1418	2.6205	−947.4032
5	0.4818	0.5182	86.1154	92.3001	2.1618	2.5709	−947.3392
6	0.4832	0.5168	86.1234	92.3096	2.1797	2.5428	−947.3112
7	0.4842	0.5158	86.1289	92.3157	2.1927	2.5252	−947.2987
8	0.4848	0.5152	86.1327	92.3198	2.2017	2.5138	−947.2932
9	0.4852	0.5148	86.1352	92.3225	2.2078	2.5063	−947.2908
10	0.4855	0.5145	86.1369	92.3243	2.2119	2.5013	−947.2897
11	0.4857	0.5143	86.1380	92.3254	2.2147	2.4981	−947.2892
12	0.4858	0.5142	86.1387	92.3262	2.2165	2.4959	−947.2890
13	0.4859	0.5141	86.1392	92.3267	2.2177	2.4945	−947.2889
14	0.4860	0.5140	86.1396	92.3271	2.2185	2.4935	−947.2889
15	0.4860	0.5140	86.1398	92.3273	2.2191	2.4929	−947.2888
16	0.4860	0.5140	86.1399	92.3275	2.2194	2.4925	−947.2888
17	0.4860	0.5140	86.1400	92.3276	2.2197	2.4922	−947.2888
18	0.4860	0.5140	86.1401	92.3276	2.2198	2.4920	−947.2888
19	0.4861	0.5139	86.1401	92.3277	2.2199	2.4919	−947.2888
20	0.4861	0.5139	86.1402	92.3277	2.2200	2.4918	−947.2888
21	0.4861	0.5139	86.1402	92.3277	2.2200	2.4918	−947.2888
22	0.4861	0.5139	86.1402	92.3277	2.2201	2.4917	−947.2888
23	0.4861	0.5139	86.1402	92.3277	2.2201	2.4917	−947.2888
24	0.4861	0.5139	86.1402	92.3278	2.2201	2.4917	−947.2888

Note: the last column, labeled "ℓ", gives the value of the log-likelihood function computed from equation (25) of Section 9.2.

Example 9.3.1 is a case of a "well-behaved" mixture likelihood. The presence of two components is obvious, and the algorithm finds the parameter values near the maximum rather quickly, although the starting values are not very good. Incidentally, Table 9.3.1 also numerically confirms a result reported in Section 4.4. In every iteration of the *EM*-algorithm the value of the likelihood function increases until a stationary point is reached. If equality of variances is assumed, convergence is typically faster. This is also the case in this example; see Exercise 2.

From the discussion at the end of Section 9.2, we know that the likelihood function has infinite peaks if the variances are not constrained. If we start the algorithm with initial parameter values near one of these peaks, then we expect the procedure to "climb the peak." This is indeed the case; see Exercise 3.

Now we return to the questions raised before Example 9.3.1. The first question was, How should the algorithm be initialized? In the univariate case, good initial values can usually be obtained by looking at the histogram, guessing some values for the means, and giving all variances the same relatively large value to make sure the algorithm will not get trapped in one of the infinite peaks. For instance, one may set the initial value of all variances equal to the sample variance of all N observations. In the multivariate case, initialization is more difficult, particularly for the covariance matrices. In the author's experience, it is usually a good idea to set the initial covariance matrices equal to the sample covariance matrix of all data. Initial values of $1/k$ for the k mixing proportions seem to be a reasonable choice, unless one has good reasons, such as previous knowledge, to choose them differently. There is no guarantee that an initialization will lead to the "right" solutions, particularly if the likelihood function has several local maxima. It is a good idea to run the algorithm several times, using distinctly different initializations. If there are two or more different final solutions, then one would choose the solution which gives the largest value of the likelihood function.

An alternative to the above strategy is to partition the data initially into k contiguous subsets and initialize the parameters as the sample statistics of the subsets. We will illustrate this strategy in Example 9.3.2.

For the stopping criterion, one may choose to quit as soon as the improvement in the value of the log-likelihood function from one step to the next is less than some small number $\epsilon > 0$. Alternatively, one may stop when the maximum absolute difference between all parameters in two successive iterations is less then ϵ. This is the criterion we used in our implementation of the *EM*-algorithm , with $\epsilon = 10^{-10}$.

Sometimes the *EM*-algorithm seems to wander around endlessly in some area of the parameter space far from a maximum, where the likelihood is relatively flat. (See Figure 9.2.2 for an illustration). In such a case, the computations may be affected by roundoff error, and convergence might never be reached. Therefore it is reasonable to impose an upper limit to the number of iterations the algorithm is allowed to perform. If this limit is reached, a different initialization should be tried.

Example 9.3.2 continuation of Example 9.3.1

Since the data take only the discrete values 82, 83, ..., 98, with no observation at 97, we partitioned the data into two groups according to a cutoff point c, and varied c from 82.5 to 96.5 in steps of 1. For the unequal variances case, the means and variances were initialized as the sample means and variances of the two subsets, and the mixing proportions were initialized as the proportion of data in the subset. For the 11 cutoffs from 84.5 to 94.5, the algorithm converged to the solution reported in Example 9.3.1, taking between 50 and 121 iterations. For the remaining four cutoffs, the procedure ended in a singularity of the likelihood, i.e., in one of the infinite peaks.

Using the same cutoffs, the algorithm was run in the equal variances version with the same initializations, except that the initial value of the common variance was set equal to the variance of the complete data set. For cutoffs from 86.5 to 92.5, the algorithm converged to the same solution reported in Example 2.8.4, taking 21 to 44 iterations. In the other cases, the calculations were aborted after 500 iterations. Consequently, having found the same solution from many different starting points, we assume that the likelihood function has a single reasonable local maximum for the case of equal variances and the case of unequal variances. □

As seen in the examples, convergence of the algorithm may be slow. Typically, if the current parameter values are near a local maximum, the Newton–Raphson algorithm and other numerical methods would converge faster. However, no other procedure matches the elegance of the *EM*-algorithm and the simplicity of its implementation.

Finally, let us add an interpretation of the equations in the *EM*-algorithm . Equation (12), representing the *E*-step, is nothing but the definition of posterior probability applied to the ith data point. The prior probabilities computed in (13) are average posterior probabilities. The mean μ_h computed in (14) is a weighted average of all N observations, where the weight of $\mathbf{y}_i$ is proportional to the posterior probability π_{hi}. This is quite natural if we think of the distribution of $\mathbf{Y}$ in the hth group as discrete with probability $\pi_{hi}/N\pi_h$ on the ith data point. That is, we can define a discrete random vector $\mathbf{Y}_h^*$ with *pdf*

$$\Pr[\mathbf{Y}_h^* = \mathbf{y}_i] = \frac{\pi_{hi}}{N\pi_h}, \qquad i = 1, \ldots, N. \tag{17}$$

In view of this, equation (14) is the mean vector of $\mathbf{Y}_h^*$. A similar interpretation holds for the covariance matrix in (15).

Exercises for Section 9.3

1. Prove equation (4)
2. Write a computer program for the *EM*-algorithm for normal mixtures. Apply it to the data in Example 9.3.1 to verify the numerical results of Table 9.3.1. Run the algorithm in the version for equal variances and verify the results reported in Example 2.8.4. If you initialize the parameters identically in both cases, which version requires more iterations?
3. Run the *EM*-algorithm for normal mixtures again on the wing length data of Example 9.3.1, using the version for unequal variances and initial parameter values $\pi_1 = 0.01, \pi_2 = 0.99, \mu_1 = 84, \mu_2 = 92, \sigma_1^2 = 0.01$, and $\sigma_2^2 = 10$. What happens?
4. Find the *EM*-algorithm for mixtures of
   (a) exponential distributions, and
   (b) Poisson distributions.
   Hint: You should find the same equations as (12) to (14), where μ_h is the mean of the hth component and f_h is an exponential density (part a) or a Poisson probability function (part b).

5. Suppose that the mixing weights π_h are known in a normal mixture with k components, but the mean vectors and the covariance matrices of the components are unknown. Find the *EM*-algorithm for this situation.

9.4 Examples

In this section we present numerical examples to illustrate finite mixture analysis and to discuss some practical aspects.

Example 9.4.1 continuation of Example 9.1.1

Prevalence of immunity against mumps in unvaccinated children. We ran the *EM*-algorithm first for equal variances, using the initial values given in the first row of Table 9.4.1, and then for unequal variances, using the parameter estimates for equal variances to initialize the computations. The final result in the case of unequal variances is shown graphically in Figure 9.4.1, confirming a rather nice fit of the normal mixture model.

In this example, recall that we are interested mostly in estimating the prevalence of immunity, i.e., the mixing proportion π_2. The numerical analysis gives estimates of $\hat{\pi}_2 = 0.7442$ in the equal variances case, and $\hat{\pi}_2 = 0.7030$ in the unequal variances case. Which one should we use? And how precise is the estimate?

There are some good reasons to prefer the equal variance solution. First, the difference in the value of the log-likelihood function between the two models is relatively small. Second, a graph of the solution for the equal variance model (Exercise 1) shows a good fit, although it is visibly worse than the solution displayed in Figure 9.4.1. Third, the large number of iterations needed to converge to the solution with unequal variances is typical for situations in which the likelihood function is relatively flat in the neighborhood of the maximum, which means that there is a relatively large set of "plausible values" for the parameters.

To assess the variability of the parameter estimates, one might attempt to use the asymptotic likelihood theory outlined in Section 4.3. However this is rather complicated, and therefore a nonparametric bootstrap analysis was performed; see Section 4.2. For $B = 1000$ samples of size $N = 385$ from the empirical distribution function given by the data in Table 9.1.1, the maximum likelihood estimates of a normal mixture model with unequal variances were computed, using the last row of Table 9.4.1 as initial values. Figure 9.4.2 shows a histogram of the bootstrap distribution of the mixing proportion $\hat{\pi}_2$. The mean and the standard deviation of this distribution are 0.6996 and 0.0392, respectively; the latter may serve as an estimate of the standard error of the maximum likelihood estimate $\hat{\pi}_2$. A nonparametric 95 percent confidence interval for π_2 can be obtained by taking the 0.025 and the 0.975 quantiles of the bootstrap distribution as lower and upper limits; this interval is (0.616, 0.770).

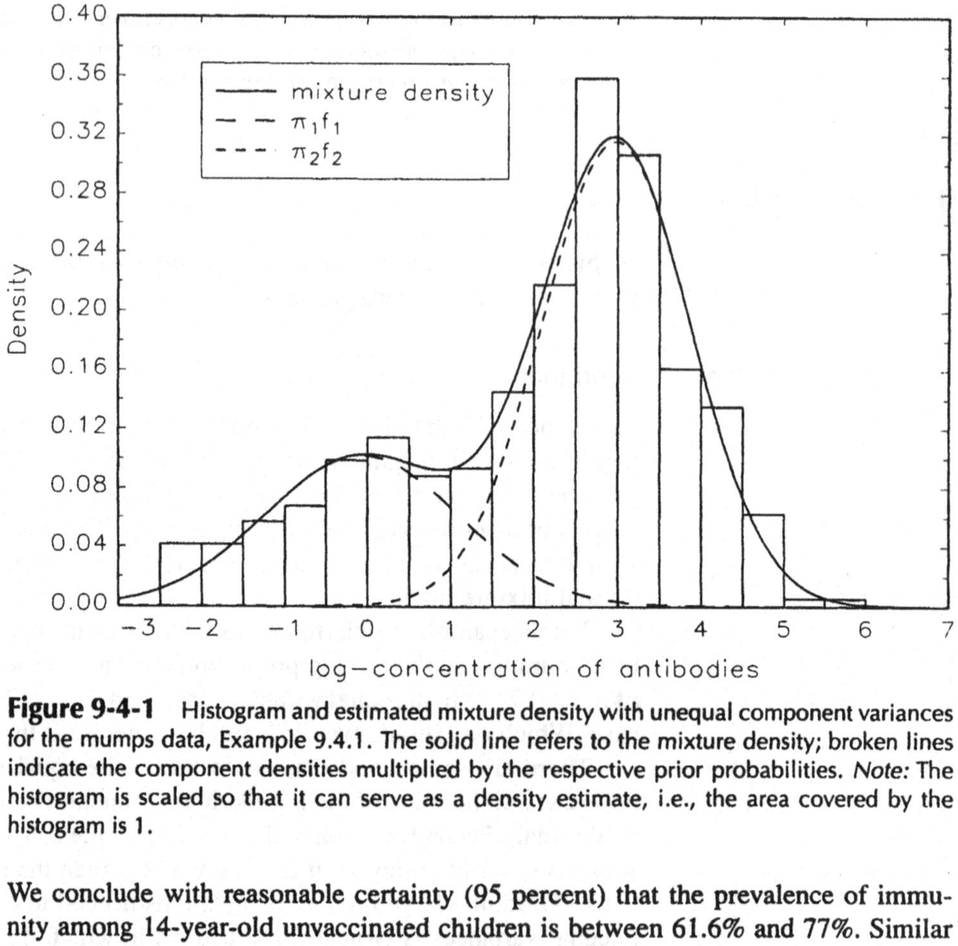

Figure 9-4-1 Histogram and estimated mixture density with unequal component variances for the mumps data, Example 9.4.1. The solid line refers to the mixture density; broken lines indicate the component densities multiplied by the respective prior probabilities. *Note:* The histogram is scaled so that it can serve as a density estimate, i.e., the area covered by the histogram is 1.

We conclude with reasonable certainty (95 percent) that the prevalence of immunity among 14-year-old unvaccinated children is between 61.6% and 77%. Similar analyses can also be performed for the other parameter estimates.

Whether or not such conclusions are reasonable depends on the validity of the normality assumptions; see the discussion at the end of Section 9.3. In the present case, we would probably have little hesitation to accept the normal model, both from experience and from the good fit of the mixture density in Figure 9.4.1. □

Table 9.4.1 Numerical results for Example 9.4.1. The initial value for σ_1 and σ_2 is the standard deviation of all observations. The value of the log-likelihood function is denoted by ℓ.

	π_1	π_2	μ_1	μ_2	σ_1	σ_2	*#iterations*	ℓ
Initialization	0.5	0.5	0	3	1.705	1.705		
mle($\sigma_1 = \sigma_2$)	0.2558	0.7442	−0.3193	2.8934	0.9668	0.9668	72	−711.45
mle($\sigma_1 \neq \sigma_2$)	0.2970	0.7030	−0.0725	2.9775	1.1631	0.8889	371	−710.02

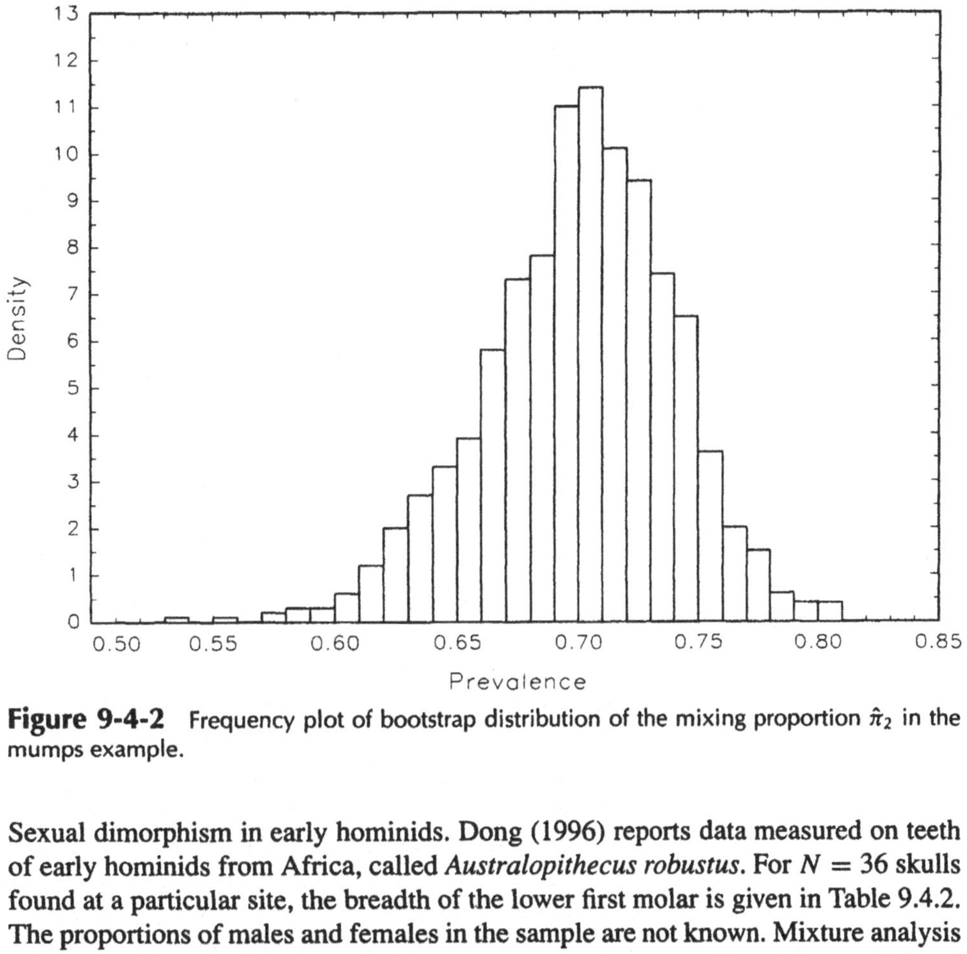

Figure 9-4-2 Frequency plot of bootstrap distribution of the mixing proportion $\hat{\pi}_2$ in the mumps example.

Example 9.4.2 Sexual dimorphism in early hominids. Dong (1996) reports data measured on teeth of early hominids from Africa, called *Australopithecus robustus*. For $N = 36$ skulls found at a particular site, the breadth of the lower first molar is given in Table 9.4.2. The proportions of males and females in the sample are not known. Mixture analysis might provide us with a reasonable estimate, assuming normality of the conditional distributions within groups, and assuming that in early hominids males tend to have larger teeth than females.

Figure 9.4.3 shows a histogram of the data, along with the estimated mixture density with the following parameters: $\hat{\pi}_1 = 0.7402$, $\hat{\pi}_2 = 0.2598$, $\hat{\mu}_1 = 13.828$, $\hat{\mu}_2 = 14.919$, $\hat{\sigma}_1 = \hat{\sigma}_2 = 0.4482$. These results are based on the *EM*-algorithm, assuming equal variances. Thus, it is reasonable to guess that about three quarters of the observations are from females and one quarter from males.

Sexual dimorphism, i.e., inhomogeneity of data due to gender differences, is a frequent justification for fitting mixtures with two components, as seen from this example and from the wing length of birds example in Section 2.8. However, in the present case, there are several difficulties: the sample is relatively small, the dimorphism is not as clear as in Example 2.8.4, and the data may be inaccurate due to decay. Moreover, even a small change in the breakpoints for the classes represented in the

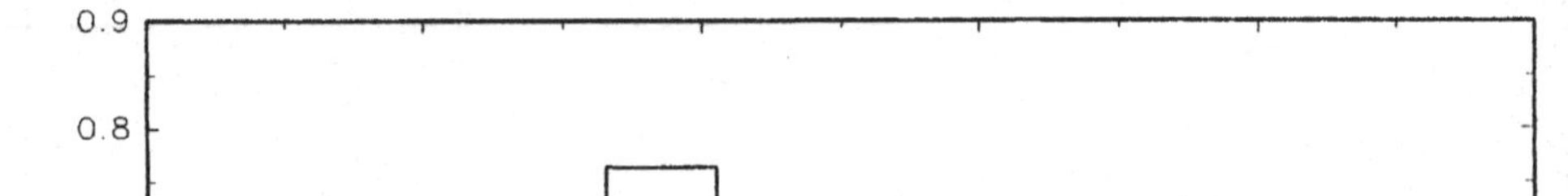

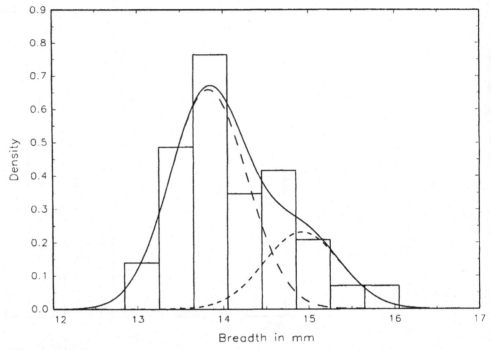

Figure 9-4-3 Histogram and estimated mixture density in Example 9.4.2.

histogram might cause the appearance of the histogram to change rather drastically, making it difficult to assess the goodness of fit of the mixture model.

All this points towards the concern that the parameter estimates obtained in this example might reflect random error more than anything else. As in Example 9.4.1, a bootstrap analysis with 1000 bootstrap samples was performed to study the variability of the parameter estimates. Details are not reported here, but a nonparametric 95 percent bootstrap confidence interval for the mixing proportion π_1 was (0.438, 0.964), showing that there is high variability in the parameter estimates.

Table 9.4.2 Breadth (in mm) of lower first molar of *Australopithecus robustus*. Data courtesy of Dr. Z. Dong, Department of Anthropology, Indiana University.

15.2	14.6	15.4	13.8	14.4	14.2
14.7	14.6	14.6	13.9	14.0	14.0
14.4	14.5	13.7	13.7	13.7	14.6
14.0	13.5	13.5	13.5	13.5	15.8
13.7	14.3	14.2	13.0	13.0	13.5
13.4	13.5	13.8	13.7	15.0	15.1

Despite all shortcomings, finite mixture analysis seems to be a powerful tool for this type of data. It is certainly more objective than commonly used methods of assessing sexual dimorphism in archeological data based on visual inspection of histograms. For further discussion, see Dong (1996). □

Example 9.4.3 Adulteration in wine production. Monetti et al. (1996) studied the chemical composition of $N = 344$ commercial samples of concentrated grape must in wine production. Table 9.4.3 gives data on four variables suitable for discovering adulterations with added sugar from plants other than grapes. For the purpose of this example, we will use the variables "myoinositol" and "D/H(I)." From experience, the two variables should follow approximately normal distributions in unadulterated samples. In our case, it was not known a priori for any of the 344 observations if they had been subject to adulteration or not. Mixture analysis serves the purposes of estimating the proportion of adulterated musts as well as establishing a classification region for acceptance or rejection of a given sample. Here, we will just focus on the estimation of the parameters of a normal mixture model.

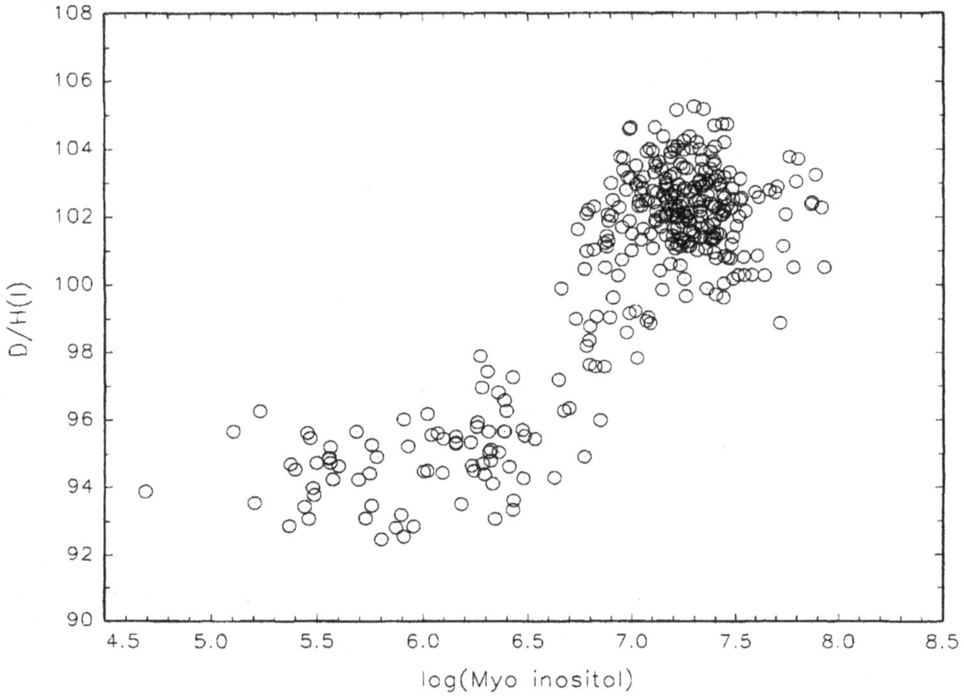

Figure 9-4-4 Scatterplot of Y_1 = log(myo-inositol) vs Y_2 = D/H(I) in Example 9.4.3.

Table 9.4.3 Sugar adulteration in wine production. Data courtesy of Dr. A. Monetti, Istituto Agrario Provinciale, 38010 S. Michele all'Adige (TN), Italy. Myo-inositol and Scylloinositol are in mg/kg sugar; D/H(I) and D/H(II) are in ppm.

Myoinositol	Scylloinositol	D/H(I)	D/H(II)
1727.94	323.53	104.74	126.22
1145.59	211.76	101.34	126.66
1532.35	242.65	103.40	130.07
164.71	32.35	95.67	125.62
108.82	10.29	93.87	124.50
236.76	35.29	95.47	126.08
1567.65	179.41	99.89	128.67
1200.00	163.24	98.88	126.45
1654.41	291.18	103.21	126.96
1748.53	285.29	103.32	127.99
1536.76	251.47	103.68	126.81
1704.41	323.53	103.18	131.83
1297.06	335.29	103.01	126.92
1545.59	311.76	105.18	131.58
1607.35	276.47	101.32	125.83
1448.53	272.06	104.38	129.37
1529.41	291.18	102.89	126.28
1689.71	270.59	104.75	130.15
1222.06	210.29	103.32	127.30
1301.47	236.76	102.94	125.34
1330.88	194.12	101.21	126.72
1558.82	255.88	103.33	126.23
1595.59	300.00	103.42	129.85
530.88	100.00	97.89	127.91
577.94	108.82	95.04	125.30
619.12	119.12	93.33	126.38
576.47	119.12	96.81	126.46
533.82	107.35	96.95	127.44
297.06	22.06	94.24	125.97
1226.47	200.00	104.65	130.25
1200.00	205.88	101.51	127.14
1351.47	238.24	103.97	129.21
1350.00	250.00	103.27	128.86
1405.88	250.00	103.45	129.27
1844.12	245.59	103.13	129.46
1092.65	164.71	101.02	129.46
1025.00	120.59	100.28	128.08
913.24	158.82	101.04	128.46
1257.35	148.53	100.42	129.62
1666.18	202.94	101.51	129.85
1494.12	192.65	102.47	130.40

Table 9.4.3 *Continued*

Myoinositol	Scylloinositol	D/H(I)	D/H(II)
216.18	35.29	94.69	126.96
270.59	23.53	94.63	125.75
258.82	33.82	94.88	127.12
239.71	23.53	93.98	127.11
263.24	33.82	94.24	125.63
258.82	35.29	94.85	126.78
241.18	27.94	93.77	126.90
260.29	30.88	95.20	127.84
244.12	20.59	94.73	127.59
186.76	19.12	96.26	127.95
1123.53	166.18	97.84	127.22
782.35	130.88	99.90	129.32
260.29	52.94	94.73	127.99
507.35	83.82	95.35	127.46
914.71	160.29	102.31	130.15
1544.12	244.12	102.91	127.66
1251.47	207.35	103.63	130.26
548.53	113.24	97.43	127.03
600.00	122.06	96.27	126.25
472.06	80.88	95.33	127.75
432.35	80.88	95.61	127.59
1214.71	227.94	103.94	129.74
1408.82	232.35	103.98	129.46
1457.35	239.71	102.74	127.52
1376.47	217.65	102.95	128.95
1629.41	227.94	103.56	130.55
1654.41	223.53	103.20	125.38
1522.06	248.53	103.99	129.83
1239.71	194.12	102.73	129.72
1498.53	244.12	104.21	129.32
1477.94	241.18	105.25	129.86
1525.00	226.47	102.03	127.68
1325.00	248.53	103.91	129.71
1194.12	214.71	104.02	129.75
1267.65	194.12	103.41	129.51
1630.88	266.18	104.09	130.21
1223.53	229.41	102.44	129.01
1598.53	275.00	103.93	130.33
233.82	23.53	95.62	127.29
1451.47	223.53	101.60	127.67
1235.29	194.12	102.17	129.65
235.29	11.76	93.07	127.94
230.88	10.29	93.42	127.24
182.35	10.29	93.53	127.96
958.82	120.59	97.58	127.66
2135.29	277.94	102.78	128.94

Table 9.4.3 *Continued*

Myoinositol	Scylloinositol	D/H(I)	D/H(II)
1705.88	279.41	104.22	131.36
2352.94	345.59	103.77	131.40
2669.12	492.65	103.24	128.13
1986.76	283.82	102.73	127.74
2611.76	372.06	102.38	129.75
792.65	129.41	96.28	128.01
594.12	85.29	96.58	129.53
557.35	86.76	95.12	127.71
472.06	72.06	95.29	128.69
411.76	66.18	94.49	127.09
541.18	102.94	94.39	127.82
552.94	111.76	95.05	128.79
442.65	27.94	94.43	126.64
941.18	148.53	96.00	127.92
569.12	108.82	93.07	125.86
592.65	85.29	95.67	126.64
607.35	98.53	94.61	127.37
755.88	123.53	94.29	126.26
2247.06	308.82	98.88	126.35
2400.00	325.00	100.52	127.69
2452.94	466.18	103.71	128.84
2433.82	335.29	103.04	129.18
872.06	170.59	94.91	125.78
385.29	23.53	92.85	126.41
307.35	26.47	93.09	128.57
444.12	69.12	95.44	127.81
472.06	42.65	95.51	128.69
536.76	116.18	94.70	127.60
363.24	27.94	93.19	125.47
330.88	29.41	92.47	126.23
2197.06	342.65	102.72	128.61
316.18	44.12	93.46	128.21
894.12	150.00	98.36	128.32
522.06	108.82	95.80	127.30
485.29	76.47	93.50	125.96
650.00	95.59	94.27	126.96
688.24	105.88	95.44	127.40
620.59	94.12	93.61	126.06
557.35	111.76	94.79	128.68
561.76	86.76	94.11	126.73
510.29	111.76	94.65	127.23
654.41	58.82	95.54	130.08
2744.12	372.06	102.28	126.90
2622.06	375.00	102.43	128.86
514.71	73.53	94.50	128.98
367.65	30.88	92.55	126.70

Table 9.4.3 *Continued*

Myoinositol	Scylloinositol	D/H(I)	D/H(II)
1129.41	152.94	102.33	126.63
2308.82	394.12	102.08	127.37
594.12	85.29	95.66	127.76
647.06	92.65	95.71	126.25
313.24	30.88	94.42	123.70
1625.00	211.76	102.31	127.53
523.53	75.00	95.94	129.23
214.71	14.71	92.86	127.82
1842.65	242.65	102.52	128.83
1189.71	161.76	99.05	127.96
1463.24	251.47	104.03	130.32
1113.24	213.24	99.23	128.21
811.76	104.41	96.35	128.06
1627.94	316.18	104.71	130.54
354.41	19.12	92.81	126.21
2020.59	357.35	102.58	129.36
1694.12	289.71	99.63	128.21
2780.88	280.88	100.52	127.26
772.06	127.94	97.19	128.84
1661.76	219.12	102.95	131.13
1855.88	279.41	102.57	130.68
998.53	151.47	99.63	130.67
1388.24	219.12	103.56	129.50
1516.18	200.00	103.02	130.52
1594.12	194.12	101.36	124.91
1339.71	201.47	104.09	131.13
1708.82	279.41	102.57	127.65
1638.24	219.12	103.09	129.17
1260.29	176.47	102.03	128.02
1388.24	201.47	101.37	126.64
1675.00	244.12	102.07	127.30
925.00	147.06	99.07	127.72
1385.29	205.88	101.13	128.60
1448.53	242.65	101.86	126.27
1352.94	242.65	102.34	126.00
1147.06	185.29	102.34	128.12
1411.76	235.29	104.25	128.67
617.65	110.29	97.26	128.38
1066.18	191.18	102.80	127.89
838.24	117.65	98.99	125.86
1301.47	250.00	102.83	126.56
882.35	139.71	102.10	127.72
1279.41	250.00	104.38	126.66
1235.29	279.41	103.50	130.40
1088.24	227.94	103.14	126.18
1588.24	264.71	103.08	127.19

Table 9.4.3 *Continued*

Myoinositol	Scylloinositol	D/H(I)	D/H(II)
895.59	139.71	97.64	126.32
919.12	139.71	97.58	126.35
1602.94	235.29	102.40	129.21
1705.88	250.00	102.17	129.71
1419.12	257.35	102.85	127.63
1367.65	235.29	101.19	124.17
1323.53	213.24	102.04	124.90
1213.24	294.12	102.78	126.07
1323.53	235.29	102.22	124.96
1176.47	198.53	98.94	127.25
1360.29	235.29	102.19	127.12
1279.41	213.24	102.64	122.91
882.35	132.35	98.20	122.26
1323.53	257.35	102.23	121.97
1323.53	227.94	103.75	125.81
970.59	154.41	101.44	125.44
1183.82	250.00	102.52	126.32
1330.88	264.71	102.57	127.40
1301.47	323.53	102.45	124.76
1522.06	279.41	101.40	127.28
1397.06	220.59	101.85	127.03
1397.06	235.29	101.33	124.67
1779.41	220.59	101.42	125.31
1147.06	235.29	102.72	125.74
1176.47	235.29	103.93	129.02
1544.12	257.35	102.47	126.30
977.94	227.94	102.08	126.52
1139.71	205.88	103.04	126.71
1411.76	205.88	102.78	126.64
1323.53	235.29	103.19	127.18
1352.94	213.24	101.07	125.19
375.00	51.47	95.23	125.18
367.65	44.12	96.03	126.89
411.76	51.47	96.18	125.34
419.12	44.12	95.56	125.95
1227.94	205.88	101.92	126.35
1117.65	294.12	102.96	124.06
1441.18	183.82	101.28	125.26
1470.59	205.88	101.55	125.05
992.65	161.76	102.04	125.25
1294.12	227.94	102.02	124.86
1411.76	220.59	102.13	125.93
970.59	147.06	101.13	124.28
1338.24	183.82	101.44	123.42
220.59	29.41	94.53	123.56
1367.65	235.29	101.69	124.28

Table 9.4.3 *Continued*

Myoinositol	Scylloinositol	D/H(I)	D/H(II)
1000.00	198.53	102.50	124.49
1169.12	250.00	102.48	124.02
1044.12	183.82	101.71	124.81
1044.12	183.82	100.75	124.28
963.24	161.76	101.23	124.45
551.47	95.59	95.66	122.26
1235.29	205.88	101.87	123.41
882.35	169.12	101.00	124.06
897.06	117.65	98.78	126.03
323.53	36.76	94.91	121.62
985.29	169.12	99.04	122.86
1294.12	198.53	101.46	123.26
889.71	191.18	102.22	124.38
404.41	44.12	94.47	124.30
1558.82	242.65	101.07	126.07
1095.59	176.47	101.51	124.88
1080.88	154.41	101.89	126.32
1661.76	205.88	102.26	125.89
1397.06	198.53	102.42	126.82
294.12	44.12	95.66	124.19
845.59	139.71	101.64	125.93
1433.82	169.12	101.14	128.14
1227.94	205.88	103.59	128.86
1051.47	205.88	103.38	127.55
1132.35	205.88	102.47	124.95
1080.88	213.24	104.60	129.92
1036.76	169.12	103.77	127.89
1286.76	191.18	102.57	126.06
1360.29	257.35	105.15	130.48
1117.65	250.00	103.52	128.22
316.18	29.41	95.27	127.85
1632.35	213.24	102.31	129.13
1073.53	213.24	103.17	127.29
1522.06	345.59	103.08	124.81
1294.12	294.12	103.14	125.53
1161.76	198.53	103.18	126.67
1426.47	257.35	103.43	126.98
992.65	191.18	103.00	125.69
1617.65	279.41	103.66	131.04
1623.53	194.12	101.41	126.18
1345.59	286.76	101.66	125.60
1367.65	191.18	102.59	127.50
1404.41	197.06	101.26	127.20
1067.65	135.29	98.60	126.03
1294.12	286.76	102.08	126.23
1382.35	294.12	102.33	126.92

Table 9.4.3 *Continued*

Myoinositol	Scylloinositol	D/H(I)	D/H(II)
1029.41	191.18	102.29	125.30
875.00	161.76	100.47	126.35
1455.88	147.06	102.13	125.81
1779.41	95.59	100.19	128.12
1352.94	294.12	102.77	127.91
1808.82	191.18	101.75	127.46
1500.00	154.41	101.02	126.64
2007.35	205.88	100.86	126.87
1360.29	264.71	101.98	125.88
1889.71	367.65	102.18	126.54
1772.06	242.65	102.85	126.62
1051.47	242.65	103.73	125.62
1154.41	264.71	102.86	126.69
1514.71	345.59	102.49	124.31
1566.18	338.24	102.98	124.87
1080.88	191.18	99.16	124.56
1419.12	235.29	99.67	125.20
1485.29	323.53	102.88	126.09
1573.53	330.88	102.67	126.73
1623.53	330.88	100.96	127.86
1544.12	323.53	102.93	123.87
1386.76	154.41	100.57	125.44
1272.06	169.12	99.86	124.23
1757.35	198.53	100.78	127.59
1441.18	250.00	101.92	126.60
1735.29	191.18	100.80	125.91
1507.35	161.76	101.83	125.62
1419.12	250.00	102.04	126.45
1625.00	227.94	101.36	125.76
1426.47	220.59	101.99	126.51
1705.88	191.18	100.85	125.97
1573.53	272.06	101.76	125.70
1551.47	235.29	101.40	125.73
1955.88	200.00	100.29	124.58
1882.35	132.35	100.29	124.91
1823.53	272.06	100.30	125.57
1698.53	455.88	102.22	123.58
1875.00	301.47	100.82	124.53
1411.76	213.24	100.17	124.56
1698.53	213.24	100.05	125.06
1698.53	367.65	102.68	123.01
2279.41	338.24	101.14	125.13
1639.71	235.29	100.79	124.65
1779.41	360.29	102.53	124.81
1213.24	242.65	101.09	125.12
1088.24	323.53	104.64	125.53

Table 9.4.3 *Continued*

Myoinositol	Scylloinositol	D/H(I)	D/H(II)
1522.06	154.41	102.19	124.99
2220.59	404.41	102.89	125.12
1617.65	272.06	101.55	124.98
1155.88	123.53	101.66	125.12
1338.24	338.24	102.08	126.96
2080.88	220.59	100.27	124.84
1323.53	272.06	100.62	123.71
1330.88	301.47	102.66	123.78
1727.94	375.00	102.35	123.89
1691.18	308.82	102.08	123.78
1367.65	338.24	104.09	127.72
1626.47	241.18	101.93	124.35
1676.47	291.18	102.97	124.77
977.94	95.59	101.30	122.42
979.41	114.71	101.89	123.98
964.71	226.47	100.52	124.69
1635.29	261.76	99.72	125.31
1763.24	279.41	101.20	125.92
1619.12	358.82	101.80	125.09
1644.12	333.82	101.52	124.59
1448.53	250.00	101.14	123.89
1757.35	338.24	102.26	124.99
1830.88	389.71	102.03	122.41
1529.41	264.71	101.68	123.20
1514.71	301.47	102.29	124.97
1477.94	213.24	101.44	124.19
1264.71	227.94	101.71	123.36

Figure 9.4.4 shows a scatterplot of $Y_1 = \log(\text{myoinositol})$ vs. $Y_2 = \text{D/H(I)}$, showing a distinct pattern that literally "asks" for a mixture distribution with two components. Using several different initializations, the *EM*-algorithm for normal mixtures with unequal covariance matrices converged to the following solution:

Mixing proportion: $\hat{\pi}_1 = 0.265, \qquad \hat{\pi}_2 = 0.735$

Means: $\hat{\mu}_1 = \begin{pmatrix} 6.113 \\ 95.299 \end{pmatrix}, \qquad \hat{\mu}_2 = \begin{pmatrix} 7.259 \\ 102.216 \end{pmatrix}$

Covariance matrices: $\hat{\psi}_1 = \begin{pmatrix} 0.250 & 0.467 \\ 0.467 & 2.456 \end{pmatrix}, \qquad \hat{\psi}_2 = \begin{pmatrix} 0.054 & 0.021 \\ 0.021 & 1.636 \end{pmatrix}$

In component 1 there is a substantial correlation of 0.597 between the two variables, whereas in component 2 they are practically uncorrelated, the estimated correlation being 0.072. The first component is associated with the adulterated samples; apparently the added sugar affects the two variables by reducing the values of both, thereby introducing a dependency. According to this analysis, the proportion of adulterated musts is estimated at roughly one-quarter.

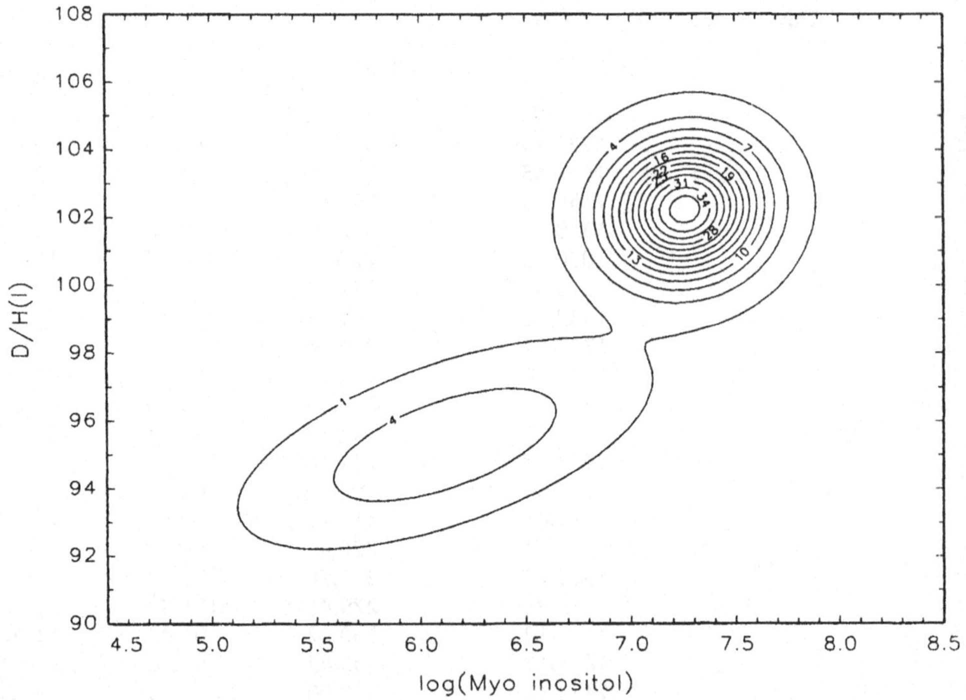

Figure 9-4-5 Estimated bivariate mixture density in Example 9.4.3. *Note:* Values of the density function are scaled by 100.

Figure 9.4.5 shows the estimated mixture density in the form of a contour plot. The two components appear quite well separated, giving strong evidence for the existence of two components.

Finally, notice that the parameter estimates found in this example are not those we would find in univariate mixture analyses of Y_1 and Y_2. That is, if we fit a univariate normal mixture to variable Y_1, we will obtain a different estimate of the mixing proportions and also different estimates of the means and variances of Y_1 in the two groups. This may be a little puzzling since marginal distributions of multivariate normal mixtures are themselves normal mixtures; see Exercise 4. Thus, if we want to graph the estimated mixture density of Y_1, it is not clear whether we should use the parameter estimates given here or estimates obtained from a univariate analysis of Y_1; see Exercise 5. □

Let us finish this section with some critical reflection on the use of normal mixtures. Mixture analysis is a modern topic in statistics, of great theoretical importance and with considerable potential for practical applications. However, it is a more difficult tool than classical methods like discriminant analysis since it requires a higher degree of statistical sophistication from the user and iterative computations not yet widely available in commercial software packages. The problem of correct understanding is the more difficult one though. To apply finite mixture analysis in a reasonable way, the user must understand that a statistical model with unknown parameters is involved, requiring careful judgment about the validity of the assumption of normality (or of some other parametric model). This is a far more challenging problem than deterministic partitioning of data into supposedly homogeneous subgroups; see the discussion at the end of Section 2.8. If we want the results of a finite normal mixture analysis to be reasonable, we must be sure about the existence and the number of components, and one must trust the normality assumptions. The importance of the normality assumptions can be seen from equation (12) in the *EM*-algorithm of Section 9.3, in which the normal density is used to compute posterior probabilities. Since the posterior probabilities are used in the computation of mean and variances, the normality assumptions affect all results. This is quite different from the standard situation of computing means and variances of samples of observations with known group membership, which can be justified without making any parametric assumptions.

Exercises for Section 9.4

1. This exercise is based on the mumps data of Examples 9.1.1 and 9.4.1.
   (a) Run the *EM*-algorithm on the data, assuming equality of variances, and verify the results displayed in Table 9.4.1.
   (b) Graph a histogram of the data, the estimated mixture density, and the two components multiplied by the mixing weights, i.e., $\pi_1 f_1$ and $\pi_2 f_2$.
   (c) Perform a bootstrap analysis as in Example 9.4.1, using $B = 1000$ bootstrap samples, but assume equality of variances. Graph the bootstrap distribution of the mixing proportion $\hat{\pi}_2$, and find a 95 percent confidence interval for π_2. Compare this interval to the one in Example 9.4.1.
   (d) Using your estimates from part (a), graph the posterior probability of being immune, given $Y =$ log concentration of antibodies as a function of Y. That is, graph the function $\pi_{2Y} = \pi_2 f_2(y)/f(y)$ for $-1 \le y \le 4$.

2. The data for this example, given in Table 9.4.4, is from the same study as Example 1.3, concerning wing length of water pipits on their way from central Europe to Africa. This time the sample consists of $N = 123$ birds caught at Tour du Valat, France.
   (a) Estimate the parameters of a normal mixture with equal variances.
   (b) Do a bootstrap analysis, and find a 95 percent confidence interval for the proportion of males (it is assumed that the component with the larger mean corresponds to males).
   (c) Fit a normal mixture with equal variances and equal mixing proportions using the result of Exercise 5 in Section 9.3.

Table 9.4.4 Wing length of 123 water pipits caught at Tour du Valat, France. From Flury, Airoldi, and Biber (1992).

wing length (mm)	84	85	86	87	88	89	90	91
# of birds	2	4	13	13	10	3	6	4
wing length (mm)	92	93	94	95	96	97	98	
# of birds	12	17	8	19	10	1	1	

3. Table 9.4.5 reports data on the breadth of the lower left second premolar of an extinct animal called a thick-mandibled giant deer. Estimate the parameters of a normal mixture with equal variances. Supposedly, the existence of two components would indicate sexual dimorphism, male animals having on the average larger teeth. Try to answer the question whether there is any evidence of bias in the sex composition, i.e., of mixing proportions different from 1/2.

4. If $\mathbf{Y}$ is a p-variate random vector that follows the distribution of a normal mixture with k components, show that every marginal distribution of dimension q $(1 \le q \le p)$ is a normal mixture with at most k components.

5. In Example 9.4.3, fit univariate normal mixture distributions with unequal variances for variables Y_1 and Y_2, and thus, show that the parameter estimates obtained this way are not identical to the mixing proportions, means, and variances estimated in the bivariate analysis. If you want to graph the estimated mixture density of (say) Y_1, should you use the estimates obtained from the bivariate analysis or those from the univariate analysis? Try to think of good arguments for both possibilities.

6. In Example 9.4.3 graph the variables "myoinositol" and "D/H(I)" in a scatterplot. Why was variable "myoinositol" transformed to logarithms? Why is a transformation to logarithms less important for variable "D/H(I)"?

7. Using all four variables in Table 9.4.3 and log-transformations for the first two, fit a four-dimensional normal mixture with two components. Use the parameter estimates from this analysis to graph an estimated univariate mixture density for the fourth

Table 9.4.5 Breadth of the lower left second premolar of the giant deer (*Megaloceros pachyosteus*). From Dong (1996).

15.0	13.5	14.0	13.0	14.0	13.0
13.0	13.0	13.0	13.0	14.0	12.4
14.0	13.0	14.0	14.0	12.0	13.0
12.4	14.0	14.5	12.0	13.0	12.0
13.0	12.0	12.0	12.0	12.0	14.0
14.0	12.5	13.0	12.0	14.0	15.0
13.0	13.0	13.0			

variable, D/H(II). Fit a univariate normal mixture to the data of variable D/H(II), and graph the solution. Which of the two solutions would you prefer?

8. In Example 9.4.3, there are theoretical reasons to believe that variables Y_1 = log(myoinositol) and Y_2 = D/H(I) are independent for unadulterated wines. Modify the *EM*-algorithm to reflect the assumption that the correlation between the two variables is zero for one of the components, and repeat the estimation procedure.

9.5 Epilogue: Normal Theory Discrimination with Partially Classified Data

In this section, we present a method that combines normal theory discriminant analysis and finite mixtures with normal components into one, more general, technique. It is relatively little known among users of discriminant analysis but has considerable potential for practical applications. Fortunately, thanks to the missing data approach to finite mixtures via the *EM*-algorithm, the method is easy to program and requires practically no additional mathematical work, as we will see shortly. The method to be discussed has no commonly accepted name in the literature except for the somewhat lengthy "Discriminant analysis with partially classified data." Since it combines discriminant analysis and mixture analysis, we will refer to it as *Discrimix*.

First, let us motivate *Discrimix* with some examples. For instance, in the wing length of birds data (Examples 1.3, 2.8.4, 9.3.1), suppose we happen to have observed the behavior of one of the birds, and we know for sure that it is male. Obviously, we would be wasting information if we decided to ignore the fact that we know the gender of this bird. But how could we use this knowledge for estimation? An intuitively reasonable answer, based on the *EM*-algorithm of Section 9.3, would be that since the group membership variable X has been measured in this case, there is no need to replace it by its expected value. That is, in each iteration of the algorithm, we will use the actual X for the bird with known gender instead of the posterior probabilities computed in equation (12) of Section 9.3. This is indeed the correct procedure, as we will see shortly.

Another example is as follows. Suppose a geneticist is interested in differentiating between identical and fraternal twins, based on external characteristics, and measures morphometric variables on, say, 100 pairs of female twins. Additional tests are performed for 20 pairs to obtain an error-free diagnosis. The tests are limited to 20 pairs because of their high cost. Suppose that of the 20 pairs, N_1 are identical twins, and N_2 are fraternal twins. In classical discriminant analysis, these would constitute the training samples, and the parameters of both groups would be estimated using only the training samples. In a subsequent stage, the classification rule obtained from the training samples would then be applied to the remaining 80 observations.

It would be unwise to assume that the data with uncertain group membership does not provide any information about the parameters. As in mixture analysis, an unclassified observation can be used for estimation if it is given a proper weight according to its posterior probability for membership in each of the groups.

Now, we proceed to the formal setup, using the same notation and terminology as in Section 9.3. Let $\begin{pmatrix} \mathbf{X}_1 \\ \mathbf{Y}_1 \end{pmatrix}, \ldots, \begin{pmatrix} \mathbf{X}_N \\ \mathbf{Y}_N \end{pmatrix}$ denote a sample of size N from the joint distribution of $\mathbf{X}$ and $\mathbf{Y}$. As before, write

$$\mathbf{X}_i = \begin{pmatrix} X_{1i} \\ \vdots \\ X_{ki} \end{pmatrix} \tag{1}$$

for the vector of indicator variables of the ith observation. Suppose that group membership is known for $N_C \le N$ of the observations. For these observations, exactly one of the X_{ij} in $\mathbf{X}_i$ is 1. For $N_U = N - N_C$ observations, X is missing. Proceeding as in Section 9.3, we see that the expectation of X_{ji} needs to be taken only for the N_U unclassified observations; for the classified ones the observed x_{ji} remains in the formula. Writing w_{ji} for the weight of the ith observation in the jth component, we use the following equations replacing (12) to (15) in the *EM*-algorithm of Section 9.3.

Step 1 (*E*-step): set

$$w_{hi} = \begin{cases} x_{hi} & \text{if } \mathbf{x}_i \text{ has been observed} \\ \dfrac{\pi_h f_h(\mathbf{y}_i)}{f(\mathbf{y}_i)} & \text{if } \mathbf{x}_i \text{ is missing.} \end{cases} \tag{2}$$

Step 2 (*M*-step): set

$$\pi_h = \frac{1}{N}\sum_{i=1}^{N} w_{hi} \qquad h = 1, \ldots, k, \tag{3}$$

$$\mu_h = \frac{1}{N\pi_h}\sum_{i=1}^{N} w_{hi}\mathbf{y}_i \qquad h = 1, \ldots, k, \tag{4}$$

and

$$\psi_h = \frac{1}{N\pi_h}\sum_{i=1}^{N} w_{hi}(\mathbf{y}_i - \mu_h)(\mathbf{y}_i - \mu_h)' \qquad h = 1, \ldots, k. \tag{5}$$

If the covariance matrices are assumed identical, then equation (5) will be replaced by

$$\psi = \frac{1}{N}\sum_{h=1}^{k}\sum_{i=1}^{N} w_{hi}(\mathbf{y}_i - \mu_h)(\mathbf{y}_i - \mu_h)' \qquad h = 1, \ldots, k. \tag{6}$$

Thus, the w_{hi} in the current setup have two meanings: They are identical to x_{hi} if group membership of the ith observation is known, or they are posterior probabilities as in the pure mixture setup. Comparing the equation system (2) to (5) with equations (4) to (6) of Section 9.3 for the case of known group membership of all observations and with (12) to (15) for the pure mixture case, we see that both are special cases. Thus, equations (2) to (6), indeed, give a more general method. Note that, if group membership is known for all N observations, no iterative computations need to be done since equations (3) to (5) give the explicit solution.

The equation system just presented is valid if all N observations are sampled from the joint distribution of $\mathbf{X}$ and $\mathbf{Y}$. However, sometimes the sampling paradigm may be different, in the sense that the number of classified observations in each group is fixed, that is, the N_C classified observations are sampled from the conditional distributions of $\mathbf{Y}$ given $X_j = 1$. Let N_h denote the number of observations from the hth group, $\sum_{h=1}^{k} N_h = N_C$. With fixed N_h and sampling from the conditional distributions, the N_h do not contain information about the prior probabilities π_h. The following equations should then replace (3) to (5): For the posterior probabilities, set

$$\pi_h = \frac{1}{N_U} \sum \pi_{hi} \qquad h = 1, \ldots, k, \tag{7}$$

where the sum extends over all unclassified observations. For the means, set

$$\mu_h = \frac{1}{N_h + N_U \pi_h} \sum_{i=1}^{N} w_{hi} \mathbf{y}_i \qquad h = 1, \ldots, k, \tag{8}$$

and, for the variances and covariances, set

$$\psi_h = \frac{1}{N_h + N_U \pi_h} \sum_{i=1}^{N} w_{hi} (\mathbf{y}_i - \mu_h)(\mathbf{y}_i - \mu_h)' \qquad h = 1, \ldots, k. \tag{9}$$

The proof of (7) to (9) is left to the reader; see Exercise 3.

Now, we will illustrate *Discrimix* with a numerical example.

Example 9.5.1: This example is based on the *Microtus* data introduced in Example 5.4.4. The data are given in Table 5.4.1. Recall that there are $N = 288$ observations on $p = 8$ variables. The variables are skull dimensions of voles of the two species *Microtus multiplex* and *M. subterraneus*. A chromosomal analysis was done for $N_C = 89$ individuals, revealing that $N_1 = 43$ observations are from *M. multiplex* and $N_2 = 46$ observations from *M. subterraneus*. In the following, we present some results originally published in Airoldi, Flury, and Salvioni (1996).

Since sampling was done from the joint distribution of $\mathbf{X}$ and $\mathbf{Y}$, we can apply equations (2) to (6). Using variables Y_1 = M1LEFT and Y_2 = FORAMEN, we ran the *EM*-algorithm in the version for equal covariance matrices, with initial values of the parameters equal to the parameter estimates obtained from the 89 classified observations alone; see Example 5.4.4. These initial parameter estimates as well as the final ones obtained from *Discrimix* are displayed in Table 9.5.1, along with the estimates of the linear discriminant function coefficients and the estimated bivariate standard distance. Figures 9.5.1 (a) and (b) show scatterplots of Y_1 vs. Y_2 for the classified and the unclassified observations, respectively, along with the classification boundary $\hat{\alpha} + \hat{\beta}'\mathbf{y} = 0$ calculated from the *Discrimix* estimates. There is little

Table 9.5.1 Parameter estimates in the *Microtus* example, using (a) ordinary discriminant analysis, i.e., only classified observation and (b) *Discrimix*. Here, $\beta = \psi^{-1}(\mu_1 - \mu_2)$, and $\alpha = \log \pi 1/\pi 2 - \frac{1}{2}(\mu_1 - \mu_2)'\psi^{-1}(\mu_1 + \mu_2)$. Group 1 is *Microtus multiplex*; group 2, *M. subterraneus*.

Parameter	Estimate using Discriminant Analysis	Estimate using *Discrimix*
π_1	0.483	0.528
π_2	0.517	0.472
μ_1	$\begin{pmatrix} 2054.5 \\ 3966.5 \end{pmatrix}$	$\begin{pmatrix} 2070.9 \\ 3958.0 \end{pmatrix}$
μ_2	$\begin{pmatrix} 1773.1 \\ 3899.0 \end{pmatrix}$	$\begin{pmatrix} 1784.1 \\ 3862.4 \end{pmatrix}$
ψ	$\begin{pmatrix} 10070 & 10245 \\ 10245 & 55149 \end{pmatrix}$	$\begin{pmatrix} 11877 & 13791 \\ 13791 & 60362 \end{pmatrix}$
$\text{corr}(Y_1, Y_2)$	0.435	0.515
α	−43.81	−37.48
β	$10^{-3}\begin{pmatrix} 32.90 \\ -4.89 \end{pmatrix}$	$10^{-3}\begin{pmatrix} 30.36 \\ -5.35 \end{pmatrix}$
Δ	2.987	2.863

difference between the discriminant analysis estimates and the *Discrimix* estimates, and the question naturally arises whether it is worth doing the iterative calculations of *Discrimix* if we are not even sure an improvement has been made. Also, the classification boundary in Figure 9.5.1 is, for practical purposes, the same as the one obtained from discriminant analysis; see Exercise 6.

Figure 9.5.2 (b) shows the joint distribution of the bootstrap estimates of Δ using discriminant analysis and *Discrimix* (the same bootstrap samples were used for both estimation procedures). It appears that, typically, the estimate from *Discrimix* is smaller than the estimate from discriminant analysis, which we may take as an indication that *Discrimix* corrects for too large values of $\hat{\Delta}$ obtained if only the classified observations are used. Thus, we have some good evidence that it is worth including the unclassified observations in the estimation procedure. Some theoretical analysis of this topic can be found in McLachlan and Basford (1988), and in O'Neill (1978).

Since the point estimates themselves are not very different, we might at least hope to find that *Discrimix* gives more stable results than discriminant analysis, in the sense of smaller standard errors of the parameter estimates. As in the examples of Section 9.4, a bootstrap analysis was performed with $B = 1000$ bootstrap replications. Since standard distance is a crucial parameter for classification, the bootstrap results

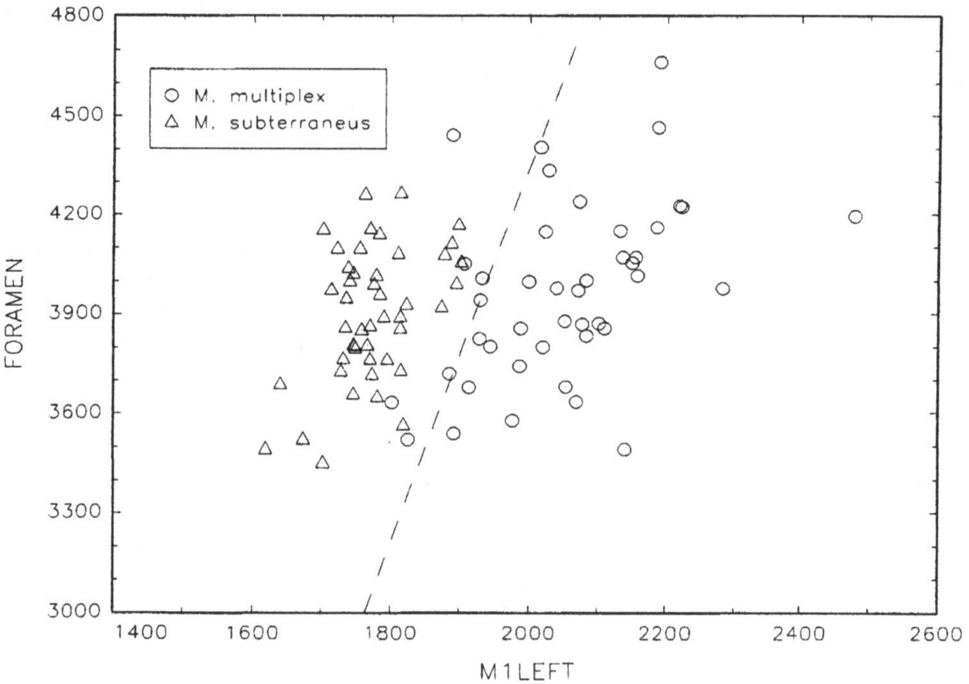

Figure 9-5-1A

for $\hat{\Delta}$ are reported here. Figure 9.5.2 (a) shows the bootstrap distribution of $\hat{\Delta}$ for both methods of estimation. A 95 percent confidence interval for Δ, using the 0.025 and 0.975 quantiles of the bootstrap distribution, is (2.494, 4.034) for discriminant analysis and (2.512, 3.371) for *Discrimix*. Thus the main effect of using *Discrimix* is that the length of the confidence interval for Δ is decreased drastically, mostly by moving the upper end down. □

Incidentally, *Discrimix* often takes care of two problems of mixture analysis almost automatically. First, if the sample covariance matrix of the classified observations is nonsingular in each group (which requires that at least $p + 1$ observations must be available from each component), then the infinite peaks in the likelihood function disappear. Second, under the same condition, the *EM*-algorithm is ideally initialized using the mean vectors and sample covariance matrices of the observations with known group membership.

Finally, a word of caution. As in pure mixture analysis, *Discrimix* is based explicitly on the normality assumptions. Thus, one needs to feel rather confident about these assumptions. In some cases a data transformation, as in Example 9.4.3, may be useful.

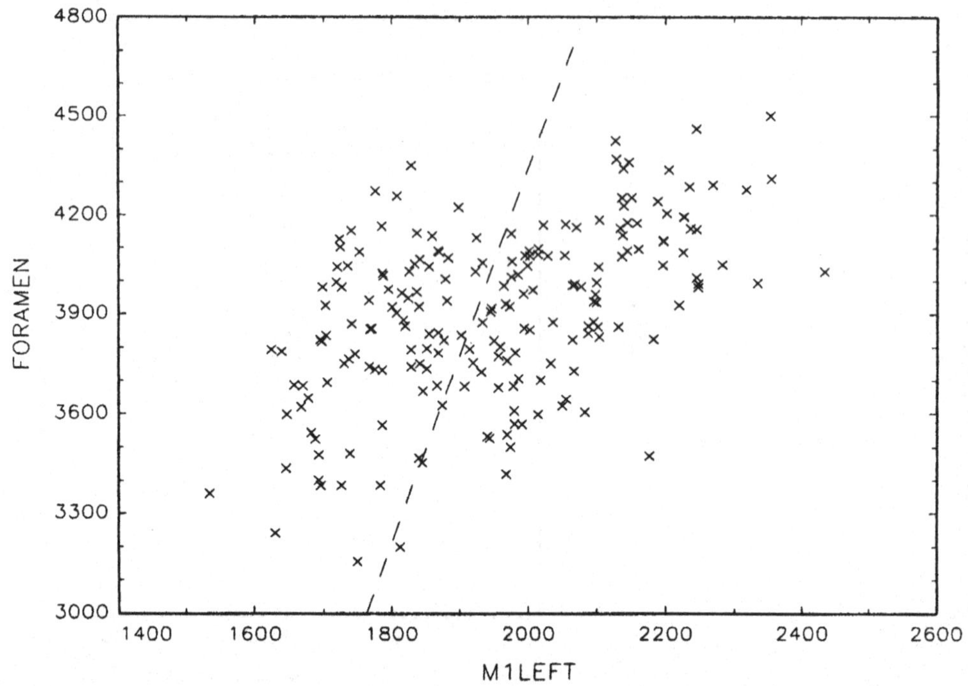

Figure 9-5-1B Scatterplot of M1LEFT vs. FORAMEN, with classification boundary based on the *Discrimix* estimates: (a) all classified observations, with circles marking *M. multiplex* ($N_1 = 43$) and triangles marking *M. subterraneus* ($N_2 = 46$); (b) all unclassified observations ($N_U = 199$). Reproduced from Airoldi et al (1995) with permission by Academic Press. Inc.

Exercises for Section 9.5

1. In the *Discrimix* setup with $p = 1$, suppose that at least two different observations with known group membership are available for each of the k components. Show that the likelihood function has no infinite peaks, even if the variances are assumed unequal.

2. In the *Discrimix* setup with a p-variate random vector $\mathbf{Y}$, suppose that the sample covariance matrix of the N_h classified observations from component h is nonsingular for $h = 1, \ldots, k$. Show that the likelihood function has no infinite peaks, even if the covariance matrices are not constrained.

3. Derive equations (7) to (9), assuming that the N_h classified observations from component h were samples from the conditional distribution of $\mathbf{Y}$, given $X_h = 1$.

4. This exercise refers to Example 9.5.1. A naive investigator might conclude that it is better to use the discriminant analysis parameter estimates instead of the *Discrimix* estimates because the former ones separate the two populations better ($\hat{\Delta} = 2.987$

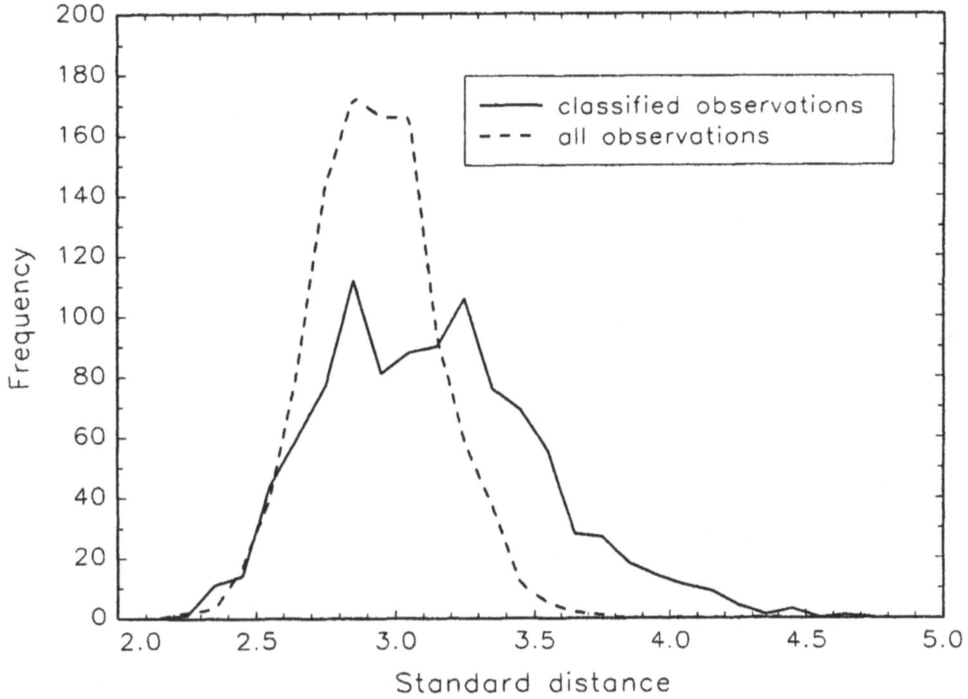

Figure 9-5-2A

for discriminant analysis and $\hat{\Delta} = 2.863$ for *Discrimix*). Is this reasoning correct? Explain.

5. This exercise refers to the same data as in Example 9.5.1 but uses the single variable Y = ROSTRUM; see Table 5.4.1.

(a) Graph the frequency distribution of Y for both samples of classified observations and for the unclassified observations, similarly to Figure 5.4.4.

(b) Compute the sample means and the pooled variance of the classified observations.

(c) Use the parameter values found in (b) as initial values, and run the *EM*-algorithm to find the *Discrimix* estimates, assuming equality of variances.

(d) Run a bootstrap analysis with $B = 1000$ bootstrap samples. Graph the bootstrap distribution of the standard distance $\hat{\Delta}$ for both methods of estimation. Find 95 percent confidence intervals for Δ using both methods of estimation and comment.

6. This exercise refers to Example 9.5.1 using only the 89 classified observations. Compute the normal theory classification boundary (assuming equality of covariance matrices), and graph it in Figure 9.5.1(a).

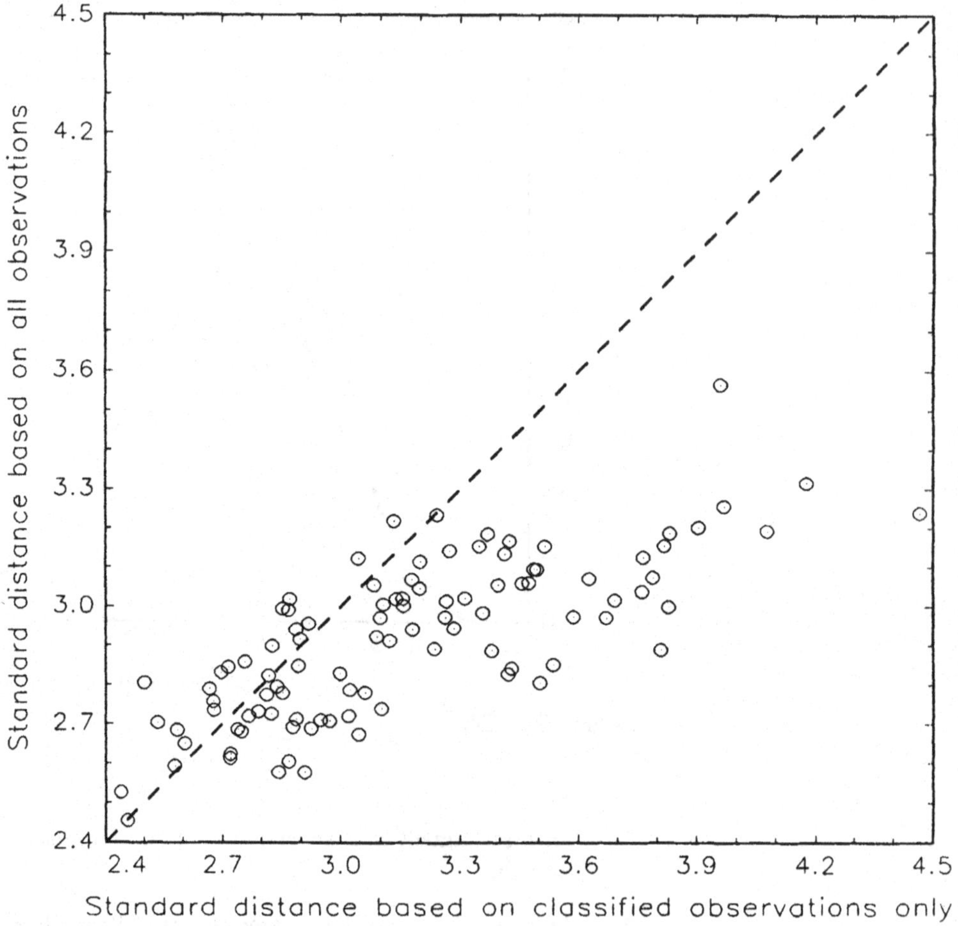

Figure 9-5-2B Bootstrap distribution of $\hat{\Delta}$ for both discriminant analysis and *Discrimix* methods of estimation: (a) frequency plots for both methods of estimation; (b) joint bootstrap distribution of both estimators. Only the first 100 bootstrap replications are shown. Reproduced from Airoldi et al (1995) with permission by Academic Press, Inc.

Suggested Further Reading

General: Everitt and Hand (1981), McLachlan and Basford (1988), Titterington et al (1985).

Section 9.3: Day (1969), Dempster et al. (1977), McLachlan and Krishnan (1997).

Section 9.5: Airoldi et al. (1996), O'Neill (1978).

Appendix: Selected Results From Matrix Algebra

A.0 Preliminaries

Mastering a fair amount of matrix algebra is a prerequisite for multivariate statistics. This appendix presents some material that the student who has taken a one–semester course on linear algebra may not be familiar with. It is assumed that the reader knows basic concepts such as matrix and vector multiplication, transposed matrices, inverses of matrices, trace, determinant, and rank of a matrix. Exercises 1 to 8 may serve as a test. The student who is not able to solve these exercises is advised to spend some time reviewing the relevant material; for instance the first six chapters of Searle (1982) or the first two chapters of Schott (1997). Standard notation used in this appendix is as follows: Capital boldface letters usually denote matrices and lowercase boldface letters denote column vectors. The transposed of a matrix $\mathbf{A}$ is written as $\mathbf{A}'$. The $p \times p$ identity matrix is denoted by $\mathbf{I}_p$. The symbols $\mathbf{O}$ and $\mathbf{0}$ denote a matrix and a vector of zeros, respectively; their dimensions are usually clear from the context. The determinant of a square matrix $\mathbf{A}$ is written as $\det(\mathbf{A})$, the trace as $\text{tr}(\mathbf{A})$, and the inverse (provided it exists) as $\mathbf{A}^{-1}$. If $a_1, \ldots, a_p$ are real numbers, then

$\text{diag}(a_1, \ldots, a_p)$ is the diagonal matrix with diagonal entries as indicated. Similarly, if $\mathbf{A}$ is a square matrix with entries a_{ij}, then $\text{diag}(\mathbf{A}) = \text{diag}(a_{11}, \ldots, a_{pp})$.

A.1 Partitioned Matrices

In many multivariate models matrices and vectors appear in partitioned form. Suppose that the matrices $\mathbf{A}$ of dimension $m \times n$ and $\mathbf{B}$ of dimension $n \times k$ are partitioned as

$$\mathbf{A} = \begin{pmatrix} \mathbf{A}_{11} & \mathbf{A}_{12} \\ \mathbf{A}_{21} & \mathbf{A}_{22} \end{pmatrix} \quad \text{and} \quad \mathbf{B} = \begin{pmatrix} \mathbf{B}_{11} & \mathbf{B}_{12} \\ \mathbf{B}_{21} & \mathbf{B}_{22} \end{pmatrix},$$

where $\mathbf{A}_{ij}$ has dimension $m_i \times n_j$, and $\mathbf{B}_{j\ell}$ has dimension $n_j \times k_\ell$. Then (Exercise 9)

$$\mathbf{AB} = \begin{pmatrix} \mathbf{A}_{11}\mathbf{B}_{11} + \mathbf{A}_{12}\mathbf{B}_{21} & \mathbf{A}_{11}\mathbf{B}_{12} + \mathbf{A}_{12}\mathbf{B}_{22} \\ \mathbf{A}_{21}\mathbf{B}_{11} + \mathbf{A}_{22}\mathbf{B}_{21} & \mathbf{A}_{21}\mathbf{B}_{12} + \mathbf{A}_{22}\mathbf{B}_{22} \end{pmatrix}. \tag{1}$$

Generalizations to any number of submatrices are straightforward as long as the dimensions of the submatrices are compatible for multiplication. In the spectral decomposition theorem (appendix A.6) we will encounter a case where two matrices $\mathbf{A}$ and $\mathbf{B}$ are partitioned by rows and columns. Suppose $\mathbf{A}$ has dimension $m \times p$, and $\mathbf{B}$ has dimension $p \times m$, partitioned as

$$\mathbf{A} = (\mathbf{a}_1 \quad \cdots \quad \mathbf{a}_p) \quad \text{and} \quad \mathbf{B} = \begin{pmatrix} \mathbf{b}'_1 \\ \vdots \\ \mathbf{b}'_p \end{pmatrix}.$$

Then

$$\mathbf{AB} = \sum_{i=1}^{m} \mathbf{a}_i \mathbf{b}'_i,$$

while

$$\mathbf{BA} = \begin{pmatrix} \mathbf{b}'_1\mathbf{a}_1 & \cdots & \mathbf{b}'_1\mathbf{a}_p \\ \vdots & \ddots & \vdots \\ \mathbf{b}'_p\mathbf{a}_1 & \cdots & \mathbf{b}'_p\mathbf{a}_p \end{pmatrix}.$$

Suppose next that a nonsingular $p \times p$ matrix $\mathbf{A}$ is partitioned as $\mathbf{A} = \begin{pmatrix} \mathbf{A}_{11} & \mathbf{A}_{12} \\ \mathbf{A}_{21} & \mathbf{A}_{22} \end{pmatrix}$, and assume that the submatrices $\mathbf{A}_{11}$ and $\mathbf{A}_{22}$ are nonsingular as well. Let $\mathbf{B} = \mathbf{A}^{-1} = \begin{pmatrix} \mathbf{B}_{11} & \mathbf{B}_{12} \\ \mathbf{B}_{21} & \mathbf{B}_{22} \end{pmatrix}$, partitioned analogously to $\mathbf{A}$. Let

$$\begin{aligned} \mathbf{A}_{11.2} &= \mathbf{A}_{11} - \mathbf{A}_{12}\mathbf{A}_{22}^{-1}\mathbf{A}_{21}, \\ \mathbf{A}_{22.1} &= \mathbf{A}_{22} - \mathbf{A}_{21}\mathbf{A}_{11}^{-1}\mathbf{A}_{12}. \end{aligned} \tag{2}$$

Then (see Exercise 10) the submatrices of $\mathbf{B}$ are

$$
\begin{aligned}
\mathbf{B}_{11} &= \mathbf{A}_{11.2}^{-1}, \\
\mathbf{B}_{12} &= -\mathbf{A}_{11}^{-1}\mathbf{A}_{12}\mathbf{A}_{22.1}^{-1} = -\mathbf{A}_{11.2}^{-1}\mathbf{A}_{12}\mathbf{A}_{22}^{-1}, \\
\mathbf{B}_{21} &= -\mathbf{A}_{22}^{-1}\mathbf{A}_{21}\mathbf{A}_{11.2}^{-1} = -\mathbf{A}_{22.1}^{-1}\mathbf{A}_{21}\mathbf{A}_{11}^{-1}, \\
\mathbf{B}_{22} &= \mathbf{A}_{22.1}^{-1}.
\end{aligned}
\tag{3}
$$

If $\mathbf{A}$ is positive definite symmetric (*pds*, see appendix A.2), then $\mathbf{A}_{11.2}$ and $\mathbf{A}_{22.1}$ are *pds* as well; in fact they are the covariance matrices of conditional distributions in the multivariate normal theory setup of Section 3.3. See Exercise 11 for the inverse of a symmetric matrix where $\mathbf{A}_{22}$ is singular.

Suppose that the submatrix $\mathbf{A}_{11}$ of dimension $k \times k$ in the partitioned matrix $\mathbf{A}$ is nonsingular, and put $\mathbf{A}_{22.1} = \mathbf{A}_{22} - \mathbf{A}_{21}\mathbf{A}_{11}^{-1}\mathbf{A}_{12}$. Let

$$
\mathbf{C}_1 = \begin{pmatrix} \mathbf{I}_k & \mathbf{O} \\ -\mathbf{A}_{21}\mathbf{A}_{11}^{-1} & \mathbf{I}_{p-k} \end{pmatrix} \quad \text{and} \quad \mathbf{C}_2 = \begin{pmatrix} \mathbf{I}_k & -\mathbf{A}_{11}^{-1}\mathbf{A}_{12} \\ \mathbf{O} & \mathbf{I}_{p-k} \end{pmatrix}.
$$

By Exercise 12, $\det(\mathbf{C}_1) = \det(\mathbf{C}_2) = 1$, and therefore $\det(\mathbf{C}_1\mathbf{A}\mathbf{C}_2) = \det(\mathbf{A})$. On the other hand $\mathbf{C}_1\mathbf{A}\mathbf{C}_2 = \begin{pmatrix} \mathbf{A}_{11} & \mathbf{O} \\ \mathbf{O} & \mathbf{A}_{22.1} \end{pmatrix}$, and therefore

$$
\det(\mathbf{A}) = \det(\mathbf{A}_{11})\det(\mathbf{A}_{22.1}); \tag{4}
$$

see Exercise 13.

A.2 Positive Definite Matrices

Let $\mathbf{A}$ denote a $p \times p$ matrix, and define a function $Q_{\mathbf{A}} : \mathbb{R}^p \to \mathbb{R}$ by

$$
Q_{\mathbf{A}}(\mathbf{x}) = \mathbf{x}'\mathbf{A}\mathbf{x} = \sum_{i=1}^{p}\sum_{j=1}^{p} a_{ij}x_i x_j .
$$

This function is called a *quadratic form*. If $\mathbf{A}$ is not symmetric, we can define a symmetric matrix $\mathbf{B} = (\mathbf{A} + \mathbf{A}')/2$; then $Q_{\mathbf{B}}(\mathbf{x}) = Q_{\mathbf{A}}(\mathbf{x})$ because $\mathbf{x}'\mathbf{A}\mathbf{x} = (\mathbf{x}'\mathbf{A}\mathbf{x})' = \mathbf{x}'\mathbf{A}'\mathbf{x}$. Thus we can always assume without loss of generality that the matrix in a quadratic form is symmetric. The quadratic form $Q_{\mathbf{A}}(\mathbf{x})$, or equivalently the matrix $\mathbf{A}$, is called *positive definite* if $\mathbf{x}'\mathbf{A}\mathbf{x} > 0$ for all $\mathbf{x} \in \mathbb{R}^p$ ($\mathbf{x} \neq \mathbf{0}$), and *positive semidefinite* if $\mathbf{x}'\mathbf{A}\mathbf{x} \geq 0$ for all $\mathbf{x} \in \mathbb{R}^p$. See Exercise 14 for examples. We will use the abbreviation *pds* for positive definite symmetric.

Suppose the partitioned matrix $\mathbf{A} = \begin{pmatrix} \mathbf{A}_{11} & \mathbf{A}_{12} \\ \mathbf{A}_{21} & \mathbf{A}_{22} \end{pmatrix}$ is *pds*. Then both $\mathbf{A}_{11}$ and $\mathbf{A}_{22}$ are *pds*; this follows by taking vectors $\mathbf{x} = \begin{pmatrix} \mathbf{x}_1 \\ \mathbf{0} \end{pmatrix}$ and $\mathbf{x} = \begin{pmatrix} \mathbf{0} \\ \mathbf{x}_2 \end{pmatrix}$, respectively.

Suppose the dimension of $\mathbf{A}_{11}$ is $k \times k$. Writing $\mathbf{C} = \begin{pmatrix} \mathbf{I}_k & \mathbf{O} \\ -\mathbf{A}_{21}\mathbf{A}_{11}^{-1} & \mathbf{I}_{p-k} \end{pmatrix}$ and noticing that $\mathbf{C}^{-1} = \begin{pmatrix} \mathbf{I}_k & \mathbf{O} \\ \mathbf{A}_{21}\mathbf{A}_{11}^{-1} & \mathbf{I}_{p-k} \end{pmatrix}$, we obtain

$$\begin{aligned} \mathbf{x}'\mathbf{A}\mathbf{x} &= \mathbf{x}'\mathbf{C}^{-1}\left[\mathbf{C}\mathbf{A}\mathbf{C}'\right](\mathbf{C}')^{-1}\mathbf{x} \\ &= \mathbf{y}'\begin{pmatrix} \mathbf{A}_{11} & \mathbf{O} \\ \mathbf{O} & \mathbf{A}_{22.1} \end{pmatrix}\mathbf{y}, \end{aligned}$$

where $\mathbf{y} = (\mathbf{C}')^{-1}\mathbf{x}$. It follows that $\mathbf{A}_{22.1}$ is *pds*, and similarly that $\mathbf{A}_{11.2}$ is *pds*.

A.3 The Cholesky Decomposition

Suppose $\mathbf{A}$ is *pds* of dimension $p \times p$. Then there exists a unique lower triangular matrix $\mathbf{C}$ with positive diagonal entries such that $\mathbf{A} = \mathbf{C}\mathbf{C}'$. This is called the Cholesky decomposition (or factorization) of a *pds* matrix. We first prove existence of such a matrix and then discuss some computational aspects.

The proof is by induction on the dimension p. For $p = 1$ the matrix $\mathbf{A}$ is a real number, and the Cholesky factorization is $\mathbf{A} = c^2$ where $c = \sqrt{\mathbf{A}}$. Assume now that $p > 1$ and the Cholesky factorization has been shown for dimension $p-1$. Partition $\mathbf{A}$ as $\begin{pmatrix} \mathbf{A}_{11} & \mathbf{a}_{12} \\ \mathbf{a}_{21} & a_{22} \end{pmatrix}$, where a_{22} is scalar. By the induction hypothesis there exists a unique lower triangular matrix $\mathbf{C}_{p-1}$ of dimension $(p-1) \times (p-1)$ with positive diagonal entries such that $\mathbf{A}_{11} = \mathbf{C}_{p-1}\mathbf{C}'_{p-1}$. Define $\mathbf{C} = \begin{pmatrix} \mathbf{C}_{p-1} & \mathbf{0} \\ \mathbf{c}' & c_{pp} \end{pmatrix}$, where $\mathbf{c} \in \mathbb{R}^{p-1}$ and $c_{pp} > 0$ have yet to be determined. We want $\mathbf{C}\mathbf{C}' = \mathbf{A}$, i.e.,

$$\begin{aligned} \mathbf{C}_{p-1}\mathbf{C}_{p-1} &= \mathbf{A}_{11}, \\ \mathbf{C}_{p-1}\mathbf{c} &= \mathbf{a}_{12}, \\ \mathbf{c}'\mathbf{c} + c_{pp}^2 &= a_{22}. \end{aligned} \tag{5}$$

Since $\mathbf{C}_{p-1}$ is nonsingular, it follows that we have to choose

$$\mathbf{c} = \mathbf{C}_{p-1}^{-1}\mathbf{a}_{12} \tag{6}$$

and

$$c_{pp} = (a_{22} - \mathbf{c}'\mathbf{c})^{1/2} = \left(a_{22} - \mathbf{a}_{21}\mathbf{A}_{11}^{-1}\mathbf{a}_{12}\right)^{1/2} = \sqrt{a_{22.1}}. \tag{7}$$

Notice that positive definiteness of $\mathbf{A}$ guarantees that $a_{22.1} > 0$; see appendix A.2.

The Cholesky decomposition is of great value in numerical analysis because it can be computed precisely without iteration. As the above proof by induction shows, we can compute the factorization by first finding the factorization of the leading 1×1 matrix (i.e., put $c_{11} = \sqrt{a_{11}}$), then the factorization of the leading 2×2 matrix, and so on. The crucial calculation in each step is solving the equation $\mathbf{C}_{p-1}\mathbf{c} = \mathbf{a}_{12}$ for $\mathbf{c}$; but since $\mathbf{C}_{p-1}$ is lower triangular this is an easy task. See Golub and Van Loan (1983) for a numerical analyst's treatment of the Cholesky factorization. Since the determinant of a triangular matrix is the product of its diagonal elements (Exercise 4), it follows that $\det(\mathbf{A}) = \prod_{i=1}^{p} c_{ii}^2$.

If $\mathbf{A} = \mathbf{C}\mathbf{C}'$ is the Cholesky factorization of a *pds* matrix A, then $\mathbf{A}^{-1} = (\mathbf{C}')^{-1}\mathbf{C}^{-1}$. Putting $\mathbf{y} = \mathbf{C}^{-1}\mathbf{x}$ it follows that $\mathbf{x}'\mathbf{A}^{-1}\mathbf{x} = \mathbf{y}'\mathbf{y} > 0$ for all $\mathbf{y} \in \mathbb{R}^p$ $(\mathbf{y} \neq \mathbf{0})$. That is if $\mathbf{A}$ is *pds*, then $\mathbf{A}^{-1}$ is *pds* as well.

A.4 Vector and Matrix Differentiation

Let $\mathbf{X}$ denote a matrix of dimension $m \times k$, and $f(\mathbf{X})$ a real–valued function of $\mathbf{X}$. Then we denote the matrix of partial derivatives of $f(\mathbf{X})$ with respect to the entries x_{ij} of $\mathbf{X}$ by

$$\frac{\partial f(\mathbf{X})}{\partial \mathbf{X}} = \begin{pmatrix} \frac{\partial f(\mathbf{X})}{\partial x_{11}} & \cdots & \frac{\partial f(\mathbf{X})}{\partial x_{1k}} \\ \vdots & & \vdots \\ \frac{\partial f(\mathbf{X})}{\partial x_{m1}} & \cdots & \frac{\partial f(\mathbf{X})}{\partial x_{mk}} \end{pmatrix}. \tag{8}$$

Thus $\partial f(\mathbf{X})/\partial \mathbf{X}$ has the same dimension as $\mathbf{X}$.

The proof of the following results (i) to (iii) is left to the reader; see Exercise 16. For result (iv) see Searle (1982, Section 12.10).

(i) If $\mathbf{x} \in \mathbb{R}^p$ and $f(\mathbf{x}) = \mathbf{a}'\mathbf{x} = \mathbf{x}'\mathbf{a}$ for some fixed vector $\mathbf{a} \in \mathbb{R}^p$, then

$$\frac{\partial f(\mathbf{x})}{\partial \mathbf{x}} = \mathbf{a}. \tag{9}$$

(ii) If $\mathbf{x} \in \mathbb{R}^p$, and $f(\mathbf{x}) = \mathbf{x}'\mathbf{A}\mathbf{x}$ where $\mathbf{A}$ is a symmetric matrix of dimension $p \times p$, then

$$\frac{\partial f(\mathbf{x})}{\partial \mathbf{x}} = 2\mathbf{A}\mathbf{x}. \tag{10}$$

(iii) If $f(\mathbf{X}) = \mathrm{tr}(\mathbf{X}\mathbf{A})$, where $\mathbf{X}$ is symmetric and $\mathbf{A}$ has the same dimension as $\mathbf{X}$, then

$$\frac{\partial f(\mathbf{x})}{\partial \mathbf{x}} = \mathbf{A} + \mathbf{A}' - \mathrm{diag}(\mathbf{A}). \tag{11}$$

(iv) If $f(\mathbf{X}) = \log\det(\mathbf{X})$, where $\mathbf{X}$ is a symmetric nonsingular matrix, then

$$\frac{\partial f(\mathbf{X})}{\partial \mathbf{X}} = \mathbf{X}^{-1} - \mathrm{diag}(\mathbf{X}^{-1}). \tag{12}$$

Next let $\mathbf{S}(\mathbf{x})$ denote a vector of m functions of $\mathbf{x} \in \mathbb{R}^p$, i.e.,

$$\mathbf{S}(\mathbf{x}) = \begin{pmatrix} S_1(\mathbf{x}) \\ \vdots \\ S_m(\mathbf{x}) \end{pmatrix}.$$

Then we can arrange the partial derivatives of the $S_j(\mathbf{x})$ with respect to the x_i in form of an $m \times p$ matrix

$$\frac{\partial \mathbf{S}(\mathbf{x})}{\partial \mathbf{x}'} = \begin{pmatrix} \frac{\partial S_1(\mathbf{x})}{\partial x_1} & \cdots & \frac{\partial S_1(\mathbf{x})}{\partial x_p} \\ \vdots & & \vdots \\ \frac{\partial S_m(\mathbf{x})}{\partial x_1} & \cdots & \frac{\partial S_m(\mathbf{x})}{\partial x_p} \end{pmatrix}. \tag{13}$$

In particular, let $f(\mathbf{x})$ be a real–valued function of $\mathbf{x} \in \mathbb{R}^p$, and put $\mathbf{S}(\mathbf{x}) = \partial f(\mathbf{x})/\partial \mathbf{x}$. Then the matrix of second derivatives of $f(\mathbf{x})$ with respect to the x_i, called the *Hessian* matrix, is

$$\frac{\partial^2 f(\mathbf{x})}{\partial \mathbf{x}\, \partial \mathbf{x}'} = \frac{\partial \mathbf{S}(\mathbf{x})}{\partial \mathbf{x}'} = \begin{pmatrix} \frac{\partial^2 f(\mathbf{x})}{\partial x_1^2} & \cdots & \frac{\partial^2 f(\mathbf{x})}{\partial x_1\, \partial x_p} \\ \vdots & \ddots & \vdots \\ \frac{\partial^2 f(\mathbf{x})}{\partial x_p\, \partial x_1} & \cdots & \frac{\partial^2 f(\mathbf{x})}{\partial x_p^2} \end{pmatrix}. \tag{14}$$

A.5 Eigenvectors and Eigenvalues

Let $\mathbf{A}$ be a matrix of dimension $p \times p$. If

$$\mathbf{A}\mathbf{x} = \lambda \mathbf{x} \tag{15}$$

for some nonzero vector $\mathbf{x} \in \mathbb{R}^p$ and some scalar λ, then $\mathbf{x}$ is called an *eigenvector* (characteristic vector, latent vector) of $\mathbf{A}$, and λ is the associated *eigenvalue* (characteristic root, latent root). Eigenvectors of $\mathbf{A}$ are vectors that are transformed into multiples of themselves when multiplied by $\mathbf{A}$. Eigenvectors are not uniquely defined: if $\mathbf{A}\mathbf{x} = \lambda\mathbf{x}$, then $\mathbf{A}(c\mathbf{x}) = \lambda(c\mathbf{x})$ for any $c \in \mathbb{R}$. Thus some normalization constraint is usually applied to eigenvectors, the most common one being $\mathbf{x}'\mathbf{x} = 1$, i.e., unit length.

Equivalently to equation (15) we can write $(\mathbf{A} - \lambda\mathbf{I}_p)\mathbf{x} = \mathbf{0}$, which implies $\det(\mathbf{A} - \lambda\mathbf{I}_p) = 0$ (otherwise the only solution to (15) would be the trivial one, $\mathbf{x} = \mathbf{0}$). The function

$$\mathcal{P}(\lambda) = \det(\mathbf{A} - \lambda\mathbf{I}_p) \tag{16}$$

is a polynomial of degree p in λ (see Exercise 18), called the *characteristic polynomial* of $\mathbf{A}$. The roots of this polynomial may all be real, all complex, or some may be real and some complex. In this book we encounter only problems where all eigenvalues are real, so we do not elaborate on the complex case. In particular, if the $p \times p$ matrix $\mathbf{A}$ is symmetric, then the characteristic polynomial can be factorized as

$$\mathcal{P}(\lambda) = \prod_{i=1}^{p} (\lambda - \lambda_i),$$

where all λ_i are real (see, e.g., Schott 1997, Theorem 3.8). That is, a symmetric matrix has p real eigenvalues. We will usually write them in algebraically decreasing order, i.e., $\lambda_1 \geq \lambda_2 \geq \cdots \geq \lambda_p$. Some of the eigenvalues may be identical; in such a case we refer to the number of times the same eigenvalue occurs as its *multiplicity*.

For symmetric matrices it is common to normalize its eigenvectors $\boldsymbol{\beta}_1, \ldots, \boldsymbol{\beta}_p$ to unit length, i.e., $\boldsymbol{\beta}_i'\boldsymbol{\beta}_i = 1, i = 1, \ldots, p$. If two eigenvalues λ_i and λ_j of a symmetric matrix $\mathbf{A}$ are distinct, then the associated eigenvectors $\boldsymbol{\beta}_i$ and $\boldsymbol{\beta}_j$ are orthogonal to each other. This can be seen as follows: Since $\mathbf{A}\boldsymbol{\beta}_i = \lambda_i\boldsymbol{\beta}_i$ and $\mathbf{A}\boldsymbol{\beta}_j = \lambda_j\boldsymbol{\beta}_j$, we have $\boldsymbol{\beta}_j'\mathbf{A}\boldsymbol{\beta}_i = \lambda_i\boldsymbol{\beta}_j'\boldsymbol{\beta}_i = \lambda_j\boldsymbol{\beta}_j'\boldsymbol{\beta}_i$, which implies $\boldsymbol{\beta}_j'\boldsymbol{\beta}_i = 0$. If $\lambda_i = \lambda_j$ then we can still choose associated eigenvectors $\boldsymbol{\beta}_i$ and $\boldsymbol{\beta}_j$ that are orthogonal to each other.

For a (not necessarily symmetric) matrix $\mathbf{A}$ with eigenvalues λ_i and associated eigenvectors $\boldsymbol{\beta}_i$, put $\boldsymbol{\Lambda} = \text{diag}(\lambda_1, \ldots, \lambda_p)$ and $\mathbf{B} = (\boldsymbol{\beta}_1 \ \cdots \ \boldsymbol{\beta}_p)$. Then the p equations $\mathbf{A}\boldsymbol{\beta}_i = \lambda_i\boldsymbol{\beta}_i$ can be written simultaneously as

$$\mathbf{AB} = \mathbf{B}\boldsymbol{\Lambda}. \tag{17}$$

A square matrix $\mathbf{A}$ is called *idempotent* if $\mathbf{A}^2 = \mathbf{A}$. Suppose $\mathbf{x}$ is an eigenvector of an idempotent matrix $\mathbf{A}$, with associated eigenvalue λ. Since $\lambda\mathbf{x} = \mathbf{Ax} = \mathbf{A}^2\mathbf{x} = \mathbf{A}(\lambda\mathbf{x}) = \lambda^2\mathbf{x}$ and eigenvectors are nonzero, it follows that $\lambda^2 = \lambda$. That is, all eigenvalues of an idempotent matrix are zero or one. See also Exercise 37.

A.6 Spectral Decomposition of Symmetric Matrices

By appendix A.5, a symmetric $p \times p$ matrix $\mathbf{A}$ has p real eigenvalues $\lambda_1 \geq \lambda_2 \geq \cdots \geq \lambda_p$ and associated eigenvectors $\boldsymbol{\beta}_1, \ldots, \boldsymbol{\beta}_p$ of unit length that can be chosen to be mutually orthogonal. Writing $\mathbf{B} = (\boldsymbol{\beta}_1 \ \cdots \ \boldsymbol{\beta}_p)$ and using orthogonality of the $\boldsymbol{\beta}_i$, we have

$$\mathbf{B}'\mathbf{B} = \mathbf{I}_p\,, \tag{18}$$

i.e., $\mathbf{B}' = \mathbf{B}^{-1}$. A $p \times p$ matrix that satisfies (18) is called an *orthogonal* matrix. It also follows from (18) that

$$\mathbf{BB}' = \sum_{i=1}^{p} \boldsymbol{\beta}_i\boldsymbol{\beta}_i' = \mathbf{I}_p\,. \tag{19}$$

Multiplying equation (17) from the left by $\mathbf{B}'$ we obtain

$$\mathbf{B}'\mathbf{A}\mathbf{B} = \Lambda = \text{diag}(\lambda_1, \ldots, \lambda_p). \tag{20}$$

Similarly, multiplying equation (17) from the right by $\mathbf{B}'$ we have

$$\begin{aligned}\mathbf{A} = \mathbf{B}\Lambda\mathbf{B}' &= (\boldsymbol{\beta}_1 \quad \cdots \quad \boldsymbol{\beta}_p) \begin{pmatrix} \lambda_1 & 0 & \cdots & 0 \\ 0 & \lambda_2 & \cdots & 0 \\ \vdots & \vdots & \ddots & \vdots \\ 0 & 0 & \cdots & \lambda_p \end{pmatrix} \begin{pmatrix} \boldsymbol{\beta}_1' \\ \vdots \\ \boldsymbol{\beta}_p' \end{pmatrix} \\ &= \sum_{i=1}^{p} \lambda_i \boldsymbol{\beta}_i \boldsymbol{\beta}_i'.\end{aligned} \tag{21}$$

This is called the *spectral decomposition* of the symmetric matrix $\mathbf{A}$. Even if all λ_i are distinct and ordered, the spectral decomposition is not unique because each column of $\mathbf{B}$ can be multiplied by -1. However, such non–uniqueness will usually not be a major concern.

Let $Q_{\mathbf{A}} = \mathbf{x}'\mathbf{A}\mathbf{x}$ be a quadratic form. Using the spectral decomposition $\mathbf{A} = \mathbf{B}\Lambda\mathbf{B}'$ we can write

$$Q_{\mathbf{A}} = \mathbf{x}'\mathbf{B}\Lambda\mathbf{B}'\mathbf{x} = \mathbf{y}'\Lambda\mathbf{y} = \sum_{i=1}^{p} \lambda_i y_i^2, \tag{22}$$

where $\mathbf{y} = (y_1 \quad \cdots \quad y_p)' = \mathbf{B}'\mathbf{x}$. It follows that $\mathbf{A}$ is positive definite exactly if all $\lambda_i > 0$, and $\mathbf{A}$ is positive semidefinite if all $\lambda_i \geq 0$; see Exercise 20.

Suppose the symmetric matrix $\mathbf{A} = \mathbf{B}\Lambda\mathbf{B}'$ has an eigenvalue of multiplicity $r > 0$. For simplicity assume that these are the first r, i.e., $\lambda_1 = \cdots = \lambda_r > \lambda_{r+1}$. Then the orthogonal matrix $\mathbf{B}$ of eigenvectors is not unique. If $\mathbf{M}$ denotes any orthogonal matrix of dimension $r \times r$, then the matrix

$$\mathbf{C} = \mathbf{B}\begin{pmatrix} \mathbf{M} & \mathbf{O} \\ \mathbf{O} & \mathbf{I}_{p-r} \end{pmatrix}$$

has the same properties as $\mathbf{B}$: it is orthogonal, and $\mathbf{A} = \mathbf{C}\Lambda\mathbf{C}'$; see Exercise 21.

The determinant of an orthogonal matrix $\mathbf{B}$ is ± 1 because $\det(\mathbf{B}'\mathbf{B}) = [\det(\mathbf{B})]^2 = 1$. Thus a symmetric matrix $\mathbf{A}$ with spectral decomposition has

$$\det(\mathbf{A}) = \det(\mathbf{B}\Lambda\mathbf{B}') = \det(\Lambda) = \prod_{i=1}^{p} \lambda_i, \tag{23}$$

and is therefore singular if and only if at least one eigenvalue is zero. For the trace of $\mathbf{A}$ we obtain

$$\text{tr}(\mathbf{A}) = \text{tr}(\mathbf{B}\Lambda\mathbf{B}') = \text{tr}(\Lambda\mathbf{B}'\mathbf{B}) = \text{tr}(\Lambda) = \sum_{i=1}^{p} \lambda_i. \tag{24}$$

A.7 The Square Root of a Positive Definite Symmetric Matrix

Suppose $\mathbf{A}$ is positive semidefinite and symmetric, with spectral decomposition $\mathbf{A} = \mathbf{B}\Lambda\mathbf{B}' = \sum_{i=1}^{p} \lambda_i \beta_i \beta_i'$. Define

$$\mathbf{A}^{1/2} = \mathbf{B}\Lambda^{1/2}\mathbf{B}' = \sum_{i=1}^{p} \lambda_i^{1/2} \beta_i \beta_i', \tag{25}$$

where $\Lambda^{1/2} = \text{diag}\left(\lambda_1^{1/2}, \ldots, \lambda_p^{1/2}\right)$. Since $\mathbf{B}$ is orthogonal, it follows that $\mathbf{A}^{1/2}\mathbf{A}^{1/2} = \mathbf{B}\Lambda^{1/2}\mathbf{B}'\mathbf{B}\Lambda^{1/2}\mathbf{B}' = \mathbf{A}$. In fact among all matrices $\mathbf{M}$ satisfying $\mathbf{A} = \mathbf{M}\mathbf{M}'$, (25) is the only symmetric one; see Exercise 23. Thus it is reasonable to call $\mathbf{A}^{1/2}$ the unique square root of $\mathbf{A}$, although some authors would call any $\mathbf{M}$ such that $\mathbf{A} = \mathbf{M}\mathbf{M}'$ a square root matrix. An example is the Cholesky decomposition of appendix A.3. Another possibility is to choose $\mathbf{M} = \mathbf{B}\Lambda^{1/2}$.

If $\mathbf{A}$ is positive definite, all eigenvalue s λ_i are strictly positive, and we can define powers of the matrix $\mathbf{A}$ by

$$\mathbf{A}^r = \mathbf{B}\Lambda^r\mathbf{B}' = \sum_{i=1}^{p} \lambda_i^r \beta_i \beta_i' \tag{26}$$

for any $r \in \mathbb{R}$. Then (Exercise 24) $\mathbf{A}^{r+s} = \mathbf{A}^r\mathbf{A}^s$, similar to powers of positive numbers. In particular $\mathbf{A}^{-1/2} = \sum_{i=1}^{p} \lambda_i^{-1/2} \beta_i \beta_i'$, with $\mathbf{A}^{-1/2}\mathbf{A}^{-1/2} = \mathbf{A}^{-1}$, and $\mathbf{A}^{-1/2} = (\mathbf{A}^{1/2})^{-1}$.

A.8 Orthogonal Projections on Lines and Planes

Let $\mathbf{a}_1, \ldots, \mathbf{a}_q$ denote q linearly independent vectors in $\mathbb{R}^p$, $1 \le q \le p$, and define the $p \times q$ matrix $\mathbf{A}$ as $\mathbf{A} = (\mathbf{a}_1 \ \cdots \ \mathbf{a}_q)$. The $\mathbf{a}_i$ span a linear subspace $\mathcal{S}$ of dimension q. It is often convenient to choose the q linearly independent vectors $\mathbf{a}_i$ such that they have unit length ($\mathbf{a}_i'\mathbf{a}_i = 1$ for all $i = 1, \ldots, q$), and such that they are mutually orthogonal, i.e., $\mathbf{a}_i'\mathbf{a}_j = 0$ for $i \neq j$. In this case $\mathbf{a}_1, \ldots, \mathbf{a}_q$ form an *orthonormal basis* of the subspace $\mathcal{S}$. For the matrix $\mathbf{A} = (\mathbf{a}_1 \ \cdots \ \mathbf{a}_q)$ we then have

$$\mathbf{A}'\mathbf{A} = \begin{pmatrix} \mathbf{a}_1'\mathbf{a}_1 & \mathbf{a}_1'\mathbf{a}_2 & \cdots & \mathbf{a}_1'\mathbf{a}_q \\ \mathbf{a}_2'\mathbf{a}_1 & \mathbf{a}_2'\mathbf{a}_2 & \cdots & \mathbf{a}_2'\mathbf{a}_q \\ \vdots & \vdots & \ddots & \vdots \\ \mathbf{a}_q'\mathbf{a}_1 & \mathbf{a}_q'\mathbf{a}_2 & \cdots & \mathbf{a}_q'\mathbf{a}_q \end{pmatrix} = \mathbf{I}_q , \tag{27}$$

and

$$\mathbf{A}\mathbf{A}' = \mathbf{a}_1\mathbf{a}_1' + \cdots + \mathbf{a}_q\mathbf{a}_q' = \sum_{i=1}^{q} \mathbf{a}_i\mathbf{a}_i'. \tag{28}$$

If $q = p$, then the matrix $\mathbf{A}$ is orthogonal of dimension $p \times p$, i.e., $\mathbf{A}' = \mathbf{A}^{-1}$.

Any vector $\mathbf{y} \in \mathbb{R}^p$ in the linear subspace $\mathcal{S}$ spanned by the q columns of $\mathbf{A}$ can be written as a linear combination of the $\mathbf{a}_i$, i.e., as $\mathbf{y} = c_1\mathbf{a}_1 + \cdots + c_q\mathbf{a}_q = \mathbf{A}\mathbf{c}$ for some vector $\mathbf{c} = (c_1, \ldots, c_q)' \in \mathbb{R}^q$. The c_i are the coordinates of $\mathbf{y}$ in the subspace $\mathcal{S}$ with respect to the basis $\mathbf{a}_1, \ldots, \mathbf{a}_q$. The subspace $\mathcal{S}$ has the representation

$$\mathcal{S} = \left\{\mathbf{y} \in \mathbb{R}^p : \mathbf{y} = \mathbf{A}\mathbf{c} \text{ for some } \mathbf{c} \in \mathbb{R}^q\right\}. \tag{29}$$

We will from now on assume that the basis is orthonormal.

Suppose we wish to find a function which maps each point $\mathbf{x} \in \mathbb{R}^p$ into the point in $\mathcal{S}$ which is closest to $\mathbf{x}$. The squared Euclidean distance between $\mathbf{x}$ and a point $\mathbf{y} = \mathbf{A}\mathbf{c}$ in $\mathcal{S}$ is

$$\begin{aligned} h(\mathbf{c}) = ||\mathbf{x} - \mathbf{A}\mathbf{c}||^2 &= (\mathbf{x} - \mathbf{A}\mathbf{c})'(\mathbf{x} - \mathbf{A}\mathbf{c}) \\ &= \mathbf{x}'\mathbf{x} - 2\mathbf{x}'\mathbf{A}\mathbf{c} + \mathbf{c}'\mathbf{c}. \end{aligned} \tag{30}$$

We need to minimize the function $h(\mathbf{c})$ with respect to $\mathbf{c} \in \mathbb{R}^q$. Using the rules for differentiation from appendix A.4 we obtain

$$\frac{\partial\, h(\mathbf{c})}{\partial\, \mathbf{c}} = -2\mathbf{A}'\mathbf{x} + 2\mathbf{c}, \tag{31}$$

which when set equal to zero, gives $\mathbf{c} = \mathbf{A}'\mathbf{x}$. That this choice of $\mathbf{c}$ leads to a unique minimum follows by forming the matrix of second derivatives (Exercise 25), or by noticing that $h(\mathbf{c})$ is a nonnegative quadratic function. An alternative proof that does not involve derivatives is given in Exercise 26.

Thus, for any $\mathbf{x} \in \mathbb{R}^q$, the $\mathbf{y} \in \mathcal{S}$ which is closest to $\mathbf{x}$ has the form $\mathbf{y} = \mathbf{A}\mathbf{c}$ for $\mathbf{c} = \mathbf{A}'\mathbf{x}$, or

$$\mathbf{y} = \mathbf{A}\mathbf{A}'\mathbf{x}. \tag{32}$$

This linear transformation maps each $\mathbf{x} \in \mathbb{R}^q$ into the nearest $\mathbf{y} \in \mathcal{S}$. We will usually write $\mathbf{Y} = \mathbf{P}\mathbf{x}$, where $\mathbf{P} = \mathbf{A}\mathbf{A}'$. The linear mapping $g : \mathbb{R}^p \to \mathcal{S}$ given by $g(\mathbf{x}) = \mathbf{P}\mathbf{x}$ is called the *orthogonal projection* onto the q–dimensional subspace $\mathcal{S}$. The matrix $\mathbf{P}$ is called a *projection matrix*. The term "projection matrix" reflects the fact that

$$g\big(g(\mathbf{x})\big) = g(\mathbf{x}) \quad \text{for all } \mathbf{x} \in \mathbb{R}^p, \tag{33}$$

or in other words, each point in $\mathcal{S}$ is mapped into itself. Equation (33) is verified by noticing that

$$\mathbf{P}^2 = (\mathbf{A}\mathbf{A}')(\mathbf{A}\mathbf{A}') = \mathbf{A}(\mathbf{A}'\mathbf{A})\mathbf{A}' = \mathbf{A}\mathbf{I}_q\mathbf{A}' = \mathbf{P}, \tag{34}$$

which implies $\mathbf{P}^2\mathbf{x} = \mathbf{P}\mathbf{x}$ for all $\mathbf{x} \in \mathbb{R}^p$. By (34), the projection matrix P is *idempotent*; see appendix A.5. By construction the projection matrix $\mathbf{P}$ is also symmetric.

The projection matrix $\mathbf{P}$ associated with the projection onto a q–dimensional linear subspace $\mathcal{S}$ is unique, although the orthonormal basis for $\mathcal{S}$ is not unique. This can be seen as follows. Let $\mathbf{a}_1, \ldots, \mathbf{a}_q$ denote an arbitrary orthonormal basis, and $\mathbf{b}_1, \ldots, \mathbf{b}_q$ a different orthonormal basis for $\mathcal{S}$. Let $\mathbf{A} = (\mathbf{a}_1 \ \cdots \ \mathbf{a}_q)$ and $\mathbf{B} = (\mathbf{b}_1 \ \cdots \ \mathbf{b}_q)$. Then there exists an orthogonal matrix $\mathbf{H}$ of dimension $q \times q$ (see Exercise 27) such that $\mathbf{B} = \mathbf{AH}$. Constructing a projection matrix using the matrix $\mathbf{B}$ then gives $\mathbf{BB}' = (\mathbf{AH})(\mathbf{AH})' = \mathbf{A}(\mathbf{H}'\mathbf{H})\mathbf{A}' = \mathbf{AA}'$. This implies that the projection matrix does not depend on the orthonormal basis chosen for the subspace $\mathcal{S}$.

Particularly simple projection matrices are those associated with projections onto subspaces spanned by some q unit basis vectors of $\mathbb{R}^p$. Let $\mathbf{e}_i$ denote the ith unit basis vector, and let $\mathcal{I}$ denote a set of q distinct integers between 1 and p, then the projection matrix associated with the projection onto the q–dimensional subspace associated with the vectors $\mathbf{e}_i$, $i \in \mathcal{I}$, is

$$\mathbf{P} = \sum_{i \in \mathcal{I}} \mathbf{e}_i \mathbf{e}_i'. \tag{35}$$

This matrix has zeros in all entries except for q diagonal entries that are equal to one.

If q orthonormal vectors $\mathbf{a}_1, \ldots, \mathbf{a}_q$ span a subspace $\mathcal{S}$ of dimension q, then we can find $p-q$ additional vectors $\mathbf{a}_{q+1}, \ldots, \mathbf{a}_p$ in $\mathbb{R}^p$, all of unit length, which are mutually orthogonal and orthogonal to $\mathbf{a}_1, \ldots, \mathbf{a}_q$. These vectors form an orthonormal basis of a $(p-q)$–dimensional subspace $\mathcal{S}^c$ orthogonal to $\mathcal{S}$. Putting $\mathbf{A} = (\mathbf{a}_1 \ \cdots \ \mathbf{a}_q)$ and $\mathbf{A}^* = (\mathbf{a}_{q+1} \ \cdots \ \mathbf{a}_p)$, the $p \times p$ matrix $(\mathbf{A} \ \ \mathbf{A}^*)$ is orthogonal because

$$(\mathbf{A} \ \ \mathbf{A}^*)'(\mathbf{A} \ \ \mathbf{A}^*) = \begin{pmatrix} \mathbf{A}'\mathbf{A} & \mathbf{A}'\mathbf{A}^* \\ (\mathbf{A}^*)'\mathbf{A} & (\mathbf{A}^*)'\mathbf{A}^* \end{pmatrix} = \begin{pmatrix} \mathbf{I}_q & \mathbf{0} \\ \mathbf{0} & \mathbf{I}_{p-q} \end{pmatrix} = \mathbf{I}_p.$$

Therefore,

$$\mathbf{I}_p = (\mathbf{A} \ \ \mathbf{A}^*)(\mathbf{A} \ \ \mathbf{A}^*)' = \mathbf{AA}' + \mathbf{A}^*(\mathbf{A}^*)' = \mathbf{P} + \mathbf{P}^*, \tag{36}$$

where $\mathbf{P} = \mathbf{AA}'$ and $\mathbf{P}^* = \mathbf{A}^*(\mathbf{A}^*)'$ are the projection matrices associated with the subspaces $\mathcal{S}$ and $\mathcal{S}^c$. From (36) it follows that the projection onto the subspace $\mathcal{S}^c$ orthogonal to $\mathcal{S}$ has projection matrix $\mathbf{I}_p - \mathbf{P}$. Every vector $\mathbf{x} \in \mathbb{R}^p$ can be decomposed uniquely as $\mathbf{x} = \mathbf{x}_1 + \mathbf{x}_2$, where $\mathbf{x}_1 = \mathbf{Px}$ and $\mathbf{x}_2 = (\mathbf{I}_p - \mathbf{P})\mathbf{x}$ are the projections onto $\mathcal{S}$ and $\mathcal{S}^c$ respectively, and $\mathbf{x}_1$ is orthogonal to $\mathbf{x}_2$ (Exercise 28).

Instead of proper linear subspaces as in (29), we will often be concerned with lines, planes, or hyperplanes, i.e., sets of the form

$$\mathcal{M} = \left\{ \mathbf{x} \in \mathbb{R}^p : \mathbf{x} = \mathbf{x}_0 + \mathbf{Ac} \text{ for some } \mathbf{c} \in \mathbb{R}^q \right\}, \tag{37}$$

where the q columns of $\mathbf{A}$ are orthonormal, and $\mathbf{x}_0 \in \mathbb{R}^p$ is an arbitrary point in the line, plane, or hyperplane. We can symbolically write $\mathcal{M} = \mathbf{x}_0 + \mathcal{S}$, where $\mathcal{S}$ is the q–dimensional subspace spanned by the columns of $\mathbf{A}$. If $\mathbf{P} = \mathbf{AA}'$ is the projection matrix associated with $\mathcal{S}$, then the orthogonal projection onto $\mathcal{M} = \mathbf{x}_0 + \mathcal{S}$ is given by the affine function

$$g(\mathbf{x}) = \mathbf{x}_0 + \mathbf{P}(\mathbf{x} - \mathbf{x}_0), \qquad \mathbf{x} \in \mathbb{R}^p. \tag{38}$$

The function $g(\mathbf{x})$ maps any $\mathbf{x} \in \mathbb{R}^p$ into $\mathcal{M}$ because the righthand side of (38) can be written as

$$\mathbf{x}_0 + \mathbf{A}\mathbf{A}'(\mathbf{x} - \mathbf{x}_0) = \mathbf{x}_0 + \mathbf{A}\mathbf{c} \ , \tag{39}$$

which has the form of the vectors in (37), with $\mathbf{c} = \mathbf{A}'(\mathbf{x} - \mathbf{x}_0)$. Moreover, the function g is a projection: for some $\mathbf{x} \in \mathbb{R}^p$, let $\mathbf{y} = g(\mathbf{x}) = \mathbf{x}_0 + \mathbf{P}(\mathbf{x} - \mathbf{x}_0)$. Then

$$\begin{aligned} g\big(g(\mathbf{x})\big) = g(\mathbf{y}) &= \mathbf{x}_0 + \mathbf{P}(\mathbf{y} - \mathbf{x}_0) \\ &= \mathbf{x}_0 + \mathbf{P}\big(\mathbf{x}_0 + \mathbf{P}(\mathbf{x} - \mathbf{x}_0) - \mathbf{x}_0\big) \\ &= \mathbf{x}_0 + \mathbf{P}^2(\mathbf{x} - \mathbf{x}_0) \\ &= \mathbf{x}_0 + \mathbf{P}(\mathbf{x} - \mathbf{x}_0) \\ &= g(\mathbf{x}) , \end{aligned}$$

as was to be shown.

A.9 Simultaneous Decomposition of Two Symmetric Matrices

Suppose $\mathbf{W}$ and $\mathbf{A}$ are two symmetric matrices of the same dimension, $\mathbf{W}$ being positive definite. Let $\mathbf{W}^{-1/2}$ be the inverse square root of $\mathbf{W}$ (see appendix A.7) and define a symmetric matrix $\mathbf{S} = \mathbf{W}^{-1/2}\mathbf{A}\mathbf{W}^{-1/2}$. Let $\mathbf{S} = \mathbf{B}\Lambda\mathbf{B}'$ denote the spectral decomposition of $\mathbf{S}$, where $\Lambda = \text{diag}(\lambda_1, \ldots, \lambda_p)$, and $\mathbf{B}$ is orthogonal. It follows that $\Lambda = \mathbf{B}'\mathbf{S}\mathbf{B} = \mathbf{B}'\mathbf{W}^{-1/2}\mathbf{A}\mathbf{W}^{-1/2}\mathbf{B}$. Putting $\Gamma = \mathbf{W}^{-1/2}\mathbf{B}$ (which is nonsingular), we have

$$\Gamma'\mathbf{W}\Gamma = \mathbf{I} \quad \text{and} \quad \Gamma'\mathbf{A}\Gamma = \Lambda. \tag{40}$$

Thus we have established existence of a nonsingular matrix Γ that diagonalizes both $\mathbf{W}$ and $\mathbf{A}$ simultaneously. Putting $\mathbf{H} = (\Gamma')^{-1}$, it follows that

$$\begin{aligned} \mathbf{W} &= \mathbf{H}\mathbf{H}', \\ \mathbf{A} &= \mathbf{H}\Lambda\mathbf{H}'. \end{aligned} \tag{41}$$

This may be called a simultaneous spectral decomposition of $\mathbf{W}$ and $\mathbf{A}$. Note however that in contrast to the spectral decomposition of a single symmetric matrix the matrix $\mathbf{H}$ is not necessarily orthogonal. With $\mathbf{H} = (\mathbf{h}_1 \ \cdots \ \mathbf{h}_p)$, the simultaneous decomposition can be written as

$$\begin{aligned} \mathbf{W} &= \sum_{i=1}^{p} \mathbf{h}_i\mathbf{h}_i' , \\ \mathbf{A} &= \sum_{i=1}^{p} \lambda_i\mathbf{h}_i\mathbf{h}_i' . \end{aligned} \tag{42}$$

From (41) it also follows that $\mathbf{W}^{-1} = (\mathbf{HH}')^{-1} = \boldsymbol{\Gamma}\boldsymbol{\Gamma}'$, and therefore

$$\mathbf{W}^{-1}\mathbf{A}\boldsymbol{\Gamma} = \boldsymbol{\Gamma}\boldsymbol{\Gamma}'\mathbf{A}\boldsymbol{\Gamma} = \boldsymbol{\Gamma}\boldsymbol{\Lambda}. \tag{43}$$

That is, the columns $\gamma_1, \ldots, \gamma_p$ of $\boldsymbol{\Gamma}$ are eigenvectors of $\mathbf{W}^{-1}\mathbf{A}$, and the diagonal entries λ_i of $\boldsymbol{\Lambda}$ are the associated eigenvalues.

Exercises for the Appendix

1. If $\mathbf{A}$ and $\mathbf{B}$ are two matrices such that both $\mathbf{AB}$ and $\mathbf{BA}$ are defined, show that $\text{tr}(\mathbf{AB}) = \text{tr}(\mathbf{BA})$.

2. If $\mathbf{A}$ is a symmetric nonsingular matrix, show that $\mathbf{A}^{-1}$ is symmetric.

3. Show that $(\mathbf{AB})' = \mathbf{B}'\mathbf{A}'$.

4. Define the determinant of a matrix. Show that if $\mathbf{A}$ is triangular (i.e., either all entries above the main diagonal or all entries below the main diagonal are zero), then the determinant is the product of the diagonal entries.

5. Show that $\det(\mathbf{A}) = \det(\mathbf{A}')$ and $\det(\mathbf{A}^{-1}) = 1/\det(\mathbf{A})$.

6. If $\mathbf{B}$ is obtained from the $p \times p$ matrix $\mathbf{A}$ by multiplying a column of $\mathbf{A}$ by a real number c, show that $\det(\mathbf{B}) = c \det(\mathbf{A})$. Use this to show that $\det(c\mathbf{A}) = c^p \det(\mathbf{A})$.

7. Suppose that the $p \times p$ matrix $\mathbf{A}$ is partitioned as

$$\mathbf{A} = \begin{pmatrix} 1 & \mathbf{0}' \\ \mathbf{a} & \mathbf{A}^* \end{pmatrix},$$

where $\mathbf{a}$ is an arbitrary column vector of dimension $p-1$, and $\mathbf{A}^*$ has dimension $(p-1) \times (p-1)$. Show that $\det(\mathbf{A}) = \det(\mathbf{A}^*)$.

8. Show that $(\mathbf{AB})^{-1} = \mathbf{B}^{-1}\mathbf{A}^{-1}$.

9. Prove equation (1).

10. Prove equation (3). *Hint*: verify that $\mathbf{AB} = \mathbf{I}_p$ and $\mathbf{BA} = \mathbf{I}_p$.

11. Suppose the symmetric nonsingular matrix $\mathbf{A}$ is partitioned as $\mathbf{A} = \begin{pmatrix} \mathbf{A}_{11} & \mathbf{A}_{12} \\ \mathbf{A}_{21} & \mathbf{O} \end{pmatrix}$. Writing the inverse in partitioned form as $\mathbf{A}^{-1} = \begin{pmatrix} \mathbf{P} & \mathbf{Q} \\ \mathbf{Q}' & \mathbf{R} \end{pmatrix}$, show that

$$\mathbf{P} = \mathbf{A}_{11}^{-1} - \mathbf{A}_{11}^{-1}\mathbf{A}_{12}\left(\mathbf{A}_{21}\mathbf{A}_{11}^{-1}\mathbf{A}_{12}\right)^{-1}\mathbf{A}_{21}\mathbf{A}_{11}^{-1},$$

$$\mathbf{Q} = \mathbf{A}_{11}^{-1}\mathbf{A}_{12}\left(\mathbf{A}_{21}\mathbf{A}_{11}^{-1}\mathbf{A}_{12}\right)^{-1},$$

$$\mathbf{R} = -\left(\mathbf{A}_{21}\mathbf{A}_{11}^{-1}\mathbf{A}_{12}\right)^{-1}.$$

12. If the $p \times p$ matrix $\mathbf{A}$ is partitioned as $\mathbf{A} = \begin{pmatrix} \mathbf{I}_q & \mathbf{O} \\ \mathbf{C} & \mathbf{I}_{p-q} \end{pmatrix}$, where $\mathbf{C}$ is an arbitrary matrix of dimension $(p-q) \times q$, show that $\det(\mathbf{A}) = 1$.

13. If the square matrix $\mathbf{A}$ is block–diagonal, i.e., $\mathbf{A} = \begin{pmatrix} \mathbf{A}_{11} & \mathbf{O} \\ \mathbf{O} & \mathbf{A}_{22} \end{pmatrix}$, show that $\det(\mathbf{A}) = \det(\mathbf{A}_{11}) \det(\mathbf{A}_{22})$.

14. Show that the matrix $\mathbf{I}_2$ is positive definite, and the matrix $\begin{pmatrix} 1 & 1 \\ 1 & 1 \end{pmatrix}$ is positive semidefinite but not positive definite.

15. Show that the symmetric $p \times p$ matrix $\mathbf{A} = \begin{pmatrix} \mathbf{A}_{11} & \mathbf{a}_{12} \\ \mathbf{a}_{21} & a_{22} \end{pmatrix}$, where $\mathbf{A}_{11}$ has dimension $(p-1) \times (p-1)$ is *pds* exactly if $\mathbf{A}_{11}$ is *pds* and $\det(\mathbf{A}) > 0$.

16. Prove equations (9) to (11).

17. Show that a triangular matrix (and in particular, a diagonal matrix) has p real eigenvalues equal to the diagonal entries.

18. Show that $\mathcal{P}(\lambda)$ of equation (16) is a polynomial of degree p in λ.

19. Find a 2×2 matrix that has only complex eigenvalues. Is it possible to construct 3×3 matrices with only complex roots?

20. Let $\mathbf{A}$ denote a symmetric matrix of dimension $p \times p$, with eigenvalues $\lambda_1, \ldots, \lambda_p$. Show that positive definiteness of $\mathbf{A}$ is equivalent to $\lambda_i > 0$ for $i = 1, \ldots, p$, and positive semidefiniteness is equivalent to $\lambda_i \geq 0$ for $i = 1, \ldots, p$.

21. Let $\mathbf{A} = \mathbf{B}\Lambda\mathbf{B}'$ be a spectral decomposition of the symmetric matrix $\mathbf{A}$, and assume that the first r eigenvalues of $\mathbf{A}$ are identical ($\lambda_1 = \cdots = \lambda_r$). Let $\mathbf{C} = \mathbf{B}\begin{pmatrix} \mathbf{M} & \mathbf{O} \\ \mathbf{O} & \mathbf{I}_{p-r} \end{pmatrix}$, where $\mathbf{M}$ is an arbitrary orthogonal matrix of dimension $r \times r$. Show that $\mathbf{C}'\mathbf{C} = \mathbf{I}_p$ and $\mathbf{C}'\mathbf{A}\mathbf{C} = \Lambda$.

22. If $\mathbf{A}$ is symmetric and nonsingular, with eigenvalues λ_i and eigenvectors $\boldsymbol{\beta}_i$, show that

$$\mathbf{A}^{-1} = \mathbf{B}\Lambda^{-1}\mathbf{B}' = \sum_{i=1}^{p} \frac{1}{\lambda_i} \boldsymbol{\beta}_i \boldsymbol{\beta}_i'.$$

23. Suppose $\mathbf{A}$ is *pds* and $\mathbf{M}$ is a symmetric matrix such that $\mathbf{A} = \mathbf{M}\mathbf{M}'$. Show that $\mathbf{M} = \mathbf{A}^{1/2}$ as defined in equation (25).

24. For a *pds* matrix $\mathbf{A}$, show that $\mathbf{A}^{r+s} = \mathbf{A}^r\mathbf{A}^s$ for any real numbers r and s.

25. For the function $h(\mathbf{c}) = ||\mathbf{x} - \mathbf{A}\mathbf{c}||^2$ in equation (30), find the matrix of second derivatives and show that h has a unique minimum.

26. Show that the function $h(\mathbf{c})$ in (30) takes a unique minimum at $\mathbf{c} = \mathbf{A}'\mathbf{x}$, without using calculus. *Hint*: write $\mathbf{x} - \mathbf{A}\mathbf{c} = (\mathbf{I} - \mathbf{P})\mathbf{x} + \mathbf{A}(\mathbf{A}'\mathbf{x} - \mathbf{c})$, where $\mathbf{P} = \mathbf{A}\mathbf{A}'$.

27. Suppose the columns of the $p \times q$ matrix $\mathbf{A}$ form an orthonormal basis of a linear subspace $\mathcal{S}$ of $\mathbb{R}^p$.

(a) If $\mathbf{H}$ is an orthogonal matrix of dimension $q \times q$, show that the columns of $\mathbf{B} = \mathbf{AH}$ form an orthonormal basis of $\mathcal{S}$.

(b) If the columns of the $p \times q$ matrix $\mathbf{B}$ form an orthonormal basis of $\mathcal{S}$, show that there exists an orthogonal matrix $\mathbf{H}$ such that $\mathbf{B} = \mathbf{AH}$. *Hint*: since both $\mathbf{A}$ and $\mathbf{B}$ have full column rank, there exists a nonsingular matrix $\mathbf{H}$ of dimension $q \times q$ such that $\mathbf{AH} = \mathbf{B}$, or equivalently, $\mathbf{AB} = \mathbf{H}^{-1}$.

28. If $\mathbf{P}$ is a projection matrix, show that for any $\mathbf{x} \in \mathbb{R}^p$ the vectors $\mathbf{x}_1 = \mathbf{Px}$ and $\mathbf{x}_2 = (\mathbf{I} - \mathbf{P})\mathbf{x}$ are orthogonal to each other.

29. Let $\mathcal{S}$ denote the one–dimensional linear subspace of $\mathbb{R}^2$ such that both coordinates are identical.

(a) Find an orthonormal basis for $\mathcal{S}$. Show that this basis is not unique.

(b) Find the projection matrix associated with $\mathcal{S}$.

(c) Graph the linear subspace $\mathcal{S}$ in $\mathbb{R}^2$, and graph the set of all $(x_1, x_2) \in \mathbb{R}^2$ that are projected into the point $(2, 2)'$.

30. Repeat parts (a) and (b) of Exercise 29 for the one–dimensional linear subspace of $\mathbb{R}^p$ in which all coordinates are equal. Show also that the projection onto $\mathcal{S}$ maps $\mathbf{x} = (x_1, \ldots, x_p)' \in \mathbb{R}^p$ in the vector $(\bar{x}, \ldots, \bar{x})'$, where $\bar{x} = (x_1 + \cdots + x_p)/p$.

31. Find an affine function which projects points $\mathbf{x} = (x_1, x_2)' \in \mathbb{R}^2$ orthogonally onto the line $x_2 = 1 + 2x_1$. Is this function unique? Graph the line $x_2 = 1 + 2x_1$, as well as the set of all points $(x_1, x_2)' \in \mathbb{R}^2$ whose image under the projection has first coordinate zero.

32. Show that $\mathbf{P} = \frac{1}{5}\begin{pmatrix} 4 & 2 \\ 2 & 1 \end{pmatrix}$ is a projection matrix.

33. If $\mathbf{P}$ is the projection matrix associated with the projection onto a linear subspace $\mathcal{S}$ of dimension q, show that $\text{rank}(\mathbf{P}) = q$.

34. Let

$$\mathbf{P} = \frac{1}{2}\begin{pmatrix} 1 & 0 & 1 & 0 \\ 0 & 1 & 0 & 1 \\ 1 & 0 & 1 & 0 \\ 0 & 1 & 0 & 1 \end{pmatrix}.$$

(a) Show that $\mathbf{P}$ is a projection matrix.

(b) Find the rank of $\mathbf{P}$.

(c) Find an orthonormal basis for the linear subspace $\mathcal{S}$ of dimension q associated with the projection matrix $\mathbf{P}$. *Note*: the solution is not unique.

(d) Using your solution from part (c), find all orthonormal bases of $\mathcal{S}$.

35. Let

$$\mathbf{P} = \frac{1}{2}\begin{pmatrix} 1 & -1 & 0 \\ -1 & 1 & 0 \\ 0 & 0 & 2 \end{pmatrix}, \qquad \mathbf{x}_0 = \begin{pmatrix} 1 \\ 1 \\ 0 \end{pmatrix}.$$

(a) Show that $\mathbf{P}$ is a projection matrix and find its rank q.

(b) Find an orthonormal basis for the linear subspace $\mathcal{S}$ of rank q associated with the projection $\mathbf{P}$.

(c) Verify that the affine function $g(\mathbf{x}) = \mathbf{x}_0 + \mathbf{P}\mathbf{x}$ is a projection in $\mathbb{R}^3$.

(d) Describe the set $\mathcal{M} \subset \mathbb{R}^3$ such that the following holds: if $\mathbf{y} \in \mathcal{M}$ then $g(\mathbf{y}) = \mathbf{y}$. That is, find a representation as in equation (18), using the matrix $\mathbf{A}$ found in part (b).

36. Find all projection matrices of dimension $p \times p$ with rank p.

37. Suppose the matrix $\mathbf{P}$ is symmetric and idempotent ($\mathbf{P}^2 = \mathbf{P}$). Show that the rank of $\mathbf{P}$ is the number of eigenvalues of $\mathbf{P}$ that are equal to 1.

Bibliography

Airoldi, J.–P., and Flury, B. 1988. An application of common principal component analysis to cranial morphometry of *Microtus californicus* and *M. ochrogaster* (Mammalia, Rodentia). *Journal of Zoology (London)* **216**, 21–36. With discussion and rejoinder pp. 41–43.

Airoldi, J.–P., Flury, B., and Salvioni, M. 1996. Discrimination between two species of Microtus using both classified and unclassified observations. *Journal of Theoretical Biology* **177**, 247–262.

Airoldi, J.–P., and Hoffmann, R.S. 1984. Age variation in voles (*Microtus californicus, M. ochrogaster*) and its significance for systematic studies. *Occasional Papers of the Museum of Natural History*. University of Kansas, Lawrence, KS, **111**, 1–45.

Aitkin, M., Anderson, D., Francis, B., and Hinde, J. 1989. *Statistical Modelling in GLIM*. Oxford: Clarendon Press.

Anderson, E. 1935. The irises of the Gaspe peninsula. *Bulletin of the American Iris Society* **59**, 2–5.

Anderson, T.W. 1963. Asymptotic theory for principal component analysis. *Annals of Mathematical Statistics* **34**, 122–148.

Anderson, T.W. 1984. *An Introduction to Multivariate Statistical Analysis*, 2nd ed. New York: Wiley.

Andersson, S. 1975. Invariant normal models, *The Annals of Statistics* **3**, 132–154.

Andrews, D.F., and Herzberg, A.M. 1985. *Data*. New York: Springer.

Bartlett, M.S., and Please, N.W. 1963. Discrimination in the case of zero mean differences. *Biometrika* **50**, 17–21.

Bensmail, H., and Celeux, G. 1996. Regularized discriminant analysis through eigenvalue decomposition. *Journal of the American Statistical Association* **91**, 1743–1748.

Bliss, C.I. 1935. The calculation of the dosage–mortality cure. *Annals of Applied Biology* **22**, 134–167.

Casella, G., and Berger, R.L. 1990. *Statistical Inference*. Belmont, CA: Duxbury Press.

Chatterjee, S., Handcock, M.S., and Simonoff, J.S. 1995. *A Casebook for a First Course in Statistics and Data Analysis*. New York: Wiley.

Chatterjee, S., and Price, B. 1977. *Regression Analysis by Example*. New York: Wiley.

Cox, D.R., and Hinkley, D.V. 1974. *Theoretical Statistics*. London: Chapman and Hall.

Cox, D.R., and Oakes, D. 1984. *Analysis of Survival Data*. London: Chapman and Hall.

Day, N.E. 1969. Estimating the components of a mixture of normal distributions. *Biometrika* **56**, 464–474.

Dempster, A.P., Laird, N.M., and Rubin, D.B. 1977. Maximum likelihood estimation from incomplete data via the *EM* algorithm (with discussion). *Journal of the Royal Statistical Society Series B*, **39**, 1–38.

Diaconis, P., and Efron, B. 1983. Computer-intensive methods in statistics. *Scientific American* **248** (May), 116–130.

Dobson, A.J. 1990. *An Introduction to Generalized Linear Models*. London: Chapman and Hall.

Dong, Z. 1996. Looking into Peking Man's subsistence — a taphonomic analysis of the middle pleistocene *Homo Erectus* site in China. Unpublished Ph.D. thesis. Department of Anthropology, Indiana University.

Edgington, E.S. 1980. *Randomization Tests*. New York: Dekker.

Edwards, A.W.F. 1984. *Likelihood*. Cambridge: Cambridge University Press.

Efron, B. 1982. *The Jackknife, the Bootstrap, and Other Resampling Plans*. Philadelphia: Society for Industrial and Applied Mathematics.

Efron, B. 1969. Students's t-test under symmetry conditions. *Journal of the American Statistical Association* **64**, 1278–1302.

Efron, B. 1975. The efficiency of logistic regression compared to normal discriminant analysis. *Journal of the American Statistical Association* **70**, 892–898.

Efron, B., and Gong, G. 1983. A leisurely look at the bootstrap, the jackknife, and other resampling plans. *The American Statistician* **37**, 36–48.

Efron, B., and Tibshirani, R. 1993. *An Introduction to the Bootstrap*. London: Chapman and Hall.

Everitt, B. S. 1993. *Cluster Analysis*. London: Edward Arnold.

Everitt, B.S., and Hand, D.J. 1981. *Finite Mixture Distributions*. London: Chapman and Hall.

Fairley, D. 1986. Cherry trees with cones? *The American Statistician* **40**, 138–139.

Fang, K., Kotz, S., and Ng, K. 1990. *Symmetric Multivariate and Related Distributions*. London: Chapman and Hall.

Finney, D.J. 1947. The estimation from original records of the relationship between dose and quantal response. *Biometrika* **34**, 320–334.

Fisher, R.A. 1936. The use of multiple measurements in taxonomic problems. *Annals of Eugenics* **7**, 179–188.

Flury, B. 1988. *Common Principal Components and Related Multivariate Models*. New York: Wiley.

Flury, B. 1995. Developments in principal component analysis: A review. In *Descriptive Multivariate Analysis*, W.J. Krzanowski, ed. Oxford: Oxford University Press, pp. 14–33.

Flury, B., Airoldi, J.–P., and Biber, J.–P. 1992. Gender identification of water pipits using mixtures of distributions. *Journal of Theoretical Biology* **158**, 465–480.

Flury, B., Nel, D.G., and Pienaar, I. 1995. Simultaneous detection of shift in means and variances. *Journal of the American Statistical Association* **90**, 1474–1481.

Flury, B., and Neuenschwander, B. 1995. Principal component models for patterned covariance matrices, with applications to canonical correlation analysis of several sets of variables. In *Descriptive Multivariate Analysis*, W.J. Krzanowski, ed. Oxford: Oxford University Press, pp. 90–112.

Flury, B., and Riedwyl, H. 1988. *Multivariate Statistics: A Practical Approach*. London: Chapman and Hall.

Flury, L., Boukai, B., and Flury, B. 1997. The discrimination subspace model. *Journal of the American Statistical Association* **92**, in press.

Frets, G.P. 1921. Heredity of head form in man. *Genetica* **3**, 193–384.

Friedman, J.H. 1989. Regularized discriminant analysis. *Journal of the American Statistical Association* **84**, 165–175.

Giri, N. 1964. On the likelihood ratio test of a normal multivariate testing problem. *Annals of Mathematical Statistics* **35**, 181–189.

Golub, G.H., and Van Loan, C.F. 1989. *Matrix Computations*. Baltimore: John Hopkins University Press.

Good, P. 1994. *Permutation Tests*. New York: Springer.

Grogan, W.L., and Wirth, W.W. 1981. A new American genus of predaceous midges related to *Palpomyia* and *Bezzia* (Diptera: Ceratopogonidae). *Proceedings of the Biological Society of Washington* **94**, 1279–1305.

Hand, D.J. 1981. *Discrimination and Classification*. New York: Wiley.

Hand, D.J. 1992. On comparing two treatments. *The American Statistician* **46**, 190–192.

Hand, D.J., and Taylor, C.C. 1987. *Multivariate Analysis of Variance and Repeated Measures*. London: Chapman and Hall.

Hartigan, J.A. 1975. *Clustering Algorithms*. New York: Wiley.

Hastie, T., and Stuetzle, W. 1989. Principal curves. *Journal of the American Statistical Association* **84**, 502–516.

Hills, M. 1982. Allometry, in *Encyclopedia of Statistical Sciences*, S. Kotz and N.L. Johnson, eds. New York: Wiley, pp. 48–54.

Hosmer, D.W., and Lemeshow, S. 1989. *Applied Logistic Regression*. New York: Wiley.

Hotelling, H. 1931. The generalization of Student's ratio. *Annals of Mathematical Statistics* **2**, 360–378.

Hotelling, H. 1933. Analysis of a complex of statistical variables into principal components. *Journal of Educational Psychology* **24**, 417–441.

Johnson, R.A., and Wichern, D.W. 1988. *Applied Multivariate Statistical Analysis*, 2nd ed. Englewood Cliffs, NJ: Prentice–Hall.

Johnson, N.L., Kotz, S., and Balakrishnan, N. 1995. *Continuous Univariate Distributions*, Vol. 2, 2nd ed. New York: Wiley.

Jolicoeur, P. 1963. The multivariate generalization of the allometry equation. *Biometrics* **19**, 497–499.

Jolicoeur, P., and Mosimann, J.E. 1960. Size and shape variation in the painted turtle: A principal component analysis. *Growth* **24**, 339–354.

Jolliffe, I.T. 1986. *Principal Component Analysis*. New York: Springer.

Kalbfleisch, J.G. 1985. *Probability and Statistical Inference*, Vol. 2, 2nd ed. New York: Springer.

Kaufman, L., and Rousseeuw, P.J. 1990. *Finding Groups in Data*. New York: Wiley.

Kaye, D.H., and Aickin, M. 1986. *Statistical Methods in Discrimination Litigation*. New York: Dekker.

Kelker, D. 1970. Distribution theory of spherical distributions and a location–scale parameter generalization. *Sankhya A* **32**, 419–430.

Khuri, A. 1993. *Advanced Calculus with Applications in Statistics*. New York: Wiley.

Klingenberg, C.P. 1996. Multivariate allometry. In *Advances in Morphometrics*, L.F. Marcus, M. Corti, A. Loy, G.J.P. Naylor, and D.E. Slice, eds. New York: Plenum Press, pp. 23–49.

Klingenberg, C.P., Neuenschwander, B.E., and Flury, B. 1996. Ontogeny and individual variation: Analysis of patterned covariance matrices with common principal components. *Systematic Biology* **45**, 135–150.

Krzanowski, W.J. 1988. *Principles of Multivariate Analysis: A User's Perspective*. Oxford: Clarendon Press.

Krzanowski, W.J., and Marriott, F.H.C. 1994. *Multivariate Analysis Part 1: Distributions, Ordination and Inference*. London: Edward Arnold.

Lachenbruch, P.A. 1975. *Discriminant Analysis*. New York: Hafner.

Lachenbruch, P.A., and Mickey, R. 1968. Estimation of error rates in discriminant analysis. *Technometrics* **10**, 1–11.

Larson, H.J. 1974. *Introduction to Probability Theory and Statistical Inference*, 2nd ed. New York: Wiley.

Lavine, M. 1991. Problems in extrapolation illustrated with space shuttle O-ring data. *Journal of the American Statistical Association* **86**, 919–922.

Levri, E.P., and Lively, C.M. 1996. The effects of size, reproductive condition, and parasitism on foraging behaviour in a freshwater snail, Potamopyrgus antipodarum. *Animal Behaviour* **51**, 891–901.

Lindsay, B.G. 1995. *Mixture Models: Theory, Geometry and Applications*. NSF-CBMS Regional Conference Series in Probability and Statistics, Vol. 5, Hayward, CA: Institute of Mathematical Statistics.

Little, R.J.A., and Rubin, D.B. 1987. *Statistical Analysis with Missing Data*. New York: Wiley.

Lubischew, A.A. 1962. On the use of discriminant functions in taxonomy. *Biometrics* **18**, 455–477.

Manly, B.F.J. 1991. *Randomization and Monte Carlo Methods in Biology*. London: Chapman and Hall.

Marais, J., Versini, G., van Wyk, C.J., and Rapp, A. 1992. Effect of region on free and bound monoterpene and C_{13}–norisoprenoid concentrations in Weisser Riesling wines. *South African Journal of Enology and Viticulture* **13**, 71–77.

Mardia, K.V., Kent, J.T., and Bibby, J.M. 1979. *Multivariate Analysis*. London: Academic Press.

McCullagh, P., and Nelder, J.A. 1989. *Generalized Linear Models*, 2nd ed. London: Chapman and Hall.

McLachlan, G.J. 1992. *Discriminant Analysis and Statistical Pattern Recognition*. New York: Wiley.

McLachlan, G.J., and Basford, K.E. 1988. *Mixture Models: Inference and Applications to Clustering*. New York: Dekker.

McLachlan, G.J., and Krishnan, T. 1997. *The EM Algorithm and Extensions*. New York: Wiley.

Monetti, A., Versini, G., Dalpiaz, G., and Reniero, F. 1996. Sugar adulterations control in concentrated rectified grape musts by finite mixture distribution analysis of the myo– and scyllo–inositol contents and the D/H methyl ratio of fermentative alcohol. *Jounal of Agricultural and Food Chiemistry* **44**, 2194–2201.

Morrison, D.F. 1976. *Multivariate Statistical Methods*, 3rd ed. New York: McGraw–Hill.

Muirhead, R.J. 1982. *Aspects of Multivariate Statistical Theory*. New York: Wiley.

Murray. G. 1977. Comment on "Maximum likelihood estimation from incomplete data via the *EM* algorithm" (by A.P. Dempster, N. Laird, and D.B. Rubin). *Journal of the Royal Statistical Society, Series B* **39**, 1–38.

Neuenschwander, B. 1991. Common principal components for dependent random vectors. Unpublished Ph.D. thesis. University of Berne (Switzerland), Department of Statistics.

O'Neill, T.J. 1978. Normal discrimination with unclassified observations. *Journal of the American Statistical Association* **73**, 821–826.

Oses, M., and Paul, J. 1997. Morphometric discrimination between two species of angels and estimation of their relative frequency. *Journal of Celestial Morphometrics* **1011**, 747–767.

Pearson, K. 1901. On lines and planes of closest fit to systems of points in space. *Philosophical Magazine Ser. 6* **2**, 559–572.

Rao, C.R. 1948. Tests of significance in multivariate analysis. *Biometrika* **35**, 58–79.

Rao, C.R. 1964. The use and interpretation of principal components in applied research. *Sankhya A* **26**, 329–358.

Rao, C.R. 1965. *Linear Statistical Inference and its Application*. New York: Wiley.

Rao, C.R. 1970. Inference on discriminant function coefficients. In *Essays in Probability and Statistics*, R.C. Bose et al., eds. Chapel Hill: University of North Carolina Press, pp. 587–602.

Ray, W.A. 1977. Prescribing of tetracycline to children less than 8 years old. *Journal of the American Medical Association* **237**, 2069–2074.

Rencher, A.C. 1995. *Methods of Multivariate Analysis*. New York: Wiley.

Ripley, B.D. 1987. *Stochastic Simulation*. New York: Wiley.

Ross, S. 1994. *A First Course in Probability*, 4th ed. New York: MacMillan College Publishing Co.

Ross, S. 1996. *Simulation*, 2nd ed. London: Academic Press.

Roy, S.N. 1953. On a heuristic method of test construction and its use in multivariate analysis. *Annals of Mathematical Statistics* **24**, 220–238.

Roy, S.N. 1957. *Some Aspects of Multivariate Analysis*. New York: Wiley.

Rubin, D.B., and Szatrowski, T.H. 1982. Finding maximum likelihood estimates of patterned covariance matrices by the *EM* algorithm. *Biometrika* **69**, 657–660.

Ryan, B.F., Joiner, B., and Ryan, T.A. 1976. *Minitab Handbook*, 2nd ed. Boston: Duxbury Press.

Schott, J.R. 1997. *Matrix Analysis for Statistics*. New York: Wiley.

Searle, S.R. 1982. *Matrix Algebra Useful for Statistics*. New York: Wiley

Seber, G.A.F. 1984. *Multivariate Obervations*. New York: Wiley.

Silvapulle, M.J. 1981. On the existence of maximum likelihood estimators for the binomial response model. *Journal of the Royal Statistical Society, Series B* **43**, 310–313.

Silvey, S.D. 1975. *Statistical Inference*. London: Chapman and Hall.

Szatrowski, T.H. 1985. Patterned covariances. In *Encyclopedia of Statistical Sciences, Vol. 6*, S. Kotz and N.L. Johnson, eds. New York: Wiley, pp. 638–641.

Takemura, A. 1985. A principal decomposition of Hotelling's T^2 statistic. In *Multivariate Analysis — VI*, P.R. Krishnaiah, ed. Amsterdam: Elsevier, pp. 583–597.

Tarpey, T., and Flury, B. 1996. Self–consistency: A fundamental concept in statistics. *Statistical Science* **11**, 229–243.

Titterington, D.M., Smith, A.F.M., and Makov, U.E. 1985. *Statistical Analysis of Finite Mixture Distributions*. New York: Wiley.

Watson, G.S. 1964. A note on maximum likelihood. *Sankhya A* **26**, 303–304.

Westfall, P.H., and Young, S.S. 1993. *Resampling–Based Multiple Testing*. New York: Wiley.

Index

Springer Texts in Statistics *(continued from page ii)*

Nguyen and Rogers: Fundamentals of Mathematical Statistics: Volume I: Probability for Statistics
Nguyen and Rogers: Fundamentals of Mathematical Statistics: Volume II: Statistical Inference
Noether: Introduction to Statistics: The Nonparametric Way
Peters: Counting for Something: Statistical Principles and Personalities
Pfeiffer: Probability for Applications
Pitman: Probability
Rawlings, Pantula and Dickey: Applied Regression Analysis
Robert: The Bayesian Choice: A Decision-Theoretic Motivation
Robert and Casella: Monte Carlo Statistical Methods
Santner and Duffy: The Statistical Analysis of Discrete Data
Saville and Wood: Statistical Methods: The Geometric Approach
Sen and Srivastava: Regression Analysis: Theory, Methods, and Applications
Shao: Mathematical Statistics
Shumway and Stoffer: Time Series Analysis and its Applications.
Terrell: Mathematical Statistics: A Unified Introduction
Whittle: Probability via Expectation, Third Edition
Zacks: Introduction to Reliability Analysis: Probability Models and Statistical Methods